DISCRETE MATHEMATICS AND ITS APPLICATIONS

Series Editor KENNETH H. ROSEN

T0229866

HANDBOOK OF MATHEMATICAL INDUCTION

THEORY AND APPLICATIONS

David S. Gunderson

University of Manitoba

Winnipeg, Canada

CRC Press

Taylor & Francis Group

Boca Raton London New York

CRC Press is an imprint of the
Taylor & Francis Group, an **informa** business

A CHAPMAN & HALL BOOK

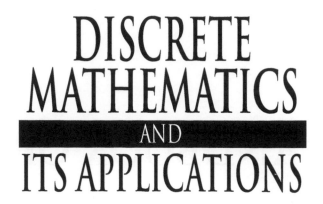

DISCRETE MATHEMATICS AND ITS APPLICATIONS

Series Editor
Kenneth H. Rosen, Ph.D.

Juergen Bierbrauer, Introduction to Coding Theory

Francine Blanchet-Sadri, Algorithmic Combinatorics on Partial Words

Richard A. Brualdi and Dragoš Cvetković, A Combinatorial Approach to Matrix Theory and Its Applications

Kun-Mao Chao and Bang Ye Wu, Spanning Trees and Optimization Problems

Charalambos A. Charalambides, Enumerative Combinatorics

Gary Chartrand and Ping Zhang, Chromatic Graph Theory

Henri Cohen, Gerhard Frey, et al., Handbook of Elliptic and Hyperelliptic Curve Cryptography

Charles J. Colbourn and Jeffrey H. Dinitz, Handbook of Combinatorial Designs, Second Edition

Martin Erickson, Pearls of Discrete Mathematics

Martin Erickson and Anthony Vazzana, Introduction to Number Theory

Steven Furino, Ying Miao, and Jianxing Yin, Frames and Resolvable Designs: Uses, Constructions, and Existence

Mark S. Gockenbach, Finite-Dimensional Linear Algebra

Randy Goldberg and Lance Riek, A Practical Handbook of Speech Coders

Jacob E. Goodman and Joseph O'Rourke, Handbook of Discrete and Computational Geometry, Second Edition

Jonathan L. Gross, Combinatorial Methods with Computer Applications

Jonathan L. Gross and Jay Yellen, Graph Theory and Its Applications, Second Edition

Jonathan L. Gross and Jay Yellen, Handbook of Graph Theory

David S. Gunderson, Handbook of Mathematical Induction: Theory and Applications

Darrel R. Hankerson, Greg A. Harris, and Peter D. Johnson, Introduction to Information Theory and Data Compression, Second Edition

Darel W. Hardy, Fred Richman, and Carol L. Walker, Applied Algebra: Codes, Ciphers, and Discrete Algorithms, Second Edition

Daryl D. Harms, Miroslav Kraetzl, Charles J. Colbourn, and John S. Devitt, Network Reliability: Experiments with a Symbolic Algebra Environment

Silvia Heubach and Toufik Mansour, Combinatorics of Compositions and Words

Leslie Hogben, Handbook of Linear Algebra

Titles (continued)

Chapman & Hall/CRC
Taylor & Francis Group
6000 Broken Sound Parkway NW, Suite 300
Boca Raton, FL 33487-2742

First issued in paperback 2016

© 2011 by Taylor and Francis Group, LLC
Chapman & Hall/CRC is an imprint of Taylor & Francis Group, an Informa business

No claim to original U.S. Government works

ISBN 13: 978-1-138-19901-9 (pbk)
ISBN 13: 978-1-4200-9364-3 (hbk)

Library of Congress Cataloging-in-Publication Data

Gunderson, David S.
    Handbook of mathematical induction : theory and applications / David S. Gunderson.
       p. cm. -- (Discrete mathematics and its applications)
    Includes bibliographical references and index.
    ISBN 978-1-4200-9364-3 (hardcover : alk. paper)
    1. Proof theory. 2. Induction (Mathematics) 3. Logic, Symbolic and mathematical. 4. Probabilities.
I. Title.

    QA9.54.G86 2010
    511.3'6--dc22
                                                      2010029756

Visit the Taylor & Francis Web site at
http://www.taylorandfrancis.com

and the CRC Press Web site at
http://www.crcpress.com

To my darling daughter, Christine.

Contents

III Solutions and hints to exercises

IV   Appendices

Foreword

The idea of mathematical induction has been with us for ages, certainly since the 16th century, but was made rigorous only in the 19th century by Augustus de Morgan who, incidentally, also introduced the term 'mathematical induction'. By now, induction is ubiquitous in mathematics and is taken for granted by every mathematician. Nevertheless, those who are getting into mathematics are likely to need much practice before induction is in their blood: The aim of this book is to speed up this process.

Proofs by induction vary a great deal. In fact, when it comes to finite structures or, more generally, *sequences* of assertions, *every* proof may be viewed as a proof by induction; when proving a particular proposition, we may as well assume that we have already proved every assertion which comes earlier in the sequence. For example, when proving the simple result that every graph with n vertices and more than $n^2/4$ edges contains a triangle, we may as well assume that this is true for graphs with fewer than n vertices. Thus, when a professor asks his class for ideas as to how to try to prove a result about finite groups and gets the suggestion *'By induction!'*, he is right to dismiss this as being unhelpful, since we are always *free* to use induction, and in some sense we are always using it. Nevertheless, it is true that in some cases induction plays a major role, while in others we hardly make any use of it. And the question is not whether to use induction but, when using it, *how* to use it.

It would be impossible for this *Handbook of Mathematical Induction* to cover all aspects of mathematical induction and its variants for infinite sets, but there is plenty of exciting material here, selected with much care, with emphasis on some of the most elegant results. This book contains all the standard exercises on induction and many more, ranging from the trifle and the trivial to the significant and challenging.

There are numerous examples from graph theory, point set topology, elementary number theory, linear algebra, analysis, probability theory, geometry, group theory, game theory, and the theory of inequalities, with results about continued fractions, logical formulae, Latin rectangles, Hankel matrices, Hilbert's affine cube, and the numbers of Fibonacci, Bernoulli, Euler, Catalan and Schröder, among others. Furthermore, the reader is guided through appropriate proofs of the theorems of Ramsey, Schur, Kneser, Hales and Jewett, Helly, Radon, Caratheodory, and many other results.

What prompts someone to write a book on mathematical induction? To share his passion for mathematics? Gunderson's passion for all of mathematics is evident. Perhaps this remarkable passion is due to the unusual road he has taken to mathematics. When I first met him, at Emory University in 1993, he was a graduate student. A rather 'mature' graduate student; as I learned later, in his youth he had flown aerobatics, and then had been a laborer and truck driver for ten years or so before starting in pure mathematics for the fun of puzzle solving. Although he has been in mathematics for over two decades, his physical prowess is still amazing: he has a penchant for ripping telephone books, and has not lost an arm-wrestling match since 1982.

This book is the first example that I know of which treats mathematical induction seriously, more than just a collection of recipes. It is sure to be an excellent student companion and instructor's guide for a host of courses.

Béla Bollobás

University of Cambridge and University of Memphis

Preface

Mathematical induction is a powerful proof technique that is generally used to prove statements involving whole numbers. Students often first encounter this technique in first or second year university courses on number theory, graph theory, or computer science. Many students report that their first exposure to mathematical induction was both scary yet simple and intriguing. In high school, formal proof techniques are rarely covered in great detail, and just the word "proof" seems daunting to many. Mathematical induction is a tool that seems quite different from anything taught in high school.

After just a few examples of proof by mathematical induction, the average student seems to gain an appreciation for the technique because the format for such a proof is straightforward and prescribed, yet the consequences are quite grand. Some students are further fascinated by the technique because of the erroneous conclusions available when the format is not followed precisely. It seems as if many students view mathematical induction as simply a necessary evil. Few beginning students in mathematics or computer science realize that all of mathematics is based on mathematical induction, nor do they realize that the foundations for the technique are of a completely different type than "scientific induction", or the "scientific method" forms of "hypothesis, testing, and conclusion" arguments used in most sciences.

In part, because of the recent explosion of knowledge in combinatorics, computing, and discrete mathematics, mathematical induction is now, more than ever, critical in education, perhaps surpassing calculus in its relevance and utility. The theory of recursion in computing science is practically the study of mathematical induction applied to algorithms. The theory of mathematical logic and model theory rests entirely on mathematical induction, as does set theory. It may be interesting to note that even in calculus, mathematical induction plays a vital role. Continuous mathematics (like calculus or analysis) uses counting numbers, dimension of a space, iterated derivatives, exponents in polynomials, or size of a matrix, and so mathematical induction might one day be taught in all junior math courses. In fact, mathematical induction is absolutely essential in linear algebra, probability theory, modelling, and analysis, to name but a few areas. Mathematical induction is a common thread that joins all of mathematics and computing science.

This book contains hundreds of examples of mathematical induction applied in a vast array of scientific areas, as well as a study of the theory and how to find and write mathematical induction proofs. The presentation here is quite unlike that of a discrete mathematics book, as theory and examples took precedence over nice pictures, charts, and chapters intended for one or two lectures.

The inception of this book

As with many books in mathematics, the incipient version of this book was a collection of notes for students. Nearly a decade ago, I put together a few pages with some standard induction problems for discrete math students. To help their writing of inductive proofs, I then provided a template and a few pages of advice on writing up induction proofs, producing a small booklet for the students that I distributed in any course requiring induction.

Since there seemed to be no readily available books on induction (most were out of print), I originally had the idea to write something small that could be universally available as a supplement to courses in discrete mathematics, including linear algebra, combinatorics, or even geometry. My first goal was to have around a hundred of the standard exercises in induction, complete with solutions. I also wanted the solutions to be written in a format that students could follow easily and reproduce. When I began to collect and write up problems for this small planned booklet, I found so many examples and major theorems that engaged me, I couldn't wait to write them down and share them with anyone who would listen. I then tried to supplement this early collection to somehow give a fair treatment to all of the mathematical sciences, including computing science.

By that time, it was too late. As many collectors do, I became obsessed with finding different kinds of inductive proofs from as many areas as possible. Even after gathering many different types of questions, I continued to add to the collection, giving more examples of some types, and also including a healthy amount of set theory and foundations—in an attempt to give a "credible" or "scholarly" representation of the theory and applications surrounding induction. In a sense, I was constructing a tribute to one of the major proof techniques in mathematics.

After the book quickly burgeoned into a few hundred pages, people (including publishers) asked me, "for whom is this book?" or "can this book be used for any course?" I could only reply that this book will work well with nearly *any* mathematics or computing science course. Then I just kept adding problems! Only when the collection began to point north of 500 pages, did Chapman & Hall/CRC suggest that I put together an encyclopedia of induction, a handbook. So, I added a few hundred more pages, sampling from as many fields I had the courage and time for. This is the product.

Who is this book for?

This book is intended for anyone who enjoys a good proof, and for those who would like to master the technique of mathematical induction. I think that nearly every student and professor of mathematics and computer science can get a little something from this book.

Students may find inductive solutions for their practice or even their homework, they may learn how to write inductive proofs, and they may discover some interesting mathematics along the way. Most topics in this book include definitions and simple theory in order to deliver the exercises, and any student perusing these might acquire new interests in areas previously unexplored.

The professor may find examples to give as exercises, test questions, or contest practice questions. Some professors and high school teachers might appreciate sections here on writing mathematical induction proofs, both for themselves and in passing along such skills to their students.

The professor or student might also use this text for definitions and references, as well as many famous theorems and their proofs. This book is designed to be a source book for everyone interested in mathematics. When I was an undergraduate, I spent all of my spare money (and more) on reference books, including collections of worked exercises, and had this book been available back then, I would have most certainly purchased it—not only to help me with induction homework, but also as a resource of popular results and mathematical tricks.

This book may enhance nearly every course in mathematics—from freshman to graduate courses. At the university, mathematical induction is taught in many different courses, including those in discrete mathematics, graph theory, theoretical computer science, set theory, logic, combinatorics, linear algebra, and math education. Other areas, including courses in computing science, engineering, analysis, statistics, modelling, game theory, and economics now use induction as a standard tool. These and many other areas are treated generously.

Structure of this book

The book is essentially divided into three parts: "theory", "applications and exercises", and "solutions". These titles aren't completely accurate, as there are exercises and solutions in the theory part and there is theory in the exercises part. The theory part also contains far more than just theory, but a more appropriate title could not be found.

In the theory part, first a brief introduction is given. The introduction is not meant to be expository nor complete in a way that some discrete mathematics books might cover mathematical induction. The formal development of natural numbers from axioms is given by mathematical induction. Many readers will want to skip this section, as it can be a little dry, but this material can be understood and appreciated by most undergraduates in their second or third year. Having basic arithmetic skills in hand, different inductive techniques are discussed: well-ordered sets, basic mathematical induction, strong induction, double induction, infinite descent, downward induction, and variants of some of these.

Chapter 4 is about mathematical induction and infinity, including an introduction to ordinals and cardinals, transfinite induction, more on well-ordering, the axiom of choice, and Zorn's lemma. The material in Chapter 4 is intended for the senior math or computer science student, and can be omitted by the inexperienced reader. One reviewer suggested that this material be moved to much later in the book; however, I feel that it fits well from a logical perspective, perhaps just not from a pedagogical one when viewed by first-year students.

There are sections on the history of induction (Section 1.8) and the present state of literature on mathematical induction (Section 1.9). Fallacies and induction (Chapter 5) and empirical induction (Chapter 6) are also surveyed. Chapters 7 and 8 on doing and writing inductive proofs are given with the intention of helping the student and perhaps providing some guidelines that a teacher might use when teaching presentation skills. Much of these two chapters are directed at the student, and so the advanced reader can safely skip these.

Part II, "Applications and exercises", contains over 750 exercises, showcasing the different levels of difficulty of an inductive proof, the variety of inductive techniques available, and the scope of results provable by mathematical induction. Topics are grouped into areas, complete with necessary definitions, theory, and notation, so each chapter is nearly independent from all others. I tried to include some famous or fundamental theorems from most major fields. In many areas, I include some very specialized problems, if only because I enjoyed them. In general, exercises are not ranked according to difficulty, so expect surprises. Many advanced topics are covered here, so there are many examples appropriate for even graduate-level courses.

The number of published mathematical induction proofs is finite; however, one might get the impression that this number is infinite! There can be no comprehensive coverage. The present collection identifies results spanning many fields, and there seems to be no end of topics that I could continue to add. It seemed that whenever I researched some mathematical induction proof, I found yet another nearby. People have joked that, by induction, I could then find infinitely many examples. At some point, I had to (at least temporarily) wrap up the project, and this is the outcome.

In part, I feel like a travel guide commissioned to write a handbook about touring Europe; after staying in Budapest for a month but only driving through Paris, the "handbook" may seem like only a biased "guidebook". I have delved deeply into specialist areas, and only glossed over some more usual topics.

If this book survives to a second edition, many more topics will be developed. For example, the theory of Turing machines or Markov processes might make worthy additions. Additive number theory, computational geometry, the theory of algorithms and recursion might be developed. I welcome suggestions for possible future inclusion.

In Part III, solutions to most exercises are given. Solutions are most often written in a strict format, making them slightly longer than what might be ordinarily found in texts (and much longer than those found in journals). The extra structure does not seem to interfere with reading the proof and, in fact, it may sometimes help. I have also attempted to eliminate as many pronouns as possible, and have avoided the royal "we" that often occurs in mathematics.

Of the over 750 exercises, over 500 have complete solutions, and many of the rest have either brief hints or references.

For some unusual exercises presented here without solutions, I have tried to provide references. Many induction exercises are now "folklore" with origins difficult to trace, so citations often just direct the reader to at least one instance of where a problem has occurred previously. Readers are invited to inform me if I have missed some key citations.

There are nearly 600 bibliography references, and results are cross referenced and indexed thoroughly. I have given over 3000 index entries to assist in quick referencing. The bibliography is also back-referenced; bold face numbers following each entry indicate where in this book the entry is cited [*].

DSG

Winnipeg, Canada

Acknowledgements

I first learned of Peano's axioms in a course on elementary mathematics from an advanced viewpoint (PMAT 340) taught to education majors by the late Eric C. Milner in 1987 at the University of Calgary. Some of the ideas related here to Peano's axioms are adapted from those notes (together with Landau's presentation in [339]).

Thanks are owed to discrete math students at Mount Royal College, Calgary, Canada, in particular Nitin Puri, for encouraging me to assemble more problems for them. Gratitude is owed to Dr. Peter Morton of MRC for observations regarding Chapter 2.

Students at the University of Manitoba provided much appreciated help. Manon Mireault researched and organized early drafts. Trevor Wares also researched and proofread. Rob Borgersen helped with research, typos, and many challenges I had with LaTeX. Karen Johannson was incredible in finding (hundreds of) typos, doing research, helping with some proofs, and making pictures.

Colleagues at the University of Manitoba have been tremendous. Dr. Sasho Kalajdzievski was especially kind and careful with his detailed comments on the first few chapters, as was Dr. Tommy Kucera on the foundations and set theory.

Drs. Julien Arino, Rob Craigen, Michelle Davidson, Michael Doob, Kirill Kopotun, Shaun Lui, Ranganathan Padmanabhan, Craig Platt, Stephanie Portet, and Grant Woods have contributed problems or helped with proofreading.

Penny Gilbert and Professor George Grätzer shared some LaTeX secrets and advice on formatting and typesetting.

Professors Ben Li (CompSci), Bill Kocay (CompSci), and Brad Johnson (Stats) were also very kind with their help. Thanks are also owed to Jenö Lehel (University of Memphis).

I also thank Dr. Ann Levey (University of Calgary, Philosophy) for support and encouragement in the early stages of this book.

Many thanks are owed to Chapman & Hall/CRC, in particular, Bob Stern and Jennifer Ahringer for their kindness, almost infinite patience, and advice. They not only accepted my proposal, but asked for a more comprehensive treatment than I had originally planned and then helped to realize it, despite my many delays in submitting a final copy. The project editor, Karen Simon, was immensely helpful and patient while guiding the final preparation. The entire manuscript was proofread by Phoebe Roth, and Kevin Craig designed the cover. Many thanks are owed for a very thoughtful external review by Dr. Kenneth Rosen; his ideas have greatly improved this book.

This manuscript was typeset by the author in LaTeX, 11pt, book style.

About the author

David S. Gunderson obtained a B.Sc. and M.Sc. from the University of Calgary, and a Ph.D. in pure mathematics from Emory University under the supervision of Vojtech Rödl in 1995. He is currently an assistant professor and head of the mathematics department at the University of Manitoba. Previous positions include postdoctoral work at the University of Bielefeld, Howard University, and McMaster University. He is an elected fellow of the Institute of Combinatorics and Its Applications, and a member of many other mathematical societies.

His research interests are primarily in combinatorics, including Ramsey theory, extremal graph theory, combinatorial geometry, combinatorial number theory, and lattice theory. As a hobby, he has made many polyhedra from wood and other mathematical models for display and teaching (see his home page at `http://home.cc.umanitoba.ca/~gunderso/`).

Part I

Theory

Chapter 1

What is mathematical induction?

Induction makes you feel guilty for getting something out of nothing, and it is artificial, but it is one of the greatest ideas of civilization.

—Herbert S. Wilf,

MAA address, Baltimore, 10 Jan. 1998.

1.1   Introduction

In the sciences and in philosophy, essentially two types of inference are used, deductive and inductive. Deductive inference is usually based on the strict rules of logic and in most settings, deductive logic is irrefutable. Inductive reasoning is the act of guessing a pattern or rule or predicting future behavior based on past experience. For example, for the average person, the sun has risen every day of that person's life; it might seem safe to then conclude that the sun will rise again tomorrow. However, one can not prove beyond a shadow of a doubt that the sun will rise tomorrow. There *may* be a certain set of circumstances that prevent the sun rising tomorrow.

Guessing a larger pattern based upon smaller patterns in observations is called *empirical induction*. (See Chapter 6 for more on empirical induction.) *Proving* that the larger pattern always holds is another matter. For example, after a number of experiments with force, one might conclude that Newton's second "law" of motion $f = ma$ holds; nobody actually proved that $f = ma$ always holds, and in fact, this "law" has recently been shown to be flawed (see nearly any modern text in physics, *e.g.*, [56, p.76]).

Another type of induction is more reliable: *Mathematical induction* is a form of reasoning that *proves*, without a doubt, some particular rule or pattern, usually infinite. The process of mathematical induction uses two steps. The first step is the "base step": some simple cases are established. The second step is called

1

the "induction step", and usually involves showing that an arbitrary large example follows logically from a slightly smaller pattern. Observations or patterns proved by mathematical induction share the veracity or assurance of those statements proved by deductive logic. The validity of a proof by mathematical induction follows from basic axioms regarding positive integers (see Chapter 2 for more on the foundations of the theory).

In its most basic form, mathematical induction, abbreviated "MI", is a proof technique used to prove the truth of statements regarding the positive integers. (The statements themselves are rarely discovered using mathematical induction.) In this chapter, mathematical induction is only briefly introduced, with later chapters spelling out a more formal presentation.

It is easy to get excited about introducing the proof technique called "mathematical induction", especially since no mathematical aptitude or training is necessary to understand the underlying concept. With only very little high school algebra (and sometimes none at all!), mathematical induction enables a student to quickly prove hundreds of fascinating results. What more can a teacher ask for—an easy to understand technique complete with an amazing array of consequences!

1.2 An informal introduction to mathematical induction

To demonstrate the claim that no mathematical sophistication is necessary to comprehend the idea of mathematical induction, let me share an anecdote. When my daughter Christine decided to keep a stray cat as a pet, the two of them soon became inseparable—until it was time to go to bed. Christine slept in the top of a set of large bunk beds, but the cat was not so eager about climbing this strange contraption we humans know as a ladder. The cat, named Jupiter, sat on the floor meowing until I lifted him to Christine's warm bed each night. (He could jump down without fear, however, via the dresser.)

So I tried to teach Jupiter how to climb the ladder. (The cat probably could climb a ladder without my help, however it seemed as if he was waiting for permission—so for the sake of this story, assume that he did not know how.) There seemed to be two separate skills that Jupiter needed to acquire. First, he was apprehensive about just getting on the ladder, so with a little guidance and much encouragement, he discovered that he could indeed get on and balance on the first rung. Second, he had to learn how to climb from one rung to the next higher rung. I put his front paws on the next step and then tickled his back feet; to escape the tickle, he brought up his hind legs to the next rung. I repeated this on the next rung; he quickly realized how to go up one more (or that it was okay to do so?), and almost immediately upon "learning" this second skill, he applied it a few more times, and a moment later was rewarded with a big hug from Christine at the top.

That's the basic idea behind what is called "the principle of mathematical induction": in order to show that one can get to any rung on a ladder, it suffices to first show that one can get on the first rung, and then show that one can climb from any rung to the next. This heuristic applies no matter how tall the ladder, or even how far up the "first" rung is; one might even consider the 0-th rung to be the floor.

1.3　Ingredients of a proof by mathematical induction

In mathematical jargon, let $S(n)$ denote a statement with one "free" variable n, where, say, $n = 1, 2, 3, \ldots$. For example, $S(n)$ might be "the cat can get on the n-th rung of the ladder" or say, "rolling n dice, there are $5n + 1$ totals possible" (see next section). To show that for every $n \geq 1$, the proposition $S(n)$ is true, the argument is often in two parts: first show that $S(1)$ is true (called the "base step"). The second part (called the "induction step") is to pick some arbitrary $k \geq 1$ and show that if $S(k)$ is true, then $S(k + 1)$ follows. In this case, $S(k)$ is called the "inductive hypothesis". Once these two parts have been shown, if one were then asked to demonstrate that $S(4)$ is true, begin with $S(1)$, then by repeating the second part three times,

$$S(1) \rightarrow S(2); \quad S(2) \rightarrow S(3); \quad S(3) \rightarrow S(4).$$

This method succeeds in reaching the truth of $S(n)$ for *any* $n \geq 1$, not just $n = 4$.

The base step above need not have been $n = 1$. Sometimes induction starts a little later. For example, the statement $S(n) : n^2 < 2^n$ is not true for $n = 1, 2, 3$, or 4, but is true for any larger $n = 5, 6, 7, \ldots$. In this case, the base step is $S(5) : 5^2 < 2^5$, which is verified by $25 < 32$. The inductive step is, for $k \geq 5$, $S(k) \rightarrow S(k + 1)$ (which is not difficult: see Exercise 159).

So the principle of mathematical induction can be restated so that the base step can be any integer (positive or negative or zero): [This is stated again formally in Chapters 2 and 3.]

> **Principle of mathematical induction**: For some fixed integer b, and for each integer $n \geq b$, let $S(n)$ be a statement involving n. If
> (i) $S(b)$ is true, and
> (ii) for any integer $k \geq b$, $S(k) \rightarrow S(k + 1)$,
> then for all $n \geq b$, the statement $S(n)$ is true.

The expression "principle of mathematical induction" is often abbreviated by "PMI", however in this text, simply "MI" is used. In the statement of the principle of mathematical induction above, (i) is the base step and (ii) is the induction step, in which $S(k)$ is the inductive hypothesis. A proof that uses mathematical induction is sometimes called simply "a proof by induction" when no confusion can arise.

For an assortment of reasons, mathematical induction proofs are, in general, easy. First, the general rule often does not need to be guessed, it is usually given. A great deal of work is often required to guess the rule, but an inductive proof starts after that hard work has been done. Another aspect of proving by mathematical induction that makes it easy is that there are usually clearly defined steps to take, and when the last step is achieved, the logic of the proof makes the answer undeniable. For some, the most challenging part of an inductive step is only in applying simple arithmetic or algebra to simplify expressions.

A proof by mathematical induction has essentially four parts:

1. Carefully describe the statement to be proved and any ranges on certain variables.

2. The base step: prove one or more base cases.

3. The inductive step: show how the truth of one statement follows from the truth of some previous statement(s).

4. State the precise conclusion that follows by mathematical induction.

For more on the structure of a proof by mathematical induction, see Chapters 2, 3; for the reader just learning how to prove by mathematical induction, see Chapter 7 for techniques and Chapter 8 for how to write up a proof by mathematical induction.

1.4   Two other ways to think of mathematical induction

Many authors compare mathematical induction to dominoes toppling in succession. If the b-th domino is tipped, (see Figure 1.1) then all successive dominoes also fall.

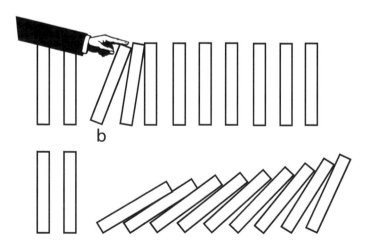

Figure 1.1: Dominoes fall successively

This comparison allows one to view mathematical induction in a slightly more general form, since all dominoes need not be in a single row for the phenomenon to work; as long as each "non-starting" domino has one "before it" which is close enough to topple it. So, in a sense, mathematical induction is not just done from any one integer to the next; induction can operate for many sequences of statements as long as for each non-initial case, there is a previous case by which one can use a rule to jump up from.

Another analogy for mathematical induction is given by Hugo Steinhaus n *Mathematical Snapshots* [508] [in the 1983 edition see page 299]: Consider a pile of envelopes, as high as one likes. Suppose that each envelope except the bottom one contains the same message "open the next envelope on the pile and follow the instructions contained therein". If someone opens the first (top) envelope, reads the message, and follows its instructions, then that person is compelled to open envelope number two of the pile. If the person decides to follow each instruction, that person then opens all the envelopes in the pile. The last envelope might contain a message "Done". This is the principle of mathematical induction applied to a finite set, perhaps called "finite induction". Of course, if the pile is infinite and each envelope is numbered with consecutive positive integers, anyone following the instructions would (if there were enough time) open all of them; such a situation is analogous to mathematical induction as it is most often used.

1.5 A simple example: Dice

Here is an example of a problem, a conjecture, and a proof of this conjecture by mathematical induction.

When rolling a single die, there are six possible outcomes: 1,2,3,4,5,6. When rolling two dice, there are 11 possible totals among two dice: 2,3, ..., 12, and for three dice, the 16 possible totals are 3,4,..., 18. After a moment of reflection, one might guess that for $n \geq 1$ dice, the number of possible totals is $5n + 1$.

Proposition 1.5.1. *The number of possible totals formed by rolling $n \geq 1$ dice is $5n + 1$.*

Proof: (By mathematical induction on n) For each positive integer n, denote the statement

$$S(n): \quad \text{When rolling } n \text{ dice, there are } 5n + 1 \text{ possible totals.}$$

So $S(1)$, $S(2)$, $S(3)$, ... form an infinite family of statements. (Using mathematical induction, all such statements are proved.)

BASE STEP: The statement $S(1)$ is already verified as there are $6 = 5(1) + 1$ outcomes.

INDUCTIVE STEP: Fix $k \geq 1$ and suppose that $S(k)$ is true (the inductive hypothesis), that is, among k dice, there are $5k + 1$ possible outcomes. To complete the inductive step, one needs only to show that the subsequent statement

$$S(k+1): \quad \text{When rolling } k+1 \text{ dice, there are } 5(k+1)+1 \text{ possible totals}$$

is also true.

Consider $k+1$ dice, say $D_1, D_2, \ldots, D_k, D_{k+1}$. Among the first k dice there are (by the inductive assumption $S(k)$) $5k+1$ possible totals. Among these totals, the smallest possible is k (where each dice shows 1), and so the lowest total possible using all $k+1$ dice is $k+1$ (when D_{k+1} also shows 1). The highest possible total for all the first k dice is $6k$ (when each of $D_1, \ldots, D_k$ show a 6). Then using D_{k+1}, each of $6k+1, 6k+2, \ldots, 6k+6$ is a new possible total. Hence, there are six new possible totals, and one old possible total (k) which no longer occurs among $k+1$ dice. Hence, there are 5 more totals possible with $k+1$ dice than with k dice, giving $5k+1+5 = 5(k+1)+1$ outcomes as desired. This completes the inductive step.

Hence, one concludes by mathematical induction that for any $n \geq 1$, the statement $S(n)$ is true. This concludes the proof of Proposition 1.5.1.   □

[The "□" indicates the end of a proof.]

1.6   Gauss and sums

It seems to be tradition in teaching induction that the first example demonstrating how well MI can work is in proving a formula for summing the first n positive integers.

There is a story about a young Carl Friedrich Gauss (1777-1855) that is often told. I first give the apocryphal version, which is an over-simplification of the supposed facts, because it so aptly creates a segue to the inductive proof. [The more historical version—which is even more unbelievable—is given after the proof of Theorem 1.6.1.]

Gauss was extremely quick as a child, and his teachers had a tough time keeping ahead of him. To keep Gauss busy, his teacher once asked him to sum the numbers from 1 to 100—to which Gauss almost immediately replied "5050". Perhaps he had discovered the following fact.

Theorem 1.6.1. *For each positive integer n,*

$$1 + 2 + 3 + \cdots + n = \frac{n(n+1)}{2}.$$

Proof of Theorem 1.6.1 by MI: Let $S(n)$ be the statement

$$S(n): \quad 1 + 2 + 3 + \cdots + n = \frac{n(n+1)}{2}.$$

BASE STEP ($n = 1$): The statement $S(1)$ says that $1 = \frac{1(2)}{2}$, which is clearly true, so $S(1)$ holds.

INDUCTIVE STEP($S(k) \to S(k+1)$): Fix some $k \geq 1$, and suppose that

$$S(k): \quad 1 + 2 + 3 + \cdots + k = \frac{k(k+1)}{2}$$

holds. (This statement is called the *inductive hypothesis*.) To complete the inductive step, it suffices to verify that the statement

$$S(k+1): \quad 1 + 2 + 3 + \cdots + k + (k+1) = \frac{(k+1)(k+2)}{2}$$

also holds. Beginning with the left-hand side of $S(k+1)$,

$$
\begin{aligned}
1 + 2 + 3 + \cdots + k + (k+1) &= (1 + 2 + 3 + \cdots + k) + (k+1) \\
&= \frac{k(k+1)}{2} + (k+1) \qquad \text{(by ind. hyp.)}, \\
&= (k+1)\left(\frac{k}{2} + 1\right), \\
&= (k+1)\left(\frac{k+2}{2}\right),
\end{aligned}
$$

which is equal to the right-hand side of $S(k+1)$. Hence $S(k) \to S(k+1)$ is proved, completing the inductive step.

Conclusion: By the principle of mathematical induction, for each $n \geq 1$, the statement $S(n)$ is true. $\qquad \square$

Many statements provable by mathematical induction are also provable in a direct manner. For example, here is one of many other proofs of the expression in Theorem 1.6.1:

Direct proof of Theorem 1.6.1: (without explicit use of MI) Write the sum $s(n) = 1 + 2 + \cdots + n$ twice, the second time with the summands in reverse order, and add:

$$
\begin{array}{ccccccccccc}
s(n) & = & 1+ & & 2 & + & 3 & + \cdots + & (n-1) & & +n \\
s(n) & = & n+ & & (n-1) & + & (n-2) & + \cdots + & 2 & & +1 \\
\hline
2s(n) & = & (n+1)+ & & (n+1) & + & (n+1) & + \cdots + & (n+1) & & +(n+1)
\end{array}
$$

The summand $(n+1)$ occurs n times, and so $2s(n) = n(n+1)$; division by 2 completes the proof. $\qquad \square$

The numbers $T_n = 1 + 2 + 3 + \cdots + n$ are called the *triangular numbers*. One reason that they are called triangular might be because if one makes a diagram with n rows of dots, starting with one dot in the first row, and in subsequent rows putting one more dot, then the dots form a triangle, and T_n is the total number of dots.

Here is an example for $n = 6$:

To compute T_n of Theorem 1.6.1, put an n by $n + 1$ box around such a triangle, and notice that T_n accounts for half of the box. See also Nelsen's wonderful little book *Proof without words* [403, p. 69], where the caption is "—"The ancient Greeks" (as cited by Martin Gardner)". Another similar "Proof without words" of the formula for T_n is given by Ian Richards [453] (also reprinted in [403, p. 70]). See also [404, p. 83]. One can also think of the triangle above as being equilateral. For other polygons, there are other "figurate numbers", for example, $n(3n-1)/2$ is a pentagonal number (the square numbers you already know). See the wonderfully illustrated [116, pp. 38ff] for more on polygonal (and polyhedral) numbers. [Polygonal numbers are also a rich source for induction problems as most are defined recursively, though few appear in this volume.]

For a moment, return to Gauss in the classroom. Expanding on the account given above, here is an excerpt from E. T. Bell's *Gauss, Prince of Mathematicians* [44] (also found in Newman's 1956 anthology [45]):

> Shortly after his seventh birthday Gauss entered his first school, a squalid relic of the Middle Ages run by a virile brute, one Büttner, ...
>
> Then, in his tenth year, Gauss was admitted to the class in arithmetic. As it was the beginning class none of the boys have heard of an arithmetical progression. It was easy then for the heroic Büttner to give out a long problem in addition whose answer he could find by a formula in a few seconds. The problem was of the following sort, $81297 + 81495 + 81693 + \cdots + 100899$, where the step from one number to the next is the same all along (here 198), and a given number of terms (here 100) are to be added.
>
> It was the custom, of the school for the boy who first got the answer to lay his slate on the table; the next laid his slate on top of the first, and so on. Büttner had barely finished stating the problem when Gauss

flung his slate on the table: "There it lies," he said—"*Liggit se*" in his pleasant dialect. Then, for the ensuing hour, while the other boys toiled, he sat with his hands folded, favored now and then by a sarcastic glance from Büttner, who imagined the youngest pupil in the class was just another blockhead. At the end of the period Büttner looked over the slates. On Gauss' slate there appeared but a single number. To the end of his days Gauss loved to tell how the one number he had written was the correct answer and how all the others were wrong.

1.7 A variety of applications

One aspect of mathematical induction is that it can be found in the proofs of a broad spectrum of results. In this section a sample is given of areas that mathematical induction is found.

Hundreds of equalities and inequalities have proofs by induction. For example, Exercise 54 asks to show the well-known formula

$$1^2 + 2^2 + 3^2 + \cdots + n^2 = \frac{n(n+1)(2n+1)}{6}.$$

Trigonometric identities also can be proved by induction, as in Exercise 124 where for any real number x and $n \geq 1$,

$$\cos^{2n}(x) + \sin^{2n}(x) \geq \frac{1}{2^{n-1}}.$$

Many such identities (or inequalities) are proved in a manner very similar to that in Theorem 1.6.1. Some inequalities have mathematical induction proofs that are not so evident. For example, in Exercise 204, induction is applied to show that any positive integer n,

$$\sqrt{2\sqrt{3\sqrt{4 \cdots \sqrt{n}}}} < 3.$$

Suppose that a sequence of numbers is defined recursively, that is, a few initial values are given, and then a formula or rule shows how to get the nth number from earlier numbers. For example, define a sequence $a_0, a_1, a_2, a_2, \ldots$ by first setting $a_0 = 3$ and let $a_1 = 3$. Then for each $n \geq 2$, define $a_n = 2a_{n-2} + a_{n-1}$, each a combination of the two previous values. Working out the first few values are $3, 3, 9, 15, 33$. There is a method by which to come up with a formula for the general term a_n; however, one might also guess that for each $n \geq 0$,

$$a_n = 2^{n+1} + (-1)^n.$$

Mathematical induction can be used to prove that this guess is correct. In the theory of recursion, mathematical induction is indispensable in proving correctness of formulas or algorithms. See Chapter 16 for mathematical induction in the theory of recursion. Many popular algorithms are analyzed here by mathematical induction.

Induction can also solve problems that have no apparent equation associated with it. For example, on a circular track, put n cars (with engines off), and among all cars, distribute just enough gas for one car to go around a track. In Exercise 590, induction is used to prove that there is a car that can make its way around a lap by collecting gas from the other cars on its way.

In an election, a votes are cast for candidate A and $b < a$ votes cast for candidate B. In Exercise 764, one counts the number of ways $a + b$ votes can be ordered so that after each vote, candidate A is winning. Similar results have an impact in game theory, probability theory, and economics.

Various forms of mathematical induction can be used to prove very general and powerful results about infinite structures. For example, a special form of mathematical induction, called Zorn's lemma, is applied in Exercise 692, to show that every vector space has a basis.

An abundance of results in discrete math and graph theory are proved by induction. For example, if a graph on n vertices has more than $n^2/4$ edges, Exercise 509 shows that the graph always contains a triangle. Problems in geometry (see Chapter 20) have surprising solutions using induction, as well.

Many basic counting principles have proofs by mathematical induction; for example, both the pigeonhole principle and the inclusion-exclusion principle have proofs by induction (see Exercises 743 and 427, respectively).

Model theory, foundations of mathematics, and computing theory are highly reliant on inductive proof techniques. Most elementary properties of arithmetic are derived using induction.

Mathematical induction is often associated with discrete counting; however, it can be used to prove many results in calculus and analysis. For example, starting with the simple product rule $(fg)' = f'g + fg'$, by induction one can prove (see Exercise 611) an extended version:

$$(f_1 f_2 \cdots f_n)' = f_1' f_2 f_3 \cdots f_n + f_1 f_2' f_3 \cdots f_n + \cdots + f_1 f_2 \cdots f_{n-1} f_n'.$$

This example hints at a theme.

Very loosely speaking, there are countless examples in mathematics where a concept is generalized or extended from one dimension to two; then from two to three; if a pattern becomes obvious in these first jumps, the pattern often describes a recursion, one that can serve as a model for an induction step taking the concept to any finite dimension required. The same is true for linear algebra and matrix theory; in fact, it might appear that most concepts in linear algebra "grow by induction" from smaller ones. See Exercises 637–668 for what might seem to be most of the major results in matrix theory, including a few applications, all proved by mathematical induction.

After only a brief perusal of the exercises in this book, one might conclude that most of mathematics is tied to induction. To many, this comes as no surprise, because counting numbers and basic rules of arithmetic and algebra are either developed or proved true using induction. Hence nearly all of discrete mathematics is based on induction, in a sense.

The first part of Chapter 2 establishes some useful notation and terminology, and the latter parts of that chapter are for those interested in the theory behind induction. To continue the introduction to mathematical induction, Chapter 3 gives examples of the many different inductive techniques and examples of each. If the reader is just beginning to learn induction and how to write proofs, I recommend also reading Chapters 7 and 8.

1.8   History of mathematical induction

I have read somewhere or other, in Dionysius of Halicarnassus, I think, that History is Philosophy teaching by examples.

—Henry St. John (Viscount Bolingbroke) (1678–1751),

On the study and use of history.

A usual (modern) development of the principle of mathematical induction begins with Peano's axioms. In this book, too, this approach is adopted. This perspective is admittedly a bit naive, since there were many other key players in establishing the present confidence held in the concept.

It is not clear who first used mathematical induction, but in Bussey's 1917 article [91], he reported that Blaise Pascal (1623–1662) recognized that an Italian named D. Franciscus Maurolycus (1494–1575) (also spelled *Francesco* or *Francesko Maurolico* or *Maurolyci*) used induction in his book [376] published in 1575. In that book, (actually, in Book I) he proved by induction that the odd numbers are formed by successively adding 2 to the first odd number, 1. Maurolycus used what is now called "induction" to prove that the sum of the first n odd numbers is n^2. These and many other ideas were learned by Blaise Pascal, in the mid 1600s, with Pascal perhaps being the first to apply induction for a formula for the sum of the first n natural numbers. In Struik's [515, p. 106] *A Concise History of Mathematics*, two works ([201] and [448]) are cited as evidence that "He [Pascal] was the first to establish a satisfying formulation of the principle of complete induction."

Maurolycus' proof of the formula for the sum of the first n numbers was non-inductive, although Georg Cantor (1845–1918) claimed that Pascal got his inductive proof from Maurolycus; Bussey refutes this claim. Cantor (Georg Ferdinand Ludwig Philip Cantor) once claimed that Pascal was the originator of mathematical induction, but later withdrew his claim after he was informed by someone named G. Vacca about Maurolycus (see [544]). So it seems, Pascal learned induction from Maurolycus.

It might be interesting to note that Bussey's article was published while Cantor was still alive. Cantor quit teaching at the University of Halle in 1905, was very ill late in life, and died in a mental hospital in Halle in 1918, so perhaps he never saw the article. Cantor is now credited with being the founder of set theory, particularly, the theory of infinite sets.

In George Pólya's (1887–1985) 1962 book *Mathematical Discovery* [435], mathematical induction is credited to Pascal as well, but in Bourbaki's *The Set Theory* [69] (1965), "Maurolico F." receives credit. [Bourbaki was not a person, but a group of sometimes 20 persons, at various times including C. Chevally, J. Delsarete, J. Dieudonne, and A. Weil—they had to retire from the group at age 50.]

It seems odd that such a simple technique was only learned in the 16th century. In fact, it would not be a surprise if Euclid (*ca.* 330–275 BC) used mathematical induction, though there does not seem to be any explicit instance of it. It might be worth noting that Euclid's result that states that there are infinitely many primes can be easily proved by induction; see Exercise 207. This has led some authors to the opinion that Euclid used, if even tacitly, induction. The debate as to whether or not Euclid knew of mathematical induction has gone as far as to interpret induction less formally. For more on Euclid and induction, see [175], [197], [541], [542], and [557]. It has been suggested [523] that Pappus (*ca.* 300AD) also knew of induction, though I have not yet seen the evidence. Even Plato might have known of the principle (see [3]).

The method of mathematical induction has been compared to the "method of exhaustion", due to Eudoxus (408–355 B.C.) [about a century before Euclid] and used by Archimedes (287–212 B.C.) in his derivation of many formulas (for areas and volumes), and his "method of equilibrium"—which often uses the method of slicing—called the *method of indivisibles* by Cavalieri (1598–1647), a technique still used in modern integral calculus. The method of exhaustion begins with an assumption that magnitudes can be divided an infinite number of times. For example, the method can be used to prove that the formula $A = \pi r^2$ for the area of a circle is correct by finding larger and larger polygons that fit inside a circle. (See [180, 11-3] for a details.) What this method has in common with mathematical induction is that a formula must first be guessed, and the proof is an iteration of (perhaps) infinitely many steps, often based on some kind of recursion depending on earlier steps. Some proofs by the method of exhaustion can be translated into proofs by induction, however the method of induction does not seem to be used explicitly by any of these masters from (nearly) ancient times.

Internet sources suggest that Ralbag (Rabbi Levi Ben Gershon) gave proofs that used induction in the 13th century. One such correspondence was from Boaz Tzaban, Bar Ilan University, Israel; another was from Ed Sandifer at Western Connecticut State University, Danbury, CT. They reported on a talk given by Shai Simonson of Stonehill College in Massachusetts, a scholar of Ben Gershon's work. It is not clear that Gershon formalized the concept, but there seems to be some agreement that he used mathematical induction. For more support on these ideas, see [444]. Many other authors report on the use of induction or inductive techniques by al-Karaji (sum of cubes formula, around 1000 A.D.), al-Haytham (sum of fourth powers), and al-Samawal (binomial theorem). [I have not yet directly seen these references, however, more information is available in [307].]

According to Quine [443, p. 243], "Mathematical induction was used and explicitly recognized by Pascal in 1654 ... and Fermat in 1659 ... But the principle of mathematical induction retained the status of an ultimate arithmetic axiom until 1879 when Frege defined the ancestral and by its means the class of natural numbers." Quine also says that "...Such inference, called *mathematical induction*, is afforded by the following metatheorem" and then uses very careful (and barely readable) logical notation to give the metatheorem.

Grimaldi reports in his textbook on discrete mathematics [238], that it was Augustus DeMorgan (1806–1871) who, in 1838, carefully described the principle and gave it its present name "mathematical induction". The reference Grimaldi gave for this fact was Bussey's paper [91], however, a quick look at Bussey's paper does not seem to confirm this. In fact, on the website *Earliest Known Uses of Some of the Words of Mathematics* [384], it is reported[1]:

> The term INDUCTION was first used in the phrase *per modum inductionis* by John Wallis in 1656 in *Arithmetica Infinitorum*. Wallis was the first person to designate a name for this process; Maurolico and Pascal used no term for it (Burton, page 440).

and

> The term MATHEMATICAL INDUCTION was introduced by Augustus de Morgan (1806-1871) in 1838 in the article *Induction (Mathematics)* that he wrote for the *Penny Cyclopedia*. De Morgan had suggested the name *successive induction* in the same article and only used the term *mathematical induction* incidentally. The expression *complete induction* attained popularity in Germany after Dedekind used it in a paper of 1887 (Burton, page 440; Boyer, page 404).

The references for the above citations are Boyer [70] and Burton [89]. See also [92] for more on the history of the name "mathematical induction". One might note that the method of mathematical induction still is occasionally referred to as "complete induction" (*e.g.*, in [556]) or "full induction."

Near the end of the 19th century, David Hilbert (1862–1943) was writing a book [269], attempting to establish geometry based not on "truths", but on axioms. Gottlob Frege (1848–1925) had been studying mathematical logic and communicated regularly with Hilbert. Much debate arose about what axioms were, what they "should" be, and what "truth" in mathematics is. (See [451] for an account of

---

[1]Used with kind permission from Jeff Miller

the discussions between Frege and Hilbert regarding axioms.) Frege was essentially trying to reduce mathematical reasoning to purely logical reasoning. For some kinds of reasoning, a "second-order" kind of logic was necessary, but Frege wanted (perhaps) to rephrase mathematical induction that did not rely on second-order logic. To this end, he used terms like "ancestors" (well, in German, he must have used "Vorfahren" or something similar) and "ancestor induction". The basic idea was to extend reasoning of the form: "Ole is an ancestor of John, and John is an ancestor of David, so Ole is an ancestor of David." [These, inasmuch as my parents tell me, are accurate statements.]

In [128] [thanks to Dr. Peter Morton for supplying this reference] Demopoulos mentions that Crispin Wright presented an argument that *Hume's principle* [the number of elements in a set S is equal to the number of elements T if and only if there is a one-to-one correspondence between elements of S and T] implies one of Peano's axioms: "...in the context of the system of second-order logic of Frege's *Bereffsschrift*, Peano's second postulate [every natural number has a successor] is derivable from Hume's principle." Demopoulos continues to mention "...that Frege discovered that, in the context of second-order logic, Hume's principle implies the infinity of the natural numbers, *Frege's theorem*." (If the reader wants another perspective, readable but confusing, on these matters, see [556].)

Ernst Zermelo (1871–1953), Richard Dedekind (1831–1916), Bourbaki, Bertrand Russell (1872–1970), and many others continued the debate regarding assumptions about the natural numbers. Concepts like "well-ordering" and "Axiom of Choice" were also introduced in an attempt to logically legitimize what students of mathematics all "know" to be "true" about natural numbers. For present purposes, assume that all the necessary groundwork has been done to establish that present assumptions (or Peano's assumptions) are reasonable. For more facts and debates regarding the history of induction, see [175], [197], [300], [523], [541], [542], and [581].

The interested reader may pursue these discussions from a model theoretic perspective as well; the mathematical logician Leon Henkin [265] examines Peano models in contrast to induction models (those with only the induction axiom). Classifying algebraic systems according to the set of axioms that generate the system, and examining which functions arise from "primitive recursion", is too deep a subject to entertain here. The reader is recommended to see some of the popular literature that is referred to in Section 2.2. The theory can get quite complex; it is hard to say what the *best* approach is.

Instead of being drawn into further discussions regarding epistemology and philosophy, this discussion is concluded with a quotation from Ernst Mach, as found in [433], regarding Jacques Bernoulli (1667–1748):

> Jacques Bernoulli's method is important to the naturalist. We find what seems to be a property A of the concept B by observing cases $C_1, C_2, C_3, \ldots,$. We learn from Bernoulli's method that we should not

attribute such a property A, found by incomplete, non-mathematical induction, to the concept B, unless we *perceive* that A is *linked* to the characteristics of B and is *independent* of the variation of the cases. As in many other points, mathematics offers here a model to natural science.

1.9 Mathematical induction in modern literature

> *One of the chiefest triumphs of modern mathematics consists in having discovered what mathematics really is.*
>
> —Bertrand Russell
>
> *International Monthly*, 1901.

In any mathematics textbook that contains a section on induction, there is usually a collection of problems, a handful of which are now used repeatedly in nearly every such text. There are perhaps about a hundred problems *in toto* that might, due to their frequency, be called "standard"; virtually all problems appearing in modern texts are adaptations of these. A few books have been devoted exclusively to induction. This chapter contains a brief overview of books specifically on induction, articles about induction, and typical books containing chapters or sections on induction, primarily from the last century; for articles concerning mathematical induction before that, see Section 1.8 on the history of mathematical induction. This overview hopefully contains most major works and a few less well-known. Aside from references given here, there are likely thousands more articles concerning induction, so parts of this review can never hope to be comprehensive. On 11 February 2009, MathSciNet showed 395 matches to "mathematical induction", 74 of which were in the title. There were 1436 titles containing simply the word "induction", most in well respected refereed journals. The number of books or articles that use induction in them is probably in the hundreds of thousands.

My own introduction to induction in high school was from *Mathematical Induction and Conic Sections* [550], a booklet excerpt from a textbook. That booklet has only a few pages on induction, but it lists 39 exercises. There have been a few other books specifically on induction, most of which I only recently became aware of, and none of which seem to be in print any more.

In 1958, a 63-page book [388] by Mitrinović on mathematical induction appeared in Serbo-Croatian, the last chapter of which contains a short history of induction. The translated title was *The Method of Mathematical Induction*. A dozen years later, the same author came out with another book [389], about half of which is reportedly devoted to problems solvable by mathematical induction, (also in Serbo-Croatian), however I have not seen either.

In 1956, I. S. Sominskii's Russian text [498] on induction was already enjoying its fourth edition. In 1959, he published *Metod Matematicheskoii Induktsii*; this was translated into English and published in 1961 as *The Method of Mathematical Induction* [499], a 57-page collection of theorems and 52 problems; most appear with helpful, complete solutions. A reviewer named N. D. Kazarinoff reviewed that book for *Math Reviews* [27 5669] and wrote "In addition to a high school training in these subjects, the reader must have good eyesight: symbols in formulas are often about the size of periods." This book has enjoyed dozens of editions in various languages, including Russian (*e.g.*, [498], 4th ed., 48 pages), German (*e.g.*, [501], 13th ed., 55 pp., [287], 120 examples, 183 pp., with two other authors), Spanish ([502], 2nd ed., 62 pp.), and Turkish (*e.g.*, [500], viii+72 pp.).

In 1964, a 55 page booklet, *Mathematical Induction* [582], by Bevan K Youse [note: there is no period after the "K" in his name] appeared, repeating many of the problems in Sominskii's book, but with a few interesting additions. Youse's book has 72 problems, most of which now commonly appear in today's texts without reference. There are only 29 complete solutions.

In 1979, the 133-page book *Induction in Geometry* [220], published in Moscow, contains inductive proofs of many difficult theorems in geometry (only a few of which are covered in this volume). This book is no longer in print and is hard to find [thanks to R. Padmanabhan for giving me his copy], but, in my opinion, well worth an effort to locate.

Another, more recent book is *Manuel d'Induction Mathématique* (Handbook of mathematical induction) by Luís Lopes [350]; this book has 100 problems complete with solutions (in French), many of which are also standard and easy; however, the author does not shy away from some really challenging solutions. The exercises occupy just over a dozen pages, with the bulk of the 127 pages being solutions.

The principles behind mathematical induction are studied in almost every logic text or set theory text (for example, in [95], [289]). There are numerous articles on mathematical induction from different points of view in logic, language, model theory, universal algebra, or philosophy (*e.g.*, [200] on predicate synthesis, [265] on model theory, [383] on formal theory of finite sets, [145] on variable free algebra and Galois connections, [111] on material implication, [139] on predicates on any well-founded set, [471] on ramified type theory as an adequate formalization of predictive methods).

More general works, like [181], [274], [400], and [556] give broad historical perspective in the modern foundations of mathematics and induction. History of mathematics texts almost always describe how induction arrived on the mathematical scene (*e.g.*, [180]) and how it relates to other areas of mathematics.

Hundreds of references have been used in assembling the collection of exercises here. Many problems using mathematical induction are now part of the folklore, but unusual problems are referenced. Here are a few kinds of books that deal explicitly with mathematical induction.

Many texts in discrete mathematics have sections on induction (*e.g.*, see [10], [8], [33], [38], [52], [55], [83], [147], [195], [222], [238], [292], [299], [355], [363], [373],

[375], [431], [462], [464], and [535]). Of these volumes, [238], [292], and [462] are very popular in North America, probably because of the colossal amount of mathematics (including induction, of course) contained in each.

Closely related are books on combinatorics, many with a prodigious array of applications of induction (for example, see [6], [77, 78], [94], [255], [266], [455], or [506]). Lovász's now classic compilation, *Combinatorial Problems and Exercises* [354] is also an abundant source of wonderful inductive proofs, many highly nontrivial. Also, for induction in advanced combinatorics, see [58].

One might be delighted to know that even some calculus books (for example, the classic book by Apostol [20], and the more modern text by Trim [534]) devote a section to induction. Books on programming cover induction, as well (see, *e.g.*, [483]). Texts that concentrate on mathematical problem solving often contain sections on induction and are a rich source of problems. In particular, Engel's book *Problem-solving Strategies* [161] contains a chapter on induction in which 39 exercises and solutions are discussed; hundreds of solutions using induction also occur throughout the book. [Some solutions are little on the brief side, but considering the plethora of problems that are actually solved, Engel's work might be considered as one of the richest sources for problem solving available today.] Three more references of this type that come to mind are [47], [124] and [461]. Such texts are an invaluable resource for mathlete training. Other works concentrate on aspects of teaching induction (*e.g.*, [194], [382], [490], and [516] to name but a few).

For anyone wanting a general insight into how to conjecture and prove mathematical statements, particularly by induction, one might be pleasantly rewarded with a look at Pólya's books [433], [434], [435]. A fairly recent collection of non-trivial problems over a broad range of fields, many of which employ induction, quickly became one of my favorites: *The Art of Mathematics: Coffee Time in Memphis* [61], by Béla Bollobás.

Leo Zippin's classic monograph *Uses of Infinity* [589] shows off induction in various settings, most notably in proving limits. In *What is Mathematics?* [120, §§1.2.1– 1.2.7, pp. 9–20] by Richard Courant and Herbert Robbins one finds a particularly easy-to-read discussion of mathematical induction. (Zippin, [589, p. 106] also refers the reader to the Courant and Robbins book.) Another, more recent delightful problem book (which has a section on induction, and various induction problems throughout) is *Winning Solutions*, by Edward Lozansky and Cecil Rousseau [357], a collection of contest problems and their solutions that might complement any library.

Some books on recreational mathematics and popular science include discussion of mathematical induction. One of the most noteworthy of these is Martin Gardner's *Penrose Tiles to Trapdoor Ciphers*, [214, Ch. 10, pp. 137–149], a chapter called "Mathematical induction and colored hats". Another, [560], discusses Penrose's non-computability of thought, consciousness, self-referencing, and discusses mathematical thinking viz-a-viz Gödel's theorem, Poincaré, and Galois, and somehow manages to tie in mathematical induction.

There has been some work on computer programs designed to produce inductive proofs (also called "automated induction"). See, for example, [28], [68] (using SPIKE), [87], [187], [303], [398], [543], [557] (using LISP), and [588]. There is a great deal more literature on this subject, as proofs by mathematical induction are central in many computer science and AI applications. An older article [139] highlights the importance of mathematical induction in termination of programs and properties of programming languages.

A special kind of mathematical induction, called "transfinite induction" (see Section 4.2) is closely related to complexity theory in [489]. The invariance theorem [whatever that is] and induction are studied in [278]. Induction and program verification and modelling are also closely related and many books and articles discuss this relation (see, *e.g.*, [302], [361], [452]).

Many texts with "finite mathematics" in the title contain sections on mathematical induction, as induction is often taught in high school and beginning university math courses. Various other subject areas (for example, number theory, algebra, and graph theory) use induction quite heavily, and some related texts contain sections on induction (*e.g.*, [150], [566]).

One can find numerous articles on induction in various popular journals, too; for example, see [82] or [265]. The article by Dragos Hrimiuc [280] is short (3 pages!) and easy to read, yet is a substantial introduction to the subject. Some are from a historical perspective (*e.g.*, [91], [175], [197], [300], [523], [541], [542], [581]). There are a variety of journal articles on induction in general (*e.g.*, [138], [175], [237] (in Spanish), [262] (in Japanese), [290] (in Chinese), and [504]).

Induction is not only applied in discrete situations. Analysis and induction are more closely related than one might think (see [155] for some classical connections). In fact, there is a kind of *continuous*, or *non-discrete* induction at play. Some of the first (and most referred to) articles in this area seem to be by Pták [441, 442] (with the Banach fixed point theorem, Banach algebra, closed graph theorem, Newton's process, and more); see also [26], [25], [27], [578], [579]. For those who can read Russian and are interested in differential equations, see [318].

Induction is ubiquitous. In fact, in any volume of a mathematics journal (popular or specialized) it seems rare *not* to find at least one proof by induction!

Incidentally, it might come as a bit of a surprise that the word "induction" does not seem to be mentioned in George Gamow's classic book *One Two Three ... Infinity*—one can be comforted, though, by the knowledge that Gamow [205, pp. 19–23] explains well two problems that are solved inductively.

Finally, there is the internet. In September 2005, a *Google* search for "mathematical induction" produced "about 2,610,000" hits! For some reason, this number dropped to 436,000 as of January 2009. Any ranking of these sites is hopeless, however, many seem to be rather well done. The sites seem to range from the very elementary to some collections of somewhat challenging problems.

Chapter 2

Foundations

The reasoning of mathematicians is founded on certain and infallible principles. Every word they use conveys a determinate idea, and by accurate definitions they excite the same ideas in the mind of the reader that were in the mind of the writer. When they have defined the terms they intend to make use of, they premise a few axioms, or self-evident principles, that every one must assent to as soon as proposed. They then take for granted certain postulates, ..., and from these plain, simple principles they have raised most astonishing speculations, and proved the extent of the human mind to be more spacious and capacious than any science.

—John Adams,

Diary.

This chapter attempts to put mathematical induction (MI) on a sound logical ground, and the principle of mathematical induction is described more formally. The usual starting point is a set of axioms called "Peano's axioms", the last of which is, essentially, the principle of mathematical induction. Using these axioms one can prove many of the basic properties of natural numbers, perhaps a reasonable place to start in mathematics.

2.1   Notation

The notation used in this text is fairly standard. If S is a set, "$x \in S$" denotes that x is an element of S. The notation "$x, y \in S$" is a common shorthand for "$x \in S$ and $y \in S$". Use "$T \subset S$" or "$T \subseteq S$" to denote that T is a *subset* of S, that is, every element of T is an element of S; in either notation, T can be equal to S. If $T \neq S$, yet $T \subseteq S$, then T is a *proper subset* of S (denoted by $T \subsetneq S$, if necessary).

Though they have yet to rigorously defined, let $\mathbb{N} = \{1, 2, 3, \ldots\}$ denote the set of natural numbers. The empty set is denoted by $\emptyset$ (this is not a computer 0).

Note: Many authors, especially combinatorists, set theorists, and those trained in the British system, include the number 0 in the natural numbers; here 0 is *not* included, and so where ever confusion can arise, different notation is used. In some schools, the set $\mathbb{W} = \{0, 1, 2, 3, \ldots\}$ is called the set of *whole numbers*, though the expression "non-negative integers" is used here. [I was taught to remember the difference by observing that the whole numbers had an extra "hole".] To avoid confusion, one might also say "positive integers" rather than "natural numbers".

There is, however, good reason to include 0 in the natural numbers (as one might witness with ordinal numbers and the Zermelo hierarchy—set theoretic interpretations of counting numbers). The tradition of natural numbers without 0 is a tradition followed in many North American schools. [I deliberated for some time on this choice of notation, and I am still not sure that I have made the correct choice; from a mathematical perspective, it seems to make more sense to include 0.]

The symbols $\mathbb{Z}$, $\mathbb{Q}$, $\mathbb{R}$, and $\mathbb{C}$ denote the sets of integers, rationals, reals, and complex numbers, respectively. The notation $\mathbb{Z}^+ = \{1, 2, 3, \ldots\}$ is often used to indicate the set of the positive integers; this notation is somewhat universal, and hence is occasionally used instead of $\mathbb{N}$ to avoid confusion (regarding the inclusion of 0). Throughout, unless otherwise noted, all variables in this text are non-negative integers. For statements p and q, use the shorthand $p \rightarrow q$ to abbreviate "if p then q", or "p implies q", and $p \Rightarrow q$ for "p logically implies q". In mathematics, one often confuses their meanings; the "$\rightarrow$" is implication in the object language, and "$\Rightarrow$" is in the metalanguage. Many mathematicians use the double arrow to mean simply "implies", perhaps to differentiate from the single arrow used for functions.

The symbol $\forall$ means "for all" and the symbol $\exists$ means "there exists"; as handy as these quantifiers are, their use is limited in this text since they tend to make simple statements unreadable to some non-mathematics students. The symbols "$\wedge$" and "$\vee$" are occasionally used to represent "and" and "or" respectively. If a paragraph is followed by "$\square$", then this indicates the end of a proof. The expression "iff" is an abbreviation for "if and only if".

2.2 Axioms

There are many statements in mathematics that are not proven, but are simply assumed to be true. For example, in Euclidean geometry, it is assumed that *for any pair of distinct points in the plane, there is a unique line that contains them.* Some people find this to be a reasonable assumption, however, might have difficulty proving such an assumption (whatever that might mean).

A statement that is assumed to be true (but not necessarily provable) is called an *axiom* or *postulate*. To state an axiom, one often requires that certain terms are accepted without meaning. For example, undefined terms might include "element", "set", "point", "line", "plane", "is incident with", and "is in between". Having established the terms, one might agree on rules of logic (where the default is usually to simply accept standard Boolean logic, with or without quantifiers, for example, and the standard connectives). With these in place, one can state an axiom, either a property of a term, (*e.g.*, "there exists something called a point") or a relation between terms (*e.g.*, "there exists a set which does not contain any elements").

A *theorem* is a statement that then follows deductively from the axioms, either directly or indirectly using other theorems. A *lemma* is a "lesser" theorem, often used to help prove a more significant theorem. (The plural for "lemma" is "lemmata" or more simply, "lemmas".) A *corollary* is a statement that is a consequence of a theorem; usually a corollary follows from a theorem in a fairly obvious way.

When speaking of the validity of a particular result, one actually only refers to whether or not the result follows from axioms. In Edmund Landau's book *Grundlagen der Analysis* [339], he begins with axioms and derives most of the foundations of arithmetic. The approach here is similar, beginning with the same set of axioms.

Any discussion in set theory, logic, geometry, number theory, or even mechanics, usually presumes that a set of axioms has been agreed upon. How would a scientist decide on a list of absolute truths (axioms) from which to develop a particular system? Of any collection of axioms a scientist might assemble, there are two properties of the collection that may be desirable:

First, insist that the list is as short as possible. Perhaps most importantly, it would not be desirable to want so many axioms that from any (or all) of the axioms, one could derive a *contradiction* (that is, both a statement and its negation). If one can not deductively derive any contradictions from a particular collection of axioms, the collection is called *consistent*, and the system that rests upon these axioms is also called consistent or *sound*.

If a particular system is sound, it might be very difficult to prove such a fact. Even proving the inconsistency of a system by exhibiting a contradiction might be an impossible task.

One reassurance of soundness is to find a *model* or *interpretation* that realizes all of the axioms. In fact, depending on your assumptions about the world, finding a model is sometimes *proof* that a set of axioms is consistent—as it is in many mathematical situations. For example, the naive image of standard Euclidean geometry seems to be a model that satisfies the postulates in Euclid's *Elements of Geometry* (written around 300 B.C.). If a collection of axioms is consistent, any subcollection is also consistent. Different models for geometries have been found that realize all but the fifth of Euclid's postulates, (*e.g*, elliptic or hyperbolic geometries). See [274, pp. 88–93] for a lively, easy to read discussion of the discoveries that led to various "non-Euclidean" geometries.

Any attempt to construct a set of consistent axioms might start by selecting a very large set of axioms, deriving some contradiction, then throwing out one possibly offending axiom, and trying again, continuing until no contradictions are derivable. It is yet another problem, however, to show that contradictions can not arise at any one stage. To support a claim that one particular axiom is consistent with a given set of axioms, one might assume its negation and try to prove a contradiction. Enough about consistency for the moment.

The second desirable property for a collection of axioms to satisfy is that the collection of axioms is large enough so as to be able to derive all truths in the system from the axioms. Such an axiomatic system is called *complete*. In Euclid's thirteen books of *Elements of Geometry* is a set of five postulates, however it seems that Euclid's postulates are not complete (see [264, p. 1636]) for what is now called "Euclidean geometry". Hilbert's set of axioms for geometry [269] arose out of efforts to find "completeness", efforts which were destined for failure as well.

There has been much discussion about what sort of minimal collection of axioms "should be" agreed upon so that one can do, say, set theory, geometry, or arithmetic. In this text, to describe the natural numbers, a set of axioms (now commonly thought to be not too problematic), Peano's axioms, is the starting point. The standard axioms of ZFC are implicitly assumed here. (See appendix IV for a list of ZFC axioms and further discussion about consistency and completeness.)

2.3   Peano's axioms

In the 19th century, Giuseppe Peano (1858–1932), a professor from the University of Turin (Italy), published (in *Formulario Matematico*, 1889), a collection of axioms for the natural numbers $\mathbb{N}$, defined here to be $\mathbb{Z}^+$.

Peano received the axioms from Dedekind in a letter, and he even recognized this in his publication, however, the term "Peano's axioms" has survived to refer to Dedekind's axioms (*e.g.*, see Pollock's book [432], though Pollock does not give references). This fact doesn't seem to be widely cited in other textbooks. Peano's axioms are generally now accepted by the mathematical community as a starting point for arithmetic.

To describe these axioms, common function notation is used: The cartesian product of sets S and T is $S \times T = \{(s,t) : s \in S, t \in T\}$. A *function* f from a domain S to T (written $f : S \longrightarrow T$) is a subset $f \subset S \times T$ so that for every $s \in S$, there is exactly one $t \in T$ so that $(s,t) \in f$. In this case, write $f(s) = t$. (See Section 18.2 for more details on functions.)

Peano's axioms are usually given as a list of five, yet one more appears in his writings, one roughly equivalent to "$\mathbb{N}$ is a class of things called numbers." (See [340, p. 1872] for a translation; many other wonderful articles regarding axioms are also found in the same collection.) His fifth axiom is really the principle that is now known as "mathematical induction".

Peano's axioms:

P1 $1 \in \mathbb{N}$.

P2 There is a function $\delta : \mathbb{N} \to \mathbb{N}$ where for each $x \in \mathbb{N}$, $\delta(x) = x' \in \mathbb{N}$ is called the successor of x.

P3 For every $x \in \mathbb{N}$, $x' \neq 1$.

P4 If $x' = y'$, then $x = y$.

P5 If $S \subset \mathbb{N}$ is such that
     (i) $1 \in S$, and
     (ii) for every $x \in \mathbb{N}$, $x \in S \to x' \in S$,
  then $S = \mathbb{N}$.

A proof employing P5 is said to be "inductive" or "is by induction". The step P5(i) is called the *base step* and P5(ii) is called the *inductive step*. Some philosophers call these two parts the *basic clause* and the *inductive clause* (for example, see [29, p. 468]). The antecedent "$x \in S$" in P5(ii) is called the *inductive hypothesis* (or sometimes *induction hypothesis*.)

2.4   Principle of mathematical induction

This section contains a very brief formulation of what is called the "principle of mathematical induction" as it is applied to various statements, instead of just for sets. Applications and various forms of this principle are discussed again in Chapter 3.

There are many forms of mathematical induction—weak, strong, and backward, to name a few. In what follows, n is a variable denoting an integer (usually nonnegative) and $S(n)$ denotes a mathematical statement with one or more occurrences of the variable n. The following is the standard presentation of mathematical induction, also called "weak mathematical induction". Observe that $\delta(x) = x' = x + 1$ is a successor function satisfying P2, P3, and P4 (it is shown in Theorem 2.5.4 that this is the only successor function on natural numbers).

Theorem 2.4.1 (Principle of Mathematical Induction (MI)).
If $S(n)$ is a statement involving n and if
     (i) $S(1)$ holds, and
     (ii) for every $k \geq 1$, $S(k)$ implies $S(k+1)$,
then for every $n \geq 1$, the statement $S(n)$ holds.

The two stages (i) and (ii) in a proof by MI are still called the *base step* (in which the *base case* is proved), and the *inductive step*, respectively. In (ii), $S(k)$ is called the *inductive hypothesis* (also called the *induction hypothesis*). Depending on the definition of the natural numbers used by different authors, the base step might also be $S(0)$.

Proof of MI from Peano's axioms: Define $A = \{n \in \mathbb{N} : S(n) \text{ is true}\}$. Then by (i), $1 \in A$. By (ii), if $k \in A$, then $k + 1 \in A$. So by P5, $A = \mathbb{N}$, proving MI.   □

2.5   Properties of natural numbers

The next few results (proved from Peano's axioms) will enable one to talk about $\mathbb{N}$ in more familiar terms.

First observe that for any successor function $\delta(x) = x'$, to each x there is a unique x', and hence $[x = y] \to [x' = y']$.

Lemma 2.5.1. *For any $x, y \in \mathbb{N}$, $[x \neq y] \to [x' \neq y']$.*

Proof: This is just the contrapositive of P4. [If a statement is of the form "if P, then Q, the contrapositive of the statement is "if not Q, then not P". The two statements are logically equivalent.]   □

Theorem 2.5.2. *If $x \in \mathbb{N}$ then $x' \neq x$.*

Proof: (By induction) Let $A = \{x \in \mathbb{N} : x' \neq x\}$.
BASE STEP: By P3, $1 \in A$.

INDUCTIVE STEP: Assume that $y \in A$, that is, $y' \neq y$. Lemma 2.5.1 then implies $(y')' \neq y'$, and so $y' \in A$.

Hence, by P5, $A = \mathbb{N}$.   □

The next result shows that predecessors are unique.

Theorem 2.5.3. *If $x \in \mathbb{N}$ and $x \neq 1$, then there is a unique y so that $x = y'$.*

Proof: (By induction) Let

$$A = \{x \in \mathbb{N} : x = 1 \text{ or there exists } y \in \mathbb{N} \text{ so that } x = y'\}.$$

BASE STEP: $1 \in A$ by definition.
INDUCTIVE STEP: Suppose that $x \in A$. Then either $x = 1$ or $x = y'$ for some $y \in \mathbb{N}$. To be shown is that $x' \in A$. If $x = 1$, then $x' \in \mathbb{N}$; if $x = y'$, then $x \in \mathbb{N}$ by P2. Hence, in any case, $x \in \mathbb{N}$, and by the definition of A, $x' \in A$.

Therefore, by P5, $A = \mathbb{N}$. Thus, for any $x \neq 1$, there is some $y \in \mathbb{N}$ so that $x = y'$. The uniqueness of y follows from P4.   □

The next theorem shows that the successor function is what one might expect it to be, namely $x' = x + 1$. In this theorem, a function is defined from $\mathbb{N} \times \mathbb{N}$ to $\mathbb{N}$, that is, it takes ordered pairs and returns natural numbers. For such a function f, it is standard to write $f(x, y)$ instead of the more proper $f((x, y))$.

Theorem 2.5.4. *There exists a unique function $f : \mathbb{N} \times \mathbb{N} \longrightarrow \mathbb{N}$ so that for all $x, y \in \mathbb{N}$,*
  (a) $f(x, 1) = x'$;
  (b) $f(x, y') = (f(x, y))'$.

Proof: There are two things to show, existence and uniqueness.

(Existence) A function from $\mathbb{N} \times \mathbb{N}$ to $\mathbb{N}$ can be described by an infinite matrix:

$$\begin{bmatrix} f(1,1) & f(1,2) & f(1,3) & f(1,4) & \cdots \\ f(2,1) & f(2,2) & f(2,3) & f(2,4) & \cdots \\ f(3,1) & f(3,2) & f(3,3) & f(3,4) & \cdots \\ f(4,1) & f(4,2) & f(4,3) & f(4,4) & \cdots \\ \vdots & \vdots & \vdots & \vdots & \end{bmatrix}$$

The idea in this existence part of the proof is to create this matrix row by row. It will suffice to show that the first row can be constructed so that (a) and (b) hold, and then to show that an arbitrary row can be constructed from a previous one.

Define B to be the set of all $x \in \mathbb{N}$ so that one can find a set of function values $\{f(x, i) : i \in \mathbb{N}\}$ so that for all y, both (a) and (b) hold (for the fixed x). To be shown is that $B = \mathbb{N}$.

BASE STEP ($x = 1$): For every $y \in \mathbb{N}$, define $f(1, y) = y'$. By definition, $f(1, 1) = 1'$, and so (a) holds with $x = 1$. Also, by definition, $f(1, y') = (y')' = (f(1, y))'$, and so (b) holds with $x = 1$.

INDUCTIVE STEP: Suppose that $x \in B$. Then $f(x, y)$ is defined for all $y \in \mathbb{N}$ so that (a) and (b) hold. Define $f(x', y) = (f(x, y))'$. Then, by definition, $f(x', 1) = (f(x, 1))' = (x')'$ and so (a) holds with x' in place of x. Also

$$\begin{aligned} f(x', y') &= (f(x, y'))' & \text{(by definition)} \\ &= ((f(x, y))')' & \text{(by (b) since } x \in B) \\ &= (f(x', y))' & \text{(by definition),} \end{aligned}$$

and so (b) holds for x'. Thus, $x' \in B$, completing the inductive step.

By induction, $B = \mathbb{N}$, finishing the existence part of the proof.

(Uniqueness) Suppose that f is defined so that for all $x, y \in \mathbb{N}$ both (a) and (b) hold and also suppose that g is a function satisfying the corresponding equalities:
  (a') $g(x, 1) = x'$, and

(b') $g(x, y') = (g(x, y))'$

Let $x \in \mathbb{N}$ be fixed and define the set $A_x = \{y \in \mathbb{N} : f(x, y) = g(x, y)\}$. Induction is used to first to show that $A_x = \mathbb{N}$.

BASE STEP: $1 \in A_x$ since $f(x, 1) = x' = g(x, 1)$.

INDUCTIVE STEP: Suppose that $y \in A_x$. that is, $f(x, y) = g(x, y)$. Then by (b) and (b'), $f(x, y') = (f(x, y))'$ and $g(x, y') = (g(x, y))'$. Hence, by P4, $f(x, y') = g(x, y')$. So $y' \in A_x$.

Hence by P5, $A_x = \mathbb{N}$. Since x was arbitrary, this completes the uniqueness part of the proof, and hence the entire proof. $\qquad\square$

The function f above is better known by its common notation, $f(x, y) = x + y$, and hence the successor function is $x' = x + 1$ (as one might expect). One can now freely use the result of the previous theorem, namely, the existence of the unique function f defined so that

(a) $f(x, 1) = x'$;
(b) $f(x, y') = (f(x, y))'$;
(c) $f(1, y) = y'$;
(d) $f(x', y) = (f(x, y))'$,

where (c) and (d) are from the way f was defined in the existence part of the proof; translating (a)–(d) into common notation using the "+" sign,

(a') $x + 1 = x'$;
(b') $x + y' = (x + y)'$;
(c') $1 + y = y'$;
(d') $x' + y = (x + y)'$.

The expression "$x + y$" is called the sum of x and y, and the process of computing $x + y$ is called addition.

Theorem 2.5.5. *Addition of natural numbers is associative, that is, for every $x, y, z \in \mathbb{N}$,*

$$(x + y) + z = x + (y + z).$$

Proof: Let x and y be fixed natural numbers and put

$$A = \{z \in \mathbb{N} : (x + y) + z = x + (y + z)\}.$$

BASE STEP: $1 \in A$ because

$$
\begin{aligned}
(x + y) + 1 &= (x + y)' && \text{(by (a'))} \\
&= x + y' && \text{(by (b'))} \\
&= x + (y + 1) && \text{(by (a')).}
\end{aligned}
$$

INDUCTIVE STEP: Suppose that $z \in A$. Then

$$
(x + y) + z' = ((x + y) + z)' \qquad \text{(by (b'))}
$$

$$
\begin{aligned}
&= (x + (y + z))' && \text{(because } z \in A) \\
&= x + (y + z)' && \text{(by (b'))} \\
&= x + (y + z') && \text{(by (b')),}
\end{aligned}
$$

and so $z' \in A$, completing the inductive step. Hence, by P5, $A = \mathbb{N}$.   □

Since addition is associative, it matters not which adjacent terms are added first, and hence parentheses are not needed.

For natural numbers $x_1, x_2, x_3, \ldots$, define inductively

$$
x_1 + x_2 + \ldots + x_n = (x_1 + x_2 + \ldots + x_{n-1}) + x_n.
$$

To abbreviate the left side, one uses so-called *sigma notation*:

$$
x_1 + x_2 + \cdots + x_n = \sum_{i=1}^{n} x_i.
$$

Such notation extends in the obvious way, for example,

$$
\sum_{j=3}^{7} y_j = y_3 + y_4 + y_5 + y_6 + y_7.
$$

For later reference, a formal definition of the sigma notation is given:

Definition 2.5.6. Let $x_1, x_2, x_3, x_4, \ldots$ be a sequence of natural numbers. Define $\sum_{i=1}^{1} x_i = x_1$ and recursively define for each $n > 1$,

$$
\sum_{i=1}^{n} x_i = \left(\sum_{i=1}^{n-1} x_i \right) + x_n.
$$

Generalizing this slightly, for any $j \in \mathbb{N}$, define $\sum_{i=j}^{j} x_i = x_j$ and recursively define for each $n > j$,

$$
\sum_{i=j}^{n} x_i = \left(\sum_{i=j}^{n-1} x_i \right) + x_n.
$$

Finally, define the sum over an empty set of indices to be zero.

According to [556], the next theorem is due to H. Grassman, from *Lehrbuch der Arithmetik*, 1861 (though I have not seen the original proof).

Theorem 2.5.7. *Addition in natural numbers is commutative, that is, for every* $x, y \in \mathbb{N}$,

$$
x + y = y + x.
$$

Proof: Let $x \in \mathbb{N}$ be fixed, and put $A = \{y \in \mathbb{N} : x + y = y + x\}$.

BASE STEP: $1 \in A$ because by (a') and (c') respectively,

$$x + 1 = x' = 1 + x.$$

INDUCTIVE STEP: Suppose that $y \in A$. Then

$$
\begin{aligned}
x + y' &= (x + y)' && \text{(by (b'))} \\
&= (y + x)' && \text{(because } y \in A) \\
&= y' + x && \text{(by (d'))},
\end{aligned}
$$

and so $y' \in A$, completing the inductive step.

Hence, by P5, $A = \mathbb{N}$, finishing the proof of the theorem.   □

Theorem 2.5.8. *For every $x, y \in \mathbb{N}$, $x + y \neq x$.*

Proof: Let A be the set of all those $x \in \mathbb{N}$ such that for every $y \in \mathbb{N}$, $x + y \neq x$.

BASE STEP: By P3 and property (c'), for every $y \in \mathbb{N}$, $1 + y \neq 1$, and so $1 \in A$.

INDUCTIVE STEP: Assume that $x \in A$, that is, x is such that for any $y \in \mathbb{N}$, $x + y \neq x$. If for some y, $x' + y = x'$ holds, then by property (b'), it follows that $(x + y)' = x'$, and so by P4, $x + y = x$, contradicting that $x \in A$; hence conclude that $x' + y \neq x'$, and thus $x' \in A$.

By P5, $A = \mathbb{N}$.   □

The next sequence of exercises establishes the properties for the operation known as "multiplication" of natural numbers; they are proved in a very similar manner to those above. The content of the exercises in this chapter are really theorems whose proofs are perhaps boring or repetitive and are not intended as the first exercises regarding induction that a student might see.

Exercise 1. *Prove that there exists a unique function $g : \mathbb{N} \times \mathbb{N} \longrightarrow \mathbb{N}$ so that for all $x, y \in \mathbb{N}$*

  (e) $g(x, 1) = x$;
  (f) $g(x, y') = (x + g(x, y))$.

Replace the notation $g(x, y)$ by $x \cdot y$, the multiplication of x and y, and then abbreviate $x \cdot y$ by xy.

Oddly enough, it helps to first prove distributivity before associativity of multiplication.

Exercise 2. *Prove that the distributive laws hold, that is, prove that for any $x, y, z \in \mathbb{N}$,*

$$x(y + z) = xy + xz,$$

and

$$(x + y)z = xz + yz.$$

Exercise 3. *Prove that the general distributive laws hold for natural numbers, that is, for $x_1, x_2, \ldots, x_n, c \in \mathbb{N}$,*

$$c \left(\sum_{i=1}^{n} x_i \right) = \sum_{i=1}^{n} c x_i.$$

Using one of the basic distributive laws, associativity comes fairly easily.

Exercise 4. *Prove that multiplication of natural numbers is associative,   that is, prove that for any $x, y, z \in \mathbb{N}$,*

$$(xy)z = x(yz).$$

Definition 2.5.9. The notation $\prod_{i=1}^{n} x_i$ is defined recursively by $\prod_{i=1}^{1} x_i = x_1$, and for $n \geq 1$,

$$\prod_{i=1}^{n+1} x_i = \left(\prod_{i=1}^{n} x_i \right) x_{n+1}.$$

Since multiplication of natural numbers is associative, if the x_i's are natural numbers, the meaning of

$$\prod_{i=1}^{n} x_i = x_1 x_2 \cdots x_n$$

is unambiguous. Finally, define the product over an empty set of vertices to be equal to one, that is,

$$\prod_{i \in \emptyset} x_i = 1.$$

Note that when all x_i's are equal, the first simple definition of exponentiation is given (for positive integers): define $x_1 = x$, and for $n > 1$, having defined x^{n-1}, define $x^n = x^{n-1} \cdot x$.

The discussions above are just a beginning to thoroughly define the real numbers, or to check all of the properties of or operations on the natural numbers. A few of these are given as exercises.

2.6   Well-ordered sets

Given a set S, a *binary relation on S* is a subset of the cartesian product $S \times S = \{(a,b) : a \in S, b \in S\}$.

A binary relation R on S is

- *reflexive* iff for every $x \in S$, $(x,x) \in R$.

- *symmetric* iff for every $x, y \in S$, $[(x,y) \in R] \to [(y,x) \in R]$.

- *antisymmetric* iff for every $x, y \in S$, $[((x,y) \in R) \wedge (x \neq y)] \to [(y,x) \notin R]$.

- *transitive* iff for every $x, y, z \in S$, $[((x,y) \in R) \wedge ((y,z) \in R)] \to [(x,z) \in R]$.

A binary relation R on S is a *partial order* if and only if R is reflexive, antisymmetric, and transitive. If R is a partial order on S, the set (S, R) is called a *partially ordered set* , abbreviated, *poset*.

One can write $x \leq_R y$ if $(x,y) \in R$, and if also $x \neq y$, write $x <_R y$ and say that x is less than y. The notation xRy is also quite common. If the relation R is implicitly understood, simply write $x \leq y$ or $x < y$, rather than $x \leq_R y$ or $x <_R y$ (or xRy). (Some texts define a partial order without reflexivity, and so a total order is then always written with "$<$" rather than "$\leq$"; such notation is often practiced regardless of whether or not reflexivity is insisted on in the definitions, since a relation without reflexive property determines precisely one with reflexivity.)

A *least element* in a partially ordered set $(P, \leq)$ is an element $x \in P$ so that for every $y \in P$, $x \leq y$. For example, the poset $\{(a,c),(b,c)\}$ has no least element (instead it has two minimal elements: a and b) but the poset $\{(x,y),(x,z)\}$ has a least element x. If a least element exists, then it is unique. For a subset $Q \subseteq P$, a *lower bound* for Q is an element $u \in P$ so that for every $q \in Q$, $u \leq q$; if u is a lower bound for Q, write $u \leq Q$. Similarly define *greatest element* and *upper bound*. A *least upper bound* for $Q \subset P$ is a least element in the set of all upper bounds; note that if a least upper bound for Q exists, it is unique. Similarly define *greatest lower bound*. Sometimes the notation $x < Q$ denotes that for every $q \in Q$, $x < Q$.

A partial order R on a set S is called a *total order* (or *linear order*) if for every $x, y \in S$, either $(x,y) \in R$ or $(y,x) \in R$ holds; in this case, the ordered set (S, R) is called a *totally ordered set*.

The standard order on $\mathbb{N}$ is often defined by $x < y$ if and only if there exists $n \in \mathbb{N}$ so that $y = x + n$. (Note that one can not yet really say in this definition "...if and only if there exists $n \geq 1$ so that...", since the order $\geq$ is being defined!) As one might expect, this standard order on $\mathbb{N}$ is indeed a linear order or total order. One first step in proving this is to show that any two elements in $\mathbb{N}$ are *comparable*, that is, for any $x, y \in \mathbb{N}$, one of $x < y$, $x = y$, or $y < x$ holds. The following "Law of Trichotomy" says precisely that.

Exercise 5 (Law of Trichotomy). *For any $x, y \in \mathbb{N}$, exactly one of $x < y$, $x = y$, or $y < x$ holds. Prove this result by induction.*

This law also confirms that since $\leq$ means $<$ or $=$, the relation $\leq$ as defined is antisymmetric. It also follows that $<$ defines a total order. Again, by induction, addition preserves order:

Exercise 6. *For any natural numbers x, y, p,*

$$x < y \quad \text{if and only if} \quad x + p < y + p.$$

There are different *kinds* of total orderings. For example, $\mathbb{N}$, the integers $\mathbb{Z}$, the rationals $\mathbb{Q}$, and the reals $\mathbb{R}$ all have no largest element. Of these, only $\mathbb{N}$ has a smallest element. Also, both $\mathbb{Q}$ and $\mathbb{R}$ are dense (between any two there is another), yet of these two, only $\mathbb{R}$ contains all its limit points.

Definition 2.6.1. A *well-ordering* on a set W is a total order $\leq$ (or $<$) on W so that for any non-empty $S \subset W$, S contains a least element. Any ordered set $(W, \leq)$ where $\leq$ is a well-ordering is called *well-ordered.*

As well [pun intended] noted in [95], the term "well-ordering" might very well be replaced with "good-ordering", because "well" in this instance is an adjective, not an adverb, however this usage has survived to become standard these days.

Peano's axioms imply that every non-empty subset of natural numbers indeed has a least element:

Theorem 2.6.2. *The standard order on $\mathbb{N}$ is a well-ordering.*

Proof: Let $S \subset \mathbb{N}$. First observe that if any least element in S exists, then it is unique, since if there were two least elements, say m_1 and m_2, then one would have both $m_1 \leq m_2$ and $m_2 \leq m_1$. Consequently, by the Law of Trichotomy, $m_1 = m_2$.

Assume that S is without a least element; to finish the proof, it suffices to show that $S = \emptyset$. Let

$$A = \{m \in \mathbb{N} : \text{no number less than } m \text{ belongs to } S\}.$$

By P3, $1 \in A$. Suppose that $k \in A$. If $n < k + 1$, then either $n < k$ (in which case $n \notin S$ since $k \in A$) or $n = k$ (in which case $n \notin S$, for if $n \in S$, then n would be least in S). In any case, such an n is not in S. Hence $k + 1 \in A$. Thus by P5, $A = \mathbb{N}$, and so $S = \emptyset$. $\square$

Exercise 7. *Let $(X, \leq)$ be a well-ordered set and let $Y \subseteq X$. Show that if $f : X \longrightarrow Y$ is an isomorphism, then for all $x \in X$, $f(x) \geq x$.*

Theorem 2.6.3. *A linearly ordered set $(W, <)$ is well-ordered if and only if there is no infinite decreasing sequence in W.*

Proof: Suppose that W is well-ordered. If $w_1 > w_2 > \ldots$ is an infinite decreasing sequence in W, put $S = \{w_1, w_2, \ldots\}$. By well-ordering, let w_k be the least element of S; then $w_k > w_{k+1} \in S$, contradicting the minimality of w_k.

Assume that W is a linearly ordered set with no infinite decreasing sequence. Fix any non-empty $S \subseteq W$, and let $t \in S$. If t is not the least element of S, pick $w_1 \in S$, $w_1 < t$. If w_1 is not the least element of S, pick $w_2 \in S$ with $w_2 < w_1$. Continue choosing successively smaller elements. Since W contains no infinite decreasing sequence, the same is true for S, so this process must stop after finitely many steps, and at that time, the least element of S is produced. $\square$

Note: Jech [289, p. 18] states that the direction "no infinite decreasing subset implies well-ordered" in Theorem 2.6.3 follows from the Axiom of Choice (see Section 4.5); however, AC does not seem to be needed in the above proof.

Definition 2.6.4. For totally ordered sets $(W_1, \leq_1)$ and $(W_2, \leq_2)$ a function $f : W_1 \to W_2$ is *order preserving* (o.p.) iff $x \leq_1 y$ implies $f(x) \leq_2 f(y)$. Well-ordered sets A and B are *similar*, written $A \sim B$, iff there is an order preserving bijection between them.

An order preserving bijection is also called a *similarity*. [Caution: "similarity" is also a term used in geometry for functions that preserve ratios of distances.] Some authors use the term "isomorphism" to describe a similarity; "isomorphism" is often used for a bijection that preserves algebraic or relational structure; in this case the structure is only the order.

Theorem 2.6.5. *2.6.3 Let $(W, <)$ be a well-ordered set and let $Y \subseteq W$. If $f : W \to Y$ is an order preserving bijection, then for all $w \in W$, $f(w) \geq w$.*

There are two ways to present this proof, one by induction, and the other by contradiction; the difference is subtle.

First proof of Theorem 2.6.3: Let $M = \{w \in W : f(w) < w\}$. If M is non-empty, pick some $m \in M$; then $f(m) < m$ and f being order preserving imply $f(f(m)) < f(m)$, so $f(m) \in M$ as well. Continue applying f, by induction, giving an infinite decreasing sequence $m, f(m), f(f(m)), \ldots$ in M. But since M is a subset of the well-ordered set W, M has a least element, so the assumption that M is non-empty must be abandoned. Thus conclude that $M = \emptyset$. $\square$

Second proof of Theorem 2.6.3: With the same notation as above, if $M \neq \emptyset$, since M is a subset of a well-ordered set, M contains a least element y_0. Then $f(y_0) < y_0$ implies that $f(f(y_0)) < f(y_0)$ and hence $f(y_0) \in M$, contradicting that y_0 is least in M. So $M = \emptyset$. $\square$

The first proof above can be thought of as a proof that uses *downward* induction to produce an infinite decreasing sequence, a sequence which contradicts a previously established fact. This very same technique is used in Fermat's method of infinite descent (see Section 3.6). Thus, many proofs by contradiction might be considered as proofs by induction.

2.7   Well-founded sets

A partial order $(P, \leq)$ is called *well-founded* if for every non-empty subset $X \subseteq P$, X contains a minimal (with respect to $\leq$). In a well-founded partial order, for every element $x \in P$, there is a well-ordered set containing both x and a minimal element of P. Just as mathematical induction is used on well-ordered sets, so too is mathematical induction valid for well-founded sets. This kind of induction might be called *generalized induction*. For example, suppose that one wants to prove a sequence of statements $P(m, n)$ that depend on two variables, say, for finitely many m. Suppose also that one knows $P(m, n) \rightarrow P(m, n+1)$. An inductive proof could start with the base cases as $P(m_i, 0)$, and from each base case, ordinary induction can be applied to reach all statements of the form $P(m_i, n)$. Generalized induction is most often used for induction on two variables, called "double induction", discussed in the next chapter. Generalized induction also includes the notion of "alternative induction", also in the next chapter.

Chapter 3

Variants of finite mathematical induction

Mathematics is either Pure or Mixed..... And as for Mixed Mathematics, I may only make this prediction, that there cannot fail to be more kinds of them, as nature grows further disclosed.

—Francis Bacon,

Advancement of Learning.

There are many forms of mathematical induction—weak, strong, and backward, to name a few. In what follows, n is a variable denoting an integer (usually non-negative) and $S(n)$ denotes a mathematical statement with one or more occurrences of the variable n.

3.1 The first principle

For convenience, the standard presentation of mathematical induction is repeated here. Sometimes this standard version of induction is called the "first principle of mathematical induction", and is also called "weak mathematical induction" (as opposed to "strong" induction, a modification appearing in Section 3.2). Recall that the notation $P \rightarrow Q$ is short for "P implies Q".

Theorem 2.4.1 [Principle of Mathematical Induction (MI)]
Let $S(n)$ be a statement involving n. If
    (i) $S(1)$ holds, and
    (ii) for every $k \geq 1$, $S(k) \rightarrow S(k+1)$,
then for every $n \geq 1$, the statement $S(n)$ holds.

By Theorem 4.5.5, an inductive step can also be accomplished indirectly by showing that the set of integers for which $S(n)$ fails has no least element, contradicting the well-ordering of $\mathbb{N}$. Such an example occurs in the following Section 3.2 on strong mathematical induction. (See also one solution to Exercise 477.)

Note that the base step in an inductive proof is essential, since, for example, if one were to attempt to prove that for any positive integer n, the statement $S(n)$: $\sum_{i=1}^{n}(2i-1) = n^2 + 5$ holds, it is not hard to show that $S(n) \to S(n+1)$, however, $S(1)$ does not even hold, and so one may not conclude that $S(n)$ holds for all $n \geq 1$. Another such statement (where n is a positive integer) is "$n^2 + 5n + 1$ is even", for which the inductive step works, but the statement is in fact never true!

The base case for MI need not be 1 (or 0); in fact, one may start at any integer. Here is a slightly more general (but equivalent) form of the principle of induction:

Theorem 3.1.1 (Principle of Mathematical Induction (MI)).
Let $S(n)$ denote a statement regarding an integer n, and let $k \in \mathbb{Z}$ be fixed. If
  (i) $S(k)$ holds, and
  (ii) for every $m \geq k$, $S(m) \to S(m+1)$,
then for every $n \geq k$, the statement $S(n)$ holds.

Proof: Let $T(n)$ be the statement $S(n + k - 1)$, and repeat the above proof, instead with T replacing every occurrence of S. Then the base case becomes $T(1) = S(1 + k - 1) = S(k)$ as desired. $\qquad\square$

3.2 Strong mathematical induction

While attempting an inductive proof, in the inductive step one often needs only the truth of $S(n)$ to prove $S(n + 1)$; sometimes a little more "power" is needed, and often this is made possible by strengthening the inductive hypothesis. The following version of mathematical induction can be viewed as contained in the principle of transfinite induction (see Section 4.2).

Theorem 3.2.1 (Strong Mathematical Induction).
Let $S(n)$ denote a statement involving an integer n. If
  (i) $S(k)$ is true and
  (ii) for every $m \geq k$, $[S(k) \wedge S(k+1) \wedge \cdots \wedge S(m)] \to S(m+1)$
then for every $n \geq k$, the statement $S(n)$ is true.

The principle of strong induction is also referred to by some as *course-of-values induction* (*e.g.*, see [42]). A few professionals use "full induction" or "complete induction" to denote strong induction; these terms have long been accepted as meaning simply "mathematical induction" (as opposed to empirical induction). [See Section 1.8 on history of induction.]

In Theorem 2.6.2, it was shown that Peano's axioms imply the well-ordering of $\mathbb{N}$. This well-ordering is used (below) to prove strong induction, and hence, to show that strong induction also follows from P5. Notice that Theorem 4.5.5 also shows that both forms of induction follow from well-ordering.

Proof of strong induction principle from weak: Assume that for some k, the statement $S(k)$ is true and for every $m \geq k$, $[S(k) \wedge S(k+1) \wedge \cdots \wedge S(m)] \rightarrow S(m+1)$. Let B be the set of all $n > m$ for which $S(n)$ is false. If $B \neq \emptyset$, $B \subset \mathbb{N}$ and so by well-ordering, B has a least element, say ℓ. By the definition of B, for every $k \leq t < \ell$, $S(t)$ is true. The premise of the inductive hypothesis is true, and so $S(\ell)$ is true, contradicting that $\ell \in B$. Hence $B = \emptyset$. □

Strong induction also implies weak induction.

Proof of weak induction from strong: Assume that strong induction holds (in particular, for $k = 1$). That is, assume that if $S(1)$ is true and for every $m \geq 1$, $[S(1) \wedge S(2) \wedge \cdots \wedge S(m)] \rightarrow S(m+1)$, then for every $n \geq 1$, $S(n)$ is true.
Observe (by truth tables, if you will), that for $m + 1$ statements p_i,

$$[p_1 \rightarrow p_2] \wedge [p_2 \rightarrow p_3] \wedge \ldots \wedge [p_m \rightarrow p_{m+1}] \Rightarrow [(p_1 \wedge p_2 \wedge \ldots \wedge p_m) \rightarrow p_{m+1}],$$

itself a result provable by induction (see Exercise 456).
Assume that the hypotheses of weak induction are true, that is, that $S(1)$ is true, and that for arbitrary t, $S(t) \rightarrow S(t+1)$. By repeated application of these recent assumptions, $S(1) \rightarrow S(2)$, $S(2) \rightarrow S(3)$, ..., $S(m) \rightarrow S(m+1)$ each hold. By the above observation, then

$$[S(1) \wedge S(2) \wedge \cdots \wedge S(m)] \rightarrow S(m+1).$$

Thus the hypotheses of strong induction are complete, and so one concludes that for every $n \geq 1$, the statement $S(n)$ is true, the consequence desired to complete the proof of weak induction. □

Hence it has been demonstrated that weak and strong forms of mathematical induction are equivalent. For remarks on this relationship, see [477].

Here is an example where strong induction is used. Recall that a prime number (or simply, a prime) is one whose only divisors are itself and 1 (and convention says that 1 is not a prime); the first few primes are

$$2, 3, 5, 7, 11, 13, 17, 19, 23, 29, 31, 37, 41, 43, 47, 53, \ldots.$$

Theorem 3.2.2. *Any positive integer $n \geq 2$ is a product of primes.*

Proof: Let $S(n)$ be the statement "n is a product of primes."
BASE STEP ($n = 2$): Since $n = 2$ is trivially a product of primes (well, actually only one prime), $S(2)$ is true.

INDUCTIVE STEP: Fix some $m \geq 2$, and assume that for every t satisfying $2 \leq t \leq m$, the statement $S(t)$ is true. To be shown is that

$$S(m+1): \quad m+1 \text{ is a product of primes,}$$

is true. If $m+1$ is prime, then $S(m+1)$ is true. If $m+1$ is not prime, then there exists r and s with $2 \leq r \leq m$ and $2 \leq s \leq m$ so that $m+1 = rs$. Since $S(r)$ is assumed to be true, r is a product of primes; similarly, by $S(s)$, s is a product of primes. Hence $m+1 = rs$ is a product of primes, and so $S(m+1)$ holds. So, in either case, $S(m+1)$ holds, completing the inductive step.

Thus, by mathematical induction, for all $n \geq 2$, the statement $S(n)$ is true. $\square$

The so-called "Fundamental theorem of arithmetic" says that any integer $n \geq 2$ is a product of primes in exactly one way, that is, the prime factorization is unique—another result provable by induction (see Exercise 206).

3.3 Downward induction

Suppose that you are trying to prove a statement $S(n)$ and a forward inductive argument is difficult for every n. Here is another strategy: first prove the statement for infinitely many n (for example, when n is a power of 2—either directly or by an inductive step of the form $S(k) \rightarrow S(2k)$, say) and then prove $S(n)$ for the gaps between. The proof for the gaps can either be by forward induction, or backward induction. For example, in the case where one has the truth of $S(n)$ for all powers of 2, one can then fill in the gaps with an inductive argument for each fixed k of the form $S(2^k + t) \rightarrow S(2^k + t - 1)$ for each t satisfying $1 \leq t \leq 2^k$.

Downward (also called "backward") inductive arguments have been around a long time; many authors, including Cauchy (1759–1857) and Weierstrass (1815–1897) (see [259, p.19]) have used them. The term "backward induction" can also be used in game theory where players reason "working backward from the last possible move in a game to anticipate each other's rational choices."[114]. What has recently become known as "downward induction" defined below might be more appropriately called "upward-downward" induction.

Downward induction: Let $S(n)$ be a statement involving n. If
  (i) $S(n)$ is true for infinitely many n, and
  (ii) for each $m \geq 2$, $S(m) \rightarrow S(m-1)$
then for every $n \geq 1$, the statement $S(n)$ is true.

Proof of downward induction from MI: Assume the hypotheses (i) and (ii) hold and let $n_1, n_2, n_3, \ldots$ be an infinite sequence so that for each $i \in \mathbb{Z}^+$, $S(n_i)$ holds. Fix some $k \in \mathbb{Z}^+$, and prove $S(k)$ holds as follows: Fix i so such that $n_{i-1} < k \leq n_i$.

For $j = 0, 1, \ldots, n_i - n_{i-1}$, define the statement $T(j) = S(n_i - j)$. It suffices to prove that $T(n_i - k) = S(k)$; this is done by induction on j.

BASE STEP $(j = 0)$: $T(j) = T(0) = S(n_i - 0) = S(n_i)$, which was assumed to be true by (i).

INDUCTIVE STEP: Suppose that for some $j \geq 0$, $T(j) = S(n_i - j)$ holds. By (ii), $T(j+1) = S(n_i - j - 1)$ holds, completing the inductive step $T(j) \to T(j+1)$.

Therefore, by MI, $T(j)$ holds for all $j \geq 0$, in particular, $T(n_i - k) = S(k)$ holds, finishing the proof of downward induction. $\qquad\square$

There are different proofs of the so-called "theorem of arithmetic and geometric means"; for example, there is one downward induction proof appearing in [259] and another simpler one also suggested there. The simpler one is presented here. Another proof follows from Jensen's inequality on convex functions—see Exercise 602 or 603—both provable by downward induction. After giving the proof by downward induction, one more simple, but tricky proof by ordinary induction due to Kong-Ming Chong [102] is presented. For other proofs prior to 1976 (mentioned in [102]) of the AM-GM inequality, see, *e.g.*, [5, pp. 200–224], [43, §5 pp. 4–5; §11 pp. 9–10], [104, p.46], [259], [135], and [397].

Theorem 3.3.1 (AM-GM inequality). *Let $a_1, \ldots, a_n$ be non-negative real numbers. Then*

$$(a_1 a_2 \cdots a_n)^{1/n} \leq \frac{a_1 + a_2 + \cdots + a_n}{n},$$

with equality holding if and only if all a_i's are equal.

Proof: Let $S(n)$ be the statement that for any $a_1, \ldots, a_n$,

$$(a_1 a_2 \cdots a_n)^{1/n} \leq \frac{a_1 + a_2 + \cdots + a_n}{n},$$

with equality holding if and only if all a_i's are equal. The first part of the proof is to show that $S(n)$ holds whenever n is a power of 2. This requires a form of strong induction, one with two base cases.

BASE STEP $n = 1$: The statement $S(1)$ reduces to $a_1 = a_1$, which is true.

BASE STEP $n = 2$: To show $S(2)$, let $a_1 = a$ and $a_2 = b$; then

$$ab = \left(\frac{a+b}{2}\right)^2 - \left(\frac{a-b}{2}\right)^2$$
$$\leq \left(\frac{a+b}{2}\right)^2,$$

with equality holding if and only if $a = b$.

UPWARD INDUCTIVE STEP $(S(k) \to S(2k))$: For some $k \geq 2$ assume that $S(k)$ holds, that is, assume that for non-negative $c_1, c_2, \ldots, c_k$,

$$c_1 c_2 \cdots c_k \leq \left(\frac{c_1 + c_2 + \cdots + c_k}{k} \right)^k,$$

with equality if and only if the c_i's are all equal. To show that $S(2k)$ follows:

$$a_1 a_2 \cdots a_k b_1 b_2 \cdots b_k$$
$$\leq \left(\frac{a_1 + \cdots + a_k}{k} \right)^k \left(\frac{b_1 + \cdots + b_k}{k} \right)^k \quad \text{(by } S(k) \text{ twice)}$$
$$= \left(\frac{(a_1 + \cdots + a_k)(b_1 + \cdots + b_k)}{k^2} \right)^k$$
$$\leq \left(\left(\frac{a_1 + \cdots + a_k + b_1 + \cdots b_k}{2} \right)^2 \frac{1}{k^2} \right)^k \quad \text{by } S(2),$$
$$= \left(\frac{a_1 + \cdots + a_k + b_1 + \cdots b_k}{2k} \right)^{2k}$$

and inequalities are strict unless all a_i's and b_j's are equal. Hence $S(k) \to S(2k)$, completing this inductive step.

By induction, for all n that are powers of 2, the statement $S(n)$ holds.

DOWNWARD INDUCTIVE STEP $(S(m) \to S(m-1))$: For some $n \geq 2$, assume that $S(m)$ holds, and let $a_1, a_2, \ldots, a_{m-1}$ be non-negative, not all equal, and put

$$A = \frac{a_1 + a_2 + \ldots + a_{m-1}}{m-1}.$$

Then

$$a_1 a_2 \cdots a_{m-1} A < \left(\frac{a_1 + a_2 + \cdots + a_{m-1} + A}{m} \right)^m \quad \text{(by } S(m)),$$
$$= \left(\frac{(n-1)A + A}{m} \right)^m$$
$$= A^m,$$

and hence $a_1 a_2 \cdots a_{m-1} < A^{m-1}$, thus showing $S(m-1)$. This completes the proof of the downward induction step, and hence the proof. $\square$

Note: Theorem 3.3.1 has a more direct proof, based on Exercise 199, the solution of which is a fairly easy inductive proof; see comments after the solution to Exercise 199.

As mentioned above, here is an outline of Chong's simple (but tricky) inductive proof of the AM-GM inequality. Suppose that the base case $n = 2$ is done, and for some $k \geq 3$, suppose that $S(k-1)$ is true, in particular, suppose that for any choice of $b_1, a_2, \ldots, a_{k-1}$ not all equal,

$$\frac{b_1 + a_2 + a_3 + \cdots + a_{k-1}}{k-1} > (b_1 a_2 a_3 \cdots a_{k-1})^{\frac{1}{k-1}}.$$

To be shown is that $S(k)$ holds (in the case when all numbers are not equal). Let $a_1 \leq a_2 \leq \cdots \leq a_n$ be not all equal, that is, $a_1 < a_n$, and let $A = (a_1 + a_2 + \cdots + a_n)/n$ be their arithmetic mean. Then $a_1 < A < a_n$, which implies that

$$A(a_1 + a_k - A) - a_1 a_k = (a_1 - A)(A - a_k) > 0,$$

and so

$$a_1 + a_k - A > \frac{a_1 a_k}{A}. \tag{3.1}$$

Let $b_1 = a_1 + a_k - A$; then

$$\frac{b_1 + a_2 + \cdots a_{k-1}}{k-1} = \frac{(\sum a_i) - A}{k-1} = \frac{kA - A}{k-1} = A.$$

Thus, by induction hypothesis,

$$A > (b_1 a_2 a_3 \cdots a_{k-1})^{\frac{1}{k-1}}$$
$$> (\frac{a_1 a_k}{A} a_2 \cdot a_k)^{\frac{1}{k-1}} \qquad \text{by eqn (3.1)}$$

which yields

$$A^{k-1} > \frac{a_1 a_2 \cdots a_k}{A},$$
$$A^k > a_1 a_2 \cdots a_k,$$

showing that $S(k)$ is true, completing the (upward) inductive step, and hence Chong's proof. □

There are other inductive proofs of the AM-GM inequality; one inductive step begins by assuming that for any a_i's satisfying $a_1 a_2 \cdots a_n = 1$ then $a_1 + \cdots + a_n \geq n$. Then assume that $b_1 b_2 \cdots b_n b_{n+1} = 1$; without loss of generality, let $b_n < 1$ and $b_{n+1} > 1$. Then $b_1 + \cdots + b_{n+1} \geq n+1$ by setting $a_1 = b_1$, $a_2 = b_2$, $\ldots$, $a_{n-1} = b_{n-1}$ but $a_n = b_n b_{n+1}$. Then by inductive hypothesis, $b_1 + b_2 + \cdots + b_{n-1} + b_n b_{n+1} \geq n$. To finish the inductive step, it suffices to show

$$b_n + b_{n+1} \geq b_n b_{n+1} + 1,$$

or

$$(1 - b_n)(1 - b_{n+1}) \leq 0,$$

which is true by the initial assumption on b_n and b_{n+1}, finishing the inductive step. □

See [146, pp. 37–40] for yet another solution by induction based on the following lemma:

Lemma 3.3.2. *For real numbers* w, x, y, *and* z, *if* $w + x = y + z$, *then the largest of the two products* wx *and* yz *is formed by the pair with the smallest difference.*

Proof: Let $w + x = y + z$. Note the following two identities:

$$(w + x)^2 - (w - x)^2 = 4wx,$$
$$(y + z)^2 - (y - z)^2 = 4yz.$$

Since $w + x = y + z$, also $(w + x)^2 = (y + z)^2$, so the left-hand side of the two identities is made largest when the second term is smallest. □

It might be interesting to note that the case $n = 2$ in Theorem 3.3.1 can be used to prove that no chord of a circle is longer than the diameter. Let three points A, B, C form a straight line segment with distances $|AB| = a$ and $|BC| = b$ units (see Figure 3.1.

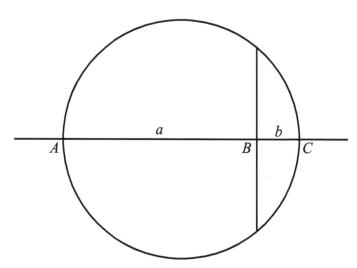

Figure 3.1: Chords are shorter than diameters

Using the line segment AC as a diameter, form the circle whose diameter is $|AC| = a + b$ units. Form a chord of that circle perpendicular to AC through B. Then with a simple application of Pythagoras' theorem, one finds that the length of that chord is $2\sqrt{ab}$. By Theorem 3.3.1, $\sqrt{ab} \leq \frac{a+b}{2}$, and so the length of the chord is not longer than the diameter of the circle.

3.4   Alternative forms of mathematical induction

There are many ways to apply inductive reasoning. For example, if $S(0)$ and $S(1)$ are true, and if for any $n \geq 0$, $S(n) \to S(n + 2)$ holds, then for all $n \geq 0$, $S(n)$ is true, since actually, two separate inductive proofs are combined in one (one for the even cases, and one for the odd cases). Here is an example of a situation where three inductive proofs are rolled into one. [Note: This very example is listed again as Exercise 311.] Other applications of this alternative form of mathematical induction appear throughout the exercises, *e.g.*, in Exercises 113 and 275.)

Theorem 3.4.1. *For any integer $n \geq 14$, n is expressible as a sum of 3's and/or 8's.*

Proof: Let $S(n)$ be the statement: n is expressible as a sum of 3's and/or 8's.

BASE CASES ($S(14), S(15), S(16)$): Since $14 = 3 + 3 + 8$, $15 = 3 + 3 + 3 + 3 + 3$, and $16 = 8 + 8$, the base steps are shown.

INDUCTIVE STEP ($S(k) \rightarrow S(k+3)$): Assume that for some $k \geq 14$, $S(k)$ holds, that this, there exist $\alpha, \beta \in \mathbb{Z}$ so that $k = \alpha \cdot 3 + \beta \cdot 8$. Then $k + 3 = \alpha \cdot 3 + \beta \cdot 8 + 3 = (\alpha + 1) \cdot 3 + \beta \cdot 8$, that is, $k + 3$ is expressible as a sum of 3's and/or 8's, showing $S(k + 3)$ holds, completing the inductive step.

By MI, for all $n \geq 14$, the statement $S(n)$ is true. (Actually, there are three separate proofs by MI rolled into one, one proving the statement for the sequence $n = 14, 17, 20, \ldots$, one for $n = 15, 18, 21, \ldots$, and another for $n = 16, 19, 22, \ldots$.) $\qquad\qquad\qquad\qquad\qquad\qquad\qquad\qquad\qquad\qquad\qquad\qquad\qquad\qquad\qquad\qquad\square$

An inductive proof might be also encountered when both $S(2)$ and $S(3)$ hold and for $k \geq 2$, $[S(k) \wedge S(k + 1)] \rightarrow S(k + 2)$, then $S(n)$ holds for all $n \geq 2$; such a proof might be classified somewhere between weak and strong induction. In Section 12.2 on Fibonacci numbers, there are many exercises where such a technique is required.

Many mathematical induction proofs use more than one base case, and such proofs can fall into the category of "generalized induction" on well-founded sets. The technique relies on the fact that in any well-founded set (a partial order with minimal elements) and a statement S about elements of that set, for each x in the set, there is an inductive argument for $S(x)$ that has as its base case one of the minimal elements. For example, consider the set $X = \{2, 3, 4, 5, 6, \ldots\}$. If X is ordered according to divisibility, the proof of any statement S about members of X might start with the base cases being a proof about each prime, the primes being the set of minimal elements in the partially ordered set X.

3.5   Double induction

A special kind of inductive argument is called "double induction"; some texts refer to double induction as an inductive step that requires, say, $S(n)$ and $S(n+1)$ to prove $S(n + 2)$. Another kind of "double induction" is where two statements involving n are proved simultaneously (see, for example, Exercise 320 or Exercise 122 where the inductive step consists of two proofs, one for each of two statements). In this section, however, "double induction", means an induction on two variables simultaneously.

Many mathematical statements involve two (or more) variables, each of which may vary independently over, say, $\mathbb{N}$. Some such statements can be proved by induction in different ways. Let $S(m, n)$ be a statement involving two positive integer variables m and n. One method to prove $S(m, n)$ for all $m \geq 1$ and $n \geq 1$ is to first prove $S(1, 1)$, then by induction prove $S(m, 1)$ for each m, and then for each fixed m_0, inductively prove $S(m_0, n)$ for all n. Here is a rather simple example of the technique.

Theorem 3.5.1. *Let positive integers m and n be given.*

$$\sum_{i=1}^{m} \sum_{j=1}^{n} (i + j) = \frac{mn(m + n + 2)}{2}.$$

Proof: Let $S(m, n)$ be the equality in the statement of the theorem.

First it is proved that for all $m \geq 1$, $S(m, 1)$ is true.

BASE STEP: The statement $S(1, 1)$ is true since $1 + 1 = 1 \cdot 1(1 + 1 + 2)/2$.

INDUCTIVE STEP (inducting on m): For some $k \geq 1$, assume that $S(k, 1)$ is true, that is, $\sum_{i=1}^{k}(i + 1) = k(k + 3)/2$. Beginning with the left-hand side of $S(k + 1, 1)$,

$$\sum_{i=1}^{k+1} (i + 1) = \left(\sum_{i=1}^{k} (i + 1) \right) + (k + 1) + 1$$

$$= \frac{k(k + 3)}{2} + k + 2 \quad \text{(by } S(k, 1))$$

$$= \frac{k^2 + 5k + 4}{2}$$

$$= \frac{(k + 1)(k + 1 + 1 + 2)}{2},$$

the right-hand side of $S(k + 1, 1)$. Hence $S(k, 1) \rightarrow S(k + 1, 1)$, and so by mathematical induction, for all $m \geq 1$, $S(m, 1)$ is true.

Fix an arbitrary m_0. Then $S(m_0, 1)$ is true and so this is a base step for proving that for all n, $S(m_0, n)$ holds.

INDUCTIVE STEP (inducting on n): This step is of the form $S(m_0, \ell) \rightarrow S(m_0, \ell+1)$. Let $\ell \geq 1$ be fixed and assume that $S(m_0, \ell)$ is true, that is,

$$\sum_{i=1}^{m_0} \sum_{j=1}^{\ell} (i + j) = \frac{m_0 \ell (m_0 + \ell + 2)}{2}.$$

Beginning with the left side of $S(m_0, \ell + 1)$,

$$\sum_{i=1}^{m_0} \sum_{j=1}^{\ell+1} (i + j) = \sum_{i=1}^{m_0} \left(\left(\sum_{j=1}^{\ell} (i + j) \right) + (i + \ell + 1) \right)$$

$$= \sum_{i=1}^{m_0} \left(\sum_{j=1}^{\ell} (i+j) \right) + \sum_{i=1}^{m_0} (i+\ell+1)$$

$$= \frac{m_0 \ell (m_0 + \ell + 2)}{2} + \sum_{i=1}^{m_0} (i + \ell + 1) \qquad \text{(by } S(m_0, \ell)),$$

$$= \frac{m_0 \ell (m_0 + \ell + 2)}{2} + \frac{m_0 (m_0 + 1)}{2} + m_0 (\ell + 1)$$

$$= \frac{m_0 \ell (m_0 + \ell + 2)}{2} + \frac{m_0 (m_0 + 1 + 2(\ell + 1))}{2}$$

$$= \frac{m_0 (\ell m_0 + \ell^2 + 2\ell + m_0 + 1 + 2(\ell + 1))}{2}$$

$$= \frac{m_0 (\ell + 1)(m_0 + \ell + 1 + 2)}{2},$$

which is the right-hand side of $S(m_0, \ell+1)$. Hence, by induction, for each fixed m_0 and all $n \geq 1$, $S(m_0, n)$ is true, completing this inductive step.

Since m_0 was arbitrary, by induction, for all $m \geq 1$ and $n \geq 1$ the statement $S(m, n)$ is proved. $\qquad \square$

Sometimes the inductive proofs contained in each stage of the double induction require multiple base cases and alternative forms of induction (or strong induction). See Exercise 304 for such a situation, where the alternative form of induction in the second stage requires two proofs by induction in the first stage.

Another way to apply a double induction argument would be to use $S(1,1)$ as a base step, then show that both $S(m,n) \to S(m+1,n)$, and $S(m,n) \to S(m,n+1)$. This would prove $S(m,n)$ for all m and n. One must be careful, however, for only the step $S(m,n) \to S(m+1, n+1)$ would not prove the statement for all m and n, only the cases where $m = n$.

A slightly trickier double induction occurs in Exercise 380, where the induction step shows $S(n-2, k-1) \wedge S(n-1, k) \to S(n, k)$. In this case, one needs to prove two families of base cases, those of the forms $S(0, k)$, $S(1, k)$ and those of the form $S(n, 0)$, $S(n, 1)$. Then, for example, to prove $S(6, 3)$, one proceeds as follows:

$$
\begin{array}{rcl}
S(0,1) \wedge S(1,2) & \to & S(2,2) \\
S(1,1) \wedge S(2,2) & \to & S(3,2) \\
S(2,1) \wedge S(3,2) & \to & S(4,2) \\
S(3,1) \wedge S(4,2) & \to & S(5,2) \\
\\
S(0,2) \wedge S(1,3) & \to & S(2,3)
\end{array}
$$

$$S(1,2) \wedge S(2,3) \quad \rightarrow \quad S(3,3)$$
$$S(2,2) \wedge S(3,3) \quad \rightarrow \quad S(4,3)$$
$$S(3,2) \wedge S(4,3) \quad \rightarrow \quad S(5,3)$$
$$S(4,2) \wedge S(5,3) \quad \rightarrow \quad S(6,3).$$

So in this situation, one fixes $k-1$ and k, inducts on n, then one repeats the process for k and $k+1$, and so on. Given that the base cases and the inductive step are proved, the interested reader can try to write up the proof formally.

3.6 Fermat's method of infinite descent

One of the more famous applications of Fermat's method of infinite descent showed that any right angle triangle with sides having rational lengths could not have integral area. His technique was to first show that the truth of the theorem follows from the special case for right triangles with integer lengths, that is, for *Pythagorean triangles*. Then he showed that if one could find such a Pythagorean triangle with integer area, one could then (using the number-theoretic properties of the lengths of the sides from the first one) produce a smaller Pythagorean triangle with the same property. From the smaller one, applying precisely the same argument, one would find yet a smaller one. By induction, one gets an infinite sequence of consecutively smaller triangles with the desired property. Clearly there is no infinite descending sequence of Pythagorean triangles (by the well-ordering of natural numbers)—a contradiction. So one must abandon the assumption that one found a Pythagorean triangle with integer area.

The above discussion hints at the possibility that there are proofs that are inductive, but not in any straightforward way. To demonstrate the beauty of Fermat's technique, the next theorem is given with a proof by infinite descent; in many respects, it duplicates the proof alluded to above for Pythagorean triangles.

Theorem 3.6.1. *The equation*

$$x^4 + y^4 = z^2 \tag{3.2}$$

has no solution in non-zero integers x, y, and z.

Proof by infinite descent: If any triple of integers (x, y, z) satisfy (3.2), then so do any of $(\pm x, \pm y, \pm z)$; thus to show the theorem, it suffices to show that are no *positive* integer solutions to (3.2).

The proof is accomplished by showing that *if* some solution x, y, z to (3.2) exists, then from that solution one can create another "smaller" solution x', y', z', where "smaller" means that $z' < z$. Since the positive integers are well-ordered, this process can not continue forever, and so one must abandon the original assumption some solution exists.

Hypothetically, suppose that x, y, z is a solution to (3.2).

Considering the equation (3.2) modulo 4, one observes that x and y can not both be odd (any odd number squared is congruent to 1 modulo 4, and since then z has to be even, z^2 is 0 modulo 4).

Let x and y be even, say $x = 2k$, $y = 2\ell$; then 16 divides z^2, and so 4 divides z, say $z = 4m$. Then $(2k)^4 + (2\ell)^4 = (4m)^2$, and division by 16 yields $k^4 + \ell^4 = m^2$, a smaller solution. Similarly, if any prime p divides both x and y, write $x = pk$, $y = p\ell$, and $z = p^2m$. Division by p^4 shows that $x' = k$, $y' = \ell$, and $z' = m$ is another smaller solution to (3.2). Hence, it suffices to assume that x and y are relatively prime.

So suppose that exactly one of x and y is even, the other odd (and x and y are relatively prime). Without loss of generality, suppose that x is even and y is odd. Then x^2 and y^2 are relatively prime, and so the triple x^2, y^2, z form a *fundamental Pythagorean triple* (a triple of positive integers a, b, c, each pair relatively prime, satisfying $a^2 + b^2 = c^2$). It is well known (e.g., see [150]) that a fundamental Pythagorean triple is a triple of the form $2mn$, $m^2 - n^2$, and $m^2 + n^2$, where m and n are relatively prime positive integers with exactly one of m, n odd.

Fix such an m and n, and write $x^2 = 2mn$, $y^2 = m^2 - n^2$, and $z = m^2 + n^2$. Since $y^2 + m^2 = n^2$, the triple y, m, n is a Pythagorean triple. Since m and n are relatively prime, so are y and m, with y odd and n even, and so y, m, n is a fundamental triple. Hence, there are relatively prime p and q, so that $m = 2pq$ and $n = p^2 + q^2$.

Since $x^2 = 2mn = 4pq(p^2 + q^2)$ and p and q are relatively prime, each of p, q, and $p^2 + q^2$ are all relatively prime and hence each must be a perfect square, say $p = \alpha^2$, $q = \beta^2$, and $p^2 + q^2 = \gamma^2$. Then $\alpha^4 + \beta^4 = \gamma^2$ with $\gamma \leq m < z$, giving a smaller solution to (3.2). $\qquad\square$

Since z^4 is a perfect square, Theorem 3.6.1 implies that $x^4 + y^4 = z^4$ has no non-zero integer solutions, a special case of what is now called "Fermat's last theorem": for each integer $n \geq 3$, the equation $x^n + y^n = z^n$ has no non-zero solutions. (This was a conjecture until Andrew Wiles *et al.* finally proved it in 1995 (see [569]). In 1753, Euler gave an incorrect proof for $n = 3$, later corrected by Gauss; both ideas were using descent, however Gauss failed to notice that unique factorization did not hold in his "proof". The case $n = 5$ was solved with infinite descent by Dirichlet (1805–1859) and Legendre (1752–1833) in 1825 (both proofs were based on a result by Sophie Germain); Dirichlet also managed $n = 14$ in 1832. Lamé settled the case $n = 7$ in 1839. [*Added note:* I forget the reference, but I recall reading that it was proved that the method of descent would not work for $n > 17$. Also, Kummer proved Fermat's last theorem for all "regular primes".]

Some authors prove that $\sqrt{2}$ is irrational by the method of infinite descent. The argument is by contradiction and begins by assuming that $\sqrt{2}$ is rational, say, $\sqrt{2} = \frac{a}{b}$ for some positive integers a and b. Squaring each side, $2 = \frac{a^2}{b^2}$ and so $2a^2 = b^2$. Then 2 divides b^2, and hence 2 divides b, so write $b = 2k$. Then replacing this in the previous equation gives $2a^2 = 4k^2$, yielding $a^2 = 2k^2$. Again, this shows that now 2 divides a, and so $a = 2\ell$, for some positive integer $\ell > 1$. Replacing a in the previous equation now gives $4\ell^2 = 2k^2$, and hence $2 = \frac{k^2}{\ell^2}$. Thus two smaller integers k and ℓ are found with $\sqrt{2} = \frac{k}{\ell}$. This process of finding smaller integers to represent $\sqrt{2}$ as a fraction can continue forever, contradicting the well-ordering of the natural numbers.

The above proof of the irrationality of $\sqrt{2}$ does not have to take the form of infinite descent if one merely assumes at the outset that a and b are relatively prime. The contradiction is then quickly arrived at since the above proof then delivers that 2 is a common factor to both a and b. See, for example, [216] or [260] for more on proving the irrationality of $\sqrt{2}$.

Exercise 8. *Using infinite descent, prove that for each positive integer n, $\sqrt{4n - 1}$ is not a rational number.*

For more results provable by descent, see (among others) Exercises 214, 222, 223, 224, and 225.

3.7   Structural induction

Computer scientists refer to mathematical induction, when applied to a recursively defined structure, as "structural induction". Apparently, the term originally came from model theory (although I cannot find the origin) where various properties of models are proved by using chains of models, and some kind of induction on each chain. The discussion here is far less serious. In the rest of mathematics, the term "structural induction" is rarely used outside of computer science applications—as a friend once said, "it's all just induction".

Assume that $\mathcal{S}$ is a class of structures (it is not important what kind of structure) with some partial order $\leq$ relating individual structures. Suppose $\mathcal{S}$ contains minimal elements, and for every structure $S \in \mathcal{S}$ there is a well-ordered set of structures beginning with a minimal element in $\mathcal{S}$ and culminating in S (in other words, $\mathcal{S}$ is well-founded). Let P be some proposition about elements of $\mathcal{S}$. Then to prove the truth of $P(S)$, it suffices to prove inductively along the chain leading to S, where each inductive step is maintained by some property of the recursion used to generate structures. Then such a proof might be called a "structural induction" proof.

The most common way in which structural induction is implemented is on recursively defined structures that have some kind of "rank"—a measure of how many recursions are necessary to construct a structure from minimal structures. The typical example to help make things clear is that of rooted trees (see Section 15.2 for terminology). The rank of a rooted tree is its height, and any finite rooted tree of height h can be constructed recursively from trees of height $h - 1$ by simply adding a new root. The inductive step for structural induction is usually proved by some simple property that follows from a recursive definition for the structure.

Structural induction is also used to prove properties with many base cases (as in generalized induction on well-founded sets) and can even be applied with transfinite induction (see Chapter 4).

Structural induction appears throughout this book. For examples using permutations, see the proof of (12.5). For examples regarding well-formed formulae, see Exercises 465, 466, 467. For examples using trees, see Exercises 483, 484, or 485. Graph theory uses structural induction frequently; as just one example, see Exercise 513, where structures are partite graphs, and r-partite graphs are constructed from $(r-1)$-partite graphs recursively. Other examples in graph theory where structural induction is used include theorems for amalgamation (see *e.g.*, Theorem 21.5.1 as a restricted form of amalgamation) because certain classes of graphs can be constructed by recursively gluing together two graphs on some common subgraph(s).

Hadamard matrices might be the structures concerned, and a simple tensor product construction creates recursively larger and larger Hadamard matrices (see Exercise 659). A similar notion is encountered when constructing latin squares recursively from latin rectangles (see Exercise 666). Functions form a large class of structures, and one can recursively define a function by its behavior on larger and larger domains (see, *e.g.*, Exercise 426). Colorings of objects are themselves functions, and so, for example, Exercise 731 is solved with structural induction. Certain classes of geometric objects can be considered as structures, in which case many exercises in Chapter 20 are by structural induction.

The instances of structural induction in this book are too numerous to list here. The index points to a few more examples of structural induction.

Chapter 4

Inductive techniques applied to the infinite

But of all other ideas, it is number, which I think furnishes us with the clearest and most distinct idea of infinity we are capable of.

—John Locke,

An essay concerning human understanding.

So far, mathematical induction has only been applied to one type of infinity, namely that of the counting numbers. In fact, mathematical induction can be performed on many other kinds of sets that have some kind of order defined on them, in particular, to sets that have a larger cardinality than that of $\mathbb{Z}^+$. These different forms of induction often depend on the axiom system decided upon. In the most common axiom systems, forms of induction for infinite sets are used to prove very powerful theorems. For example, the fact every vector space has a basis is easily proved by one of these forms.

4.1 More on well-ordered sets

Theorem 4.1.1. *There is at most one order-preserving bijection between any two well-ordered sets.*

Proof: Let $(A, <)$ and $(B, \prec)$ be well-ordered sets. Suppose that both f and g are order-preserving bijections from a A onto B. Then $g^{-1} \circ f$ is an order preserving bijection from A to itself. By Theorem 2.6.3, for all $a \in A$, $a \leq g^{-1}(f(a))$, and applying g to each side, $g(a) \leq f(a)$. Similarly, applying $f^{-1} \circ g$ to A, for each $a \in A$, $f(a) \leq g(a)$. Combining these two facts shows that for all $a \in A$, $f(a) = g(a)$. $\square$

Definition 4.1.2. For a well-ordered set $(W, <)$ and $t \in W$, define the *initial segment* of $(W, <)$ up to t by

$$\text{seg}_{(W,<)}(t) = \{w \in W : w < t\}.$$

Define a *closed initial segment* by

$$\overline{\text{seg}(t)} = \{w \in W : w \le t\} = \text{seg}(t) \cup \{t\}.$$

When no confusion can arise, the notations $\text{seg}_W(t)$, $\text{seg}_<(t)$, or $\text{seg}(t)$ denote $\text{seg}_{(W,<)}(t)$. If a subset S of W satisfies $a \in S, b \le a \Rightarrow b \in S$, then either S is an initial segment of W or $S = W$.

A closed initial segment $\overline{\text{seg}(t)}$ is also an initial segment, for if ℓ is the least element of $W \backslash \text{seg}(t)$, then $\overline{\text{seg}(t)} = \text{seg}(\ell)$. However, an initial segment need not be closed; for example, consider the well-ordered set $X = \omega + 1 = \{0, 1, 2, 3, \ldots, \omega\}$. Then $\text{seg}_X(\omega) = \omega$, which is not a closed initial segment in X.

A well-ordered set is similar (or isomorphic) to the collection of all its initial segments:

Theorem 4.1.3. *Let $(W, <)$ be a well-ordered set, and put $S = \{\text{seg}(w) : w \in W\}$. Then $(W, <) \sim (S, \subset)$.*

Proof outline: It is not difficult to verify that the function $f(x) = \text{seg}(x)$ is the desired order preserving bijection. $\qquad \square$

Lemma 4.1.4. *Let $(P, <_1)$ and $(Q, <_2)$ be well-ordered sets with $a, b \in P$ and $s, t \in Q$ and let $g : \text{seg}_P(a) \to \text{seg}_Q(s)$ and $h : \text{seg}_P(b) \to \text{seg}_Q(t)$ be order preserving bijections. If $a < b$, then $h \mid_{\text{seg}_P(a)} = g$.*

Proof: Let $a < b$. Suppose the conclusion fails, that is, suppose $h \mid_{\text{seg}_P(a)} \neq g$. Because the set of all those $y \in \overline{\text{seg}_P(a)}$ with that $g(y) \neq h(y)$ is a subset of a well-ordered set, fix the least element $y_0 \in \text{seg}_P(a)$ such that $g(y_0) \neq h(y_0)$. There are a number of ways to derive a contradiction. If $g(y_0) < h(y_0)$, then for every $x, z \in \text{seg}(a)_P$ with $x < y_0 < z$, since h is order-preserving, $h(x) = g(x) < g(y_0) < h(y_0) < h(z)$ shows that $g(y_0)$ is not in the range of h, contradicting h being onto an initial segment. Similarly, $h(y_0) < g(y_0)$ implies that g is not onto an initial segment. $\qquad \square$

Theorem 4.1.5. *Let $(W, <)$ be a well-ordered set. For any $w \in W$, there is no order preserving bijection from W to $\text{seg}(w)$.*

Proof: If $f : W \to \text{seg}(w)$ is any *function*, then $f(w) < w$, so by Theorem 2.6.3, such a function can not be an order preserving bijection. $\qquad \square$

Exercise 9. *Show that an arbitrary union of initial segments in a well-ordered set $(W, <)$ is either another initial segment of W or is W itself. Similarly, the union of closed initial segments will always be an initial segment, the closure of an initial segment, or W itself.*

4.2   Transfinite induction

The principle of mathematical induction, as seen so far, applies to only sets that have a well-ordering identifiable with the well-ordered set $\mathbb{N}$. In short, transfinite induction works just like the principle of mathematical induction, however applies to any well-ordered set, particularly, to infinite ordinals other than ω. In fact, transfinite induction is a generalization of "strong induction" (see Theorem 3.2.1).

Principle of transfinite induction: Let A be a subset of a well-ordered set X with x_0 being the least element of X. If
   (i) $x_0 \in A$, and
   (ii) for every $x \in X$, $[\text{seg}_X(x) \subseteq A] \to [x \in A]$
then $A = X$.

In fact, in the statement of the principle of transfinite induction, one can even dispense with part (i), since $\emptyset \subset A$ and if (ii) holds, $\emptyset = \text{seg}(x_0)$ implies that $x_0 \in A$.
Proof of the transfinite induction principle: Suppose that it fails, that is, suppose the condition (ii) holds, but $A \neq X$. Put $C = X \backslash A$. Since X is well-ordered and $C \subset X$, C has a least element, say $c \notin A$. Then $\text{seg}_X(c) \subseteq A$, and by (ii), $c \in A$, a contradiction.   □

Transfinite induction is suited to proving theorems about initial segments of well-ordered sets. The same principle can be adapted to other statements, however, caution is needed regarding what axioms are being used (see [160] and the comments at the end of Section 4.4). Loosely speaking, if some process or construction is based on transfinite induction over a well-ordered set, the process is called *transfinite recursion*. For example, if W is a well-ordered set, one can define a function f on W to any set X by defining inductively the map f, by defining each $f(w) \in X$ according to how f is defined before w. By transfinite induction, the resulting map on all of W is again a function.

Lemma 4.2.1. *For any two well-ordered sets P and Q, either they are similar or one is similar to an initial segment of the other.*

There are many proofs of Lemma 4.2.1; the proof below can be found in, *e.g.*, [160]. Given any function g and set A contained in the domain of g, the shorthand $g[A] = \{g(a) : a \in A\}$ is used for the image of A under g; the function g restricted to A is denoted by $g|_A$. If either of P or Q is finite, then the smaller one is similar to an initial segment of the other; hence any proof need only be applied when P and Q are infinite. One may interpret the idea in the proof given below as an attempt to construct (inductively) an order-preserving injection f from P onto an initial segment of Q; if this process fails for some $P' \subseteq P$, then f takes P' onto Q (then f^{-1} is a bijection from Q onto P').

Proof of Lemma 4.2.1: Let P and Q be well-ordered sets. Fix some symbol x not in Q. By transfinite recursion, define the function $f : P \to Q \cup \{x\}$ by

$$f(p) = \begin{cases} \text{the least element of } Q \setminus f[\text{seg}(p)], & \text{if } Q \setminus f[\text{seg}(p)] \neq \emptyset \\ x, & \text{otherwise.} \end{cases}$$

Set $Q' = Q \cap f[P] = f[P] \setminus \{x\}$ and $P' = f^{-1}[Q] = f^{-1}[Q']$ and define $f' : P' \to Q'$ by $f' = f|_{P'}$. Then f' is onto Q'.

To see that f' is one-to-one, let $a, b \in P'$ with $a < b$; since $a \in \text{seg}(b)$ and $f(b) \in Q \setminus f[\text{seg}(b)] \subseteq Q \setminus \{f(a)\}$, $f(b) \neq f(a)$.

To see that f' is order preserving, let $a, b \in P'$ with $a \leq b$. Then since $f[\text{seg}(a)] \subseteq f[\text{seg}(b)]$, $Q \setminus f[\text{seg}(a)] \supseteq Q \setminus f[\text{seg}(b)]$ and so $f(a) = f'(a) \leq f'(b) = f(b)$.

Claim: Either $P' = P$ or $Q' = Q$.

Proof of claim: If $P' \neq P$, then there is $a \in P$ so that $f(a) \notin Q$, in which case $f(a) = x$; this means that $Q \subseteq f[\text{seg}(a)] \subseteq f[P]$ and hence $Q' = Q$, proving the claim.

Also, P' is an initial segment of P or $P' = P$ and Q' is an initial segment of Q or $Q' = Q$. If $a, b \in P$ with $a \leq b$ and $b \in P'$, then $\emptyset \neq Q \setminus f[\text{seg } b] \subseteq Q \setminus f[\text{seg}(a)]$. Thus $f(a) \in Q$ and hence $a \in P'$. Given $z, w \in Q$ with $z < w$ and $w \in Q'$, let $p \in P$ be such that $f(p) = w$. Then w is the least element of $Q \setminus f[\text{seg } p]$ and since $z < w$, $z \notin Q \setminus f[\text{seg}(p)]$ and so $z \in f[\text{seg}(p)] \subseteq f[P]$ and hence $z \in Q'$.

Therefore, f' is an order-preserving bijection either from P onto an initial segment of Q or else from an initial segment of P onto Q. $\qquad\square$

Transfinite induction can be applied with any well-founded sets (not just well-ordered), including in proofs by structural induction (the term "structural induction" likely originated in model theory); see Section 3.7.

4.3 Cardinals

This section is a very brief introduction to cardinals to establish some terminology. If there is an injection from a set A into a set B, write $|A| \leq |B|$. If there is a bijection from A to B, write $|A| = |B|$, and say that A and B have the same *cardinality*, or are *equinumerous*. This definition is due to Cantor.

To define a cardinal, one needs to give an interpretation for $|A|$. Given two sets A and B, if there is a bijection from A to B, write $A \approx B$. It is easily seen that the relation $\approx$ is an equivalence relation on the collection of all sets. (Note that one does not say "an equivalence relation on the set of all sets", for this leads to Russell's paradox.) Although the following definition leaves open just what an element of an equivalence class is, it is convenient:

Definition 4.3.1. A *cardinal number*, or simply a *cardinal*, is an equivalence class for $\approx$.

Denote the cardinal number containing A by $|A|$. Then two sets have the same cardinality iff $|A| = |B|$.

A set A is called *countable* iff A is either finite or is equinumerous with $\mathbb{N}$, and *uncountable* otherwise. Standard proofs show that $\mathbb{Z}$ and $\mathbb{Q}$ are countable, yet $\mathbb{R}$ is uncountable. Even the set of algebraic numbers in $\mathbb{R}$ is countable. It is known that a union of countably many countable sets is again countable (see Exercise 434).

Cardinal numbers are well-ordered, so induction is often carried out on cardinalities of sets.

Cardinal numbers say something about the "size" of a set. The cardinal number (or *cardinality* of) for a well-ordered set says something about its size; to differentiate between well-ordered sets of the same size, something called *ordinals* are introduced.

4.4 Ordinals

There are different ways to define "ordinals", all equivalent. Recall that for well-ordered sets A and B (in fact, for any linearly ordered sets) A is similar to B, written $A \sim B$, if and only if there exists an order preserving bijection $f : A \to B$.

Definition 4.4.1. An *ordinal* is an equivalence class under $\sim$.

Ordinals are sometimes called *ordinal numbers*. Different ordinals have different "shape". If α is an ordinal and $A \in \alpha$, then A is said to be of type α. An ordinal α is often identified with any set of type α. For two ordinals α and β, say that α precedes β if and only if there exist $A \in \alpha$ and $B \in \beta$ so that $A \subseteq B$. The order on ordinals is given by $\alpha < \beta$ iff α precedes β. Thus, with an abuse of notation, $\alpha < \beta$ can be written $\alpha \subsetneq \beta$, or simply $\alpha \subset \beta$.

For an ordinal β, one can identify each element $\alpha \in \beta$ with its initial segment $\text{seg}(\alpha)$, the set of predecessors of α. Given this identification, some define an *ordinal*, to be a well-ordered set $(X, <)$ with the property that every element $\alpha \in X$ is equal to its initial segment. One can (and some do) *define* natural numbers (and 0) as ordinals: Put $0 = \emptyset$; $1 = \{\emptyset\}=\{0\}$; $2 = \{\emptyset, \{\emptyset\}\} = \{0, 1\}$; $3 = \{\emptyset, \{\emptyset\}, \{\emptyset, \{\emptyset\}\}\} = \{0, 1, 2\}$, and in general, $n = \{0, 1, \ldots, n-1\}$.

The ordinal number $\omega = \{0, 1, 2, \ldots\}$ (with the usual well-order) is the first infinite ordinal, which is really the set of all natural numbers together with 0. (This is a "good" reason why some texts use 0 in the definition of natural numbers—so that they can identify ω with $\mathbb{N}$.) In ordinal arithmetic, if any one of $\alpha < \beta$, $\alpha \in \beta$, or $\alpha \subset \beta$ hold, then all hold.

An ordinal $\beta \neq \emptyset$ is one of two types:

- β is called a *limit ordinal* if $\beta = \cup_{\alpha < \beta}\alpha$, and

- β is called a *successor ordinal* if $\beta = \alpha \cup \{\alpha\}$ for some ordinal α (in this case, β is the smallest ordinal larger than α, sometimes denoted by $\beta = \alpha + 1$ or α^+).

Recall that two sets have the same *cardinality* (think "size") iff there is a bijection between them (see Exercises 430 and 431). A *cardinal number* is an ordinal β whose every initial segment has a different cardinality than β. Every infinite cardinal number is a limit ordinal.

Here are (without proof) a sequence of lemmas that can be used to prove the subsequent theorem. Some of these facts have proofs that rely on facts already proved for well-ordered sets in general; most proofs are simple, and can be considered as exercises. (For details, see [95, pp. 42–43].) The subsequent theorem is used later to give a simple proof by transfinite induction, a proof that could otherwise be very complicated.

Lemma 4.4.2. *Every initial segment of an ordinal is again an ordinal.*

Lemma 4.4.3. *If α and γ are ordinals with $\alpha \subset \gamma$, then α is an initial segment of γ.*

Lemma 4.4.4. *For any distinct ordinals α and β, one is an initial segment of the other. Thus any collection of distinct ordinals is linearly ordered by inclusion. This order is indeed a well-order.*

Lemma 4.4.5. *The union of a set of ordinals is again an ordinal.*

Theorem 4.4.6 (Burali-Forti paradox). *The collections of all ordinals is not a set.*

Proof outline: Let C be the collection of all ordinals. Then C itself is an ordinal, greater than each of its members, a contradiction. □

Suppose that $P(\alpha)$ is a statement involving an ordinal α, perhaps infinite. If $P(\alpha_0)$ holds and $P(\alpha) \to P(\alpha + 1)$ then by transfinite induction, one can conclude that $P(\beta)$ holds from α_0 up to any ordinal below the next limit ordinal. As is, however, one can not 'jump' to the limit ordinal.

For a limit ordinal β the statement $P(\beta)$ is proved by showing that for every $\alpha \in \beta$, $P(\alpha)$ holds. This allows one to prove P "across limit ordinals". Ordinary induction can be thought of as 'pushing up' from n to $n + 1$; transfinite induction can be considered more as 'reaching down and pulling up'. Proofs by transfinite induction are often divided into three cases, one for the base case, one for successor ordinals, and one for limit ordinals. In many instances, only two steps are required, since the process for limit ordinals usually works for successor ordinals, too. In fact, by the comment above, often only one case is necessary.

There are many proofs in set theory that rely on constructing functions by transfinite induction. One might find different proofs in the literature for the same result, both using some kind of transfinite induction, some relatively short, some very long (and horrid). Thanks to an explanation by Prof. Kucera [331] the reason for difference in complexity lies in a subtlety not usually a concern for *non-foundationalists* (translation: mere mortals). Here is a very rough account of that subtlety: In the development of set theory, some authors prefer proofs that don't invoke a "replacement axiom" (see Appendix IV for statement) unless necessary. Avoiding unnecessary assumptions can make a proof very difficult.

As Kucera commented, the replacement axioms are not necessary if transfinite induction is used to construct functions from ordinals to ordinals, however for functions from ordinals to the class of all sets, often the extra axioms are required. (See [160, pp. 178–9] for more intelligent discussion on this matter.) Others implicitly assume these axioms (or assume them at the onset, many pages before the proof in question). For example, (these abbreviations are defined below) compare proofs of AC implying WO, [160, Thm 6M] or [347, p. 182] with [289, Thm 15], or proofs of AC implying ZL, [416, p.531] with [289, p. 40].)

4.5   Axiom of choice and its equivalent forms

In set theory, one begins not with Peano's axioms, but with axioms that apply to sets in general, not just $\mathbb{N}$. (See Appendix IV for such a collection of axioms.) The most famous of such axioms is the "Axiom of Choice" (AC). There are, in fact, many axioms that have been shown to be equivalent to the Axiom of Choice. In this section, a few of these forms are given together with a sequence of proofs showing them all to be equivalent. In any such sequence, it seems that there is always at least one step that is difficult, especially if one restricts the tools available. For more on such equivalences, the reader might look at the reference standards by Herman and Jean Rubin [472, 473]; see also a more recent book [281] by P. Howard and Jean Rubin on consequences of the Axiom of Choice.

To state the Axiom of Choice, a definition is helpful. If $\mathcal{F}$ is a family of sets, a choice function for $\mathcal{F}$ is a function

$$\gamma : \mathcal{F} \to \cup_{F \in \mathcal{F}} F$$

so that for every $F \in \mathcal{F}$, $\gamma(F) \in F$. So a choice function picks an element from every set in the family.

> **Axiom of Choice (AC):** If $\mathcal{F}$ is a non-empty family of non-empty sets, then $\mathcal{F}$ has a choice function.

A standard example (some say it goes back to Bertrand Russell (1872–1970)) used to demonstrate what AC says is: among infinitely many pairs of shoes, it would be easy to pick one shoe from each pair—pick the left one of each pair. If, however, there were infinitely many pairs of socks, AC guarantees that there is still a choice function that picks one sock from each pair. This doesn't seem very surprising, and in fact, most would argue that this goes without saying. The subtlety might lie in the fact that the Axiom of Choice says that all these socks can be picked *at once* even though there is no way to differentiate between socks of a pair.

It might be interesting to note that the following apparently weaker "Axiom of Choice" is indeed equivalent to the Axiom of Choice:

> **Zermelo's Postulate:** For every non-empty family of disjoint non-empty sets $\mathcal{S}$ there exists a choice function $f : \mathcal{S} \to \cup_{S \in \mathcal{S}} S$.

Theorem 4.5.1. *The Axiom of Choice is equivalent to Zermelo's Postulate.*

Proof: The Axiom of Choice clearly implies Zermelo's Postulate, so it suffices to prove only the other direction. Assume that Zermelo's Postulate is true and let $\mathcal{S} = \{S_i : i \in I\}$ be a non-empty family of sets, not necessarily disjoint. From $\mathcal{S}$, create a disjoint family as follows. For each $i \in I$, set $S_i^* = S_i \times \{i\} = \{(s, i) : s \in S_i\}$. Then $\mathcal{S}^* = \{S_i^* : i \in I\}$ is a disjoint family of non-empty sets. By Zermelo's postulate, fix a choice function $f* : \mathcal{S}^* \to \cup_{i \in I} S_i^*$, and for each $i \in I$, set $f * (S_i^*) = (s_i, i)$. Then the function $f : \mathcal{S} \to \cup_{i \in I} S_i$ defined by $f(S_i) = s_i$ is a choice function for $\mathcal{S}$. □

Note: In the above proof, the fact was used that the family of sets was indexed. If a given family of sets is not indexed, how can one create an index set for this family? One has to look more closely at what at an indexed set is. An *indexing* of a family of sets $\mathcal{F}$ by a set I is a bijection

$$\eta : I \to \mathcal{F}.$$

In this case, write $\eta(i) = F_i$ and $\mathcal{F} = \{F_i\}_{i \in I}$. If one chooses η to be the identity function on $\mathcal{F}$, a family of sets can itself act as the index set! Hence, any family of sets can be indexed.

The Axiom of Choice can be stated for indexed families of sets, but is often done using product notation. Since product notation can be a bit confusing for infinite products, the reader might be forgiven for erring on the side of being too pedantic in the following explanation.

Recall that the cartesian product of two sets is written

$$X_1 \times X_2 = \{(a, b) : a \in X_1, b \in X_2\}.$$

To generalize this to a product of infinitely many sets, reinterpret the product $X_1 \times X_2$ as follows: Each $(a, b) \in X_1 \times X_2$ can be considered as the image of a function, $\gamma : \{X_1, X_2\} \to X_1 \cup X_2$, where $\gamma(X_1) = a \in X_1$ and $\gamma(X_2) = b \in X_2$. As a trivial example, if $X_1 = \{5, 10\}$ and $X_2 = \{3, 5\}$, since there are four ordered pairs in $X_1 \times X_2$, there are four different functions, say γ, δ, ϕ, ξ, to be considered:

$$\gamma(X_1) = 5, \qquad \gamma(X_2) = 3;$$
$$\delta(X_1) = 5, \qquad \delta(X_2) = 5;$$
$$\phi(X_1) = 10, \qquad \phi(X_2) = 3;$$

$$\xi(X_1) = 10, \qquad \xi(X_2) = 5.$$

Notice that the meaning of the function γ is clear without the order mattering: "$\gamma(X_1) = 5$ and $\gamma(X_2) = 3$" means precisely the same thing as "$\gamma(X_2) = 3$ and $\gamma(X_1) = 5$". The meaning of the ordered pair $(3,5)$ captures this if one remembers that the 3 came from X_1 and the 5 came from X_2. Then

$$X_1 \times X_2 = \{f : \{X_1, X_2\} \to X_1 \cup X_2 : f(X_1) \in X_1, f(X_2) \in X_2\}.$$

Dealing with indices only, write

$$X_1 \times X_2 = \{f : \{1, 2\} \to X_1 \cup X_2 : f(1) \in X_1, f(2) \in X_2\}.$$

If the sets were indexed by something other than numbers, the meaning of their product would then not depend on order at all:

$$X_\clubsuit \times X_\triangle = \{f : \{\clubsuit, \triangle\} \to X_\clubsuit \cup X_\triangle : f(\clubsuit) \in X_\clubsuit, f(\triangle) \in X_\triangle\}.$$

Recall that $A \times B \neq B \times A$, because the first is ordered pairs of the form (a, b) while the second consists of ordered pairs of the form (b, a). The difference is only in the order in which one writes them down. In fact, either would be fine, if only one had some way of knowing which of the ordered pair came from which set. Usually, the first coordinate is to mean that the element came from the first set listed in the collection A, B. If there are infinitely many sets, however, and no order imposed on the list, then what does one do? The answer is simple: go back to the function interpretation of the product.

Definition 4.5.2. For a family of indexed sets $\{F_i\}_{i \in I}$, define the infinite product

$$\prod_{i \in I} F_i = \left\{ f : I \to \bigcup_{i \in I} F_i : \text{ for each } i \in I, f(i) \in F_i \right\}.$$

Any function $f : I \to \cup_{i \in I} F_i$ for which each $f(i) \in F_i$, in fact determines a choice function. Thus, the Axiom of Choice can be restated as follows:

> **Axiom of Choice (indexed version):** Let $\{A_i\}_{i \in I}$ be a family of non-empty sets. Then $\prod_{i \in I} A_i \neq \emptyset$.

Another axiom that is often a starting point in set theory is called the *well-ordering principle*. By Theorem 2.6.2, the natural numbers can be (or are) well-ordered; can any set be well-ordered? No one has been able to *prove* otherwise, so the following might seem like a reasonable axiom:

> **Well-ordering principle (WO):** Any set can be well-ordered.

In 1904, Zermelo wrote in a letter to Hilbert that the Axiom of Choice implied the well-ordering principle (see [586]). However, Eves [181, p. 297] says that after Zermelo proved the well-ordering principle, it was Emile Borel who was searching for a flaw in Zermelo's proof, and discovered that it relied on the Axiom of Choice, and pointed out that the Axiom of Choice (AC) is equivalent to the well-ordering principle (WO) (that is, if AC is true, then WO is true, and if WO is true, then AC is true). A modified proof of AC implying WO was then published by Zermelo [587] in 1908. Some authors call the well-ordering principle the *well-ordering theorem* (since it can be derived it from the Axiom of Choice). The reverse direction is easy:

Theorem 4.5.3. *The well-ordering principle implies the Axiom of Choice.*

Proof: Suppose that WO holds and that $\mathcal{F}$ is a family of non-empty sets. Since WO holds, each $F \in \mathcal{F}$ can be well-ordered. Since every well-ordered set F contains a minimal element, say, $\min F$, then the function $f : \mathcal{F} \to \cup_{F \in \mathcal{F}} F$ defined by $f(F) = \min F$ is a choice function. □

The other direction (AC implying WO) is not as simple. Two proofs are given here. The first is an adaptation of that found in [347, p. 182] combined with notes on a lecture given by R. Aharoni, (at University of Calgary, 1986). This proof apparently does not rely on replacement axioms. A second proof, occupying only one paragraph, is from Jech [289, p. 39] and is vastly simpler, relying on the stronger form of transfinite induction.

Theorem 4.5.4. *The Axiom of Choice implies the well-ordering principle.*

First proof of Theorem 4.5.4: Let X be a set let $f : 2^X \backslash \{\emptyset\} \to X$ be a choice function.

Look at pairs of the form $(W, <)$, where $W \subseteq X$ and $<$ is a well-ordering of W. Define a pair $(W, <)$ to be *f-compatible* iff for every $t \in W$,

$$f(X \backslash \text{seg}_{(W,<)}(t)) = t.$$

Such f-compatible sets exist by the following: let $x_0 = f(X)$, $x_1 = f(X \backslash \{x_0\})$, and $x_2 = f(X \backslash \{x_0, x_1\})$. It is not difficult to verify that $W = \{x_0, x_1, x_2\}$ with the ordering $x_0 < x_1 < x_2$ is indeed f-compatible.

[*Comment:* If $(W, <)$ is f-compatible, then $(W, <)$ was created according to the rule: choose a next element to be $f(X \backslash \text{elements chosen so far})$].

Fact 0: For any f-compatible sets $(W_1, <_1)$ and $(W_2, <_2)$, either they are equal or one is an initial segment of the other.

Proof of Fact 0: Let $(W_1, <_1)$ and $(W_2, <_2)$ be f-compatible. Since $(W_1, <_1)$ and $(W_2, <_2)$ are well-ordered sets, by Lemma 4.2.1, either they are similar or one is similar to an initial segment of the other. Without loss, let $\alpha : W_1 \to W_2$ be a similarity from W_1 onto either W_2 or an initial segment of W_2.

Let $W_1^* = \{w \in W_1 : \alpha(w) \neq w\}$, everything that α moves. If $W_1^* = \emptyset$, then either $W_1 = W_2$ or W_1 is an initial segment of W_2 (since α is either a similarity onto W_2 or onto an initial segment of W_2). Thus, Fact 0 is proven if it can be shown that $W_1^* = \emptyset$.

Suppose, in hopes of contradiction, that $W_1^* \neq \emptyset$, and let t_0 be the least element of W_1^*. Then

$$t_0 = f(X\backslash \text{seg}_{(W_1,<_1)}(t_0)) = f(X\backslash \text{seg}_{(W_2,<_2)}(\alpha(t_0))) = \alpha(t_0),$$

contradicting $t_0 \in W_1^*$. Hence $W_1^* = \emptyset$, proving Fact 0.

So by Fact 0, without loss assume that $(W_1, <_1)$ and $(W_2, <_2)$ are such that W_1 is an initial segment of W_2 (or $W_1 = W_2$). If $a, b \in W_1 \subset W_2$ then $a <_1 b$ iff $a <_2 b$ (that is, the orders are "compatible" for f-compatible well-orderings).

Let $V = \{x \in X : \text{for some } f\text{-compatible w.o. } (W, <), x \in W\}$. Define $(V, \prec) = \cup\{(W, <) : (W, <) \text{ is } f\text{-compatible}, W \subset X\}$.

By the compatibility of the f-compatible well-orderings, the next fact follows:

Fact 1: $(V, \prec)$ is a totally ordered set.

Fact 2: $(V, \prec)$ is a well-ordered set.

Proof of Fact 2: Let $T \subset V$, $T \neq \emptyset$, and let $t \in T$. Then for some well-ordered set $(W, <)$, $t \in W$. Since W is well-ordered, $W \cap T$ has a least element, call it x.

[*Aside*: If $x \in V$ and W is a w.o. such that $x \in W$, then for any $y \in V$, if $y \prec x$, then $y \in W$ because for some w.o. W', $y \in W'$, but one of W or W' is an initial segment of the other, so $y \in W \cap W'$.]

Thus for any $s \in T$, if $s \prec x$, then $s \in W$. But then $s \prec x$ and $s \in W \cap T$, contradicting x being the least in $W \cap T$. Therefore, x is the least element of T, proving Fact 2.

Fact 3: $(V, \prec)$ is f-compatible.

Proof of Fact 3: Pick $v \in V$ with $(W_1, <_1)$ an f-compatible set such that $v \in W_1$.

The next thing to show is that $\text{seg}_{(W_1,<_1)}(v) = \text{seg}_{(V,\prec)}(v)$. Since $(W_1, <_1) \subseteq (V, \prec)$, it follows that $\text{seg}_{(W_1,<_1)}(v) \subseteq \text{seg}_{(V,\prec)}(v)$. Let $x \in \text{seg}_{(V,\prec)}(v)$; then for any $(W_2, <_2)$ with $V \in W_2$, (by the aside above) $x \in W_2$, so $x \in \text{seg}_{W_1,<_1}(v)$. Therefore, $\text{seg}_{(W_1,<_1)}(v) = \text{seg}_{(V,\prec)}(v)$.

So $f(X\backslash \text{seg}_{(V,\prec)}(v)) = f(X\backslash \text{seg}_{W_1,<_1}(v)) = v$ since W_1 is f-compatible.

Fact 4: $V = X$.

Proof of Fact 4: Suppose not, that is, suppose that $X\backslash V \neq \emptyset$, and since f is a choice function, put $z = f(X\backslash V)$. Put $V' = V \cup \{z\}$ and extend the order $\prec$ to $\prec'$ by defining for every $v \in V$, $v \prec' z$. Then $\text{seg}_{(V',\prec')}(z) = V$ and so $f(\text{seg}_{(V',\prec')}(z)) = f(X\backslash V) = z$. So $(V', \prec')$ is f-compatible, and so $z \in V' \subset V$

(since V was the union of all f-compatible sets). Then $z = f(X \backslash V) \notin X \backslash V$, contradicting that f is a choice function. Hence no such z exists, and so $V = X$, finishing the proof of Fact 4.

Therefore, X is well-ordered. □

Second proof of Theorem 4.5.4: (Based on [289, p. 39].) Let A be a set. To well-order A, it suffices to construct a transfinite sequence $\langle a_\alpha : \alpha < \theta \rangle$ that enumerates A. Such a sequence can be found by induction, using a choice function f for the family S of all nonempty subsets of A. Let $a_0 = f(A)$, and

$$a_\alpha = f(A \backslash \{a_\xi : \xi < \alpha\})$$

if $A \backslash \{a_\xi : \xi < \alpha\}$ is non-empty. Let θ be the least ordinal such that $A = \{a_\xi : \xi < \theta\}$. Then $\langle a_\alpha : \alpha < \theta \rangle$ enumerates A. □

Instead of assuming Peano's axioms for the natural numbers, one could take as an axiom that the natural numbers are well-ordered and then *derive* that P5 holds.

Theorem 4.5.5 (WO→P5). *Assuming that the usual order on $\mathbb{N}$ is a well-ordering, then P5 holds.*

Proof: Assume that $\mathbb{N}$ is well-ordered and assume the hypothesis of P5 holds, that is, that $S \neq \emptyset$ is a non-empty set of natural numbers with $1 \in S$ and satisfying $(x \in S) \to (x' \in S)$. Let $T = \{t \in S : t \notin \mathbb{N}\}$. To show that P5 holds, one must show that $S = \mathbb{N}$, that is, that $T = \emptyset$.

In hopes of a contradiction, suppose that $T \neq \emptyset$. By well-ordering, T contains a least element, say $t_0 \in T$. Since $1 \in S$, $t_0 \neq 1$. Since t_0 is the least element in T, $t_0 - 1 \notin T$, and so $t_0 - 1 \in \mathbb{N}$. By the hypothesis of P5, then $(t_0 - 1)' = t_0 \in \mathbb{N}$, contradicting that $t_0 \in T$. So one must abandon the assumption that $T \neq \emptyset$ and conclude that $T = \emptyset$ and hence $S = \mathbb{N}$, thereby showing that P5 holds. □

For an article (in Spanish) on the equivalence between WO and P5, see [237]. It is interesting to note that there has been some controversy regarding the implication W0 to P5. In a review [*Math. Reviews* 2002k:03003] of an article "Is the least integer principle equivalent to the principle of mathematical induction?" [140], the reviewer (Victor V. Pambuccian) writes: "... the purported equivalence may have been erroneously read into an article by Pieri, in which he proposes an axiom system which differs from the one proposed by Padoa (a variant of Peano's) not only in replacing PI [P5] with LEP [WO], but in the other axioms as well, such as in adding an axiom requiring that there is at most one number which is not a successor." [I have not seen the article, so I can not comment more here, but it might be interesting to investigate the matter more.]

One more axiom, also equivalent to the Axiom of Choice, is called "Zorn's lemma" (named after Max Zorn (1906–1993)). It has many forms.

> **Zorn's lemma (version 1):** In a partial order $(P, \leq)$, if every totally ordered subset of P has an upper bound in P, then P contains at least one maximal element.

Note: Stating Zorn's lemma without the phrase "in P" can allow for misinterpretation.

Some authors suggest that Zorn's lemma is more appropriately called *Kuratowski's Lemma* as Kuratowski published the statement in 1922 [333] whereas, Zorn published in 1935 [590, statement (42)]. Jech [289, p.40] avoids this controversy by calling it the "Kuratowski–Zorn Lemma". Hausdorff and Brouwer also stated "Zorn's lemma" before Zorn did.

Another version of Zorn's lemma is often used in application. Recall that for a set X, a *chain* of subsets of X is a totally ordered sequence of sets, ordered by containment. For example, $\{4\}$, $\{4, 7\}$, $\{4, 5, 7\}$, $\{4, 5, 6, 7\}$ is a chain. If $\mathcal{C}$ is a chain, then the notation $\cup\mathcal{C}$ denotes the union of all sets in $\mathcal{C}$. If $\mathcal{F}$ is a family of sets, a *maximal* element of $\mathcal{F}$ is a set $S \in \mathcal{F}$ so that for every $T \in \mathcal{F}$, if $S \subset T$ then $S = T$. Note, maximal elements might not be unique (for example, the family of three sets $\{2\}, \{2, 4\}, \{2, 5\}$ has two maximal elements).

> **Zorn's lemma (version 2):** Let $\mathcal{F}$ be a family of subsets of a set X with the property that for every chain $\mathcal{C} \subseteq \mathcal{F}$, $\cup\mathcal{C} \in \mathcal{F}$. Then $\mathcal{F}$ has at least one maximal element.

The first version of Zorn's lemma easily implies the second:

Exercise 10. *Prove that version 2 of Zorn's lemma follows from version 1.*

On the other hand, somewhat surprisingly, the second version also implies the first!

Theorem 4.5.6. *The two versions of Zorn's lemma are equivalent.*

Proof: As one direction is left as an exercise above, assume that version 2 holds, and let $(P, <)$ be a poset so that every totally ordered subset of P has an upper bound in P. Let X be the set of all chains (totally ordered subsets) in P. Order X by inclusion and now consider chains in X. If $\mathcal{C}$ is a chain in $(X, \subseteq)$, then

$$\cup\mathcal{C} = \bigcup_{C \in \mathcal{C}} C$$

is also an element of X (another chain in $(P, <)$). Thus, by version 2, $(X, \subset)$ contains a maximal chain $\mathcal{F}$.

Then $F = \cup\mathcal{F}$ is a maximal totally ordered subset of $(P, <)$; by assumption, F has an upper bound x in P. Then x is a maximal element of P (for if not, there is another element $y \in P$ with $F < y$, in which case $F \cup y$ together with $\mathcal{F}$ form a larger chain, contradicting the maximality of $\mathcal{F}$). $\square$

The utility of Zorn's lemma is demonstrated in many well-known results, many of them occurring as exercises in this book. These results include Tychonoff's product theorem for compact spaces (see Exercise 449), the Hahn–Banach theorem (see Exercise 607), the existence and uniqueness of the algebraic closure of a field, the existence of maximal ideals in rings with 1 (see Exercise 682), and that every vector space has a basis (see Exercise 692). See also Exercise 592.

Theorem 4.5.7. *Zorn's lemma (version 1) implies the Axiom of Choice.*

Proof outline: Let $\mathcal{S}$ be a family of non-empty sets. Let P be the set of all choice functions on subsets of $\mathcal{S}$. The set P is ordered by restriction (or inclusion) in the natural way: for two subsets T_1 and T_2 of $\mathcal{S}$, if $T_1 \subset T_\in$ and f is a choice function for T_2, then the restriction of f to T_1 is a choice function g for T_1. In this case, $f \subseteq g$ (as f both functions are sets of ordered pairs). In this manner, $P(, \subseteq)$ is a partially ordered set. For any chain $\mathcal{C}$ in $(P, \subseteq)$, the union

$$\bigcup_{f \in \mathcal{C}} f$$

is a choice function for the union of the domains, so Zorn's lemma applies, yielding a maximal $h \in P$.

Observe that h is a choice function for $\mathcal{S}$, for if it were not, there is some $S^* \in \mathcal{S}$ for which h is not a choice function. Extending h to a function h^* defined by $h^*(S) = h(S)$ for any S in the domain of h, and selecting any $x \in S^*$, put $h^*(S^*) = x$. Thus h is a proper subset of h^*, contradicting the maximality of h. $\square$

Theorem 4.5.8. *Zorn's lemma implies the Well-Ordering Principle.*

Proof: Let X be a set, and assume that Zorn's lemma is true. To show is that X can be well-ordered, that is, that there exists a well-ordering of X. Define the set

$$S = \{(W, \leq) : W \subseteq X, (W, \leq) \text{ is a well-ordered set}\}.$$

Since each well-ordered $(W, \leq)$ is a subset of $X \times X$, consider the partially ordered set $(S, \subseteq)$, where the order is containment. Note that if $(W_1, \leq_1) \subset (W_2, \leq_2)$, then $W_1 \subseteq W_2$ and the orders agree on W_1.

Observe that S is non-empty because the trivial ordering defined by equality, $(X, =) = \{(x, x) : x \in X\}$, is indeed a well-ordered set.

Let $\mathcal{C} = \{(W_i, \leq_i) : i \in I\}$ be a chain in $(S, \subseteq)$. The next claim is that $\cup_{i \in I}(W_i)$, the union of the chain is again a well-ordered set. The sets W_i are nested and so any subset $Y \subset \cup_{i \in I}(W_i)$ is contained entirely within some W_i, and so has a least element (each W_i is well-ordered); this proves the claim. Hence the union of any chain in S is again an element of S.

Since the conditions for Zorn's lemma are satisfied, S has a maximal element, say $(M, \preceq)$ (where $\preceq$ is a well-ordering of some subset M). The next claim is that $X = M$. If otherwise, that is, if $X \neq M$, then there exists $y \in X$ that is not in M. Then one can create a larger well-ordered set $(M \cup \{y\}, \preceq^*)$ defined by $y \preceq^* x$ for all $x \in M$. But then $(M, \preceq) \subseteq (M \cup \{y\}, \preceq^*)$, contrary to $(M, \preceq)$ being maximal. This finishes the claim that $X = M$. Hence $\preceq$ is a well-ordering of X. □

Since ZL $\Rightarrow$ WO $\Rightarrow$ AC has been shown, to complete the demonstration that all three are equivalent, it suffices to show that AC $\Rightarrow$ ZL. This is accomplished by giving a principle from which Zorn's lemma easily follows, and proving that principle using the Axiom of Choice.

> **Hausdorff's maximality principle:** Every partially ordered set has a maximal totally ordered subset (chain).

The Hausdorff maximality principle is sometimes called the "Hausdorff-Birkhoff maximality principle", a special version of which is called the "high-chain principle" (a *high chain* is a chain with no proper upper bound, so maximal chains are high chains), but the proofs below are attempted without this extra terminology.

Theorem 4.5.9. *Hausdorff's maximality principle implies Zorn's lemma.*

Proof: Let $(P, \leq)$ be a poset, and assume that every chain in P has an upper bound in P. By Hausdorff's maximality principle, let $\mathcal{C}$ be a maximal chain in P. By assumption, $\mathcal{C}$ has an upper bound in P, say $u \in P$. Observe that u is a maximal element in P, for if there were to exist an element v with $u < v$, then $\mathcal{C} \cup \{v\}$ would be a larger chain containing $\mathcal{C}$, contrary to $\mathcal{C}$ being maximal. □

For the reader interested in Hausdorff's maximality principle, its applications, and connection to inductive proofs, see [155], a short article in the *American Mathematical Monthly*. The Heine-Borel theorem and the uniform continuity of continuous functions are examples discussed.

The proof of Zorn's lemma from the Axiom of Choice is now completed by showing that the Axiom of Choice implies the maximality principle.

Theorem 4.5.10. *The Axiom of Choice implies the Hausdorff maximality principle.*

Three proofs are given, each significantly different from the others. The first proof is reminiscent of the (difficult) proof of Theorem 4.5.4 and is adapted from [416, pp. 529–532], without invoking terms like "high chain". The correlation between "f-compatible" sets from the proof of Theorem 4.5.4 and "f-chains" below will soon be apparent. Apparently, this proof has evolved from Zermelo's first (1904) proof of the well-ordering principle, Hellmuth Kneser's proof [319] of Zorn's lemma, and a (one page!) proof outline by Weston [567].

Proof of Theorem 4.5.10: The outline of the proof is given, with some details to be filled in by the reader. Let M be a partially ordered set. For any chain (totally ordered subset) C in M, let $\hat{C}$ denote the set of all proper upper bounds for C, (often called the roof of C), that is, all upper bounds for C not including any upper bound contained in C.

Step 1: By AC, let f be a choice function for the family of all non-empty $\hat{C}$'s, that is, for any chain C in M, if $\hat{C} \neq \emptyset$, then $f(\hat{C}) \in \hat{C}$. Call a chain K in M an f-*chain* iff for any subchain $C \subset K$ satisfying $\hat{C} \cap K \neq \emptyset$, then $f(\hat{C})$ is the least element of $\hat{C} \cap K$. That is, if C is a chain in K with proper upper bound in K, then $f(\hat{C})$ is the least of these proper upper bounds.

Step 2: Show that if K is a chain and $C \subseteq K$, then $\hat{K} = \hat{C}$ is equivalent to $\hat{C} \cap K = \emptyset$.

Step 3: Show that if $\mathcal{K}$ is an f-chain with $\hat{K} \neq \emptyset$, then $K^* = K \cup f(\hat{K})$ is an f-chain.

Proof of Step 3: First observe that K^* is indeed a chain with greatest element $f(\hat{K})$ and $K^* \not\subseteq K$. Assume that $C \subseteq K^*$ with $\hat{C} \cap K^* \neq \emptyset$ and let $s \in \hat{C} \cap K^*$. Then for every $c \in C$, $c < s \leq f(\hat{K})$, and so $f(\hat{K}) \notin C$ and $C \subseteq K$.

If $\hat{C} = \hat{K}$, then $f(\hat{C}) = f(\hat{K})$ and $\hat{C} \cap K^* = \hat{K} \cap K^* = f(\hat{K})$. If $\hat{C} \neq \hat{K}$, it follows from Step 2 that $f(\hat{C}) \in K$ and $f(\hat{C} \leq \hat{C} \cap K$.

Step 4: Show that if K and L are f-chains, then $L \subseteq K \cup \hat{K}$ and $K \subseteq L \cup \hat{L}$.

Step 5: If K and L are f-chains, then either $K \subseteq L$ or $L \subseteq K$.

Step 6: Show that the union of an arbitrary set of f-chains is again an f-chain.

Step 7: Let V be the union of all f-chains. By Step 6, V is an f-chain, and so by Step 3, $\hat{V} = \emptyset$. $\square$

Two more ways to show that AC implies Zorn's lemma are more direct.

Theorem 4.5.11. *The Axiom of Choice implies Zorn's lemma.*

Two outlines of proof are presented here, the first from Thomas Jech, the second from Peter Cameron. Both proofs follow the same general philosophy, however they are distinct in their approach. Both assume AC.

First proof of Theorem 4.5.11: (Paraphrased from [289, p. 40]) Let $(P, <)$ be a partially ordered set so that every chain has an upper bound. The idea is by using a choice function for the non-empty subsets of P, construct a chain in P that leads to a maximal element of P.

Let, by transfinite induction, $a_\alpha \in P$ be such that for every $\xi < \alpha$ (if there are any) $a_\xi < a_\alpha$. If $\alpha > 0$ is a limit ordinal, then

$$C_\alpha = \{a_\xi : \xi < \alpha\}$$

is a chain in P and a_α exists by assumption. "Eventually" there is a θ such that there is no $a_{\theta+1} \in P$ with $a_{\theta+1} \in P$. Thus a_θ is a maximal element of P.   □

Second proof of Theorem 4.5.11: (See [95, p. 119].) Let $(P, <)$ be a partially ordered set in which every chain has an upper bound. Suppose, in hopes of contradiction, that P has no maximal element. Attempt to construct a function by transfinite induction from ordinals to P. Let f be a choice function for non-empty subsets of P.

Set $h(0) = f(P)$. Since every chain has an upper bound, and $\emptyset$ is a chain, it has an upper bound, so $P \neq \emptyset$, and thus $h(0)$ is defined. If $h(\alpha)$ has been defined, $\{x \in P : x > h(\alpha)\}$ is non-empty, for otherwise, $h(\alpha)$ would be a maximal element of P. Hence, for each α, let α^+ be the successor to α, and put

$$h((\alpha^+)) = f(\{x \in P : x > h(\alpha))\}).$$

Finally, if λ is a limit ordinal, observe that $\lambda = \{\gamma : \gamma < \lambda\}$ and so can be considered as a chain, and since for $\alpha < \beta$, $h(\alpha) < h(\beta)$, $\{h(\gamma) : \gamma < \lambda\}$ is a chain C_γ in P. Then set

$$h(\lambda) = f(\text{the set of all upper bounds for } C_\gamma).$$

By transfinite induction, h is a function from the class of all ordinals into X. However, h is 1:1, and since the class of all ordinals is not a set, this is a contradiction.   □

See Theorem 13.2.2 for a countable version of Zorn's lemma for measurable sets.

Chapter 5

Paradoxes and sophisms from induction

As lightning clears the air of impalpable vapours, so an incisive para-
dox frees the human intelligence from the lethargic influence of latent
and unsuspected assumptions. Paradox is the slayer of Prejudice.

—J. J. Sylvester,

On a lady's fan, etc.

In [589] is a quotation from Rostrand's *Cyrano de Bergerac*, describing a "jocular version of mathematical induction":

I stand on a platform holding a strong magnet which I hurl upwards.
The platform follows. I catch the magnet and hurl it up again, the
platform following, and repeating this is in stages, I ascend to the moon.

Mathematical induction can be used to prove both simple mundane results and truly fantastic constructs. Occasionally, however, inductive reasoning leads to controversial conclusions. Sometimes this is because the inductive reasoning itself is faulty, or at times, deliberately deceitful! (See [86] for a few remarkable examples.) There might be, however, some bizarre results that mathematical induction "proves", yet no error in reasoning can be found. In this chapter are introduced a few standard inductive arguments that yield questionable results. Can you tell which type each is? Some of the conclusions below are due to faulty reasoning, some may lead to unsolved paradoxes.

5.1 Trouble with the language?

5.1.1 Richard's paradox

There are a couple of paradoxes arising from induction that have become quite famous, the first of which mentioned here is called "Richard's paradox", given by the French mathematician Jules Richard in 1905.

Statement: "Every natural number is definable by an English expression of less than thirty syllables."

Proof (?) by strong induction: Let $S(n)$ be the statement "n is definable by an English expression of less than thirty syllables."

BASE STEP: $n = 1$ is definable as "the least natural number", an expression with less than thirty syllables, and so $S(1)$ holds.

INDUCTION STEP: Let $k \geq 1$ be fixed and assume that $S(1), S(2), \ldots, S(k-1)$ hold, that is, every number less than k is definable by an English expression of less than 30 syllables. If k is not so definable, then k is "the least natural number that is not definable by an English expression of less than thirty syllables"—an expression of 29 syllables, and so is definable after all. This contradiction proves the inductive step.

Hence by mathematical induction, $S(n)$ is true for all n, and so the statement of theorem is true.   □

Exercise 11. *Decide whether or not the result in Richard's paradox is true, and if it is not, find the error (if any) in the given inductive proof.*

Richard's paradox is very similar to something called "Berry's least integer paradox", given by a British librarian, G. G. Berry in 1908. (For a more thorough discussion, see Nicholas Falletta's book *The Paradoxicon* [185, p. 49].) Here's the paradox: Since the set of all natural numbers is well-ordered, the set of all integers n describable by the expression "n is not nameable in fewer than 19 syllables" has a least element, say n^*. (According to [185], Bertrand Russell claimed this number is 111,777.) But n^* is then described by the expression "the least integer not nameable in fewer than nineteen syllables" has eighteen syllables, a contradiction.

5.1.2 Paradox of the unexpected exam

Suppose that a professor announces to a class that "there will be an exam in the next week, and that the exam will be unexpected".

The students agree that the exam can not be on the following Friday—because if by Thursday night they still have not yet had the exam, then the exam *must* occur on Friday. In this case, the exam would be expected.

So, the exam could only occur on one of Monday, Tuesday, Wednesday, or Thursday. If by Wednesday night, the exam has still not yet taken place, the exam could then only occur on Thursday, and it would be expected—so now they have argued that the exam could not occur on either Thursday or Friday! Continuing inductively, the exam could not take place on Wednesday, it could not take place on Tuesday, and so it must occur on Monday—again an expected scenario. So the class concludes that they will not have an exam the next week and that the professor was deceitful.

Exercise 12. *The professor gives the exam on Tuesday, and the students were surprised. Where did the inductive reasoning go awry?*

5.2 Fuzzy definitions

Another kind of paradox comes when examining certain definitions that are not really precise.

5.2.1 No crowds allowed

How many people does it take to form a crowd? If a certain group of people do not already form a crowd, it is unlikely that the addition of just one more person would create a crowd. Continuing one person at a time, one could prove by induction that no crowds ever assemble.

5.2.2 Nobody is rich

Reasoning similar to that used for no crowds could be used to show that nobody could ever be rich, since the addition of one penny to your bank account would not ever transform you from being "not rich" to "rich". Thus, by induction, one could prove that nobody is rich! This reasoning can also "show" that there are no heaps of sand.

5.2.3 Everyone is bald

Certainly, a person with no hair is called "bald". However, if a person has only a single hair, it is likely that most would consider that person to bald, too. Adding a single hair to someone's head would not change one from being bald to "not bald". Therefore, by induction, everyone is bald.

5.3 Missed a case?

Many attempts at an inductive proof fail because the base case is missing. A classic example (*e.g.*, see [355, p. 29]) is the following "proof" that for every positive integer

n, the number $n(n+1)$ is an odd number: since $n(n+1) = (n-1)n + 2n$, if $(n-1)n$ is odd, then certainly so is $n(n+1)$, and so the inductive step flies. However, when checking any value for n, one sees that $n(n+1)$ is always even. The error is that no base case was proved. Sometimes, however, it is not the base case that is missing, but some other case.

5.3.1 All is for naught

Here is a clever example that has appeared in many books (*e.g.*, see [373]).

Statement: "Every non-negative integer is equal to 0."

False Proof: For each non-negative integer n, let $S(n)$ be the statement "$n = 0$". Certainly the base case $S(0)$ is true. So fix some $k \geq 0$ and assume that $S(0), \ldots, S(k)$ are true. To prove that $S(k+1)$ is true, notice that $S(k)$ says $k = 0$ and $S(1)$ says $1 = 0$, hence $k + 1 = 0 + 0 = 0$, proving $S(k+1)$. This concludes the inductive step, and hence the proof by strong induction.

Exercise 13. *Why is this reasoning faulty?*

5.3.2 All horses are the same color

Here is one that has appeared in a number of places; it is apparently due to Pólya.

Statement: *All horses are the same color.*

False proof: The "proof" is by induction. The base case is that one horse is the same color as itself, which is clearly true. For some fixed $k \geq 1$, assume that any k horses are the same color. Examine a remuda of $k + 1$ horses, say H_1, $H_2, \ldots, H_k, H_{k+1}$. By induction hypothesis, $H_1, \ldots, H_k$ are the same color, say roan. Also by induction hypothesis, the horses $H_2, H_3, \ldots, H_k, H_{k+1}$, are of the same color, and since H_2 is roan, so are all the others. Hence, all $k + 1$ horses are the same color, completing the inductive step, and hence the proof.

Exercise 14. *Find the flaw in the reasoning that "proves" that all horses have the same color.*

5.3.3 Non-parallel lines go through one point

No discussion of bizarre conclusions from inductive reasoning would be complete without this old classic (see, *e.g.* [355, 2.1.13 p.30]).

Statement $S(n)$: For any collection of n lines in the plane, if no two are parallel, then all lines intersect in one point.

Proof? For $n = 1$ the statement is plainly true, as it is for $n = 2$ (since no two lines are parallel). Let $k \geq 2$ and assume $S(k)$, that is, that any collection of k non-parallel lines intersect in a single point. To prove $S(k+1)$, it suffices to show that any $k + 1$ non-parallel lines intersect in a point. Let $\ell_1, \ell_2, \ldots, \ell_{k+1}$ be lines, no two of which are parallel. By the induction hypothesis $S(k)$, the first k lines $\ell_1, \ldots, \ell_k$ intersect in some point X. Again by $S(k)$, the last k lines $\ell_2, \ldots, \ell_{k+1}$ intersect in some point Y. The point X is on lines ℓ_2 and ℓ_3, and so is Y. Since two lines intersect in a unique point, $X = Y$, which is the intersection of all lines, concluding the proof of $S(k + 1)$.

By mathematical induction, for all n, the statement $S(n)$ is true.

Exercise 15. *Find the flaw in the reasoning above.*

5.4 More deceit?

5.4.1 A new formula for triangular numbers

Recall that in Section 1.6, the sum of the first n positive integers was called the triangular number T_n, and by Theorem 1.6.1, $T_n = n(n-1)/2$.

Problem: prove for that for all positive integers n, the assertion

$$A(n): \quad \sum_{i=1}^{n} i = \left(n + \frac{1}{2}\right)^2 / 2.$$

Bogus solution: $A(1)$ is true, so assume that for some $k \geq 1$, $A(k)$ is true. Then

$$
\begin{aligned}
\sum_{i=1}^{k+1} i &= \left(\sum_{i=1}^{k} i\right) + (k+1) \\
&= \frac{(k + \frac{1}{2})^2}{2} + k + 1 \qquad\qquad \text{(by } A(k)\text{)} \\
&= \frac{(n + 1 + \frac{1}{2})^2}{2} \qquad\qquad \text{(by algebra)}
\end{aligned}
$$

proves $A(k+1)$ and hence the inductive step. Hence, for all $n \geq 1$, $A(n)$ gives a new formula for the sum of the first n positive integers.

Exercise 16. *Find the error in the above bogus solution.*

The next two examples are quoted from [36], where it is cited that they appeared a few years earlier in *Mathematical Gazette* **72**. The first one is rather standard (*e.g.*, see [180, pp. 450–451]), however the second one might raise an eyebrow! Can you find the flaw in each?

5.4.2 All positive integers are equal

Here is a "proof" by induction that all positive integers are equal. The first step is to prove by induction that for each $n \geq 1$ the statement $S(n)$: if n is the maximum of two positive integers a and b, $(a, b \in \mathbb{Z}^+)$ then $a = b$.

BASE STEP: When $n = 1$, $\max\{a, b\} = 1$, and $a, b \in \mathbb{Z}^+$, then $a = b = 1$, so $S(1)$ holds.

INDUCTIVE STEP: Let $k \in \mathbb{Z}^+$, and suppose that $S(k)$ holds. Suppose that two positive integers c, d satisfy $\max\{c, d\} = k + 1$. Then $\max\{c - 1, d - 1\} = k$, and so by $S(k)$, $c - 1 = d - 1$, and so $c = d$. This completes the inductive step $S(k) \to S(k+1)$.

By MI, one concludes that for all $n \geq 1$, $S(n)$ is true.

Once this is achieved, then, for any two positive integers x and y, taking n to be their maximum, one concludes that $x = y$.

Exercise 17. *Find the flaw in the above reasoning that all positive integers are equal.*

5.4.3 Four weighings suffice

An old popular puzzle concerns 12 coins, one of which is counterfeit and has a different weight from the others; using a balance scale (and no extra weights), the counterfeit coin can be identified with three weighings. In general, if m coins are given, one of which is counterfeit, what is the minimum number of weighings required to identify the fake? This question is answered in Exercise 586.

This puzzle has a variant that is much easier to solve (see Exercise 585); if the counterfeit coin is known to be lighter than the rest, three weighings can locate the counterfeit coin from among 27 coins. It seems reasonable that as the number of coins goes up, so does the number of weighings required to spot the bogus coin, so one might be suspicious of the following claim:

Statement: For any $m \geq 2$, if exactly one of m coins is counterfeit and weighs less than the rest, then the light coin can be identified with at most four weighings on a balance scale.

Proof (?): Base step: If there are only two coins, only one weighing is required. Induction step: Suppose that the result is true for $m \geq 2$ coins, and consider $m + 1$ coins, only one of which is lighter. Lay any one coin aside and apply the induction hypothesis to the remaining m coins. If the light coin is not determined from among these m coins in four weighings, then the coin set aside is the counterfeit, so the result is true for $m + 1$ coins, completing the inductive step. By MI, the statement above is true for any number $m \geq 2$ of coins.

Exercise 18. *What is wrong with the "proof" for the above coin weighing statement?*

In Martin Gardner's article "Mathematical induction and colored hats" [214] more paradoxes concerning induction are entertained. Two articles Gardner references are the first chapter of [348] and [378], an article on paradoxes. (The problem of the colored hats is introduced in Section 17.4 in this volume.)

Chapter 6

Empirical induction

What is the good of drawing conclusions from experience?

—G. C. Lichtenberg, 18th century.

6.1 Introduction

The above quotation was found in a daily bridge column by Phillip Alder (*Calgary Sun*, 25 February, 2001); the column was entitled "Don't jump to conclusions". (Alder says that Lichtenberg was an 18th century German physicist and philosopher.)

Every cow that I have seen has four legs, and so it would be easy for me to conclude that all cows have four legs. Such reasoning is called *empirical induction*—empirical evidence suggesting a pattern that holds in all cases. (Come to think of it, I have seen a variety of cow with no legs—it's called "ground beef".) Okay, perhaps a better example is that since the sun has risen every day this century, it will rise again tomorrow, and hence the expression "is as certain as the sun rising tomorrow." Quoting Martin Gardner [214, p. 137], Charles Sanders Peirce once wrote "I like that phrase, for its great moderation because it is infinitely far from certain that the sun will rise tomorrow." Gardner continues: "There is not a single truth of science, Peirce said, on which he would 'bet more than about a million of millions to one.' "

Exercise 19. *Give an example of a statement $S(n)$ that is true only for $n = 1$ to $n = 1,000,000$, but fails at $n = 1,000,001$.*

When working on a problem, one often gathers information about small cases, and based on this empirical evidence, one might spot what seems to be a pattern; it is "empirical induction" that leads one to believe, at least in part, that the same pattern always holds in more general situations. This guess at the pattern can then sometimes be proved directly, or by mathematical induction. Both empirical induction and mathematical induction are types of what one might call "inductive reasoning". Coming up with the guess is usually done by empirical induction, and proving it is sometimes done by mathematical induction. The difference between the two types of induction is highlighted in this chapter. It seems that experience is the only teacher of how to use empirical induction in formulating guesses, and so no explicit training is given here on how to guess patterns.

There are many examples through the ages of famous mathematicians or other scientists making incorrect guesses based on patterns. In [245] and [247], Richard Guy exhibits a collection of patterns in sequences from which it would be easy to "conclude" a general rule, but many times, incorrectly. Such patterns exhibit something Guy has called "The Strong Law of Small Numbers", roughly to be interpreted as "there aren't enough small numbers to fit all perceived patterns."

Richard Guy has toured the world giving many wonderful lectures based on this theme and has been the subject of many articles and interviews (for example, see [427]). Later in this section and throughout this text are included a few of the more famous examples found Guy's lectures and articles. In [247], Richard Guy also cites Leonhard Euler (1707–1783) as one of the earlier discoverers of The Strong Law of Small Numbers, , which Euler called

"exemplum memorable inductionis fallacis."

The word "induction" has been used even in mathematical literature with different meanings. An interesting quotation due to Neils Henrik Abel (1802–1829) was given by Lakatos [336, p. 133]:

In a letter to Hansteen dated 29 March 1826, Abel characterized "miserable Eulerian induction" as a method that leads to false and unfounded generalizations and he asks what the reason is for such procedures having in fact led to so *few* calamities. His answer is

To my mind the reason is that in analysis one is largely concerned with functions that can be represented by power-series. As soon as other functions enter—and this happens but rarely—then [induction] does not work any more and an infinite number of incorrect theorems arise from these false conclusions, one leading to others. I have investigated several of these and I was lucky enough to solve the problem...

It is noteworthy to see that in the above quotation, "theorems" can be incorrect! Also in the above quotation, it is not exactly clear what "Eulerian induction" is, but certainly this can't have the same meaning as what is now known as "mathematical induction." If one reads Pólya (*e.g.*, [433, p. 90ff]), it becomes immediately clear that "Eulerian induction" is so named because of Euler's techniques, reasoning, and presentation of reasons that led him to his discoveries. Pólya writes:

Yet Euler seems to me almost unique in one respect: he takes pains to present the relevant inductive evidence carefully, in detail, in good order.

He presents convincingly but honestly, as a genuine scientist should do. His presentation is "the candid exposition of the ideas that led him to those discoveries" and has a distinctive charm.

Pólya's sentiments are implicitly shared in [499], where Leonhard Euler was called "one of the first St. Petersburg Academicians".

Later, in [336], Lakatos goes on to contrast the deductivist approach in mathematics with to the inductivist style of science in general. He claims that deductive reasoning stifles independent and critical thought (see pp. 142–143). The present concept of mathematical induction really does typify deductivism, but one might benefit from viewing a MI proof as a final stage in some creative mathematical thinking. The first stages in finding a theorem usually consist of a different type of induction, namely, collecting data, seeing general patterns, making conjecture after conjecture (as in most science), and then finally trying to prove some of the conjectures.

The reader is cautioned that most books with "induction" in the title (even some math texts!) are philosophical discussions about *inference*, not mathematical induction. For example, in *Foundations of Geometry and Induction* [413] by Jean Nicod (a student of Bertrand Russell) one finds a definition of induction:

> **Definition of induction**—What sort of inference is induction? It is defined in current times by the logical form of its premises and its conclusion by saying that it is a passage from the individual to the universal.

Nicod then later says "...perfect induction does not concern us here." What he really examines is how one establishes probabilities concerning generalizing from the individual to the universal. Nicod also notes that probability is different from certainty not only in degree, but in nature. One must be careful even when a probability is 1. If an integer taken at random from a given set X satisfies a certain property with probability 1, this still does not guarantee that *every* number from X satisfies the property; for example, a random natural number in base 10 has at least 3 digits with probability 1, however there are 99 numbers that do not have 3 digits.

In many of the hard sciences (and perhaps many of the soft, too), empirical induction is the only way to guess "the rule". Mathematical induction differs in that it can be used to prove (or sometimes disprove) the rule once it is conjectured. Here are some situations where it is easy to guess the rule, yet ultimately, it is not obvious how to prove the rule. Some of the examples in this chapter have been reported by Guy in [245] (also see [247]).

6.2   Guess the pattern?

Many intelligence tests give a pattern of numbers and ask to provide the next number in the sequence. If you find too ingenious a rule, you won't get the answer that was intended. Your rule could have been a polynomial! For any finite sequence of integers, there are infinitely many polynomials that produce the given sequence; the next number in the sequence could really be anything, depending on "the rule" you find. As an easy example [564, p. 123], consider the polynomial

$$p(x) = 4x^3 - 18x^2 + 32x - 15.$$

One can check that $p(1) = 3$, $p(2) = 9$, $p(3) = 27$, and $p(4) = 81$. Is it reasonable to guess what $p(5)$ is? Would you guess $3^5 = 243$, or the correct answer 195?

6.3   A pattern in primes?

This example is mentioned in the delightful book *Hidden Connections, Double Meanings* by David Wells [564, p. 122]. All primes except 3 are either one more than a multiple of 3 or one less. Call these "more-primes" and "less-primes", respectively. Of the primes less than 100, there are (two) more less-primes than more-primes. This property persists through the hundreds of thousands. It might be reasonable to conjecture that this property is true forever, however, it fails at the plus-prime 608,981,813,029, where the plus-primes then dominate for a while. It has been proved that the lead changes an infinite number of times, thereby destroying any hope of a conjecture either way.

As an added note, Chebychev once conjectured that primes of the form $4k + 3$ eventually outnumber those of the form $4k + 1$; however, it has been proven that the lead again changes hands infinitely often. For more on this and the distribution of primes, see [199].

6.4   A sequence of integers?

Let $s_0 = 1$ and for $n \geq 0$, let

$$s_{n+1} = \frac{1 + s_0^2 + s_1^2 + \cdots s_n^2}{n + 1}.$$

For example, $s_1 = 2$, $s_2 = 3$, $s_3 = 5$, $s_4 = 10$, $s_5 = 28$, $s_6 = 154$, $s_7 = 3520$, $s_8 = 1,551,880$, $s_9 = 267,593,772,160$, and

$$s_{10} = 7,160,642,690,122,633,501,504.$$

Exercise 20. *Is s_n always an integer?*

6.5 Sequences with only primes?

Pierre de Fermat (1601–1665), along with Marin Mersenne (1588–1648) and a host of others, worked extensively searching for primes among integers of the form $2^k + 1$ or $2^k - 1$. If k is even and greater than 2, the number $2^k - 1$ is not prime, since $2^{2m} - 1 = (2^m - 1)(2^m + 1)$. For $k = 2, 3, 5, 7$, the expression $2^k - 1$ is prime, however $2^{11} = 2047 = 23 \cdot 89$, which spoils the conjecture that if p is prime, then $2^p - 1$ is prime.

For $t \geq 0$, define the *Fermat numbers* $F_t = 2^{2^t} + 1$. Then $F_0 = 3$, $F_1 = 5$, $F_2 = 17$, $F_3 = 257$, and $F_4 = 65537$, all prime numbers. Fermat conjectured that for every non-negative integer t, F_t is prime, but Euler proved this to be false.

Theorem 6.5.1 (Euler). $F_5 = 2^{32} + 1$ *is not prime.*

Proof: Put $a = 2^7$ and $b = 5$. Then $a - b^3 = 128 - 125 = 3$ and $1 + ab - b^4 = 1 + (a - b^3)b = 1 + 3b = 16 = 2^4$. Hence

$$
\begin{aligned}
2^{2^5} + 1 &= 2^{32} + 1 = (2^8)^4 + 1 = (2a)^4 + 1 = 2^4 a^4 + 1 \\
&= (1 + ab - b^4)a^4 + 1 \\
&= (1 + ab)a^4 - (a^4 b^4 - 1) \\
&= (1 + ab)a^4 - (a^2 b^2 - 1)(a^2 b^2 + 1) \\
&= (1 + ab)a^4 - (ab + 1)(ab - 1)(a^2 b^2 + 1) \\
&= (1 + ab)[a^4 - (ab - 1)(a^2 b^2 + 1)],
\end{aligned}
$$

and so $1 + ab = 1 + 2^7 \cdot 5 = 641$ is a divisor of F_5. $\square$

Over a century later, Landry proved that F_6 is not prime either! Since then, it has been shown that for $5 \leq n \leq 21$, F_n is composite. For further references, see, e.g., [456, pp. 214–215].

Fermat numbers do, however, share one property (provable by induction):

Exercise 21. *Prove that for every $t \geq 2$, the last digit of the Fermat number F_t is 7.*

The next property of Fermat numbers was mentioned in *Proofs from the Book* [7], and although it says nothing about producing primes, it has an amazing connection to the number of primes being infinite. The following statement follows the convention that an empty product is 1.

Exercise 22. *Prove that for $n = 0, 1, 2, \ldots$,*

$$
F_n = \prod_{i=0}^{n-1} F_i + 2. \tag{6.1}
$$

From Equation (6.1), it follows that the Fermat numbers are relatively prime (if any two had a common divisor, it would also have to divide 2, but Fermat numbers are odd). As pointed out in [7], it then follows easily that there are infinitely many primes (as there are infinitely many Fermat numbers). (Compare with Exercise 207, a variation of Euclid's proof.) Note that a similar divisibility situation to that in Euclid's proof (with a product and something small added) occurs in the consequence of Exercise 6.1.

When addressing the next question, a table of primes might be used to verify the first few values; one might write a small computer program to check larger values if a proof by induction is not immediately apparent.

Exercise 23. *Are all of the numbers in the infinite sequence*

$$31, 331, 3331, 33331, 333331, 3333331, 33333331, 333333331, \ldots$$

prime?

Another classic example is a remarkable polynomial discovered by Leonhard Euler that generates a long sequence of primes. For $n \geq 0$, define $f(n) = n^2 - n + 41$. One notices that $f(0) = 41$, $f(1) = 41$, $f(2) = 43$, $f(3) = 47$, are all prime numbers.

Checking more values, $f(4) = 53$, $f(5) = 61$, $f(6) = 71$, $f(7) = 83$, $f(8) = 97$, $f(9) = 113$, $f(10) = 131$, $f(11) = 151$, $f(12) = 173$, $f(13) = 197$, $f(14) = 223$, $f(15) = 251$, $f(16) = 281$, $f(17) = 313$—all primes! Given the empirical evidence, one might make the following guess:

Conjecture: For each $n \geq 0$, $f(n) = n^2 - n + 41$ is prime.

The reader can check the next twenty values, and still get primes! Can you prove the conjecture in general? It seems that an inductive proof is required, however, primes are curious creatures; knowing the first n primes, there is no known way to predict what the $(n+1)$-st prime is.

Exercise 24. *Determine whether or not for every $n \geq 1$, $f(n) = n^2 - n + 41$ is prime.*

The history of prime-producing polynomials is quite rich; the interested reader might look in [150] or [247] for a start. See also [215] for interesting discussion and references.

6.6　Divisibility

Leibniz observed that for any positive integer n, 3 divides $n^3 - n$ (see Exercise 243), 5 divides $n^5 - n$ (Exercise 252), and 7 divides $n^7 - n$, and for a short time, thought that if t is odd, then t divides $n^t - n$, until he noticed that with $n = 2$ and $t = 9$, $2^9 - 2 = 510$, which is not divisible by 9.

According to [220], the Soviet mathematician D. A. Grave once conjectured that for any prime p, that p^2 divides $2^{p-1} - 1$. This conjecture may seem reasonable since it is true for all primes less than 1000.

Exercise 25. *Find the first prime p that proves Grave's conjecture false.*

6.7 Never a square?

In attempting to solve the next exercise, one might be inclined to invoke a computer search.

Exercise 26. *Define $f(n) = 991n^2 + 1$. Decide whether or not for each $n \geq 1$, $f(n)$ is never a perfect square.*

6.8 Goldbach's conjecture

Christian Goldbach (1690–1764) conjectured in 1742 that every even number greater than 2 is the sum of two primes. For example, $4 = 2 + 2$, $6 = 3 + 3$, $8 = 3 + 5$, and $16 = 3 + 13$. Some even numbers are the sum of two primes in more than one way, for example,

$$20 = 3 + 17 = 7 + 13.$$

(See [24] for an early work on how many ways an even number can be the sum of two primes.) To this day, Goldbach's conjecture has not been resolved, though it has been verified for all even numbers up to 1.615×10^{12} in 1988 [133]. For progress on Goldbach's conjecture up to the late 1940s, see [288]; popular, inviting discussion and more facts can also be found in more recent works, for example, [120], [153], [248], [428].

In 1752, Goldbach also conjectured that any odd natural number greater than 1 is either a prime, or a perfect square, or can be written as the sum of a prime and twice a square. For example, the first non-prime non-square is 15, which is $7 + 2 \cdot 4$, and the next is $21 = 13 + 2 \times 4$. Calculations for the first few thousand cases might have very well caused some mathematicians to search for a proof, perhaps an inductive proof. However, all such efforts were doomed to fail. The first value for which this breaks down is 5777. The following exercise can be resolved with much less effort.

Exercise 27. *Can every odd natural number greater than 3 be written as the sum of a prime and a power of 2? For example, $5 = 3+2$, $7 = 5+2 = 3+4$, $9 = 5+4 = 7+2$.*

6.9 Cutting the cake

Mark n dots on the edge of a circle, and then connect all dots with straight chords as in Figure 6.1; this cuts the circle into various regions. Given n dots, what is the maximum number of regions the circle can be cut into? For $n = 1$ dot, there are no chords, and hence only 1 region. The numbers of regions for the first five cases are 1,2,4,8,16.

Figure 6.1: Cutting the cake

By empirical induction, one might pose the following guess for the number of regions:

Conjecture: The maximum number of regions in a circle created by n dots joined by chords is 2^{n-1}.

Apparently (see [564, pp. 119–120]), the disruption in the pattern was discovered by the mathematician Leo Moser.

Exercise 28. *Show that this conjecture is false by checking the case $n = 6$.*

Exercise 29. *Show that by cutting a cake between every pair of n dots on its circumference, the maximum number of regions formed is*

$$r(n) = \frac{1}{24}(n^4 - 6n^3 + 24n^2 - 18n + 24).$$

6.10 Sums of hex numbers

Start with a penny on a table, then surround this penny with six others to form a hexagon (see Figure 6.2). Again, surround this hexagon with 12 more pennies to form yet a larger hexagon, now with a total of 19 pennies.

Continue this process and get a sequence of so-called *hex numbers*:

$$1, 7, 19, 37, 61, 91, 127, 169, \ldots.$$

Adding these hex numbers cumulatively, the partial sums are

$$1 = 1,$$
$$1 + 7 = 8,$$
$$1 + 7 + 19 = 27,$$
$$1 + 7 + 19 + 37 = 64,$$
$$1 + 7 + 19 + 37 + 61 = 125,$$
$$1 + 7 + 19 + 37 + 61 + 91 = 216,$$

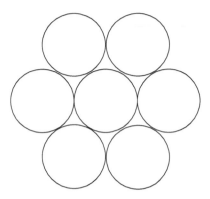

Figure 6.2: Pennies and the second hex number

$$1 + 7 + 19 + 37 + 61 + 91 + 127 = 343,$$
$$1 + 7 + 19 + 37 + 61 + 91 + 127 + 169 = 512,$$

$$\vdots$$

each of which is a perfect cube, $1^3, 2^3, 3^3, 4^3, 5^3, 6^3, 7^3, 8^3, \ldots$. Will this always happen? Perhaps one could begin by finding a general formula for the n-th hex number, and then show that the total of the first n hex numbers is the desired difference of cubes?

Exercise 30. *Prove or disprove that the difference of consecutive cubes is always a hex number.*

6.11 Factoring $x^n - 1$

The polynomial $p(x) = x^n - 1$ occurs in mathematics in a number of different contexts. For example, if r is a real number with $|r| < 1$, one can ask what the infinite geometric series

$$1 + r + r^2 + r^3 + \cdots$$

converges to, if indeed it does. Treating r like a variable, one begins by noticing (multiply it out!) that for any positive integer n,

$$(r^{n-1} + r^{n-2} + \cdots + r^2 + r + 1)(r - 1) = r^n - 1,$$

and so

$$1 + r + r^2 + \ldots + r^{n-1} = \frac{r^n - 1}{r - 1}.$$

As $n \to \infty$, this last equation says

$$1 + r + r^2 + r^3 + \cdots = \lim_{n \to \infty} \frac{r^n - 1}{r - 1},$$

and since $|r| < 1$, $\lim_{n \to} r^n = 0$, and so

$$1 + r + r^2 + r^3 + \cdots = \frac{-1}{r-1} = \frac{1}{1-r}.$$

Other common occurrences of the polynomial $p(x) = x^n - 1$ are in the study of power series, in finding roots of unity in complex numbers, in number theory, in field theory, and even in cryptography.

   The polynomial $p(x) = x^n - 1$ appears in many mathematical calculations, and hence it might be interesting to look at $p(x)$ a little more closely. Can one factor this polynomial into polynomials with integer coefficients? The answer is "of course", as has already been seen:

$$x^n - 1 = (x-1)(x^{n-1} + x^{n-2} + \ldots + x^2 + x + 1).$$

What if one asks further to have $p(x)$ factored into polynomials each with integer coefficients and each polynomial is as small as possible (irreducible, not having further factorization into polynomials with integer coefficients)? For example,

$$
\begin{aligned}
x^2 - 1 &= (x-1)(x+1) \\
x^3 - 1 &= (x-1)(x^2+x+1) \\
x^4 - 1 &= (x-1)(x+1)(x^2+1) \\
x^5 - 1 &= (x-1)(x^4+x^3+x^2+x+1) \\
x^6 - 1 &= (x-1)(x+1)(x^2-x+1)(x^2+x+1).
\end{aligned}
$$

In each of these five factorizations, all coefficients are ± 1 or 0. In 1938, the Soviet mathematician N. G. Chebotarëv [100] (also spelled "Tschebotareff") asked if this always holds. It wasn't until 1941 that Ivanov [285] published an answer to the question. It turns out that for all $n < 105$, the coefficients are indeed 0 or ± 1, but when $n = 105$, this fails. One irreducible factor of $x^{105} - 1$ has degree 48 and coefficients 1, 1, 1, 0, 0, -1, -1, $-\mathbf{2}$, -1, -1, 0, 0, 1, 1, 1, 1, 1, 1, 0, 0, -1, 0, -1, 0, -1, 0, -1, 0, 0, 1, 1, 1, 1, 1, 1, 0, 0, -1, -1, $-\mathbf{2}$, -1, -1, 0, 0, 1, 1, 1, (where the first 1 is the coefficient of x^{48}). In fact, what Ivanov proved was the following (from the review written by J. A. Shohat in *Math. Reviews* 3,164a): Let X_m be the irreducible factor of the polynomial $x^m - 1$ whose zeros are the primitive m-th root of unity. If $m = pq$ is the product of two distinct (odd) primes, then the coefficients in X_m have only values -1,0,1. If $m = pqr$ $(p < q < r, p+q > r)$ is the product of three distinct primes then in X_m, the coefficient of x^r is -2. [Note that $105 = 3 \cdot 5 \cdot 7$.] If m has sufficiently many distinct primes in its factorization, then coefficients in X_m can attain arbitrarily large absolute values.

6.12   Goodstein sequences

Hercules had a fight with the Hydra, and every time Hercules chopped off one of the heads of the Hydra, two more grew back. Can Hercules ever kill the Hydra? It seems not, for if the fight goes on long enough, Hydra will eventually have millions of heads, a monster that even Hercules might not be able to conquer. In the Hercules–Hydra fight, only simple subtraction of 1 and addition of 2 is required to calculate the number of heads at each step. However, what hides behind that "mathematics" is yet another monster, and this monster is defeated by well-ordering. [This delightful paradox was first shown to me by Ron Aharoni over 20 years ago, and only recently did I find its name—thanks to KR!]

Consider the sequence: $1^1 = 1$, $2^2 = 4$, $3^3 = 27$, $4^4 = 256$, $5^5 = 3125$, $6^6 = 46656$, ... The terms are growing faster than any simple exponential sequence. The situation is much more drastic with exponents stacked three high:

$$2^{2^2} = 16, \ 3^{3^3} = 7625597484987, \ 4^{4^4} > 10^{154}, \ 5^{5^5} > 10^{2184}, 6^{6^6} > 10^{36305}, \ldots$$

One can only imagine how fast such expressions grow if the exponents are stacked even higher.

Expressions of the above form can be used to write large numbers using only small digits. For example, using only 1's and 2's, one can express

$$300 = 2^{2^{2+1}} + 2^{2^2+1} + 2^{2+1} + 2^2.$$

Using only digits at most 3,

$$300 = 3^{3+2} + 2 \cdot 3^3 + 3.$$

In general, if n is a large integer and $b \geq 2$ is smaller, a simple algorithm produces such a form. First write $n = qb^x + r$, where $q < b$, x is the largest power possible, and $r < b^x$ is the remainder upon dividing n by b^x. Now apply the same decomposition to each of x and r, and continue until all the exponents are are at most b. For example, with $b = 4$,

$$3205 = 3 \cdot 4^5 + 133 = 3 \cdot 4^{4+1} + 2 \cdot 4^3 + 5 = 3 \cdot 4^{4+1} + 2 \cdot 4^3 + 4^1 + 1.$$

Such a representation is called the *hereditary base b representation of n*. When $b = \omega$, the first infinite ordinal, such a representation is called the *Cantor normal form* (see [160]) of an ordinal.

For each $n, b \in \mathbb{Z}^+$, $b \geq 2$, define $B_b(n)$ to be the positive integer obtained by replacing each "b" with "$b+1$" in the hereditary base b representation of n. For example,

$$B_3(300) = 4^{4+2} + 2 \cdot 4^4 + 4 = 4612.$$

Beginning with any number n, define the *Goodstein sequence* $n_0, n_1, n_2, \ldots$ recursively by setting $n_0 = n$ (written in hereditary base 2), and for each $k = 0, 1, 2, \ldots$, if $n_k > 0$, then define

$$n_{k+1} = B_{k+2}(n_k) - 1;$$

if some $n_k = 0$, the sequence terminates. So the sequence begins with a base 2 form, and at each stage, the base is increased and then 1 is subtracted. The following example for $n = 266$ is given in [67, p. 36] [and many other places on-line]:

$$266 = n_0 = 2^{2^{2+1}} + 2^{2+1} + 2^1;$$
$$n_1 = B_2(n_0) - 1$$
$$= 3^{3^{3+1}} + 3^{3+1} + 3^1 - 1$$
$$= 3^{3^{3+1}} + 3^{3+1} + 2;$$
$$n_2 = B_3(n_1) - 1$$
$$= 4^{4^{4+1}} + 4^{4+1} + 1;$$
$$n_3 = B_4(n_2) - 1$$
$$= 5^{5^{5+1}} + 5^{5+1};$$
$$n_4 = B_5(n_3) - 1$$
$$= 6^{6^{6+1}} + 6^{6+1} - 1;$$
$$= 6^{6^{6+1}} + 5 \cdot 6^6 + 5 \cdot 6^5 + \cdots + 5 \cdot 6 + 5;$$
$$n_5 = B_6(n_4) - 1$$
$$= 7^{7^{7+1}} + 5 \cdot 7^7 + 5 \cdot 7^5 + \cdots + 5 \cdot 7 + 4;$$
$$\vdots$$

Despite the rapid growth of this sequence, it actually terminates at 0. Indeed, this is true for any Goodstein sequence, which was proved by R. L. Goodstein [223] in 1944. Most people might agree that this result is not to be believed, because the growth of the sequence seems to vastly outweigh subtracting just 1 each time.

Here is a proof sketch: Given any hereditary base 2 representation, replace all 2's with ω's. This new ordinal is larger than each term in the sequence, and since an ordinal is a well-ordered set, subtracting 1 from the ordinal number can only be done a finite number of times.   □

In a sense, this proof might seem like a cheat since one had to "go through infinity", whereas every term in the sequence is finite. In fact, in 1982, Kirby and Paris [312] showed that any proof of Goodstein's theorem indeed had to go outside of Peano arithmetic. (The Kirby–Paris result seems very similar to the Paris–Harrington theorem in Ramsey theory—see [231].)

The calculations reported on in [67] are amazing. For the above Goodstein sequence starting with $n = 266$, for $k = 3(2^{402,653,211} - 1)$ (which is roughly $10^{121,210,695}$), $n_k = 0$, and the sequence terminates.

Chapter 7

How to prove by induction

A good proof is one that makes us wiser.

—Yu I. Manin,

A course in mathematical logic.

When learning to prove theorems by mathematical induction, there are usually two challenges. First one must find the proof idea (or understand one from the literature). The second concern is how to present the proof formally. This chapter is concerned with the discovery or understanding process. Some tips include how a stronger result might be easier to prove, or how induction can be used to prove limits. Reading this chapter might implicitly help one's written proof, as well. The next chapter focusses more on aspects of how to present the written proof, complete with a template for writing an inductive proof, and information on notation.

For more on the thinking that surrounds the discovery and writing of an inductive proof, see Pólya [433].

7.1 Tips on proving by induction

Here are some tips that might help while trying to prove a statement $S(n)$ using induction.

1. A problem that says "... for all $n \geq 0$..." says the same as "... for each $n \geq 0$...". Sometimes, a problem might ask to prove "...for any $n \geq 0$..."—this does not mean that you can pick your favorite n and solve only that case; it really means that you must show the problem for any *arbitrary* n, that is, for each possible n.

2. Work out a few examples with actual numbers in $S(n)$ and confirm the truth of the statement for yourself. This process helps one to see how an inductive proof

might go. Furthermore, it is *very easy* to copy the question down incorrectly (heaven knows, I have done it dozens of times while writing this book!), and then get frustrated trying to prove something that is not even true! Needless to say (so I say it anyway) a few minutes of simple checking can often prevent a big headache (or cure it).

3. Work backwards. When proving the inductive step of an equality (or inequality), what one often does in practice is to put the left side of $S(k)$ at the top of a page, and put $S(k+1)$ at the bottom. If one gets stuck going down, one might start at the other end and try working back. For example, in the solution to Exercise 61, one is faced with trying to derive a sequence of equalities of the form

$$1^3 + 3^3 + 5^3 + \cdots + (2k-1)^3 + (2(k+1)-1)^3$$
$$= 1^3 + 3^3 + 5^3 + \cdots + (2k-1)^3 + (2k+1)^3$$
$$= k^2(2k^2 - 1) + (2k+1)^3 \qquad \text{(by } E(k))$$
$$= \vdots$$
$$= (k+1)^2(2(k+1)^2 - 1).$$

It looks pretty daunting, so the next step might be to write

$$= \vdots$$
$$= k^2(2k^2 - 1) + (2k+1)^3 \qquad \text{(by } E(k))$$
$$= 2k^4 - k^2 + 8k^3 + 12k^2 + 6k + 1$$
$$= \vdots$$
$$= (k^2 + 2k + 1)(2k^2 + 4k + 1)$$
$$= (k+1)^2(2(k+1)^2 - 1),$$

working from each end. When you get to the middle and the expressions are the same, you know that you have it!

Note: It is easy to sometimes fool yourself with this method, so be honest with yourself about every step. To write up a proof with this method and then pull some magic trick in the middle might go unnoticed by your instructor, but if it is noticed, it will say more about your work habits than a simple note to the instructor that says that you can't quite bridge the gap. (Many instructors would prefer to see an inductive proof in proper format with an admission of difficulty in the middle rather than a poorly formatted proof or an outright "fudge".)

4. There are usually many different solutions to one problem, so don't panic if yours does not agree precisely with the solution given. For example, there

might be completely different sequences of algebraic manipulation that will prove a particular equality. In fact, it is often very instructive to deliberately seek a different solution.

5. Use only simple algebra in the steps so that any reader (including yourself) can clearly follow what has happened. It is much better to err on the side of showing too many operations than too few. The only steps one should feel comfortable in omitting are those that can be verified with the simplest of high school algebra in a very few steps. For example, adding fractions first requires common denominators—showing this intermediate step is often helpful but not required. Just don't force your readers to go off and do a page of calculations just to verify one equality.

To ease simplification of a huge expression, it is often helpful to put all common factors "out front" first, then simplify the smaller inner factor. For example, in the inductive step of Exercise 54, one is faced with simplifying

$$\frac{k(k+1)(2k+1)}{6} + (k+1)^2.$$

One could proceed like

$$\frac{k(k+1)(2k+1)}{6} + (k+1)^2 = \frac{k(k+1)(2k+1) + 6(k+1)^2}{6}$$

$$= \frac{2k^3 + 3k^2 + k + 6k^2 + 12k + 6}{6}$$

$$= \frac{2k^3 + 9k^2 + 13k + 6}{6}$$

$$\vdots$$

and then one would have to factor a cubic polynomial—sometimes not very much fun. However, if one factors out the term $(k+1)$ first, things get a bit simpler, having only to factor a quadratic:

$$\frac{k(k+1)(2k+1)}{6} + (k+1)^2 = (k+1)\left[\frac{k(2k+1)}{6} + (k+1)\right]$$

$$= (k+1)\frac{k(2k+1) + 6(k+1)}{6}$$

$$= (k+1)\frac{2k^2 + k + 6k + 6}{6}$$

$$= (k+1)\frac{2k^2 + 7k + 6}{6}$$

$$= (k+1)\frac{(k+2)(2k+3)}{6}.$$

This idea of separating common factors from the rest of the calculations can be extremely helpful for more complicated expressions, especially when adding terms each having many factors.

6. The base step can be vacuously true (as in Exercise 684, the base case is true because it never applies) but it is good practice to explicitly mention this fact in an inductive proof. Many writers might say something like "since for $n = 0$ this is clearly true, so assume $n > 0$." This is often the only clue that you are reading an inductive proof.

7. If an inductive proof in some text ends up confusing you, try rewriting it in a formal style, using your own variables, clearly identifying the statement, the parameters, and both the hypothesis and conclusion of the inductive step. For example, when I first discovered the result in Exercise 684, I tried reading the proof in some text, and got lost—it said that the result was "elementary", and so I felt a bit stupid. The notation was a little different from that I was used to, and perhaps that threw me. Only after rewriting it carefully did I agree that it was indeed fairly elementary!

8. Sometimes in the inductive step, there seems to be no way to relate an expression to that found in the inductive hypothesis. One trick is to add and subtract the same term. For example, in Exercise 279, in the inductive step, one has to show that $6^{k+2} + 7^{2k+1}$ is divisible by 43, based on the hypothesis that $6^{k+1} + 7^{2k-1}$ is divisible by 43. To connect the two expressions, one can add and subtract the expression $6 \cdot 7^{2k-1}$ as follows:

$$
\begin{aligned}
6^{k+2} + 7^{2k+1} &= 6^{k+2} + 6 \cdot 7^{2k-1} - 6 \cdot 7^{2k-1} + 7^{2k+1} \\
&= 6(6^{k+1} + 7^{2k-1}) + (-6 + 7^2)7^{2k-1}.
\end{aligned}
$$

In the last line, the first of the two expressions in parentheses is divisible by 43 by the induction hypothesis, and in the second expression, a 43 serendipitously drops out, making the whole expression divisible by 43.

Don't feel bad if you miss this trick—for it is only that, a trick. After a few math courses, you might witness this trick only a few times, but it is nevertheless worth remembering. In this particular case, the "trick" can take a different form by writing

$$6^{k+2} + 7^{2k+1} = 6 \cdot 6^{k+1} + 49 \cdot 7^{2k-1} = 6(6^{k+1} + 7^{2k-1}) + 43 \cdot 7^{2k-1},$$

essentially the same idea as above.

9. Recheck your inductive step with the first case. For example, if $S(n)$ is true for $n \geq 1$, recheck your inductive step to see that it indeed proves $S(1) \rightarrow S(2)$.

Often, one has to subtract 1 in a denominator, say, making an expression nonsensical if $k = 1$. Once you have checked it for the first jump, try another, say, for $n = 4$ to $n = 5$. Checking the inductive step that gets you from $n = 1$ to $n = 2$, say, often prevents errors of the type mentioned Chapter 5 on paradoxes in this volume.

10. In the inductive step $S(k) \rightarrow S(k+1)$, try to do it in one long sequence of equalities (or inequalities). If one has to manipulate $S(k)$ to get $S(k+1)$, be aware of the rules that preserve equality. For example, taking square roots of each side does not necessarily preserve equality. Here is a silly example I was taught in high school that shows how dangerous it is to go from equation to equation, rather than using one long string of equalities (or inequalities):

$$-20 = -20 \qquad \qquad \text{(obviously true)}$$

$$16 - 36 = 25 - 45 \qquad \qquad \text{(rewrite each side)}$$

$$16 - 36 + \frac{81}{4} = 25 - 45 + \frac{81}{4} \qquad \qquad \text{(add } \tfrac{81}{4} \text{ to each side)}$$

$$\left(4 - \frac{9}{2}\right)^2 = \left(5 - \frac{9}{2}\right)^2 \qquad \qquad \text{(factor)}$$

$$4 - \frac{9}{2} = 5 - \frac{9}{2} \qquad \qquad \text{(take square roots)}$$

$$4 = 5 \qquad \qquad \text{(add } \tfrac{9}{2} \text{ to each side)}$$

11. When proving the inductive step, one often gets stuck, not seeing how to get to the last line in a sequence of equalities or inequalities. Often a simple observation must be made, but one that requires a separate proof. Figure out this step, usually working backwards, and then put this observation, usually with proof, *before* you start the string of inequalities so that you can simply refer to it when needed, streamlining the presentation. For example, in the solution to Exercise 188, one soon finds that an inequality like $2\sqrt{k} + \frac{1}{\sqrt{k+1}} \leq 2\sqrt{k+1}$ would be very handy. To check this, one might first investigate by multiplying by $\sqrt{k+1}$, squaring, and then standing back and staring—the actual proof is then done in reverse, starting with the obvious $4k^2 + 4k < 4k^2 + 4k + 1$.

12. If the statement to be proved has the variable n in it, use a different variable, say k, for the inductive step. The reason is that in an inductive step, the k is fixed, whereas n could be thought to be varying. [This comment is echoed in [194].] Some authors use the same variable for both, but this can easily lead to confusion. For example, if $S(n)$ is stated, the inductive step could be $S(k) \rightarrow S(k+1)$—some express this by "Assume that the statement is true for $n = k$ (inductive hypothesis); to be proved is the statement for $n = k+1$." For the beginner, I suggest to stay away from such a format. Professional

mathematicians often use more concise shortcuts; *e.g.*, in [58, p. 20], the following is used: "The case $n = 1$ being trivial, we assume that $n > 1$ and that the assertion holds for all smaller values of n." This is particularly poor style for a novice; such shorthand might be reserved for only those with years of experience with induction.

In this volume, many different variables are used so that the student thinks more about the proof than the letters on the page; it helps one's problem solving ability to be flexible in notation. Standard variables for inductive proofs are usually those reserved for integers, like m, n, p, q, i, j, k, M, and N, but you are certainly not restricted to these. Some authors prefer lower case Greek letters like α, β, γ and δ, and these are recommended when working with ordinals. Try to select variables that might remind the reader (or author) as to their meaning.

13. Be suspicious! When reading a proof in some book (including this one) and something doesn't seem quite right, don't just blindly copy it down and hope that sense can be made of it later. Everyone makes mistakes, including professors and textbook authors. Ask your instructor! Convince yourself whole-heartedly that each step is justified—you learn a great deal more this way.

14. Fight the urge to read the solution after only a few minutes of effort. If after scratching your head for a day, maybe take a peek, get an idea, then try again without the solution in front of you. If you must gain the idea from a published solution, try to rewrite the proof in your own style, perhaps using new variables. If you are submitting your work, and you have discovered the solution in some text, cite your sources! It might be considered *academic dishonesty* by some instructors to find a solution somewhere, use it, and not tell anyone of the source.

7.2 Proving more can be easier

An interesting feature regarding proofs is that it sometimes makes a proof easier if one strengthens the original statement! This is particularly true for some statements provable by induction; such a technique is sometimes called "loading the inductive hypothesis", or "inductive loading". Pólya [434, p. 121] calls this the inventor's paradox—an inventor might be more successful in trying a more ambitious project.

For example, if one were to try to prove the statement "the sum of the first n cubes is a perfect square", one might have trouble finding the proof, however, if one strengthens the statement to "the sum of the first n cubes is $\left[\frac{n(n+1)}{2}\right]^2$", the proof is straightforward (see Exercise 56).

In his wonderful book *Problem-solving Strategies*, Engel [161, p. 180, 7.16] gives the following as an exercise (see Exercise 192 in this volume): prove that for every

$n \geq 1$,

$$\frac{1}{2} \cdot \frac{3}{4} \cdot \frac{5}{6} \cdots \frac{2n-1}{2n} \leq \frac{1}{\sqrt{3n+1}}.$$

Engel then asks the reader to try and prove by a weaker result, namely

$$\frac{1}{2} \cdot \frac{3}{4} \cdot \frac{5}{6} \cdots \frac{2n-1}{2n} \leq \frac{1}{\sqrt{3n}}.$$

Though the second inequality is not as tight, it is much harder to prove by induction (try it!). (This example also occurs in a number of other texts; *e.g.*, see [462].)

Another example (regarding Fibonacci numbers) is found in the proof of Exercise 332 in this volume; there it helps to actually prove two statements simultaneously, the truth of which imply the one result asked for. Perhaps an even more bizarre example (also with Fibonacci numbers) occurs in Exercise 365, a very simple looking result which seems impossible to prove without first proving a more general statement, that of Exercise 352. (The reader is invited to try and prove Exercise 365 first! In fact, when I was writing an earlier draft of this book, I tried to do the innocuous looking one before the more complicated looking one—and got stumped; the more complicated one is really quite easy, and the other follows *directly*.) In Exercise 381, one is asked to prove that a particular sum is a Fibonacci number; if one first guesses as to *which* Fibonacci number is arrived at, then one has a better chance of proving the result. The similar situation arises in Exercise 577, counting ways to place dominoes, where a more precise count is easy to prove by induction, and the proof yields a result stronger than what was asked for! Without making the extra assumption in the inductive step, it is not apparent how one would solve the question. See also the solution of Exercise 515, where a far stronger claim is easier to prove inductively. Again, proving a more precise result is often easier than proving a weaker statement.

Here is yet another example from Pólya [433, pp. 119, 243, Ex. 12] that requires only freshman calculus:

Define a sequence of functions $f_0, f_1, f_2, \ldots$ recursively by

$$f_0(x) = \frac{1}{1-x},$$

and for $n \geq 0$, define

$$f_{n+1}(x) = x\frac{d}{dx}[f_n(x)].$$

The goal is to prove (by induction) that for each $n \geq 0$, the statement

$$S(n): \quad \text{The numerator of } f_n(x) \text{ is a polynomial.}$$

For the moment, ignore the fact that the statement is meaningless, because one has placed no constraints on the denominator, or made any claims about f being a rational function, but the intent might be made clear. [Thanks to Sasho Kalajdzievski for this observation.] Perhaps $S(n)$ was to mean that f is naturally written as a ratio of two simple looking expressions, and the expression in the numerator is a polynomial, as opposed to, say, an exponential function or a square root. One might naively attempt a proof as follows:

BASE STEP: Since the numerator of $f_0(x)$ is 1, $S(0)$ is clearly true.

INDUCTIVE STEP: $(S(k) \to S(k+1))$. For some fixed $k \geq 0$, assume that $S(k)$ is true, that is, assume that the numerator of $f_k(x)$ is a polynomial, say $p(x)$. To be proved is that $S(k+1)$ holds, that is, it remains to prove that the numerator of $f_{k+1}(x)$ is also a polynomial.

All one has at hand is that $f_{k+1}(x) = x\frac{d}{dx}[f_k(x)]$, and that the numerator of $f_k(x)$ is a polynomial. How can one conclude *anything* about $f_{k+1}(x)$? In fact, one can not—not without more information. One needs to strengthen the inductive hypothesis. So, instead, examine the following statement:

$T(n)$: The denominator of $f_n(x)$ is $(1-x)^{n+1}$ and the numerator of $f_n(x)$ is a polynomial of degree n having constant term 0 and with all other coefficients being positive integers.

The statement $T(0)$ is still true, and for each n, the statement $T(n)$ is stronger than $S(n)$, that is, $T(n) \to S(n)$. So instead of proving $S(k) \to S(k+1)$, the following is proved:

INDUCTIVE STEP $(T(k) \to T(k+1))$: For some fixed $k \geq 0$, assume that $T(k)$ is true, that is, there are positive integers $a_1, \ldots, a_k$ so that

$$f_k(x) = \frac{a_1 x + a_2 x + \cdots + a_k x^k}{(1-x)^{k+1}}.$$

Then putting $p(x) = a_1 x + a_2 x + \cdots a_k x^k$,

$$
\begin{aligned}
f_{k+1}(x) &= x\frac{d}{dx}\left[\frac{p(x)}{(1-x)^{k+1}}\right] \\
&= x\left(\frac{p'(x)(1-x)^{k+1} - p(x)(k+1)(1-x)^k(-1)}{(1-x)^{k+1})^2}\right) \\
&= \frac{x[p'(x)(1-x) + p(x)(k+1)]}{(1-x)^{k+2}} \\
&= \frac{x[a_1 + 2a_2 x + \cdots + ka_k x^{k-1}](1-x) + xp(x)(k+1)}{(1-x)^{k+2}} \\
&= \frac{(a_1 x + 2a_2 x^2 + \cdots + ka_k x^k)(1-x) + xp(x)(1+k)}{(1-x)^{k+2}},
\end{aligned}
$$

the numerator of which is

$$a_1 x + (2a_2 + ka_1)x^2 + (3a_3 + (k-1)a_2)x^3 + \cdots + (ka_k + 2a_{k-1})x^k + a_k x^{k+1}.$$

Thus f_{k+1} satisfies $T(k+1)$, completing the inductive step $T(k) \to T(k+1)$.

Hence by MI, for all $n \geq 0$, $T(n)$ is true, and so for all $n \geq 0$, the original statement $S(n)$ is true.   □

7.3   Proving limits by induction

Mathematical induction can be an invaluable tool in evaluating the long term behavior of a sequence or a series. Induction can be used to prove if a sequence or series has a limit, and often, some information about a limit when it exists. Many sequences are defined inductively, and so occasionally, an inductive proof of some property of a sequence is fairly simple.

A sequence is an ordered list, and some sequences have terms that tend to a particular value L as one goes further down the sequence. An infinite sequence $s_1, s_2, s_3, \ldots$ is said to *converge to a limit* L iff for any small real number number $\epsilon > 0$, there is an integer $N = N(\epsilon)$, so that for all $n \geq N$, the nth term s_n is within ϵ of L; in this case, write $\lim_{s \to \infty} s_n = L$. If no such L exists, say that the sequence *diverges* or is *divergent*. Two divergent sequences are 0, 1, 0, 1, 0, 1, ..., and 2, 4, 6, 8, ...

For example, the sequence 1.1, 1.11, 1.111, ..., tends to the value 10/9, because as nearly every child knows, dividing 10 by 9 gives the infinite decimal 1.11111.... Induction can be used to come to the same conclusion in a rather indirect way. In this example, it hardly seems worth the work, but one way is to find an expression for the nth term in the sequence,

$$a_n = 1 + \frac{10^{n-1} + 10^{n-2} + \ldots + 10 + 1}{10^n}$$

that will clearly reveal the same conclusion. Express the nth term of the sequence by

$$a_n = 1 + \frac{10^n - 1}{9 \cdot 10^n}$$

and simplify to get

$$a_n = 1 + \frac{1}{9}\left(1 - \frac{1}{10^n}\right). \tag{7.1}$$

The expression (7.1) is easy to verify using mathematical induction, as is the relation $a_{n+1} = a_n + 10^{-n}$. It then follows that $\lim_{n \to \infty} a_n = \frac{10}{9}$. In fact, from (7.1), one observes that each a_n is strictly less than 10/9, an observation that is also seen by looking at the sequence directly. When a given sequence is not so simple, induction can often be used to prove that a sequence is bounded above (or below) by some number used as a guess for any putative limit.

The following theorem is a standard result (which follows from the completeness of the real numbers, or the Bolzano–Weirstrass theorem, one version of which that says any bounded infinite sequence of real numbers has a convergent subsequence) that is most useful in analyzing "monotonic" sequences. Recall that a sequence $s_1, s_2, s_3, \ldots$ of real numbers is called *non-decreasing* iff for each $i = 1, 2, 3, \ldots$, $s_i \leq s_{i+1}$ or *non-increasing* iff for each $i = 1, 2, 3, \ldots$, $s_i \geq s_{i+1}$.

Theorem 7.3.1. *If $s_1, s_2, \ldots$ is a non-decreasing sequence of real numbers bounded above by a real number U (i.e., for each $i = 1, 2, \ldots$, $s_i \leq U$), then the sequence converges, and converges to a value that is at most U. The analogous result is true for non-increasing sequences bounded from below.*

As an example, for each $n = 1, 2, \ldots$, define

$$a_n = \left(1 + \frac{1}{n}\right)^n.$$

To see that the sequence $\{a_n\}$ converges, by Theorem 7.3.1 it suffices to show this sequence is increasing and bounded above. One way to see that this sequence is increasing is to check the derivative of the function f given by $f(x) = (1 + x^{-1})^x$. Here is the outline of an inductive proof that $\{a_n\}$ is increasing: expand $(1 + 1/n)^n$ by the binomial theorem (see Exercises 103 or 104), and expand $(1 + 1/(n+1))^{n+1}$ also. For $k = 0, 1, \ldots, n$, compare the $(k+1)$-th terms of each. The $(k+1)$ term of the second is greater than equal to the first iff $(n+1-k)(n+1)^{k-1} \leq n^k$, and this is provable by induction on k. So corresponding terms in the expansion get bigger and the second expansion has an additional term, so $a_n \leq a_{n+1}$. To see that the sequence is bounded from above, prove by induction that for each $k = 0, 1, \ldots, n$

$$\frac{n!}{(n-k)!} \leq n^k,$$

and then (see also Exercise 182 for another proof)

$$(1 + \frac{1}{n})^n = \sum_{k=0}^{n} \binom{n}{k} (\frac{1}{n})^k \leq \sum_{k=0}^{n} \frac{1}{k!} \leq 1 + \sum_{k=0}^{n} 2^{-k} \leq 3.$$

So the sequence is bounded above by 3. In fact, the limit of the sequence is e, roughly 2.71828....

Knowing that a particular sequence has a limit (say, by use of Theorem 7.3.1) can sometimes reveal precisely what the limit is. For example, let $a_1 = \sqrt{2}$ and for each $n \geq 1$, define $a_{n+1} = \sqrt{2 + a_n}$. It can be shown that the sequence $\{a_n\}$ is increasing and bounded above by 2, so by Theorem 7.3.1, the sequence has a limit L. Using standard properties of limits (of continuous functions)

$$L = \lim_{n\to\infty} a_n = \lim_{n\to\infty} a_{n+1} = \lim_{n\to\infty} \sqrt{2 + a_n} = \sqrt{2 + \lim_{n\to\infty} a_n} = \sqrt{2 + L}$$

and so $L = \sqrt{2 + L}$, from which it follows that $L^2 - L - 2 = 0$. The roots of this quadratic are $L = -1$ and $L = 2$, and since $L > 0$, $L = 2$ is the desired limit.

As another example occurring in Exercise 546, for any real number $c \in (0,1]$ define the sequence $s_1, s_2, s_3, \ldots$ recursively by $s_1 = c/2$, and for each $n \geq 1$,

$$s_{n+1} = \frac{s_n^2 + c}{2}.$$

Mathematical induction is used to show that the sequence is strictly increasing and strictly bounded above by 1, that is, for each $n \geq 1$, $s_n < s_{n+1} < 1$. Then one can conclude that $\lim_{n \to \infty} s_n$ exists and is at most 1. Throughout analysis, induction is used to prove that certain sequences are monotonic and bounded.

Induction can also help to prove that a complicated sequence can be compared with some known simple sequence (see Exercises 561 and 559 and many others in Section 16.3 and elsewhere throughout this book). Occasionally, a complicated looking sequence can be bounded above and below by two convergent sequences, thereby restricting the limit of the complicated sequence, either precisely, or to some small interval. As a trivial example, in Exercise 191, it is shown that

$$\frac{1}{2n} \leq \frac{1 \cdot 3 \cdot 5 \cdots (2n-1)}{2 \cdot 4 \cdot 6 \cdots (2n)} \leq \frac{1}{\sqrt{n+1}}.$$

Viewing this as a comparison of three sequences, the center sequence is then forced to converge to 0 (by what some call "the squeeze theorem").

An *infinite series* is a sum of the form

$$\sum_{i=1}^{\infty} a_i = a_1 + a_2 + a_3 + \cdots .$$

Loosely speaking, an infinite series is said to *converge* if the series sums to a single finite number. To have infinitely many numbers adding up to a finite number might be counterintuitive, but the following standard example might help.

Consider a square with side length 1. As in Figure 7.1, cut it in half, cut one of the remaining halves in half, cut one of the remaining quarters in half, and so on. Measuring areas of all (infinitely many) pieces gives an intuitive proof of

$$1 = \frac{1}{2} + \frac{1}{4} + \frac{1}{8} + \cdots .$$

Using summation notation, this reads

$$\sum_{n=1}^{\infty} \frac{1}{2^{n-1}} = 1$$

or

$$\sum_{m=2}^{\infty} \frac{1}{2^m} = 1. \tag{7.2}$$

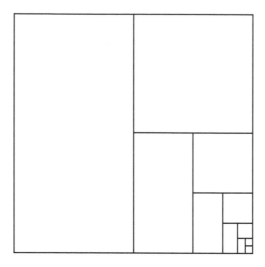

Figure 7.1: The series $\frac{1}{2} + \frac{1}{4} + \frac{1}{8} + \cdots$

At the n-th step, by construction, the remaining area is

$$1 - \left(\frac{1}{2} + \frac{1}{4} + \frac{1}{8} + \cdots + \frac{1}{2^n} \right) = \frac{1}{2^n}, \tag{7.3}$$

a result also provable by induction without too much difficulty (see Exercise 49 for the general formula for a geometric series). In particular, equation (7.3) implies that

$$\frac{1}{2} + \frac{1}{4} + \frac{1}{8} + \cdots + \frac{1}{2^n} = 1 - \frac{1}{2^n}, \tag{7.4}$$

and so in the limit,

$$\sum_{i=1}^{\infty} \frac{1}{2^i} = \lim_{n \to \infty} \left(1 - \frac{1}{2^n} \right) = 1.$$

Zeno's paradox (that finite distance can be made up by an infinite number of steps) is soon resolved by equation (7.2).

If one is faced with an arbitrary infinite series, say $\sum_{n=1}^{\infty} b_n$, what does it mean for it to converge? By definition, this series converges iff the *sequence* of *partial sums*

$$\begin{aligned}
s_1 &= b_1 \\
s_2 &= b_1 + b_2 \\
s_3 &= b_1 + b_2 + b_3 \\
&\vdots \\
s_m &= b_1 + b_2 + \ldots + b_m
\end{aligned}$$

$$\vdots$$

converges. If one can find a closed form for each of the partial sums (as was done in equation (7.2) above), then one might be able to evaluate the sum of the series because then

$$\sum_{n=1}^{\infty} b_n = \lim_{n \to \infty} s_n$$

might be obvious. Even if one can not find an explicit form for each partial sum, one might be able to prove that each partial sum is bounded above by some number or expression.

Notice, that if each term in a series is positive, then the partial sums are increasing. In particular, Theorem 7.3.1 can often be applied to the sequence of partial sums. If each term in a series is positive or zero, then the partial sums are non-decreasing, and so if the partial sums are all at most U, then the above theorem shows that the partial sums have a limit, that is, the series converges, to a value at most U.

For example, in Exercise 180, the inequality to be shown is that for each $n > 1$,

$$1 + \frac{1}{4} + \frac{1}{9} + \cdots + \frac{1}{n^2} \le 2 - \frac{1}{n}.$$

Once this has been shown, one can then conclude that the series $\sum_{j=1}^{\infty} \frac{1}{n^2}$ converges, and converges to a value that is at most

$$\lim_{j \to \infty} \left(2 - \frac{1}{j} \right) = 2.$$

In fact, the sum of the reciprocals of the squares converges to the value $\pi^2/6$, a rather unexpected value, but it is at most 2 as promised. A statement involving limits is similarly implicit in Exercise 181 and others.

There are many other ways to apply mathematical induction regarding limits. In some cases, finding an expression for a partial sum by induction then gives way to seeing that a series diverges (sums to infinity, or does not sum to any single finite number). If the partial sums grow larger than any given n, the series *diverges* (see Exercise 395 for such an example).

Sequences that are defined recursively often don't easily lend themselves to detailed analysis unless some intermediary observations can be shown, and such observations are often proved by induction. Examine the sequence $x_1, x_2, x_3, \ldots$ defined recursively by $x_1 = 2$, and for $n = 1, 2, 3, \ldots$, $x_{n+1} = \frac{1}{2}(x_n + 6)$. This sequence converges to 6, and this is shown below by first proving that the sequence is increasing, and then by showing each term is bounded above by 6. These two facts show (by Theorem 7.3.1) that the sequence converges to something at most 6. Finally, knowing the limit exists, a simple computation shows that it is indeed 6. Computing a few terms,

$$x_1 = 2, x_2 = 4, x_3 = 5, x_4 = 5.5, x_5 = 5.75, x_6 = 5.875, x_7 = 5.9375, \ldots,$$

so the result seems reasonable. To show that the sequence is increasing is done by induction. Let $I(n)$ be the statement that $x_n < x_{n+1}$.

BASE STEP ($I(1)$: The statement $I(1)$ is true since $x_1 = 2 < 4 = x_2$.

INDUCTION STEP ($I(k) \rightarrow I(k+1)$): Suppose that for some fixed $k \geq 1$, $I(k)$ is true, that is, $x_k < x_{k+1}$. Next, it is shown that $I(k+1): x_{k+1} < x_{k+2}$ follows:

$$
\begin{aligned}
x_{k+2} &= \frac{1}{2}(x_{k+1} + 6) \\
&> \frac{1}{2}(x_k + 6) &&\text{(which follows by } I(k)) \\
&= x_{k+1} &&\text{(by def'n)}.
\end{aligned}
$$

This completes the inductive step.

By mathematical induction, for each $n \geq 1$, $I(n)$ is true, and so the sequence is increasing.

To show boundedness: For each $n \geq 1$, let $B(n)$ be the statement that $x_n < 6$. Since $x_1 = 2 < 6$, the base case $B(0)$ is true. For some $k \geq 1$, suppose that $B(k)$ is true. Then

$$
\begin{aligned}
x_{k+1} = \frac{1}{2}(x_k + 6) &< \frac{1}{2}(6 + 6) \quad \text{(by } B(k)) \\
&= 6,
\end{aligned}
$$

shows $B(k+1)$ is also true, completing the inductive step. Therefore, by induction, each term of the sequence is bounded above by 6. Hence, the sequence converges to a limit L that is at most 6. Using the fact (which only holds for convergent sequences) that

$$\lim_{n \to \infty} x_{n+1} = L = \lim_{n \to \infty} x_n,$$

one has

$$L = \lim_{n \to \infty} x_{n+1} = \lim_{n \to \infty} \frac{1}{2}(x_n + 6) = \frac{1}{2}(\lim_{n \to \infty} x_n + 6) = \frac{1}{2}(L + 6);$$

solving the equation $L = \frac{1}{2}(L + 6)$ gives $L = 6$. This concludes the example.

Some other interesting applications of induction to limits can be found in examining interesting expressions such as

$$\sqrt{2 + \sqrt{2 + \sqrt{2 + \ldots}}} = 2.$$

(see Exercise 203, for one). Results for continued fractions are also often obtained using induction.

There are many exercises in this volume that can be used to derive certain limits, sometimes in a less than obvious manner. (See the index under "limits, proving by induction" for many other exercises.)

7.4 Which kind of induction is preferable?

Determining which kind of mathematical induction to use is a difficult problem. There are no general rules. When is it reasonable to use the first principle of mathematical induction instead of strong induction? For which kinds of problems does well-ordering help; when is it feasible to attempt a proof by contradiction and downward induction? What kind of inductive step is preferable? If a problem has two variables, how do you decide upon which to induct? How many base cases need be employed? None of these questions seem to have firm answers.

7.4.1 When is induction needed?

How does someone decide whether or not to attempt an inductive solution to a problem? A few mathematicians may not believe in induction at all, and are only willing to accept theorems that have a direct proof. Some mathematicians suggest (in good fun) that an inductive proof is used only as a last resort, for if one really knows the subject, induction is not required. Others insist (or joke) that only real mathematicians use induction, and they use it often. [Names of individuals in each class are suppressed.]

Some scholars examine theorems that have been proved using the axiom of choice (AC) and work very hard to find proofs that don't rely on AC. There is a large school of thought that prefers proofs that don't rely on contradiction, for the law of the excluded middle is forbidden to them; for these people, inductive proofs that use infinite descent might be troublesome. [See, *e.g.*, [316, p. 332–340] for a discussion of 3-valued logic.]

For many, whether or not to use induction boils down to simply a matter of individual taste, ignoring the deep philosophical questions regarding axiomatic assumptions. Sometimes, a proof by mathematical induction seems just "beautiful", and sometimes not. To decide among which of all proofs for a result is more beautiful, or more appropriate, is very personal.

There is the opinion that induction is used excessively. One writer [71] said in a letter to the editor of *MAA Focus*, "...mathematical induction tends to be over-used as a proof technique." "...I am sure that I am not alone in feeling that induction should generally be avoided. ... If we come across an identity for which the only known proof relies on induction, then it's our job to gain a better understanding of that identity until a more conceptual proof is found." Perhaps the word "should" in the above quotation is inappropriate, since induction is often taught via, at first, simple examples (perhaps those that can be verified independent of induction). [Personally, I object to the word "should" when it is used as an imperative; if an instance of the word "should" is a true imperative, it would be nice to see the rest of the tacitly assumed "if you want... then you should (must?)..." sentence, so that it can be verified logically.] Perhaps one reason that induction seems to be overused is that in teaching induction, often simple statements are used, statements that otherwise have simple direct proofs. For example, for $n \geq 7$, to prove that $n^2 - 5x - 6 > 0$ can be done by induction, however a simpler proof is by factoring: $n^2 - 5x - 6 = (n - 6)(n + 1)$, each factor of which is positive for $n \geq 7$. The inductive proof might be discovered first, however the latter argument is certainly more efficient.

On the other hand, in a response to the letter referred to above, Stockmeyer [514] replies "We can certainly construct proofs of combinatorial identities, such as his example—$1+2+3+\cdots+n = n(n+1)/2$—that hide the induction from our students. As mathematicians, though, we should [sic] keep in mind that with identities of this type induction is always present, at least in the background." Stockmeyer continues "We should not be surprised, then, when induction turns out to be a natural proof technique for identities that sum over the positive integers." Stockmeyer has a point: since the counting numbers are defined recursively, and many operations in math (like addition of integers) are defined recursively, and confirmed inductively, induction is almost always at work. One might take this reasoning a bit further and argue that induction is actually alive in *any* mathematical statement. [The term "mathematical statement" is used here deliberately without definition.]

Induction certainly suffers from the weakness that one already needs to "know" (or guess) the desired result before induction can be applied; only in certain situations can induction be used to discover, say, a particular identity. Finding a particular identity might be done without induction, but for more complicated problems, one often guesses at a formula via non-inductive techniques, whereas induction may provide the easiest proof.

Sometimes a statement (like the sum mentioned above, see Theorem 1.6.1) with an inductive proof has a simple direct proof. Some prefer a direct proof—if one is at all available. For example, some might say that since the formula for the sum of squares (Exercise 54) has many "non-inductive" proofs, induction "should" never be used to prove the formula! One advantage of an inductive proof is that one never has to remember the "trick" behind some direct proof. The formula for the sum of the first n squares can be proved in a way analogous to the Gauss proof for the sum of the first n integers, and similarly so can a formula for the sum of the first cubes (see Exercise 56) be proved "directly", however such proofs soon become more involved for more complicated sums.

Another simple example is in showing that for any positive integer n, the number $n^2 - n$ is even. Since $n^2 - n = n(n - 1)$, and one of n or $n - 1$ is even, the result is nearly immediate. An inductive proof of this fact is also available. Is $n^3 - n$ always divisible by 3? Is there a pattern? An advantage of MI is that perhaps just one technique can be executed repeatedly with little variation, producing a host of results, and even though each individual result may have a "cute" proof, there is no need to remember all the cute proofs. Mathematical induction often requires little thought when applied in such situations. There is an old adage that says something like "why remember all of the details, when you can just remember where to find them?" In a sense, mathematical induction is like a place where one can find many proofs, and so induction is like having the proofs forever at hand.

Sometimes a proof by upward induction can be turned into a proof using well-ordering and downward induction (descent). Some people feel rather strongly about which way is preferable. In fact, it may be true that a vast majority of inductive proofs can be given using a minimal counterexample, and to give such a proof might be pedagogically interesting. Some problems only seem to have a proof by contradiction and downward induction (see, *e.g.*, Exercise 201), so discovering *any* proof might entail trying a number of techniques. Again, it may seem to be a matter of taste as to which kind of inductive proof need be attempted, or taught, but one can imagine that there are situations where one method seems preferable to another.

It seems that not all statements provable by induction also have direct (using only deductive logic and no induction) proofs (like Theorem 1.6.1), though proving this claim might be difficult! It's very likely that there are mathematical statements for which only an inductive proof is known.

There is probably no good answer to "When is induction needed?". As Charles Caleb Colton said in 1825, "... for the greatest fool may ask more than the wisest man can answer". This next exercise might indeed ask more than can be answered, but the questions in it might make for interesting discussion.

Exercise 31. *Does there exist a mathematical truth that does not have a (meaningful) proof by induction? If someone handed you such a truth, how could you guarantee that no inductive proof exists? Does there exist a property provable by induction, but with no other kind of proof? Again, how would one show that no direct proof exists? Can one characterize those mathematical truths for which no inductive proof exists, or can one characterize those statements that have inductive proofs, but fail to have any other proof?*

Perhaps, one might be able to more safely approximate an answer to "Should I try an inductive proof for this problem?" There are some obvious earmarks to a problem that might easily be solved using induction. Finding a particular formula is often not done by induction, however once a correct formula is guessed, mathematical induction is often a natural choice for a proof technique. If the statement of the problem contains only one variable, and that variable is meant to only hold integer values, then induction might be a reasonable choice for a proof technique. If the problem involves a process of steps that can be matched to the positive integers, then induction might be efficacious. It seems that many problems involving recursive definitions (like those in logic, set theory, combinatorics, or computer science) might be candidates for an inductive solution.

Some might say that induction should be one of the first choices of proof technique. Often one gets lucky, even when there are different choices for variables to induct on. For example, one can prove the handshaking lemma (Lemma 15.1.1) by either inducting on the number of vertices in a graph or on the number of edges in a graph; the first of these two proofs is slightly more difficult, so sometimes it pays to try different approaches.

Mathematical induction is often a more powerful technique than one might expect. Even if one attempts an inductive proof and fails, invaluable familiarity with the problem may be gained, knowledge that might very well lead to a direct proof. It may seem that attempting a proof of a conjecture by induction is rarely a (total) waste of time. Induction is an invaluable technique to any student of mathematics and is one of the most powerful tools in the hands of any working mathematician. Practicing induction by proving known results is often how one first learns the method.

7.4.2 Which kind of induction to use?

There are many theorems having two or more different kinds of inductive proof. For example (see Exercise 477), to prove that a tree on n vertices has $n - 1$ edges, one can use either the first principle of mathematical induction, or one can use strong mathematical induction—or one can use well-ordering. The choice of which kind of induction to use may depend upon the idea behind the proof. In the inductive step of one such proof, one assumes the existence of a leaf in a tree (by Lemma 15.2.1), and deleting this leaf gives a tree with one fewer vertex and one fewer edge. Then the induction hypothesis need only be applied to the remaining tree on $n - 1$ vertices, giving $n - 2$ edges; together with the edge deleted, this gives $n - 1$ edges in all.

On the other hand, if one did not think to delete a leaf vertex, one could, by Lemma 15.2.3, delete any edge and get two smaller trees. To apply any induction hypothesis to the smaller trees, one must assume that the the statement holds for *all* smaller trees, that is, one must use strong induction. See the solutions for yet another proof, one using well-ordering and contradiction.

Another example of a theorem that has various proofs is that in the statement of Exercise 515, showing that a tournament has a king. One inductive proof is slightly tricky, however a proof by strong induction is remarkably straightforward! [Thanks to Liji Huang for reminding me of this example.]

In Exercise 214, one is asked to show that if s and t are relatively prime (nonzero) integers so that st is a perfect square, then both s and t are perfect squares. A fairly simple proof comes to mind involving unique factorization, however two very different proofs are available by some form of induction; a proof by strong induction and a proof by infinite descent are given in the solutions. Both seem relatively natural and painless, but there are likely fans of both proofs.

Often one can not tell that strong induction is needed until after failing in an attempt to prove the inductive step using only one inductive hypothesis. It is for this reason that some authors tend to use strong induction for *every* induction proof, for then it matters not that the extra inductive hypotheses are superfluous.

Proving as many (small) cases as you can without any induction serves many purposes. Doing so might make it obvious as to which kind of proof will work. For example, doing just the first two cases directly might reveal that a general proof of the n-th statement might depend on whether n is even or odd. This might suggest two base cases, and an inductive step of the form $S(k) \to S(k+2)$ to prove all cases.

Some results involving more than one variable often require some kind of double induction (see Section 3.5), where it is often easy to identify which variable to induct on first. Sometimes a variable occupies essentially two different roles, and so perhaps a more general statement involving two variables is easier to prove than the one with a single variable. See Exercise 330 for a problem where, at first, it is not clear as to which variable one might try to induct on.

Making decisions about what other types of inductive proof to try is challenging. About all that one can hope is that after many failed attempts, one gains a better sense of when to use, for example, Hausdorff's maximality principle instead of transfinite induction, or when to use infinite descent over induction. Studying the many famous proofs might be the best advice toward learning which kind of inductive proof to try first. Consequently, many failed attempts may be invaluable in any attempt to gain a "feel" for which kind of inductive proof is most apt.

Comment: Professor Farahat once told me that pages of seemingly wasted work (mistakes, preliminary calculations, failed proof attempts, and clumsy proof exposition) are never really wasted; "they build character". Of course, he might have meant "character" in the sense that hard-working people have solid character, but he probably also intended "character" to mean "mathematical character", the wisdom to choose appropriate efficient notation, the *mathematical maturity* [pardon the trite phrase] to select proof techniques that are elegant, and the faith that hard work will produce many answers. His wisdom can be especially appropriate when trying to find (or write or teach) inductive proofs.

Chapter 8

The written MI proof

We may always depend upon it that algebra, which cannot be translated into good English and sound common sense, is bad algebra.

—William Kinston Clifford (1845–1879),

Common sense in the exact sciences.

This chapter is directed at the student, especially the student just learning how to write an inductive proof. Many teachers might also benefit from guidelines presented here when experimenting in how to teach MI writing. Some comments given here may also be useful to professional mathematicians, however most professionals can afford to be slightly relaxed in their presentation because they may expect that most of their audience can reconstruct a formally written proof from their outline.

In this chapter, the student is given a template to follow for writing up an inductive proof. Other aspects of the written proof are given that not only may help with style, but with organization and logical presentation.

Many mathematicians have different ideas as to what is "well written math", and this collection of comments is an attempt to capture the best from many of my teachers. When I was being taught how to write mathematics, at first I resented all the red ink on my assignments—but have since appreciated the incredible work my professors put into my education. Three professors, E. C. Milner, H. K. Farahat, and N. Sauer, stand out as profound influences on my written word (but don't blame them for any idiosyncracies here).

The style of an inductive proof has certain necessary parts, and when learning to write such proofs, I have found that a very strict format is helpful. In my opinion, only after writing up many inductive proofs should the student attempt to abbreviate this style. Most proofs in this text are written keeping in mind this strict format; only a few are written up in more conversational style.

Before getting into what an "ideal" inductive proof might look like, let me tell a brief story. I once gave a one-hour lecture on induction to a group of keen freshman, going over the theory and a few examples. At the end of the lecture, all nodded their heads politely that they understood, and so I announced that, in next class, I would give a quiz—just one proof by induction. The outcome was extremely disappointing. So, I developed a plan. I told the class that I would give them yet another quiz, but I offered them a template most inductive proofs "should" follow. I promised that any submitted solution that followed the template PRECISELY would earn 9 out of 10 marks. To gain full marks, the student needed only to manage the algebraic manipulation proving the inductive step. My goal was to first eliminate many common logical mistakes by teaching them how to format and present their proofs.

8.1 A template

Here is some of the rationale behind the template (the actual template then follows). Suppose that a particular statement regarding n is to be proved for $n \geq 3$.

1. Define the statement that needs to be proved. For example: "For each $n \geq 3$, let $S(n)$ be the statement ..." . If there is more than one variable, be careful of quantification; for example, the expression

   For each $n \geq 3$ let $S(n)$ be the statement that for all $m \leq n$ ...

   is different from

   For each $n \geq 3$ and all $m \leq n$, let $S(n)$ be the statement that ...

   In the second expression, the lower bound for m is not stated, and it is not clear whether or not $S(n)$ depends on the particular value of m, so perhaps something like

   For each $n \geq 3$ and each m satisfying $1 \leq m \leq n$, let $S(m,n)$ be the statement...

   is better. It might help to also identify in advance for which variables a particular sentence even makes sense, later restricting the variable to the cases that are being proved.

2. State the range of n for which the statement is to be proved: For example: "To be proved is that for each integer $n \geq 3$, the statement $S(n)$ is true."

3. Base step: Write the words "Base step" and verify that the base case is true (giving reasons if it is not trivial). For example:

   BASE STEP: $S(3)$ says ... which is true.

4. Inductive step: Write out the words "INDUCTIVE STEP:"

5. State the inductive hypothesis. For simple mathematical induction, this will read like: For some fixed $k \geq 3$, assume that $S(k)$ is true. [Writing out precisely what $S(k)$ says is usually an excellent idea.] For strong induction, this will read something like: "For some fixed $k \geq 3$, assume that $S(3)$, $S(4)$, ..., $S(k)$ are all true," or "For some fixed $k \geq 3$, assume that for $3 \leq j \leq k$, $S(j)$ is true." Labelling the inductive hypothesis with the words "inductive hypothesis" (or "IH") is often a useful practice for the novice.

6. State what needs to be proved, namely $S(k+1)$. It is highly recommended that one writes out $S(k+1)$ specifically so that one sees the required form of the conclusion in the inductive step.

7. Prove $S(k+1)$. If $S(n)$ is an equality (or inequality), it is best (see comments in Section 7.1) to start with one side of $S(k+1)$, and via a sequence of equalities (or inequalities), derive the other side. At the point where the inductive hypothesis is used, this should be mentioned either as a side comment "by $S(k)$", "by induction hypothesis", or even by putting the initials "IH" over the relevant equal sign. For example, in the solution to Exercise 245, the induction hypothesis $S(k)$ is that there exists an integer m so that $2^{2k} - 1 = 3m$. The equalities

$$2^{2(k+1)} - 1 = 4 \cdot 2^{2k} - 1 = 4(2^{2k} - 1) + 3 \stackrel{IH}{=} 4(3m) + 3 = 3(4m + 1)$$

are then used. The above equations could have been written

$$
\begin{aligned}
2^{2(k+1)} - 1 &= 4 \cdot 2^{2k} - 1 \\
&= 4(2^{2k} - 1) + 3 \\
&\stackrel{IH}{=} 4(3m) + 3 \\
&= 3(4m + 1),
\end{aligned}
$$

or in the manner most commonly used:

$$
\begin{aligned}
2^{2(k+1)} - 1 &= 4 \cdot 2^{2k} - 1 \\
&= 4(2^{2k} - 1) + 3 \\
&= 4(3m) + 3 \qquad\qquad \text{(by } S(k)\text{)} \\
&= 3(4m + 1).
\end{aligned}
$$

8. Mention when the inductive step is done. For example, one might write "... completing the inductive step $S(k) \rightarrow S(k + 1)$.", or simply "This completes the inductive step."

9. State the conclusion: "Therefore by mathematical induction, for all $n \geq 3$, $S(n)$ is true. □", using the symbol "□" to denote that the entire proof is complete. Some mathematicians prefer to quantify variables before they are used, as in "... for all $n \geq 3$, $S(n)$ is true." This is a good practice, as it reads more logically, however remember to insert a comma (because "$n \geq 3 \ S(n)$" might be meaningless) or an extra phrase like "... for $n \geq 3$, the statement $S(n)$ holds."

The template I gave was the following, with the instructions to change certain letters and numbers as appropriate:

Problem: Prove that for all $n \geq 3$, ... holds.

Solution: For any integer $n \geq 3$, let $S(n)$ denote the statement...
BASE STEP $(n = 3)$: $S(3)$ says ... which is true because...

INDUCTIVE STEP $S(k) \rightarrow S(k+1)$: Fix some $k \geq 3$. Assume that

$$S(k): \quad \text{(write out what } S(k) \text{ says)}$$

holds. [$S(k)$ is called the inductive hypothesis.] To be proved is that

$$S(k+1): \quad \text{(write out what } S(k+1) \text{ says)}$$

follows. Beginning with the left side of $S(k+1)$,

$$
\begin{aligned}
\text{LHS of } S(k+1) \ &= \ \text{simplify or rearrange} \\
&= \ \cdots \\
&= \ \cdots \\
&= \ \cdots \qquad\qquad (\text{by } S(k)) \\
&= \ \cdots \\
&= \ \text{RHS of } S(k+1),
\end{aligned}
$$

one arrives at the right side of $S(k+1)$, thereby showing $S(k+1)$ is also true, completing the inductive step.

CONCLUSION: By mathematical induction, it is proved that for all $n \geq 3$, the statement $S(n)$ is true. □

Note: In the above template, if the proof is by strong induction, the induction hypothesis should be replaced with "assume that for each j, $3 \leq j \leq k$,

$$S(j): \quad \text{(write out what } S(j) \text{ says)}$$

holds." Also, in the sequence of equations, at the point where the induction hypothesis is invoked, either write "by IH" or mention which statements of the IH are used (*e.g.*, by $S(4)$ and $S(k)$).

Amazingly, in an attempt to simply memorize the format of an inductive proof (for which my students received healthy marks), the students also seemed to discover what was wrong with their previous formats. Also incredibly, students began to ask for more induction problems to practice on! (That was the genesis of this book.) The result was that nearly every student seemed to look forward to cracking inductive proofs on exams (believe it or not!).

8.2   Improving the flow

In some sense, there are two languages for mathematics. There is the language of *doing* mathematics, and there is the language for *writing* mathematics as a formal record of logical implications.

When doing mathematics, one often tries a few cases, drafts a dozen or so diagrams with dots, arrows, and sausages all over the place, and makes a few guesses. Details are often worked out from the desired result, that is, backwards, and variables are changed a few times depending on how confused the writer gets. Only after all the mathematics has been done, can it be written.

Unfortunately, in mathematics journals, the output that readers see is often sterile, uninspiring, a bit on the terse side, usually without superfluous or auxiliary observations, and with a strict adherence to notational consistency. A proof might start out "Let $\delta = \epsilon^3/2$," with no insight as to why such a choice was made. Some authors like to develop a proof, showing why it is required to use such a δ, say, and do the proof from the bottom up. It takes years of practise to come to a balance between the two styles that is acceptable to both the writer and reader.

There are many ways to make an inductive proof read more "smoothly". Some of these ways include rearranging details and careful use of language. In the next section are a few comments on notation, proper use of which may improve presentation. Again, please be reminded that many comments are intended for the novice, and that much of what follows are my personal opinion or style, not rigid rules. Other styles can be equally effective.

8.2.1   Using other results in a proof

If some auxiliary fact is used in proving an inductive step, state the fact (and prove it, if necessary) before starting the induction; this streamlines the proof, as you can simply reference it when needed in the body of the proof. (This topic was briefly discussed in Section 7.1. where Exercise 188 was mentioned.) This means that sometimes you will have to rewrite the proof two or three times. Here is a simple example (from the solution of Exercise 236), which uses only basic number theory:

For $n \geq 1$, let $S(n)$ be the statement that $3^n + 7^n \equiv 2 \pmod 8$ [the notation $x \equiv 2 \pmod 8$ means that x is 2 more than a multiple of 8].

What happens when one tries a proof of the inductive step $S(k) \rightarrow S(k+1)$? Fix $k \geq 1$ and assume $S(k)$ is true. In a first attempt to prove $S(k+1)$,

$$
\begin{aligned}
3^{k+1} + 7^{k+1} &= 3 \cdot 3^k + 7 \cdot 7^k \\
&\equiv 3 \cdot 3^k - 7^k \pmod 8 \\
&\equiv 4 \cdot 3^k - (3^k + 7^k) \pmod 8 \\
&\equiv 4 \cdot 3^k - 2 \pmod 8 \qquad\qquad \text{(by } S(k)\text{)}
\end{aligned}
$$

one gets stuck. Notice that if $4 \cdot 3^k \equiv 4 \pmod 8$ were to hold, the proof becomes simple. In fact, upon a moment of reflection, one sees that 3^k is always an odd number (for $k \geq 1$), and multiplying any odd number by 4, say $(2b+1)4 = 8b+4$, gives precisely 4 modulo 8. Put this observation first, and then write the proof, citing this observation in the appropriate place. One might even comment as to why a particular observation may be needed later. The proof then reads more smoothly. (See the solution for Exercise 236 for the final outcome.)

8.2.2 Clearly, it's trivial!

Particularly in the base step of an inductive proof, there is a tendency to want to say something like "$S(1)$ clearly holds", or "which is obviously true". In general, try to avoid such phrases. There are many examples in mathematical literature that an author uses such a phrase, only to find out later that some special case violates the claim. Eric Temple Bell (a famous mathematician and mathematics historian) once said

" 'Obvious' is the most dangerous word in Mathematics."

Another interesting relevant quotation is from Pólya [433]:

"The advanced reader who skips parts that appear too elementary may miss more than the less advanced reader who skips parts that appear too complex."

In other words, there is often immense value in checking what appear to be simple details. As a rule, if something is obvious, or trivial, then it doesn't take very much effort to give a rigorous proof—so do so.

Also, if one person has a certain picture in their head, indeed an observation can be trivial, but not everyone will have that same picture, so it might help the reader to give at least a strong indication as to why something is true. Once you have seen why something is true, it is easy to say that "it is trivial". There is a story (perhaps apocryphal) about a professor giving a lecture and in the middle, said "this is trivial". He then scratched his head, standing there speechless for a minute. To the surprise of the audience, without saying anything, he left the room. The crowd remained seated, waiting patiently for some news. After 20 minutes, he came back, and announced "yes, it's indeed trivial." [At least five versions of this story persist—I cannot locate the source.]

The famous physicist Richard P. Feynman wrote in *Surely You're Joking, Mr. Feynman* [189, p. 70]:

> We decided that 'trivial' means 'proved'. So we joked with mathematicians: 'We have a new theorem—that mathematicians can only prove trivial theorems, because every theorem that is proved is trivial'.

8.2.3   Pronouns

Try to avoid using too many pronouns or vague descriptors. For example, in the middle of your proof, if you write "...then it is an even number..." the term "it" might be meant to indicate any of a long list of things. Generally, any occurrence of words like "it" or "this" will point to the very last thing written, so be careful to check what "it" might mean. Also, if you want to refer to a formula or equation already mentioned, one can go back, display that formula on its own line, mark it with a star or a number, and then later say something like "... and so by equation (*) above, one has...". This saves one having to repeat the same formula again, yet there is no ambiguity as to which equation you are now referring to.

8.2.4   Footnotes

In mathematical writing, it was once common practice to use footnotes, especially for comments and references. Footnotes can be used to provide a comment, one which would interrupt the flow if included in the text. Footnotes were often used for bibliographic references, saving the reader from flipping to the back of the book. Mathematical typesetting has evolved a great deal in the past few years, and it seems that footnotes are now on the way out in math; in other sciences, footnotes are still rather common. Today, in mathematics it is now customary to refer the reader to a bibliography by use of labels. For comments that are an aside, footnotes are still efficacious, but putting such comments in square brackets also works.

8.2.5   We, let's, our, will, now, must

In mathematical writing, it is common to encounter the "royal we", as in "We see that...", or "we have ...." or "Let's calculate...", or "in our assumption that ...". In most cases, there is no need for such, and some might think that such usages are downright silly. Once a reader gets used to "we wish to prove...", it is rather difficult to write while avoiding such phrases.

Some authors believe that using "we" makes the mathematics more personal, that its use makes the reader feel invited to a cooperative process with the author.

Students emulate that which they see in the literature, and so "we" is now commonly found in homework, too. [This might be appropriate for joint submissions where the authors say "We researched this problem...", in first person.] If one attempts to write mathematics without the word "we", a writer is often forced to switch from a narrative about how to proceed, to concentrating on the logic behind the mathematics. I encourage younger mathematicians to try writing without "we"; the mathematics then tends to stand out. After all, the royal "we" is just a tradition, and in my opinion, an unnecessary one. [For consistency, I have removed most occurrences of "we" in this book, but doing so was a real challenge.]

There is only one situation where I can see using the word "we", and that is when two or more authors are passing along thoughts in a first person manner. Some authors write in the first person (*e.g.*, "I once proved that...", or "Erdős once told me ...") whereas some find such conversational writing a bit too informal; I think that first person prose in mathematics is sometimes refreshing.

Some authors insist on not using the word "will". For example, the phrase "the proof will be presented in Section 2..." might be written as "the proof is presented in Section 2". One reason for eliminating such a "will" is that word can be taken to mean that something in the future is about to happen—or as an order ("you will soon see..."). [Thanks to Ted Bisztriczky for the suggestion.] If an author wants to indicate what will be done in an subsequent volume, "will" might be replaced with "plan to", quite possibly a more reliable phrase. In a similar vein, sentences beginning "Now, let x be ..." can be avoided, especially since the word "Now" can often be tacitly assumed, and it is very easy to overuse it.

There is one more word that a writer *must* be careful about: "must". This word smells more of imperative than of proof. If one wants to skip some logic, one might say "this must be true". Such a phrase can mean any number of things; first, it could mean "this is true", and supporting arguments have been given. On the other hand, it could mean "it is probably true, yet I don't have a proof handy". When refereeing a journal article, a professional mathematician might raise an eyebrow if the word "must" is used anywhere; "must" often points to where errors are hiding.

8.3   Using notation and abbreviations

Here are a few points on notation that might be worthwhile to know. These points refer not only to inductive proofs, but written mathematics in general.

Perhaps one of the most useful things to keep in mind is that when using some particular notation, it is always safe to describe it first. For example, in calculus, one often denotes $\frac{dy}{dx}$ by simply y', however, if given the expression

$$y^2 = s^2 - 3^x,$$

the term y' might not be so meaningful; if one is asked to find $\frac{dy}{ds}$, one might start by saying "let $y' = \frac{dy}{ds}$ and $x' = \frac{dx}{ds}$" and then calculating might begin "upon differentiating with respect to s, one finds

$$2y \cdot y' = 2s - 3^x \ln(3)x',$$

and so ..." With the extra words about notation at the beginning, there is little room for doubt about what later expressions mean.

On the other hand, there is much notation that is standard and has well accepted meaning in any circumstances. Unfortunately, some students have learned to use certain notations without remembering what they really mean. It is often helpful to read out loud that which you have written in symbols; this technique may help the author to find better notation.

Symbols that are commonly misused are "$\Rightarrow$" and "$\rightarrow$". Let P and Q be (complete) sentences that can be either true or false. The expression "$P \rightarrow Q$" is notation for "P implies Q" or "if P then Q". The statement $P \rightarrow Q$ is an *implication*. The expression $P \Rightarrow Q$ means that P *logically* implies Q, that is, the truth of $P \rightarrow Q$ follows directly from the rules of the language. Many mathematicians tend to use the logical implication arrow $\Rightarrow$ when they mean only $\rightarrow$; however, this abuse of the notation is often handy when there many other arrows, like those used in limits. In lectures, it is often tacitly assumed that "$\Rightarrow$" means only "implies". Sometimes the expression "$P \rightarrow Q$" is replaced with "P therefore Q". The word "therefore" (often capitalized to indicate the beginning of a new sentence) used too many times in a row can appear boring, so the words "hence" or "thus" are often used. The word "whence" means "from which" and its use is not generally preferred (however, I like to use it once in a while). "From whence" is improper, as is any sentence beginning "Whence..."; "whence" is not a fancy form of "hence".

If both $P \rightarrow Q$ and $Q \rightarrow P$, then write $P \leftrightarrow Q$, often expressed as "P if and only if Q" (where the "P if Q" part is the implication $Q \rightarrow P$). The expression "if and only if" is often abbreviated "iff".

Two pieces of notation that seem to creep into (and take over!) solutions on math exams or homework are "$\therefore$" and "$\because$" shorthand for "therefore" and "because" respectively. My guess is that in university homework assignments, such notation is misused over half of the time. It seems that many students have learned to write $\therefore$ beside nearly every expression, and include no real reasons. Students copy their teachers, and since it seemed cool for their teacher to use these symbols, it is cool for them too. In my own classes, I have since forbidden the use of these symbols, requiring my students to instead use words (and complete sentences) to convey what "therefore" means. (The proper use of implication arrows is encouraged, also.) Misused, these notations confuse the reader; overused, they offer no real help to the reader. [Also, their usage seems to be becoming passé—maybe that is only wishful thinking.]

When practical, I encourage students to use the *words* instead of symbols, at least until proper usage of notation is learned. "Longhand" has an added benefit of forcing the student to think while writing, with the goal of communicating their ideas, rather than writing something that merely looks "technically fancy". The central reason for learning how to write up proofs is to learn how to communicate ideas clearly. Often, an added benefit to writing things out in longhand is that doing so forces you to organize your thoughts more.

It is incredible how many exams and assignments are turned in for grading that do not contain a single word, only equations, dots, arrows, and charts with numbers. One marked difference between high school and university is that in high school, students seem to get away with writing only a record of steps processed, whereas in university, most often, the student is expected to give coherent and complete reasoning. Someone once told me "good mathematics is good prose," perhaps a thought worthwhile remembering. Most math texts are written so that *everything* is a part of a complete sentence (look for periods at the end of equations, even in this book). The process by which a student learns to write in this manner is often painful, but the rewards are incredible.

As in ordinary prose, abbreviations are useful, but incorrect usage can drastically alter meaning in mathematics—or at least make reading clumsy. Here are a few more abbreviations that are often misused. The expression "Q.E.D." is short for "quod erat demonstratum", literally, "that which was to be demonstrated", and often appears at the end of a proof. This does *not* mean that one can put "QED" at the end of every proof—only use it if the phrase "that which was to be demonstrated" would make sense in its place. ["QED" is also short for "quantum electrodynamics", as in Richard Feynman's book *QED The strange theory of light and matter* [190], but I digress.] Similarly, "Q.E.F" is short for "quod erat faciendum", meaning "that which was to be done". (This is sometimes used when, for example, a construction of a promised object is accomplished, so a loose translation might be "that which was to be made"; in any case, QEF is rather archaic and is seldom used in mathematical works lately.) One standard way to indicate the end a proof these days is with "□".

Two other commonly misused abbreviations are "*e.g.*", an abbreviation for *exempli gratia*, ("for the sake of example"), and "*i.e.*", short for *id est* ("that is"). Misuse of these abbreviations only detracts from a beautiful proof. One particularly bad place to use "*i.e.*" is in any proof with complex numbers (since $i = \sqrt{-1}$) or exponential functions (where $e = 2.718...$).

Avoid using "*etc.*" in mathematical proofs, *especially in proofs by induction*, since it can to lead to ambiguity in what is really intended. After all, "*etc.*" is only an abbreviation for *et cetera*, meaning, "and the rest", a notion that can usually be made precise in mathematics.

The notation "s.t." is used in place of "such that", or "so that", and is often safely employed in a proof with no variables named s or t. Another notation for this notion is "$\cdot \ni \cdot$", but this is not so common (some professors use this notation without the dots on either side, but this can be confused with "contains as an element"). Finally, the notation "$\forall$" means "for all", and "$\exists$" means "there exists", and are examples of what are called "quantifiers". The sentence "$\forall x \exists y \cdot \ni \cdot y > x$

then says that for every x there is a y so that y is larger than x.

Requiring complete sentences in proofs often helps the student organize thoughts; appropriate notation helps one to express these thoughts efficiently. Acquiring both of these skills is often quite painful. Learning to write induction proofs properly can often be a gentle introduction to these arts, as the format and notation are usually straightforward.

Part II

Applications and exercises

In open statements in this part, n is an integer, usually $n \geq 1$ unless otherwise specified. For the most part, all necessary notation and definitions used in a question are given in the same section as the question, so many problems here can be understood without much previous knowledge. With only a few exceptions, nearly all exercises here can be solved by using some form of induction; nearly all exercises requiring an infinite form of induction or not requiring induction are clearly identified.

Among these exercises one can find an amazing array of classic theorems. Most exercises can be done with only simple algebra; only a few require calculus or more sophisticated machinery. Most questions have solutions, hints, or references in Part III.

Chapter 9

Identities

> *The business of concrete mathematics is to discover the equations which express the mathematical laws of the phenomenon under consideration; and these equations are the starting-point of the calculus, which must obtain from them certain quantities by means of others.*
>
> —Comte,
>
> *Positive philosophy.*

Identities are tools of every working mathematician. Standard identities involving simple sums, products, fractions, or exponents are used by every student; it may be amazing to see how many of these can be proved by induction. Some of the more useful identities involve binomial coefficients or trigonometry, and many of these also can be proved by mathematical induction. Exercises below are often roughly grouped by category; however, many exercises could easily fall into more than one category.

Many of the identities below have "proofs without words", that is, a pictorial representation. The interested reader can find many of these in Roger B. Nelsen's two delightful books [403], [404], available from the MAA. A few specific references to these "picture proofs" are mentioned below. Many identities regarding integers also have combinatorial proofs; in fact, many of these identities form the heart of combinatorics. Riordan's *Combinatorial Identities* [454] is a rich handbook for those studying identities that come from counting in different ways. Many of the identities in that text are provable by induction, so Riordan's text might be a great place to start for references or other resources.

9.1 Arithmetic progressions

For the first two exercises, let O_n denote the nth non-negative odd number and E_n denote the nth even number. So $O_1 = 1, O_2 = 3, O_3 = 5, \ldots$, and $E_1 = 0, E_2 =$

$2, E_3 = 4, \ldots$. The first two theorems have been attributed to Maurolycus (see, for example, [91], where the reference is [376]).

This next exercise has a direct proof, and proving it inductively is not nearly as straightforward as one might think.

Exercise 32. *Prove that for each positive integer* n,

$$O_n + 2 = O_{n+1}.$$

Exercise 33. *Use the result in Exercise 32 and mathematical induction to prove that for each positive integer* n,

$$n + (n - 1) = O_n.$$

Although I haven't seen the following referenced, it's very likely that the identity in the next exercise was also used by Maurolycus. Its proof is nearly identical to that of Exercise 32.

Exercise 34. *Prove that for each positive integer* n, $E_{n+1} = E_n + 2$.

Exercise 35. *Define* $T_n = 1 + 2 + 3 + \cdots + n$. *Use induction to prove that for* $n \geq 1$,

$$T_n = \frac{n(n + 1)}{2}.$$

In Section 1.6, the numbers T_n are called *triangular numbers*.

Recalling Definition 2.5.6, the summation notation is $\sum_{i=a}^{b} x_i = x_a + x_{a+1} + \ldots + x_b$.

Exercise 36. *Prove that for any* $1 \leq m < n$,

$$\sum_{i=m+1}^{n} i = \frac{(n - m)(n + m + 1)}{2}.$$

A special case of Exercise 36 is the following (see [404, p. 85] for a pictorial representation).

Exercise 37. *Prove that for every positive integer* n,

$$n^2 + (n^2 + 1) + (n^2 + 2) + \cdots + (n^2 + n) = (n^2 + n + 1) + \cdots + (n^2 + 2n) = (2n + 1)T_n.$$

The solution to the next exercise was given by Maurolycus, and Bussey [91] commented that the proof given was "a clear case of a complete induction proof."

Exercise 38. *Prove that for $n \geq 1$,*

$$1 + 3 + 5 + \cdots + (2n - 1) = n^2.$$

Exercise 39. *Prove that for $n \geq 1$,*

$$2 + 4 + 6 + \cdots + 2n = n(n + 1).$$

Exercise 40. *Prove that for $n \geq 1$,*

$$2 + 5 + 8 + \cdots + (3n - 1) = \frac{n(3n + 1)}{2}.$$

Exercise 41. *Show that for $n \geq 1$,*

$$3 + 11 + 19 + \cdots + (8n - 5) = 4n^2 - n.$$

Exercise 42. *Prove that for $n \geq 1$,*

$$\sum_{i=1}^{n}(3i - 1) = \left(\sum_{i=1}^{n} i\right) + n^2.$$

Exercise 43. *Prove that for $n \geq 1$,*

$$5 + 9 + 13 + \cdots + (4n + 1) = n(2n + 3).$$

Exercise 44. *Prove that for $n \geq 1$,*

$$(2n + 1) + (2n + 3) + (2n + 5) + \cdots + (4n - 1) = 3n^2.$$

The next exercise generalizes nearly every exercise in this section so far.

Exercise 45 (Summing arithmetic progressions). *Let a and d be fixed real numbers. Prove that for each $n \geq 1$,*

$$a + (a + d) + (a + 2d) + \cdots + (a + (n - 1)d) = \frac{n}{2}[2a + (n - 1)d].$$

As in Section 1.6, for each $n \geq 1$, the n-th triangular number is $T_n = 1 + 2 + \cdots + n$.

Exercise 46. *Prove by induction that for each $n \geq 2$,*

$$T_n^2 - T_{n-1}^2 = n^3.$$

9.2   Sums of finite geometric series and related series

Exercise 47. *Prove that for every $n \geq 1$,*

$$1 + 2 + 2^2 + 2^3 + \cdots + 2^n = 2^{n+1} - 1.$$

Exercise 48. *Prove that for each $n \geq 1$,*

$$1 + 3 + 3^2 + 3^3 + \cdots + 3^{n-1} = \frac{3^n - 1}{2}.$$

The next exercise generalizes Exercises 47 and 48:

Exercise 49 (Summing a geometric series). *Let a and r be real numbers with $a \neq 0$ and $r \neq 1$. Prove that for each integer $n \geq 1$,*

$$a + ar + ar^2 + \cdots + ar^n = a\frac{r^{n+1} - 1}{r - 1}.$$

For the particular geometric series with $r = a = \frac{1}{2}$, Exercise 49 gives equation (7.3) in Section 7.3.

Exercise 50. *Prove that for each natural number $n \geq 1$,*

$$1 \cdot 2^1 + 2 \cdot 2^2 + 3 \cdot 2^3 + \cdots + n \cdot 2^n = 2 + (n - 1)2^{n+1}.$$

Here is the same problem stated slightly differently (and with its own solution).

Exercise 51. *Prove that for every $n \geq 1$,*

$$1 + 2 \cdot 2 + 3 \cdot 2^2 + \cdots + n2^{n-1} = (n - 1)2^n + 1.$$

Exercise 52. *Prove that for every $n \geq 1$,*

$$1 + 2 \cdot 3 + 3 \cdot 3^2 + \cdots + n3^{n-1} = \frac{3^n(2n - 1) + 1}{4}.$$

Note: Results in Exercises 50, 51, and 52 can all be found by taking the derivative of the identity in Exercise 49 and, if necessary, shifting exponents by multiplying each side by the appropriate term.

Exercise 53. *Generalize Exercises 50, 51, and 52 to powers of an arbitrary $k \geq 2$, and prove your answer by induction.*

9.3 Power sums, sums of a single power

Exercise 54. *Prove that for $n \geq 1$,*

$$1^2 + 2^2 + 3^2 + \cdots + n^2 = \frac{n(n+1)(2n+1)}{6}.$$

Exercise 55. *Prove that for every $n \geq 1$,*

$$(1 + 2 + \cdots + n)(2n+1) = 3(1^2 + 2^2 + \cdots + n^2).$$

Exercise 56. *For $n \geq 1$,*

$$1^3 + 2^3 + 3^3 + \cdots + n^3 = \left[\frac{n(n+1)}{2}\right]^2.$$

By virtue of Exercise 35, the equality in Exercise 56 can also be stated as

$$(1 + 2 + \ldots + n)^2 = 1^3 + 2^3 + 3^3 + \cdots + n^3,$$

which was the form in which it appeared in the 1990 Canadian Mathematical Olympiad (a solution appears in [367], for example).

For integers $k \geq 0$ and $n > 0$, define

$$S_k(n) = \sum_{i=1}^{n} i^k = 1^k + 2^k + 3^k + \cdots + n^k. \tag{9.1}$$

For any $m \geq 0$ define $S_m(0) = 0$.

Note: many authors (*e.g.* [230]) define $S_k(n) = 0^k + 1^k + 2^k + 3^k + \cdots + (n-1)^k$, which can lead to a great deal of confusion when researching such sums.

Identities in Theorem 1.6.1 and in Exercises 54 and 56 say

$$S_1(n) = \frac{n(n+1)}{2}$$
$$S_2(n) = \frac{n(n+1)(2n+1)}{6}$$
$$S_3(n) = \left[\frac{n(n+1)}{2}\right]^2.$$

The formula for $S_1(n)$ was derived independently, and then proved by induction. The expressions for $S_2(n)$ and $S_3(n)$ were simply given, and an inductive proof was used to check each. Nichomachus (*ca.* 100 A.D.) knew of the expression for $S_3(n)$. How were they found? Here are a few more (each of which also has an inductive proof, for those with the energy):

$$S_4(n) = \sum_{i=1}^{n} i^4 = \frac{1}{30} n(n+1)(2n+1)(3n^2 + 3n - 1)$$

$$S_5(n) = \sum_{i=1}^{n} i^5 = \frac{1}{12}n^2(n+1)^2(2n^2+2n-1)$$

$$S_6(n) = \sum_{i=1}^{n} i^6 = \frac{1}{42}n(n+1)(2n+1)(3n^4+6n^3-n^2-3n+1)$$

$$S_7(n) = \sum_{i=1}^{n} i^7 = \frac{1}{24}n^2(n+1)^2(3n^4+6n^3-n^2-4n+2)$$

$$S_8(n) = \sum_{i=1}^{n} i^8 = \frac{1}{90}n(n+1)(2n+1)(5n^6+15n^5+5n^4-15n^3-n^2+9n-3)$$

$$S_9(n) = \sum_{i=1}^{n} i^9 = \frac{1}{20}n^2(n+1)^2(2n^6+6n^5+n^4-8n^3+n^2+6n-3)$$

$$S_{10}(n) = \sum_{i=1}^{n} i^{10} = \frac{n}{66}(6n^{11}+33n^9+55n^8-66n^6+66n^4-33n^2+5).$$

Each $S_m(n)$ above is a polynomial in n of degree $m+1$ with 0 constant term. A general expression for $S_m(n)$, called "Faulhaber's formula", is given in Section 9.6.3; each is indeed a polynomial with properties shared by those listed above. (One development of Faulhaber's formula relies heavily on binomial coefficients and induction, so it appears in a later section.)

Exercise 57. *Show that for each $n \geq 1$,*

$$1^2 + 3^2 + 5^2 + \cdots + (2n-1)^2 = \frac{n(2n-1)(2n+1)}{3}.$$

Exercise 58. *Prove that for each $n \geq 1$,*

$$2^2 + 4^2 + 6^2 + \cdots + (2n)^2 = \frac{2n(n+1)(2n+1)}{3}.$$

Exercise 59. *Show by induction that for each $n \geq 1$,*

$$1 - 4 + 9 - 16 + \cdots + (-1)^{n+1}n^2 = (-1)^{n+1}(1+2+3+\ldots+n).$$

Exercise 60. *Show that for every $n \geq 1$,*

$$n^2 - (n-1)^2 + (n-2)^2 + \cdots + (-1)^{n-1}(1)^2 = \frac{n(n+1)}{2}.$$

Note: this equality follows directly from that in Exercise 59 by Theorem 1.6.1 and division by $(-1)^{n+1}$, however this equality is to be proved by induction without Exercise 59 or Theorem 1.6.1.

Exercise 61. *Prove that for each $n \geq 1$,*

$$1^3 + 3^3 + 5^3 + \cdots + (2n-1)^3 = n^2(2n^2-1).$$

9.4 Products and sums of products

Recall that the definition of the factorial function is recursive: $0! = 1$ and for $n \geq 1$, $n! = n \cdot (n-1)!$. The first exercise in this section is not really a sum unless the trivial sum is counted as such.

Exercise 62. *For each $n \geq 1$, prove that*

$$2 \cdot 6 \cdot 10 \cdots (4n-2) = \frac{(2n)!}{n!}.$$

Exercise 63. *Prove that for $n \geq 1$,*

$$1 \cdot 2 + 2 \cdot 3 + 3 \cdot 4 + \cdots + n(n+1) = \frac{n(n+1)(n+2)}{3}.$$

Exercise 64. *Show by induction that for each $n \geq 2$,*

$$1 \cdot 3 + 2 \cdot 4 + 3 \cdot 5 + \cdots + (n-1)(n+1) = \frac{(n-1)(n)(2n+5)}{6}.$$

Exercise 65. *Prove that for $n \geq 1$,*

$$1 \cdot 2 \cdot 3 + 2 \cdot 3 \cdot 4 + 3 \cdot 4 \cdot 5 + \cdots + n(n+1)(n+2) = \frac{1}{4}n(n+1)(n+2)(n+3).$$

Exercise 66. *Prove that for each $n \geq 1$,*

$$\sum_{j=1}^{n} j(j+1)(j+2)(j+3) = \frac{n(n+1)(n+2)(n+3)(n+4)}{5}.$$

The next exercise generalizes Exercises 63, 65 and 66.

Exercise 67. *For a fixed $k \in \mathbb{Z}^+$, show that for each natural number $n \geq 1$,*

$$\sum_{j=1}^{n} j(j+1) \cdots (j+k-1) = \frac{(k+n)!}{(k+1) \cdot (n-1)!}.$$

Exercise 68. *Prove that for each $n \geq 1$,*

$$\sum_{k=1}^{n} (2k-1)(2k+1)(2k+3) = n(2n^3 + 8n^2 + 7n - 2).$$

Exercise 69. *Prove that for any $n \in \mathbb{Z}^+$,*

$$\sum_{i=1}^{n} i^2 2^i = 2^{n+1}(n^2 - 2n + 3) - 6.$$

One might want to compare the next exercise to Exercise 593.

Exercise 70. *Prove that for $n \geq 0$,*

$$0 \cdot 0! + 1 \cdot 1! + 2 \cdot 2! + 3 \cdot 3! + \cdots + n \cdot n! = (n+1)! - 1.$$

9.5   Sums or products of fractions

Exercise 71. *Prove that for every $n \geq 1$,*

$$\frac{1}{1 \cdot 2} + \frac{1}{2 \cdot 3} + \frac{1}{3 \cdot 4} + \cdots + \frac{1}{n(n+1)} = \frac{n}{n+1}.$$

Exercise 72. *Show that for each $n \geq 1$,*

$$\sum_{i=0}^{n-1} \frac{1}{(n+i)(n+i+1)} = \frac{1}{2n}.$$

Exercise 73. *Prove that for each $n \geq 1$,*

$$\frac{1}{1 \cdot 2 \cdot 3} + \frac{1}{2 \cdot 3 \cdot 4} + \frac{1}{3 \cdot 4 \cdot 5} + \cdots + \frac{1}{n(n+1)(n+2)} = \frac{n(n+3)}{4(n+1)(n+2)}.$$

Exercise 74. *Prove that for each $n \geq 1$,*

$$1 - \frac{1}{2} + \frac{1}{3} - \frac{1}{4} + \cdots + \frac{1}{2n-1} - \frac{1}{2n} = \frac{1}{n+1} + \frac{1}{n+2} + \cdots + \frac{1}{2n}.$$

Exercise 75. *Show that for each $n \geq 1$,*

$$\frac{1}{1 \cdot 3} + \frac{1}{3 \cdot 5} + \frac{1}{5 \cdot 7} + \cdots + \frac{1}{(2n-1)(2n+1)} = \frac{n}{2n+1}.$$

Related to Exercise 75 is the following:

Exercise 76. *Prove that for each $n \in \mathbb{Z}^+$,*

$$\sum_{i=2}^{n} \frac{1}{i^2 - 1} = \frac{(n-1)(3n+2)}{4n(n+1)}.$$

Exercise 77. *Show that for every $n \geq 1$,*

$$\frac{1}{1 \cdot 3} + \frac{1}{2 \cdot 4} + \frac{1}{3 \cdot 5} + \cdots + \frac{1}{n(n+2)} = \frac{n(3n+5)}{4(n+1)(n+2)}.$$

Exercise 78. *Prove that for $n \geq 1$,*

$$\frac{1}{1 \cdot 5} + \frac{1}{5 \cdot 9} + \frac{1}{9 \cdot 13} + \cdots + \frac{1}{(4n-3)(4n+1)} = \frac{n}{4n+1}.$$

Exercise 79. *Show that for $n \geq 1$,*

$$\frac{1^2}{1 \cdot 3} + \frac{2^2}{3 \cdot 5} + \frac{3^2}{5 \cdot 7} + \cdots + \frac{n^2}{(2n-1)(2n+1)} = \frac{n(n+1)}{2(2n+1)}.$$

Exercise 80. *Use Exercise 71 to prove that for $n \geq 1$,*

$$\sum_{k=1}^{n} \frac{1}{k^2 + 3k + 2} = \frac{n}{2(n+2)},$$

then give an inductive proof (that does not rely on Exercise 71).

Exercise 81. *Prove by induction on n that*

$$\sum_{i=1}^{n} \frac{i}{1 + i^2 + i^4} = \frac{n(n+1)}{2(n^2 + n + 1)}.$$

Exercise 82. *Prove that for each $n \geq 1$,*

$$\frac{1}{1 \cdot 4} + \frac{1}{4 \cdot 7} + \frac{1}{7 \cdot 10} + \cdots + \frac{1}{(3n-2)(3n+1)} = \frac{n}{3n+1}.$$

Exercise 83. *Prove that for each $n \in \mathbb{Z}^+$,*

$$\sum_{m=1}^{n} \frac{m+4}{m(m+1)(m+2)} = \frac{n(3n+7)}{2(n+1)(n+2)}.$$

Exercise 84. *Prove that for each $n \geq 2$,*

$$\left(1 - \frac{1}{2}\right)\left(1 - \frac{1}{3}\right)\left(1 - \frac{1}{4}\right) \cdots \left(1 - \frac{1}{n}\right) = \frac{1}{n}.$$

Exercise 85. *Prove that for each $n \geq 2$,*

$$\left(1 - \frac{1}{4}\right)\left(1 - \frac{1}{9}\right)\left(1 - \frac{1}{16}\right) \cdots \left(1 - \frac{1}{n^2}\right) = \frac{n+1}{2n}.$$

Exercise 86. *For $n \geq 3$, suppose that $a_1, a_2, \ldots, a_n$ are positive integers so that all the quotients*

$$p_1 = \frac{a_n + a_2}{a_1}, \quad p_2 = \frac{a_1 + a_3}{a_2}, \quad \ldots, \quad p_n = \frac{a_{n-1} + a_1}{a_n}$$

are integers. Show that

$$p_1 + p_2 + \cdots + p_n \leq 3n - 1.$$

9.6 Identities with binomial coefficients

Recall the definition of binomial coefficients: for integers $0 \leq k \leq n$, "n choose k" is defined by

$$\binom{n}{k} = \frac{n!}{k!(n-k)!}.$$

Binomial coefficients are so-called because they are the coefficients in the binomial theorem (see Exercise 104). In other texts this same number is represented by various other notations, including $_nC_k$, C_k^n, and $C(n,k)$. If $n < k$, or if $n < 0$, then by convention put $\binom{n}{k} = 0$. One might note that $\binom{n}{k}$ is defined for other choices of n and k, (for example, when k is rational) but these situations are described only as needed. As it is, the number $\binom{n}{k}$ counts the number of different collections of k objects chosen from a set of n distinguishable elements (see Exercise 419).

Certain recursions regarding binomial coefficients are often useful, especially in inductive proofs.

Lemma 9.6.1. *Let s and m be non-negative integers. If $0 \leq s \leq m-1$, then*

$$\binom{m}{s} = \frac{m}{m-s}\binom{m-1}{s} \tag{9.2}$$

If $1 \leq s \leq m$,

$$\binom{m}{s} = \frac{m-s+1}{s}\binom{m}{s-1}, \tag{9.3}$$

and

$$\binom{m}{s} = \frac{m}{s}\binom{m-1}{s-1}. \tag{9.4}$$

Proof:

$$\frac{m!}{(m-s)!s!} = \frac{m}{m-s} \cdot \frac{(m-1)!}{(m-1-s)!s!}$$

$$= \frac{m}{m-s}\binom{m-1}{s},$$

$$\frac{m!}{(m-s)!s!} = \frac{m-s+1}{s} \cdot \frac{m!}{(m-s+1)!(s-1)!}$$

$$= \frac{m-s+1}{s} \cdot \frac{m!}{(m-(s-1))!(s-1)!}$$

$$= \frac{m-s+1}{s}\binom{m}{s-1},$$

$$\frac{m!}{(m-s)!s!} = \frac{m}{s} \cdot \frac{(m-1)!}{(m-s)!(s-1)!}$$

$$= \frac{m}{s} \cdot \frac{(m-1)!}{(m-1-(s-1))!(s-1)!}$$

$$= \frac{m}{s}\binom{m-1}{s-1}.$$

Exercise 87. *Show by induction that for each $n \geq 0$,*

$$\sum_{j=0}^{n}\binom{n+j}{j}\frac{1}{2^j} = 2^n.$$

Exercise 88. *For each $n \geq 1$, show that*

$$\binom{2}{2} + \binom{3}{2} + \binom{4}{2} + \cdots + \binom{n+1}{2} = \frac{n(n+1)(n+2)}{6}.$$

Exercise 89. *Prove that for every $n \geq 2$,*

$$\sum_{i=1}^{n}\frac{1}{\binom{i+1}{2}} = 2 - \frac{2}{n+1},$$

and conclude that

$$\sum_{i=1}^{\infty}\frac{1}{\binom{i+1}{2}} = 2.$$

The next exercise asks to prove "Pascal's identity", one of the most useful identities in combinatorics and discrete mathematics. (It's very likely that the original proof was not inductive *per se*, but instead based on properties of Pascal's triangle.) There are (at least) two standard simple proofs of the identity, however it may also be proved by induction. (The proof by induction is not nearly so elegant as the counting proof, but Lemma 9.6.1 helps.) In the solution to this exercise, one proof of each kind is presented; see the comments following Exercise 94 for yet another (non-inductive) proof.

Exercise 90 (Pascal's identity). *Prove that for any fixed $r \geq 1$, and all $n \geq r$,*

$$\binom{n+1}{r} = \binom{n}{r} + \binom{n}{r-1}.$$

The next exercise has a direct counting proof, and another very simple proof invoking the binomial theorem (see Exercise 104); however, as is the case with many identities involving binomial coefficients, the next exercise can also be solved by induction (where at least one proof uses Pascal's identity).

Exercise 91. *For any $n \geq 0$, prove that*

$$\sum_{i=0}^{n}\binom{n}{i} = 2^n.$$

The next result was also proved by Pascal, probably *circa* 1659 (see [91], where the reference is "Consequence XII, Vol. III, p.248 of [424]"). The result has a very simple direct proof using the definition of binomial coefficients, namely

$$\frac{\binom{n}{k}}{\binom{n}{k+1}} = \frac{\frac{n!}{k!(n-k)!}}{\frac{n!}{(k+1)!(n-k-1)!}} = \frac{n!(k+1)!(n-k-1)!}{n!k!(n-k)!} = \frac{k+1}{n-k}.$$

Exercise 92 (Pascal). *Using Lemma 9.6.1, prove by mathematical induction that for each $n \geq 2$ and all k satisfying $1 \leq k \leq n-1$,*

$$\frac{\binom{n}{k}}{\binom{n}{k+1}} = \frac{k+1}{n-k}.$$

The next exercise asks to reprove the result from Exercise 92, again using induction, but in a different (more cumbersome) way.

Exercise 93 (Pascal). *Using Pascal's identity, prove by mathematical induction that for each $n \geq 2$ and all k satisfying $1 \leq k \leq n-1$,*

$$\frac{\binom{n}{k}}{\binom{n}{k+1}} = \frac{k+1}{n-k}.$$

In the following exercise, it might help to imagine an $m \times n$ rectangular array of city blocks, m blocks east to west, and n blocks north to south. All streets are one-way either northbound or eastbound. The goal is to count the number of ways to drive from the southwest corner to the northeast corner.

Exercise 94. *Consider the integer lattice grid $[0, m] \times [0, n]$, points in the plane with integral coordinates (x, y) where $0 \leq x \leq m$ and $0 \leq y \leq n$. Prove that the number of walks from $(0,0)$ to (m, n) on the grid that go up and/or to the right is*

$$\binom{m+n}{m} = \binom{m+n}{n}.$$

Though not an inductive exercise, one can use the result from Exercise 94 to give a new proof of Pascal's identity. The idea is to separate the paths from $(0,0)$ to (n, m), into two groups, those going through the point $(m, n-1)$ and those which do not. The number of paths from $(0,0)$ to (m, n) that pass through $(m, n-1)$ is (by Exercise 94) $\binom{m+n-1}{m}$. All other paths must go through $(m-1, n)$ and the number of such paths is (again by Exercise 94) $\binom{m-1+n}{m-1}$. Putting $p = m + n$, then $p \geq m$ and the total number of paths is

$$\binom{p}{m} = \binom{p-1}{m} + \binom{p-1}{m-1},$$

which is Pascal's identity.   □

The next wonderful identity is often attributed to Euler, but is also called "Vandermonde's convolution". This identity is stated next in terms of a theorem, together with a simple counting proof; an exercise is to find a purely inductive proof.

Theorem 9.6.2 (Euler). *For any non-negative integers m and n and p,*

$$\sum_{i=0}^{p} \binom{m}{i}\binom{n}{p-i} = \binom{m+n}{p}.$$

Proof: Let X be a set with $m+n$ elements, and fix a partition $X = L \cup R$, where $|L| = m$ and $|R| = n$. If a subset of X with p elements intersects L in i elements, then it intersects R in $p-i$ elements. This can occur in $\binom{m}{i}\binom{n}{p-i}$ ways. Note that if $i > m$, then $\binom{m}{i} = 0$ or if $p - i > n$, then $\binom{n}{p-i} = 0$. Summing over all i finishes the proof.   □

Exercise 95. *Find a purely inductive proof of Theorem 9.6.2.*

A special case of Theorem 9.6.2 leads to what is sometimes called "Lagrange's identity" (after Joseph Louis Lagrange (1736–1813), one of the most powerful mathematicians of his time, rivalling even Euler). [This result is special to me, as it was first shown to me by Paul Erdős when I was in grad school; this beautiful simple result had somehow escaped my attention until then.]

Corollary 9.6.3 (Lagrange). *For each non-negative integer k,*

$$\binom{2k}{k} = \sum_{i=0}^{k} \binom{k}{i}^{2}.$$

Proof: By Theorem 9.6.2 with $m = n = p = k$,

$$\binom{2k}{k} = \sum_{i=0}^{k} \binom{k}{i}\binom{k}{k-i}.$$

It is easy to check by merely writing out the definitions for corresponding binomial coefficients that they are symmetric, that is, for each $i = 0, \ldots, k$,

$$\binom{k}{i} = \binom{k}{k-i},$$

and so the result follows directly.   □

Exercise 96. *Discuss why any inductive proof of Corollary 9.6.3 might be more complicated than the inductive proof of Theorem 9.6.2, if indeed one can be found.*

In the next two exercises, one must first decide on the ranges for m.

Exercise 97. *Prove that for each $n \geq 0$,*

$$\sum_{i=0}^{n} \binom{m+i}{m} = \binom{m+n+1}{m+1}.$$

Exercise 98. *Show that for each $n \geq 0$,*

$$\sum_{i=0}^{n} \binom{m+i}{i} = \binom{m+n+1}{n}.$$

Exercise 99. *Prove that for any $1 \leq m \leq n$,*

$$\sum_{i=0}^{n-m} (-1)^i \binom{n}{m+i} = \binom{n-1}{m-1}.$$

Note: Here is an interesting consequence of Exercise 99; this consequence can be proved in other ways.

$$\sum_{i=0}^{n} (-1)^i \binom{n}{i} = \binom{n}{0} + \sum_{i=1}^{n} (-1)^i \binom{n}{i}$$

$$= \binom{n}{0} - \sum_{j=0}^{n-1} (-1)^j \binom{n}{1+j}$$

$$= \binom{n}{0} - \binom{n-1}{0} \qquad \text{(with } m = 1 \text{ in Ex. 99)}$$

$$= 0,$$

and so for $n \geq 1$,

$$\sum_{i=0}^{n} (-1)^i \binom{n}{i} = 0. \tag{9.5}$$

Exercise 100. *Find a proof by induction of the identity (for $n \geq 1$)*

$$\sum_{i=0}^{n} (-1)^i \binom{n}{i} = 0$$

that does not use Exercise 99. You may, however, use Pascal's identity.

Another interesting identity follows from equation (9.5):

$$\sum_{0 \le i \le j \le n} \binom{n}{j}\binom{j}{i}(-1)^i = 1.$$

The proof actually follows fairly easily from equation (9.5), if the left side is rewritten as

$$\sum_{j=0}^{n} \binom{n}{j} \sum_{i=0}^{j} (-1)^i \binom{j}{i},$$

and j is used in place of n in equation (9.5), giving 0 for each of the outer summands except for when $j = 0$.

The "well-known" identity in the following exercise relates to something called the "Euler characteristic" (see [315] for details). Conventions dictate that $\binom{-1}{0} = 1 = \binom{n}{0}$ and for $i > n$, $\binom{n}{i} = 0$.

Exercise 101. *Prove that for $m \ge 0$ and any $n \ge 0$,*

$$\sum_{i=0}^{m} (-1)^i \binom{n}{i} = (-1)^m \binom{n-1}{m}.$$

Exercise 102. *Prove that for every $n \ge 0$,*

$$\sum_{i=0}^{n} \binom{i}{2}\binom{i}{5} = \binom{n+1}{2}\binom{n+1}{6} + \binom{n+2}{8} - n\binom{n+2}{7}.$$

Exercise 103 (Binomial theorem, simple case). *Prove that for each $n \ge 1$,*

$$(1+x)^n = \binom{n}{0} + \binom{n}{1}x + \binom{n}{2}x^2 + \cdots + \binom{n}{n}x^n.$$

The following more general version of the binomial theorem is sometimes called "Newton's binomial theorem" named after Isaac Newton (1642–1727) (*e.g.*, in [499]), but is most often referred to as simply "the binomial theorem".

Exercise 104 (Binomial theorem). *Give an inductive proof of the binomial theorem: for each $n \ge 1$,*

$$(x+y)^n = \binom{n}{0}x^n + \binom{n}{1}x^{n-1}y + \binom{n}{2}x^{n-2}y^2 + \cdots + \binom{n}{n-1}xy^{n-1} + \binom{n}{n}y^n.$$

In summation notation, the binomial theorem can be written

$$(x+y)^n = \sum_{j=0}^{n} \binom{n}{j}x^{n-j}y^j. \tag{9.6}$$

Observe that $(1 - 1)^n = 0$, and expanding (9.6) (with $x = 1$ and $y = -1$), equation (9.5) is obtained, that is, $\sum_{i=0}^{n}(-1)^n \binom{n}{i} = 0$, offering another solution to Exercise 100.

Exercise 105. *Show that differentiating the simple binomial theorem (Exercise 103) with respect to x gives*

$$\sum_{i=0}^{n} \binom{n}{i} x^i = nx(1+x)^{n-1},$$

then give an inductive proof of this equality.

The identities called "difference of squares", $x^2 - y^2 = (x - y)(x + y)$, and "difference of cubes", $x^3 - y^3 = (x - y)(x^2 + xy + y^2)$, are special cases of a more general equality:

Exercise 106. *Prove that for every positive integer $n \geq 1$,*

$$x^n - y^n = (x - y) \left(\sum_{\nu=1}^{n} x^{n-\nu} y^{\nu-1} \right).$$

The idea in Exercise 105 of differentiating a known equality is very powerful. For example, by the binomial theorem,

$$(x - 1)^n = \sum_{k=0}^{n} \binom{n}{k} x^n (-1)^{n-k};$$

differentiating with respect to x yields

$$n(x - 1)^{n-1} = \sum_{k=0}^{n} \binom{n}{k} k x^{k-1} (-1)^{n-k}.$$

If $n \geq 2$, using $x = 1$ gives a fairly remarkable identity,

$$0 = \sum_{k=0}^{n} \binom{n}{k} (-1)^{n-k} k, \tag{9.7}$$

which can be used as a base case for proving a family of (perhaps surprising) equalities:

Exercise 107. *Induct on j to show that for every $1 \leq j < n$,*

$$\sum_{k=0}^{n} \binom{n}{k} (-1)^{n-k} k^j = 0.$$

A similar identity to that in Exercise 107, but for $j \geq n$, appears when counting surjective functions in Exercise 595.

For many more identities involving binomial coefficients, see, *e.g.*, [225] or [454].

9.6.1 Abel identities

The next two identities are attributed to the Norwegian Neils Henrik Abel (1802–1829) [1], perhaps most famous for showing that fifth degree equations are not, in general, solvable by radicals.

Exercise 108 (Abel identity 1)**.** *For any $a \in \mathbb{R}$ and each $n \geq 1$, show that*

$$(x + y)^n = \sum_{k=0}^{n} \binom{n}{k} x(x - ka)^{k-1}(y + ka)^{n-k}.$$

Exercise 109 (Abel identity 2)**.** *Prove that for each $n \geq 1$,*

$$(x + y + n)^n = \sum_{k=0}^{n} \binom{n}{k} x(x + k)^{k-1}(y + n - k)^{n-k}.$$

Vandermonde's convolution (Theorem 9.6.2) is the special case of $b = 0$, $a = m$, and $c = n$ in Rothe's formula [468], given in 1793: For $n \geq 1$, and any $a, b \in \mathbb{C}$,

$$\sum_{k=0}^{n} \frac{a}{a + bk} \binom{a + bk}{k} \frac{c}{c + b(n - k)} \binom{c + b(n - k)}{n - k} = \frac{a + c}{a + c + bn} \binom{a + c + bn}{n}.$$

If one is energetic, one *might* be able to prove Rothe's formula by induction as well. Rothe's equality is a special case of yet an even more general equality proved by Hagen [251] in 1891:

$$\sum_{k=0}^{n} (p + qk) \frac{a}{a + bk} \binom{a + bk}{k} \frac{c}{c + b(n - k)} \binom{c + b(n - k)}{n - k}$$

$$= \frac{p(a + c) + aqn}{a + c + bn} \binom{a + c + bn}{n}.$$

Perhaps p and q are meant to be positive integers, so *maybe* with even more effort, one can prove Hagen's formula by induction as well. In [324], Knuth shows how the identities of Rothe, Hagen, and Abel, all follow from the binomial theorem and Vandermonde's convolution. [Note: I have not seen if these proofs are by induction.]

9.6.2 Bernoulli numbers

Jakob Bernoulli (1654–1705) (and independently, perhaps earlier, Seki Takakazu; see [558]) studied the following numbers:

Definition 9.6.4. For $n \geq 0$, define the *Bernoulli numbers* B_n recursively: set $B_0 = 1$, and for $n \geq 1$,

$$B_n = \frac{-1}{n + 1} \sum_{j=0}^{n-1} \binom{n + 1}{j} B_j.$$

The first few Bernoulli numbers are $B_0 = 1$, $B_1 = -\frac{1}{2}$, $B_2 = \frac{1}{6}$, $B_3 = 0$, $B_4 = \frac{-1}{30}$, $B_5 = 0$, $B_6 = \frac{1}{42}$. (For $n \geq 1$, $B_{2n+1} = 0$.) Bernoulli numbers do not seem to have a simple description. One property of Bernoulli numbers needed below is nearly immediate from the definition:

$$\sum_{j=0}^{n} \binom{n+1}{j} B_j = 0. \tag{9.8}$$

For the reader who knows about exponential generating functions,

$$\frac{x}{e^x - 1} = \sum_{n=0}^{\infty} B_n \frac{x^n}{n!},$$

and by induction, one can check that this definition agrees with the above definition (e.g., see [558]).

The interested reader might try to prove by induction the *Carlitz identity*: for any non-negative integers m and n,

$$(-1)^m \sum_{i=0}^{m} \binom{m}{i} B_{n+i} = (-1)^n \sum_{j=0}^{n} \binom{n}{j} B_{m+j}.$$

9.6.3   Faulhaber's formula for power sums

For integers $k \geq 0$ and $n \geq 1$, define

$$S_m(n) = 1^m + 2^m + \cdots + n^m.$$

For $m \geq 0$, one can also define $S_m(0) = 0$. As observed following Exercise 56, for $m \leq 12$, each $S_m(n)$ is a polynomial in n of degree $m + 1$ with 0 constant term (so n divides $S_m(n)$). In fact, for any $m \geq 0$, this fact is true, and is provable by strong induction on m. The induction can be based on the following trick:

Lemma 9.6.5. *For integers $m \geq 0$ and $n > 0$,*

$$(n+1)^{m+1} = \sum_{j=0}^{m} \binom{m+1}{j} S_j(n). \tag{9.9}$$

Proof: There are two proofs, both relying on the binomial theorem (Exercise 104). Here is the first: for $n \geq 0$ and $m \geq 0$,

$$S_{m+1}(n) + (n+1)^{m+1} = \sum_{i=1}^{n+1} i^{m+1}$$

$$= \sum_{k=0}^{n} (k+1)^{m+1} \qquad \text{(replace } i = k+1)$$

$$= \sum_{k=0}^{n} \sum_{j=0}^{m+1} \binom{m+1}{j} k^j \qquad \text{(binomial thm)}$$

$$= \sum_{j=0}^{m+1} \sum_{k=0}^{n} \binom{m+1}{j} k^j$$

$$= \sum_{j=0}^{m+1} \binom{m+1}{j} \sum_{k=0}^{n} k^j$$

$$= \sum_{j=0}^{m+1} \binom{m+1}{j} S_j(n).$$

Subtracting $S_{m+1}(n)$ gives equation (9.9). $\qquad\qquad \square$

The second proof of (9.9) uses a collapsing sum:

$$(n+1)^{m+1} = \sum_{k=0}^{n} ((n-k+1)^{m+1} - (n-k)^{m+1})$$

$$= \sum_{k=0}^{n} \sum_{j=0}^{m} \binom{m+1}{j} (n-k)^j$$

$$= \sum_{j=0}^{m} \sum_{k=0}^{n} \binom{m+1}{j} (n-k)^j$$

$$= \sum_{j=0}^{m} \binom{m+1}{j} \sum_{k=0}^{n} (n-k)^j$$

$$= \sum_{j=0}^{m} \binom{m+1}{j} S_j(n).$$

$\qquad\qquad \square$

Exercise 110. *Prove by strong induction on m that each $S_m(n)$ is a polynomial of degree $m+1$ in n with constant term 0.*

As noted in the solution to Exercise 110, the following is a consequence of Lemma 9.6.5:

Corollary 9.6.6. *For positive integers m, n,*

$$(m+1)S_m(n) = (n+1)^{m+1} - \sum_{j=0}^{m-1} \binom{m+1}{j} S_j(n). \qquad (9.10)$$

Is there a way to calculate the coefficients of the polynomial $S_k(n)$? Since the degree of $S_k(n)$ is $k + 1$, any $k + 2$ values uniquely determine the polynomial and this is seen easily with a little linear algebra. For example, if one sets $S_2(n) = an^3 + bn^2 + cn + d$, the values $S_2(1) = 1$, $S_2(2) = 5$, $S_2(3) = 14$, $S_2(4) = 30$ give the system

$$a + b + c + d = 1$$
$$8a + 4b + 2c + d = 5$$
$$27a + 9b + 3c + d = 14$$
$$64a + 16b + 4d + d = 30.$$

which has a unique solution $a = \frac{1}{3}$, $b = \frac{1}{2}$, $c = \frac{1}{6}$, $d = 0$. In general, the above technique works since the coefficient matrix used is a Vandermonde matrix and so is invertible (see Exercise 661). However, this technique does not yield any simple formula for the coefficients. See [157] for more related matrix methods. Another way to find a polynomial of degree $n - 1$ that fits n values is to use the *Lagrange interpolation formula*: if $f : \mathbb{R} \to \mathbb{R}$ is a function and $x_1, \ldots, x_n$ are distinct, then

$$p(x) = \sum_{i=1}^{n} \frac{\prod_{j \neq i}(x - x_j)}{\prod_{j \neq i}(x_i - x_j)} f(x_i).$$

is a degree $n - 1$ polynomial so that for each i, $p(x_i) = f(x_i)$.

The following expression for $S_k(n)$ is now eponymous with Johann Faulhaber (1580–1635), who published a form in the 1631 edition of *Academiae Algebrae* [186] (though the form here is due to Bernoulli). See [116, p. 106] and [325] for more references.

Theorem 9.6.7 (Faulhaber's formula). *For integers $k \geq 0$ and $n \geq 1$,*

$$S_k(n) = \frac{1}{k+1} \sum_{j=0}^{k} \binom{k+1}{j} B_j(n + 1)^{k+1-j}.$$

No solution is given for the following exercise; solving it uses some tricky sums involving binomial coefficients; the reader is recommended to see [230]—however, be warned that their notation is different than that used here.

Exercise 111. *By induction on k, prove Theorem 9.6.7.*

9.7   Gaussian coefficients

This next definition, similar to that of the binomial coefficient, relies upon the versatile notion of a finite-dimensional vector space over a finite field (see Sections 19.4 and 19.5 for definitions). The notation and terminology varies in the literature.

For integers $0 \le k \le n$ and a prime power q, the *Gaussian coefficient*

$$\begin{bmatrix} n \\ k \end{bmatrix}_q$$

is the number of k-dimensional subspaces of the n-dimensional vector space over the field $\mathrm{GF}(q)$. These coefficients are sometimes called the q-analogues of binomial coefficients (and this can be justified by examining limits as $q \to 1$ below). It is known that

$$\begin{bmatrix} n \\ k \end{bmatrix}_q = \frac{(q^n - 1)(q^{n-1} - 1) \cdots (q^{n-k+1} - 1)}{(q^k - 1)(q^{k-1} - 1) \cdots (q - 1)},$$

and from this, one can also directly prove a q-analogue of Pascal's identity

$$\begin{bmatrix} n+1 \\ k \end{bmatrix}_q = \begin{bmatrix} n-1 \\ k \end{bmatrix}_q + q^k \begin{bmatrix} n \\ k \end{bmatrix}_q.$$

Using the last equality, the statement in the next exercise is (relatively easily) proved by induction on n (as in [94, p. 127]).

In Definition 2.5.9, product notation for natural numbers was introduced; the same notation applies to real numbers and polynomials since multiplication is associative: $\prod_{i=1}^n x_i = x_1 x_2 \cdots x_n$. [See Exercise 4 for proof that multiplication of positive integers is associative.]

Exercise 112 (q-binomial theorem). *Prove that for $n \ge 1$,*

$$\prod_{i=0}^{n-1} (1 + q^i t) = \sum_{k=0}^{n} q^{k(k-1)/2} \begin{bmatrix} n \\ k \end{bmatrix}_q t^k.$$

9.8 Trigonometry identities

Recall the three main identities in trigonometry:

$$\sin(\alpha + \beta) = \sin(\alpha)\cos(\beta) + \cos(\alpha)\sin(\beta) \tag{9.11}$$
$$\cos(\alpha + \beta) = \cos(\alpha)\cos(\beta) - \sin(\alpha)\sin(\beta) \tag{9.12}$$
$$sin^2(\theta) + \cos^2(\theta) = 1. \tag{9.13}$$

From these, all other standard trigonometric identities follow. For example, a standard trigonometric identity is:

$$\tan(a + b) = \frac{\tan(a) + \tan(b)}{1 - \tan(a)\tan(b)} \tag{9.14}$$

Here is one derivation of equation 9.14:

$$\tan(a + b) = \frac{\sin(a + b)}{\cos(a + b)}$$

$$= \frac{\sin(a)\cos(b) + \cos(a)\sin(b)}{\cos(a)\cos(b) - \sin(a)\sin(b)}$$

$$= \frac{\frac{\sin(a)\cos(b)}{\cos(a)\cos(b)} + \frac{\cos(a)\sin(b)}{\cos(a)\cos(b)}}{1 - \frac{\sin(a)\sin(b)}{\cos(a)\cos(b)}}$$

$$= \frac{\frac{\sin(a)}{\cos(a)} + \frac{\sin(b)}{\cos(b)}}{1 - \frac{\sin(a)}{\cos(a)}\frac{\sin(b)}{\cos(b)}}$$

$$= \frac{\tan(a) + \tan(b)}{1 - \tan(a)\tan(b)}$$

Remembering that $\tan(-b) = -\tan(b)$,

$$\tan(a - b) = \frac{\tan(a) - \tan(b)}{1 + \tan(a)\tan(b)}. \tag{9.15}$$

Various other identities are required in the following exercises, and the reader will benefit most by *deriving* each at least once. Complete solutions using induction are given for all exercises in this section, though some may have direct proofs.

Exercise 113. *Show that for every positive integer n,*

$$\cos(n\pi) = (-1)^n.$$

Exercise 114. *Let $x \in \mathbb{R}$ be fixed. Show that for each $n \geq 1$,*

$$|\sin(nx)| \leq n|\sin(x)|.$$

The result in the next exercise is named for Abraham De Moivre (1667–1754), a friend of Isaac Newton. (In Newman's anthology *The world of mathematics*, his name is spelled "Demoivre".) De Moivre, as mentioned in [4, p. 155] "emigrated from France to England to escape religious prosecution, and made his living as a coffee-house consultant to students of mathematics."

Recall that i is a (complex) number satisfying $i^2 = -1$.

Exercise 115 (De Moivre's Theorem). *Prove that for each $n \geq 1$,*

$$[\cos(\theta) + i\sin(\theta)]^n = \cos(n\theta) + i\sin(n\theta).$$

In [550] it is explained how some of the these next statements aid in high speed computing.

Exercise 116. *Prove that for any $n \geq 1$ and any angle θ,*

$$\sin(\theta + n\pi) = (-1)^n \sin(\theta).$$

Exercise 117. *Prove that for any $n \geq 1$ and any angle θ,*

$$\cos(\theta + n\pi) = (-1)^n \cos(\theta).$$

Exercise 118. *Prove that for each $n \geq 1$, and any angle θ that is not a multiple of 2π,*

$$\sin\theta + \sin(2\theta) + \cdots + \sin(n\theta) = \frac{\sin(\frac{n+1}{2}\theta)\sin(\frac{n\theta}{2})}{\sin(\theta/2)}.$$

Exercise 119. *Prove that for $n \geq 1$, and any angle θ that is not a multiple of 2π,*

$$\cos\theta + \cos(2\theta) + \cdots + \cos(n\theta) - \frac{\cos(\frac{n+1}{2}\theta)\sin(\frac{n\theta}{2})}{\sin(\theta/2)}.$$

Exercise 120. *Prove that for each $n \geq 1$, and any angle θ that is not a multiple of π,*

$$\sin(\theta) + \sin(3\theta) + \cdots + \sin((2n-1)\theta) = \frac{\sin^2(n\theta)}{\sin(\theta)}.$$

Exercise 121. *Prove that for each $n \geq 1$, and any angle θ that is not a multiple of π,*

$$\cos\theta + \cos(3\theta) + \cdots + \cos((2n-1)\theta) = \frac{\sin(2n\theta)}{2\sin(\theta)}.$$

Exercise 122. *Fix some θ that is not a multiple of π. Let $s_0 = 0$, $s_1 = 1$, and for $n \geq 2$, recursively define*

$$s_n = 2\cos(\theta)s_{n-1} - s_{n-2}.$$

If θ is not a multiple of π, prove that for each $n \geq 0$,

$$s_n = \frac{\sin(n\theta)}{\sin(\theta)},$$

and for each $n \geq 1$,

$$\cos(n\theta) = \cos(\theta)s_n - s_{n-1}.$$

The next exercise uses the same recursion as in Exercise 122, however with different initial values.

Exercise 123. *Let θ be any fixed angle. Define a sequence of real numbers recursively by $s_1 = \cos(\theta)$, $s_2 = \cos(2\theta)$ and for $n > 2$, define*

$$s_n = 2\cos(\theta)s_{n-1} - s_{n-2}.$$

Prove for every $n \geq 1$, that $s_n = \cos(n\theta)$.

The following was a contest question in [429]; the solution follows with help from another exercise given in Chapter 10.

Exercise 124. *Let x be any real number. Prove that for any $n \geq 1$,*

$$\cos^{2n}(x) + \sin^{2n}(x) \geq \frac{1}{2^{n-1}}.$$

Exercise 125. *Prove that for $n \geq 0$,*

$$\cos(\alpha) \cos(2\alpha) \cos(4\alpha) \cdots \cos(2^n \alpha) = \frac{\sin(2^{n+1}\alpha)}{2^{n+1} \sin(\alpha)}.$$

Exercise 126. *Prove that for $n \geq 1$, and any angle t that is not a multiple of 2π,*

$$\frac{1}{2} + \cos(t) + \cos(2t) + \cdots + \cos(nt) = \frac{\sin((2n+1)t/2)}{2 \sin(t/2)}.$$

The expression

$$D_n(t) = \frac{1}{2} + \sum_{j=1}^{n} \cos(jt)$$

is called the *Dirichlet kernel,* arising in the theory of convex functions and Fourier series (see, *e.g.,* [154], p.64). The next exercise regards the average of Dirichlet kernels, and was developed by Lipot Fejér, perhaps appearing first in R. Courant's 1937 book *Differential and Integral Calculus,* 2nd. ed., [although I have not yet been able to verify this]. It is also interesting to note that among Fejér's students was Paul Erdős, and that Fejér's advisor was Schwarz (as in "Cauchy–Schwarz inequality").

Exercise 127. *Using the notation from Exercise 126 put*

$$K_N(t) = \frac{1}{N+1} \sum_{n=0}^{N} D_n(t).$$

$K_N(t)$ is called the Fejér kernel (see, e.g., [154, p.64]). Note that when $n = 0$, the sum in $D_n(t)$ is empty, so $D_0(t) = 1/2$. Prove that for $N \geq 0$,

$$K_N(t) = \frac{\sin^2\left((N+1)t/2\right)}{2(N+1) \sin^2(t/2)}.$$

Exercise 128. *Prove that for each $n \geq 1$ and any real number x that is not a multiple of 2π,*

$$\sin x + 2\sin(2x) + 3\sin(3x) + \cdots + n\sin(nx)$$
$$= \frac{(n+1)\sin(nx) - n\sin((n+1)x)}{4\sin^2(x/2)}.$$

Exercise 129. *Prove that for each $n \geq 1$ and any real number x that is not a multiple of 2π,*

$$\cos(x) + 2\cos(2x) + 3\cos(3x) + \cdots + n\cos(nx)$$
$$= \frac{(n+1)\cos(nx) - n\cos((n+1)x) - 1}{4\sin^2(x/2)}.$$

Exercise 130. *For any x that is not a multiple of π, prove that for each $n \geq 1$,*

$$\frac{1}{2}\tan\left(\frac{x}{2}\right) + \frac{1}{2^2}\tan\left(\frac{x}{2^2}\right) + \cdots + \frac{1}{2^n}\tan\left(\frac{x}{2^n}\right) = \frac{1}{2^n}\cot(\frac{x}{2^n}) - \cot(x).$$

Exercise 131. *Prove that for each $n \geq 1$,*

$$\cot^{-1}(3) + \cot^{-1}(5) + \cdots + \cot^{-1}(2n+1)$$
$$= \tan^{-1}(2) + \tan^{-1}\left(\frac{3}{2}\right) + \cdots + \tan^{-1}\left(\frac{n+1}{n}\right) - n\tan^{-1}(1).$$

Exercise 132. *Use mathematical induction to show that for $n \geq 2$, there exist constants $a_0, a_1, \ldots, a_n,$ and $b_0, b_1, \ldots, b_n$ so that*

$$\sin^n(x) = \sum_{r=0}^{n}\left(a_r\cos(rx) + b_r\sin(rx)\right).$$

Exercise 133. *Let x and α be real numbers so that $x + \frac{1}{x} = 2\cos(\alpha)$. Prove that for every $n \geq 1$,*

$$x^n + \frac{1}{x^n} = 2cos(n\alpha).$$

9.9   Miscellaneous identities

Exercise 134. *Prove that for any $n \geq 1$ and non-negative real numbers $x_1, \ldots, x_n,$ if $x_1 + \cdots + x_n = 0$, then $x_1 = \cdots = x_n = 0$.*

Exercise 135. *For $x, b \in \mathbb{R}^+$ with $b \neq 1$, prove that for all $n \in \mathbb{Z}^+$,*

$$\log_b(x^n) = n\log_b(x).$$

Exercise 136. *For any real numbers $a_1, a_2, \ldots, a_n$ and $b_1, b_2, \ldots, b_n,$*

$$\sum_{i=1}^{n}(a_i + b_i) = \sum_{i=1}^{n}a_i + \sum_{j=1}^{n}b_j.$$

Exercise 137 (Telescoping sum). *Prove that if $a_1, a_2, a_3, \ldots,$ are real numbers, then for each positive integer $n \geq 1$,*

$$\sum_{i=1}^{n}(a_i - a_{i+1}) = a_1 - a_{n+1}.$$

Exercise 138. *Prove that for any $n \geq 1$,*

$$\sum_{j=1}^{n}(3j^2 - j + 2) = n(n^2 + n + 2).$$

In [366], José Nieto, Unversidad del Zulia, Venezuela, posed the problem of showing that there are infinitely many $a > b > c > d > 1$ with $a!d! = b!c!$. The equality in the next exercise exhibits such families of four numbers. The equality is trivial to verify, but with a little more work, one can prove it inductively as well.

Exercise 139. *Prove by induction that for every $n \geq 3$,*

$$(n^2 + n)!(n - 1)! = (n^2 + n - 1)!(n + 1)!.$$

Here is an elegant result that seems related to Exercise 139:

Lemma 9.9.1. *Show that any rational number can be expressed as a quotient, where each of the numerator and denominator is a product of factorials of prime numbers.*

A question in the 2009 Putnam exam (question B1) asked to prove Lemma 9.9.1, and the example given in the question was

$$\frac{10}{9} = \frac{5!2!}{3!3!3!}.$$

At the time of writing this exercise, the official solutions to the Putnam have not yet been released (usually appearing in the *American Mathematical Monthly* in the fall), however one easy solution is to use induction:

Exercise 140. *Prove Lemma 9.9.1.*

In the next few exercises, x represents an indeterminate or variable (representing, for example, a real number).

Exercise 141. *Prove by induction that for $n \geq 1$,*

$$\frac{1}{x(x + 1)} + \frac{1}{(x + 1)(x + 2)} + \cdots + \frac{1}{(x + n - 1)(x + n)} = \frac{n}{x(x + n)}.$$

Exercise 142. *Prove that for $n \geq 1$,*

$$\frac{1}{1 + x} + \frac{2}{1 + x^2} + \frac{4}{1 + x^4} + \cdots + \frac{2^n}{1 + x^{2^n}} = \frac{1}{x - 1} + \frac{2^{n+1}}{1 - x^{2^{n+1}}}.$$

Exercise 143. *Prove that for any non-integer x and for every $n \geq 1$,*

$$1 - \frac{x}{1!} + \frac{x(x - 1)}{2!} - \cdots + (-1)^n \frac{x(x - 1) \cdots (x - n + 1)}{n!}$$

$$= (-1)^n \frac{(x - 1)(x - 2) \cdots (x - n)}{n!}.$$

Exercise 144. *Show by induction that for any $n \geq 0$,*

$$(1)(3)(5) \cdots (2n + 1) = \frac{(2n + 1)!}{2^n n!}.$$

Exercise 145. *Show that for each non-negative integer n,*

$$1 + n(n+1)(n+2)(n+3)$$

is a perfect square. Hint: Work out the first five or six cases, conjecture an inequality, and prove it by induction.

This next exercise contains a surprisingly elegant result. Recall that the notation $S \subseteq [n]$ means that S is a set of positive integers all at most n, and that $\prod_{s \in S} s$ is the product of all elements in S. (See the example given after the exercise.)

Exercise 146. *Show that for each $n \geq 1$,*

$$\sum_{\emptyset \neq S \subseteq [n]} \frac{1}{\prod_{s \in S} s} = n.$$

For example, when $n = 3$, the possible subsets in Exercise 146 are

$$\{1\}, \{2\}, \{3\}, \{1,2\}, \{1,3\}, \{2,3\}, \{1,2,3\},$$

and so the righthand side of the equality in Exercise 146 becomes

$$\frac{1}{1} + \frac{1}{2} + \frac{1}{3} + \frac{1}{2} + \frac{1}{3} + \frac{1}{6} + \frac{1}{6} = 3.$$

Exercise 147. *For any fixed real number x that is not an integer, and for all $n \geq 1$,*

$$\frac{n}{x} + \frac{n(n-1)}{x(x-1)} + \frac{n(n-1)(n-2)}{x(x-1)(x-2)} + \cdots = \frac{n}{x-n+1}.$$

Exercise 148. *For any given $n \geq 1$, consider all the subsets of $\{1,2,\ldots,n\}$ that do not contain two consecutive numbers. For example, when $n = 4$, the sets are $\{1\}$, $\{2\}$, $\{3\}$, $\{4\}$, $\{1,3\}$, $\{1,4\}$, and $\{2,4\}$. Prove that the sum of the squares of the products in each set is $(n+1)! - 1$. (For example, when $n = 4$, the number is $1^2 + 2^2 + 3^2 + 4^2 + 3^2 + 4^2 + 8^2 = 119 = 5! - 1$.)*

Exercise 149. *Let positive integers $x_1,\ldots,x_n$ and $y_1,\ldots,y_m$ be given so that $x_1 + \cdots + x_n = y_1 + \cdots + y_m \leq mn$. Prove that there are proper subsets $I \subset \{1,\ldots,n\}$ and $J \subset \{1,\ldots,m\}$ so that $\sum_{i \in I} x_i = \sum_{j \in J} y_j$.*

Chapter 10

Inequalities

It is from this absolute difference and tranquility of the mind, that mathematical speculations derive some of their most considerable advantages; ...All proportions, every arrangement of quantity, is alike the understanding, because the same truths result to it from all; from greater and lesser, from equality and inequality.

—Edmund Burke,

On the sublime and beautiful.

When proving inequalities, it might serve well to be reminded of transitivity. For example, if $a = b \leq c \leq d = e$, then $a \leq e$. Furthermore, if $a = b < c = d \leq e = f$, then both $a < f$ and $a \leq f$ are true. Also when proving inequalities, the reader is reminded that many inequalities often fail for some or all integers less than some threshold value.

Proving inequalities is usually tantamount to comparing functions; comparing polynomial functions with exponential functions gives some sense as to which functions are "larger". Let $p(x)$ be any polynomial and let $f(x) = 2^x$. It is well known, that eventually, f will "dominate" p, that is, there is a smallest x_0 so that if $x \geq x_0$, then $f(x) > p(x)$. Many of these problems ask to prove this notion in particular settings. Recall that the relative "sizes" of some functions might be captured in increasing order as follows: $log_b(n)$, $\sqrt{n}$, n, $p(n)$, 2^n, $n!$, n^n, and so on. Many of these relationships are demonstrated in the exercises.

Exercise 150. *Prove by induction that if x and y are positive real numbers with $x < y$, then for each $n \geq 1$, $x^n < y^n$.*

Note: another simple inequality based on the above idea is also presented in Exercise 698.

Exercise 151. *Prove that for $n \geq 6$, $4n < n^2 - 7$.*

Exercise 152. *Show that if $n \geq 3$, then $2n + 1 < n^2$.*

Exercise 153. *Prove that for $n \geq 2$, $4n^2 > n + 11$.*

Exercise 154. *Prove by induction that if $n \geq 3$ then $2n < 2^n$.*

Exercise 155. *Prove by induction that for each $n \geq 2$, $1 + 2^n < 3^n$.*

Exercise 156. *Prove by induction that if $n \geq 2$, then $n + 1 < 2^n$.*

Exercise 157. *Prove by induction that for each $n \geq 1$,*

$$\left(1 + \frac{1}{3}\right)^n \geq 1 + \frac{n}{3}.$$

Exercise 158. *Prove by induction that for $n \geq 4$, $2^n < n!$.*

The next problem was commented on in Chapter 1.

Exercise 159. *Prove by induction that if $n \geq 5$, then $n^2 < 2^n$.*

Exercise 160. *Prove by induction that for any integer $\ell \geq 6$,*

$$6\ell + 6 < 2^\ell.$$

Exercise 161. *Use Exercise 160 and mathematical induction to show that for any integer $k \geq 10$,*
$$3k^2 + 3k + 1 < 2^k.$$

Exercise 162. *Use Exercise 161 to prove inductively that for $n \geq 10$,*

$$n^3 < 2^n.$$

Exercise 163. *Use Exercise 160 and mathematical induction to show that for $k \geq 4$,*

$$3k^2 + 3k + 1 < 2(3^k).$$

(Note: in order to apply Exercise 160, the cases $k = 4, 5$ have to be handled separately.)

Exercise 164. *Use Exercise 163 to prove by induction that for $n \geq 4$,*

$$n^3 < 3^n.$$

(Although this result follows indirectly from Exercises 150 and 162 this exercise asks for yet another solution.)

Exercise 165. *Prove by induction that if $n > 6$, then $3^n < n!$.*

Exercise 166. *Prove by induction that for each positive integer k,*

$$k! \geq \left(\frac{k}{e}\right)^k.$$

Exercise 167. *Prove by induction that for $n \geq 1$, $n^2 \geq 2n - 1$.*

Exercise 168. *Prove by induction that if $n \geq 1$, then $2n + 1 \leq 3^n$.*

Exercise 169. *Prove by induction that for $n \geq 3$,*

$$n^n \geq (n + 1)!.$$

Exercise 170. *Prove by induction that for $n \geq 3$,*

$$n^{n+1} > (n + 1)^n.$$

Exercise 171. *Prove by induction that for $n \geq 5$,*

$$(n + 1)! > 2^{n+3}.$$

Exercise 172. *Prove by induction that if $n \geq 3$ then $(n!)^2 > n^n$.*

Exercise 173. *Prove by induction that for $n \geq 2$,*

$$1 \cdot 3 \cdot 5 \cdots (2n - 1) < n^n.$$

Exercise 174. *Prove by induction that for $n \geq 5$,*

$$(2n)! < (n!)^2 4^{n-1}.$$

Exercise 175. *Prove by induction that if $n \geq 2$, then $\frac{4^n}{n+1} < \frac{(2n)!}{(n!)^2}$.*

A statement $S(n)$ is said to be true for *sufficiently large n* iff there exists an n_0 so that for every $n \geq n_0$, $S(n)$ holds; to prove that $S(n)$ is true for sufficiently large n, one need not name the least such n_0, only show that such an n_0 exists. If one can find any suitable n_0, and $S(n)$ is defined only for natural numbers, then, by the well-ordering of $\mathbb{N}$, a least such n_0 exists.

Exercise 176. *Prove by induction that for n sufficiently large, $n! > 4^n$.*

Exercise 177. *Prove that for each positive integer n, $\ln(n) < n$ holds.*

Similar inequalities are sometimes hard to prove by induction. For example, for $n > 16$,
$$\log_2(n) < \sqrt{n},$$
is true, and a base step of $n = 17$ is true (check with a calculator, and find $\log_2(17) = 4.0874...$, and $\sqrt{17} = 4.123...$); however, to find an inductive step might be challenging. On the other hand, calculus easily proves this inequality since both sides are equal for $n = 16$, and derivatives show that $\log_2(x)$ grows more slowly than does $\sqrt{x}$ for $x > 16$.

Exercise 178. *Let $b \in \mathbb{R}$ with $b > 1$. Prove that for each $n \geq 2$, if k is the number of prime factors in n, then $\log_b(n) \geq k \log_b(2)$.*

Exercise 179. *Prove by induction that for $n \geq 1$,*

$$1 + 2 + 3 + \cdots + n < \frac{1}{8}(2n + 1)^2.$$

Exercise 180. *Prove that for each $n \geq 1$,*

$$1 + \frac{1}{4} + \frac{1}{9} + \cdots + \frac{1}{n^2} \leq 2 - \frac{1}{n}.$$

What does this result imply regarding the corresponding infinite series $1 + \frac{1}{4} + \frac{1}{9} + \cdots$?

Exercise 181. *Prove that for every $n > 1$,*

$$2\left(1 + \frac{1}{8} + \frac{1}{27} + \cdots + \frac{1}{n^3}\right) < 3 - \frac{1}{n^2}.$$

Exercise 182. *Prove by induction that for $n \geq 4$,*

$$1 + \frac{1}{1!} + \frac{1}{2!} + \frac{1}{3!} + \cdots + \frac{1}{n!} < 3 - \frac{1}{n}.$$

Exercise 183. *Prove by induction that for $n \geq 1$,*

$$\frac{1}{1!} + \frac{2}{3!} + \frac{3}{5!} + \cdots + \frac{n}{(2n-1)!} \leq 2 - \frac{1}{(2n)!}.$$

Exercise 184. *Prove that for $n \geq 1$,*

$$\frac{1}{2!} + \frac{2}{3!} + \frac{3}{4!} + \cdots + \frac{n}{(n+1)!} \leq 1 - \frac{1}{(n+1)!}.$$

Exercise 185. *Prove by induction that for $n \geq 2$,*

$$\frac{1}{n+1} + \frac{1}{n+2} + \cdots + \frac{1}{2n} > \frac{13}{24}.$$

Exercise 186. *Prove that for each $n \geq 2$,*

$$2^n < \binom{2n}{n} < 4^n.$$

Exercise 187. *Prove that for each $n \geq 1$,*

$$\binom{2n}{n} \geq \frac{2^{2n}}{2n}.$$

Exercise 188. *Prove that for* $n \geq 1$,

$$\frac{1}{\sqrt{1}} + \frac{1}{\sqrt{2}} + \frac{1}{\sqrt{3}} + \cdots + \frac{1}{\sqrt{n}} \leq 2\sqrt{n} - 1.$$

Exercise 189. *Prove that for* $n \geq 1$,

$$\frac{1}{\sqrt{1}} + \frac{1}{\sqrt{2}} + \frac{1}{\sqrt{3}} + \cdots + \frac{1}{\sqrt{n}} \geq \sqrt{n}.$$

Exercise 190. *Prove that for* $n \geq 1$,

$$2 + \frac{1}{\sqrt{1}} + \frac{1}{\sqrt{2}} + \frac{1}{\sqrt{3}} + \cdots + \frac{1}{\sqrt{n}} > 2\sqrt{n+1}.$$

Exercise 191. *Prove that for each* $n \geq 1$,

$$\frac{1}{2n} \leq \frac{1 \cdot 3 \cdot 5 \cdots (2n-1)}{2 \cdot 4 \cdot 6 \cdots (2n)} \leq \frac{1}{\sqrt{n+1}}.$$

Exercise 192. *Prove by induction that for* $n \geq 1$,

$$\frac{1}{2} \cdot \frac{3}{4} \cdot \frac{5}{6} \cdots \frac{2n-1}{2n} \leq \frac{1}{\sqrt{3n+1}}.$$

In solutions to some of the following exercises, the following common observation (which is actually a special case of the statement of Exercise 150) might come in handy:

Lemma 10.0.2. *If* $a, b \in \mathbb{R}$ *are both non-negative, then* $a \leq b$, *if and only if* $a^2 \leq b^2$.

Proof of Lemma 10.0.2: Let $a \geq 0$ and $b \geq 0$. If $a = b$, the result is trivial, so suppose that $a \neq b$, and if either $a = 0$ or $b = 0$, then again the result is obvious, so assume that both a and b are non-zero. If $a < b$ then $b^2 = (a + (b - a))^2 = a^2 + 2a(b - a) + (b - a)^2 > a^2$, since both $a(b - a)$ and $(b - a)^2$ are positive. If $a^2 \leq b^2$, then $0 \leq b^2 - a^2 = (b - a)(b + a)$, and since $b + a > 0$, it follows that $b - a > 0$, that is, $b > a$. $\square$

Exercise 193 (Triangle inequality). *Prove that for each* $n \geq 1$ *and for real numbers* $x_1, x_2, \ldots, x_n$,

$$|x_1 + x_2 + \cdots + x_n| \leq |x_1| + |x_2| + \cdots + |x_n|.$$

Exercise 194. *Prove that for* $n \in \mathbb{N}$, *and for any non-negative real numbers* x *and* y,

$$\left(\frac{x+y}{2}\right)^n \leq \frac{x^n + y^n}{2}.$$

Exercise 195. *Prove that for each $n \geq 1$, if $s_1, s_2, \ldots, s_n$ are positive integers, then*

$$(s_1 + s_2 + \cdots + s_n)^2 \leq s_1^3 + s_2^3 + \cdots + s_n^3.$$

Exercise 196. *For non-negative real numbers $x_1, x_2, \ldots, x_n$, prove that*

$$\frac{x_1^2 + x_2^2 + \cdots + x_n^2}{n} x_1 x_2 \cdots x_n \leq \left(\frac{x_1 + x_2 + \cdots + x_n}{n} \right)^{n+2}.$$

Exercise 197. *Let x be a positive real number. Prove that for each $n \geq 1$,*

$$x^n + x^{n-2} + x^{n-4} + \cdots + \frac{1}{x^{n-4}} + \frac{1}{x^{n-2}} + \frac{1}{x^n} \geq n+1.$$

There are many proofs of the following exercise, named for Jacques Bernoulli (1654–1705), that do not require induction; for example, one can use the binomial theorem.

Exercise 198 (Bernoulli's inequality). *For any non-zero real number x with $x > -1$, prove by induction that for each $n \geq 2$,*

$$(1 + x)^n > 1 + nx.$$

Exercise 199. *Prove that if $n \geq 2$ and $a_1, a_2, \ldots, a_n \in \mathbb{R}^+$ are so that $a_1 a_2 \cdots a_n = 1$, then*

$$a_1 + a_2 + \cdots + a_n \geq n.$$

The *geometric mean* of n positive numbers $a_1, a_2, \ldots, a_n$ is $(a_1 a_2 \cdots a_n)^{1/n}$. The *harmonic mean* of $a_1, a_2, \ldots, a_n$ is

$$\frac{n}{\frac{1}{a_1} + \frac{1}{a_2} + \cdots + \frac{1}{a_n}}.$$

This next exercise relates the geometric and harmonic means. The proof asked for does not use induction, but instead the AM-GM inequality. Since the AM-GM inequality was proved by induction in Theorem 3.3.1, the result is still, technically, one provable by induction.

Exercise 200 (GM-HM inequality). *Let $a_1, a_2, \ldots, a_n$ be positive real numbers. Using the AM-GM inequality, (not induction) prove the GM-HM inequality:*

$$(a_1 a_2 \cdots a_n)^{1/n} \geq \frac{n}{\frac{1}{a_1} + \frac{1}{a_2} + \cdots + \frac{1}{a_n}}.$$

Using downward induction, one can solve this next problem.

Exercise 201. *Let* $1 \leq a_1 \leq a_2 \leq \cdots \leq a_n$ *be positive integers. Prove that if*

$$\frac{1}{a_1} + \frac{1}{a_2} + \cdots + \frac{1}{a_n} = 1,$$

then $a_n < 2^{n!}$.

Exercise 202. *Let* $a_1, a_2, a_3, \ldots$ *be a sequence of real numbers, each* $a_i \geq 1$. *Prove that*

$$\sum_{i=1}^{n} \frac{1}{1 + a_i} \geq \frac{n}{1 + (a_1 a_2 \cdots a_n)^{1/n}}.$$

Prove that equality holds if and only if $a_1 = a_2 = \cdots = a_n$.

The next three exercises give results that might be helpful in evaluating certain limits.

Exercise 203. *Prove that*

$$\sqrt{2 + \sqrt{2 + \sqrt{2 + \sqrt{2 + \ldots}}}} = 2.$$

Exercise 204. *Prove that for any positive integer* n,

$$\sqrt{2\sqrt{3\sqrt{4 \cdots \sqrt{n}}}} < 3.$$

A more general form of Exercise 204 is the following:

Exercise 205. *Prove that for every* $n \in \mathbb{Z}^+$ *and every non-negative real number* a,

$$\sqrt{a + 1 + \sqrt{a + 2 + \cdots + \sqrt{a + n}}} < a + 3.$$

The reader interested in "nested square root" problems can find more problems and references in the 2008 article [272].

There are a number of other famous inequalities one could list here, but they occur in other sections. For example, inequalities that often arise in vector spaces are given in Section 19.5. Perhaps the most popular of these is the Cauchy–Schwarz inequality (see Theorem 19.5.2):

For each $n \geq 1$ and real numbers $a_1, a_2, \ldots, a_n, b_1, b_2, \ldots, b_n$, then

$$(a_1^2 + a_2^2 + \cdots + a_n^2) \cdot (b_1^2 + b_2^2 + \cdots + b_n^2) \geq (a_1 b_1 + a_2 b_2 + \cdots + a_n b_n)^2.$$

For a generalization of the triangle inequality, called Minkowski's inequality, see Exercise 694. For Hölder's inequality for vectors, see Theorem 19.5.4. For Minkowski's inequality for p-norms, see Exercise 695.

Chapter 11

Number theory

Mathematics is the Queen of Sciences, and number theory, the Queen of Mathematics. She often condescends to render service to astronomy and other natural sciences, but under all circumstances the first place is her due.

—Carl Friedrich Gauss (1777–1855).

The area of mathematics called "number theory" might be loosely described as the study of arithmetic properties of integers, usually those concerning divisibility, prime factorization, congruences, and representations of a natural number as sums of others. See virtually any text on number theory (for example, [19]) for a better idea of what other topics might be considered as number theoretic.

In this chapter, many of the most popular aspects of number theory are considered, together with a few others, for example, the decomposition of integers into sums of fractions. For many more examples of proof by mathematical induction in advanced additive number theory, the books by Nathanson [401] and Tao and Vu [527] are highly recommended.

11.1   Primes

Recall that a prime number is a natural number $n \geq 2$ with only two (positive integer) divisors, n itself, and 1. The next exercise asks to prove what is called "The fundamental theorem of arithmetic", often abbreviated by "FTOA".

Exercise 206 (FTOA). *Prove that every integer $n \geq 2$ can be written uniquely as a product of powers of primes (also called prime powers), that is, for each $n \geq 2$, there exist unique primes $p_1, \ldots, p_k$ and positive integers $\alpha_1, \ldots, \alpha_k$ so that*

$$n = p_1^{\alpha_1} p_2^{\alpha_2} \cdots p_k^{\alpha_k}.$$

Euclid proved that there are infinitely many primes. The standard proof is by contradiction: suppose that there are only finitely many primes, say $p_1, p_2, \ldots, p_n$. Since the number $x = p_1 p_2 \cdots p_n + 1$ is not divisible by any of the primes $p_1, \ldots, p_n$, then x itself must be either prime or a product of primes not listed. This contradicts that $p_1, \ldots, p_n$ are the only primes.   □

There has been debate as to whether or not Euclid knew of mathematical induction, a debate that has gone as far as to interpret induction less formally. (See, for example, [175], [197], [541], [542], [557].) The next exercise addresses Euclid's tacit use of induction.

Exercise 207. *Rewrite the above proof for Euclid's theorem as a proof by induction.*

Compare Exercise 6.1 for another proof of Euclid's theorem, which is based on a simple equation that is proved by induction.

For any positive integer n, let $\sigma(n)$ denote the sum of all positive divisors of n. For example, if p is a prime and m is some positive integer, then $\sigma(p^m) = 1 + p + p^2 + \cdots + p^m = (p^{m+1} - 1)/(p - 1)$.

Exercise 208. *Prove that if $n = p_1^{e_1} p_2^{e_2} \cdots p_k^{e_k}$ is the prime power decomposition of n, then*

$$\sigma(n) = \sigma(p_1^{e_1})\sigma(p_2^{e_2}) \cdots \sigma(p_k^{e_k}).$$

(See Exercise 213 for an extension of Exercise 208.)

Exercise 209 (Division lemma)**.** *Use well-ordering to prove that for any natural numbers m, n, there exist unique integers $q \geq 0$ and r with $0 \leq r < m$ so that $n = qm + r$. (The numbers q and r are called the* quotient *and* remainder*, respectively, upon dividing n by m.) Hint: examine the set $\{a \in \mathbb{Z} : n + am \geq 0\}$.*

Two non-zero integers m and n are said to be *relatively prime* if they share no common factors other than 1. For example, 40 and 27 are relatively prime, however 40 and 28 are not (they share a common factor of 4).

Exercise 210. *Let $m_1, \ldots, m_n$ be pairwise relatively prime natural numbers. Prove that if y is a natural number so that for each $i = 1, 2, \ldots, n$, m_i divides y, then so also the product $m_1 m_2 \cdots m_n$ divides y.*

The greatest common divisor of two natural numbers x and y is denoted $\gcd(x, y)$ and their least common multiple is denoted by $\operatorname{lcm}(x, y)$. For example, $\gcd(4, 6) = 2$ and $\operatorname{lcm}(4, 6) = 12$. Positive integers n and m are relatively prime if and only if $\gcd(m, n) = 1$. The greatest common divisor and least common multiple are related by $mn = \gcd(m, n) \cdot \operatorname{lcm}(m, n)$.

There are two minor technicalities regarding the $\gcd(m, n)$ notation when m and n are allowed to be any integers. Since for any positive integer a, a 'divides' 0, say that $\gcd(0, n) = n$. If either m or n are negative integers, the greatest common factor is non-negative, and so $\gcd(m, n) = \gcd(|m|, |n|)$. The results below are used primarily when m and n are non-negative, and since using negative integers adds little information, generally, non-negative integers are used—the interested reader can make the generalizations to all integers.

For any positive integers m and n, how does one know that $\gcd(m,n)$ exists and is unique? One simple way to see this is to write out the prime factorizations of each (which are unique by the FTOA), and see which factors are common to both. For example, if $m = 1680 = 2^4 \cdot 3 \cdot 5 \cdot 7$ and $n = 3528 = 2^3 \cdot 3^2 \cdot 7^2$, then 2^3, 3 and 7 are the only factors common to each, so $\gcd(m,n) = 2^3 \cdot 3 \cdot 7 = 168$. One can calculate $\gcd(m,n)$ without factoring—by something called the *Euclidean division algorithm* (see below). Proving that it works can be done by induction. The following simple lemma is at the heart of that inductive proof.

Lemma 11.1.1. *For positive integers m and n, if q and r are integers satisfying $q \geq 0$, $0 \leq r < m$ and $n = mq + r$, then $\gcd(m,r) = \gcd(m,n)$.*

Proof: Let $d = \gcd(m,n)$. Since d divides two of the three terms in $n = mq + r$, it must divide the third, namely r. So d divides both m and r. Next, let c be *any* common divisor of m and r. Again, since $n = mq + r$, c divides n; hence c is a common divisor of m and n, and so divides $\gcd(m,n) = d$. Thus $c \leq d$. So indeed, d is the greatest common divisor of m and r. $\qquad\square$

The following central result has a proof by induction, which is given as Exercise 211 below:

Theorem 11.1.2 (Euclidean division algorithm). *Given natural numbers m and n, n not a multiple of m, if one applies the division lemma repeatedly producing $k \geq 2$ quotients $q_1, q_2, \ldots, q_k$ and remainders $r_1, r_2, \ldots, r_k$, where*

$$\begin{aligned}
n &= q_1 m + r_1 & (0 < r_1 < m) \\
m &= q_2 r_1 + r_2 & (0 < r_2 < r_1) \\
r_1 &= q_3 r_2 + r_3 & (0 < r_3 < r_2) \\
r_2 &= q_4 r_3 + r_4 & (0 < r_4 < r_3) \\
&\ \ \vdots & \\
r_{k-3} &= q_k r_{k-2} + r_{k-1} & (0 < r_{k-1} < r_{k-2}) \\
r_{k-2} &= q_k r_{k-1} + r_k & (0 = r_k),
\end{aligned}$$

then $r_{k-1} = \gcd(m,n)$.

So, in the above example, with $m = 1680$ and $n = 3528$, applying the Euclidean division algorithm, the gcd is arrived at rather quickly:

$$\begin{aligned}
3528 &= 2 \cdot 1680 + 168 \\
1680 &= 10 \cdot 168 + 0,
\end{aligned}$$

and so by Theorem 11.1.2, $\gcd(1680, 3528) = 168$.

Perhaps one more example is instructive: Let $m = 192$ and $n = 513$. Then

$$513 = 2 \cdot 192 + 129 \tag{11.1}$$
$$192 = 1 \cdot 129 + 63 \tag{11.2}$$
$$129 = 2 \cdot 63 + 3 \tag{11.3}$$
$$63 = 21 \cdot 3 + 0.$$

So $\gcd(192, 513) = 3$.

Exercise 211. *Prove the Euclidean division algorithm by induction. Hint: use Lemma 11.1.1.*

One powerful consequence of the Euclidean division algorithm is the following simple result that underpins much of number theory.

Lemma 11.1.3 (Bezout's Lemma). *Let a and b be positive integers, and put $d = \gcd(a, b)$. Then there exist integers k and ℓ so that*

$$d = ka + \ell b.$$

In particular, if a and b are relatively prime, then there exist integers k and ℓ so that $1 = ka + \ell b$.

Rather than give a formal proof, here is an example of how it works with the above $m = 219$, $n = 513$ and $d = 3$; let us just say that Bezout's Lemma is proved by *undoing* the Euclidean division algorithm, actually enabling one to compute the desired k and ℓ:

$$
\begin{aligned}
3 &= 129 - 2 \cdot 63 && \text{(by eq'n (11.3))} \\
&= 129 - 2(192 - 1 \cdot 129) && \text{(by eq'n (11.2))} \\
&= 3 \cdot 129 - 2 \cdot 192 \\
&= 3 \cdot (513 - 2 \cdot 192) - 2 \cdot 192 && \text{(by eq'n (11.1))} \\
&= 3 \cdot 513 - 8 \cdot 192,
\end{aligned}
$$

and so $k = 3$ and $\ell = -8$ in the statement of Bezout's Lemma. This seemingly trivial result has rather grand consequences in the theory of numbers and, for example, in the study of finite fields.

For more regarding the Euclidean division algorithm, and bounds on the number steps the algorithm takes, see Exercises 377, 378, and 379.

Next are a few other standard results in number theory that are provable by induction.

Exercise 212. *Prove that for $n \geq 2$, the product of all primes at most n is at most 2^{2n}.*

Let $d(n)$ denote the number of positive divisors of n, and let $\sigma(n)$ be the sum of these divisors. For example, if p is a prime, then $d(p) = 2$ and $\sigma(p) = p + 1$. The next exercise contains, in part, the result of Exercise 208.

Exercise 213. *Prove that for each $n \geq 2$, if $n = p_1^{\alpha_1} \cdots p_s^{\alpha_s}$ is the prime factorization, then*

$$d(n) = (\alpha_1 + 1)(\alpha_2 + 1) \cdots (\alpha_s + 1), \tag{11.4}$$

and

$$\sigma(n) = \frac{p_1^{\alpha_1+1} - 1}{p_1 - 1} \cdot \frac{p_2^{\alpha_2+1} - 1}{p_2 - 1} \cdots \frac{p_s^{\alpha_s+1} - 1}{p_s - 1}. \tag{11.5}$$

In number theory, a function $f : \mathbb{Z}^+ \to \mathbb{Z}^+$ is said to be *completely multiplicative* if and only if for any m, n, $f(mn) = f(m)f(n)$, and f is called *multiplicative* if and only if for any relatively prime m, n, $f(mn) = f(m)f(n)$. As a consequence of the second part of Exercise 213, σ is multiplicative. (This fact is also implicitly contained in the solution of Exercise 208.)

An integer k is called a *perfect square* (or simply, a square) if for some integer ℓ, $k = \ell^2$. This next exercise has at least three different proofs, one of which is by strong induction, and another is by infinite descent.

Exercise 214. *Prove by induction that if s and t are relatively prime natural numbers, and st is a perfect square, then both s and t are perfect squares. Hint: use strong induction, but not on either s or t. Then give a proof of this result by infinite descent.*

Exercise 215. *Prove that if a, b, and $q = \frac{a^2+b^2}{ab+1}$ are non-negative integers, then $q = (\gcd(a,b))^2$. Hint: induct on ab.*

Exercise 216. *For $n \geq 1$ let $S_n = \sum \frac{1}{xy} = 1$ where the sum is taken over all $x, y \leq n$ with $\gcd(x, y) = 1$, and $x + y > n$. Prove that for $S_n = 1$.*

Exercise 217. *Let $n \geq 0$ be a non-negative integer, p be a prime, s be the sum of p-ary digits needed to represent n in base p, and let m be the largest integer so that p^m divides $n!$. Prove by induction on n that*

$$m = \frac{n - s}{p - 1}.$$

Exercise 218. *Prove that for each positive integer n,*

$$2^{2n-1} + 4^{2n-1} + 9^{2n-1}$$

is not a perfect square.

Exercise 219. *Prove that for each positive integer n,*

$$8^{2^n} - 5^{2^n}$$

is not a perfect square.

Exercise 220. *Show by infinite descent that there are no non-trivial integer solutions to*

$$a^2 + b^2 = 3(c^2 + d^2).$$

Hint: first show that if $a^2 + b^2$ is divisible by 3, then so are each of a and b.

Another application of descent occurs in the next exercise.

Exercise 221. *Prove that the equation*

$$a^2 + b^2 + c^2 + d^2 = 2abcd \qquad (11.6)$$

has no solutions in positive integers.

The result in the next exercise shows (with a little work) that the area of a Pythagorean triangle is never a perfect square a negative answer to Bachet's problem).

Exercise 222. *Adapt the above proof of Theorem 3.6.1 by infinite descent to show that $x^4 - y^4 = z^2$ has no non-trivial (none of x, y, z are zero) integer solutions.*

Exercise 223. *Use infinite descent to show that the equation $x^4 - 4y^4 = z^2$ has no non-trivial integer solutions.*

Exercise 224. *Use infinite descent to prove that $x^4 + 2y^4 = z^2$ has no non-trivial integer solutions.*

The equation in the next exercise was studied by Brahmagupta (*circa* 600 AD) and Bhaskara (*circa* 1100 AD), but is presently (and mistakenly) named after John Pell.

Exercise 225. *Let $N > 1$ be a non-square integer. Prove (by induction) that the equation (called* Pell's equation*)*

$$x^2 - Ny^2 = 1 \qquad (11.7)$$

has infinitely many solutions (x_n, y_n) in positive integers. Furthermore, prove that if x_1 and y_1 are the values for which $x + \sqrt{N}y$ is least, then x_n, y_n satisfy

$$x_n + \sqrt{N}y_n = (x_1 + \sqrt{N}y_1)^n.$$

The x_n and y_n can be explicitly computed by

$$x_n = \frac{1}{2}[(x_1 + \sqrt{N}y_1)^n + (x_1 - \sqrt{N}y_1)^n],$$

$$y_n = \frac{1}{2\sqrt{N}}[(x_1 + \sqrt{N}y_1)^n - (x_1 - \sqrt{N}y_1)^n],$$

or recursively by

$$x_{n+1} = x_1 x_n + N y_1 y_n,$$

$$y_{n+1} = y_1 x_n + x_1 y_n.$$

11.2 Congruences

For integers a and b and for a positive integer m, the notation

$$a \equiv b \pmod{m}$$

means that $a - b$ is divisible by m, which is equivalent to saying that there exists $k \in \mathbb{Z}$ so that $a = b + km$. The notation above is read "a is congruent to b modulo m." So, for example, $8 \equiv 2 \pmod 3$, $16 \equiv 0 \pmod 4$, and $32 \equiv 2 \pmod 5$. These equations are sometimes called "congruences". For a fixed n, if $0 \le m < n$, all those integers congruent to m modulo n form what is called a congruence class, denoted $[m]$ or $[m]_n$. For example, when $n = 5$,

$$[1] = \{\ldots, -9, -4, 1, 6, 11, 16, \ldots\}.$$

When trying to prove an equation, it sometimes helps to first try and prove it modulo some n (see, for example, Exercise 232) though the truth of a statement modulo n does not guarantee the truth of the statement in general.

Congruences yield many surprises. One (though not provable by induction) is a spectacular example from [245]: $2^n \equiv 3 \pmod n$ for the first time when $n = 4700063497$.

A useful fact regarding congruences is that if $a \equiv b \pmod n$, then for any $c \in \mathbb{Z}$, $ac \equiv bc \pmod n$. This has an easy proof (left to the reader) using the definition of congruence. Another powerful fact is an immediate consequence of Bezout's lemma:

Lemma 11.2.1. *If* $\gcd(a, n) = 1$*, then there exists* $b \in \{1, 2, \ldots, n - 1\}$ *so that* $ab \equiv 1 \pmod n$*.*

Define Euler's totient function, $\phi : \mathbb{Z}^+ \to \mathbb{Z}^+$, by

$$\phi(n) = |\{m \in \{1, \ldots, n-1\} : \gcd(m, n) = 1\}|,$$

the number of positive integers less than n relatively prime to n. For example, $\phi(2) = 1$, $\phi(5) = 1 + 2 + 3 + 4 = 10$, and $\phi(6) = 1 + 5 = 6$. Apparently (see [230, p. 132]) the totient function was so named by James Joseph Sylvester, "a British mathematician who liked to invent new words" in his 1883 paper [520] on Farey fractions.

As is well-known, Euler's totient function is multiplicative, that is, for any relatively prime pair a, b, $\phi(ab) = \phi(a)\phi(b)$. It is also easy to see that for each prime p and each $k \in \mathbb{Z}^+$,

$$\phi(p^k) = p^k - p^{k-1} = p^k\left(1 - \frac{1}{p}\right),$$

and so a simple proof by induction shows that

$$\phi(n) = n \prod_{\substack{p \,|\, n \\ p \text{ prime}}} \left(1 - \frac{1}{p}\right).$$

Another result that is often needed is Euler's theorem (see nearly any book on number theory for the proof; one particularly easy and popular proof is in [357], for example). Its proof is generally not by induction.

Theorem 11.2.2 (Euler's theorem [177]). *If* $\gcd(a, m) = 1$, *then*

$$a^{\phi(m)} \equiv 1 \pmod{m}.$$

Exercise 226. *Prove by induction that for any fixed positive integer n, the sequence*

$$2, 2^2, 2^{2^2}, 2^{2^{2^2}}, \ldots, \pmod{n}$$

is eventually constant.

Exercise 227. *Prove by induction that for every $k \in \mathbb{Z}^+$, and every odd number $n \in \mathbb{Z}^+$,*

$$(n+1)^k \equiv n+1 \pmod{2n}.$$

The result in the next exercise is sometimes called "Fermat's little theorem", and can be proved by induction (although the now standard proof is not inductive—see [150], for example).

Exercise 228 (Fermat's little theorem). *Prove that if p is a prime and a is an integer, then $a^p - a$ is divisible by p.*

Another formulation of Fermat's little theorem is that if p is a prime and p does not divide a, then $a^{p-1} \equiv 1 \pmod{p}$. This follows from the result in Exercise 228 since $a^p - a$ being divisible by p is equivalent to $a^p \equiv a \pmod{p}$; if p does not divide a, then a is relatively prime to p, and so, (by Lemma 11.2.1) modulo p, a has an inverse—simply multiply this last equation by this inverse. Note also that Fermat's little theorem is a special case of Euler's theorem.

Exercise 229. *Prove that for each $n \geq 0$, and prime number p, and for any polynomial $f(x) = a_n x^n + a_{n-1} x^{n-1} + \cdots + a_1 x + a_0$ where each $a_i \in \mathbb{Z}$, if a_n is not divisible by p, then $f(x) \equiv 0 \pmod{p}$ has at most n distinct solutions modulo p.*

Exercise 230. *Prove by induction that for any $n \geq 1$, the number $(16)^n$ always ends in a 6 (in standard decimal representation, of course). Restating this in terms of congruences, this says that $(16)^n \equiv 6 \pmod{10}$.*

Exercise 231. *Prove that for all $n \geq 1$, $10^n \equiv (-1)^n \pmod{11}$.*

A famous unsolved conjecture in number theory is called Goldbach's conjecture; it states that each even number greater than 2 is the sum of exactly two prime numbers (in at least one way). For example, $16 = 5 + 11$, and $16 = 3 + 13$. This conjecture essentially dates back to a letter Christian Goldbach wrote to Euler in 1742, and is still unsolved. (Goldbach was a tutor to the royal family in Moscow in 1728.) The conjecture has recently been verified for all even n up to 4×10^{14} by Richstein (see [428]). Here is a *finite* (*i.e.*, modular) version of Goldbach's conjecture.

Exercise 232. *Prove that for any positive integers m and n, there exist positive integers a and b relatively prime to $2m$ so that $2n \equiv a + b \pmod{2m}$.*

The next result is called the *Chinese Remainder Theorem* because it was apparently discovered by Sun Zi (4th century), the author of *Sunzi suanjing* (*Master Sun's Mathematical manual*), a manual of arithmetic, part of a required course for Chinese civil servants. A general procedure for solving systems of linear congruences was published by Qin Jiushao (1201–1261) in *Shushu jiuzhang* (Mathematical treatises in nine sections), in 1247.

Theorem 11.2.3 (Chinese Remainder Theorem). *For $n \geq 1$, let*

$$m_1, m_2, \ldots, m_n \in \{2, 3, 4, \ldots\}$$

be pairwise relatively prime, and let $a_1, a_2, \ldots, a_n$ be arbitrary integers. Then there exists x so that

$$x \equiv a_1 \pmod{m_1},$$
$$x \equiv a_2 \pmod{m_2},$$
$$\vdots$$
$$x \equiv a_n \pmod{m_n},$$

and x' is another solution to the above n equations if and only if

$$x' \equiv x \pmod{m_1 m_2 \cdots m_n}.$$

There are different proofs of the Chinese Remainder Theorem; however, seldom is a proof given by induction. The reader might be surprised to see just how easy the proof is when it is presented as a standard induction argument. Only the notation seems cumbersome, but even this problem is easily overcome by first proving the cases $n = 1, 2$, just to get the idea (the arbitrary inductive step is virtually identical to proving the case $n = 2$ from $n = 1$).

Exercise 233. *Prove the Chinese Remainder Theorem by induction on n. Hint: Use Bezout's Lemma with $d = 1$ for relatively prime pairs. You might also need the result from Exercise 210.*

Virtually all the many problems from Section 11.3 on divisibility can be couched in terms of congruences, so only a few problems of that type are given here using congruence notation.

Exercise 234. *Prove that if $n \in \mathbb{Z}^+$, then $2^n + 3^n \equiv 5^n \pmod{6}$.*

Exercise 235. *Prove that if $n \in \mathbb{Z}^+$, then $16^n \equiv 1 - 10n \pmod{25}$.*

Exercise 236. *Prove that if $n \geq 1$, then $3^n + 7^n \equiv 2 \pmod{8}$.*

Exercise 237. *Prove that if $n \geq 0$, then $10^n \equiv (-1)^n \pmod{11}$.*

11.3   Divisibility

The notation $a \mid b$ means a divides b (for example, $4 \mid 12$; since 5 does not divide 12, one writes $5 \nmid 12$). In divisibility problems, one might find it handy to first observe that if $n \mid a$ and $n \mid b$, then $n \mid (a + b)$, $n \mid (a - b)$, and $n \mid ab$. Many of the following problems have direct proofs using congruences (or even more straightforward techniques), but they also can be proved by induction.

Exercise 238. *Prove that for any integer $n \geq 2$, the product of n odd numbers is also odd.*

Exercise 239. *Prove that if n is odd, the sum of n odd numbers is odd.*

The following lemma is proved using the pigeonhole principle (see Exercise 743), however it also has an inductive proof.

Theorem 11.3.1. *Let n be a positive integer. If $n + 1$ distinct numbers are chosen from $\{1, 2, \ldots, 2n\}$, then one of these numbers divides another.*

Proof: Fix $n \geq 1$, and let $x_1, \ldots, x_{n+1}$ be numbers chosen from $\{1, 2, \ldots, 2n\}$. For each $i = 1, \ldots, k + 1$, write $x_i = 2^{\alpha_i} m_i$, where m_i is odd. Since $\{m_1, \ldots, m_{n+1}\} \subseteq \{1, 3, \ldots, 2n - 1\}$, by the pigeonhole principle, there exists $i \neq j$ so that $m_i = m_j$. Since one of 2^{α_i} and 2^{α_j} divides the other, one of x_i or x_j divides the other. $\square$

One might think that an inductive proof of Theorem 11.3.1 is elementary, and after discovering a simple trick, it is elementary!

Exercise 240. *Give an inductive proof of Theorem 11.3.1.*

The following exercise generalizes Exercise 240; (the case $r = 1$ is Theorem 11.3.1):

Exercise 241. *Prove by induction that for any positive integers r and n, if $T \subset \{1, 2, \ldots, 2^r n\}$ has $|T| = (2^r - 1)n + 1$ elements, then there exist $r + 1$ numbers $t_{i_1} < t_{i_2} < \cdots < t_{i_{r+1}}$ in T so that for each $j = 1 \ldots r$, t_{i_j} divides $t_{i_{j+1}}$.*

The next problem has a one line direct proof (do you see it?); however, the proof asked for is by induction.

Exercise 242. *Prove by induction that for every $n \geq 1$, $n(n + 1)$ is even, that is, $2 \mid n(n + 1)$.*

The next result was observed by Leibniz.

Exercise 243. *Prove by induction that for every non-negative integer n, 3 divides $n^3 - n$.*

Exercise 244. *Prove that for every $n \geq 1$, $3 \mid (n^3 + 2n)$.*

Exercise 245. *Prove that if $n \geq 1$, then $3 \mid (2^{2n} - 1)$.*

Exercise 246. *Prove n that for every $n \geq 1$, $3 \mid (2^{2n+1} + 1)$.*

Exercise 247. *Prove that if $n \geq 1$ then $3 \mid (5^{2n} - 1)$.*

Exercise 248. *Prove that for every $n \geq 1$,*

$$3 \mid \left(\frac{10^n + 5}{3} + 4^{n+2} \right).$$

Exercise 249. *Prove that if n is a non-negative integer, then $3 \mid (7^n + 2)$.*

Exercise 250. *Prove that for every $n \geq 1$, $4 \mid n^2(n+1)^2$. (A direct proof is trivial; can you see it?)*

Exercise 251. *Prove that for every $n \geq 1$, $4 \mid (6 \cdot 7^n - 2 \cdot 3^n)$.*

The following popular exercise was also known by Leibniz.

Exercise 252. *Prove that if $n \geq 1$ then $5 \mid (n^5 - n)$.*

Exercise 253. *Prove that for all $n \geq 1$, $(3^{2n} + 4^{n+1})$ is divisible by 5.*

Exercise 254. *Prove that for any natural number n, $8^n - 3^n$ is divisible by 5.*

Exercise 255. *Prove that for every $n \geq 1$, $6 \mid (n^3 - n)$.*

Exercise 256. *Prove that for each $n \geq 1$, $6 \mid (7^n - 1)$.*

Exercise 257. *Prove that for every $n \geq 1$, 6 divides $n^3 + 5n$.*

Exercise 258. *Prove that for every $n \geq 0$, $6 \mid n(n + 1)(n + 2)$.*

Exercise 259. *Prove by induction that for every $n \geq 0$, $n(n-1)(2n-1)$ is divisible by 6.*

The result in the following was also known to Leibniz.

Exercise 260. *By induction, prove that for every $n \geq 0$, 7 divides $n^7 - n$.*

Exercise 261. *Prove that for every $n \geq 1$, $7 \mid (2^{n+2} + 3^{2n+1})$.*

Exercise 262. *Prove by induction that for every $n \geq 1$, $7 \mid (11^n - 4^n)$.*

Exercise 263. *Prove by induction that for every $n \geq 1$, $7 \mid (23^{3n} - 1)$.*

For yet another result concerning divisibility by 7, see Exercise 344 (involving Fibonacci numbers).

Exercise 264. *Prove that for every $n \geq 1$, $3^{2n} - 1$ is divisible by 8.*

Exercise 265. *Prove that for every $n \geq 1$, 8 divides $3^n + 7^n - 2$.*

Exercise 266. *Prove that for every $n \geq 1$,*

$$8 \mid (5^{n+1} + 2 \cdot 3^n + 1).$$

Exercise 267. *Prove that for every $n \geq 1$,*

$$9 \mid (n^3 + (n+1)^3 + (n+2)^3).$$

Exercise 268. *Prove that for every $n \geq 1$, $4^n + 6n - 1$ is divisible by 9.*

Exercise 269. *Prove that for every $n \geq 1$, $n^5 - n$ is a multiple of 10.*

For an exercise for divisibility by 13, see Exercise 351 (involving Fibonacci numbers).

Exercise 270. *Prove that for every $n \geq 1$,*

$$15 \mid (4(47)^{4n} + 3(17)^{4n} - 7).$$

Exercise 271. *Prove that for every $n \geq 0$, $15 \mid (2^{4n} - 1)$.*

Exercise 272. *Prove that for every $n \geq 1$, 16 divides $5^n - 4n - 1$.*

Exercise 273. *Prove that for every $n \geq 0$,*

$$17 \mid (3 \cdot 5^{2n+1} + 2^{3n+1}).$$

Exercise 274. *Prove that all numbers of the form*

$$12008, 120308, 1203308, 12033308, 120333308, \ldots$$

are divisible by 19.

The next exercise has a remarkably elegant solution if the given hint is followed; without following the hint, it is very challenging. [Note: this is a special case of Exercise 295 part (c).]

Exercise 275. *Prove that for every positive integer n,*

$$2^{2^n} + 3^{2^n} + 5^{2^n}$$

is divisible by 19. Hint: induct from k to $k + 2$.

Exercise 276. *Prove that for every $n \geq 1$,*

$$21 \mid (4^{n+1} + 5^{2n-1}).$$

Exercise 277. *Prove by induction that for every odd positive integer n,*

$$24 \mid n(n^2 - 1).$$

Exercise 278. *Let f be a function satisfying $f(1) = f(2) = 1$, and for $n \geq 3$,*

$$f(n) = 3f(n-2) + 3f(n-1) + 1.$$

Prove that for all positive integers n, $f(3n) + f(3n+1)$ is divisible by 32. Hint: Show that f has a period of 12 modulo 32.

Exercise 279. *Prove that for every $n \geq 1$,*

$$43 \mid (6^{n+1} + 7^{2n-1}).$$

Exercise 280. *Prove that all numbers of the form 1007, 10017, 100117, 1001117, 10011117, ... are divisible by 53.*

Exercise 281. *Prove that for every $n \geq 1$,*

$$57 \mid (7^{n+2} + 8^{2n+1}).$$

Exercise 282. *Prove that for every $n \geq 0$,*

$$64 \mid (3^{4n+1} + 10 \cdot 3^{2n} - 13).$$

Exercise 283. *Prove by mathematical induction that for any positive integer n, the number $3^{2n+2} - 8n - 9$ is divisible by 64.*

Exercise 284. *Prove by mathematical induction that for any positive integer n, $9^n - 8n - 1$ is divisible by 64.*

Exercise 285. *Prove that for every $n \geq 0$,*

$$73 \mid (8^{n+2} + 9^{2n+1}).$$

Exercise 286. *Prove that if $n \geq 1$, then $80 \mid (3^{4n} - 1)$.*

Exercise 287. *Prove that for every $n \geq 0$,*

$$133 \mid (11^{n+2} + 12^{2n+1}).$$

Exercise 288. *Prove that for every $n \geq 1$,*

$$576 \mid (5^{2n+2} - 24n - 25).$$

Exercise 289. *If n is an arbitrary natural number, prove by mathematical induction that $7^{2n} - 48n - 1$ is divisible by 2304.*

The next exercise uses the ceiling function. For any real number x, define $\lceil x \rceil$ to be the least integer not less than x. For example, $\lceil 3.2 \rceil = 4$, $\lceil 4 \rceil = 4$, and $\lceil -4.8 \rceil = -4$.

Exercise 290. *Prove that for every $n \geq 1$,*

$$2^{n+1} \mid \lceil (1 + \sqrt{3})^{2n} \rceil.$$

Exercise 291. *Prove that if $n \geq 0$, then $2^{3^n} + 1$ is divisible by 3^{n+1}.*

Exercise 292. *Prove by induction that the polynomial $x^{2n} - y^{2n}$ is divisible by $x^2 - y^2$.*

Exercise 293. *Prove by induction that the polynomial $x^{2n+1} + y^{2n+1}$ is divisible by $x + y$.*

Exercise 294. *Prove by induction that for any $n \geq 1$, $\binom{n+1}{2}$ divides $\sum_{j=1}^{n} j^5$. [Note: No satisfactory solution has been found, yet.]*

In the following exercise, part (c) generalizes Exercise 275.

Exercise 295. *Let a, b, and c be positive integers with $a + b = c$, and let d be an odd factor of $a^2 + b^2 + c^2$. Prove that for all positive integers n:*
(a) $a^{6n-4} + b^{6n-4} + c^{6n-4}$ is divisible by d.
(b) $a^{6n-2} + b^{6n-2} + c^{6n-2}$ is divisible by d^2.
(c) $a^{2^n} + b^{2^n} + c^{2^n}$ is divisible by d.
(d) $a^{4^n} + b^{4^n} + c^{4^n}$ is divisible by d^2.
For parts (c) and (d), assuming that d is not divisible by 3 makes the problem slightly easier.

Exercise 296. *Suppose that A, B, and C are positive integers so that $BC \mid (A - B - C)$. Prove that for every $n \geq 1$, $BC \mid (A^n - B^n - C^n)$.*

Exercise 297. *Let p be a prime and $a_1, a_2, \ldots, a_n, \ldots$ be positive integers each larger than one. Prove by induction that if $p \mid (a_1 \cdot a_2 \cdot \ldots \cdot a_n)$, then p divides one of the a_i's.*

Exercise 298. *Prove that for each $n \geq 1$, there is an n-digit number N (in standard decimal representation), where each digit is a 1 or a 2, so that N is divisible by 2^n.*

Exercise 299. *Prove that for every $n \geq 1$, $\dfrac{(2n)!}{2^n}$ is an integer. This exercise has a very easy direct proof, but prove it by induction anyway.*

Exercise 300. *Prove by induction that for every $n \geq 1$, $\dfrac{(2n)!}{n! 2^n}$ is an integer.*

Exercise 301. *Prove that if $n \geq 1$, then $\frac{n^5}{5} + \frac{n^3}{3} + \frac{7n}{15}$ is an integer.*

Exercise 302. *For each $n \geq 1$, prove that $\frac{n^7}{7} + \frac{n^3}{3} + \frac{11n}{21}$ is an integer.*

Exercise 303. *Prove that if n is a positive integer, so is $(n^3 + 6n^2 + 2n)/3$.*

Exercise 304. *Suppose that x and y are non-zero real numbers for which $x + \frac{1}{x}$, $y + \frac{1}{y}$, and $xy + \frac{1}{xy}$ are integers. Prove that for all integers m and n, $x^m y^n + \frac{1}{x^m y^n}$ is an integer.*

Exercise 305. *Prove that for every $n \in \mathbb{Z}^+$, $(2 + \sqrt{3})^n + (2 - \sqrt{3})^n$ is an integer.*

Exercise 306. *Let n be a power of 2. Prove that for any $n \geq 1$, in any set of $2n - 1$ positive integers, there is a subset of n of these integers whose sum is divisible by n.*

Exercise 307. *Prove that for each $n \geq 1$, $2^{2^n} - 1$ is divisible by at least n distinct primes.*

The next few problems are due to José Espinosa [176]. [They are reproduced here with kind permission.]

Exercise 308. *Let p be a prime of the form $4k + 3$. Prove that for each $n \geq 1$,*

$$\sum_{i=1}^{2k+1} i^{2^n}$$

is divisible by p.

Exercise 309. *Let p be a prime of the form $6k + 5$. Prove that for all $n \geq 0$,*

$$\sum_{i=1}^{3k+2} i^{2 \cdot 3^n}$$

is divisible by p.

Exercise 310. *Let p be a prime of the form $4k + 1$. Prove that for all $n \geq 0$,*

$$\sum_{i=1}^{2k} i^{4n+2}$$

is divisible by p.

The reader is invited to see Espinosa's website [176] for many more challenging induction problems regarding divisibility. For example, Espinosa's Problem 13 asks to show that for each $n \geq 0$,

$$1 + 2^{4n+2} + 3^{4n+2} + 4^{4n+2} + 5^{4n+2} + 6^{4n+2}$$

is divisible by 13.

11.4   Numbers expressible as sums

Let x and y be positive integers, and say that a positive integer k is *expressible as
a sum of x's and y's* iff there exist non-negative integers α, β so that $k = \alpha x + \beta y$.
Many of the following statements have multiple base cases. The first result has
already been proved as Theorem 3.4.1 in case a hint as to the techniques is needed.

Exercise 311. *Prove that any integer $n \geq 14$, can be written as a sum of 3's and
8's.*

Exercise 312. *Prove that any integer $n \geq 8$ is expressible as a sum of 3's and 5's.*

Exercise 313. *Prove that any positive integer n, where $n \notin \{1, 3\}$, can be written
as a sum of 2's and 5's.*

Exercise 314. *Prove that any integer $n \geq 24$ can be written as a sum of 5's and
7's.*

Exercise 315. *Prove that any integer $n \geq 64$ can be expressed as a sum of 5's and
17's.*

11.5   Egyptian fractions

Ancient records show that at a time, some Egyptians used only fractions of the form
$\frac{1}{q}$ (called *unit fractions*).

In 1858, Scot Henry Rhind bought a 3200 year-old papyrus written by a scribe
named Ahmose, and Ahmose wrote that he copied it from work some 400 years previous. (Ahmose is also referred to as "the priest Ahmes, who lived before 1700 B.C."
in Turnbull's article [538].) According to Newman (see his commentary just before
[412]) this papyrus is the second oldest mathematical document known in existence,
as of 1956, anyway. (The oldest is an Egyptian papyrus called the Golenischev, now
in Moscow, also from the same dynasty.) The Rhind papyrus is now a roll 13 inches
high and almost 18 feet long. See [412] for a detailed discussion of its contents,
along with a photo of part of the papyrus. It was found alongside a leather scroll,
both now in the British Museum. (See [42] for a photo of the leather scroll.)

According to Turnbull, the Rhind papyrus was called "directions for knowing all
dark things". The papyrus holds a collection of problems in geometry and arithmetic. For example, it seems to indicate that the Egyptians of the Twelfth Dynasty
(2000–1788 B.C.) knew the formula for the volume of a truncated pyramid. Much
in the papyrus is concerned with expressing fractions as sums of unit fractions such
as
$$\frac{2}{29} = \frac{1}{24} + \frac{1}{58} + \frac{1}{174} + \frac{1}{232},$$
and they did it without + signs! Problem 23 of the papyrus asks how to complete
$$1 = \frac{1}{3} + \frac{1}{4} + \frac{1}{8} + \frac{1}{10} + \frac{1}{30} + \frac{1}{45} + ??$$

using as few as possible unit fractions. The answer given was

$$?? = \frac{1}{9} + \frac{1}{40}.$$

For more on such decompositions and history, together with further references, see [72].

For positive integers p and q, if there are positive integers $n_1 < n_2 < \cdots < n_k$, so that

$$\frac{p}{q} = \frac{1}{n_1} + \frac{1}{n_2} + \cdots + \frac{1}{n_k},$$

then the fraction $\frac{p}{q}$ is said to be written in Egyptian form. If a fraction is in Egyptian form, its representation is not necessarily unique. For example,

$$\begin{aligned}
\frac{5}{6} &= \frac{1}{2} + \frac{1}{3} \\
&= \frac{1}{2} + \frac{1}{4} + \frac{1}{12} \\
&= \frac{1}{3} + \frac{1}{4} + \frac{1}{6} + \frac{1}{12}.
\end{aligned}$$

In [510], Ian Stewart examines a problem regarding the division of camels among sons, and discusses the number of representations of a fraction in nearly Egyptian form. For example, Stewart lists the 14 solutions of "Mustapha's" equation

$$\frac{1}{a} + \frac{1}{b} + \frac{1}{c} + \frac{1}{d} = 1$$

subject to $a \leq b \leq c \leq d$.

Exercise 316. *Prove that if a fraction can be written in Egyptian form, there are infinitely many representations of the fraction in Egyptian form. Hint: Express $\frac{1}{n(n-1)}$ as a sum of two unit fractions.*

Exercise 317. *Prove by induction that for every pair of positive integers p and q with $1 \leq p < q$, the fraction $\frac{p}{q} < 1$ can be written in Egyptian form. Hint: Use a greedy algorithm, finding the largest unit fractions possible.*

Exercise 318. *Prove by induction that for every pair of positive integers p and q, the fraction $\frac{p}{q}$ can be written in Egyptian form, that is, every positive fraction can be written in Egyptian form.*

See also Exercise 201 for a problem related to Egyptian fractions.

11.6   Farey fractions

All variables here represent non-negative integers. For each $n \geq 1$, the set $\mathcal{F}_n$ of Farey fractions is the set of reduced proper fractions $0 \leq \frac{p}{q} \leq 1$ where $0 \leq p \leq q \leq n$. For example,

$$\mathcal{F}_4 = \left\{ \frac{0}{1}, \frac{1}{4}, \frac{1}{3}, \frac{1}{2}, \frac{2}{3}, \frac{3}{4}, \frac{1}{1} \right\}.$$

Farey fractions are usually written as a sequence, and so the set $\mathcal{F}_n$ is often called a *Farey sequence*, and the list of all Farey sequences is sometimes called the *Farey series*.

Exercise 319. *Prove that for each $n \geq 2$, $\mathcal{F}_n$ has an odd number of elements.*

Say that $\frac{a}{b} < \frac{c}{d}$ are adjacent in $\mathcal{F}_n$ if no other $\frac{e}{f}$ exists in $\mathcal{F}_n$ with $\frac{a}{b} < \frac{e}{f} < \frac{c}{d}$. Here is a useful fact that has an easy proof (not inductive).

Lemma 11.6.1. *If $\frac{a}{b} < \frac{c}{d}$, then for every $h, k \in \mathbb{Z}^+$,*

$$\frac{a}{b} < \frac{ha + kc}{hb + kd} < \frac{c}{d}.$$

Another useful fact that has a fairly direct proof is the following.

Lemma 11.6.2. *If $\frac{a}{b} < \frac{c}{d}$ with $bc - ad = 1$, and $\frac{e}{f}$ satisfies $\frac{a}{b} < \frac{e}{f} < \frac{c}{d}$, then there exists $h, k > 0$, so that $e = ah + ck$ and $f = bh + dk$.*

To prove the above lemma, solve the system of equations $e = ah + ck$ and $f = bh + dk$ for the unknowns h and k, and show that they are indeed positive integers.

This next exercise is proved by induction on n (lemmas above may be used.)

Exercise 320. *Prove that if $\frac{a}{b} < \frac{c}{d}$ are reduced fractions adjacent in $\mathcal{F}_n$, then both*
  $S_1(n)$: $bc - ad = 1$, *and*
  $S_2(n)$: *If the reduced fraction $\frac{e}{f} \in \mathcal{F}_{n+1}$ separates $\frac{a}{b}$ and $\frac{c}{d}$, that is, if $\frac{a}{b} < \frac{e}{f} < \frac{c}{d}$, then $e = a + c$ and $f = b + d$.*

As a consequence of Exercise 320, if $\frac{a}{b} < \frac{c}{d}$ are adjacent Farey fractions, then

$$\frac{c}{d} = \frac{a}{b} + \frac{1}{bd}$$

gives a recursive way to generate Egyptian forms for fractions.

11.7 Continued fractions

In some circles, the most famous book on continued fractions is the classic *Continued Fractions* [311] by Aleksandr Ya. Khinchin (1894–1959); this little book has been recently reprinted by Dover. Among the many other possible resources, I recommend *Number Theory with Computer Applications* [332] by Kumanduri and Romero; the authors discuss history, theory, and applications of continued fractions. The introduction given here is naive and very sporadic, touching on only selected basics. For example, the deep connections between continued fractions and approximation theory, transcendental numbers, generating functions, hypergeometric functions, or solutions to quadratics are not discussed here.

According to the references above, continued fractions in the modern sense were apparent in the work of Rafal Bombelli (1526–1573), whose techniques were used by Christiaan Huygens (1629–1695) in the construction of a mechanical planetarium (of the first six planets). Similar methods were used by the Indian astronomer Aryabhata. See [73] for more on the history of continued fractions. According to the references mentioned, the basic properties of continued fractions given below were developed by Euler and Lagrange.

For present purposes, a *continued fraction* is an expression of the form (where each of the a_i's and b_i's are real numbers):

$$a_0 + \cfrac{b_1}{a_1 + \cfrac{b_2}{a_2 + \cfrac{b_3}{a_3 + \cfrac{b_4}{a_4 + \cdots}}}},$$

either stopping after finitely many steps or continuing on to infinity. Implicit in the above form is that each of $a_1, a_2, \ldots$ are non-zero. In what follows, the a_i's and b_i's are most often positive integer values, with perhaps a_0 being non-positive, and if any, the last a_n perhaps being a positive real. A continued fraction is called *finite* iff its expansion is finite, and *infinite* otherwise. A finite continued fraction can be evaluated from the bottom up, *i.e.*, by first calculating a bottom level $a_{n-1} + \frac{b_n}{a_n}$, producing a continued fraction with one fewer level. Simplification is repeated recursively on each new last level. Any *finite* continued fraction is a real number (since there are only finitely many simple operations to evaluate a finite continued fraction). When all the a_i's and b_i's are rational numbers, such a finite continued fraction is again a rational (see Exercise 321 below).

Do infinite continued fractions have meaning? Let C be an infinite continued fraction with coefficients a_i's and b_i's as above. For each $k \geq 0$, define the k-*convergent* (denoted C_k) of C to be the finite continued fraction formed by truncating the expression after a_k and b_k:

$$C_k = a_0 + \cfrac{b_1}{a_1 + \cfrac{b_2}{a_2 + \cfrac{b_3}{a_3 + \cfrac{b_4}{a_4 + \cfrac{\ddots}{a_{k-1} + \cfrac{b_{k-1}}{\quad \cfrac{b_k}{a_k}}}}}}},$$

[A k-convergent is also sometimes called an *initial k-segment*.] Since each C_k is finite, each is a real number. The infinite continued fraction C is said to *converge* to some limit $L \in \mathbb{R}$ iff $\lim_{k \to \infty} C_k = L$.

In some very special circumstances, an infinite continued fraction can be given meaning by calculating convergents, or simply by its algebraic properties. For example, if one writes

$$C = 1 + \cfrac{1}{1 + \cfrac{1}{1 + \cfrac{1}{1 + \cfrac{1}{1 + \cdots}}}},$$

the first few convergents are $C_1 = 2$, $C_2 = \frac{3}{2}$, $C_3 = \frac{5}{3}$, $C_4 = \frac{8}{5}$. The numbers used in these convergents look familiar. In Section 12.2, the Fibonacci numbers are defined recursively by $F_0 = 0$ and $F_1 = 1$, and for $n \geq 2$, $F_n = F_{n-2} + F_{n-1}$; the first few are 0,1,1,2,3,5,8,..., One soon guesses (and proves by induction) that the convergents are ratios of consecutive Fibonacci numbers (see Exercise 322 below), which tend to $\tau = \frac{1+\sqrt{5}}{2}$, the golden ratio. Another way to confirm this result is to observe that, as expressions, $C = 1 + \frac{1}{C}$, so if x is a convergent of C, $x = 1 + \frac{1}{x}$. Upon solving, $x^2 - x - 1 = 0$, the only non-negative root of which is indeed τ.

One might conclude that if C is a continued fraction whose coefficients eventually repeat with a fixed period, the same idea as above can be used to find a quadratic polynomial, one of whose roots is the limit of the convergents of C. The reader is invited to try this with

$$\sqrt{7} = [2, 1, 1, 1, 4, 1, 1, 1, 4, 1, 1, 1, 4, \ldots].$$

In fact, a continued fraction C is eventually periodic iff C is a quadratic irrational. See, *e.g.* [332, §11.4]. It was Lagrange that proved the expansion of any quadratic irrational is eventually periodic.

If an infinite continued fraction is denoted by, say, the letter C, sometimes C is interpreted as a real number, and until it is established that C is a real number, C is taken to denote only the continued fraction *expression*.

By a construction described below, for any real number x, there is a unique continued fraction converging to x that has, in the continued fraction expansion, all the b_i's equal to 1 and all the a_i's are integers with all but perhaps a_0 being positive. For any continued fractions with all b_i's being 1, there is a convenient one-line notation:

$$[a_0; a_1, a_2, \ldots, a_n] = a_0 + \cfrac{1}{a_1 + \cfrac{1}{a_2 + \cfrac{1}{a_3 + \cfrac{1}{a_4 + \cfrac{\ddots}{\cfrac{1}{a_{n-1} + \cfrac{1}{a_n}}}}}}},$$

and a corresponding infinite continued fraction of this form is $[a_0; a_1, a_2, a_3, \ldots]$.

For a continued fraction of the form $C = [a_0; a_1, a_2, \ldots]$, (or $[a_0; a_1, \ldots, a_n]$), for each $k = 0, 1, 2, \ldots$ (or $k = 0, 1, \ldots n$), the *k-convergent* of C is $C_k = [a_0; a_1, \ldots, a_k]$. Similarly define a *k-remainder* to be $[a_k, a_{k+1}, \ldots]$. Observe that k-convergents are themselves finite continued fractions (hence are real numbers) and k-remainders of a continued fraction are again continued fractions, perhaps infinite. Many theorems can be proved by using remainders rather than convergents; however, most proofs below rely on convergents.

A *simple continued fraction* is a continued fraction of the form $[a_0, a_1, a_2, \ldots, a_n]$ (or $[a_0, a_1, a_2, \ldots]$) where $a_0 \in \mathbb{Z}$ and all remaining a_i's are *positive* integers. Before showing how every real number can be represented by a simple continued fraction, it is instructive to first consider expansions of rational numbers.

11.7.1 Finite continued fractions

By the Euclidean division algorithm, (see Exercise 211) every rational number has a unique representation as a finite simple continued fraction. For example,

$$\frac{119}{35} = 3 + \frac{14}{35} = 3 + \frac{1}{\frac{35}{14}} = 3 + \frac{1}{2 + \frac{7}{14}} = 3 + \frac{1}{2 + \frac{1}{2}},$$

and so $\frac{119}{35} = [3; 2, 2]$.

A finite continued fraction $[a_0, a_1, \ldots, a_n]$ is evaluated with finitely many operations, so when the a_i's are rational, the number $[a_0; a_1, \ldots, a_n]$ is rational; this claim is also justified by a (nearly trivial) inductive argument based on the following observation: For each $n \geq 1$,

$$[a_0, a_1, \ldots, a_n] = [a_0, a_1, \ldots, a_{n-1} + \frac{1}{a_n}]. \tag{11.8}$$

Exercise 321. *Use equation (11.8) and induction to prove that if a real number x has a finite continued fraction expansion $x = [a_0, a_1, \ldots, a_n]$ where all a_i's are rational, then x is rational. Conclude that a number is rational iff its continued fraction expansion is finite.*

Oddly enough, the proof suggested in Exercise 321 fails for simple (integer) continued fractions; one seems forced to prove the theorem for a class of finite continued fractions whose last a_n is allowed to be rational. This is an instance of "proving more is easier". [Perhaps it is for this reason that some authors (*e.g.*, [332]) define a "simple" continued fraction to be of the form $[a_0, a_1, a_2, \ldots, a_n]$ where a_0 is an integer, each of $a_1, \ldots, a_{n-1}$ are positive integers and a_n is allowed to be any rational (or real).]

Let $C = [a_0, a_1, a_2, \ldots]$ be a continued fraction with all a_i's rational, and for each $i \geq 1$, $a_i > 0$. In much of what follows, proofs do not depend on $a_1, a_2, \ldots$ being positive; however, for simplicity assume that from now on, all such a_i's are positive, and that $a_0 \geq 0$. For each $k \geq 1$, the k-convergent of C is a positive rational number, so write

$$C_k = \frac{p_k}{q_k}, \tag{11.9}$$

where p_k and q_k are relatively prime positive integers. The 0-th convergent $C_0 = a_0 = \lfloor x \rfloor$ can also be considered as a fraction over 1.

Exercise 322. *Let F_n denote the n-th Fibonacci number ($F_0 = 0$, $F_1 = 1$, and for $n \geq 2$, $F_n = F_{n-2} + F_{n-1}$; see Section 12.2). Prove that for $n \geq 2$, the continued fraction expansion of $\frac{F_n}{F_{n-1}}$ is of the form $[1, 1, \ldots, 1]$, where there are $n - 1$ ones. It follows that any simple continued fraction of the form $[1, 1, \ldots, 1]$ is a ratio of consecutive Fibonacci numbers.*

Theorem 11.7.1. *Let $C = [a_0, a_1, a_2, \ldots]$ be a rational continued fraction and for each $k \geq 0$, let $C_k = \frac{p_k}{q_k}$. Then for $k \geq 2$,*

$$p_k = a_k p_{k-1} + p_{k-2} \tag{11.10}$$

$$q_k = a_k q_{k-1} + q_{k-2}. \tag{11.11}$$

Exercise 323. *Prove equations (11.10) and (11.11) by induction on k. [Hint: To start the inductive step, apply equation (11.8).]*

Corollary 11.7.2. *Let $C = [a_0, a_1, \ldots, a_n]$ be a rational continued fraction with, for $0 \leq k \leq n$, k-convergent $C_k = \frac{p_k}{q_k}$. Then*

$$\text{for } k \geq 1, \ p_k q_{k-1} - p_{k-1} q_k \ = \ (-1)^{k-1} \tag{11.12}$$

$$\text{for } k \geq 2, \; p_k q_{k-2} - p_{k-2} q_k \;=\; (-1)^l a_k. \tag{11.13}$$

Exercise 324. *Prove equation (11.12) of Corollary 11.7.2.*

Exercise 325. *Prove equation (11.13) of Corollary 11.7.2.*

Multiplying the equations (11.12) and (11.12) by $1/(q_{k-1}q_k)$ gives the following useful result.

Corollary 11.7.3. *Let $C = [a_0, a_1, \ldots,]$ be a rational continued fraction (finite or not), where for each $k = 0, 1, 2, \ldots$, the k-convergent is $C_k = \frac{p_k}{q_k}$. Then*

$$\text{for } 1 \leq k \leq n, \quad C_k - C_{k-1} \;=\; \frac{(-1)^{k-1}}{q_{k-1}q_k}; \tag{11.14}$$

$$\text{for } 2 \leq k \leq n, \quad C_k - C_{k-2} \;=\; \frac{a_k(-1)^k}{q_{k-2}q_k}. \tag{11.15}$$

Lemma 11.7.4. *Let a_0 be rational, and for each $i = 1, 2, \ldots$, let $a_i \geq 1$ be rational, and form the continued fraction $C = [a_0, a_1, a_2, \ldots]$ (finite or not), and for each $k \geq 0$, let $C_k = [a_0, a_1, \ldots, a_k] = \frac{p_k}{q_k}$. Then the sequence $1 = q_0, q_1, q_2, q_3, \ldots$ is strictly increasing.*

Exercise 326. *Prove Lemma 11.7.4.*

In Lemma 11.7.4, denominators of the convergents are increasing, but more can be said about just how fast.

Exercise 327. *Let a_0 be rational, and for each $i = 1, 2, \ldots$, let $a_i \geq 1$ be a rational, and form the continued fraction $C = [a_0, a_1, a_2, \ldots]$ (finite or not), and for each $k \geq 0$, let $C_k = [a_0, a_1, \ldots, a_k] = \frac{p_k}{q_k}$. Prove by induction that for each $k \geq 2$,*

$$q_k \geq 2^{k/2}.$$

The following exercise is somewhat novel in that continued fractions are turned "inside out".

Exercise 328. *Let $[a_0, a_1, a_2, \ldots]$ be a continued fraction with all a_i's rational, and for each $k \geq 0$, $C_k = [a_0, a_1, \ldots, a_k] = \frac{p_k}{q_k}$. Prove by induction that for each $k \geq 1$,*

$$\frac{q_k}{q_{k-1}} = [a_k, a_{k-1}, \ldots, a_1].$$

11.7.2   Infinite continued fractions

Theorem 11.7.5. *Let $a_0, a_1, a_2, \ldots$ be an infinite sequence of rational numbers where, for $i \geq 1$, $a_i \geq 1$. Then $C = [a_0, a_1, a_2, \ldots]$ converges.*

Proof: By equation (11.14) of the previous section, the convergents of C satisfy $C_1 = C_0 + \frac{1}{q_1}$, $C_2 = C_1 - \frac{1}{q_1 q_2} = C_0 + \frac{1}{q_1} - \frac{1}{q_1 q_2} > C_0$, since the q_i's are increasing by Lemma 11.7.4. Continuing in the same manner, $C_3 = C_2 + \frac{1}{q_2 q_3} < C_2 + \frac{1}{q_1 q_2} = C_1$, so $C_0 < C_2 < C_3 < C_1$. By equation (11.15) it is easier to see that the even convergents are increasing, $C_0 < C_2 < C_4 < \cdots$, and the odd convergents are decreasing, $C_1 > C_3 > C_5 > \cdots$, and by equation (11.14), each C_{i+2} is between C_i and C_{i+1}. Thus, the convergents form the partial sums of an alternating decreasing series,

$$C_0 + (C_1 - C_0) + (C_2 - C_1) + \ldots,$$

which converges to some limit; call this limit C.   □

Notice that, from the above proof, each convergent C_k is within $\frac{1}{q_{k-1} q_k}$ of C. It might be interesting to know (see [311] for details) that an infinite continued fraction of the form $[a_0, a_1, a_2, \ldots]$ converges iff the series $\sum_{i=1}^{\infty} a_i$ diverges (so just forcing the a_i's to be at least one is a very weak condition).

Corollary 11.7.6. *If $C = [a_0, a_1, a_2, \ldots]$ is an infinite simple continued fraction (where $a_0 \in \mathbb{Z}$ and for each $i \geq 1$, $a_i \in \mathbb{Z}^+$), then the continued fraction C converges to an irrational real number.*

Proof: Any infinite *simple* continued fraction satisfies the hypothesis of Theorem 11.7.5, and so converges to some $x \in \mathbb{R}$. By Exercise 321, x is irrational.   □

Recall that $\lfloor x \rfloor$ denotes the largest integer not greater than a real number x, often called the *integer part* or *floor* of x. For any real number x, write $x = \lfloor x \rfloor + [[x]]$, where $[[x]] \in [0, 1)$ is often called the *fractional part* of x..

To each irrational number x, there is a "canonical" *simple* infinite continued fraction $C(x) = [a_0, a_1, a_2, \ldots]$ that converges to x; this representation of x is recursively defined in the same manner that the Euclidean division algorithm yields a finite expansion for a rational—at each stage, break off the integer part, and invert the fractional part. Here is the process:

Fix an irrational x. By Exercise 321, no finite simple continued fraction defines x, confirming that the process about to be defined does not stop. Define recursively a sequence $x_0, x_1, x_2, \ldots$ of real numbers, by $x_0 = x$, and for each $i \geq 1$,

$$x_i = \frac{1}{[[x_{i-1}]]}.$$

For each $i \geq 1$, (by induction) $[[x_{i-1}]]$ is an irrational number between 0 and 1, and so each of $x_1, x_2, \ldots$ is a real number larger than one (thus the recursion never stops—or explodes with some division by zero). Using these x_i's, the first few steps in an expansion of x are:

$$x = \lfloor x_0 \rfloor + [[x_0]]$$
$$= \lfloor x_0 \rfloor + \frac{1}{x_1}$$
$$= \lfloor x_0 \rfloor + \frac{1}{\lfloor x_1 \rfloor + [[x_1]]}$$
$$= \lfloor x_0 \rfloor + \frac{1}{\lfloor x_1 \rfloor + \dfrac{1}{x_2}}$$
$$= \lfloor x_0 \rfloor + \frac{1}{\lfloor x_1 \rfloor + \dfrac{1}{\lfloor x_2 \rfloor + [[x_2]]}}$$
$$= \lfloor x_0 \rfloor + \frac{1}{\lfloor x_1 \rfloor + \dfrac{1}{\lfloor x_2 \rfloor + \dfrac{1}{x_3}}}$$
$$\vdots$$

For each $i = 0, 1, 2, \ldots$, put $a_i = \lfloor x_i \rfloor$. Then for each $n \geq 1$,

$$x = [a_0, a_1, \ldots, a_{n-1}, x_n].$$

Then for each $n \geq 0$,

$$x_n = a_n + \frac{1}{x_{n+1}}. \tag{11.16}$$

Define the *canonical representation* of an irrational x to be the infinite simple continued fraction $C = [a_0, a_1, a_2, \ldots]$ defined above. By Corollary 11.7.6, C converges to an irrational number y. Since the $n - 1$ first k-convergents of both $x = [a_0, a_1, \ldots, a_{n-1}, x_n]$ and $C = [a_0, a_1, \ldots, a_{n-1}, a_n, \ldots]$ are the same, it follows (as in the proof of Theorem 11.7.5) that $x = y$. Hence any real number x has a unique simple continued fraction expansion, and that expansion is found by the algorithm above.

As a penultimate note of interest, the continued expansion of e, the base of the natural logarithm, is

$$e = [2, 1, 2, 1, 1, 4, 1, 1, 6, 1, 1, 8, 1, 1, 10, \ldots].$$

In the 1760s, Johann Heinrich Lambert (1728–1777) proved that the pattern above continues on forever, hence first proving that e is irrational (Legendre is sometimes credited) the proof (see *e.g.*, [332, pp. 276–8]) is another (beautiful?) application of induction.

In this section, the b_i's have been virtually ignored. For an intriguing introduction to more general continued fractions (even with complex numbers), starting with a famous continued fraction by John Wallis (1616–1703) for $\pi/4$, and containing many more inductive proofs regarding continued fractions, I highly recommend the reader to look at *The Number π* [182, pp. 65–78]. [This book is an incredible introduction to most undergraduate mathematics, including geometry and analysis. The irrationality of π is another goal the authors use to introduce continued fractions used in approximation theory, which then, through Liouville's work, lead to showing π is trancscendental. Also, all theorems given in this section are implicitly contained in that book.]

Chapter 12

Sequences

As yet a child, nor yet a fool to fame,
I lisp'd in numbers, for the numbers came.

—Alexander Pope (1688–1744),

Epistle to Dr. Arbuthnot.

Sequences seem to fascinate the young mathematician in all of us. In Chapters 9 and 10, most of the equalities (or inequalities) can be seen as a comparison of two sequences. In this section, more relevant are properties of elements of a particular sequence, rather than simply seeing if two sequences are comparable.

It is helpful to make clear what is meant by the term "sequence". If you were asked to explain to a child what "a sequence of numbers" is, what comes to mind? Would you tell the child that it is merely a list of numbers, ordered so that one comes after another? This reply might seem satisfactory in most instances, but how would you respond if a precocious child then asked whether or not such a list can go on forever, or, say, if it must have a starting point? Can there be infinitely many numbers between two numbers in the list? If two numbers in a sequence are interchanged, is the new sequence the same as the old? Must you be able to describe each number in the list, that is to say, for example, if one wanted to know what the 1001-th number in the sequence is, must you be able to calculate it without first calculating the previous 1000 numbers? You *could* respond to all of these questions by giving the following definition to the child:

Definition 12.0.7. A finite sequence of elements from a set X is a function, where for some $n \in \mathbb{Z}^+$,
$$f : \{1, 2, \ldots, n\} \longrightarrow X.$$
An infinite sequence of elements from a set X is a function
$$f : \mathbb{Z}^+ \longrightarrow X.$$

With this definition, elements of a sequence are $f(1), f(2), f(3), \ldots$ and the sequence can be called simply "f". In practice, however, one often writes a sequence by listing the elements in the range of f in the order determined by the natural numbers. For example, if $X = \{a, b, 5, x, y\}$, and f is defined by $f(1) = 3$, $f(2) = a$, $f(3) = a$, and $f(4) = y$, then the sequence f is written as $3, a, a, y$. By definition, the sequence $3, a, y, a$ is different from $y, a, a, 4$, so order is important. Often, instead of naming a sequence by f, use subscripted variables, like $x_1, x_2, x_3, \ldots$, where $x_1 = f(1)$, $x_2 = f(2)$, and so on.

In the literature, there is not any one standard way to denote sequences. For example, an infinite sequence $s_1, s_2, s_3, \ldots$, can be denoted by $(s_i)_{i \in \mathbb{Z}^+}$ (or simply (s_i)); by $\{s_i\}_{i \in \mathbb{Z}^+}$ (or simply $\{s_i\}$); by s (where s is viewed as a function and for each i, $s_i = s(i)$); by a boldface $\mathbf{s}$ (especially for finite sequences—as in vector notation); or by $\langle s_i \rangle_{i \in \mathbb{Z}^+}$. Even though set braces $\{\ \}$ are often used, and ordinarily "sets" are collections with no order imposed, this notation for sequences is very popular. Some authors prefer the angle bracket notation $\langle\ \rangle$, however angle brackets are also reserved for something called "inner products" and a few other notions. Sometimes sequences are indexed starting at 0, in which case a sequence $s_0, s_1, s_2, \ldots$ can be denoted by $(s_i)_{i \in \omega}$. The notation varies throughout the exercises, but I hope that no confusion arises.

Some mathematicians might mean "sequence" to include sequences that are infinite in both directions, that is, functions of the form

$$f : \mathbb{Z} \longrightarrow X,$$

however here, ordinarily only sequences with a first element are used.

Perhaps the most famous of all sequences are those whose adjacent elements are equally far apart. For fixed integers a, d, and ℓ, a sequence of the form $a, a + d, a + 2d, a + 3d, \ldots, a + (\ell - 1)d$ is called an *arithmetic progression of length ℓ* with difference d. For example, $2, 5, 8, 11, \ldots, 62$ is an arithmetic progression of length 21.

12.1 Difference sequences

Given some sequence $s = \{s_i\}_{i \in \mathbb{Z}^+}$, define the *difference sequence* $\Delta s = \{s_{i+1} - s_i\}_{i \in \mathbb{Z}^+}$. The *second difference* sequence $\Delta^2 s$ is found by taking the differences between consecutive elements in the difference sequence. In general, one finds the p-th difference sequence $\Delta^p s$ by taking the difference sequence for the previous difference sequence $\Delta^{p-1} s$. To be precise, let $x = x_1, x_2, x_3, \ldots$ be a sequence. Then

$$
\begin{aligned}
\Delta x &= x_2 - x_1, x_3 - x_2, x_4 - x_3, \ldots \\
\Delta^2 x &= x_3 - 2x_2 + x_1, x_4 - 2x_3 + x_2, x_5 - 2x_4 + x_3, \ldots \\
\Delta^3 x &= x_4 - 3x_3 + 3x_2 - x_1, x_5 - 3x_4 + 3x_3 - x_2, x_6 - 3x_5 + 3x_4 - x_3, \ldots \\
\Delta^4 x &= x_5 - 4x_4 + 6x_3 - 4x_2 + x_1, x_6 - 4x_5 + 6x_4 - 4x_3 + x_2, \ldots
\end{aligned}
$$

Exercise 329. *If $x = x_1, x_2, x_3, \ldots$, is a sequence of real numbers, prove by induction on k that the n-th term in the k-th difference sequence $\Delta^k x$ is*

$$\sum_{i=0}^{k} (-1)^i \binom{k}{i} x_{n+k-i}.$$

While playing with Fermat's last theorem (*i.e.*, for $n > 2$, the equation $x^n + y^n = z^n$ has no non-trivial solutions in integers) one might be interested in differences of consecutive "like powers". Some amazing patterns evolve if one repeatedly takes differences. Examine the sequence of squares

$$s = 1, 4, 9, 16, \ldots, n^2, \ldots;$$

the difference sequence is

$$\Delta s = 3, 5, 7, 9, 11, \ldots,$$

an arithmetic progression. If one again takes differences, the second difference sequence is the constant sequence

$$\Delta^2 s = 2, 2, 2, 2, \ldots.$$

Examine the sequence of the cubes,

$$s = 1, 8, 27, 64, 125, 216, 343, 512, 729, 1000, \ldots.$$

The first difference sequence is

$$\Delta s = 7, 19, 37, 61, 91, 127, 169, 217, 271, \ldots,$$

and the second difference sequence is

$$\Delta^2 s = 12, 18, 24, 30, 36, 42, 48, 54, \ldots;$$

the third differences are all $6 = 3 \cdot 2$. If one tries the same procedure for fourth powers, the fourth differences are all equal to $24 = 4 \cdot 3 \cdot 2$ (try it!). Similarly, the fifth differences of fifth powers are all $120 = 5 \cdot 4 \cdot 3 \cdot 2$.

Exercise 330. *For each positive integer k, guess an expression for the k-th difference of the sequence $1^k, 2^k, 3^k, 4^k, \ldots$, and prove your result by induction.*

The following exercise, in a sense, is concerned with differences, but not in the same way as described above.

Exercise 331. *Suppose that a sequence of positive integers $a_1, a_2, \ldots, a_d$ satisfies, for each $i = 1, 2, \ldots, d - 2$,*

$$a_{i+2} = |a_{j+1} - a_j|.$$

Prove that if $a_1 \leq a_2$, then $d \leq \frac{3}{2}(a_2 + 1)$. Hint: induct on a_2.

12.2   Fibonacci numbers

The next few statements use the *Fibonacci numbers*, which are defined by $F_0 = 0$, $F_1 = 1$, and for $n \geq 2$, the recursion

$$F_n = F_{n-2} + F_{n-1}.$$

So $F_2 = F_0 + F_1 = 0 + 1 = 1$, $F_3 = F_1 + F_2 = 1 + 1 = 2$, $F_4 = F_2 + F_3 = 1 + 2 = 3$, $F_5 = F_3 + F_4 = 2 + 3 = 5$, and so on. Here are the first 31 Fibonacci numbers:

n	0	1	2	3	4	5	6	7	8	9	10	11	12	13	14
F_n	0	1	1	2	3	5	8	13	21	34	55	89	144	233	377

n	15	16	17	18	19	20	21	22	23
F_n	610	987	1597	2584	4181	6765	10946	17711	28657

n	24	25	26	27	28	29	30
F_n	46368	75025	121393	196418	317811	514229	832040

The Fibonacci numbers were named after Fibonacci, also called "Leonardo of Pisa" (1175–1230), a mathematician and merchant from Italy. ("Fibonacci" means "son of good fortune".) The sequence occurred in his book *Il Liber Abaci*, published in Pisa in 1202 A.D. A second edition appeared in 1228 and, according to [589], it is the one that survived for many centuries. Apparently, the sequence arose from studying rabbits.

Suppose that rabbits are fertile by the age of one month, (some say that this period should be two months, but even so, the development is analogous) and that one begins with a matching pair (one male, one female) of mature, or fertile rabbits. Every matching pair of rabbits produce two rabbits after one month, let's say one male and one female. At the end of one month, there is one new pair. At the end of two months, the first pair produces another pair, and the second pair matures, giving 3 pairs, 2 of which are mature. At the end of three months, the first 2 pairs produce a pair each, giving a total of 5 pairs, three of which are mature. In Exercise 332, one is asked to complete this analysis to see that indeed one arrives at the number of rabbits at the end of each month to be a Fibonacci number.

When proving results regarding Fibonacci numbers, a form of strong induction is often useful. In particular, the inductive step in many proofs is of the form $[S(k-1) \wedge S(k)] \rightarrow S(k+1)$; note that in such instances, two base cases are often required. The first exercise has a solution that is just slightly underhanded.

Exercise 332. *Prove by induction that after $k \geq 0$ months, in the above description there are F_{k+2} pairs of rabbits. What result do you get if rabbits mature only after two months instead of one month?*

Exercise 333. *Show that if $n > 5$, then $F_n > \left(\frac{3}{2}\right)^{n-1}$.*

Exercise 334. *Prove that if $n \geq 1$, then $F_n \leq \left(\frac{7}{4}\right)^{n-1}$.*

Exercise 335. *Prove that if $n \geq 1$, then $F_n \leq \left(\frac{18}{11}\right)^{n-1}$.*

Exercise 336. *Prove by induction that if $n \geq 0$, then $F_n \leq \left(\frac{5}{3}\right)^{n-1}$.*

Exercise 337. *Give an inductive proof (one whose inductive step actually uses the inductive hypothesis) of the fact that for each $n \geq 0$,*

$$F_n + 2F_{n+1} = F_{n+3}.$$

Exercise 338. *Prove by induction that (for non-negative integers n) F_n is an even number if and only if n is divisible by 3.*

Lemma 12.2.1. *If $k \geq 0$, then $F_{4k+4} = 2F_{4k} + 3F_{4k+1}$.*

For example, $F_8 = 2F_4 + 3F_2$ since $21 = 2 \cdot 3 + 3 \cdot 5$ and $F_8 = 21$, $F_4 = 3$, and $F_5 = 5$. The general proof is non-inductive and only relies on the recursive definition of Fibonacci numbers:

Proof of Lemma 12.2.1:

$$
\begin{aligned}
F_{4k+4} &= F_{4k+2} + F_{4k+3} \\
&= F_{4k} + F_{4k+1} + F_{4k+1} + F_{4k+2} \\
&= F_{4k} + F_{4k+1} + F_{4k+1} + F_{4k} + F_{4k+1} \\
&= 2F_{4k} + 3F_{4k+1}.
\end{aligned}
$$

$\square$

An inductive proof is also available for a slight generalization of the equality in Lemma 12.2.1:

Exercise 339. *Prove by induction on m that for every $m \geq 0$,*

$$F_{m+4} = 2F_m + 3F_{m+1}.$$

Note that Lemma 12.2.1 is the special case $m = 4k$ in Exercise 339.

Exercise 340. *Use Lemma 12.2.1 or Exercise 339 to prove by induction that for each $i \geq 0$, F_{4i} is divisible by 3.*

Exercise 341. *Beginning with F_0, prove that every fifth Fibonacci number is divisible by 5.*

The result in the next exercise has a simple, albeit cumbersome, direct proof; it also has an inductive proof.

Exercise 342. *Prove by induction that if $n \geq 0$, then $F_{n+8} = 7F_{n+4} - F_n$.*

Exercise 343. *Use Exercise 342 to prove by induction that every 8-th Fibonacci number (beginning at F_0) is divisible by 7.*

Exercise 344. *Prove that for all positive integers n,*

$$1 + 2^{2n} + 3^{2n} + 2[(-1)^{F_n} + 1]$$

is divisible by 7.

Exercise 345. *Prove by induction that for $n \geq 0$,*

$$F_{n+10} = 11F_{n+5} + F_n.$$

Exercise 346. *Using Exercise 345, prove that every 10th Fibonacci number (starting with F_0) is divisible by 11, that is, for every $i \geq 0$, F_{10i} is divisible by 11.*

Exercise 347. *Prove that for all non-negative integers n, $F_{5n+3} + F_{5n+4}^2$ is divisible by 11.*

Exercise 348. *Prove that for every non-negative integer n, F_{12n} is divisible by 6, 8, 9, and 12.*

The following result has a number of different proofs, one that is asked for in Exercise 350:

Lemma 12.2.2. *For each $n \geq 0$,*

$$F_{n+15} = 10F_{n+10} + 10F_{n+5} + 10F_{n+1} + 7F_n.$$

Before giving the proof of Lemma 12.2.2, here is a verification that the above equality for $n = 0$ (this is not a base step to any inductive proof, but merely a check to see if it works): $F_{15} = 610$ and $10F_{n+10} + 10F_{n+5} + 10F_{n+1} + 7F_n = 10(55) + 10(5) + 10(1) + 7(0)$, which is 610.

Proof of Lemma 12.2.2:

$$
\begin{aligned}
&F_{n+15} \\
&= 11F_{n+10} + F_{n+5} && \text{(by Exercise 345)} \\
&= 10F_{n+10} + F_{n+10} + F_{n+5} \\
&= 10F_{n+10} + (11F_{n+5} + F_n) + F_{n+5} && \text{(by Exercise 345)} \\
&= 10F_{n+10} + 10F_{n+5} + 2F_{n+5} + F_n \\
&= 10F_{n+10} + 10F_{n+5} + 2(3F_n + 5F_{n+1}) + F_n && \text{(see Ex. 341 sol'n)} \\
&= 10F_{n+10} + 10F_{n+5} + 10F_{n+1} + 7F_n.
\end{aligned}
$$

$\square$

Exercise 349. *Give an inductive proof of the equality in Lemma 12.2.2.*

Exercise 350. *Relying on previous exercises, give a simple proof that, beginning with F_0, every fifteenth Fibonacci number is divisible by 10. Then, using Lemma 12.2.2, find an inductive proof of this result. Verify that the among $F_0, F_1, \ldots, F_{30}$, the only Fibonacci numbers that are divisible by 10 are those of the form F_{15k}.*

Exercise 351. *Prove that for every $n > 0$,*

$$2(2^{2n} + 5^{2n} + 6^{2n}) + 3(-1)^{n+1}[(-1)^{F_n} + 1]$$

is divisible by 13.

The next exercise asks to prove a simple but powerful identity, one that generalizes some of the results stated above and facilitates the proofs of more elegant results below.

Exercise 352. *Prove that for all integers m and n satisfying $n > m > 1$,*

$$F_{n-m+1}F_m + F_{n-m}F_{m-1} = F_n.$$

Which of the above exercises does this result generalize?

Exercise 353. *Use the result from Exercise 352 to prove by induction that for any positive integers m and n, F_m divides F_{nm}.*

It might be interesting to note that Exercise 352 (or Exercise 376 below) implies the following remarkable fact:

Corollary 12.2.3. *If F_n is prime, and $n > 4$, then n is prime.*

Unfortunately, the implication in Corollary 12.2.3 does not go the other way. For example, $F_{19} = 4181 = 113 \cdot 37$ is not prime. The first few prime Fibonacci numbers occur when $n = 3, 4, 5, 7, 11, 13, 17, 23, 29, 43, 47, 83, 131, 137, 359, 431, 433, 449, 509, 569, 571,\ldots$ (See [148] for more discussion.)

The following identity, known as "Cassini's identity", was presented by the Italian astronomer Giovanni Domenico Cassini (1625–1712) to the Royal Academy (Paris) in 1680, but these proceedings [97] were published only in 1733 (under the name Jean Dominique Cassini). In 1671, Cassini moved from Bologna to Paris and became the first royal astronomer of France, and so sometimes is considered to be French. His son, Jacques Cassini (1677–1756), grandson, César-Francois Cassini, and great grandson Jacque Dominique Cassini (1748–1845) all were French scientists. Giovanni is also known for the *Cassinian curve*, the locus of points whose product of distances from two fixed points is constant.

Cassini's identity is also often called "Simson's identity", since it was independently discovered by the Scottish geometer Robert Simson (1687–1768) in 1753 [493] (see also, *e.g.* [122, pp. 165–168], and [123, p. 41]). Perhaps the identity should be known as "Kepler's identity", as, according to Graham, Knuth, and Patashnik [230, p. 292], Johannes Kepler (1571–1630) knew of it in 1608. [They offer [310] as evidence of this.]

Exercise 354 (Cassini's identity). *Prove, using mathematical induction, that for every $n \geq 1$,*

$$F_{n-1}F_{n+1} = F_n^2 + (-1)^n.$$

Note that Cassini's identity can be written as

$$F_{n-1}^2 + F_{n-1}F_n - F_n^2 = (-1)^n,$$

and so by multiplying by -1 and taking absolute values, it takes the form $|m^2 - \ell m - \ell^2| = 1$. A remarkable fact is that this equation essentially implies that ℓ and m are adjacent Fibonacci numbers, giving a kind of converse to Exercise 354.

Exercise 355. *Prove that if ℓ and m are integers such that $|m^2 - \ell m - \ell^2| = 1$, then there is an integer n so that $\ell = \pm F_n$ and $m = \pm F_{n+1}$.*

Cassini's identity can now be viewed as a special case ($r = 1$) of a more general result, known as "Catalan's identity":

Exercise 356 (Catalan's identity). *Prove that for $1 \leq r \leq n$,*

$$F_n^2 - F_{n-r}F_{n+r} = (-1)^{n-r}F_r^2.$$

Related to Catalan's identity is a result sometimes known as "d'Ocagne's identity":

Exercise 357 (d'Ocagne's identity). *Prove that for $0 \leq m \leq n$,*

$$F_nF_{m+1} - F_mF_{n+1} = (-1)^m F_{n-m}.$$

Note that with $m = 1$ in d'Ocagne's identity, the original Fibonacci recurrence relation $F_n - F_{n+1} = -F_{n-1}$ is recovered.

The next exercise contains a result that Yuri Matijasevich [370] found and used in his (negative) answer to Hilbert's tenth problem: *Is there an algorithm to solve an arbitrary diophantine equation?*. For more on the solution to Hilbert's tenth problem, see the book by Matiyasevich [371], or websites [591] (with an article by YM) and [592]. [*Note:* sometimes the "j" is replaced by "y" in "Matijasevich".]

Exercise 358 (Matijasevich's lemma). *Prove that for $n > 2$, F_m is a multiple of F_n^2 if and only if m is a multiple of nF_n.*

Exercise 359. *Prove that for every $n \geq 0$,*

$$F_0 + F_1 + F_2 + \cdots + F_n = F_{n+2} - 1.$$

Exercise 360. *Show that for every $n \geq 0$,*

$$F_0^2 + F_1^2 + F_2^2 + \cdots + F_n^2 = F_nF_{n+1}.$$

Exercise 361. *Show that for each $n \geq 1$,*

$$F_1 + F_3 + \cdots + F_{2n-1} = F_{2n}.$$

Exercise 362. *Prove that for any $n \geq 1$,*

$$F_0 + F_2 + F_4 + \cdots + F_{2n} = F_{2n+1} - 1.$$

Exercise 363. *Show that for every $n \geq 1$,*

$$F_1 F_2 + F_2 F_3 + \cdots + F_{2n-1} F_{2n} = F_{2n}^2.$$

Exercise 364. *Prove that for every $n \geq 1$,*

$$F_1 F_2 + F_2 F_3 + \cdots + F_{2n} F_{2n+1} = F_{2n+1}^2 - 1.$$

For use in Exercise 365, the following fact is presented:

Lemma 12.2.4. *For every $m \geq 1$,*

$$F_{2m} = F_m(F_m + 2F_{m-1}).$$

Proof: Using $n = 2m$ in the result of Exercise 352,

$$
\begin{aligned}
F_{2m} &= F_{m+1}F_m + F_m F_{m-1} \\
&= F_m(F_{m+1} + F_{m-1}) \\
&= F_m(F_{m-1} + F_m + F_{m-1}).
\end{aligned}
$$

$\square$

The equality in the next exercise is merely a special case of the more general statement given in Exercise 352, though it seems difficult to prove it without first proving the more general statement!

Exercise 365. *Using Lemma 12.2.4, give an inductive proof that for every $n \geq 0$,*

$$F_n^2 + F_{n+1}^2 = F_{2n+1}.$$

Verify that this equality also follows directly (no induction) from Exercise 352 using $n = 2m + 1$ and then replacing all m's by n.

Another elegant equality follows from Lemma 12.2.4, though it may be difficult to prove by induction (without using some consequence of Exercise 352):

$$F_{2n} = F_{n+1}^2 - F_{n-1}^2.$$

This follows directly from Lemma 12.2.4, since

$$F_{n+1}^2 - F_{n-1}^2 = (F_{n+1} - F_{n-1})(F_{n+1} + F_{n-1}) = F_n(F_n + 2F_{n-1}).$$

Exercise 366. *Prove that for all $n \geq 0$,*

$$F_n^2 + F_{n+1}^2 + F_{n+2}^2 + F_{n+3}^2 = 3F_{2n+3}.$$

Exercise 367. *Prove that for every $n \geq 1$,*

$$\frac{F_0}{2} + \frac{F_1}{4} + \frac{F_2}{8} + \cdots + \frac{F_{n-1}}{2^n} = 1 - \frac{F_{n+2}}{2^n}.$$

Exercise 368. *Prove by induction that for $n \geq 1$,*

$$\begin{bmatrix} 1 & 1 \\ 1 & 0 \end{bmatrix}^n = \begin{bmatrix} F_{n+1} & F_n \\ F_n & F_{n-1} \end{bmatrix}.$$

One might notice that the ratio of consecutive Fibonacci numbers seems to approach something near to 1.6 (or 8/5). For the moment, suppose that this ratio indeed approaches some number in the limit, that is, suppose that there is some number t so that

$$\lim_{n \to \infty} \frac{F_{n+1}}{F_n} = t.$$

Beginning with the recursion $F_{n+1} = F_{n-1} + F_n$, dividing throughout by F_n, and taking limits gives

$$\lim_{n \to \infty} \frac{F_{n+1}}{F_n} = 1 + \lim_{n \to \infty} \frac{F_{n-1}}{F_n}.$$

Since $\lim_{n \to \infty} \frac{F_{n+1}}{F_n} = \lim_{n \to \infty} \frac{F_n}{F_{n-1}}$, and $\frac{F_n}{F_{n+1}} = \frac{1}{F_{n+1}/F_n}$, the above equation becomes $t = 1 + \frac{1}{t}$, and so

$$t^2 = t + 1. \tag{12.1}$$

The roots of this equation are $\phi_1 = \frac{1+\sqrt{5}}{2}$ and $\phi_2 = \frac{1-\sqrt{5}}{2}$. These two numbers arise in a closed form for Fibonacci numbers. This closed form formula was discovered by Daniel Bernoulli in 1728; however, it was Jacques Philippe Marie Binet's [53] rediscovery in 1843 that makes the form eponymous with him.

Exercise 369 (Binet's formula, Fibonacci numbers). *Give an inductive proof of the fact that for every $n \geq 0$,*

$$F_n = \frac{1}{\sqrt{5}} \left[\left(\frac{1+\sqrt{5}}{2} \right)^n - \left(\frac{1-\sqrt{5}}{2} \right)^n \right].$$

With the notation above,

$$F_n = \frac{1}{\sqrt{5}} [\phi_1^n - \phi_2^n].$$

Using Binet's formula, without too much difficulty, one can now prove, for example, that the ratio $\frac{F_{n+1}}{F_n}$ tends to the golden ratio $\frac{1+\sqrt{5}}{2}$, approximately 1.618034.

Binet's formula can also be used to prove some interesting identities directly, some of which are difficult to see using induction. For example, here is one due to the Italian mathematician Ernesto Cesaro (1859–1906) given around 1888.

Lemma 12.2.5. *For $n \geq 0$,*

$$\binom{n}{0} F_0 + \binom{n}{1} F_1 + \cdots + \binom{n}{n} F_n = F_{2n}.$$

Proof:

$$
\begin{aligned}
\sum_{n=0}^{n} \binom{n}{i} F_i &= \sum_{n=0}^{n} \binom{n}{i} \frac{1}{\sqrt{5}} [\phi_1^i - \phi_2^i] \\
&= \frac{1}{\sqrt{5}} \left[\sum_{n=0}^{n} \binom{n}{i} \phi_1^i - \sum_{n=0}^{n} \binom{n}{i} \phi_2^i \right] \\
&= \frac{1}{\sqrt{5}} [(1 + \phi_1)^n - (1 + \phi_2)^n] \\
&= \frac{1}{\sqrt{5}} [(\phi_1^2)^n - (\phi_2^2)^n] \qquad\qquad \text{(by eqn (12.1))} \\
&= F_{2n}.
\end{aligned}
$$

$\square$

Similarly, one can prove that $\sum_{k=0}^{n} \binom{n}{k} 2^k F_k = F_{3n}$ (see [277, pp. 109–110]). The next exercise generalizes Cesaro's result.

Exercise 370. *Prove that for any non-negative integers m and n,*

$$\binom{n}{0} F_m + \binom{n}{1} F_{m+1} + \cdots + \binom{n}{n} F_{m+n} = F_{m+2n}.$$

Hint: Use Lemma 12.2.5 for the case $m = 0$, prove a similar result for $m = 1$, and then use these both as base cases for an inductive proof.

Exercise 371. *Prove that for $n \geq 1$,*

$$\binom{2n}{0} + \binom{2n-1}{1} + \binom{2n-2}{2} + \cdots + \binom{n}{n} = F_{2n+1}$$

and

$$\binom{2n+1}{0} + \binom{2n}{1} + \binom{2n-1}{2} + \cdots + \binom{n+1}{n} = F_{2n+2}.$$

Hint: To prove just one of these identities by induction might be far more difficult than to prove them together.

The result of the following exercise is often attributed to Edouard Zeckendorf (see [585] and [345]) and can be used to establish a unique representation number system:

Exercise 372 (Zeckendorf's theorem). *Prove that every $n \in \mathbb{Z}^+$ is a unique sum of distinct non-consecutive Fibonacci numbers.*

Exercise 373. *The number of binary strings of length $n \geq 1$ not containing "11" as a substring is F_{n+2}.*

Exercise 374. *Let a_n be the number of binary strings of length n that do not have two 1's at distance 2 apart. Find a_n in terms of Fibonacci numbers.*

The next exercise uses the "floor function": for any real number x, the number $\lfloor x \rfloor$ is the largest integer less than or equal to x. For example, $\lfloor \pi \rfloor = 3$, $\lfloor 7 \rfloor = 7$, and $\lfloor -1.2 \rfloor = -2$. Exercise 375 appears implicitly as Example 40 in [247]; another way to interpret the exercise is to draw Pascal's triangle shifted so that $\binom{n}{r}$ appears in the n-th row and $(n+r)$-th column of an array, and then calculate the sums of the columns:

	0	1	2	3	4	5	6	7	8	
0	1									
1		1	1							
2			1	2	1					
3				1	3	3	1			
4					1	4	6	4	1	
5						1	5	10	10	...
6							1	6	15	...
7								1	7	...
8									1	...
	1	1	2	3	5	8	13	21	34	...

Exercise 375. *Prove that for each $n \geq 1$,*

$$F_{n+2} = \sum_{i=0}^{\lfloor (n+1)/2 \rfloor} \binom{n+1-i}{i}.$$

The next exercise can be solved by the Euclidean division algorithm, not really mathematical induction; however, since the division algorithm follows from well-ordering, the result can be considered as related. In any case, the exercise reveals a very interesting property of Fibonacci numbers.

Exercise 376. *Prove that for $n \geq 1$, adjacent Fibonacci numbers F_n and F_{n+1} are relatively prime. Furthermore, if $c = \gcd(a, b)$, then $\gcd(F_a, F_b) = F_c$.*

The result in Exercise 353 said that F_{mn} is a multiple of F_n. A converse of this statement is also true:

Lemma 12.2.6. *For $m > 2$, if F_n is divisible by F_m, then n is divisible by m.*

Proof: Let $m > 2$, and suppose that F_m divides F_n. Then F_m divides $\gcd(F_m, F_n)$ and by Exercise 376, $\gcd(F_m, F_n) = F_{\gcd(m,n)}$, so F_m divides $F_{\gcd(m,n)}$. Thus $m \leq \gcd(m, n)$, which occurs only when $m = \gcd(m, n)$, that is, when m divides n. $\qquad\square$

The next three exercises are steps toward a result proved by Gabriel Lamé (1795–1870) in 1845; this result is apparently (see [195]) historically important as the first practical application of the Fibonacci numbers. (These exercises are also proved in [363], pp. 66–67; I have included proof outlines of the last two in the Hints and Solutions section.) Lamé is also known for his solution of Fermat's Last Theorem for $n = 7$ in 1839.

Exercise 377. *If m and n are positive integers with $m < n$, and if the Euclidean division algorithm computes $\gcd(m, n)$ in k steps, then $m \geq F_{k+1}$.*

The following exercise has a proof that follows from the previous one by simple algebra and Binet's formula for F_{k+1} (not by induction).

Exercise 378. *If m and n are positive integers with $m < n$, the Euclidean division algorithm will compute $\gcd(m, n)$ in no more than $\log_\phi(m) + 1$ steps, where $\phi = \frac{1+\sqrt{5}}{2}$ is the golden ratio.*

Exercise 379 (Lamé's Theorem). *If m and n are positive integers with $m < n$, and N is the number of digits in the decimal expansion of m, then the Euclidean division algorithm will compute $\gcd(m, n)$ in no more than $5N$ steps.*

For the next exercise, the expression "k-subset of a set" means a subset with k elements (with no order imposed and no element repeated).

Exercise 380. *Let $a_{n,k}$ be the number of k-subsets of $\{1, 2, \ldots, n\}$ that do not contain a pair of consecutive integers. Prove that*

$$a_{n,k} = \binom{n - k + 1}{k},$$

and that

$$\sum_{k \geq 0} a_{n,k} = F_{n+2}.$$

Exercise 381. *Show that $\sum_k \binom{n}{k} F_{m+k}$ is always a Fibonacci number.*

The next clever result was first communicated to me in 1997 by Louis Shapiro, at Howard University. It is based on Kirchoff's law: if two resistors with resistance R_1 and R_2 are put in parallel, the net resistance is $\frac{R_1 \times R_2}{R_1 + R_2}$.

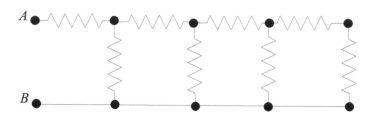

Figure 12.1: The Fibonacci electrical circuit $FC(4)$

Exercise 382. *For each $n = 1, 2, 3, \ldots$ define an electrical circuit $FC(n)$ with $2n$ resistors, each with resistance 1Ω (one ohm), as in Figure 12.1 (where there are $n = 4$ resistors across the top).*

Prove that for each n, the net resistance between points A and B in $FC(n)$ is

$$\frac{F_{2n+1}}{F_{2n}}.$$

Exercise 383. *Recursively define a sequence $f_1, f_2, f_3, \ldots$ of rational functions by $f_1(x) = \frac{1}{1+x}$, and for $n \geq 1$, $f_{n+1}(x) = \frac{1}{1+f_n(x)}$. Prove that for $n \geq 2$,*

$$f_n(x) = \frac{F_n + F_{n-1}x}{F_{n+1} + F_n x}.$$

Exercise 384. *Prove by induction that for integers $m \geq 2$ and $n \geq 0$,*

$$F_{n-1} - mF_n \equiv (-1)^n m^n \pmod{m^2 + m + 1}.$$

Another popular exercise regarding Fibonacci numbers is one using dominoes (Exercise 577). See also Exercises 679 and 680 for Fibonacci numbers arising from counting permutations. Fibonacci numbers also arise in the study of continued fractions (see Exercise 322). See the journal *Fibonacci Quarterly* for many more properties of Fibonacci numbers. See also Conway and Guy's book *The Book of Numbers* [116].

12.3   Lucas numbers

Closely related to Fibonacci numbers are Lucas numbers, defined by $L_0 = 2$, $L_1 = 1$, and for $n \geq 2$, the same recursion as in Fibonacci numbers:

$$L_n = L_{n-2} + L_{n-1}.$$

Leonhard Euler (1707–1783) wrote about the Lucas numbers in his *Introductio in Analysin Infinitorum*, published in 1748.

Exercise 385. *Prove that if $n \geq 1$, then $L_n = F_{n-1} + F_{n+1}$.*

Exercise 386. *Prove that if $n \geq 0$, then $F_n + L_n = 2F_{n+1}$.*

Exercise 387. *Prove that for $n \geq 1$,*

$$L_{n-1}L_{n+1} - L_n^2 = 5(-1)^{n+1}.$$

Exercise 388. *Prove that for $n \geq 1$, $\displaystyle\sum_{i=1}^{n} L_i = L_{n+2} - 3$.*

Exercise 389. *Prove that if $n \geq 1$, then $L_n + 2L_{n+1} = L_{n+3}$.*

Exercise 390. *Prove that for $n \geq 1$,*

$$L_1^2 + L_2^2 + \cdots + L_n^2 = L_n L_{n+1} - 2.$$

Instead of the equality in Exercise 390, one could instead prove

$$L_0^2 + L_1^2 + L_2^2 + \cdots + L_n^2 = L_n L_{n+1} + 2,$$

another common form for this identity.

Exercise 391. *Prove that for $n \geq 1$,*

$$2\sum_{i=0}^{n} L_{3i} = 1 + L_{3n+2}.$$

Exercise 392. *Prove that if $n \geq 1$, then $5F_{n+2} = L_{n+4} - L_n$.*

The next formula is called the *Binet Formula* for L_n, (after Jacques Philippe Marie Binet) even though it was given earlier by Euler in 1765.

Exercise 393 (Binet's formula, Lucas numbers). *Prove that for each $n \geq 1$,*

$$L_n = \left(\frac{1 + \sqrt{5}}{2}\right)^n + \left(\frac{1 - \sqrt{5}}{2}\right)^n.$$

Exercise 394. *Prove that for $n \geq 1$, $F_{2n} = F_n L_n$.*

12.4   Harmonic numbers

For each $m \geq 1$, the m-th Harmonic number is defined to be

$$H_m = 1 + \frac{1}{2} + \frac{1}{3} + \cdots + \frac{1}{m}.$$

There are many statements about harmonic numbers and most seem provable by induction; only a few are given here. One of the first properties of the harmonic numbers that a student is likely to encounter is that the harmonic series

$$1 + \frac{1}{2} + \frac{1}{3} + \frac{1}{4} + \cdots$$

diverges. This is formalized in the next exercise.

Exercise 395. *Prove by induction that for any $n \in \mathbb{Z}^+$, there exists an m so that $H_m \geq n$. Conclude that as $m \to \infty$, $H_m \to \infty$.*

Exercise 396. *Prove that for $n \geq 1$,*

$$H_n = \sum_{i=1}^{n} \frac{(-1)^{i-1} \binom{n}{i}}{i}$$

Exercise 397. *Prove that for each $n \geq 1$,*

$$H_1 + H_2 + \cdots + H_n = (n+1)H_n - n.$$

Exercise 398. *Prove that for $n \geq 1$,*

$$1 + \frac{n}{2} \leq H_{2^n} \leq 1 + n.$$

Exercise 399. *Prove that for $n \geq 1$,*

$$H_1 + 2H_2 + 3H_3 + \cdots + nH_n = \frac{n(n+1)}{2}H_{n+1} - \frac{n(n+1)}{4}.$$

Exercise 400. *Prove that for $n \geq 2$,*

$$\sum_{i=2}^{n} \frac{1}{i(i-1)}H_i = 2 - \frac{1}{n+1} - \frac{H_{n+1}}{n}.$$

Exercise 401. *Prove that for every $n \geq 1$ and $1 \leq m \leq n$,*

$$\sum_{i=1}^{n} \binom{i}{m}H_i = \binom{n+1}{m+1}\left[H_{n+1} - \frac{1}{m+1}\right].$$

The next exercise is the subject of a rich discussion in [106] and [88, pp. 90–99]. For generalizations and more references, see the recent articles [425] or [426].

Exercise 402. *Suppose that n identical planks, each of length L, are stacked flat on top of each other, upper planks shifted to the right as in Figure 12.2, how much*

Figure 12.2: Overhanging planks, $n = 4$

further to the right can the top plank span past the end of the bottom plank without the stack falling over? Prove that this maximum span is

$$\frac{L}{2}\left[1 + \frac{1}{2} + \frac{1}{3} + \cdots + \frac{1}{n-1}\right].$$

Notice that by Exercise 395, the span above can be made arbitrarily large if only enough planks are used—it may seem highly non-intuitive that with enough 1-meter planks, one could reach an overhang sufficient to cross the Pacific Ocean! [How many planks would this take? How high is such a stack?]

12.5 Catalan numbers

12.5.1 Introduction

The *Catalan numbers* $C_0, C_1, C_2, \ldots$ form an increasing sequence of integers beginning

$$1, 1, 2, 5, 14, 42, 132, 429, 1430, 4862, 16796, 58786, 208012, 742900, \ldots.$$

and have many different definitions and interpretations. It may be interesting to try and discover the rule for this sequence; two definitions are given below, one in terms of a short formula, and the other, in terms of counting the number of ways to perform some combinatorial feat. Each definition has its own advantages.

The Catalan numbers arise naturally in a variety of contexts, including formal logic, combinatorial geometry, probability, and computing science. It is not uncommon to see definitions for Catalan numbers using different combinatorial interpretations; for example, Stanley [506, Vol 2, p. 219] lists 66 of these, with many more occurring throughout the text and exercises, and 9 more advanced algebraic interpretations [p. 231]. Only a few of the possibly hundreds of representations are given below.

On 15 September, 1988, Richard K. Guy gave a lecture at the University of Calgary titled "The ubiquitous Catalan numbers"; some of his observations appear here. See Guy's article [247] for many more, along with many references. The interested reader can spend a lifetime investigating Catalan numbers; some general sources include [14], [30], [78], [93], [105], [116], [158], [210], [213, Ch. 20], [224], [230], [270], [276], [277, 146–150], [377], [458], [474], [494], [552], and [563]. In many of these sources listed, two common areas where Catalan numbers occur are in triangulating polygons, and in counting certain trees (both of these ideas are briefly examined below). For more detailed information on polygon division and Catalan numbers, see papers [146, 21–27], and [244]. For treatments regarding specifically trees, see [103], [131], [314], and [506, Vol. 2].

12.5.2 Catalan numbers defined by a formula

One succinct way to define the Catalan numbers is as follows:

Definition 12.5.1. For each $n = 0, 1, 2, \ldots$, define the Catalan number

$$C_n = \frac{1}{n+1}\binom{2n}{n}.$$

Recall that binomial coefficients have two common definitions, one is a formula, and the second is a combinatorial definition; from the second, it is clear that the binomial coefficients are integers, but the from the formula, it takes a bit of work to show this. The situation is similar with Catalan numbers. As defined above, each Catalan number is indeed an integer; according to John H. Conway (as quoted in [247]), a more general observation is that

$$\frac{\gcd(m, n)(m + n - 1)!}{m! n!}$$

is an integer because

$$\frac{m(m + n - 1)!}{m! n!} = \binom{m + n - 1}{m - 1}$$

and

$$\frac{n(m + n - 1)!}{m! n!} = \binom{m + n - 1}{n - 1},$$

are both integers, and by the Euclidean division algorithm, $\gcd(m, n)$ is a linear combination of m and n. (Use $m = n + 1$, where $\gcd(n + 1, n) = 1$.) For further remarks on divisibility and Catalan numbers, see [15]. From any of the many combinatorial interpretations (a few are given below), it is easier to see that Catalan numbers are indeed integers.

With only a little algebra, Catalan numbers can take on many different forms; three common forms are

$$C_n = \frac{1}{n + 1}\binom{2n}{n} = \frac{1}{2n + 1}\binom{2n + 1}{n} = \binom{2n}{n} - \binom{2n}{n - 1}.$$

12.5.3 C_n as a number of ways to compute a product

If x, y, z are real numbers, the expression "xyz", written in this specific order can be parenthesized in two ways, $(xy)z$ or $x(yz)$. In the real numbers, multiplication is both commutative [$xy = yx$] and associative [$(xy)z = x(yz)$], so order and parentheses do not matter when working out a product given by symbols in some order. However, since multiplication is defined as a product of only two elements, parentheses are needed to indicate the order of operations.

It is a straightforward induction that shows a product $x_1 x_2 \cdots x_n$ of real numbers is performed by $n - 1$ multiplications. Each multiplication requires two factors, and if $n > 2$, $n - 2$ pairs of matching left-right parentheses are required. For example, when $n = 4$,

$$((x_1 x_2)x_3)x_4, \quad (x_1(x_2 x_3))x_4, \quad (x_1 x_2)(x_3 x_4), \quad x_1((x_2 x_3)x_4), \quad x_1(x_2(x_3 x_4))$$

are the five different ways to multiply out $x_1 x_2 x_3 x_4$ in that order. In the third of the above five expressions, it does not matter which of $x_1 x_2$ or $x_3 x_4$ is evaluated first; precisely the same operations are performed.

Parentheses are inserted in matching left-right pairs subject to certain rules. Every time a multiplication is executed, it is of the form AB where each of A or B is a single x_i, or is a product previously worked out that is surrounded by a matching pair of parentheses. The result of AB is then indicated by putting parenthesis around it, as in (AB), unless it is the last multiplication, in which case parentheses are not required. If matching pairs of parentheses are considered as an open interval, any two intervals are either nested or are disjoint.

For $n \geq 3$, a brief proof by induction shows that when multiplying $x_1 x_2 \cdots x_n$, exactly $n-2$ matching pairs of parentheses are inserted (recursively) and $n-1$ pairs of factors are multiplied.

Here is a second definition for Catalan numbers (compare with Definition 12.5.1 above):

Definition 12.5.2. Define $C_0 = 1$ and for $n \geq 1$, define C_n to be the number of ways to evaluate a product of $n+1$ elements in a given order, in other words, the number of ways to properly parenthesize a product of $n+1$ elements by using $n-1$ pairs of parentheses.

Checking the first few values, $C_0 = 1$, and $C_1 = 1$ since no parentheses are required to multiply two elements. Also, by above, $C_2 = 2$, and $C_3 = 4$, so these numbers agree with with Definition 12.5.1. In Section 12.5.4, it is shown that indeed these two definitions are equivalent. Before showing this, a popular recursion for Catalan numbers is given.

In evaluating a product of n elements as in Definition 12.5.2, the last multiplication occurs between two terms (and the number of ways to evaluate each term is again a Catalan number) and so the following recursion for the Catalan numbers comes directly from Definition 12.5.2. Let $C_0 = C_1 = 1$ and when $n \geq 1$, having defined $C_0, C_1, \ldots, C_n$, define C_{n+1} by

$$C_{n+1} = C_1 C_n + C_2 C_{n-1} + C_2 C_{n-1} + \cdots C_n C_1. \tag{12.2}$$

This formula was discovered by Euler and rediscovered by Segner (1704–1777) while solving the problem of counting the number of ways to triangulate an n-gon (see Exercise 405 below) and is sometimes called "Segner's recurrence relation" for Catalan numbers.

12.5.4   The definitions are equivalent

Proving that Definition 12.5.2 or equation (12.2) follows from Definition 12.5.1 seems rather difficult; generating functions can be used to prove that this recursive definition yields Definition 12.5.1. However, one can easily show (as in [286]) that the two definitions are equivalent by considering a related problem provable by induction.

For each positive integer n, let P_n be the number of ways to evaluate a product of n elements in *any order* (using $n-1$ multiplications); so $P_1 = 1$, $P_2 = 2$. The possible products of three elements a, b, c are

$$(ab)c, a(bc), (ac)b, a(cb), (ba)c, b(ac), (bc)a, b(ca), (ca)b, c(ab), (cb)a, c(ba),$$

and so $P_3 = 12$.

Theorem 12.5.3. *For each $n \geq 0$, the number of ways to multiply $n + 1$ elements in any order is*

$$P_{n+1} = \frac{(2n)!}{n!}.$$

Proof: The proof is by induction on n; for each $n \geq 0$, let $S(n)$ be the statement that $P_{n+1} = \frac{(2n)!}{n!}$. Let $x_1, x_2, x_3, \ldots$ be given elements to be multiplied.

BASE STEP: When $n = 0$, $P_1 = 1 = \frac{0!}{0!}$, so $S(0)$ is true. To further check when $n = 1$, there are two products ($x_1 x_2$ and $x_2 x_1$), so $P_2 = 2$. since $2 = \frac{2!}{1!}$, $S(1)$ is also true.

INDUCTION STEP: Fix $k \geq 1$ and assume that

$$S(k-1) : \quad P_k = \frac{(2(k-1))!}{(k-1)!}$$

is true. Consider $k + 1$ elements $x_1, \ldots, x_k, x_{k+1}$. Fix some product of $x_1, \ldots, x_m$, which contains $m - 1$ pairs of factors. Then x_{m+1} can be attached to either end of this product, or on either side of two existing factors. For example, for the fixed product $x_1(x_2 x_3)$, the number x_4 can be attached at either end: $x_4(x_1(x_2 x_3))$ or $(x_1(x_2 x_3))x_4$. Looking at the factors x_1 and $x_2 x_3$, insertion of x_4 can be before or after either:

$$(x_4 x_1)(x_2 x_3), \quad (x_1 x_4)(x_2 x_3), \quad x_1(x_4(x_2 x_3)), \quad x_1((x_2 x_3)x_4)),$$

and looking at the factors x_2 and x_3, get

$$x_1((x_4 x_2)x_3), \quad x_1((x_2 x_4)x_3), \quad x_1(x_2(x_4 x_3)), \quad x_1(x_2(x_3 x_4)).$$

In all, for any fixed product of $x_1, \ldots, x_k$, there are $2 + 4(k-1) = 4k - 2$ products created this way. Since every product of $k + 1$ elements arises in such a way, the recursion $P_{k+1} = (4k - 2)P_k$ holds. Then

$$
\begin{aligned}
P_{k+1} &= (4k - 2)P_k \\
&= (4k - 2)\frac{(2(k-1))!}{(k-1)!} \qquad \text{(by } S(k-1)) \\
&= \frac{2(2k-1)(2k-2)!}{(k-1)!} \\
&= \frac{2(2k-1)!}{(k-1)!}
\end{aligned}
$$

$$= \frac{2(2k-1)!}{(k-1)!}$$
$$= \frac{2k(2k-1)!}{k(k-1)!}$$
$$= \frac{(2k)!}{k!},$$

verifying $S(k)$, completing the inductive step.

Therefore, by MI, for each $n \geq 0$, the number of ways to multiply $n+1$ elements in any order is $P_{n+1} = \frac{(2n)!}{n!}$. $\qquad\square$

Since there are $(n+1)!$ ways to order $n+1$ elements, it follows from Definition 12.5.2 that $P_{n+1} = (n+1)!C_n$, and so, by Theorem 12.5.3,

$$C_n = \frac{1}{(n+1)!} P_{n+1} = \frac{1}{(n+1)!} \cdot \frac{(2n)!}{n!} = \frac{1}{n+1}\binom{2n}{n},$$

thereby proving that the definitions agree.

A recursion for Catalan numbers also follows directly from the equalities $P_{n+1} = (n+1)!C_n$ and $P_{k+1} = (4k-2)P_k$ derived above, namely,

$$C_n = \frac{4n-2}{n+1} C_{n-1}.$$

In retrospect, this recursion is also easy to prove from Definition 12.5.1.

12.5.5 Some occurrences of Catalan numbers

A point $(x,y) \in \mathbb{R}^2$ in the cartesian plane is called a *lattice point* iff both x and y are integers. The set of all lattice points

$$\mathbb{Z}^2 = \{(a,b) : a, b \in \mathbb{Z}\}.$$

is sometimes called the *integer lattice*. [The term "lattice" has another mathematical meaning: a partially ordered set with meets and joins.] A *walk* on a lattice is a sequence of lattice points, consecutive points differing by one of the vectors $(0,1)$, $(0,-1)$ $(1,0)$, or $(-1,0)$. In the next exercise, consider walks on the integer lattice, or *lattice paths* that move only up or to the right, that is, each step in the walk is of the form $(0,1)$ or $(1,0)$.

Exercise 403. *Prove by induction that the number of lattice paths from $(0,0)$ to (n,n) that move only up and to the right is $\binom{2n}{n}$.*

One way to view Exercise 403 is to tilt the grid by $3\pi/4$ clockwise, and then ask how many downward paths there are from the top to the bottom. (This approach was developed in [125], along with other interesting [to me, at least] approaches to identifying sequences constructed recursively.) Using this tilting idea, a downward path is obtained by choosing n down-left segments, and n down-right segments, among $2n$ segments in total. There are many other combinatorial proofs of the result in Exercise 403.

This next exercise also has a direct combinatorial proof, but an inductive proof is possible. (No solution is given.)

Exercise 404. *Show that the number of walks on the integer lattice from $(0,0)$ to (n,n) without crossing $x = y$ and moving only to the right or upward is C_n. [Touching the line $x = y$ is okay.]*

Adding a diagonal to a square divides the square into two triangles; this process may be called "triangulating the square". There are two different ways to triangulate a square. A convex pentagon can be triangulated in exactly five ways (see Figure 12.3).

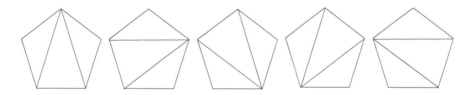

Figure 12.3: Five ways to triangulate a pentagon

Exercise 405. *Show that for $n \geq 1$, there are C_n ways to triangulate a convex $(n+2)$-gon. (See also Exercise 710.)*

Two of many relations between Catalan numbers and trees are also given in Exercise 481 (on rooted plane trees) and Exercise 482 (on full binary trees) Rooted plane trees and parenthesizing are related. For another exercise surrounding Catalan numbers and matrices, see Exercise 668. Catalan numbers also count the number of ways votes can be cast so that the eventual winner is ahead at all times (see the "ballot problem" in Exercises 764, and 765 and comments following them).

12.6   Schröder numbers

Schröder numbers are named after F. W. K. Ernst Schröder (1841–1902), a German mathematician and logician. Let S_n denote the nth Schröder number, defined to be the number of lattice paths from $(0,0)$ to (n,n) where each step is of the form $(0,1)$, $(1,0)$, or $(1,1)$ [that is, each step is one up, one to the right, or diagonally up and to the right] and contains no points above the line $x = y$. Define $S_0 = 1$. The first few are $S_1 = 2$, $S_2 = 6$, $S_3 = 22$, and $S_4 = 90$.

Exercise 406. *Prove that the Schröder numbers satisfy the recurrence*

$$S_n = S_{n-1} + \sum_{k=0}^{n-1} S_k S_{n-1-k}. \tag{12.3}$$

For more information on Schröder numbers, the interested reader might see [395], [459], [479], [505], [517]. The reader might be aware that there are both large and small Schröder numbers, and variants of each; if crossing the line $x = y$ is allowed, the Delannoy numbers are the analog to the Schröder numbers (see, for example, [394] or [552]).

12.7 Eulerian numbers

Another family of numbers, called the "Eulerian numbers", (due to Euler [178]) is implicitly tied to counting kinds of permutations, and they are often defined in that manner. In the literature, I have found at least *six* ways to define Eulerian numbers, and some of these ways are conflicting. In different sources, there are different meanings for the same word, leading to even more confusion.

The presentation here is somewhat standard, however with a bit more notation so that one can compare different definitions for the Eulerian numbers. For most of the ways to define Eulerian numbers, permutations are used. Although the discussion surrounding permutations here is self-contained, one can also see Section 19.2.2.

12.7.1 Ascents, descents, rises, falls

Recall that a *permutation* on a set X is a bijection $\sigma : X \to X$. Any permutation σ on an ordered n-element set can be seen as a permutation on $\{1, \ldots, n\}$, and so can be written

$$\sigma = (\sigma(1), \sigma(2), \ldots, \sigma(n)).$$

An *ascent* in a permutation σ on $[n] = \{1, \ldots, n\}$ is a consecutive pair $(\sigma(i), \sigma(i+1))$ with $\sigma(i) < \sigma(i+1)$, and a *descent* is an adjacent pair of entries $(\sigma(i), \sigma(i+1))$ with $\sigma(i) > \sigma(i+1)$. An *ascending run* (sometimes called a *rise*) in a permutation is a maximal consecutive sequence of entries whose every pair is an ascent. Similarly define a *descending run* (or *fall*). For example, the permutation $(2, 3, 4, 7, 5, 1, 6)$ has ascents $(2, 3)$, $(3, 4)$, $(4, 7)$, and $(1, 6)$; descents $(7, 5)$ and $(5, 1)$; rises $(2, 3, 4, 7)$ and $(1, 6)$; and only only one fall $(7, 5, 1)$.

For integers n and k, $0 \le k < n$ define

$$N_a(n, k) = \text{ the number of permutations on } [n] \text{ with exactly } k \text{ ascents;}$$
$$N_d(n, k) = \text{ the number of permutations on } [n] \text{ with exactly } k \text{ descents;}$$
$$N_r(n, k) = \text{ the number of permutations on } [n] \text{ with exactly } k \text{ rises;}$$
$$N_f(n, k) = \text{ the number of permutations on } [n] \text{ with exactly } k \text{ falls.}$$

By symmetry, $N_a(n, k) = N_d(n, k)$ and $N_r(n, k) = N_f(n, k)$. Since a permutation with k rises (ascending runs) has precisely $k - 1$ descents, $N_r(n, k) = N_d(n, k - 1)$. Similarly, $N_f(n, k) = N_a(n, k-1)$. Note that the number of ascents plus the number of descents in a permutation of length n is always $n-1$, so $N_a(n, k) = N_d(n, n-1-k)$, and then by symmetry, $N_a(n, k) = N_a(n, n - 1 - k)$. Concluding some of the relationships so far,

$$N_a(n, k) = N_d(n, k) = N_r(n, k + 1) = N_f(n, k + 1)$$
$$= N_a(n, n - 1 - k) = N_d(n, n - 1 - k) = N_r(n, n - k) = N_f(n, n - k).$$

How does one actually compute these numbers? Since there is only one permutation of each length with no ascents, for every $n \geq 1$, $N_a(n, 0) = 1$. Similarly, $N_a(n, n - 1) = 1$. Also, $N_a(2, 1) = 1$, $N_a(3, 1) = 4$; $N_a(3, 2) = 1$; $N_a(4, 0) = 1$; and $N_a(4, 1)$ is already a bit clumsy to compute directly. Of the 24 permutations on $\{1, 2, 3, 4\}$, only the permutations (deleting brackets and commas) 1432, 2143, 2431, 3142, 3214, 3241, 3421, 4132, 4213, 4231, 4312 have exactly one ascent, so $N_a(4, 1) = 11$. The permutations with exactly two ascents are all others except 1234 and 4321, another 11, so $N_a(4, 2) = 11$, and $N_a(4, 3) = 1$.

To develop a recursion formula for $N_a(n, k)$, consider a recursive construction for permutations. Any permutation σ on $\{1, 2, \ldots, n\}$ can be constructed by inserting n in a permutation τ on $\{1, 2, \ldots, n - 1\}$. For example, the permutation $(1, 2, 4, 5, 3)$ is created by inserting the 5 in between the third and fourth entry of $(1, 2, 4, 3)$. Insertion of n can be done in one of two ways: (i) at the beginning or inside an ascent, keeping the same number of ascents, or (ii) at the end or inside a descent, increasing the number of ascents by one.

There are $(k + 1)N_a(n - 1, k)$ permutations of length n with k ascents created in way (i). Since a permutation of length $n - 1$ with $k - 1$ ascents has $n - k - 1$ descents, there are $n - k$ positions described by (ii), and so there are $(n - k - 1)N_a(n - 1, k - 1)$ permutations formed by (ii). In all, for each $k \in \{1, \ldots, n - 1\}$,

$$N_a(n, k) = (n - k)N_a(n - 1, k - 1) + (k + 1)N_a(n - 1, k). \tag{12.4}$$

To verify this $N_a(4, 2) = 11$ and $(4 - 2)N_a(3, 1) + (3)N_a(3, 2) = 2 \cdot 4 + 3 \cdot 1 = 11$.

12.7.2 Definitions for Eulerian numbers

Mathematicians essentially give two different definitions for "the" Eulerian numbers; some seem to agree on $N_a(n, k)$ but most agree on $N_r(n, k) = N_a(n, k - 1)$; it often takes some work to understand what is meant by some authors. Bóna [62, p. 5] uses $A(n, k) = N_d(n, k - 1)$ as the definition of an Eulerian number, and so $N_a n, k - 1$ is his definition. Aigner [6, p. 123] uses $W_{n,k} = N_d(n, k)$, and defines the Eulerian numbers to be $A_{n,k} = W_{n,k-1}$, which, in present notation is $N_d(n, k - 1) = N_a(n, k - 1)$. Graham, Knuth and Patashnik [230, pp. 267–271] define Eulerian numbers to be $\langle \begin{smallmatrix} n \\ k \end{smallmatrix} \rangle = N_a(n, k)$. Weisstein [562] defines Eulerian numbers to be

$\langle \begin{smallmatrix} n \\ k \end{smallmatrix} \rangle$, "the number of permutation runs of length n with $k \leq n$", which is an obvious typo for many reasons (the k is not identified, it is not the runs that are counted, and "run" *could* mean ascending or descending). [This may have been corrected in the later edition.] Conway and Guy [116] define Eulerian numbers to be "the total number of arrangements of $1, 2, \ldots, n$ in which there are just $k - 1$ rises ... the Eulerian number, $A(n, k)$." One might guess because of the "A" that they meant "ascents" and not "rises"; this is confirmed only after they cite an explicit expression for them, and so Conway and Guy also use $N_a(n, k - 1)$ for Eulerian numbers.

Other authors first derive some polynomial and define its coefficients to be Eulerian numbers (actually, the number $W_{n,k}$ from Aigner arises in this way, too). For example, Stanley [506, Vol. 1] lets $d(\pi)$ be the number of descents in π, and defines the *Eulerian polynomial*

$$A_n(x) = \sum_{\text{all } \pi \text{ on } \{1, \ldots, n\}} x^{1+d(\pi)},$$

and defines $A(n, k)$ to be the k-th coefficient of $A(n)$, in which case, $A(n, k) = N_a(n, k - 1)$.

The definition here goes with the majority, even though the other definition is often easier to work with. [It may be that the standout, Graham, Knuth and Patashnik, actually looked at Euler's original work, and are consistent with that—I don't know, as I have never looked at the actual manuscript.]

Definition 12.7.1. For integers n and k with $0 < k \leq n$, define the *Eulerian number* $E_{n,k}$ to be the number of permutations of $\{1, 2, \ldots, n\}$ with precisely $k - 1$ ascents (or equivalently, with k rises) as defined above.

In present notation,

$$N_a(n, k) = E_{n,k+1}.$$

Only values for $1 \leq k \leq n$ make sense, and based on the values for $N_a(n, k)$ computed above, $E_{1,1} = 1$; $E_{2,1} = 1$; $E_{2,2} = 1$, $E_{3,1} = 1$, $E_{3,2} = 4$; $E_{3,3} = 1$; $E_{4,1} = 1$; $E_{4,2} = E_{4,3} = 11$, and $E_{4,4} = 1$.

Perhaps to avoid further confusion, authors might call ascents "up-steps", and rises "upward runs", or some such, because in English (according to the O.E.D.), an "ascent" can mean an upward slope or an instance of ascending, as in a climb to the top of a mountain, and a "rise" can be taken to mean "the vertical height of a step". [I am not suggesting that there needs to be an "uprising" to settle this, but perhaps "steps" can be taken to correct things.]

12.7.3 Eulerian number exercises

The recursion equation (12.4) for $N_a(n,k)$ says (with the present definition of Eulerian numbers):

$$E_{n,k+1} = (n-k)E_{n-1,k} + (k+1)E_{n-1,k+1}. \tag{12.5}$$

Find a similar equation to (12.5) and use it solve the following.

Exercise 407. *Show (by induction) that for all $1 \le k \le m$, the Eulerian numbers $E_{m,k}$ can be defined recursively by: for each positive integer j, set $E_{j,1} = 1$ and $E_{j,j} = 1$. For each $2 \le k \le m-1$, put*

$$E_{m,k} = (m-k+1)E_{m-1,k-1} + kE_{m-1,k}. \tag{12.6}$$

Exercise 408. *Prove that for each fixed $m \ge 1$,*

$$\sum_{j=1}^{m} E_{m,j} = m!.$$

The next identity is named after Julius Daniel Theodor Worpitzky, who proved it in 1883. [577].

Exercise 409 (Worpitzky's identity). *For any fixed $m \in \mathbb{Z}^+$ and for each $n \ge 1$, show that*

$$m^n = \sum_{i=1}^{n} E_{n,i} \binom{m+i-1}{n}.$$

For example,

$$m^3 = \binom{m}{3} + 4\binom{m+1}{3} + \binom{m+2}{3} \text{ and}$$

$$m^4 = \binom{m}{4} + 11\binom{m+1}{4} + 11\binom{m+2}{4} + \binom{m+3}{4}.$$

The following exercise apparently has many different proofs, one of which involves counting and the inclusion-exclusion principle (see [62, pp. 8–9]). I think that a direct approach with induction is also possible, although I have not yet worked out a solution:

Exercise 410. *Prove that the Eulerian numbers are given explicitly by*

$$E_{n,k} = \sum_{j=0}^{k-1} (-1)^j \binom{n+1}{j} (k-j)^n.$$

Just as a check,

$$\sum_{j=0}^{2} (-1)^j \binom{4+1}{j} (2-j)^4 = \binom{5}{0} 2^4 - \binom{5}{1} = 1 \cdot 16 - 5 = 11,$$

which is correct.

12.8   Euler numbers

Another collection of numbers is named after Euler (see [232]): for $n \geq 0$, the *Euler number E_n* is the number of permutations of the integers $1, 2, \ldots, n$ that first rise, then alternately fall and rise between consecutive elements. Call a permutation $\sigma = (\sigma(1), \ldots, \sigma(n))$ *alternating* if either

$$\sigma(1) < \sigma(2) > \sigma(3) < \sigma(4) > \cdots ,$$

or

$$\sigma(1) > \sigma(2) < \sigma(3) > \sigma(4) < \cdots .$$

For an alternating permutation σ, if σ begins with an ascent, $\sigma(1) < \sigma(2)$ then say σ is *up-alternating* and say σ is *down-alternating* if σ starts off with a descent, $\sigma(1) > \sigma(2)$. If σ is an up-alternating permutation on $\{1, 2, \ldots, n\}$, then turning its graph upsidedown (by setting $\tau(i) = n + 1 - \sigma(i)$, also called the *complement permutation*) produces a down-alternating permutation, (and similarly, a down gives an up) so the number of each are the same. Thus the definition for the Euler number E_n can be restated as either the number of up-alternating permutations on $\{1, \ldots, n\}$, or the number of down-alternating permutations on $\{1, \ldots, n\}$.

For example, for $n = 3$, the only permutations that are up-alternating are 132 and 231, so $E_3 = 2$. Turning each of these upside down gives 312 and 213, both down-alternating. Defining the term $E_0 = 1$, the first ten Euler numbers are 1,1,1,2,5,16,61,272,1385,7936, as found in sequence # 587 in [497]. According to Grassl [232], Euler numbers were first studied by D. André in 1879 (though the reference cites a paper from 1871, so I don't know precisely which date is accurate); many other (perhaps) useful references occur in Grassl's paper. The Euler numbers satisfy a recursion:

Exercise 411. *Prove that $E_0 = E_1 = 1$ and that for $n \geq 1$,*

$$E_{n+1} = \frac{1}{2} \sum_{i=0}^{n} \binom{n}{i} E_i E_{n-i}.$$

The even Euler numbers occur in coefficients of the Maclaurin expansion for secant:

$$\sec(x) = E_0 + E_2 \frac{x^2}{2!} + E_4 \frac{x^4}{4!} + E_6 \frac{x^6}{6!} + \cdots ,$$

and so the even Euler numbers are also called *secant numbers*. Similarly, the odd Euler numbers E_{2n+1} occur in the coefficients of the Maclaurin expansion for $\tan(x)$, and so are called *tangent numbers*.

12.9 Stirling numbers of the second kind

For $0 \leq k \leq n$, the Stirling number of the second kind $S_{n,k}$ is the number of partitions of an n-element set into k non-empty parts. Restating the definition in another way, $S_{n,k}$ is the number of ways to put n distinguishable elements into k indistinguishable cells. These numbers are named after James Stirling (1692–1770).

One can easily verify that $S_{0,0} = 1$, $S_{1,0} = 0$, $S_{1,1} = 1$, $S_{2,0} = 0$, $S_{2,1} = 1$, $S_{2,2} = 1$, and for example with $n = 5$, $S_{5,0} = 0$, $S_{5,1} = 1$, $S_{5,2} = 15$, $S_{5,3} = 25$, $S_{5,4} = 10$, and $S_{5,5} = 1$. In general, $S(n,n) = 1$, and for $k > n$, $S(n,k) = 0$. By a simple combinatorial argument, Stirling numbers of the second kind enjoy the following recursion:

Lemma 12.9.1. *For $n \geq 1$, $k \geq 1$:*

$$S_{n,k} = S_{n-1,k-1} + kS_{n-1,k}. \tag{12.7}$$

Proof: Let $X = \{x_1, \ldots, x_n\}$ be any set of n elements. Focus on any one element of X, say x_n, and put $X' = \{x_1, \ldots, x_{n-1}\}$.

Identify two kinds of a partition of X into k non-empty parts: when x_n is alone in its own part, or when x_n is in a part with other elements.

For the first case, start with a partition of X' into $k - 1$ parts (which can be done in $S_{n-1,k-1}$ ways), and add $\{x_n\}$ as the k-th part. For the second case, begin with a partition of X' into k parts, (which can be done in $S_{n-1,k}$ ways) and add x_n to any one of these k parts. $\square$

In practice, one often uses the recursion in (12.7) only when $1 \leq k < n$, however the formula continues to hold when $k \geq n$: when $k = n$, the second summand on the right is zero, so the recursion reads $S_{n,n} = S_{n-1,n-1}$, which is true because each side is 1; when $k > n$, each side of (12.7) is 0.

Exercise 412. *Using the above recursion in (12.7), prove by induction on n that for any variable x,*

$$x^n = \sum_{k=1}^{n} S_{n,k}x(x-1)\cdots(x-k+1).$$

A standard inclusion-exclusion argument (see Exercise 427) yields an explicit formula for the Stirling numbers of the second kind, however, with a little effort this formula can also be proved directly (so to speak) by induction:

Exercise 413. *Prove that for positive integers n and k,*

$$S_{n,k} = \frac{1}{k!}\sum_{i=1}^{k}(-1)^{k-i}\binom{k}{i}i^n.$$

Exercise 414. *For a fixed n, let $f(n) = \max\{S_{n,0}, S_{n,1}, \ldots, S_{n,n}\}$ and let $M(n) = \max\{k : S_{n,k} = f(n)\}$. Prove by induction on n that the sequence*

$$S_{n,0}, S_{n,1}, S_{n,2}, \ldots, S_{n,n}$$

is unimodal of one of the following forms:

$$S_{n,0} < S_{n,1} < \cdots < S_{n,M(n)} > S_{n,M(n)+1} > \cdots > S_{n,n}$$

or

$$S_{n,0} < S_{n,1} < \cdots < S_{n,M(n)-1} = S_{n,M(n)} > \cdots > S_{n,n},$$

where either $M(n) = M(n-1)$ or $M(n) = M(n-1)+1$.

The result in the next exercise can be thought of as an exponential generating function for the Stirling numbers of the second kind.

Exercise 415. *Let x denote a real variable. Prove by induction on k that for $k \geq 1$,*

$$\sum_{n=k}^{\infty} S(n,k)\frac{x^n}{n!} = \frac{1}{k!}(e^x - 1)^k.$$

Hint: taking a derivative might help.

For a relationship between surjective functions and Stirling numbers of the second kind, see Exercise 595.

Chapter 13

Sets

> *The essence of mathematics lies in its freedom.*
>
> —Georg Cantor,
>
> *Mathematische Annalen.*

The study of sets can be taken to include most of mathematics, however this chapter is restricted to exercises regarding "basic" properties of sets, some set theory *per se*, posets and lattices, topology, and Ramsey theory.

13.1   Properties of sets

Exercise 416. *Use strong induction to show that in every set of n natural numbers, there is a greatest.*

One might see the following exercise couched in any number of notations. For a set X, the collection of all subsets of X is often called the *power set* of X, denoted $\mathcal{P}(X)$, 2^X, or sometimes $^X 2$.

Exercise 417. *Prove by induction that number of subsets of a k-element set is 2^k.*

Exercise 418. *Fix a set A with $|A| = m < \infty$ elements. Prove by induction on n that for any set B with $|B| = n \geq 1$ elements, the cartesian product*

$$A \times B = \{(a, b) : a \in A, b \in B\}$$

is a set with $|A \times B| = m \cdot n$ elements.

Exercise 419. *For $0 \leq k \leq n$, let the definition of the binomial coefficient "n choose k" be*

$$\binom{n}{k} = \frac{n!}{(n-k)!k!}.$$

Use this definition (and Pascal's identity—see Exercise 90) to prove inductively that the number of different k-element sets that can be chosen from an n-element set is indeed $\binom{n}{k}$, thereby justifying the terminology.

Using Exercise 417, the result in the next exercise has a fairly obvious proof; however, an inductive proof is also available.

Exercise 420. *Give an inductive proof that any set with n elements has $2^{n-1} - 1$ partitions into two non-empty subsets.*

For each $n \geq 1$, if $A_1, A_2, \ldots, A_{n+1} \subseteq U$, the intersection

$$A_1 \cap A_2 \cap \cdots \cap A_n \cap A_{n+1}$$

is defined recursively to be

$$(A_1 \cap A_2 \cap \cdots \cap A_n) \cap A_{n+1}.$$

Exercise 421. *For $1 \leq r < n$,*

$$(A_1 \cap A_2 \cap \cdots \cap A_r) \cap (A_{r+1} \cap \cdots \cap A_n) = A_1 \cap A_2 \cap \cdots \cap A_n.$$

For sets A and B, let $A \backslash B = \{a \in A : a \notin B\}$ denote set substraction, the set of those elements in A but not in B. Other texts also use $A - B$ to denote this, however such subtraction also has other meanings, so the backslash is preferred here. For subsets of a universal set U, the complement of B in U is denoted by $\overline{B} = U \backslash B$. Some texts also use B^c denote the complement of B. Just like the sigma notation abbreviates sums, the big cup is used to denote unions; for example,

$$\bigcup_{i=3}^{5} A_i = A_3 \cup A_4 \cup A_5.$$

The analogous notation is used for intersections.

The results in the next two exercises are often called the "extended DeMorgan's laws".

Exercise 422. *Prove that for every $k \geq 1$, if $A_1, A_2, \ldots, A_k$ are subsets of a universal set U, then*

$$\overline{\bigcap_{i=1}^{k} A_i} = \bigcup_{i=1}^{k} \overline{A_i}.$$

Exercise 423. *Prove that for each $k \geq 1$, if $A_1, A_2, \ldots, A_k$ are subsets of a universal set U, then*

$$\overline{\bigcup_{i=1}^{k} A_i} = \bigcap_{i=1}^{k} \overline{A_i}.$$

Exercise 424. *Prove that for every $n \geq 1$, if $A_1, A_2, \ldots, A_n$ and B are subsets of U, then*

$$\left(\bigcup_{k=1}^{n} A_i \right) \cap B = \bigcup_{k=1}^{n} (A_k \cap B).$$

Exercise 425. *Prove that for each $n \geq 1$, if $A_1, A_2, \ldots, A_n, B \subseteq U$, then*

$$\left(\bigcup_{k=1}^{n} A_i \right) \setminus B = \bigcup_{k=1}^{n} (A_k \setminus B).$$

If A and B are two sets, let

$$A \triangle B = (A \cup B) \setminus (A \cap B)$$

denote the *symmetric difference* of the sets A and B; in other words, the symmetric difference of two sets consists of elements that are in precisely one of the sets (not both).

Exercise 426. *Let X be a set and recursively define functions $D_1, D_2, \ldots$ by the following: for any subset $S \subset X$, define $D_1(S) = S$, and for any and subsets $S_1, S_2, \ldots, S_{n+1}$ of X, define*

$$D_{n+1}(S_1, S_2, \ldots, S_n, S_{n+1}) = D_n(S_1, S_2, \ldots, S_n) \triangle S_{n+1}.$$

Prove that for any subsets $S_1, S_2, \ldots, S_n$, the set $D_n(S_1, S_2, \ldots, S_n)$ is precisely the set of all elements of X that are contained in an odd number of the sets $S_1, \ldots, S_n$.

The next exercise is to prove the so-called "inclusion-exclusion principle", also called the "sieve formula". The standard proof is by counting, however an inductive proof is possible. As an example of the principle, if S is a set, and both $A \subset S$ and $B \subset S$, then the number of elements in S that are not in either A or B is

$$|S \setminus (A \cup B)| = |S| - |A| - |B| + |A \cap B|.$$

Exercise 427 (Inclusion-exclusion principle, (IE)). *Fix $n \geq 1$ and a finite universe set S. Let $A_1, A_2, \ldots, A_n$ be subsets of S. For any subset $B \subset S$ use $\overline{B} = S \setminus B$ to denote the complement with respect to S, and let $|A|$ denote the cardinality of a set A (the number of elements in A). Prove that*

$$|\overline{A_1} \cap \overline{A_2} \cap \cdots \cap \overline{A_n}|$$

$$= |S| + \sum_{m=1}^{n} (-1)^m \sum_{|K|=m} \left| \bigcap_{i \in K} A_i \right|$$

$$= |S| - \sum_{i=1}^{n} |A_i| + \sum_{1 \leq i < j \leq n} |A_i \cap A_j| - \sum_{1 \leq i < j < k \leq n} |A_i \cap A_j \cap A_k| +$$

$$\ldots + (-1)^n |A_1 \cap A_2 \cap \cdots \cap A_n|.$$

Exercise 428. *Let R be a transitive binary relation on an infinite set X and let $x_1, x_2, x_3, \ldots$ be a sequence of elements in X satisfying for each $i \in \mathbb{Z}^+$, $(x_i, x_{i+1}) \in R$. Prove that for each $n \in \mathbb{Z}^+$, $(x_1, x_{n+1}) \in R$. Then prove that for each $j, n \in \mathbb{Z}^+$ that $(x_j, x_{n+j}) \in R$, hence $x_1, x_2, \ldots,$ is a totally ordered sequence.*

John Wilder Tukey (1915–?) was a co-inventor of the fast Fourier transform. He also coined the term "bit" to mean a binary digit.

Exercise 429. *Using Zorn's lemma, prove Tukey's Lemma: Let $\mathcal{F}$ be a family of subsets of X with the property that if $F \in \mathcal{F}$ if and only if every finite subset of F is in $\mathcal{F}$. Then $\mathcal{F}$ has a maximal member.*

Two sets, A and B, are said to be of the same cardinality if there is a bijection between them, and this is denoted by $|A| = |B|$.

Exercise 430. *For $m, n \in \mathbb{Z}^+$, if $\{0, 1, \ldots, n-1\}$ and $\{0, 1, \ldots, m-1\}$ have the same cardinality, then $m = n$.*

If A and B are sets, write $|A| \le |B|$ if there exists an injection from A into B.

Exercise 431 (Cantor–Bernstein–Schröder Theorem). *Prove that for any sets A and B, if $|A| \le |B|$ and $|B| \le |A|$, then $|A| = |B|$.*

A set A is said to be *Peano finite* if and only if there exists an n so that there is a bijection between A and $n = \{0, 1, 2, \ldots, n-1\}$. (By Exercise 430, if there exists such an n, it is unique.) A set A is said to be *Peano infinite* if and only if A is not Peano finite. In 1882, Dedekind proposed to Cantor another definition of infinite (see, *e.g.*, [284] for more details): Dedekind called a set A *infinite* if there is a bijection A and some proper subset of A; a set satisfying Dedekind's definition is said to be *Dedekind infinite*, (and *Dedekind finite* if it does not).

Exercise 432. *Prove by induction that if a set is Peano finite, then it is Dedekind finite.*

Exercise 433. *Assuming the Axiom of Choice, (or Well-Ordering Principle) prove that if a set is Peano infinite, then it is Dedekind infinite.*

Exercise 434. *Assuming the Axiom of Choice, prove that a countable union of countable sets is again countable.*

The next theorem contains a popular result that, in 1965, Béla Bollobás originally proved by induction. Other proofs that do not explicitly rely on induction have since been given, *e.g.*, by Katona [306] in 1974 (see, *e.g.*, [58] and [539] for more references and related results).

Definition 13.1.1. Two families of distinct sets $A_1, \ldots, A_k$ and $B_1, \ldots, B_k$ are a *cross-intersecting* family of sets if and only if for each $i = 1, \ldots, k$,

$$A_i \cap B_i = \emptyset, \tag{13.1}$$

and for every $j \neq i$

$$A_i \cap B_j \neq \emptyset. \tag{13.2}$$

Theorem 13.1.2 (Bollobás [57])**.** *For $k \geq 1$, if $A_1, \ldots, A_k$ and $B_1, \ldots, B_k$ form a cross-intersecting family, then*

$$\sum_{i=1}^{k} \binom{|A_i| + |B_i|}{|A_i|}^{-1} \leq 1. \tag{13.3}$$

Setting $X = \cup_{i=1}^{k} A_i \cup B_i$, it was furthermore shown (using induction) that equality in (13.3) holds if and only if there is an a and b so that $|X| = a + b$, the A_i's are all a-element subsets of X, the B_i's are all the b-element subsets (and so $k = \binom{a+b}{a}$ and each $A_i \cup B_i = X$).

Perhaps the most popular consequence of Theorem 13.1.2 is when all the A_i's are of the same size, and all the B_i's are of the same size.

Corollary 13.1.3. *Let $A_1, \ldots, A_k, B_1, \ldots, B_k$ be a cross-intersecting family where for each i, $|A_i| = a$, $|B_i| = b$. Then $k \leq \binom{a+b}{a}$.*

One remarkable feature of both results is that the number of elements in any ground set is irrelevant, and so it might be some surprise that inductive proofs of both can be made by inducting on the order of the ground set.

In a sense, proving Theorem 13.1.2 by induction is easier than an inductive proof of the simpler Corollary 13.1.3 because one does not need to worry about the sizes of the sets (proving more is easier).

Exercise 435. *Prove Theorem 13.1.2 by induction on the number of elements in the ground set.*

To provide context for the next exercise, a famous conjecture by Martin Kneser [321] is discussed.

Conjecture 13.1.4 (M. Kneser, 1955)**.** *For each $n \geq 1$ and $k \geq 0$, if the n-subsets of a $(2n + k)$-set are partitioned into $k + 1$ classes, then some class contains two disjoint n-sets.*

This was proved by Lovász in 1978 [353], and a simpler proof was given by Bárányi [35] later the same year; both proofs use Borsuk's theorem. See [235] for another simple proof (again, using Borsuk's theorem).

Kneser's conjecture with $n = 2$ says that if the pairs of a $(k + 4)$-set are partitioned into $k+1$ classes, then some class contains at least two disjoint pairs. Thus if the pairs of an $(k+4)$-set are partitioned into r classes so that no two such disjoint pairs exist in any one class, then $r \geq k+2$. This case was shown by Kneser in 1956 [322], where in the next exercise, n now plays the role of $k+4$.

Exercise 436. *Prove by induction on $n \geq 3$ that if the pairs of an n-set are partitioned into k classes so that every two pairs in the same class share a vertex, then $k \geq n-2$.*

In the next exercise, some special terminology is used [perhaps unnecessarily so, but it helps to make the statement of the exercise and the proof briefer, and it is used in many other related problems].

For $1 \leq q \leq p$, a family of sets is said to have the (p,q) *property* if among every p sets in the family, there are q with non-empty intersection. For the solution of the next exercise, the following observation can be helpful:

Lemma 13.1.5. *Let $2 \leq q \leq p$ and let $\mathcal{F}$ be a family of sets that has the (p,q) property; then for each $r = 1, 2, \ldots, q-1$, $\mathcal{F}$ has the $(p-r, q-r)$ property.*

Proof: Fix r, and suppose that $p-r$ sets are chosen. Extend this family to any p sets by adding r "new sets" (from $\mathcal{F}$ of course). By the (p,q) property, some q of these p sets have non-empty intersection, and so any $q-r$ of these q sets have non-empty intersection; in particular, when omitting the r new sets at least some $q-r$ sets from the original $p-r$ sets have non-empty intersection, proving the lemma.   □

The next exercise looks simple, however, even with the help of Lemma 13.1.5 and the hint, solving it might be challenging:

Exercise 437. *Let $2 \leq q \leq p$ be integers, and suppose that a collection of (at least p) closed segments of the real line has the (p,q) property. Prove that the segments can be partitioned into $p-q+1$ classes so that sets in each class have non-empty intersection. Hint: Induct on $p-q$.*

For the next exercise, some special terminology is used. Let X be a set and let $\mathcal{H} \subseteq \mathcal{P}(X)$ be a family of subsets of X. For any $Y \subset X$, let

$$\mathcal{H}|_Y = \{H \cap Y : H \in \mathcal{H}\},$$

called the restriction of $\mathcal{H}$ to Y. A set $Y \subset X$ is called *shattered* by $\mathcal{H}$ if and only if $\mathcal{H}|_Y = \mathcal{P}(Y)$, that is, for every subset $W \subset Y$, there exists $H \in \mathcal{H}$ such that $H \cap Y = W$. Define the *VC-dimension* of $\mathcal{H}$ to be

$$\text{VC-dim}(\mathcal{H}) = \sup_{Y \subset X} \{|Y| : Y \text{ is shattered}\}.$$

The notion of shattered sets is often expressed in terms of hypegraphs, and has been used in set theory, combinatorics, combinatorial geometry, and probability. For more on VC dimension, see, for example, [372] or [421].

Observe that if $\mathcal{G} \subset \mathcal{H}$, any set shattered by $\mathcal{G}$ is certainly shattered by $\mathcal{H}$ and so VC-dim$(\mathcal{G}) \leq$ VC-dim$(\mathcal{H})$.

For example, (from [372, p. 238]), let $X = \mathbb{R}^2$ and let $\mathcal{H}$ be the set of all closed half-planes. Perhaps surprisingly (by reasoning given below) VC-dim$(\mathcal{H}) = 3$. To see this, any set of three points in general position is shattered, but no four-point set is shattered because

- if three of the four points are in a row, the middle point cannot be separated by a half-plane;

- if four points are convex, a pair of points at opposite corners can be not be separated; and

- if one point lies in the convex hull of the others, then that point can not be separated.

Below, only finite sets X (and hence finite families) are considered. Note that if $\mathcal{H}$ is the empty family, no set is shattered, in which case VC-dim$(\mathcal{H}) = 0$. To eliminate complete trivialities, assume that X is non-empty.

In the early 1970s, the following theorem was proved independently by Sauer [475], Shelah [487], and Vapnik and Chervonenkis [551].

Theorem 13.1.6. *Let X be a set with $n \geq 1$ elements, and let $\mathcal{H} \subseteq \mathcal{P}(X)$. If VC-dim $= d$, then*

$$|\mathcal{H}| \leq \binom{n}{0} + \binom{n}{1} + \cdots \binom{n}{d},$$

and this bound is best possible.

To see that the bound in Theorem 13.1.6 is best possible, let $|X| = n$ and let $\mathcal{H}$ be the family of subsets of X that have at most d elements, that is, let $\mathcal{H} = [X]^{\leq d}$. Then any d-set is shattered, but any $(d+1)$-subset is not, and equality holds.

Exercise 438. *Using mathematical induction, prove the inequality in Theorem 13.1.6.*

13.2 Posets and lattices

See Section 2.6 for some of the terminology used here.

Recall that a poset $(P, \leq)$ is a set P together with a binary relation $\leq$ that is reflexive, antisymmetric, and transitive. Let X be any set with $|X| = \ell$. The power set $\mathcal{P}(X)$ (which has 2^ℓ elements) together with the partial order $\subseteq$ is a poset, sometimes called the ℓ-*dimensional boolean lattice* $B(\ell) = (\mathcal{P}(X), \subseteq)$.

A *chain* is a family $\mathcal{C}$ of sets so that for any two distinct elements $A, B \in \mathcal{C}$, either $A \subset B$ or $B \subset A$. A chain in a poset $(P, \leq)$ is a set $C \subset P$ so that for any

$a, b \in C$, one of $a \leq b$ or $b \leq a$ holds. A subset $A \subset P$ is an *antichain* if no two elements in A are comparable.

Perhaps the most famous result on antichains was published by R. P. Dilworth [141] in 1950, and is now eponymous with Dilworth, even though it was, according to Tverberg [540], discovered earlier by T. Gallai in 1936.

Exercise 439 (Dilworth's theorem, finite). *Let $(P, \leq)$ be a finite poset. Prove that the minimum number of disjoint chains necessary to cover P is equal to the maximum number of elements in an antichain. Hint: Induct on $|P|$.*

Dilworth's theorem is perhaps the beginning of what is now called "dimension theory" for posets, and has far-reaching consequences in many aspects of combinatorics and set theory.

Dilworth's theorem is also valid for infinite sets:

Theorem 13.2.1 (Dilworth's theorem, infinite). *Let P be poset (not necessarily finite) whose largest antichain contains $\alpha < \infty$ elements. Then P can be covered by a union of α chains.*

Exercise 440. *(Challenging) Prove Theorem 13.2.1. Hint: Use a compactness argument (perhaps using Tychonoff's theorem, Theorem 13.3.4 below).*

The claim in the following exercise is due to Mirsky [387], and is a dual to Dilworth's theorem.

Exercise 441. *If S is a poset with no chains of length greater than m, then S can be covered by at most m antichains. Hint: induct on m. For $m > 1$, let M be the set of all maximal elements in S. Then $S \setminus M$ has no chains of length greater than $m - 1$, and M is an antichain.*

For a set X with n elements, a chain $\mathcal{C} \subseteq \mathcal{P}(X)$ is *convex* iff whenever $A \subset B \subset C$ and $A, C \in \mathcal{C}$, then $B \in \mathcal{C}$. A chain $\mathcal{C} \subseteq \mathcal{P}(X)$ is *symmetric* if for every $C \in \mathcal{C}$ there exists $C' \in \mathcal{C}$ so that for some $i \geq 0$, $\{|C|, |C'|\} = \{\lceil \ell/2 \rceil + i, \lfloor \ell/2 \rfloor - i\}$.

For any set X, a *symmetric chain decomposition* of $\mathcal{P}(X)$ is a partition

$$\mathcal{P}(X) = \mathcal{C}_1 \cup \mathcal{C}_2 \cup \cdots \cup \mathcal{C}_s,$$

into disjoint symmetric convex chains. The goal of Exercise 442 below is to show that for any finite X, such a decomposition always exists. There are a number of methods by which a symmetric chain decomposition can be found or constructed (see, for example, [234]); however, one of the easiest ways to construct one is by induction. Note that since each chain contains precisely one set with $\lfloor |X|/2 \rfloor$ elements, any symmetric chain decomposition has $\binom{|X|}{\lfloor |X|/2 \rfloor}\}$ chains.

As an example, it is convenient to use $X = [n] = \{1, 2, \ldots, n\}$, where sets are written without commas or set brackets, and chains are vertical. For $n = 4$:

$$
\begin{array}{cccccc}
1234 & & & & & \\
123 & 124 & 234 & & 134 & \\
12 & 14 & 23 & 24 & 13 & 34 \\
1 & 4 & 2 & & 3 & \\
\emptyset & & & & &
\end{array}
$$

It might be enlightening for the reader to find such a decomposition for $n = 5$ before examining the following example (a recursion mentioned in the subsequent solution was used):

$$
\begin{array}{cccccccccc}
12345 & & & & & & & & & \\
1234 & 1235 & 1245 & & 2345 & & & 1345 & & \\
123 & 125 & 124 & 145 & 234 & 235 & 245 & 134 & 135 & 345 \\
12 & 15 & 14 & 45 & 23 & 25 & 24 & 13 & 35 & 34 \\
1 & 5 & 4 & & 2 & & & 3 & & \\
\emptyset & & & & & & & & &
\end{array}
$$

Exercise 442. *For any finite set X, prove that there is a partition of $(\mathcal{P}(X), \subseteq)$ into disjoint symmetric convex chains.*

This next section introduces a version of a countable Zorn's lemma applied to measurable sets, described to me by Steve Kalikow [297], a form of which appears in his new book [298]. [Kalikow suspects that this result has been known for decades, however I could not find another source.]

Let (X, Ω, μ) be a finite measure space, that is, X is a set, $\Omega \subseteq \mathcal{P}(X)$ is a σ-algebra (closed under finite intersections, countable unions), and $\mu : \Omega \to \mathbb{R}^+ \cup \{0\}$ is a measure with $\mu(X) < \infty$.

Let $\Theta \subseteq \Omega$ be an arbitrary subset of measurable sets. Define a partial order $\precsim$ on Θ as follows: For $A, B \in \Theta$, define $A \precsim B$ if and only if $\mu(A \backslash B) = 0$. If both $A \precsim B$ and $B \precsim A$, then $\mu(A \triangle B) = 0$ (where $X \triangle Y = (X \cup Y) \backslash (Y \cap X)$). So define the relation $\approx$ by $A \approx B$ if and only if $\mu(A \triangle B) = 0$.

It is nearly trivial to check that $\precsim$ is indeed a partial order; reflexivity and antisymmetry are built into the definition, and transitivity follows since if $A, B, C \in \Theta$ satisfy $A \precsim B$ and $B \precsim C$, then $\mu(A \backslash C) = \mu(A \backslash B) + \mu(B \backslash C) = 0 + 0 = 0$.

Theorem 13.2.2 (Countable Zorn's lemma for measurable sets). *As defined above, let $P = (\Theta, \precsim)$ have the property that any countable chain in P has an upper bound in P, then P contains a maximal element.*

Proof: Define a chain (by the axiom of choice or the principle of MI)

$$
\mathcal{A} = A_1 \precsim A_2 \precsim A_3 \precsim \cdots,
$$

in P as follows: Choose $A_1 \in \Theta$ arbitrarily, and put

$$a_1 = \sup\{\mu(B) : B \in \Theta, A_1 \precsim B\}.$$

For $i \in \mathbb{Z}^+$, having defined $A_i \in \Theta$ and a_i, choose $A_{i+1} \in \Theta$ so that

$$\mu(A_{i+1}) \geq \frac{a_i + \mu(A_i)}{2},$$

and define

$$a_{i+1} = \sup\{\mu(B) : B \in \Theta, A_{i+1} \precsim B\}.$$

So each A_{i+1} sits at least halfway (in measure) up from A_i to any potential set in Θ which is above (in P) A_i. The chain $\mathcal{A}$ is at most countable. By hypothesis, let $U \in \Theta$ be an upper bound for the chain $\mathcal{A}$, and put $A = \cup_{i=1}^{\infty} A_i$. Note that for each n, $A_n \subseteq A$ and so $\mu(A_n \backslash A) = 0$. Thus, $\mu(A_n \backslash U) = \mu(A_n \backslash A) + \mu(A \backslash U) = 0 + 0 = 0$ shows that $A_n \precsim U$.

Claim: $\mu(U \backslash A) = 0$ (and so $U \cong A$).

Exercise 443. *Prove the claim and complete the proof of Theorem 13.2.2.*

13.3 Topology

The first few exercises ask to prove some standard results regarding the real line $\mathbb{R}$.

Exercise 444. *Prove that for open intervals (a_1, b_1), $(a_2, b_2), \ldots, (a_n, b_n)$ of real numbers, if for each i, j, $(a_i, b_i) \cap (a_j, b_j) \neq \emptyset$, then*

$$\bigcap_{i=1}^{n} (a_i, b_i) \neq \emptyset.$$

For $A \subseteq \mathbb{R}$, a point $a \in \mathbb{R}$ is a called a *limit point* of A if there exists a sequence $\{x_n\}$ so that $x_n \to a$ as $n \to \infty$.

Exercise 445. *Using the Axiom of Choice, prove that if for every $\epsilon > 0$, there exists $x \in A$ such that $|x - a| < \epsilon$, then a is a limit point of A.*

Exercise 446. *Let X and Y be sets, and let $\mathcal{F}$ be a collection of functions from some subset of X into Y. Define a partial order $\preceq$ on $\mathcal{F}$ by $f \preceq g$ if g extends f (i.e., $dom(f) \subseteq dom(g)$, and for all $x \in dom(f)$, $f(x) = g(x)$). Prove that the union of any chain in $(\mathcal{F}, \preceq)$ is again a function (whose domain is the union of domains of f's in the chain and range is the union of the ranges).*

The next few paragraphs give some basic definitions in topology so that a very few major results may be stated, and perhaps proved by induction.

Definition 13.3.1. A *topological space* is a pair $(X, \mathcal{T})$, where X is a set, and $\mathcal{T} \subseteq \mathcal{P}(X)$ is a collection of subsets satisfying (i) $X \in \mathcal{T}$, (ii) $\emptyset \in \mathcal{T}$, (iii) the union of any elements in $\mathcal{T}$ is an element of $\mathcal{T}$, and (iv) the intersection of finitely many elements of $\mathcal{T}$ is an element of $\mathcal{T}$. The elements of $\mathcal{T}$ are called *open sets*.

A topological space $(X, \mathcal{T})$ is often referred to by simply X, when it is clear what $\mathcal{T}$ is; $\mathcal{T}$ is called a *topology* on X. If $\mathcal{T}$ and $\mathcal{U}$ are topologies on X, then $\mathcal{T}$ is said to be *coarser* than $\mathcal{U}$ iff $\mathcal{T} \subseteq \mathcal{U}$.

A set $S \subset X$ is *closed* iff the complement $X \setminus S$ is an open set. For a set S, the closure of S, is denoted $\overline{S}$ (the smallest closed set containing S, found by intersecting all closed supersets of S).

If X is a topological space, an *open cover* of a subset $A \subseteq X$ is a collection of open sets whose union contains A. A subset A of a topological space is *compact* iff for any open cover of A, there exists a finite subcollection of the open cover that is an open cover of A ("any open cover contains a finite subcover"). A family of sets has the *finite intersection property* (FIP) iff any finite subcollection has non-empty intersection.

The next lemma gives another way to look at compactness.

Lemma 13.3.2. *A topological space X is compact iff for every family $\{A_i : i \in I\} \subset \mathcal{P}(X)$ with the FIP, $\cap_{i \in I} \overline{A_i} \neq \emptyset$.*

Proof: $(\Rightarrow)$ Let X be compact, and let $\{A_i : i \in I\} \subset \mathcal{P}(X)$ be a family with FIP. For each $i \in I$, define $U_i = X \setminus \overline{A_i}$. If $F \subseteq I$ is finite,

$$\cup_{j \in F} U_j = \cup_{j \in F} X \setminus \overline{A_j} = X \setminus (\cap_{j \in F} \overline{A_j}),$$

and by FIP, the family $\{U_j : j \in F\}$ does not cover X. Since X is compact and the family $\{U_i : i \in I\}$ has no finite subcover, the family $\{U_i : i \in I\}$ is not a cover, that is, $\cup_{i \in I} U_i \neq X$, and it follows that $\cap_{i \in I} A \neq \emptyset$.

$(\Leftarrow)$ Suppose that every family of sets with FIP has a non-empty intersection. To show that X is compact, let $\{Y_i : i \in I\}$ be an open cover of X and for each $i \in I$, put $W_i = X \setminus Y_i$. Then

$$\emptyset = \cap_{i \in I} W_i = \cap_{i \in I} \overline{W_i},$$

so the family $\{W_i : i \in I\}$ does not have FIP; therefore, there exists some finite $F \subseteq I$ so that $\cap_{j \in F} W_i = \emptyset$, and so $\{Y_i : i \in F\}$ is a finite subcover. $\qquad\square$

Exercise 447. *Use Zorn's lemma to show that for a set X, there exists a maximal (with respect to inclusion) family of subsets of X with FIP.*

Given any family $\{X_i : i \in I\}$ of sets, let $X = \prod_{i \in I} X_i$ denote their cartesian product. Recall that (see Section 4.5, indexed version of AC) the axiom of choice is equivalent to saying that any product of non-empty sets is again non-empty. A point in X is of the form $\mathbf{x} = (x_i \in X_i : i \in I)$. For $j \in I$, the j-th projection $\pi_j : X \to X_j$ is defined by $\pi_j(\mathbf{x}) = x_j$. If one assumes the well-ordering principle, consider I to be well-ordered, and so elements of the product space are "generalized" sequences (ordered tuples, perhaps infinite). A function f on a topological space is called *continuous* iff for every open set O, $f^{-1}(O)$ is again open.

Definition 13.3.3. Let $\{(X_i, \mathcal{T}) : i \in I\}$ be a family of topological spaces, and let $X = \prod_{i \in I} X_i$. The product topology $\mathcal{T}$ on X is coarsest topology on X so that all projections $\pi_i : X \to X_i$ are continuous.

Each X_i in the product topology denoted above is called a *coordinate space* (or *factor space*—but some authors reserve this term to mean "quotient space", a term not defined here).

Theorem 13.3.4 (Tychonoff's theorem). *(Assume AC.) If for each $i \in I$, X_i is a compact topological space, then $X = \prod_{i \in I} X_i$ is compact in the product topology.*

The next exercise asks to prove a finite version version of Tychonoff's theorem, but it may be helpful to review four facts about compact spaces (all of which have easy proofs):

1. A space X is compact iff every open cover of X with basic open sets has a finite subcover.
2. The set

$$\{U \times V : U \text{ is open in } X \text{ and } V \text{ is open in } Y\}$$

forms a basis for the product topology on $X \times Y$.
3. Compactness is preserved by isomorphism.
4. Associativity of products: $X_1 \times \cdots \times X_{n-1} \times X_n$ is isomorphic to $(X_1 \times \cdots \times X_{n-1}) \times X_n$.

Exercise 448. *Prove the finite version of Tychonoff's theorem: For any $n \geq 1$, if $X_1, \ldots, X_n$ are compact topological spaces, then the space $X_1 \times X_2 \times \cdots \times X_n$ is compact.*

The following exercise is quite ambitious, and might be directed to only those individuals with much experience in topology:

Exercise 449. *Assuming AC (and Zorn's lemma), prove Tychonoff's theorem (Theorem 13.3.4). Hint: There are many proofs, one using Lemma 13.3.2, another using Exercise 446.*

The following exercise has a fairly elementary proof, often left as an exercise in topology texts (*e.g.*, see [571]).

Exercise 450. *Prove that if Tychonoff's theorem is rewritten to say that any non-empty product of compact spaces is compact, then this version implies the Axiom of Choice.*

It is well known that the reals have the *Archimedean property*, that is, for every $x \in \mathbb{R}$, there exists an $n \in \mathbb{N}$ so that $x < n$. (Note: there are similar properties that are also referred to by "the Archimedean property", so don't take this as a standard definition.) Here's an easy proof: Suppose the contrary, that there is an x so that for every $n \in \mathbb{N}$, $n \leq x$ holds. Then $\mathbb{N}$ is bounded from above by x and so has a least upper bound s. Then $s \geq 2$ implies $s - 1 \geq 1$, and similarly, $s - 1 \geq n$ for all $n \in \mathbb{N}$, showing that $s - 1$ is an upper bound, and so s is not the least upper bound.

Exercise 451. *Show that $\mathbb{Q}$ is dense in $\mathbb{R}$, that is, between any two distinct real numbers, there exists a rational number. Hint: Well-ordering.*

Recall that a function $f : \mathbb{R} \to \mathbb{R}$ is *continuous* at $a \in \mathbb{R}$ if and only $\lim_{x \to a} f(x) = f(a)$. The limit $\lim_{x \to a} f(x)$ exists only if both one-sided limits $\lim_{x \to a^-} f(x)$ and $\lim_{x \to a^+} f(x)$ exist, are finite, and agree. Equivalently, a function f is continuous at a point a iff for any $\epsilon > 0$, there exists $\delta > 0$ so that for every $x \in \mathbb{R}$, if $|x - a| < \delta$ then $|f(x) - f(a)| < \epsilon$.

Exercise 452. *Let $f : \mathbb{R} \to \mathbb{R}$ be a function and let $a \in \mathbb{R}$ be fixed. Use the Axiom of Choice to prove that if for every sequence $\{x_n\}$ that converges to a, the sequence $\{f(x_n)\}$ converges to $f(a)$, then f is continuous.*

13.4   Ultrafilters

For a set S, recall that $\mathcal{P}(S) = 2^S = \{X : S \subseteq X\}$ denotes the power set of S. For $A \subset S$, let $\overline{A} = S \backslash A$ denote the complement of A (with respect to S). In this section, occasionally families of sets, and families of families of sets are used, so the usual notation for elements, sets, and families of sets (*e.g.* $x \in X \in \mathcal{X}$) is often abandoned.

A collection $\mathcal{F} \subset \mathcal{P}(S)$ is called a *filter on S* iff
(i) $S \in \mathcal{F}$,
(ii) $A \in \mathcal{F}$ and $A \subset B \subseteq S$ imply $B \in \mathcal{F}$, and
(iii) $A \in \mathcal{F}$ and $B \in \mathcal{F}$ imply $A \cap B \in \mathcal{F}$.

If moreover, $\emptyset \notin \mathcal{F}$, then $\mathcal{F}$ is called a *proper* filter. A trivial filter on a set S consists only of the set S. From now on, only non-trivial filters are considered. Note that when $\mathcal{F}$ is non-empty, (ii) implies (i), so often condition (i) is not mentioned explicitly, but instead filters are restricted to non-empty collections.

There are three typical (or standard) examples of filters; the reader is invited to prove the simple exercises verifying that each is indeed a filter:

1. Let S be a set and fix some $C \subseteq S$. Then $\mathcal{F} = \{A \subseteq S : C \subseteq A\}$ is a filter, called a *principal filter*. (A *principal filter* $\mathcal{F}$ is one that satisfies $\cap_{A \in \mathcal{F}} A \neq \emptyset$.)

2. For fixed infinite set S, the family $\mathcal{F} = \{A \subseteq S : \overline{A} \text{ is finite}\}$ is a filter, called the *Frechet filter* on S. (A set whose complement is finite is often called *co-finite*.)

3. Let $\mathcal{M} = \{T \subset S : |T| < \infty\}$ be the family of all finite subsets of S. For any $A \in \mathcal{M}$, define $\hat{A} = \{M \in \mathcal{M} : M \supseteq A\}$. Then

$$\mathcal{F} = \{\mathcal{A} \subseteq \mathcal{M} : \exists A \text{ s.t. } \hat{A} \subseteq \mathcal{A}\}$$

is a filter on $\mathcal{M}$.

Another example of a filter can be found by intersecting filters.

Lemma 13.4.1. *Let $\{\mathcal{F}_i : i \in I\}$ be a collection of filters on S. Then $\mathcal{F} = \cap_{i \in I} \mathcal{F}_i$ is a filter.*

Proof: (Trivial) (a) For each $i \in I$, $S \in \mathcal{F}_i$, and so $S \in \mathcal{F}$.

(b) If $A \in \mathcal{F}$ and $A \subset B$, then for each $i \in I$, $B \in \mathcal{F}_i$, so $B \in \mathcal{F}$.

(c) If $A, B \in \mathcal{F}$, then for each $i \in I$, $A \cap B \in \mathcal{F}_i$; hence $A \cap B \in \mathcal{F}$. □

Notation: For $\mathcal{E} \subseteq 2^S$, define

$$\mathcal{F}_{\mathcal{E}} = \cap\{\mathcal{F} \supseteq \mathcal{E} : \mathcal{F} \text{ is a filter on } S\},$$

called the filter *generated by* $\mathcal{E}$ (which is the smallest filter containing $\mathcal{E}$). By Lemma 13.4.1, $\mathcal{F}_{\mathcal{E}}$ is a filter. When is $\mathcal{F}_{\mathcal{E}}$ a proper filter? Recall that a collection $E \subseteq 2^S$ is said to have the *finite intersection property* (FIP) iff the intersection of finitely many elements in E is not empty.

Theorem 13.4.2. *For any $\mathcal{E} \subseteq 2^S$, $\mathcal{F}_{\mathcal{E}}$ is a proper filter on S iff $\mathcal{E}$ has FIP.*

Proof: One direction is given here; the other direction is an exercise.

($\Leftarrow$) Let $\mathcal{E}$ have FIP, and put

$$\mathcal{F} = \{A \subset S : \exists B_1, B_2, \dots, B_n \in \mathcal{E} \text{ s.t. } B_1 \cap B_2 \cap \cdots \cap B_n \subseteq A\}.$$

Claim: $\mathcal{F}$ is a filter. Clearly $S \in \mathcal{F}$. Also, if $A \in \mathcal{F}$ and $B \supset A$, then $B \in \mathcal{F}$ holds. For $B_1, \dots, B_n \in E$, and $D_1, \dots, D_n \in E$, if $A \subset B_1 \cap \cdots B_n$ and $C \subset D_1 \cap \cdots \cap D_n$, then $A \cap B \supset B_1 \cap \cdots \cap B_n \cap D_1 \cap \cdots D_n$. Thus $\mathcal{F}$ is a filter, and by FIP, $\mathcal{F}$ is a proper filter. Also, every filter containing $\mathcal{E}$ must also contain $\mathcal{F}$, since a filter having elements $B_i \in \mathcal{E}$ has elements of $\mathcal{F}$ by properties (ii) and (iii). So $\mathcal{F} = \mathcal{F}_{\mathcal{E}}$.

The proof of the other direction is a simple induction.

Exercise 453. *Prove the remaining direction of Lemma 13.4.2.*

A filter $\mathcal{F}$ on a set S is called an *ultrafilter* on S iff for all $A \subseteq S$, either $A \in \mathcal{F}$ or $\overline{A} \in S$ and not both. (Hence an ultrafilter is a proper filter.) A filter $\mathcal{F}$ on S is called a *maximal* filter on S iff $\mathcal{F}$ is proper and whenever $\mathcal{F}' \supsetneq \mathcal{F}$ is a filter, then $\mathcal{F}' = 2^S$. The following well-known result has a fairly straightforward proof using Theorem 13.4.2.

Theorem 13.4.3. *A filter $\mathcal{F}$ is maximal if and only if $\mathcal{F}$ is an ultrafilter.*

The following theorem is one of the central theorems for ultrafilters:

Theorem 13.4.4. *Every proper filter $\mathcal{F}$ on a set S can be extended to an ultrafilter on S.*

It suffices to show that $\mathcal{F}$ can be extended to a maximal filter. The proof below relies on Zorn's lemma, but

Exercise 454. *Prove Theorem 13.4.4 by Zorn's lemma.*

For (at least) countable cases, a more constructive enumeration is possible without AC. Theorem 13.4.4 has another form (which appears, *e.g.*, in [271, Thm 6.5]) which says that any family with FIP can be extended to an ultrafilter; this result follows directly from Theorems 13.4.2 and 13.4.4.

The following is a simple, but useful, property of ultrafilters.

Exercise 455. *Let $\mathcal{F}$ be an ultrafilter on X, $k \in \mathbb{Z}^+$, and let $X = X_1 \cup \cdots \cup X_k$ be a partition of X. Then for precisely one $i \in [1, k]$, $X_i \in \mathcal{F}$.*

The theory of ultrafilters (and ultraproducts) is closely related to semigroup theory, topology, dynamical systems, Ramsey theory, and model theory. Unfortunately, these many exciting pursuits are not covered here.

Chapter 14

Logic and language

Logic, properly used, does not shackle thought. It gives freedom, and above all, boldness.

—Alfred North Whitehead (1861–1947),

The Organization of Thought.

14.1 Sentential logic

In this section, it is assumed that the reader is familiar with basic logic and can verify truth of statements by truth tables. For example, $p \wedge q \rightarrow p$ is a fairly simple true statement. Many more exercises could be given in the section, many duplicating those given for boolean algebras, sets with unions and intersections, or even associative laws for many binary relations like integer addition. The rightarrow $\rightarrow$ is implication, whereas $\Rightarrow$ is logical implication. The symbols $\wedge$ and $\vee$ are short for "AND" and "OR" respectively, and $\neg p$ is the negation of p. Only a few solutions are given to the following exercises as most are rather routine.

For $n \geq 2$, define recursively the *conjunction* of $n+1$ statements $p_1, \ldots, p_n, p_{n+1}$ by

$$p_1 \wedge \cdots \wedge p_n \wedge p_{n+1} \Leftrightarrow (p_1 \wedge \cdots \wedge p_n) \wedge p_{n+1}.$$

Exercise 456. *Prove by induction that for statements $p_1, \ldots, p_{m+1}$,*

$$[p_1 \rightarrow p_2] \wedge [p_2 \rightarrow p_3] \wedge \ldots \wedge [p_m \rightarrow p_{m+1}] \Rightarrow [(p_1 \wedge p_2 \wedge \ldots \wedge p_m) \rightarrow p_{m+1}].$$

Exercise 457. *Prove that for every $m \geq 1$, and statements $q_1, \ldots, q_m$ and q,*

$$(q_1 \wedge \cdots q_m) \wedge q \Leftrightarrow q_1 \wedge \cdots q_m \wedge q$$

and

$$q \wedge (q_1 \wedge \cdots \wedge q_m) \Leftrightarrow q \wedge q_1 \wedge \cdots \wedge q_m.$$

For sentences p_1, p_2, p_3, it is easily checked by truth tables that

$$(p_1 \wedge p_2) \wedge p_3 \Leftrightarrow p_1 \wedge p_2 \wedge p_3, \tag{14.1}$$

and similarly,

$$p_1 \wedge (p_2 \wedge p_3) \Leftrightarrow p_1 \wedge p_2 \wedge p_3. \tag{14.2}$$

Exercise 458. *Prove inductively that for $n \geq 3$ and $1 \leq r < n$,*

$$(p_1 \wedge \cdots \wedge p_r) \wedge (p_{r+1} \wedge \cdots \wedge p_n) \Leftrightarrow p_1 \wedge \cdots \wedge p_r \wedge p_{r+1} \wedge \cdots \wedge p_n.$$

Exercise 459. *Give a recursive definition for the disjunction of $n + 1$ statements $p_1, p_2, \ldots, p_n, p_{n+1}$.*

Exercise 460. *Prove inductively that for $n \geq 3$ and $1 < r \leq n$,*

$$(p_1 \vee \cdots \vee p_r) \vee (p_{r+1} \vee \cdots \vee p_n) \Leftrightarrow p_1 \vee \cdots \vee p_r \vee p_{r+1} \vee \cdots \vee p_n.$$

Exercise 461. *Use the result from Exercise 458 to show that for $n \geq 2$ and statements $p, q_1, q_2, \ldots, q_n$,*

$$p \vee (q_1 \wedge q_2 \wedge \cdots \wedge q_n) \Leftrightarrow (p \vee q_1) \wedge (p \vee q_2) \wedge \cdots \wedge (p \vee q_n).$$

The next two exercises state the sentential equivalents of the extended DeMorgan's laws (cf. Exercises 422 and 423).

Exercise 462. *Prove that for $n \geq 2$ and statements $p_1, p_2, \ldots, p_n$,*

$$\neg(p_1 \vee p_2 \vee \cdots \vee p_n) \Leftrightarrow \neg p_1 \wedge \neg p_2 \wedge \cdots \wedge \neg p_n.$$

Exercise 463. *Prove that for $n \geq 2$ and statements $p_1, p_2, \ldots, p_n$,*

$$\neg(p_1 \wedge p_2 \wedge \cdots \wedge p_n) \Leftrightarrow \neg p_1 \vee \neg p_2 \vee \cdots \vee \neg p_n.$$

Exercise 464. *For each $n \geq 2$, let $q(x_1, x_2, \ldots, x_n)$ be defined inductively by*

$$q(x_1, x_2) = (x_1 \rightarrow x_2),$$

and for $k \geq 2$,

$$q(x_1, x_2, \ldots, x_k, x_{k+1}) = q(x_1, x_2, \ldots, x_k) \wedge (x_k \rightarrow x_{k+1}).$$

Prove that for every $n \geq 2$, $q(x_1, x_2, \ldots, x_n) \rightarrow (x_1 \rightarrow x_n)$.

14.2 Equational logic

Only a comment or two is made here. When L is a set, a partial order on L is a binary relation $\leq$ which is symmetric, antisymmetric, and transitive; in other words, a partial order on L is a relation $\leq$ that satisfies the three axioms

P1: $\forall x \in L,\ x \leq x$;

P2: $\forall x \in L,\ \forall y \in L,\ [x \leq y\ \&\ y \leq x] \Rightarrow x = y$;

P3: $\forall x \in L,\ \forall y \in L, \forall z \in L,\ [x \leq y\ \&\ y \leq z] \Rightarrow x \leq z$.

If one adds either of the least upper bound or greatest lower bound axioms,

P4: $\forall x \in L,\ \forall y \in L,\ \exists s \in L$ such that $[x \leq s\ \&\ y \leq s]$ and $\forall z[x \leq z\ \&\ y \leq z] \Rightarrow s \leq z$ (l.u.b. axiom);

P5: $\forall x \in L,\ \forall y \in L,\ \exists p \in L$ such that $[p \leq x\ \&\ p \leq y]$ and $\forall z[z \leq x\ \&\ z \leq y] \Rightarrow z \leq$ (g.l.b. axiom);

then the partial order is called either a *join semilattice* or a *meet semilattice* (respectively). For each x and y, the elements s and p, if they exist, are unique, enabling one to define an "algebra" with two operations, join $\vee$, where $x \vee y = s$, and meet $\wedge$, where $x \wedge y = p$. If all P1–P5 hold, the structure is called a *lattice*. Such a system of axioms was proposed by Charles Sanders Peirce in the early 1880s, and corrected by Oystein Ore in 1935. There have been various axiom systems for lattices.

Given any algebra, what is the fewest number of axioms that define it? Certain classes of axioms for semilattices can be reduced (by induction) to smaller classes. For more on such results, please see Padmanabhan and Rudeanu's new book *Axioms for Lattices and Boolean Algebras* [423]. In particular, see Theorem 1.1.2, a result due to Padmanabhan and Wolk, and on page 117, where if a finite lattice satisfies a certain axiom system, then any lattice in its variety does, too.

14.3 Well-formed formulae

Consider a propositional language built only on negation and conjunction. Let $A = \{a_1, a_2, a_3, \ldots\}$ be a set of (atomic) sentence symbols. The elements of A are often referred to as *atoms* or *primitive statements*. Define W to be the smallest (see below for meaning of "smallest") set of expressions satisfying the following recursive definition:

(i) $A \subseteq W$, and
(ii) If $p \in W$ and $q \in W$, then both $(\neg p) \in W$ and $(p \wedge q) \in W$.
(To say "smallest" means that if any other set W' satisfies both (i) and (ii), then $W \subset W'$; one could define W as the intersection of all sets satisfying (i) and (ii),

but that definition might introduce more questions than it answers. The definition suggests that W is unique, and the reader may assume this.)

The set W is called a set of *well-formed formulae* in the language. Induction on the length of well-formed formulae can be considered structural induction (see Section 3.7 and Chapter 16 for more on structural induction).

Exercise 465. *Show that any formula $x \in W$ has an even number of parentheses.*

Exercise 466. *Show that the set of all expressions (constructed from atomic sentence symbols with $\neg$ and $\wedge$ and parentheses) with an even number of parenthesis is exactly W.*

Let T_0 denote the tautology, and F_0 denote the contradiction. [Interestingly, the tautology seems to have been introduced by Wittgenstein only in 1921.] Extend the above language to include

$$A' = A \cup \{T_0, F_0\}, \neg, \wedge, \vee, \rightarrow, \leftrightarrow, (,).$$

Let WFF denote the set of well-formed formulae defined as the "smallest" set that satisfies the following recursive definition:

(i') $A' \subseteq WFF$, and

(ii') If $p \in WFF$ and $q \in WFF$, then $(\neg p) \in WFF$, $(p \wedge q) \in WFF$, $(p \vee q) \in WFF$, $(p \rightarrow q) \in WFF$, and $(p \leftrightarrow q) \in WFF$.

Exercise 467. *Show that any well-formed formula $x \in WWF$ has an even number of parentheses.*

For more on parenthesizing, see Section 12.5, where it is shown that the number of ways parentheses can be placed in a non-associative product is a "Catalan number"; in particular, the result applies to repeated occurrences of $\vee$ (or to $\wedge$).

14.4 Language

If Σ is an alphabet, then Σ^* is the set of all finite strings formed by letters from Σ (including λ, the empty string). Subsets of Σ^* are called languages. If C and D are languages, then CD is the language formed by concatenating strings, first one from C, then one from D.

Exercise 468. *Prove that if $A \subseteq B \subseteq \Sigma^*$, then for all $n \geq 1$, $A^n \subseteq B^n$.*

The following exercise deals with a purely combinatorial question about words that has a surprisingly elegant looking solution. Let A be a finite alphabet. For $n \geq 1$, let $(a_1, a_2, \ldots, a_n) \in A^n$ be denoted by simply $a_1 a_2 \cdots a_n$, called a word, or string, of length n (or simply, an n-string). For $m < n$, a word $\mathbf{b} = b_1 b_2 \cdots b_m$ is a substring of $\mathbf{a} = a_1 a_2 \cdots a_n$ if the letters of $\mathbf{b}$ all occur in proper order in $\mathbf{a}$—they need not appear consecutively in **a**. For example, let $A = \{a, b, c\}$, and $\mathbf{s} = cc$. The words in A^3 that contain **s** as a substring are

$$ccc, acc, bcc, cca, ccb, cac, cbc.$$

The result in the following exercise is a general formula for counting the number of words containing some fixed substring was given by Chvátal and Sankoff [109]. [No solution is given.]

Exercise 469. *For any alphabet A with k letters, and any m-string $\mathbf{s} \in A^m$, for each $n \geq m$, define $F(k, n, \mathbf{s})$ to be the number of n-strings in A^n that contain $\mathbf{s}$ as a substring. Prove by induction on n that*

$$F(k, n, \mathbf{s}) = \sum_{i=m}^{n} \binom{n}{j} (k-1)^{n-i}.$$

In particular, this number does not depend on the content in $\mathbf{s}$, only its length.

Note that the example above with $n = 3$, $k = 3$, and $m - 3$, gives the correct answer, 7, as given by Exercise 469. For another exercise on words, see Exercise 373.

For more combinatorics on words, see, *e.g.*, the standard [351].

Chapter 15

Graphs

The time has now come when graph theory should be part of the education of every serious student of mathematics and computer science, both for its own sake and to enhance the appreciation of mathematics as a whole.

—back cover of *Modern Graph Theory* [59]

Perhaps thousands of theorems in graph theory have been proved using induction. The selection given here is rather arbitrary and is a bit sparse, but contains at least a few standard highlights. For references and more examples, the interested reader can consult virtually any modern graph theory text (*e.g.*, see [59], [64], [137], [226], or [566]).

For those not familiar with the language of graph theory, here is a brief introduction.

15.1   Graph theory basics

For a given set S and $k \in \mathbb{Z}^+$, let the collection of k-sets in S be defined by

$$[S]^k = \{T \subset S : |T| = k\}.$$

A *graph* (also called a "simple graph") is a pair (V, E) where V is a set of elements called vertices and $E \subseteq [V]^2$ is a collection of distinct unordered pairs of distinct elements in V; elements of E are called *edges*.

A *multigraph* is a pair $G = (V, E)$ where E is allowed to contain a pair $\{x, y\}$ more than once (called multi-edges) and pairs of the form $\{x, x\}$, called *loops*, are also allowed. Most definitions and problems below apply to multigraphs, however, usually the graph under consideration is simple (no multi-edges, no loops).

A *hypergraph* is an ordered pair $G = (V, E)$ where each *hyperedge* $e \in E$ is a subset of V. If all hyperedges of G contain the same number, say k, of vertices, the hypergraph is called *k-uniform*; 2-uniform hypergraphs are just graphs.

The graph on n vertices with all possible edges is denoted by K_n, a *complete graph*. [Should K_9 be called "the dog graph"?] The notation mK_n denotes m vertex disjoint copies of K_n.

Exercise 470. *Prove that any graph G on n vertices has at most $\binom{n}{2}$ edges.*

If the graph is named G, then write $V = V(G)$ and $E = E(G)$. A graph H is a (weak) subgraph of G if $V(H) \subseteq V(G)$ and $E(H) \subseteq (E(G) \cap [V(H)]^2)$ and is an *induced subgraph* if $E(H) = (E(G) \cap [V(H)]^2)$.

For any vertex $x \in V(G)$, the *degree* of x is the number of edges containing x. (In a multigraph, loops count twice.) The degree of a vertex x in G is denoted by $d_G(x)$ or $\deg_G(x)$, or when clear, simply by $d(x)$ or $\deg(x)$. A vertex x with $\deg(x) = 0$ is said to be an *isolated vertex*. Since each edge of a graph contributes to counting two degrees, one proof of the following is direct:

Lemma 15.1.1 (Handshaking lemma). *For any graph G,*

$$\sum_{x \in V(G)} \deg(x) = 2|E(G)|.$$

(The handshaking lemma is true also for multigraphs, where loops count 2 toward the degree of vertex.)

Exercise 471 (Handshaking lemma). *By induction, prove the handshaking lemma. Hint: inducting on the number of vertices is possible, but inducting on the number of edges is nearly trivial.*

The result in the next exercise is often called the "handshake problem", not to be confused with the handshaking lemma (Lemma 15.1.1).

Exercise 472 (The handshake problem). *At a party with n couples, (including a host and hostess) people shake hands subject to the following conditions: no couple shakes hands; no pair shakes hands more than once; and besides the host, all people shake a different number of hands. Prove that the hostess shakes hands with precisely $n - 1$ people.*

The smallest degree in a graph G is denoted by $\delta(G)$ and the largest degree by $\Delta(G)$.

A *walk* of length m in a graph G is sequence of vertices (not necessarily distinct) $w_0, w_1, w_2, \ldots, w_m$ so that for each $i = 0, \ldots, m - 1$, $(w_i, w_{i+1}) \in E(G)$. A *closed walk* is a walk with $w_m = w_0$; a walk is open if it is not closed. A *trail* is a walk with no edge repeated and a *path* is an (open) walk with no vertex repeated (and hence no edge repeated). A path with k edges (and hence $k + 1$ vertices) is denoted by P_k, and is said to be of length k. [Note: this terminology varies in the literature, as some use P_{k+1} to denote a path with k edges.]

A *leaf* in a graph is a vertex of degree 1; the edge incident to a leaf is called a *pendant edge*.

A *cycle* is a closed walk with no vertex repeated (except the first and last in the walk, of course). A cycle with k vertices is denoted by C_k. A graph with no cycles is called *acyclic*. (Note: in some older texts, the word "circuit" is used to denote a cycle; however, it is now standard that a circuit is a closed trail, that is, a trail with initial and terminal vertex the same.)

The following lemma is useful:

Lemma 15.1.2. *For a graph G, if $\delta(G) = k$, then G contains a path of length k; furthermore, if $k \geq 2$, then G contains a cycle of length $k + 1$.*

Proof: Let $k \geq 1$ and suppose that G is a graph with $\delta(G) = k$. Let P be a longest path in G, between, say, u and v. Since P is maximal, every neighbor of u is in P; since u has at least k neighbors, P has at least k vertices besides u, that is, P has length at least k. If $k \geq 2$, the edge to the farthest neighbor of u in P completes a cycle of length at least $k + 1$. $\qquad\square$

A graph is *connected* if there is a path between every pair of vertices. A *connected component* of a graph is a maximal connected subgraph.

A graph G is called *Eulerian* if there is a closed trail (called an *Eulerian circuit*) containing all the edges of G. That is, a graph is Eulerian iff there exists a closed walk passing through every edge of the graph precisely once (and returning to the original start point). [Note: if G has isolated vertices, and an Eulerian circuit in the remainder of G, then G is still called Eulerian, however, some texts insist that G is connected.]

Exercise 473. *Prove that a graph G is Eulerian iff G is connected (up to isolated vertices) and every vertex has even degree. Hint: use Lemma 15.1.2.*

See material surrounding Exercise 514 for the related result on digraphs.

Exercise 474. *Let G be a graph with n vertices. Prove that if G has at least n edges, then G contains a cycle.*

Exercise 475 (Erdős–Gallai, 1959 [169]). *Let $3 \leq c \leq n$. If G is a graph with more than $(c - 1)(n - 1)/2$ edges, then the length of the longest cycle in G is at least c. Hint: Fix c and induct on n.*

The *distance* between vertices v and w in a connected component of a graph G is the length of the shortest walk between them; this distance is denoted $d_G(v, w)$, or simply $d(v, w)$ when clear. If v and w lie in different components of G, one can say $d(u, v)$ is undefined or is ∞, depending on the application.

Exercise 476. *Let v be a vertex in a connected graph G. Prove that the sum of the distances from v to all other vertices of B, $\sum_{w \in V(G)} d(v, w)$, is at most $\binom{n}{2}$.*

15.2   Trees and forests

A *tree* is a connected acyclic graph. A *forest* is an acyclic graph; so connected components of a forest are trees. The next few lemmata demonstrate some basic properties of trees and forests.

Lemma 15.2.1. *Every tree contains at least one leaf.*

Proof: Let T be a tree. If every vertex has degree at least two, then by Lemma 15.1.2, T contains a cycle, contrary to T being a tree, so there exists at least one vertex with degree either 0 or 1. Since T is connected, there are no isolated vertices, that is, T contains no vertices of degree 0. Hence, T contains at least one vertex of degree 1. $\qquad\square$

Lemma 15.2.2. *Between any two distinct vertices in a tree, there is a unique path.*

Proof outline: If two vertices x and y are connected by two different paths, a cycle results. $\qquad\square$

A *bridge* in a connected graph is an edge whose removal disconnects the graph.

Lemma 15.2.3. *Every edge in a tree is a bridge, and removal of any edge creates two trees.*

Proof: Suppose that T is a tree. By Lemma 15.2.2, for any $x, y \in V(T)$, there is precisely one path joining x and y. If $\{x, y\} \in E(T)$, there is no other $x - y$ path than $\{x, y\}$ itself, hence its removal produces a graph with no $x - y$ path, that is, a disconnected graph. Let $X \subseteq E(T)$ be the set of those vertices connected to x via a path that does not use y, and let Y be those vertices connected to y by a path not going through x. After removing the edge $\{x, y\}$, the graph induced by X is still connected and contains no cycles, so is a tree. Similarly, the vertex set Y induces a tree. $\qquad\square$

The following exercise characterizes trees on n vertices.

Exercise 477. *For a graph on $n \geq 1$ vertices, prove that the following statements A, B, C are equivalent:*

  A: G is connected and acyclic (i.e., G is a tree);

  B: G is connected and has $n - 1$ edges;

  C: G is acyclic and has $n - 1$ edges.

[Hint: The proof of A implies B can be done in at least three ways, each inductive.]

Lemma 15.2.4. *Every tree has at least two leaves.*

Proof: A tree is connected, so if a tree T has more than one vertex, $\delta(T) \geq 1$. For a tree T on n vertices, by Exercises 471 and 477, $\sum_{x \in V(T)} d(x) = 2|E(T)| = 2(n-1)$, and if T were to have fewer than two leaves, then this sum would be at least $2n - 1$. $\qquad\square$

Exercise 478. *Prove that if G is a graph with minimum degree $\delta(G) = k$, then G contains every tree with k edges as a (weak) subgraph.*

The *diameter* of a graph G is the maximum distance between vertices in G, that is, $\mathrm{diam}(G) = \max_{v,w \in V(G)} d(v,w)$. The *eccentricity* of a vertex v, denoted $\epsilon(v)$, is $\max_{w \in V(G)} d(v,w)$. The *radius* of G is the minimum eccentricity of any vertex in G, and the *center* of G is the subgraph induced by those vertices of minimum eccentricity. The result in the next exercise is sometimes called *Jordan's lemma*.

Exercise 479. *Prove that the center of a tree is either a vertex or a single edge. Hint: Use induction on the number of vertices in a tree.*

In 1972, Horn proved [279] three theorems regarding trees. The notation and names might have changed since then, but they can be roughly described as follows. Let T be a tree and let $T_1, \ldots, T_n$ be a collection of subtrees of T.

1. A recursive algorithm is proved to find a set P of vertices in T, of least cardinality, containing at least one point from each T_i.

2. An algorithm is proved to find a collection $T_1', \ldots, T_n'$ of subtrees of T with each $T_i \subset T_i'$ and $T_i' \cap T_{i+1}' \neq \emptyset$ and $\sum_{i=1}^n |V(T_i')|$ is minimized.

3. If each pair of T_i's have a common vertex, then there is a vertex common to all.

Exercise 480. *Which of the above three results by Horn has an inductive proof?*

A *rooted tree* is a tree with one vertex specified as its root. Rooted trees are often drawn with the root at the bottom and branches extending upward with no edges crossing. The *rank* of a vertex in a rooted tree is its distance from the root. In a rooted tree, a *descendant* of a vertex v is a neighbor x of v with $\mathrm{rank}(x) = \mathrm{rank}(v)+1$. The *height* of a rooted tree is the maximum rank of its vertices.

A *plane tree* is a rooted tree together with its drawing having a left-right orientation. In Figure 15.1 are depicted the five plane trees with 3 edges.

Recall that the Catalan numbers are defined in two ways, in Definition 12.5.1 by $C_n = \frac{1}{n+1}\binom{2n}{n}$, and in Definition 12.5.2, by the number of ways to parenthesize a product of $n+1$ elements in a fixed order. The result in the next exercise follows almost directly from Definition 12.5.2, if interpreted properly. A separate proof by induction is also possible.

Figure 15.1: The $C_3 = 5$ plane trees with three edges

Exercise 481. *Prove that the number of (rooted) plane trees with n edges is the Catalan number C_n.*

A *binary tree* is a plane rooted tree where each vertex has at most two descendants, and if there are two descendants, they are designated "left" or "right". (If a vertex has a single descendant, it is not considered left or right.) In Figure 15.1 are depicted the five binary trees with precisely three edges.

A *full binary tree* is one where every non-leaf has precisely two descendants. See Figure 15.2 for two examples of two full plane binary trees of height 3, that without left-right orientation, are otherwise isomorphic.

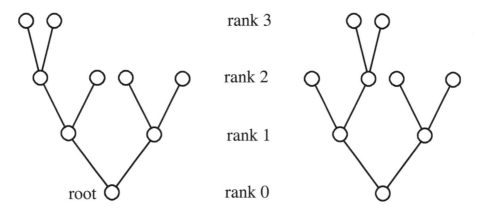

Figure 15.2: Two different full plane binary trees of height 3

The next exercise is a bit more challenging (and is given without solution; see the references given in Section 12.5).

Exercise 482. *Prove that the number of full binary trees with n internal nodes is the Catalan number C_n.*

A plane binary tree is called *complete* if and only if every leaf has rank equal to the height of the tree. [Caution: some authors use "full" to mean "'complete".] So a complete plane binary tree has 1 vertex of rank 0, 2 vertices of rank 1, 4 vertices of rank 2, 8 vertices with rank 3, and so on. It is fairly clear that a full binary tree of height h has $1 + 2 + \cdots + 2^h$ vertices, and by Exercise 47, the total number of vertices is $2^{h+1} - 1$. Another proof is suggested in the next exercise, and although it is nearly trivial, it uses a technique quite common in induction for graphs.

Exercise 483. *Induct on h to show that if T is a complete plane binary tree with height h, then $|V(T)| = 2^{h+1} - 1$.*

Virtually the same proof as for Exercise 483 solves the following (however there is much simpler proof based on the fact that every binary tree is contained in a complete binary tree):

Exercise 484. *Use induction on h to show that if T is a plane binary tree with height h, then $|V(T)| \leq 2^{h+1} - 1$.*

If a rooted tree T has vertices labelled $0, 1, 2 \ldots, n$, where 0 is the root and every path starting at 0 consists of an increasing sequence of vertices, then T is called *increasing*. For example, ignoring left-right orientation, there are six increasing trees on vertices $\{0, 1, 2, 3\}$, as listed in Figure 15.3.

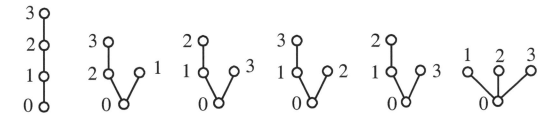

Figure 15.3: The six increasing trees on $\{0, 1, 2, 3\}$

To prove the next exercise, instead of deleting the root as in Exercise 483 (or Exercise 484), one deletes (or attaches) a leaf.

Exercise 485. *Prove by induction that the number of increasing rooted trees on $\{0, 1, \ldots, n\}$ is n!. [Consider two trees with precisely the same edge set to be equal, regardless of what orientation is used to draw them.]*

Recall from Section 12.7.2 that the Eulerian number $E_{n,k}$ is the number of permutations on $\{1, 2, \ldots, n\}$ that have precisely k ascents (adjacent pairs $(\sigma(i), \sigma(i+1))$ with $\sigma(i) < \sigma(i+1)$).

Exercise 486. *Prove that the number of increasing trees on $n + 1$ vertices with precisely k leaves (endpoints) is the Eulerian number $E_{n,k}$.*

See [506, Vol 1, pp. 24–25] for other exercises on increasing trees.

15.3 Minimum spanning trees

Recall that for a graph G and a subgraph $H \subset G$, the subgraph H is said to be a *spanning subgraph* if and only if $|V(G)| = |V(H)|$; in other words, a spanning subgraph contains all of the vertices of G and some of the edges in G. A spanning tree of G is any spanning subgraph that happens to be a tree (or any tree that happens to be a spanning subgraph of G).

Let G be a graph, and let $w : E(G) \to [0, \infty)$ be called a *weight function*, assigning non-negative real numbers to the edges in G. [One can and does also entertain vertex weights, but such is not used here.] For example, one might use a graph G to indicate a certain network of highways between cities; if there is a highway joining two particular cities, the corresponding edge in B might be weighted with mileage, or cost of transport. One might say then that such a graph is weighted by w. The weight of any subgraph of an edge-weighted graph is the total weight of all its edges. In notation, if H is a subgraph of a weighted graph G, define

$$w(H) = \sum_{e \in E(H)} w(e).$$

Given an edge-weighted graph G, a *minimum* spanning tree (MST) (or *minimum weight* spanning tree) of G is a spanning tree T of G with minimum weight, where the minimum is taken over all spanning trees of G. Note that MST's are not, in general unique, but always exist (the set of weights for all spanning trees is finite, and so has a minimum, perhaps not a unique minimum, however). The *minimum spanning tree problem* is to find a MST. See [229] for a history of the problem and [529] for additional references. The MST problem is also known as the *minimum connector problem*. Is there an effective procedure to find a MST? There are two very simple algorithms that produce MSTs called "Prim's algorithm" (due to Prim, [439] in 1957) and "Kruskal's algorithm" (due to Kruskal [330] in 1956). Both algorithms are simple to describe, greedy algorithms (at each step, take the "least", somehow, option available). These algorithms were discovered independently and earlier by other authors, but somehow it is these names that survive.

Briefly, Prim's algorithm "grows" a tree starting at any vertex by attaching recursively available edges of least weight without forming a cycle.

Prim's algorithm:
INPUT: a non-empty graph $G = (V, E)$ on $|V| = n$ vertices with weighted edges (given by a weight function $w : E \to [0, \infty)$).

BASE STEP: Pick $x_1 \in V$ arbitrarily. Set $V_1 = \{x\}$ and G_1 to be the graph consisting of just x.

RECURSIVE STEP: Suppose that a subgraph $G_i = (V_i, E_i)$ has been formed, has been formed by the previous step. Among all edges using having one vertex in V_i, select an edge $\{x, y\}$ (say, with $x \in V_i$) with least weight so that the addition of $\{x, y\}$

does not form a cycle together with edges of G_i (so $y \notin V_i$). Set $V_{i+1} = V_i \cup \{y\}$, $E_{i+1} = E_i \cup \{x, y\}$, and $G_{i+1} = (V_{i+1}, E_{i+1})$.

Terminate the algorithm only when $V_n = V$, that is, after $n - 1$ recursive steps. OUTPUT: $G_n = (V, E_n)$.

Theorem 15.3.1. *The output of Prim's algorithm is a minimum weight spanning tree.*

Exercise 487. *Prove Theorem 15.3.1.*

The other easy algorithm for the MST problem is called "Kruskal's algorithm." Informally, instead of taking the least weight edge attaching to the last tree in the construction, Kruskal's algorithm merely selects the minimum weight edge still available that does not form a cycle. The graph at each stage is not necessarily a tree. In a sense, instead of growing a tree, one is throwing a tree, just throwing in the cheapest edges possible.

Kruskal's algorithm:

INPUT: a non-empty graph $G = (V, E)$ on $|V| = n$ vertices with a weight function $w : E \rightarrow [0, \infty)$, and edges labelled $e_1, e_2, \ldots, e_{|E|}$, so that $w(e_1) \leq w(e_2) \leq \cdots \leq w(e_{|E|})$ weights are increasing order.

BASE STEP: Pick e_1 the least weighted edge, and set G_1 to be the graph consisting of just e_1 on vertex set V.

RECURSIVE STEP: Suppose that a subgraph $G_i = (V, E_i)$ has been formed, has been formed by the previous step. Select an edge $e \in E \setminus E_i$ of least weight that does not does not form a cycle if added to G_i, and set $E_{i+1} = E \cup \{e\}$ and $G_{i+1} = (V, E_{i+1})$.

Terminate the algorithm after $n - 1$ recursive steps. OUTPUT: $G_n = (V, E_n)$.

Exercise 488. *Prove that when Kruskal's algorithm terminates, G_n is a minimum spanning tree. Hint: see the proof of Prim's algorithm (the solution to Exercise 487) for a possible methodology.*

15.4 Connectivity, walks

If a graph G is connected, a set $S \subseteq V(G)$ of vertices is called a *cutset* if the removal of S (and all edges using these vertices) disconnects the graph. A graph is called *k-connected* iff every cutset contains at least k vertices.

Exercise 489 (Whitney's theorem). *Prove that a graph with at least 3 vertices is 2-connected if and only every pair of vertices are connected by two disjoint paths (vertex disjoint, except for endpoints, of course; in other words, every pair of vertices lies on a common cycle).*

For the following exercise, it is assumed that the reader is familiar with matrices and how to multiply them—see Chapter 19 for details. The *adjacency matrix* of a graph G on vertices $v_1, v_2, \ldots, v_n$ is an $n \times n$ 0-1 matrix $A = [a_{ij}]$ with $a_{ij} = 1$ if and only if $(v_i, v_j) \in E(G)$.

Exercise 490. *If $A = [a_{ij}]$ is the adjacency matrix of a graph G on vertices $v_1, \ldots, v_n$, then the (i, j)-entry of A^k is the number of walks of length k from v_i to v_j.*

All binary strings of length n can be viewed as the numbers from 0 to $2^n - 1$ in their binary expansion. Another way to write all such binary strings is the n-fold cartesian product of the set $\{0, 1\}$,

$$B_n = \{0, 1\}^n = \{(\epsilon_0, \epsilon_1, \ldots, \epsilon_{n-1}) : \forall i, \epsilon_i \in \{0, 1\}\}.$$

The n-dimensional unit cube (also called the unit n-cube) has B_n as its vertices, and the graph of the unit cube is defined to be the graph with vertex set B_n, where two vertices are connected by an edge if and only if their binary representations differ in exactly one coordinate. A *hamiltonian cycle* in a graph is a cycle containing all vertices of the graph (each vertex contained precisely once). A *hamiltonian path* is a path containing all vertices precisely once. Compare this definition with that of a Gray code (*cf.* Exercise 564) and see that a hamiltonian path in the n-cube is a Gray code. A graph containing a hamiltonian cycle is called *hamiltonian*.

Exercise 491. *For every $n \geq 2$, prove that the graph of the n-dimensional unit cube is hamiltonian.*

Exercise 492. *For any $n \geq 2$, prove that the graph of the n-dimensional unit cube is n-connected.*

The next theorem, due to G. A. Dirac [143] in 1952, is one of the more central results regarding hamiltonicity. Recall that $\delta(G)$ is the minimum degree of vertices in G.

Theorem 15.4.1 (Dirac's theorem). *Let G be a simple graph on $n \geq 3$ vertices. If $\delta(G) \geq n/2$, then G is hamiltonian.*

Dirac's theorem has at least three simple proofs (see [63, pp. 21–22]), however, Bondy [63, pp. 23] mentions that no inductive proofs are known. It is also mentioned that Woodall constructed a bogus proof to "illustrate the potential pitfalls" of inductive reasoning. [I have not seen Woodall's demonstration, but it can be found in [576], an article with the delightful title "Reductio ad absurdum?".]

Exercise 493. *(Unsolved!) Find an inductive proof of Dirac's theorem.*

15.5 Matchings

A *matching* in a graph G is a collection of edges $M \subseteq E(G)$, so that M is an independent set of edges (*i.e.*, no two edges share a vertex). The empty set can be considered as an "empty matching", though such is seldom used. A *perfect matching* is a matching that uses all vertices (so a graph with a perfect matching must have an even number of vertices). For a definitive work on matching theory, see [356].

Exercise 494. *Prove by induction that the number of perfect matchings in K_{2n} is $\frac{(2n)!}{2^n(n!)}$.*

If $X \subset V(G)$ and $Y \subset V(G)$, where $X \cap Y = \emptyset$, a matching M in G *matches X to Y* if for every $x \in X$, there exists a $y \in Y$ so that $\{x, y\} \in M$. (Such a matching is often called *X-saturated*.)

Recall that for $x \in V(G)$, $N_G(x)$ is the neighborhood of x; when clear, only $N(x)$ is written. For any $S \subset V(G)$, define $N(S) = \cup_{s \in S} N(s)$.

An obvious necessary condition for a bipartite graph $G = (X \cup Y, E)$ to have a matching from X into Y is

$$\forall S \subseteq X, \ |N(S)| \geq |S|. \tag{15.1}$$

It turns out that condition (15.1) is also sufficient! This was given by Philip Hall in 1935 [256] in what is now often called "Hall's marriage theorem". The sets X and Y in the marriage theorem below are often considered as men and women respectively, and an edge between some $x \in X$ and $y \in Y$ can be taken to indicate that x and y find each other acceptable as mates.

Theorem 15.5.1 (Philip Hall's matching theorem). *A bipartite graph $G = (X \cup Y, E)$ has a matching from X into Y if and only if*

$$\forall S \subseteq X, |N(S)| \geq |S|.$$

Exercise 495. *Prove Hall's marriage theorem (Theorem 15.5.1) by inducting on $|X|$.*

Hall's marriage theorem can be interpreted as follows. Let $G = (X, Y, E)$ be a bipartite graph where $|X| \leq |Y|$. Define X to be a set of men and Y to be a set of women, and call a pair $(x, y) \in X \times Y$ *acceptable* if and only if $(x, y) \in E$. An acceptable man-woman pair may be thought of as pair each of whom find the thought of marriage to the other to be "acceptable". Hall's theorem then says that if the pattern of acceptability is appropriate, a perfect matching pairs up all of the men with acceptable partners, what one might call an "acceptable" marriage, although using graph theory vernacular, the marriage is a "perfect" matching when $|X| = |Y|$.

15.6   Stable marriages

As above, let X denote a set of men, and let Y denote a set of women. For now, consider only the case where men and women are equinumerous ($|X| = |Y|$). A perfect matching $M \subseteq X \times Y = \{(m,w) : m \in X, w \in Y\}$ is called a *marriage*, and an edge $(m,w) \in M$ is called a married couple, or a couple married by M. It is convenient to drop the parentheses and comma and denote a married couple by simply mw.

Suppose that each person ranks the members of the opposite sex, and in the case of a tie, breaks it arbitrarily; in other words, each man provides a linear order on the women, and each woman provides a linear order on the men. (So, in contrast to marriages in the last section, all members of the opposite sex are acceptable, just some are "more acceptable" than others.)

How does one decide what marriage is best? What can "best" mean? Consider the following example (which extends an example from [326]), with men m_1, m_2, m_3, women w_1, w_2, w_3, and preference lists:

Man	first	second	third		Woman	first	second	third
m_1	w_2	w_1	w_3		w_1	m_1	m_3	m_2
m_2	w_1	w_3	w_2		w_2	m_3	m_1	m_2
m_3	w_1	w_2	w_3		w_3	m_2	m_1	m_3

If $M_0 = (m_1w_1, m_2w_2, m_3w_3)$ is a marriage, the last two couples might want to divorce and swap partners, because in doing so, each of the four people m_2, m_3, w_2, w_3 would get a mate higher on their list. So in trying to create a "good" marriage, one might want to avoid having two couples, each man preferring the other's wife, and each woman preferring the other's husband. If there exists a pair of couples where each of the two men and two women prefer the mate of the other, one might say that this pair of couples is *unstable*. Some authors (see, *e.g.*, [528]) say that "a marriage is unstable" if and only if there exists an unstable *pair* of couples (and *stable* otherwise). In swapping among an unstable pair of couples, nobody gets hurt. The standard definition of a stable marriage precludes even more:

Definition 15.6.1. For a given marriage M, a man-woman pair $(m,w) \notin M$ is called *unstable* iff m prefers w to his mate in M and w prefers m to her mate in M. A marriage is called *unstable* iff there exists an unstable pair, and *stable* otherwise.

For example, the marriage $\{m_1w_1, m_2w_3, m_3w_2\}$ is stable since each woman gets their first pick, and hence there are no unstable pairs. When two women have the same first choice and two men have the same first choice, another strategy to find a stable marriage is needed.

Can an unstable marriage be made stable by marrying unstable pairs (and marrying the remaining partners)? In each of the following marriages, an unstable pair is identified in boldface, and the subsequent marriage is created by marrying the marked unstable pair:

$$M_1 = \{m_1 w_3, m_2 \mathbf{w_1}, \mathbf{m_3} w_2\}$$
$$M_2 = \{\mathbf{m_1} w_3, m_2 w_2, m_3 \mathbf{w_1}\}$$
$$M_3 = \{\mathbf{m_1} w_1, m_2 \mathbf{w_2}, m_3 w_3\}$$
$$M_4 = \{m_1 \mathbf{w_2}, m_2 w_1, \mathbf{m_3} w_3\}$$
$$M_5 = \{m_1 w_3, m_2 \mathbf{w_1}, \mathbf{m_3} w_2\} = M_1,$$

arriving back at the original marriage M_1.

So it seems that marrying unstable pairs may not be a good strategy to find a stable marriage (recall that a stable marriage is a marriage of all men and women). Is there a strategy to find a stable marriage? The answer to this question is "yes", and the proof is in an algorithm that finds a stable marriage.

Theorem 15.6.2 (Stable marriage theorem). *For any collection of n men and n women, each person with a ranking of all members of the opposite sex, there exists a stable marriage marrying all men and women.*

An algorithm that proves the stable marriage theorem is called the *Gale–Shapley algorithm*, due to David Gale and Lloyd Stowell Shapley in 1962 [204]. First the algorithm is described (often called a "deferred acceptance algorithm"), and the proof that it works is left as an exercise.

Gale–Shapley algorithm: Men propose to women in rounds.

ROUND 1: Each man proposes to the favorite woman on his list. If every woman receives a proposal, then stop, and have each woman accept that proposal; this produces a perfect marriage (and easily seen to be stable, because all men got their first choice). If some woman has not been proposed to, proceed to the next round. If any woman receives one or more proposals, she gives her favorite of these a "maybe" and rejects all others.

Suppose that $j > 1$ and that the algorithm has not terminated in round $j - 1$.

ROUND j: Every man rejected in Round $j - 1$ proposes to the next woman on his list. Again, any woman receiving more than one proposal keeps the highest proposer (among all rounds) as a "maybe" and rejects all others. Any man with "maybe" status can be rejected later if some more preferred man proposes to her on any later round and she upgrades.

At the end of each round, if every woman has received at least one proposal, (in the course of all rounds) terminate the algorithm and marry each woman to her "maybe".

Exercise 496. *Prove that the Gale–Shapley algorithm terminates and produces a stable marriage.*

It might be interesting to note that the Gale–Shapley algorithm can be extended to the case where the number of men and women are different (see, *e.g.*, [380]), and to the case where some men or women have equal preferences for some other women or men. Another (solved) problem asks to find a stable marriage where each person gets a partner as high on his/her list as possible, a "minimum regret" problem.

One hint given in [462, Prob. 24, p. 293] for the next exercise is to use strong induction.

Exercise 497. *Prove that the stable marriage provided by the Gale–Shapley algorithm is optimal for men, that is, in any other stable marriage any man could not get a woman preferable to the one he was paired up with.*

The book *Insights into Game Theory* [241] has an easy-to-read chapter with many examples regarding stable matchings and the Gale–Shapley algorithm. The small (74 pages + *xiii*) book, *Stable Marriages and its Relation to other Combinatorial Problems: an Introduction to the Mathematical Analysis of Algorithms* by Knuth [326] is a much more theoretical examination of the problem, providing many higher level connections to other problems and abstractions (including probability, hashing, the coupon collector problem, the shortest path, and data structures). Knuth's book also contains some key references.

The stable matching problem is also associated with the problem of university acceptance algorithms. Briefly, if certain schools have a quota, they might decide on whom they accept by first ranking them. Similarly, prospective students rank the universities they would like to attend. Some variants of the Gale–Shapley algorithm are presently applied in placement of college students (*e.g.*, the National Resident Matching Program). In fact, the title of the original Gale–Shapley paper is "College admissions and stability of marriage", but a version of their algorithm placing medical students had been in use since 1952. The algorithm is also in use in other countries for student placement (Singapore, for example). For more on placing students, see also [2], [9], [188], [381], and [495, pp. 245–246].

15.7 Graph coloring

A graph G is k-colorable if there is an assignment of k colors to the vertices of G, so that no two vertices connected by an edge receive the same color. For example, the triangle K_3 is 3-colorable but not 2-colorable; in fact, any cycle with an odd number of vertices is 3-colorable but not 2-colorable. The least number of colors c for which a graph G is c-colorable is called the *chromatic number* of G, usually denoted $\chi(G)$. For this next exercise, recall that $\Delta(G)$ is the maximum degree of vertices in G.

Exercise 498. *Show that for any simple graph G, $\chi(G) \leq \Delta(G) + 1$. Hint: Induct on $|V(G)|$, not on $\Delta(G)$.*

A stronger result than Exercise 498 was proved by R. L. Brooks [75] in 1941, namely that if G is neither the complete graph K_{k+1} on $k + 1$ vertices nor a cycle with an odd number of vertices, then $\chi(G) \leq \Delta(G)$. Essentially, the proof of Brooks' theorem is also by induction on $\Delta(G)$, however one needs to separate cases depending on whether or not G has a cut-vertex (a vertex which upon removal disconnects the graph). This proof is slightly beyond the scope of this book, so it is recommended that the reader looks at any almost any text on graph theory, *e.g.*, [566, pp. 197–199].

In the next exercise, the notation $\overline{G}$ denotes the complement of the graph G, that is, $V(G) = V(\overline{G})$ and edges [non-edges] in G are replace by non-edges [edges, resp.] in $\overline{G}$.

Exercise 499. *Prove that for any graph G,*

$$\chi(G) + \chi(\overline{G}) \leq |V(G)| + 1.$$

Hint: Induct on $|V(G)|$.

For a set X and a positive integer r, a coloring $\Delta : S \to [r]$ is said to be onto iff for each $i \in [r]$, $\Delta^{-1}(i) \neq \emptyset$. A coloring of X is said to be a *rainbow coloring* iff for all $s, t \in X$, $s \neq t$ implies $\Delta(s) \neq \Delta(t)$. Rainbow colorings are often called *injective*.

The result in the following exercise appears to be part of folklore:

Exercise 500. *Prove that for $n \geq 3$, if $\Delta : E(K_n) \to [n]$ is an onto n-coloring, then there exists a triangle that is rainbow colored.*

The following theorem is often called the *de Bruijn-Erdős compactness theorem* [80], and is intimately related to model theory.

Theorem 15.7.1. *An infinite graph G is k-colorable iff every finite subgraph of G is also k-colorable.*

Exercise 501. *Prove Theorem 15.7.1 using Zorn's lemma.*

For more exercises regarding chromatic number, see Chapter 21.

15.8 Planar graphs

A graph is called *planar* if it can be drawn in the Euclidean plane with no edges crossing. A *face* is a connected region of the plane not crossing any edges; the outside infinite region is counted as a face.

Although it is somewhat obvious that any tree is planar, one formal proof of this fact is by induction. [Recall that a *plane tree* is a tree together with a planar drawing; this next exercise shows that every tree can be drawn as a plane tree.]

Exercise 502. *Prove that any tree is planar.*

The next theorem might be considered the most important theorem regarding planar graphs and has many proofs.

Exercise 503 (Euler's formula for planar graphs). *If G is a connected planar graph with $v = |V(G)|$ vertices, $e = |E(G)|$ edges, and f faces (in some planar drawing), then*

$$v - e + f = 2 \quad \text{(Euler's formula)}.$$

Note that Euler's formula implies that any planar drawing of a graph has the same number of faces.

Exercise 504. *Show by induction that if a planar graph has n components, then*

$$v - e + f = n + 1.$$

The following basic facts about planar graphs might be helpful in later exercises.

Lemma 15.8.1. *If G is a planar graph on v vertices with e edges, then*

$$e \leq 3v - 6. \tag{15.2}$$

Proof: Let G be a planar graph on v vertices, with e edges and f faces and fix some planar drawing of G. Let k be the number of edge-face incidences, that is, count pairs (edge, face), where the edge forms a border of the corresponding face. Since each edge is adjacent to at most two faces, $k \leq 2e$. On the other hand, since every face has at least 3 edges, $3f \leq k$. Thus

$$3f \leq 2e. \tag{15.3}$$

Multiplying Euler's formula by 3 gives $3v + 3f = 3e + 6$, and so (15.3) yields $3v + 2e \geq 3e + 6$, whence the result follows. $\qquad \Box$

Lemma 15.8.2. *Every planar graph contains a vertex of degree at most 5; that is, if G is planar, then $\delta(G) \leq 5$.*

Proof: Let G be planar; it suffices to consider only when G is connected. If every vertex of G has degree at least 6, then $2e = \sum_{x \in V(G)} \deg(x) \geq 6v$, and so $e \geq 3v$, which violates Lemma 15.8.1. $\qquad \Box$

Using Lemma 15.8.2, the next exercise has an almost immediate solution by induction on the number of vertices of a planar graph.

Exercise 505. *Prove that every planar graph is 6-colorable.*

If $G = (V, E)$ is a planar graph, it has a *dual*, or *planar dual*. If G has faces $F_1, F_2, \ldots, F_f$, then the dual of G, denoted $G^* = (V^*, E^*)$, is defined with vertices $V^* = \{v_1^*, \ldots, v_f^*\}$, and edge set defined by $\{v_i^*, v_j^*\} \in E^*$ if and only if faces F_i and F_j share a common border. Note that G^* might not be simple, that is, it might have multiple edges or loops, but that it is still planar. To color the regions of a map is equivalent to coloring the vertices of its dual.

The next exercise has a proof also by induction on the number of vertices, but is considerably more complicated than the 6-colorable case. It is generally attributed separately to Kempe and Heawood, and is a precursor to the famous "four color conjecture" for map colorings which said that any map with contiguous countries could be colored with four colors so that neighboring countries were colored differently. The four color conjecture is now a theorem, first proved by Appel, Haken, and Koch (see [21], [22] or see [573] for a description of the problem and its history) with a rather difficult proof involving hundreds of cases; it was recently proved again by Neil Robertson, Daniel Sanders, Paul Seymour, and Robin Thomas [457], apparently in a more elegant and efficient fashion; both proofs were accomplished with the aid of computers. Another substantial (but earlier) reference for the four color theorem (4CT) is a book by Oystein Ore [418].

Exercise 506. *Prove that every planar graph is 5-colorable.*

Exercise 507. *Prove that if G is a planar graph, its faces can be properly 2-colored (that is, its dual is 2-colorable) if and only if every vertex in G has even degree.*

15.9   Extremal graph theory

An area of graph theory, called *extremal graph theory*, is concerned with thresholds at which certain substructures or properties occur. For example, if a graph on n vertices has n edges, the graph contains a cycle, and there are examples (trees) that have one less edge and no cycle. So the threshold for cycles is n edges.

In the following exercise, the graph $K_1 + K_{n-1}$ is an n-vertex graph with two components, an isolated vertex and a complete graph.

Exercise 508. *Prove that the maximum number of edges in a disconnected simple n-vertex graph is $\binom{n-1}{2}$, with equality only for $K_1 + K_{n-1}$.*

The following exercise is usually one of the first studied in extremal graph theory.

Exercise 509 (Mantel's theorem [362]). *If a simple graph G on $n \geq 3$ vertices has more than $\frac{n^2}{4}$ edges, then G contains a triangle.*

The energetic reader might notice that in fact two triangles are formed under the condition of Exercise 509, (try to prove it!). The addition of one more edge actually forces two triangles tied together:

Exercise 510. *Prove that if a simple graph G on $n \geq 5$ vertices has more than $\frac{n^2}{4} + 1$ edges, then G contains two triangles joined at a single vertex. Hint: an inductive proof has one case that requires special treatment.*

Exercise 511. *Prove that a graph on n vertices with more than $\lfloor n^2/3 \rfloor$ edges contains the graph of a tetrahedron (a K_4).*

Both Mantel's theorem (Exercise 509) and the result in Exercise 511 are special cases of a much stronger theorem proved by Paul Turán (1910–1976). [Some say that he proved this while in a concentration camp, although I can not recall the source for this story.] Turán was also known for his work in number theory and analysis. To state this theorem, a few definitions help.

For positive integers n, k, let $T(n,k)$ be the complete k-partite graph on n vertices whose partite sets have sizes that are as nearly equal as possible. Denote the number of edges in $T(n,k)$ by $|E(T(n,k))| = t(n,k)$. If $n = qk + r$, where q and r are non-negative integers with $0 \leq r < k$, then r of the partite sets in $T(n,k)$ have $q + 1 = \lceil n/k \rceil$ vertices, and the remaining $k - r$ have $q = \lfloor n/k \rfloor$ vertices (see Figure 15.4).

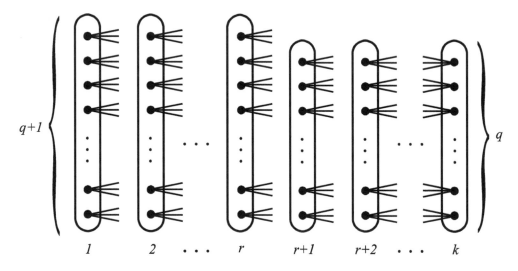

Figure 15.4: The Turan graph $T(n,k)$

Hence

$$t(n,k) = \binom{r}{2}(q+1)^2 + r(k-r)(q+1)q + \binom{k-r}{2}q^2.$$

Theorem 15.9.1 (Turán [536, 537]). *Let $k \geq 1$. Then $ex(n; K_{k+1}) = t(n,k)$ and $T(n,k)$ is the unique extremal K_{k+1}-free graph.*

Exercise 512. *Prove Turán's theorem by induction on n. Hint: One proof has an inductive step that leaps from m to $m + k$ vertices.*

Erdős [165] has related results for certain multipartite hypergraphs. The following theorem is due to Erdős [166]:

Theorem 15.9.2. *Let $G = (V, E)$ be a graph containing no K_{r+1}. Then there exists an r-partite graph H on vertex set V so that for each $z \in V$, $d_G(z) \leq d_H(z)$. If G is not a complete r-partite graph, then there exists at least one z for which $d_G(z) < d_H(z)$.*

Exercise 513. *Prove Theorem 15.9.2 by induction on r.*

The observant reader may notice that Turán's theorem (Theorem 15.9.1) follows from Theorem 15.9.2.

15.10 Digraphs and tournaments

The term *digraph* denotes a graph with directed edges. To be precise, a digraph $D = (V, E)$ is a set V of vertices, and a collection E of *ordered pairs* from V. Elements of E are called *directed edges*, or sometimes *arcs*. Directed edges are usually drawn with an arrow; if $(x, y) \in E$, the arrow usually goes from x to y, in which case say that x *dominates* y.

A digraph D is called *simple* iff for every $x, y \subset V$, precisely one of (x, y) or (y, x) is in E, any ordered pair occurs as a directed edge at most once, and there are no loops, that is, arcs of the form (x, x). So a simple digraph D can be obtained from a simple graph $G = (V, E)$ by simply *orienting* each edge in E, that is, for each $\{x, y\} \in E$, choosing either (x, y) or (y, x) to be a directed edge in D. Another way to say this is that D is an orientation of some graph G. A path in a digraph must follow the arrows, so one might say, for example, that there is a directed path from v_4 to v_1, but there might not be one from v_1 to v_4.

If $x \in V$ is a vertex of some digraph D, then the outdegree of x, denoted $d^+(x)$, is the number of edges of the form (x, z), that is, in the case of a simple digraph,

$$d^+(x) = |\{z \in V : (x, z) \in E(D)\}|.$$

Similarly define the *indegree* $d^-(x)$.

A digraph D is called *Eulerian* if there is a closed (directed) trail (called an *Eulerian circuit*) containing all the edges of D. The following version of Euler's theorem was found by Good [221] in 1946.

Exercise 514. *Prove that a digraph is Eulerian if and only if the underlying (undirected) graph has one non-trivial component and for each vertex v, $d^+(v) = d^-(v)$. Hint: If D is a digraph with minimum outdegree at least 1, then D contains a directed cycle; then use strong induction on the number of directed edges.*

In graph theory, a *tournament* is an orientation of a complete graph, that is, a tournament T on vertex set V is a collection $E(T) \subset V \times V$ of ordered pairs so that for every pair of distinct vertices $a, b \in V$, precisely one of $(a, b) \in E(T)$ or $(b, a) \in E(T)$ holds. The reason for the name is that if in a real (round-robin) tournament with n players where every player meets every other in a match precisely once, the winner of each match can be recorded by an appropriate arrow.

If T is a tournament, a fixed vertex $v \in V(T)$ is a *king* iff for every other vertex $y \in V(T) \backslash \{v\}$, there is a directed path from v to y with at most two edges. So a king can be thought of as a person in a tournament so that for every other player, the king either beat that player, or beat someone who beat that player. There are at least three proofs of the result in the following exercise, two of which are by induction.

Exercise 515. *Prove that every tournament on finitely many vertices has a king.*

Exercise 516. *Let $p_1, p_2, \ldots, p_n$ be integers with $0 \le p_1 \le p_2 \le \cdots \le p_n$, and for each $k = 1, \ldots, n$, denote the partial sums by $s_k = \sum_{i=1}^{k} p_i$. Prove that there exists a tournament with outdegrees $p_1, \ldots, p_n$ if and only if for each $k < n$, $s_k \ge \binom{k}{2}$ and $s_n = \binom{n}{2}$. Hint: Use induction on $\sum_{k=1}^{n} \left(s_k - \binom{k}{2} \right)$.*

The result in the following exercise is known as *Redei's theorem*, published by L. Redei in 1934 [449].

Exercise 517. *A tournament among n players is held, where every pair meets. Prove that there is a listing of all the players $a, b, c, \ldots$, so that a beats b, b beats c, and so on, continuing until the last player.*

In graph theory parlance, Exercise 517 says that every finite tournament has a directed hamiltonian path; such a listing of players is also called a *ranking*.

Exercise 518. *Prove that for any $n \ge 3$, if a tournament has a (directed) cycle on n vertices, then it contains a directed cycle on three vertices.*

A related exercise occurred as a Putnam problem in 1958, with inductive solution published in the *Monthly* [90]:

Exercise 519. *For $n \ge 3$, let T be a tournament on vertices $x_1, \ldots, x_n$ and for each $i = 1, \ldots, n$, let $d^+(x_i)$ denote the outdegree of x_i (the number of players that x_i beat). Prove that there exists a directed cycle on three vertices if and only if*

$$\sum_{i=1}^{n} (d^+(x_i))^2 < \frac{(n-1)n(2n-1)}{6}.$$

15.11   Geometric graphs

This short section is an introduction to an area of combinatorial geometry that concerns graphs, each graph considered together with a drawing of that graph in the plane. For many more combinatorial geometry problems with solutions by induction (along with countless other results that might seem beautiful and amazing), see, *e.g.*, the comprehensive textbooks [372] or [421].

A *geometric graph* is a collection of points in the plane in general position together with line segments (edges) drawn between pairs of points; a geometric graph on n points is complete if each of the $\binom{n}{2}$ pairs of points is joined by an edge. Note that because the points are in general position (no three on a line), each edge contains only two of the given points. More simply put, a geometric graph is a graph that has been drawn in the plane so that the structure is completely visible.

The results in the next two exercises were published with elegant solutions by Károlyi, Pach, and Tóth in 1997 [304, Thm. 1.1]. Analogous results for convex patterns of points were known by Bialostocki and Dierker, and appeared in an exercise in *Combinatorial Geometry* [421, 14.9, p. 240] by Pach and Agarwal. (See also [304] for more details.) The results for ordinary graphs analogous to those in the next two exercises were known some time earlier (again, see [304] for details), the first being trivial.

Recall that a tree is a connected acyclic graph; a *plane tree* is a tree drawn in the plane with no crossing edges.

Exercise 520. *Prove by strong induction on $n \geq 2$ that if the edges of a complete geometric graph on n vertices are 2-colored, there exists a plane spanning tree that is monochromatic.*

The result of the next exercise was also proved by induction in [304]; no solution is given here, but the proof idea is similar to that from Exercise 520.

Exercise 521. *Prove that if the edges of a complete geometric graph on $3n - 1$ vertices are 2-colored, there exist n pairwise disjoint edges of the same color.*

Chapter 16

Recursion and algorithms

recursive *adj. See* RECURSIVE.

—Stan Kelly-Bootie,

the Devil's DP Dictionary.

Many common operations in mathematics are defined recursively, that is, some initial values are given, and then some rule is repeated successively. For example, in Definition 2.5.9, the product of n numbers is defined by successively multiplying a number by the previous product.

In a proof by mathematical induction, suppose that some base step $S(1)$ is proved, and for any $k \geq 1$, the induction step $S(k) \to S(k+1)$ is also proved. The induction step can be thought of as a rule that generates a true $S(k+1)$ from a true $S(k)$. In the inductive step, the proof of the statement $S(k+1)$ usually depends, in a critical way, on the truth of some previous statements, either the base case, or from the output of some previous application of the rule. Then for any large n, the truth of $S(n)$ follows by beginning at $S(1)$, and applying the rule $n-1$ times. Applying the rule just once might be called "inducting" and the application of the rule many times might be called "induction". Analogously, recursion is the repeated application of a recurrence relation.

In this chapter, only some elementary aspects of recursion are considered. For more on pure recursion theory, as applied to model theory and language, including the theory of recursive invariance, recursively enumerable sets, recursive functions and decidability questions, see, *e.g.*, [460].

Many exercises elsewhere in this text could also fall under the heading "recursion"; for example, see Exercises 122, and 123 (trig identities), Exercise 427 (inclusion-exclusion), Exercises 595, 596 (counting functions), and Exercise 719 (counting triangles).

16.1   Recursively defined operations

Many operations are really defined recursively from binary operations. For example, exponentiation can be defined recursively by defining $x^2 = x \cdot x$ and for $n \geq 3$, $x^n = (x^{n-1}) \cdot x$. Another way to define exponentiation is to first show that multiplication is associative and thus an expression of the form $x_1 \cdot x_2 \cdots x_n$ is meaningful; proving associativity can often be done by induction (see Exercise 4 for associativity for multiplication of natural numbers). The definition of $n!$ can be given in two ways. One way is $n! = \prod_{i=1}^{n} i$, and another way is recursively: $0! = 1$, and for $n \geq 1$, $n! = n(n-1)!$. Keeping in mind this recursive nature of many definitions, one can prove results extending those at the end of Section 2.3. Here are a couple (presented without solution).

Exercise 522. *For $n \geq 3$, $1 \leq r < n$, and real numbers $x_1, x_2, \ldots, x_r, x_{r+1}, \ldots, x_n$, prove that*

$$(x_1 + x_2 + \cdots + x_r) + (x_{r+1} + \cdots + x_n) = x_1 + x_2 + \cdots + x_r + x_{r+1} + \cdots + x_n.$$

Exercise 523. *For $n \geq 3$, $1 \leq r < n$, and real numbers $x_1, x_2, \ldots, x_r, x_{r+1}, \ldots, x_n$, prove that*

$$(x_1 x_2 \cdots x_r)(x_{r+1} \cdots x_n) = x_1 x_2 \cdots x_r x_{r+1} \cdots x_n.$$

16.2   Recursively defined sets

This book is filled with recursively defined sets, elements of which may be, for example, numbers, vectors, functions, geometric shapes, edges (in the graph sense) or words. It doesn't seem practical to even attempt to list the various examples here. For more on recursively defined sets (and induction), see, *e.g.*, [464] or [462]. Some recursively defined structures are discussed later in this chapter. In the following Section 16.3, recursively defined sequences are examined in more detail.

Instead, here are slightly different examples of recursively defined sets and one sequence. In the first example, one can not enumerate the set nor ascertain too many properties of the set:

Define the set A recursively by

(i) $1 \in A$,

(ii) if $n \in A$, then $2n \in A$, and

(iii) if n is odd and $3n + 1 \in A$ then $n \in A$.

Question 16.2.1. *Is A the set of positive integers?*

As of this writing, Question 16.2.1 is still notoriously unsolved (see, *e.g.*, [464, p. 395]). Question 16.2.1 is more famously known as the "$3n + 1$ problem" (or the

Collatz conjecture, the Ulam conjecture, or....), but is often stated in terms of sequences, as follows:

To test whether or not some fixed $n \in \mathbb{Z}^+$ is in A, recursively (or inductively) define the sequence of positive integers $s_1, s_2, s_3, \ldots$ by setting $n = s_1$, and for $k \geq 1$, having defined s_k, define

$$s_{k+1} = \begin{cases} \frac{s_k}{2} & \text{if } s_k \text{ is even,} \\ 3s_k + 1 & \text{if } s_k \text{ is odd.} \end{cases}$$

For example, starting with $n = 9$, the sequence is

$$9, 28, 14, 7, 22, 11, 34, 17, 52, 26, 13, 40, 20, 10, 5, 16, 8, 4, 2, 1, 4, 2, 1, 4, 2, 1, \ldots$$

Since by (i), $1 \in A$, the above sequence shows that $9 \in A$. So the $3n + 1$ problem is to show whether or not for every starting number n, the corresponding sequence eventually hits 1 (in finitely many steps). There is an immense amount of literature on this problem; one starting point may be, for example, [335]. [I think that the preponderance of opinions is that Question 16.2.1 has an affirmative answer.]

This next example might be of interest to Ramsey theorists or combinatorial number theorists.

Exercise 524. *Define sets S_n and T_n as follows: Let $S_1 = \{1\} = T_1$. For $n > 1$, recursively define*

$$\begin{aligned} S_n &= \{3^{n-2} + x : x \in T_{n-1}\}, \\ T_n &= S_n \cup T_{n-1}. \end{aligned}$$

Prove that T_n does not contain any 3-term arithmetic progression and that $|T_n| = 2^{n-1}$, with the largest element in T_n being $(3^{n-1} + 1)/2$.

16.3 Recursively defined sequences

A sequence $s_0, s_1, s_2, s_3, \ldots$, perhaps infinite, is called *recursively defined* if and only if for some $i \geq 1$, some initial values, say $s_0, s_1, s_2, \ldots, s_{i-1}$ are defined individually, and for every $n \geq i$, s_n is defined by a rule which depends only on previous values $s_0, s_1, \ldots, s_{n-1}$. This "rule" can be viewed as a function of previous values, sometimes depending on more and more inputs, and is sometimes called a *recurrence relation* (or simply, a *recurrence*) and the process of executing the recurrence relation repeatedly is called *recursion*; however, it is common to use these expressions interchangeably.

Note: Doubly indexed sequences like the Stirling numbers of the second kind (and many others in this book, too many to list) also can be defined recursively (see equation (12.7)), however, only linearly ordered sequences are discussed here.

For example, consider the sequence of positive integers $a_0, a_1, a_2, a_3, \ldots$ defined by $a_0 = 2$, and having defined $a_0, \ldots, a_n$, let a_{n+1} be the product $a_0 \cdot a_1 \cdot a_2 \cdots \cdots a_n$. One can work out the first few terms:

$$a_0 = 2, a_1 = a_0 = 2, a_2 = a_0 a_1 = 2 \cdot 2 = 4, a_3 = a_0 a_1 a_2 = 2 \cdot 2 \cdot 4 = 16.$$

Is there a simple general formula for a_n that depends only on n? Working out a few more terms,

$$a_4 = a_0 a_1 a_2 a_3 = 2^1 \cdot 2^1 \cdot 2^1 \cdot 2^2 \cdot 2^4 = 2^{1+1+2+4} = 2^8,$$

and

$$a_5 = a_0 a_1 a_2 a_3 a_4 = \cdot 2^1 \cdot 2^1 \cdot 2^2 \cdot 2^4 \cdot 2^8 = 2^{1+1+2+4+8} = 2^{16}.$$

It seems like for $n \geq 1$ a general formula for a_n might be

$$a_n = 2^{2^{n-1}}, \tag{16.1}$$

which has been verified for $n = 1, 2, 3, 4, 5$.

Exercise 525. *Use Exercise 47 to give a simple inductive proof that in the sequence defined above, for each $n \geq 1$, equation (16.1) is true.*

Is there an expression for a_n that applies also for $n = 0$? If so, it is not clear how to find such an expression.

The above formula was found by "inspection"; are there formal rules about how to find such a formula (provided one exists)? In general, recursively defined sequences do not necessarily have a "closed form", a general formula for a_n that uses finitely many simple operations, including addition, multiplication, exponentiation, and inverses. (See Section 18.5 for discussion of *primitive* recursive functions, ones that can be written in closed form). For some recurrences (for example, see the Ackermann function, Section 18.6), the only "adequate" way to describe the sequence is by the recursion itself. Most recurrence relations here, however, yield a closed form for the nth term in the resulting sequence.

Given some initial values and some recurrence, finding a general formula for the nth term of the sequence is called "solving the recurrence". For some functions sometimes an exact value for each term in a recursively defined sequence is not required, but only some loose bound or approximation. This is especially true in the study of *complexity* (the study of run-times of algorithms; see Section 16.6 below).

Just as in the theory of differential equations , or difference equations, there is a variety of specialized techniques to solve various kinds of recurrences. [I am told that many of the ways to solve recurrences came directly from differential equations.]

In many of the problems below, the recurrence has been solved and it only remains to verify the solution by induction. In general, mathematical induction is well suited to verifying solutions to recursion, but is virtually useless in solving recursions (except for proving parts of some general methods for solving them). The general theory of how each was solved is not included here (see nearly any book on combinatorics).

In the recursion given above for a_n, for terms later in the sequence, an increasing number of previous terms are used. This chapter focusses on recursions where some maximum number of previous terms is required.

Perhaps some of the easiest recurrences involve a dependence on only the one previous term. For example, if $P = P_0$, and for $n \geq 1$, define $P_n = (1.07)P_{n-1}$, after executing the recurrence a couple of times, the solution is clear, as in this next exercise.

Exercise 526. *Let r denote an interest rate expressed as a decimal (e.g., $7\% = .07$). If an initial principle P is deposited in an account that pays interest r compounded annually, prove by mathematical induction that the amount $A(n)$ in the account at the end of $n \geq 0$ years is*

$$A(n) = P(1+r)^n.$$

A recursively defined sequence $\{x_i\}_{i=1}^{\infty}$ is called *recursive of order p* if for some $p \geq 1$, the initial values $x_1, x_2, x_3, \ldots, x_p$ are given, and there exists function $f : \mathbb{R}^p \to \mathbb{R}$ of p variables (that uses all variables) so that for each $n > p$, the recurrence relation (or briefly, the recurrence)

$$x_n = f(x_{n-p}, \ldots, x_{n-1})$$

holds; one also defines such a recurrence relation to be of order p. [Notice that, for convenience of exposition, the sequence $x_1, x_2, \ldots$ started with index 1, whereas in many situations, calculations are easier if the sequence begins with index 0.]

16.3.1 Linear homogeneous recurrences of order 2

A recurrence f of order p is called *linear* (with constant coefficients) if and only if there exist non-zero constants $c_0, c_1, \ldots, c_p$ so that

$$f(y_1, y_2, \ldots, y_p) = c_0 + c_1 y_1 + c_2 y_2 + \cdots + c_p y_p;$$

A linear recurrence of finite order is said to be *homogeneous* if $c_0 = 0$.

For example, since Fibonacci numbers are defined by $F_n = F_{n-2} + F_{n-1}$, the recursion is of order 2, linear (with both $c_1 = c_2 = 1$), and homogeneous. Similarly, the recursion behind the Lucas numbers is also linear homogeneous of order 2. A recurrence relation of order 1 given by $s_n = s_{n-1} + 3^n$ is not linear because 3^n is not a constant. Catalan numbers can be defined by the recursion $C_{n+1} = C_0 C_n + C_1 C_{n-1} + \cdots C_n C_0$, which is not linear. The recurrence (12.3) for Schröder numbers is also not linear.

Questions regarding recurrences can often be answered without solving the recursion.

Exercise 527. *Define the sequence $a_0, a_1, a_2, \ldots,$ by $a_1 = a_2 = 1$, and for $n \geq 2$, define $a_{n+1} = 3a_n + a_{n-1}$. Prove by induction that for each $n = 1, 2, 3, \ldots,$ $\gcd(a_n, a_{n+1}) = 1$, that is, consecutive elements are relatively prime.*

16.3.2 Method of characteristic roots

Linear homogeneous recurrences are "easy to solve", that is, there is a method, called the "method of characteristic roots" (a simple case of which is shown below), by which a closed form for the nth term can be found. Any closed form that arises from this method can usually be verified by induction rather easily, since the given recurrence is the key to the inductive step. Thus choosing arbitrary linear homogeneous recurrences of order 2 provides an endless source of exercises of the form "here is a recurrence, and a general form for the nth term; prove this form by induction."

The method of characteristic roots also has analogues for many higher order linear homogeneous recursions, however only the case $p = 2$ is discussed here in detail. Also, generating functions can be used to solve linear homogeneous recurrences of higher order, but these methods are not presented here. See many texts on combinatorics for these additional techniques and results (*e.g.*, [455]).

Suppose that $a_0, a_1, a_2, \ldots$ is a recursively defined sequence, where initial values a_0 and a_1 are defined, and for non-zero constants c_1 and c_2, consider the linear homogeneous recurrence of order 2 given by, for $n \geq 2$,

$$a_n = c_1 a_{n-2} + c_2 a_{n-1}. \tag{16.2}$$

The trick is to first "guess" that there may be a number $x \in R$ so that the sequence $1, x, x^2, x^3, \ldots$ satisfies the recurrence (16.2), ignoring the initial values. Such a sequence is called a *general solution* to the recurrence. For the moment, suppose that such an x exists. Then for any $n > 2$,

$$x^n = c_1 x^{n-2} + c_2 x^{n-1}. \tag{16.3}$$

Further assuming that $x \neq 0$, cancelling x^{n-2} from each side of the last equation gives

$$x^2 = c_1 + c_2 x,$$

or

$$x^2 - c_2 x - c_1 = 0. \tag{16.4}$$

Equation (16.4) is called the *characteristic equation* for the recurrence (16.2), and roots of this equation are called *characteristic roots*. [The case for $p > 2$ is handled similarly.] By the quadratic equation, the characteristic equation has two (perhaps complex) roots, say α and β, not necessarily distinct.

Some critical observations enable the next step in the method, the first few of which are easy to confirm.

Observation 1: If α is a root of the characteristic equation (16.4), then the sequence $1, \alpha, \alpha^2, \alpha^3, \ldots$ is a solution to the recurrence (16.2). So both sequences $\{\alpha^n\}_{n=0}^{\infty}$ and $\{\beta^n\}_{n=0}^{\infty}$ satisfy the recurrence.

Observation 2: If two sequences $\{s_n\}_{n=0}^{\infty}$ and $\{t_n\}_{n=0}^{\infty}$ both satisfy the recurrence (16.2), then the sequence $\{s_n + t_n\}_{n=0}^{\infty}$ also does.

Observation 3: If a sequence $\{s_i\}_{i=0}^{\infty}$ satisfies (16.2), then for any constant $A \in \mathbb{C}$, so does the sequence $\{As_i\}_{i=1}^{\infty}$.

Combining Observations 1, 2, and 3:

Observation 4: For any (constant) complex numbers A and B, a general solution to the recurrence (16.2) is $\{A\alpha^n + B\beta^n\}_{n=0}^{\infty}$.

Can one find suitable A and B so that the initial values a_1 and a_2 are the first two terms of the above sequence? The answer depends on α and β (and hence on c_1 and c_2). In some cases, one more observation is required:

Observation 5: Let α be a root of (16.4). Then the sequence $\{n\alpha^n\}_{n=0}^{\infty}$ is a solution to the recurrence (16.2).

Proof of Observation 5: There are many ways to show this; however, the simplest way might be just to take derivatives (with respect to x) of each side of (16.3), yielding

$$nx^{n-1} = c_1(n-2)x^{n-3} + c_2(n-1)x^{n-2},$$

and then multiplication through by x gives

$$nx^n = c_1(n-2)x^{n-2} + c_2(n-1)x^{n-1},$$

from which Observation 5 now follows directly. $\qquad\square$

The method now breaks down into cases:

Case 1: Let $\alpha \neq \beta$ be real roots of (16.4). Replacing $n = 0$ and $n = 1$ respectively in $a_n = \{A\alpha^n + B\beta^n\}_{n=0}^{\infty}$. gives $a_1 = A + B$ and $a_2 = A\alpha + B\beta$, two equations in A and B that have solutions since $\alpha \neq \beta$. Thus for each $n \geq 0$, $a_n = A\alpha^n + B\beta^n$ is a specific solution to the recurrence that agrees also with the initial values.

Case 2: Let $\alpha = \beta$ be a repeated root of (16.4). Then it suffices to find A such that for each $n \geq 0$, $a_n = A\alpha^n$. Using the two initial values a_0 and a_1, seek A such that both $a_0 = A$ and $a_1 = A\alpha$. Unless $\alpha = \frac{a_1}{a_0}$, these equations do not have a simultaneous solution for A. What if there is no such A?

Combining the Observations 1, 2, 3, and 5, if α is a solution to (16.3), then for any constants A and B, the sequence $\{A\alpha^n + Bn\alpha^n\}_{n=0}^{\infty}$ is also a general solution to (16.3). To find the particular solution, use $n = 0$ and $n = 1$ respectively in

$$a_n = A\alpha^n + Bn\alpha^n,$$

get $a_0 = A$, and $a_1 = A\alpha + B\alpha$, in which case one can solve for $B = \frac{a_1}{\alpha} - a_0$.

Case 3: Let α and β be complex numbers (where $\alpha \neq \beta$). This case proceeds nearly identically to that of Case 1, however with some extra work at the end. Polar coordinates and DeMoivre's formula are used.

From the quadratic formula, it follows that these two roots are complex conjugates (that is, if $\alpha = c + di$, then $\beta = c - di$). Recall that $|\alpha| = \sqrt{c^2 + d^2} = |\beta|$. Let $\theta \in [0, 2\pi)$ be such that $\cos\theta = c$ and $\sin\theta = d$. Then $\alpha = |\alpha|(\cos\theta + i\sin\theta)$ and $\beta = |\alpha|(\cos\theta - i\sin\theta)$.

Continuing from Case 1, seek A and B so that for every $n \geq 0$, the particular solution to the recurrence is

$$a_n = A\alpha^n + B\beta^n,$$

the difference now is that A and B may be complex. Replacing the above expressions for α and β, and using DeMoivre's theorem (see Exercise 115) from the second to third line below:

$$
\begin{aligned}
a_n &= A(|\alpha|(\cos\theta + i\sin\theta))^n + B(|\alpha|(\cos\theta - i\sin\theta))^n \\
&= |\alpha|^n[A(\cos\theta + i\sin\theta)^n + B(\cos\theta - i\sin\theta)^n] \\
&= |\alpha|^n[A(\cos(n\theta) + i\sin(n\theta)) + B(\cos(n\theta) - i\sin(n\theta))] \\
&= |\alpha|^n[A(\cos(n\theta) + i\sin(n\theta)) + B(\cos(n\theta) - i\sin(n\theta))] \\
&= |\alpha|^n[(A + B)(\cos(n\theta) + i(A - B)\sin(n\theta)].
\end{aligned}
$$

Using $n = 0$, $n = 1$, and the initial values a_0, a_1 respectively,

$$
\begin{aligned}
a_0 &= A + B; \\
a_1 &= |\alpha|[(A + B)\cos(\theta) + i(A - B)\sin(\theta).]
\end{aligned}
$$

Reducing the equation for a_1 by

$$a_1 = c(A + B) + id(A - B) = (c + id)A + (c - id)B = \alpha A + \beta B,$$

and since $\alpha \neq \beta$ and each is non-zero, such a solution for A and B exists. In fact, the solutions are found by simple reduction to be $B = \frac{a_1 - a_0\alpha}{\beta - \alpha}$ and $A = a_0 - B$. Having found the desired A and B, the solution to the recurrence is complete, and is $a_n = A\alpha^n + B\beta^n$. In the more convenient form above, the solution is

$$a_n = |\alpha|^n[(A + B)(\cos(n\theta) + i(A - B)\sin(n\theta)],$$

where $A + B = a_0$ and $A - B = a_0 - 2(\frac{a_1 - a_0\alpha}{\beta - \alpha})$. $\qquad\square$

If one works out the various solutions for A and B needed above, the summary of outcomes of the method of characteristic roots can be expressed in a three part theorem:

Theorem 16.3.1 (2nd order linear homogeneous recurrence solutions). *Let a recursively defined sequence $a_0, a_1, a_2, a_3, \ldots$ have initial values a_0 and a_1 given, and for non-zero constants c_1 and c_2, for $n \geq 2$, let $a_n = c_1 a_{n-2} + c_2 a_{n-1}$. Suppose that α and β are the roots of the associated characteristic polynomial.*

1. *If α and β are distinct and real, a general solution is of the form $a_n = A\alpha^n + B\beta^n$. To find A and B, solve the system of two equations found for $n = 0$ and $n = 1$ (and the two initial values), giving*

$$A = a_0 - \frac{\alpha}{\beta - \alpha}\left(\frac{a_1}{a_0} - a_0\right), \quad B = \frac{\alpha}{\beta - \alpha}\left(\frac{a_1}{a_0} - a_0\right).$$

2. *If $\alpha = \beta$, then the solution is $a_n = a_0\alpha^n + \left(\frac{a_1}{\alpha} - a_0\right)n\alpha^n$.*

3. *If α and β are distinct and complex, a general solution is still of the form $a_n = A\alpha^n + B\beta^n$, where A and B are described in Case 1. If $\alpha = (|\alpha|, \theta)$ in polar coordinates, an easier form of the solution to work with is*

$$a_n = |\alpha|^n[a_0(\cos(n\theta) + iL\sin(n\theta)],$$

where $L = a_0 - 2\left(\frac{a_1 - a_0\alpha}{\beta - \alpha}\right)$.

16.3.3   Applying the method of characteristic roots

Exercise 528. *Recursively define a sequence by $a_0 = 3$, $a_1 = 3$, and for $n \geq 2$, $a_n = 2a_{n-2} + a_{n-1}$. First prove by induction that every a_i is odd. Use the method of characteristic roots to solve this recursion, and then prove the result by induction.*

Exercise 529. *The recursive definition of the Fibonacci numbers is $F_0 = 0$, $F_1 = 1$, and for $n \geq 1$, $F_n = F_{n-2} + F_{n-1}$. Solve this recursion and obtain Binet's formula for F_n,*

$$F_n = \frac{1}{\sqrt{5}}\left[\left(\frac{1 + \sqrt{5}}{2}\right)^n - \left(\frac{1 - \sqrt{5}}{2}\right)^n\right].$$

(which was to be proved by induction in Exercise 369).

Exercise 530. *Solve the recurrence given by $a_0 = 2$, $a_1 = 5$, and for $n \geq 2$, $a_n = -16a_{n-2} - 8a_{n-1}$.*

Exercise 531. : *Let $a_0 = 1$, $a_1 = 2$, and for $n \geq 3$, define $a_n = -2a_{n-2} + 2a_{n-1}$. First, prove by induction that all but the first term are even. Then solve the recursion.*

Exercise 532. *Define the sequence $a_1, a_2, a_3, \ldots$ by $a_1 = 1$, $a_2 = 3$ and for each $k \geq 2$,*

$$a_{k+1} = 3a_k - 2a_{k-1}.$$

Prove that for all $n \geq 1$, $a_n = 2^n - 1$.

Exercise 533. *For a constant $b \in \mathbb{R}^+$, define a sequence by $a_1 = b$, $a_2 = 0$, and for $n \geq 3$, $a_n = ba_{n-2} - b^2a_{n-1}$. Solve this recursion, and then prove the solution by induction.*

Exercise 534. *Put $a_0 = a_1 = 1$ and for $n \geq 1$, define $a_{n+1} = a_n + 2a_{n-1}$. Prove by induction that for each $n \geq 1$,*

$$a_n = \frac{2^{n+1} + (-1)^n}{3}. \tag{16.5}$$

Confirm this solution by the method of characteristic roots.

Exercise 535. *Define $a_1, a_2, a_3, \ldots$ by $a_1 = 2$, $a_2 = 3$ and for each $k \geq 2$,*

$$a_{k+1} = 3a_k - 2a_{k-1}.$$

Show by the method of characteristic roots that for all $n \geq 1$, $a_n = 2^{n-1} + 1$. Confirm this by mathematical induction.

16.3.4 Linear homogeneous recurrences of higher order

The method of characteristic roots also applies to solving linear homogeneous recurrences of order $p > 2$. Repeating the method of characteristic roots given above for $p = 2$, the following generalization is obtained (given here without proof): Suppose that some characteristic equation of degree $p \geq 2$ has roots $\alpha_1, \ldots, \alpha_r$ each respectively occurring with multiplicities $m_1, \ldots, m_r$ (where $m_1 + \cdots + m_r = p$). Then for each $i = 1, \ldots, r$ a particular solution to the given recursion is of the form

$$A_1 \alpha_1^n + A_2 n \alpha_1^n + A_3 n^2 \alpha^n + \cdots + A_{m_i} n^{m_i - 1} \alpha_i^n,$$

and so a general solution can be found by forming linear combinations of the r expressions above.

In this section, only a few examples are given, and solving recurrences of higher order is not asked for.

Exercise 536. *Let $x_1 = x_2 = 1$, $x_3 = 4$, and for $n \geq 1$, define*

$$x_{n+3} = 2x_{n+2} + 2x_{n+1} - x_n.$$

Using mathematical induction, prove that for each $n \geq 1$, x_n is a perfect square.

Exercise 537. *Define the sequence of integers $s_0, s_1, s_2, \ldots$, by $s_0 = s_1 = s_2 = 1$ and for $n \geq 3$,*

$$s_n = s_{n-1} + s_{n-3};$$

prove by induction that for all $n \geq 0$, $s_{n+2} \geq (\sqrt{2})^n$.

Exercise 538. *Define the sequence $s_0 = 1$, $s_1 = 2$, $s_2 = 3$, and for $n \geq 3$,*

$$s_n = s_{n-3} + s_{n-2} + s_{n-1}.$$

Prove by induction that for $n \geq 0$, $s_n \leq 3^n$.

Exercise 539. *Define the sequence* $s_1, s_2, s_3, s_4, \ldots$ *by* $s_1 = s_2 = s_3 = 1$, *and for* $n \geq 1$,

$$s_{n+3} = s_n + s_{n+1} + s_{n+2}.$$

Prove by induction that for each $n \geq 1$, $s_n < 2^n$.

Exercise 540. *Define the sequence* $s_1, s_2, s_3, s_4, \ldots$ *by* $s_1 = 2$, $s_2 = 4$, $s_3 = 7$, *and for* $n \geq 1$,

$$s_{n+3} = s_n + s_{n+1} + s_{n+2}.$$

Prove that s_n *is the number of binary strings of length* n *that do not contain the substring "000". [A non-trivial formula for* s_n *exists and has a proof by induction, however both the discovery and proof of the formula are more challenging exercises.]*

16.3.5 Non-homogeneous recurrences

Non-homogeneous recurrences are common; for example, the recurrence $f(n) = 2f(n-1) + 1$ is found in Exercise 562 below (Towers of Hanoi). There are many specific techniques for solving various non-homogeneous recurrences; however, that theory is not covered here.

Exercise 541. *Suppose that a function* f *satisfies* $f(1) = f(2) = 1$, *and for all* $n \geq 3$, $f(n) = f(n-1) + 2f(n-2) + 1$. *Using mathematical induction, prove that for any positive integer* n,

$$f(n) = 2^{n-1} - \frac{(-1)^n + 1}{2}.$$

Exercise 542. *Let* a *and* b *be fixed real numbers. Define the sequence* $s_0, s_1, s_2, \ldots,$ *of real numbers by* $s_0 = a$ *and*

$$s_n = 2s_{n-1} + b.$$

Prove that for each $n \geq 1$, $s_n = 2^n a + (2^n - 1)b$.

The next theorem includes the result from Exercise 542. For more on difference equations, see [159].

Theorem 16.3.2. *Consider the first-order linear nonhomogeneous difference equation defined for* $t = 0, 1, 2, \ldots$ *by*

$$x_{t+1} = a_t x_t + b_t.$$

If an initial value x_0 *is known, then the solution is unique and is given by*

$$x_t = \left[\prod_{i=0}^{t-1} a_i \right] x_0 + b_{t-1} + \sum_{i=0}^{t-2} \left[\prod_{r=i+1}^{t-1} a_r \right] b_i.$$

In particular,

- If $x_{t+1} = ax_t + b_t$, then

$$x_t = a^t x_0 + \sum_{i=0}^{t-1} a^{t-i-1} b(i).$$

- If $x_{t+1} = ax_t + b$, then

$$x_t = \begin{cases} a^t x_0 + b \left[\dfrac{a^t - 1}{a - 1} \right] & a \neq 1 \\ x_0 + bt & a = 1. \end{cases}$$

Exercise 543. *Prove Theorem 16.3.2 by mathematical induction.*

Exercise 544. *Let $a_1, a_2, a_3, \ldots$ be a sequence of positive integers satisfying*
(1) $a_{2n} = a_n + n$, and
(2) if a_n is prime, then so is n.
Prove that for each $n \geq 1$, $a_n = n$.

16.3.6 Finding recurrences

A problem that a computer scientist might run into is opposite in nature to solving a recurrence. Given a formula for the n-th term in a sequence, how can one arrive at a recursive definition? In general, the methods are quite *ad hoc*, varying according to the individual situation, and only one simple example is considered here.

Suppose the sequence $s_1, s_2, s_3, \ldots$ is defined by

$$s_n = \frac{1}{4n}$$

(so $s_1 = \frac{1}{4}$, $s_2 = \frac{1}{8}$, and so on). Can one give a purely recursive definition for s_n? The reciprocals $t_n = \frac{1}{s_n} = 4n$ form an arithmetic progression with difference 4 and satisfy

$$t_n = t_{n-1} + 4 = t_{n-1} + (t_{n-1} - t_{n-2}) = 2t_{n-1} - t_{n-2}.$$

It now follows that

$$\begin{aligned} s_n &= \frac{1}{t_n} \\ &= \frac{1}{2t_{n-1} - t_{n-2}} \\ &= \frac{1}{\frac{2}{s_{n-1}} - \frac{1}{s_{n-2}}} \\ &= \frac{1}{\frac{2s_{n-2} - s_{n-1}}{s_{n-1} s_{n-2}}} \end{aligned}$$

$$= \frac{s_{n-1}s_{n-2}}{2s_{n-2} - s_{n-1}}.$$

Hence, the recursive definition for the sequence $\{\frac{1}{4n}\}$ is

$$s_1 = \frac{1}{4}, \quad s_2 = \frac{1}{8}, \quad \text{and for } n \geq 3, \quad s_n = \frac{s_{n-1}s_{n-2}}{2s_{n-2} - s_{n-1}},$$

nastier looking than the closed form formula $s_n = \frac{1}{4n}$. Ideally, a recursive definition first gives some initial values for the sequence, and then defines subsequent elements in terms of only previously defined elements. In [437, soln to 17-2, p.202], a "recursive definition" for this sequence was given by

$$s_{n+1} = \frac{2s_n s_{n+2}}{s_n + s_{n+2}},$$

and $s_1 = \frac{1}{4}$, $s_2 = \frac{1}{8}$. This definition fails to be an ideal recursive definition in two ways: it is not precisely recursive (one would have to solve for s_{n+2} first, of course—try recursively finding s_3, for example, using their formula), and initial values are given *after* the general formula for recursion.

To verify that indeed the correct (non-linear) recursive definition was found, one can use induction:

Exercise 545. *Define the sequence of integers $s_1, s_2, s_3, \ldots$, recursively by $s_1 = \frac{1}{4}, s_2 = \frac{1}{8}$, and for $n \geq 2$,*

$$s_n = \frac{s_{n-1}s_{n-2}}{2s_{n-2} - s_{n-1}}.$$

Prove by induction that $s_n = \frac{1}{4n}$.

16.3.7 Non-linear recurrence

Exercise 546. *Suppose that c is a real number where $0 < c \leq 1$. Define recursively the sequence $s_1, s_2, s_3, \ldots$ by $s_1 = c/2$, and for each $n \geq 1$, define*

$$s_{n+1} = \frac{s_n^2 + c}{2}.$$

Prove that the sequence is strictly increasing and strictly bounded above by 1, that is, for each $n \geq 1$, $s_n < s_{n+1} < 1$. [Then one can conclude that $\lim_{n\to\infty} s_n$ exists and is at most 1.]

Exercise 547. *Define the sequence $a_1, a_2, a_3, \ldots$ by $a_1 = 1$ and for each $n \geq 1$,*

$$a_{n+1} = \sqrt{a_n + 5}.$$

Prove that for all $n \geq 1$, both $a_n < 3$ and $a_{n+1} > a_n$. Since the sequence is bounded above and increasing, it has a limit; find it.

Exercise 548. *Define the sequence* $a_1, a_2, a_3, \ldots$ *by* $a_1 = 1$ *and for each* $n \geq 1$,

$$a_{n+1} = 1 + \sqrt{a_n + 5}.$$

Prove that for all $n \geq 1$, $a_n > 4 - \frac{4}{n}$.

Exercise 549. *Define the sequence* $a_1, a_2, a_3, \ldots$ *by* $a_1 = 1$ *and for each* $n \geq 1$,

$$a_{n+1} = \sqrt{2a_n + 1}.$$

Prove that for all $n \geq 1$, *both* $a_n < 4$ *and* $a_{n+1} > a_n$. *Decide if the sequence converges or diverges, and if it converges, find the limit.*

Exercise 550. *Define the sequence* $a_1, a_2, a_3, \ldots$ *by* $a_1 = 4$ *and for each* $n \geq 1$,

$$a_{n+1} = \sqrt{a_n + 2}.$$

Prove that for all $n \geq 1$, *both* $a_n > 2$ *and* $a_{n+1} < a_n$. *Find* $\lim_{n \to \infty} a_n$ *or show that it does not exist.*

Exercise 551. *Define the sequence* $a_1, a_2, a_3, \ldots$ *by* $a_1 = 2$ *and for each* $n \geq 1$,

$$a_{n+1} = 1 + \sqrt{a_n + 5}.$$

Prove that for all $n \geq 1$, *both* $a_n < 4$ *and* $a_{n+1} > a_n$. *Then find* $\lim_{n \to \infty} a_n$ *if it exists.*

Exercise 552. *Let* $x_1 = 1$, *and for* $n \geq 1$, *define* $x_{n+1} = 1 + n/x_n$. *Show that*

$$\sqrt{n} \leq x_n \leq \sqrt{n} + 1.$$

Exercise 553. *Let* $s_1, s_2, s_3, \ldots$, *be an integer sequence recursively defined by* $s_1 = 0$ *and for* $n \geq 2$,

$$s_n = 1 + s_{\lfloor n/2 \rfloor}.$$

Prove that for every $n \in \mathbb{Z}^+$, $s_n = \lfloor \log_2(n) \rfloor$.

Exercise 554. *Let* $s_1, s_2, s_3, \ldots$ *be recursively defined by* $s_1 = 3$, *and for* $n \geq 1$,

$$s_{n+1} = \begin{cases} \text{the smallest odd integer} \geq s_n^{(n+1)/n} & \text{if } n+1 \text{ is odd} \\ \text{the smallest even integer} \geq s_n^{(n+1)/n} & \text{if } n+1 \text{ is even} \end{cases}.$$

Prove that $\{s_n\}$ *is an increasing sequence of positive integers satisfying*

$$s_n^{(n+1)/n} \leq s_{n+1} \leq s_n^{(n+1)/n} + 2,$$

and prove by induction that for each n, $s_n \geq 3^n$.

Exercise 555. *Let $\alpha \neq \beta$ be reals. Define the sequence $u_1, u_2, u_3, \ldots$ by*

$$u_1 = \frac{\alpha^2 - \beta^2}{\alpha - \beta}, \quad u_2 = \frac{\alpha^3 - \beta^3}{\alpha - \beta},$$

and for $k > 2$, define $u_k = (\alpha + \beta)u_{k-1} - \alpha\beta u_{k-2}$. Prove that for each $n \geq 1$,

$$u_n = \frac{\alpha^{n+1} - \beta^{n+1}}{\alpha - \beta}.$$

Exercise 556. *Let $a_0 = 9$, and for each $n \geq 0$, define $a_{n+1} = a_n^3(3a_n + 4)$. Show that the decimal representation of a_n ends in 2^n nines.*

Exercise 557. *Put $f(x) = \frac{x+1}{x-1}$, and let $y_1 = f(x)$. Then put*

$$y_2 = f(y_1) = \frac{\frac{x+1}{x-1} + 1}{\frac{x+1}{x-1} - 1}$$

and for each $n \geq 1$, put $y_{n+1} = f(y_n)$. Prove that for each odd n, $y_n = \frac{x+1}{x-1}$ and for each even n, $y_n = x$.

For the next exercise, the reader is reminded of the product notation (see Definition 2.5.9)

$$\prod_{i=1}^{n} x_i = x_1 \cdot x_2 \cdot \cdots \cdot x_n.$$

Exercise 558. *Define the sequence $t_1, t_2, t_3, \ldots$ by $t_1 = 2$ and for $n \geq 1$, define*

$$t_{n+1} = 1 + \prod_{i=1}^{n} t_i.$$

Prove by induction that for each $n \geq 1$, $t_n > n$ and

$$\sum_{i=1}^{n} \frac{1}{t_i} = 1 - \frac{1}{t_n(t_n - 1)}.$$

Exercise 559. *Let $x_0 = 1$, and for $n \geq 0$, define $x_{n+1} = x_n + \frac{1}{x_n}$. Prove that as $n \to \infty$, $x_n \to \infty$ and that*

$$\sqrt{2n} < x_n < \sqrt{2n + \frac{1}{2}\log n}.$$

The sequence in the next exercise is used by many pocket calculators to calculate square roots, and is based on an expression of the form $x = f(x)$, where $f(x) = \frac{1}{2}(x + c/x)$. [Do you see the relation between this exercise and Newton's method? See nearly any calculus text for Newton's method.]

Exercise 560. *Let $c > 0$ be a fixed real number. Define a sequence by selecting any $x_0 > 0$, and for each non-negative integer n, define*

$$x_{n+1} = \frac{1}{2}(x_n + c/x_n).$$

Prove that if x_n is an approximation for $\sqrt{c}$ correct to t decimal places, then x_{n+1} is correct to at least $2t - 1$ places. Furthermore, $\lim_{n\to\infty} x_n = \sqrt{c}$, regardless of the choice for $x_0 > 0$.

For example, in the above exercise, for $c = 2$, if $x_0 = 1.5$ is chosen, then $x_1 \sim 1.416666...$, $x_2 = 1.414215686274...$, and $x_3 = 1.414213562375...$, compare favorably with

$$\sqrt{2} = 1.414213562373... \,.$$

Exercise 561. *Let $0 < b < a$ be positive reals. Put $a_1 = \frac{a+b}{2}$ and $b_1 = \sqrt{ab}$, the arithmetic and geometric means, respectively. For each $n \geq 1$, recursively define $a_{n+1} = \frac{a_n+b_n}{2}$ and $b_{n+1} = \sqrt{a_n b_n}$. Using induction, prove that*

$$b_n < b_{n+1} < a_{n+1} < a_n,$$

and show that both $\lim_{n\to\infty} a_n$ and $\lim_{n\to\infty} b_n$ exist, and are equal.

Gauss called the limit in Exercise 561 the *arithmetic-geometric mean* of a and b. (See [513, p. 703].)

16.3.8 Towers of Hanoi

Many computer science students are taught the puzzle or game called "The Towers of Hanoi", usually in the context of recursive programming. Though some say that the Towers of Hanoi puzzle was invented in India, it was apparently produced by Édouard Lucas in 1883. It appeared in [358] (see [292, p. 229] or [42, p. 48]; also see [11] or[207]). [I once read that it was while Catalan was playing with this puzzle that he discovered Definition 12.5.2 (the one using parentheses) for what are now called Catalan numbers.]

Here is how the game works. There are n discs of increasing diameter, each with a hole in the center and three pegs on which discs can be positioned. Start with n discs on the first peg, the discs in increasing size with the smallest disc on the top and the largest on the bottom (see Figure 16.1). The goal is to move all n discs to another peg using the following rules: Discs are moved from peg to peg one at a time. At any stage, no larger disc can sit on any smaller one. In the original puzzle, there were $n = 8$ disks.

Exercise 562 (Towers of Hanoi). *Using all three pegs in the Towers of Hanoi game, show that the number of moves required to move all n discs from the first peg to another peg is $2^n - 1$.*

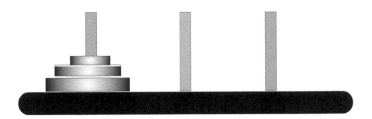

Figure 16.1: Towers of Hanoi, with $n = 3$ disks

A myth originally accompanied the puzzle: monks of Brahma were to transfer a tower of 64 discs, and when this task was complete, the world would end. Well, the time necessary for the

$$2^{64} - 1 = 18,446,744,073,709,551,615$$

moves is certainly a long time—over 58 billion centuries at one move per second! For other interesting references or renditions of the stories regarding the Towers of Hanoi and the towers of Brahma, see [12], [99], [149], [193], [205, 20–23] [208], [209], [305, pp. 169–171], [329, §3.12.4, pp. 91–93] [436], [478], [552], and [574].

The following variation of the Towers of Hanoi problem was discovered by S. Althoen and his students [16], published in 2009. He says that his students inadvertently assumed two extra rules, thereby changing the game. [It would be surprising to me if this variation had not been previously discovered because this variation and its solution are so elegant.] In addition to the usual rules (move disks one at a time, a larger disk can never be placed upon a smaller one), further insist that n disks start on peg 1, the final position to have all disks on peg 3, and disks can be moved only between adjacent pegs. With some simple experimentation, when $n = 1$, 2 moves ($1 \to 2 \to 3$) are required, and when $n = 2$, the following 8-move sequence is optimal (and virtually forced): $1 \to 2$, $2 \to 3$ $1 \to 2$, $2 \gets 3$, $1 \gets 2$, $2 \to 3$, $1 \to 2$, and $2 \to 3$. When $n = 3$, it turns out that 26 moves are required. Before looking below, can you guess the pattern (beginning $2, 8, 26, \ldots$) and prove it?

Exercise 563 (Hanoi revisited). *Solve the game of Towers of Hanoi with n disks with three pegs A, B, C, given in a row, subject to two additional rules: disks start on peg A, and must finish on peg C, and disks are moved only between adjacent pegs. Prove that $3^n - 1$ is the minimum number of moves required to complete this altered Towers of Hanoi game.*

16.4 Loop invariants and algorithms

The term "loop invariant" is frequently used in computer science. Some say that a *loop invariant* is a relationship among variables that is maintained in all iterations of a loop. In fact, it is probably fair to say that a "loop invariant" is the same as "a statement to be proved by induction", denoted throughout this book as $S(n)$, or some such. According to [119], the notion of a loop invariant is first due to due R. W. Floyd. [The basic reference for algorithms is [323].] The use of loop invariants in programming is the use of (finite) mathematical induction.

The importance of induction to computing scientists and engineers can not be overstated. It seems, however, that many engineering schools give very limited instruction in formal mathematical induction. In [302] is a report on how mathematical induction is the basis for "verification and validation (V&V) in modeling". They examined universities and colleges that taught courses in Modeling and Simulation (M&S) and observed how few are teaching V&V and mathematical induction, yet how valuable M&S is today. Here is a quote from that paper: "Upon contacting ten other universities, none were found that introduce any Formal technique [incl. MI] into their classrooms, ... This is an oversight that must be corrected, otherwise universities will be producing not engineers of technology, but users of technology."

The basic idea is that in most computer programs, there are **while** loops, and upon execution of each pass through a loop, some desired property continues to hold. This "property" could just be a number staying the same (invariant), or it could be the truth of a particular claim being maintained. Proving that some property continues to hold after all passes through the loop (and so outputs a correct result) is often easily done by mathematical induction. Many recursive procedures (algorithms) can be complicated, and in many cases, looking at an algorithm as an inductive machine makes clear whether or not the procedure outputs a correct result.

The textbooks [119], and [496], and [570] have been recommended to me for the modern study of algorithms. In the first, there are loop invariants for search algorithms, sorting algorithms, spanning-tree algorithms, merging, exponentiation, the simplex algorithm, and many more. Only a few basics are given here.

Some computer scientists say that checking loop invariants is *like* induction, and others say that this is just applying induction on the number of iterations of a loop. Since **while** loops are to be executed only finitely many times, induction here is finite. Some refer to the three stages of an induction proof, base case, inductive step, and conclusion, as *initialization*, *maintenance*, and *termination*, respectively. Aside from terminology, computer scientists are doing induction all the time.

Suppose at each iteration $i = 0, 1, \ldots, n$ through a loop in a program, there are variables u_i, v_i, w_i, and $S(i)$ is some statement about these three variables. In the initialization step, one checks that $S(0)$ is true prior to the iteration of the first loop. At the maintenance step, one shows that for any k, if $S(k)$ is true, then $S(k + 1)$ is true. In the termination step, the loop invariant S gives a property, namely $S(n)$ that is useful.

Examine the following pseudocode:

1. $x := 0$
2. $y := 0$

3. INPUT: a
4. **while** $y \neq a$
4a. $x \leftarrow x + a$
4b. $y \leftarrow y + 1$
5. OUTPUT: x

What does it do? Working out a few steps, it seems to give $x = a^2$. To prove this, for each $i = 0, 1, 2, \ldots$, let x_i, y_i be the values of x and y after the i-th iteration of the **while** loop. Then for each $i = 1, 2, \ldots$, by line 4a, $x_i = x_{i-1} + a$. Is there some property of x_i that remains true throughout the program? It appears as if for each i that $x_i = ia$, which, if true, after the a-th iteration, $x_a = a \cdot a$ as desired. Here is a formal proof of the loop invariant $S(i) : x_i = ia$. It is helpful to notice (also by an inductive proof, if need be) that $y_i = i$.

INITIALIZATION: For $i = 0$, $S(0)$ states $x_0 = 0 \cdot a = 0$, which is correct.

MAINTENANCE: Fix $0 \leq k < a$, and assume that $S(k)$ is true. Then at the end of the k-th iteration, $x_k = ka$ and $y_k = k$. Since $k < a$, $y_k \neq a$, so the while loop is executed the $k + 1$ time; then upon execution of 4a,

$$
\begin{aligned}
x_{k+1} &= x_k + a \\
&= ka + a \qquad \text{(by } S(k)) \\
&= (k+1)a.
\end{aligned}
$$

Hence $S(k + 1)$ is true.

TERMINATION: By mathematical induction, as long as the loop executes, the statement $S(i)$ remains true. However, when $i = a$, $y_a = a$, and so the **while** loop fails to be executed, and the program terminates. At that time, (by $S(a)$), $x = a^2$. Thus the program outputs the correct output, and so the program is the squaring function. □

This book has a number of algorithms, each with a proof of correctness by induction. For example, some algorithms analyzed or used in this book are:

- Euclidean division algorithm for finding gcd's (Exercise 211);

- greedy algorithm for Zeckendorf's theorem (Exercise 372);

- Prim's algorithm for finding minimum spanning trees (Exercise 487);

- Kruskal's algorithm for minimum spanning trees (Exercise 488);

- greedy coloring algorithm for graphs (Exercise 498);

- Gale–Shapley algorithm for finding stable marriages (Exercise 496);

- detecting a counterfeit coin (see Section 17.5).

- soldiers in a circle (see Exercise 588)

- greedy algorithm for producing Egyptian fraction representation (Exercise 317).

Not only can induction be used to check programs, but it can be used in determining the number of steps a program takes, also called the *running time*, or *complexity* of the algorithm. Counting steps in a recursive proof can often be done inductively—especially when a procedure calls itself and there are nested loops that complicate the picture.

One challenge in applying induction to programs, algorithms, or recursive procedures, is to select the proper variables upon which to induct. Another challenge is to separate the steps sufficiently so it is clear what properties are required of a given structure before a recursive step will succeed. Such dilemmas are usually solved after going through some loops manually a few times, keeping careful track of all variables. For example, when reading some of the standard proofs of Prim's algorithm, it is not immediately clear what magic property makes the algorithm work; it just seemed to work. [The solution to Exercise 487 is written so that it is clear what property is being preserved from step to step.]

16.5 Data structures

Computers are used to store information as well as to compute. Depending on how information is stored, modifying sets of data (like inserting, deleting, or sorting) or locating and retrieving data can be accomplished at different speeds. As a trivial example, if one million numbers are stored in a simple list, finding the largest of these numbers can be very easy if the list is already sorted in increasing or decreasing order. For some applications, it may be convenient to put one million numbers in a 1000×1000 array (a matrix). For other data sets, trees can be used to model relationships between individual entries. Loosely speaking a *data structure* is a way to store data in order to enable modification or retrieval. Many data structures are created or defined recursively, and so mathematical induction is a natural tool to analyze such structures; when induction is applied to a recursively defined structure, it is sometimes called *structural induction*, especially in computing science.

A few of the well-studied data structures are linked lists, rooted trees, arrays, heaps, red-blue trees, stacks, hash tables, and dictionaries. Data structures are often defined recursively; for example, one can imagine that a large binary tree can be created by adding a new root adjoining the roots of two smaller trees. It then seems like mathematical induction might be a natural choice to prove properties of various data structures. In the vernacular of computer science, these properties are proved by *structural* induction, which may be considered as an abbreviation for "mathematical induction on recursively defined structures".

Rooted trees have already been examined in Section 15.2 and elsewhere in this text. For example, Exercise 502 asks to show (the nearly obvious) that all trees are planar; plane binary trees and and the height of full binary trees are discussed in Exercises 482, 483, and 484; increasing trees occur in Exercises 486 and 485. Arrays are implicitly discussed in Chapter 19.

This section contains only a few other examples; to give any reasonably broad overview of data structures would take another chapter. For a more accurate picture of data structures, the reader may want to consult *e.g.*, [334] (with Java), [346], [529] or any of many more recent titles on data structures. In 1980, Musser published on proving inductive properties of abstract data types [398]. See also nearly any book on discrete mathematics, algorithms (*e.g.*,[119] or [496]) for more on data structures.

16.5.1   Gray codes

Let B_n be the set of binary strings of length n. A Gray code for B_n is a listing of the 2^n strings in B_n so that any two adjacent strings differ in exactly one position. Compare this with the definition of a hamiltonian circuit in the unit n-cube (see Exercise 491).

Exercise 564. *For every $n \geq 1$, prove that there is a Gray code for B_n by finding a recursive construction.*

Gray codes are named after F. Gray, who published a paper [233] where these codes are developed and applied in computing. For other information on Gray codes, see [51], [553], or [555]; for an unexpected application in combinatorial (polytope) geometry, see [54].

16.5.2   The hypercube

The next exercise concerns a graph called the *n-dimensional hypercube* Q_n, also called the n-cube graph or the binary n-cube graph. The n-cube consists of vertices

$$V = \{(\epsilon_1, \epsilon_2, \ldots, \epsilon_n) : \forall i = 1, \ldots, n, \epsilon_i \in \{0, 1\}\},$$

and for each $d = 1, \ldots, n$, d-dimensional facets, each facet a collection of vertices determined by fixing $n - d$ coordinates, and letting the remaining d coordinates vary over $\{0, 1\}$. For example, the vertices of Q_3 are

$$V = \{(0,0,0), (0,0,1), (0,1,0), (0,1,1), (1,0,0), (1,0,1), (1,1,0), (1,1,1)\},$$

and the 2-dimensional facets are:

$$\{(0,0,0), (0,0,1), (0,1,0), (0,1,1)\}, \qquad (\text{fixing } (0, *, *));$$
$$\{(1,0,0), (1,0,1), (1,1,0), (1,1,1)\}, \qquad (\text{fixing } (1, *, *));$$
$$\{(0,0,0), (0,0,1), (1,0,0), (1,0,1)\}, \qquad (\text{fixing } (*, 0, *));$$

$$\{(0,1,0),(0,1,1),(1,1,0),(1,1,1)\}, \qquad (\text{fixing } (*,1,*));$$
$$\{(0,0,0),(0,1,0),(1,0,0),(1,1,0)\}, \qquad (\text{fixing } (*,*,0));$$
$$\{(0,0,1),(0,1,1),(1,0,1),(1,1,1)\}, \qquad (\text{fixing } (*,*,1)),$$

each corresponding to a face of the cube. One-dimensional facets of Q_3 are called "edges" and one can consider vertices as 0-dimensional facets. Ordinarily, Q_n simply refers to the simple graph formed by the vertices and the edges alone. See Exercises 491 (Q_n is hamiltonian), 492 (Q_n is n-connected) and 564 (Gray codes) for other properties of the graph Q_n.

Exercise 565. *Let v_i^n denote the number of i-dimensional facets of Q_n. If either $i > n$ or $i < 0$ holds, put $v_i^n = 0$. Prove that for each $n \geq 1$, $v_0^n = 2^n$ and $v_1^n = n2^{n-1}$. Give an inductive proof that for general i,*

$$v_i^n = 2^{n-i}\binom{n}{i}.$$

16.5.3   Red-black trees

Red-black trees on n vertices are a class of binary trees with height $O(\log n)$, nearly optimal.

Definition 16.5.1. A *red-black tree* is a binary search tree T with nodes (vertices) 2-colored, say red and black, so that

1. Every node is either red or black.

2. The root is black.

3. Every leaf node is black.

4. If a node is red, then both its children are black.

5. For each node, all paths from that node to descendant leaves contain the same number of black nodes.

Figure 16.5.3 shows a red-black tree based on [119, p. 275].

The *height* of a node v in a red-black tree is the length of a longest path from v to any leaf. (For trees or other partial orders drawn with a root at the bottom, often the height of a node is defined to be the length of the path to the root.) Define the *height* of a red-black tree to be the height of the root. To confirm ideas, the height of the red-black tree in Figure 16.5.3 is 6.

By property 5 above, for any node v, the number of black nodes on any descending path to a leaf is the same. For each node v in a red-black tree, define the *black height* of v, denoted bh(v), to be the number of black nodes, not including v, on any descending path from v to a leaf. So all leaves have black height 0. Define the black height of a red-black tree to be the black height of the root. The black height of the tree in Figure 16.5.3 is 3.

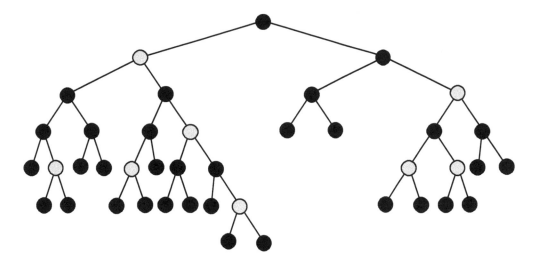

Figure 16.2: A red-black tree with 20 internal vertices and 21 leaves

Lemma 16.5.2. *A red-black tree with n internal nodes has height at most*

$$2\log_2(n+1).$$

Proof: The proof is in two parts: first a claim (involving black height) is made and proved inductively, and then some calculations follow the claim.

Claim: For any red-black tree T and any node v of T, the subtree rooted at v contains at least $2^{\mathrm{bh}(v)} - 1$ internal nodes.

Proof of Claim: This is achieved by induction on the height of a node. [This is another example of structural induction; the usual formal style is partially abandoned here. This proof could have easily been given as an exercise—if the hint to induct on height was given.]

BASE STEP: If the height of v is 0, then v is a leaf (with $\mathrm{bh}(v) = 0$). In this case, the subtree rooted at v contains no internal nodes, and $2^{\mathrm{bh}(x)} - 1 = 2^0 - 1 = 0$.

INDUCTION STEP: Let $r > 0$ and suppose that for all nodes w of height less than r, the subtree rooted at w contains at least $2^{\mathrm{bh}(w)} - 1$ internal nodes. Let v have height $r > 0$; then v is an internal node with 2 children. If v is black, then each of its children has black height $\mathrm{bh}(v) - 1$; if v is red, each of these children has black height $\mathrm{bh}(v)$. By the induction hypothesis (where each of these two children play the role of w above), the subtrees rooted at these two children each have at least $2^{\mathrm{bh}(v)-1} - 1$ internal children. In all, the tree rooted at v has at least

$$2(2^{\mathrm{bh}(v)-1} - 1) + 1 = 2^{\mathrm{bh}(v)} - 2 + 1 = 2^{\mathrm{bh}(v)} - 1$$

internal nodes, satisfying the claim for any vertex at height r, thereby completing the inductive step.

By mathematical induction on height, for all v, the Claim is true. $\qquad\square$

Returning to the proof of Lemma 16.5.2, let T be a red-black tree of height h and with n internal nodes.

Because of property 4, where v is the root of the tree, along any downward path from the root to a leaf, for every red, there is a black, and so the number of non-root blacks is at least the number of reds on that same path. Hence the black height of the root of T is at least $h/2$. By the claim above, since the tree rooted at v is T itself, $n \geq 2^{h/2} - 1$. Upon solving for h,

$$h \leq 2\log_2(n+1),$$

as desired, completing the proof of Lemma 16.5.2. $\qquad\square$

16.6 Complexity

Recall that complexity is the "run-time" of an algorithm. Usually n is the size of the input, and $f(n)$ is the number of operations performed by executing the algorithm upon such an input of size n.

16.6.1 Landau notation

Often one is not concerned with finding precisely values for some function $f : \mathbb{Z}^+ \to \mathbb{Z}^+$, but only the "order of magnitude" of values $f(n)$. For example, in the study of run times for algorithms, one may care only that, for "large n", $f(n)$ behaves more like n^2 or like $n\log_2(n)$. To compare functions and their asymptotic rates of growth, certain notation is helpful (due to Landau).

For two functions f and g, write $f = o(g)$ [read "f is little oh of g"] or $f(n) = o(g(n))$ if and only if $\lim_{n\to\infty} \frac{f(n)}{g(n)} = 0$. For example, $\ln(x) = o(x^2)$, and $\frac{1}{n} = o(1)$. Hence, the notation $f(n) = (1 + o(1))g(n)$ means that f and g are approximately equal for large n, that is, $\lim_{n\to\infty} \frac{f(n)}{g(n)} = 1$, in which case one often writes $f \sim g$.

For two functions f and g, if there is a constant $C > 0$ and some n_0 so that for every $n \geq n_0$, $f(n) \leq C \cdot g(n)$, say that f is *big Oh of* g. One often writes $f(n) = O(g(n))$, even though this is really an abuse of notation since $O(g(n))$ is really the class of functions that are big Oh of g, and so one might more properly write $f(n) \in O(g(n))$. For example, $3n(\ln(n)+1) \in O(n^2)$, or $n^2+1 = O(3n^2+14n)$.

Turning the big oh notation inside out, define $f = \Omega(g)$ if and only if $g = O(f)$. If both $f = O(g)$ and $f = \Omega(g)$, write $f = \Theta(g)$; this essentially describes the situation where f and g satisfy $cg \leq f \leq Cg$ for some constants c and C.

In analysis of algorithms, another notation is convenient: for a positive integer k and positive reals x, b with $b > 1$, let $\log_b^k(x)$ denote the iterated logarithm, *i.e.*, $\log_b(x) = \log_b^1(x)$ and for $k > 1$, $\log_b^k(x) = \log_b(\log_b^{k-1}(x))$.

16.6.2   The master theorem

In many "divide and conquer" algorithms, one encounters a recurrence of the form $f(n) = 2f(n/2) + h(n)$, where h is usually a linear function of n. For example, when $n = 2^k$, suppose that $f(n)$ is the number of comparisons necessary to find the maximum and minimum elements of a set of n distinct real numbers. Split the set of numbers into two groups, find the max and min in each group, thereby using $2f(n/2)$ comparisons. To find the global max and mins, two more comparisons are necessary. So $f(1) = 0$, $f(2) = 1$, and for larger n, the recurrence $f(n) = 2f(n/2) + 2$ is found. How does one solve such a recurrence? With a little experimentation, one discovers that $f(4) = 2$, $f(4) = 2f(2) + 2 = 4$, $f(8) = 2f(4) + 2 = 10$, $f(16) = 22$, $f(32) = 46$. It seems as if $f(n) \in O(n)$. In fact, this is true and follows from a special case of a more general "master theorem" explained below.

For many such recurrences, it is nearly hopeless to arrive at a closed form; however, an order of magnitude can be calculated in some cases. Some divide and conquer algorithms have simple run time (the basis for binary searches):

Exercise 566. *If $A_n = \{a_1, a_2, \ldots, a_{2^n}\} \subset \mathbb{R}$ is a set of 2^n real numbers in increasing order (i.e., if $i < j$ then $a_i < a_j$) and $x \in \mathbb{R}$, then the number of comparisons necessary to determine whether or not $x \in A_n$ is $n + 1$. Hint: divide and conquer.*

The following, called "the master theorem", covers many cases of divide and conquer, even when the number of parts b a job is divided into is larger than two. The proof of this master theorem has a proof which is largely based on induction (beginning with the case where n is a power of b, and then a lot of careful detail to handle cases where n/b is not an integer). There are many variations of this theorem in the literature, many more adapted for special cases. See, *e.g.*, [119, pp. 73–84] (with complete proof), [346, p. 32], or [464, pp. 244ff] (for the case $a = b = 2$ with a nice induction proof).

Theorem 16.6.1 (Master theorem). *Let $a, b \in \mathbb{Z}^+$ where $b > 1$, and let some function $h : \mathbb{Z}^+ \to \mathbb{Z}^+$ be given. If f is defined by the recurrence, $f(1) = 1$, and for $n \geq b$,*

$$f(n) = af(n/b) + h(n),$$

where $f(n/b)$ is taken to mean either $f(\lfloor n/b \rfloor)$ or $f(\lceil n/b \rceil)$, then

1. *If there exists a constant $\epsilon > 0$ so that $h \in O(n^{\log_b(a) - \epsilon})$, then $f(n) = \Theta(n^{\log_b(a)})$.*

2. *If there exists a constant $K \geq 0$ so that $h \in \theta(n^{\log_b(a)} \log_2^K n)$, then $f(n) = \theta(n^{\log_b(a)} \log_2^{K+1} n)$.*

3. *If there exists a constant $\epsilon > 0$ so that $h(n) = \Omega(n^{\log_b(a) + \epsilon})$, and if for some constant $c < 1$ and sufficiently large n, $a \cdot h(n/b) \leq c \cdot h(n)$, then $f(n) = \Theta(h(n))$.*

In the problem above for finding both the maximum and minimum in a collection of 2^k distinct real numbers, use $a = b = 2$, $h(n) \leq 2$, and $\epsilon = 1$, Theorem 16.6.1 confirms that $f(n) = \theta(n)$.

An inductive proof for the case $a = b = 2$ is similar to that of a proof by upward and downward induction as in Section 3.3, first solving the cases when n is a power of 2, and then filling in the steps going downward—however induction does not seem to be needed to fill in the gaps, only meticulous handling of floor and ceiling functions (which take a few pages).

16.6.3   Closest pair of points

The next example is a computational geometry problem brilliantly solved with a divide and conquer approach as given by Shamos and Hoey [484] in 1975. The given case here is only for points on the plane, but a similar result can be obtained for points in any metric space.

Problem: Given n points in the plane find a pair of points whose distance is closest.

One way to find such a closest pair of points is to simply compute and list each of the distances for all $\binom{n}{2} = O(n^2)$ pairs, and then in a single pass (with $n - 1$ comparisons in a "take the best so far" algorithm), find a closest pair. [One may assume that no distance is repeated, and so only one such pair is found.] If computing a distance counts as an operation and if comparison is an operation, then the total number of operations in this "brute force" algorithm is $O(n^2) + n - 1 = O(n^2)$. However, with a simple divide and conquer approach, an algorithm uses only $O(n \log_2 n)$ steps.

Here is a brief explanation of the algorithm from [484] (also see [119] for the many details overlooked here.) Describe the procedure with run time $f(n)$ as follows: Put $f(1) = 0$, $f(2) = 0$, and $f(3) = 3$, Let $n \geq 3$, and suppose that for all $m < n$, the procedure has been defined for m points, taking $f(m)$ steps. Let X be a set of n points in the plane, and without loss of generality, suppose that all x-coordinates are different (if they are not, one can rotate them a bit, or use non-vertical lines in the algorithm). Let $X = X_L \cup X_R$ be a partition of X into equal or almost equal parts determined by some vertical line ℓ. Now call the procedure twice, once for each of X_L and X_R. Thus the closest pair in X_L and and the closest pair in X_R are found in $f(\lfloor n/2 \rfloor) + f(\lceil n/2 \rceil)$ steps. For simplicity, assume that n is even. Then the two pairs are found in $2f(n/2)$ steps. Pairs with one point on each side of ℓ may have smaller distance than in either of X_L or X_R, so such pairs must be checked. In all, there are $n^2/4$ pairs, but they need not all be checked. The remarkable observation is that for any point near ℓ, only six neighboring points need to be checked (essentially because if a point is at minimum distance to another, at most six other points can fit in a circle having the previous minimum distance as its radius)! So at worst, $6n/2 = 3n$

more pairs need to be checked. Thus $f(n) \leq 2f(n/2) + 3n$. By the master theorem, $f(n) \in O(n \log_2(n))$. □

See Exercise 700 for another exercise regarding distances in the plane.

Many other algorithms, like those finding a convex hull of points (see, *e.g.* [438] or [119, 947–957]), or other computational geometry problems (see, *e.g.* [48], [420]) can be analyzed in a similar "divide and conquer" method.

The following is the key behind "mergesort", where a linear order is found by splitting the input into two halves, sorting each, then gluing them together. The following two exercises might be considered typical applications of structural induction.

Exercise 567. *Prove by induction that the number of comparisons needed to merge two disjoint sorted sets with k and ℓ distinct elements respectively is at most $k+\ell+1$.*

Exercise 568. *Prove that the number of comparisons needed to place a set of 2^n different real numbers in order is bounded above by $n2^n$.*

Chapter 17

Games and recreations

One day, when I was doing well in class and had finished my lessons, I was sitting there trying to analyze the game of tic-tac-toe... The teacher came along and snatched my papers on which I had been doodling... She did not realize that analyzing tic-tac-toe can lead into dozens of non-trivial mathematical questions.

—Martin Gardner,

Math. Intell., 1997.

17.1   Introduction to game theory

Games have been played by humans for centuries, and until recently, the study of games seems to have concentrated on some board games and games of chance. Presently, the study of games concentrates more on strategy and maximizing payoffs or minimizing losses, especially in business ventures or tactical situations. Early in the 20th century, Emile Borel is often credited as starting the new theory of games. In 1928, John von Neumann [554] published his first paper on game theory, with the more extensive treatise *Theory of Games and Economic Behavior* [393] co-authored with the economist Oskar Morgenstern appearing in 1944. Another soon to be classic *The Compleat Strategyst* [572] (primarily on two-person zero sum games), appeared in 1954 and 1966. Other "modern" texts on game theory include [41], [359], [379], and [447] (to name but a very few). Game theory today is now extensively studied in economics, operations research, military, political and other social settings, with the foundational work of von Neumann and John Nash standing out. The book [530] gives an account of game theory in terms of hypergraphs.

Game theory is now a very broad subject, addressing many questions regarding (to list just a few) winning strategies, expected winnings, minimizing losses and maximizing gains, zero-sums games, perfect and non-perfect information games, continuous or infinite games, equilibriums and saddle points, n-person games, games with no natural outcome, voting theory, and tree games. Much of modern game theory is represented with and analyzed by matrix theory, probability (for games of chance), and linear programming—mathematical induction seems to be only occasionally used explicitly in such discussions regarding game theory.

The paper [113] discusses extensively backward induction in game theory, including the centipede game, NIM, the Prisoner's dilemma, the chain-store game (a game with real world business implications), and essentially, the colored hats puzzle (with muddy children, instead). The "backward induction paradox" and many topics mentioned in that paper are not developed fully here. A second paper about induction in game theory is [127], about backward induction in mortgage analysis, with reference to Monte Carlo analysis in finance. The paper [301] addresses "inductive game theory" with respect to Nash equilibrium and other topics. In [530, 103–108] is an inductive proof of a major result due to Snevily regarding Chvátal points in a simple game (details go beyond the intended scope of this section).

The following sections on games only include the analysis of a small class of games where induction is highlighted. For many more examples of induction applied to games, see [462]. For a general introduction to games that has many carefully worked simple examples regarding matching algorithms, social justice issues (including voting theory and Arrow's theorem), cooperative games, and many others, see [241].

In many simple finite two-person games, analysis of strategy can invoke inductive reasoning; after all, a player's next move often depends on the position that arose from previous moves. In *On Numbers and Games* [115], John Conway developed a notation for such games that is inductively defined.

17.2 Tree games

17.2.1 Definitions and terminology

Definition 17.2.1. A *tree game* is a two-player (White and Black) game where the players take alternate turns making a move, where

(i) at each turn the next player has only finitely many possible moves,

(ii) each player knows the moves of the opponent (the game is a "perfect information game"), and

(iii) the game takes at most a predetermined number of moves, and at or before the end of that number of moves, the outcome (one player wins or there is a draw) is determined.

Note: Some authors (*e.g.*, see [533]) insist that there are no draws in their definition of a tree game.

For example, tic-tac-toe is a tree game, but poker is not. Bridge is a tree game, where a player is really a team of two players. Chess can be considered as a tree game, for under many systems of rules, if a position is repeated three times, a draw is declared. (Many chess games are timed, or have a limit of say, 52 moves.) The famous compilation *Winning Ways* [49] (the older version has two volumes, a more recent edition has four volumes) is perhaps the bible of two-person tree games; see also [246]. Backgammon, Hex, and NIM are all tree games. Rock-scissors-paper is not a tree game as players do not alternate moves. For topics covered here and more, the delightful book *Excursions into Mathematics* [42, Ch. 5] contains a very easy to read introduction to game theory, in particular, tree games. Similar comments can be made about [533], although it is probably harder to find outside of the Chicago area.

Tree games can be represented by a rooted tree (see Section 15.2 for relevant terminology): the vertices of the tree are the possible positions of the game (where the root is a starting position) and there is a directed edge from position p_1 to p_2 iff there is a proper move from p_1 to p_2. [To get an actual tree structure, some positions may be repeated; also, a game may have many starting positions, so one null position might serve as a root for the tree—or could one call these "forest games"?] Condition (i) above ensures that the associated tree is locally finite (for each vertex v, there are only a finite number of positions p_i so that $v \rightarrow p_i$). Condition (iii) above guarantees that the tree has only finite height (the length of its longest branch).

A *strategy* for playing a game is a set of rules that determine the moves of a player, that is, if $\mathcal{S}$ is the set of possible positions of a game, then a strategy is a function $\psi : \mathcal{S} \rightarrow \mathcal{S}$.

Convention dictates that White is the player who moves first. If it is possible for White to win no matter what Black then moves, one says that "White to win" is a *natural outcome*, or that White has a winning strategy. Similarly, define the natural outcome "Black to win". If both players can prevent the other from winning, the natural outcome is a draw. If $\mathcal{G}$ is a game, then $\mathcal{G}'$ is the game with the roles of Black and White reversed; $\mathcal{G}'$ is called an inverted game, the inverse of $\mathcal{G}$. If $\mathcal{G}$ is a game with a natural outcome, then $\mathcal{G}'$ also has a natural outcome.

A position of the game is called an N-position if the next player to move can force a win, and a position is called a P-position if the previous player can force a win. An N-position is often called a *winning position* and a P-position is called a *losing position*. These definitions avoid one having to say that a position is a winning or losing position for which player.

Remark: A position is an N-position iff every proper move results in a P-position, and a position is a P-position iff every proper move then results in an N-position.

Exercise 569. *Prove by induction on the length of the longest sequence of moves that every tree game has a natural outcome.*

Exercise 570 (December 31 game). *Two players alternately select from among the 365 dates in a calendar year. On any move, a player can increase the month, or* *the day, but not both. The starting position is January 1, and the player naming December 31 is the winner. The first player can name any day in January, or the first of any month. Derive a winning strategy for the first player.*

17.2.2 The game of NIM

The game of NIM is played by two players. To begin, some stones are presented in various piles. A move consists of one player removing any positive number of stones from any one pile. The players alternately make a move. The winner is the player who picks up the last stone(s). For example, if there are only two piles each consisting of two stones and the first player removes both stones from the first pile, the second player wins by removing both stones from the second pile. If however, the first player removes only one stone from the first pile, the second player can force a win by removing only one stone from the second pile (because the first player can not remove both remaining stones on the next move).

Consider p piles to be in some order, and at some point in the game, suppose that for each i, let n_i denote the number of stones in the i-th pile. Such a position is denoted by $(n_1, n_2, \ldots, n_p)$. In the above example with starting position $(2,2)$, the position $(1,1)$ is a losing position (a P-position); the position $(0,0)$ is also a P-position.

The game with starting position $(1,3,5,7)$ is now popular; a variant of it was played in the movie "Last year at Marienbad" (see [42, p. 341]). The word "NIM" apparently comes from the German word "nehmen", which means "to take". [In Germany, take-out food is ordered "nehmen mit", to "take with".]

Exercise 571. *Prove that the position $(1, n, n+1)$ is a losing position in NIM if and only if n is even.*

The secret to finding a winning strategy in NIM is in evaluating what are called "NIM-sums". For a position $(n_1, n_2, \ldots, n_t)$, define $\sigma_i = \sigma_i(n_1, \ldots, n_t)$ to be sum of the i-th binary digits of n_j $(1 \leq j \leq t)$. The following is well-known:

Theorem 17.2.2. *A position $(n_1, n_2, \ldots, n_t)$ in NIM is a P-position iff for each $j = 1, \ldots, t$, the NIM-sum σ_i is even.*

So a winning strategy for a game of NIM is to remove stones from one pile that makes all NIM-sums even. One might notice that if a position already has all NIM-sums even, then it is impossible to remove any stones without destroying this pattern. The example $(1,3,5,7)$ is a losing position (P-position) because 1=1, 3=11, 5=101, and 7=111; the NIM-sums are (from left to right) 2, 2, and 4 respectively.

As another example, $(41, 58, 26, 9)$ is a P-position: 41=101001, 58=111010, 26=11010, and 9=1001. The sums of the six columns are (from left to right) 2,2,2,0,2,2, respectively.

On the other hand, the position $(150, 37, 93, 106)$ is an N-position:

$$150 = 10010110, 37 = 100101, 93 = 1011101, \text{ and } 106 = 1101010,$$

and the column sums are 1,2,2,2,2,3,2,2 respectively (so bad columns correspond to the digits in the 2^7 column and the 2^2 column). In fact, there is only one winning move, removing $132 = 128 + 4 = 2^7 + 2^2$ from the first pile:

$$150 = 10010110 \rightarrow 00010010 = 18.$$

The following exercise is given without solution, however the same idea is used in Exercise 573, which does have a solution.

Exercise 572. *Prove Theorem 17.2.2. Hint: Induct on the sum of the n_i's.*

One can extend the game of NIM to NIM(k), where the rules are the same as NIM, except that a player may remove stones from up to k different piles. (So NIM is the same as NIM(1).) Call a position in NIM(k) *satisfactory* if for every i, $\sigma_i \equiv 0$ (mod $k+1$).

The following exercise is non-trivial.

Exercise 573. *Prove by induction on $n = n_1 + \cdots + n_t$ that a position $(n_1, \ldots, n_t)$ in NIM(2) is a losing position (P-position) if and only if it is satisfactory.*

For example, the position $(7, 8, 9, 10)$ is an N-position for NIM(2); the move to $(1, 8, 9, 9)$ creates a P-position.

The reader might have already observed that Theorem 17.2.2 says that a position in NIM is a P-position iff for every i, $\sigma_i \equiv 0$ (mod 2), and so is a satisfactory position. The interested might guess that NIM and NIM(2) are special cases of the following exercise (given without solution).

Exercise 574. *Prove that for each $k \geq 1$, a position in NIM(k) is a P-position iff it is satisfactory.*

17.2.3 Chess

In chess, a knight's move consists in moving two squares vertically and one horizontally, or or two squares horizontally and one vertically.

Exercise 575. *Suppose that a knight sits on a chessboard, infinite in every direction. Let $f(n)$ denote the number of squares the knight could reach after precisely n moves. Observe that from Figure 17.2.3, $f(0) = 1$ and $f(1) = 8$. Prove that $f(2) = 33$, and for $n \geq 3$, $f(n) = 7n^2 + 4n + 1$.*

Exercise 576. *Let an infinite chessboard have squares labelled (m, n), where m and n are positive integers. Show that by starting at $(1, 1)$, a knight can reach any square on the chessboard in a finite number of moves.*

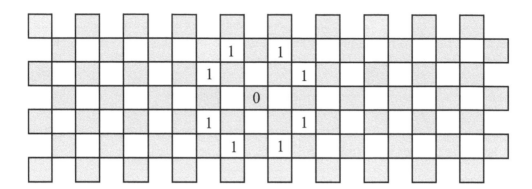

Figure 17.1: Squares reachable by knight after 0 or 1 moves

17.3   Tiling with dominoes and trominoes

The next few problems fall under what are known as *tiling* problems. A *polyomino* of order s is a shape consisting of s identical squares attached by edges. Solomon W. Golomb introduced polyominoes in 1954 [218], and wrote a popular book [219] on the topic. For the interested reader, a standard modern reference is George Martin's book [365] on the subject. (See also [292, p. 45], for further references.)

A domino is a shape formed by joining two squares of same size along an edge. (In actuality, dominoes have dots on them, but the dots will not be used here.) A standard problem regarding dominoes is found in Exercise 577, related to Fibonacci numbers.

Exercise 577. *Show that the number of ways to place dominoes in a $2 \times n$ array is a Fibonacci number.*

There are only two polyominoes of order 3, called *trominoes*, namely three squares in a row, and three forming the shape of an L. There are many tiling problems for polyominoes in general: however, only a small sample of problems for L-shaped trominoes is given here.

To gain familiarity with the general idea, the reader might first try to take three L-shaped trominoes and form a perfect 3×3 square—it can't be done. How about a 3×5 rectangle? A 5×5 board with a missing square next to a corner can not be covered by L-shaped trominoes; however, such a board with a corner missing can be tiled.

The next exercise is rather easy, and does not really need induction:

Exercise 578. *For any m and n that are multiples of 2 and 3 respectively, an $m \times n$ checkerboard (with no squares missing) can be covered with L-shaped trominoes.*

Exercise 579. *For any $n \geq 2$, a $6 \times n$ checkerboard can be covered by L-shaped trominoes.*

Exercise 580. *For each $n \in \mathbb{Z}^+$, a $2^n \times 2^n$ checkerboard with any one square removed can be covered with L-shaped trominoes.*

The next two exercises are far more challenging; solutions are only outlined.

Exercise 581. *For any even number $n \geq 14$ that is not a multiple of 3, an $n \times n$ checkerboard with any one square removed can be covered by L-shaped trominoes.*

Exercise 582. *For $n > 5$, every $n \times n$ checkerboard with a square missing can be covered with L-shaped trominoes if n is odd and 3 divides $n^2 - 1$.*

17.4 Dirty faces, cheating wives, muddy children, and colored hats

Problems and puzzles in this section may be roughly described to be of the type

I know that you know that I know that you know that I know ...

The remarkable feature of the puzzles in this section is that information is gained by people repeatedly saying "I don't know." Some authors might loosely classify the problems in this section as "knowledge propagation" problems, or "common knowledge" problems, falling into a modern area of epistemology some have recently called "reasoning about knowledge" (see [184]).

The variety of problems of this type now includes puzzles regarding (to name but a few) dirty faces, holding cards up to one's forehead, cheating wives, muddy children, and colored hats. (See [511] and other references below for examples.) Some of these problems are simply a restatement of others; some enjoy subtle differences. Only a very brief chronology of four such problems is given here, the first three of which are essentially the same problem, and the fourth a slight strengthening of the first three.

17.4.1 A parlor game with sooty fingers

According to [66], the history of the following problem goes back to at least 1832, from a game where people pinch their neighbor on the face, and some of the people have soot on their fingers. If two people end up with a smudge on their faces, everybody laughs, but each of the two with a smudge think that the rest are laughing at the other. Also according to [66], a version with three people was mentioned in the mathematical literature in 1935.

One of the popular variants appeared in Littlewood's 1953 *A Mathematician's Miscellany* [348, p.3] (or [349, p. 25]). Implicit in Littlewood's story are the following rules or assumptions:

- Three intelligent and honest women A, B, and C, are riding together on a train.

- Some or all of the women have a dirty face, each woman not knowing the condition of their own face.

- If a woman sees dirt on the face of another, she begins to laugh.

- Any woman who is laughing stops laughing if and when she (correctly) deduces that she also has a dirty face.

The original statement of the problem starts out by saying that all three women have dirty faces, and hence all begin to laugh; can each deduce the condition of their own face?

Can A deduce that her face is dirty? She might argue as follows: Suppose that her face is not dirty. Then B ought to know that B has a dirty face because C is laughing at someone other than A. Since B is still laughing, A deduces that her face must be dirty.

Littlewood points out that an extension to n ladies, all laughing, is possible and is provable by induction. How would such an argument go? The induction essentially proceeds on the number of dirty ladies—and some kind of time-step for each logical conclusion: suppose that ladies $A_1, \ldots, A_n$ all have dirty faces. Then A_1 can conclude her face is dirty when the remaining $n - 1$ women have not, after a reasonable amount of time, yet concluded what their own conditions are.

What is a "reasonable" amount of time? The next two variations of the dirty face puzzle have since clarified how long is reasonable by incorporating into the puzzle steps that can count the decisions and inferences being made.

17.4.2   Unfaithful wives

In 1958, George Gamow and Marvin Stern [206, pp. 20–23] published the problem (or puzzle, or paradox) of the "unfaithful wives".

The unfaithful wives problem highlights that the reasoning by the women in the train not only must take place in steps, but also that the reasoning also works when not all of the faces are dirty. [The analysis for dirty faces on the train also works when not all women are dirty.]

In the original version, Sultan Ibn-al-Kuz knows that 41 wives in a city are unfaithful, including the wife of his vizier, and decides that something needs to be done. The sultan proclaims that if a husband can correctly deduce that his wife is cheating, he should shoot her. The vizier was not so intelligent as the rest of the citizens in the city, and only 40 of these wives were discovered and shot. The version here only has 40 cheating wives. The assumptions are:

- A city has $n \geq 40$ married couples, and a Sultan who rules the city; the Sultan is not married.

- Precisely 40 of the wives are cheating on their husbands.

- The Sultan announces that at least one wife is cheating, but does not share precisely how many.

- If a particular wife is cheating, everybody in the city except her husband is aware of the infidelity.

- Husbands never find out about their wives' activities through personal communication.

- All husbands are honest, intelligent, and think logically.

- If a husband of a cheating wife can logically (and correctly) deduce that his wife has been stepping out, then he makes such a deduction, and by an order from the Sultan of the city, shoots his wife.

Two more assumptions are implicit in the problem, but for clarity are given explicitly:

- All shootings are to occur at roughly the same time each evening.

- All husbands can hear precisely how many shots, if any, are fired each evening.

On the day the proclamation was announced (day 1), no shots are fired (and all citizens "hear" no shots). Up to and including day 39, no shots are fired, however, on the 40th day, there were 40 shots, all of the cheating wives being shot (and no more). [The version from [206] actually started on day 0, and on day 40, forty wives were shot, but one wife escaped execution because her husband was incapable of the logic required.]

What is the logic behind the unfaithful wives problem? Rather than totally expounding on such a morbid, insensitive, and politically incorrect puzzle, only a brief discussion is given here. The "muddy children" puzzle in the next section has precisely the same ingredients, and so more of the logic is examined then.

Rather than forty wives cheating, suppose that only one wife, W is cheating on husband H. On day 1, all but H know that W is cheating, and H knows that no other wife is cheating, but because of the announcement by the Sultan, knows that at least one wife is unfaithful, so it must be his wife.

The next simplest case is when two wives are cheating, say W_1 and W_2, married to H_1 and H_2 respectively. On day 1, both H_1 and H_2 know about each other's wife, but not about their own wives. At the end of the first day, no shots are fired, and so each concludes that at least two have been unfaithful (otherwise the scenario is like in the last paragraph, and someone would have been shot). Since, for example, H_1 knows that W_2 is cheating and at most one more is cheating (his own wife, W_1), H_1 deduces that his wife is cheating; H_2 arrives at a similar conclusion, so on day 2, both W_1 and W_2 are shot.

An inductive argument then shows how it was that on the fortieth day, all cheating women were shot. See the next section (Exercise 583) for the proof.

17.4.3 The muddy children puzzle

A more sensitive variant of the cheating wives puzzle, which is "mathematically identical", has recently become a standard presentation. This variant is called the "muddy children puzzle", perhaps first formulated by Barwise [39] in 1981. A slight variation of this puzzle is as follows:

- $n \geq 1$ children are in a playground, all honest and very clever.

- Each child may or may not have a muddy forehead; each child cannot see mud on their own forehead, but can clearly see all other foreheads.

- There are $k \geq 0$ children with a muddy forehead.

- A teacher proclaims publicly that at least one child has a muddy forehead, that is, the teacher informs the group of children that $k \geq 1$. The teacher does not state the precise value of k.

- Once each minute, the teacher announces the same statement: "if any of you know for certain that your forehead is muddy, all at once, raise your hand—now".

When $n = 1$, then $k = 1$ and so the one child knows before the first announcement by the teacher.

When $n = 2$ and $k = 1$, the muddy child sees no mud on the other, and so can raise a hand upon the first announcement; the non-muddy child sees mud on the other, and cannot yet tell the state of his/her own forehead. When $n = 2$ and $k = 2$, each of the two children see a muddy forehead, and so upon the first announcement, neither can raise a hand. Upon the second announcement, each may conclude properly that their own forehead is muddy; each argues that if only one had a muddy forehead, one would see no muddy forehead and so that child would have been able to deduce mud on his/her forehead before the first announcement.

The general claim is that after $k - 1$ announcements, nobody has yet put up a hand, and after the kth announcement, only the k muddy children raise a hand. However simple this claim is, in order to prove it by induction, it is convenient to state the claim in much more detail:

Proposition 17.4.1. *If $n \geq 1$ children play the above game, where precisely $k \geq 1$ of the children have muddy foreheads ($1 \leq k \leq n$), then after each of the first $k - 1$ repetitions of the teacher's announcement, no child raises a hand, but only after the kth announcement, all of the k muddy children raise their hand (and no others do) by using reasoning based on seeing only $k - 1$ other muddy faces. Furthermore, after the $(k + 1)$st announcement, all children raise their hand, the non-muddy children basing their conclusions upon seeing k muddy faces (and those children raising their hands upon the kth announcement).*

The solution to the following exercise is given:

Exercise 583. *Prove Proposition 17.4.1 by induction on k.*

17.4.4 Colored hats

In Martin Gardner's book *Penrose Tiles to Trapdoor Ciphers* [214, pp. 138–149], the dirty faces (or muddy children) puzzle is taken to yet another level.

Three men, A, B, and C are in a dark room, and someone puts on each head either a red hat or a black hat. The lights are turned on, and each can see the hats of the other two men, but not his own. Any man seeing a red hat must raise his hand. The first person to deduce the color of his own hat is the winner.

If all three hats are red, everyone raises their hand, and the fastest thinker of the three, say C, could deduce the color of his own hat as follows: if C's hat were to have been black, A would know that since B sees a red hat, it must be A's, so A would quickly announce that his hat was red, but he hasn't done so yet, so C deduces that his hat is red.

Consider the case with four men, all with red hats, and suppose that D is even quicker than the rest. He reasons as follows: "The other three have red hats. If I have a black hat, the remaining three have red hats, and they now are in the situation above. So if I have a black hat, the fastest of the remaining three, say C again, will deduce his hat is red, but he doesn't, so mine must be red."

Arguing by induction, if n men have red hats, the fastest thinker, after waiting an appropriate time, would deduce that his hat is red. Of course, the fastest thinker would have to know how long it would take C to make the first deduction, and failing that, how fast it would take D to deduce his color, and so on, so in this solution, one assumes that the fastest thinker also knows how fast the others are.

To eliminate vague aspects of the aforementioned inductive solution, suppose that there are n men seated in a column, one behind another, so that each can only see the color of hats on men in front of them. There only $n - 1$ black hats and n red hats (and the men know this). Each of the men are asked, in order (from the back of the column), if they know the color of their own hat. Assume that each is capable of making the deductive reasoning as above (and they are honest). For the case of three men, all with red hats, the man in the back, A, answers "no", because he sees two red hats, and knows that one remains, perhaps on his head or not. B sees one red hat and similarly can not deduce his color, so replies "no". Then C deduces that, because A said "no", there is at least one red hat in front of C, and if B were to have seen only a black hat, B would know that his was red; so B's negative response indicates that B saw a red hat, and C answers "yes".

Exercise 584. *Extend this last scenario to n men, (n red hats, n − 1 black hats, and all receive red) and prove that the man sitting at the front can be blind and still deduce the that his hat is red.*

Returning to the original scenario with three men A, B, and C, each being able to see the other two. If questioned in order, A will say "no", B will say "no" and then C will say "yes". In fact, before A is questioned, both B and C know that A will say "no", so to ask A provides no useful information. However, if the questioning begins with B (skipping A), it seems that C can not make his deduction. Is this a paradox? Not really, since when A is asked, C does not know that B knows A will say "no".

17.4.5   More related puzzles and references

See Gardner's book [214] for a more detailed and lively discussion of induction and colored hats. Gardner also expands on a wonderful paradox regarding cards with consecutive integers on opposite sides, taken from Littlewood's *Mathematician's Miscellany* [349, p. 26], a "monstrous hypothesis" attributed to the physicist Erwin Schrödinger. [Note: There is a mistake in Gardner's bibliography; the title he gave Littlewood's book was *A Mathematician's Apology*, the title of a book by G. H. Hardy (1877–1947).] Gardner also cites [81], [144], [211, prob. 87], [378], and [417], (among others) as references for further reading.

Terence Tao recently put a version of the muddy children puzzle (called the "blue-eyed islander puzzle") on his blog [526]. The comments posted might make very interesting reading, as many readers seem to try very hard to find a flaw in the language used to present the puzzle.

A recent (2009) article by E. Brown and J. Tanton [76] is an entertaining introduction to the colored hats problem (and many of its variants). The article [66] also has many other puzzles related to the muddy children puzzle, including the "consecutive integer game" (briefly mentioned above) the "arithmetic mean game" (by David Silverman), the "sum game" (by Andy Liu), and the powerful "Conway–Paterson game" (due to John Horton Conway and Mike Paterson), which generalizes the muddy children puzzle.

See [548] or [258] for on-line articles for more on "reasoning about knowledge". As mentioned above, the book *Reasoning About Knowledge* [184] describes many applications of the kind of reasoning shown above to game theory, economics, and communication theory.

17.5   Detecting a counterfeit coin

One is given a collection of $m \geq 3$ coins that all look the same, but one of which is known to be counterfeit and has a weight slightly different from that of a genuine coin. Using only a balance scale (with two pans and equal arm lengths), how many weighings are necessary to find the coin? This problem can be stated in another way:

Question 17.5.1. *For an integer $n \geq 1$, what is the maximum number of coins from among which a (single) counterfeit can be detected by at most n weighings.*

Since the actual weight of the counterfeit coin is unknown, but close to the original, the only kind of useful weighing is with the same number of coins are put in each pan (call them left and right). Note that for two coins, there is essentially only one weighing, and such a weighing does not reveal which is counterfeit, only which is heavier. It follows then that in Question 17.5.1 only $n \geq 2$ and $m \geq 3$ need be examined.

A fairly simple argument shows that a counterfeit coin can be found from among at most $m = 3$ coins with at most $n = 2$ weighings: Suppose that one of three coins C_1, C_2, and C_3 is counterfeit. Setting one coin aside, say C_1, put C_2 in the left pan and C_3 in the right. If the scale balances, then neither of C_2 or C_3 is counterfeit, and so C_1 is the fake (found after only one weighing). If the scale tilts to one side, then C_1 is not counterfeit, and one more weighing with C_1 against C_2 gives enough information to to say which is the counterfeit (and whether or not it is heavy or light).

With a bit of work, one can see that two weighings are not sufficient to determine a counterfeit from among $m \geq 4$ coins, but three weighings are. So the answer to Question 17.5.1 for $n = 2$ is $m = 3$ coins.

The next case, for three weighings, is already a bit complicated. It turns out that from among at most 12 coins, a counterfeit can be found with three weighings. There are essentially two different sequences of weighings, one sequence being contingent upon outcomes, and another sequence prescribed in advance.

For the contingency method, label 12 coins A, B, C, ..., J, and K. Follow the weighings in the chart below, depending on the outcomes; follow the left branch if the left side is heavy, the center branch if the scales balance, and the right branch if the right side is heavy.

Notice that there are only 24 possible outcomes, as three positions for the final weighing are impossible (due to information gained from first two weighings).

Examine more closely the second weighing; some information from the first is carried down. In the first case, where ABCD is heavier than EFGH, then I, J, K, and L are all genuine, so in the first weighing of the second row, I can be taken as a genuine (or test) coin. Furthermore, A, B, and C are not light, and E and F are not heavy.

For the second case in the second row (when ABCD–EFGH balances), eight coins are known to be genuine, leaving only I, J, K, and L, as unknown. Any one of the first coins, in this case, A, can be used as a reference (or test) coin. The third case in the second row is symmetric with the first case.

These two patterns (3 not light, 2 not heavy, and 1 test coin, or, 4 unknown coins and a test coin) are actually early cases of some special variants of Question 17.5.1. To give a solution to Question 17.5.1, certain special cases or variants of the question are shown first (each by induction). Some variants of the counterfeit coin

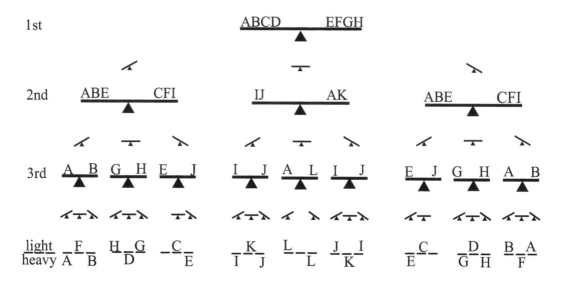

Figure 17.2: Three weighings to find counterfeit among 12 coins

problem may now seem natural:

- The weight of the counterfeit coin is known (to be, say, lighter).

- One marked coin is known to be genuine (a reference, or labelled test coin, is one of the coins).

- All coins are either white or black; if the counterfeit is white, it is light, and if the counterfeit is black, it is heavy.

- Identify the n weighings in advance (not using information from one weighing to decide what to weigh next). [Suppose that some algorithm successfully identifies a counterfeit coin from among m coins using n weighings. If this algorithm has steps that depend on previous weighings, such an algorithm is called a *contingency* algorithm. If a sequence of n weighings can be prescribed in advance, then this sequence is called a *prescriptive* algorithm. So this variant is to insist upon a prescriptive algorithm.]

- At most one coin is counterfeit (as opposed to exactly one).

- After the counterfeit coin is identified, one must identify whether it is heavier or lighter than a genuine coin.

- There are two (or more) counterfeit coins.

- There are any number less than $m/2$ counterfeit coins (there may be none).

The answer to the first variant (and including the sixth variant) is surprisingly simple:

Exercise 585. *Suppose that a counterfeit coin is always lighter than a genuine coin. Use induction to prove that for each $n \in \mathbb{Z}^+$, 3^n is the greatest number of coins for which a single counterfeit coin can identified by (at most) n weighings. Hint: Partition 3^n coins into three equal groups.*

The information that a counterfeit is always lighter is rather powerful; losing this information reduces the number of coins (to about half—see below) that work for n weighings, but this number is still 3^n for the black-white variant above:

Lemma 17.5.2. *Suppose that all coins are either white or black; if the counterfeit is white, it is light, and if the counterfeit is black, it is heavy. Furthermore, assume that the number of white coins and the number of black coins differ by at most one. Under these conditions, for each $n \in \mathbb{Z}^+$, a counterfeit coin can be identified and classified among 3^n coins by (at most) n weighings.*

Proof outline: For each $n \in \mathbb{Z}^+$, let $C(n)$ be the statement of the lemma.

BASE STEP: For $n = 1$, consider $3^1 = 3$ coins, and without loss of generality, suppose that two are black and one is white. Put a black on each pan and set the white aside; if the scale balances, the white coin is counterfeit, and since it is white, is lighter than a genuine coin. If the scales tips, say the left side down, then the black on the left is heavy (the black on the right can not be light). In any case, the counterfeit coin is identified from among three coins and classified as heavy or light, so $C(1)$ holds.

INDUCTION IDEA: To see how the induction works, it is helpful to look next at the case $n = 3$, assuming, for the moment, that $C(2)$ has been shown. Consider $3^3 = 27$ coins, with, say, 14 black, and 13 white. Partition these 27 coins into three groups, say G_1 with 4 black and 5 white, and G_2 and G_3 each with 5 black and 4 white. Put G_2 on the left, weighed against G_3 on the right. If the scale balances, the counterfeit coin is in group G_1 and so $C(2)$ applies to a set of 9 coins, 4 black and 5 white. If the left pan goes down, either one of the 5 black from G_2 is heavy, or one of the 4 white from G_3 is light; the induction hypothesis $C(2)$ now applies to these $3^2 = 9$ coins. Similarly, if the scale tips to the right, the counterfeit is among the 4 white in G_2 or the 5 black in G_3, and again $C(2)$ applies.

Once the inductive step is proved (see Exercise 586 below), the result is true by mathematical induction. $\qquad\square$

Exercise 586. *Finish the inductive step in the proof of Lemma 17.5.2.*

Two points are to be made: First, Lemma 17.5.2 is still true for any distribution of black and white (see [364]). Second, Lemma 17.5.2 is optimal; if more than 3^n coins are used, then at least $n + 1$ weighings are required (no matter what the distribution of black and white).

The following discussion might make the second assertion believable. Since each weighing yields one of L, R, or B(alance), and in n weighings, there are 3^n (theoretically) possible outcomes, one could not expect to have gathered enough information when more coins are used.

In fact, when one coin is counterfeit, a sequence of n balances does not help to find and classify the coin, so at most $3^n - 1$ outcomes make sense. When the relative weight of the counterfeit is required, the number of different outcomes must be at least twice the number m of coins (each could be light or heavy), $3^n - 1 \geq 2m$. In fact, m must be slightly smaller, but before presenting the answer to Question 17.5.1, another variant is solved, where one coin is known as a test-coin.

Lemma 17.5.3. *Fix $n \in \mathbb{Z}^+$, and consider a collection of coins, exactly one of which is counterfeit, and one genuine coin is marked "test-coin". Then $m = \frac{3^n - 1}{2}$ is the maximum number of coins so that n weighings identifies the counterfeit coin and finds if it is light or heavy.*

The following gives an answer to Question 17.5.1:

Theorem 17.5.4. *For $n \in \mathbb{Z}^+$, $\frac{3^n - 3}{2}$ is the maximum number of coins, one of which is counterfeit, so that n weighings identifies the coin and finds if it is light or heavy. If one needs only locate a counterfeit coin, and not classify it as light or heavy, then one more coin can be used, giving $\frac{3^n - 1}{2}$ coins (and this number is optimal).*

17.6 More recreations

In this section, only four puzzles are given, the first two of which have solutions that involve iteration.

17.6.1 Pennies in boxes

Online sources say that this next problem is first found in [98, pp. 27–28], where there are many other applications and games that are solved by iterative processes.

Exercise 587. *Suppose that N pennies are distributed among n boxes labelled $B_1, B_2, \ldots, B_n$. If box B_i and B_j have p_i and p_j pennies respectively, with $p_i \leq p_j$, an allowable move is to transfer p_i pennies from B_j to B_i (doubling the number of pennies in the less full box). Prove that no matter what the original distribution of pennies is, it is possible after only finitely many allowable moves, to transfer all pennies into one or two boxes, and if the total number N of pennies is a power of 2, then all pennies can be transferred to just one box.*

17.6.2 Josephus problem

The next problem is often called the "Josephus" problem, or the "Flavius" problem. Josephus ben Matthias was a Jewish "priest" who became an unwilling soldier in the

Jewish rebellion against the Romans (in fact, he was made commander of Galilee). The Jews were defeated in 66 A.D. and Josephus was taken prisoner under Vespasion Flavius. Two years later, Vespasion was named emperor, and subsequently, Josephus was released, took the emperor's family name (calling himself Josephus Flavius) and then became a soldier and historian for the Romans. Much of the history he writes about, he himself took part in.

The situation regarding the capture of Josephus is the basis of an intriguing [to some, at least] mathematical problem. Facing ultimate defeat, Josephus's men agreed that it would be better to commit suicide rather than to be captured and enslaved by the Romans; Josephus was of another mind—his men could not convince Josephus to join the mass suicide. In fact, some passages of Josephus's writings seem to indicate a certain animosity on behalf of his soldiers and they wanted him dead. Josephus seemed to agree to join in if there were some kind of rule that would determine how they would all die, a rule which perhaps appeared random, however "allowed" himself (and one co-conspirator) to survive. In his own words [295, p. 787, 3.8.7, line 388]:

> "And now" said he, "since it is resolved among you that you will die, come on, let us commit our mutual deaths to determination by lot. He whom the lot falls to first, let him be killed by him that hath the second lot, and thus fortune shall make its progress through us all: nor shall any of us perish by his own right hand, for it would be unfair if, when the rest are gone, somebody should repent and save himself."

Josephus did not say how lots were arranged—in his biography, he just continues (line 391) "yet was he [Josephus] with another [co-conspirator] left to the last, whether he must say it happened so by chance, or whether by providence of God;...", that he made a pact with another last survivor [whom I called the "co-conspirator" above, though he might have been innocent] and then the story moves on. [The order of execution might have been "everyone else first"!]

From this story, somehow a mathematical question arose; I have not located precisely who first invented this question, but somehow the story above has evolved (into many similar stories). Perhaps someone first asked "if the pattern of suicides is as follows, ..., where should Josephus place himself in the pattern to survive?" At least in the biographical-historical records of Josephus, there does not seem to be any evidence of Josephus applying mathematics to the problem.

So one story goes (see [267, pp. 121–126]) Josephus and 39 others remained in a cave, and to avoid their ultimate defeat and surrender, they formed a suicide pact. The soldiers were to stand around in a circle, say numbered 1–40. Beginning with soldier number 7, every seventh surviving soldier in rotation was to be killed. In one version, Flavius did not want to die, and so calculated where he must stand to be the last person standing. (The soldier to be killed was to be killed by the next soldier in the sequence, but for present purposes, it suffices to assume that each commits suicide.) In order, soldiers numbered 7, 14, 21, 28, 35, 2, 10, 18, 27, 36, 4, 13, 25, 32, 1, 12, 24, 34, 6, 19, 31, 5, 20, 37, 11, 29, 8, 26, 9, 33, 17, 3, 40, 39, 15, 22, 38, 16, 25, 30 are to be killed, so Josephus stood in position 30. [Note: in [267], they claim position 24 is the right one; however, this must be because the soldier numbered 1 was the first to go.] One could ask a more general question: what if Josephus had a co-conspirator and both were to be spared? Where should the two of them stand to be in the last two positions standing?

In another version (see [230, 8–20]) there were 41 soldiers in all, and every third soldier is killed, in which case Josephus and his friend should then stand in positions 16 and 31 if to survive. For more history of this problem, further references, and other interpretations and generalizations, see [267] and [481].

For each $n \geq 2$, and $q \geq 2$ (it is possible that $q \geq n$) define $J(n,q)$ to be the last surviving position when n soldiers are in a circle and beginning with position q, every q-th remaining soldier is killed. Apparently (see [267]), there is no closed formula for $J(n,q)$, however there is one for $J(n,2)$ (given as an exercise below). Before looking at the result in the exercise below, here is a chart for some small values of n; can you spot the pattern?

n	2	3	4	5	6	7	8	9	10	11	12	13	14
$J(n,2)$	1	3	1	3	5	7	1	3	5	7	9	11	13

An expression for $J(n,2)$ is remarkably easy to compute (see Exercise 588 below), and one proof for this expression relies on an idea contained following more general lemma.

Lemma 17.6.1. *For positive integers n and q (where $q \geq n$ is allowed),*

$$J(n+1,q) = J(n,q) + q \pmod{n+1}. \tag{17.1}$$

Proof: Suppose that $n+1$ soldiers are in a circle, named $1, 2, \ldots, n+1$. The first soldier to die is q modulo $n+1$. The remaining n soldiers are

$$1, 2, \ldots, q-1, q+1, \ldots, n+1 \pmod{n+1},$$

and starting with $q+1$, the soldiers

$$q+1, q+2, \ldots, n, n+1, 1, \ldots, q-1,$$

form a new "game" with $J(n,q) + q$ being the last to go. $\qquad\square$

Note that when $q = 2$, equation (17.1) says that

$$J(n+1,2) = J(n,2) + 2 \pmod{n+1},$$

a key inductive step to finding a formula for $J(n,2)$.

Exercise 588. *Prove that if $n = 2^a + t$, where $t < 2^a$, then $J(n,2) = 2t+1$. Hint: First prove the result by downward induction when n is a power of 2, and then apply upward induction between powers of 2.*

17.6.3 The gossip problem

Consider a collection of n people, each person with a secret unknown to all others. People phone each other sharing all information known at the time of the phone call. Conference calls are not allowed. What is the strategy that allows for the fewest number of phone calls that result in everybody knowing all the secrets?

For only one person, no phone calls are required; for two people, only one call is necessary; for three people, three phone calls are required. With four people, say A,B,C, and D, four calls are sufficient, say A-B, C-D, then A-C and B-D.

One obvious way to communicate all secrets is to first order the people, say, $P_1, P_2, \ldots, P_n$. If P_1 calls P_2, then P_2 calls P_3, and so on until P_{n-1} calls P_n, then P_n is the only person knowing all the secrets. Then P_n knows everything, as does P_{n-1}, so P_n makes $n-2$ calls to inform the first $n-2$, giving $2n-3$ calls in all. However, one can do slightly better.

In advance, the people are divided into four groups—to set ideas, portion the people into roughly quarters, however the same proof works when the four groups are lopsided. One person from each group is identified as a leader. Each leader places phone calls to all other members of his/her group; the four leaders in total placing $n-4$ phone calls. Four phone calls between the leaders then share all secrets, and each leader then returns to call everyone in their group. In all, there are $(n-4) + 4 + (n-4) = 2n - 4$ phone calls.

Exercise 589. *Show by induction that at most $2n-4$ calls among n people are required for the gossip problem.*

17.6.4 Cars on a circular track

Exercise 590. *There are n identical cars (with engines off) stationed around a circular track. Only enough gas for one car to complete a lap is distributed among all cars. Show that there is a car that can make its way around a lap by collecting gas from the other cars on its way.*

Chapter 18

Relations and functions

That flower of modern mathematical thought—the notion of a function.

—Thomas J. McCormack,

On the nature of scientific law and scientific explanation.

18.1 Binary relations

For more definitions relating to this section, see Section 2.6. Let A and B be sets. A *binary relation* from A to B is a subset of the ordered pairs $A \times B = \{(x, y) : x \in A, y \in B\}$.

If R is a binary relation from A to B, and S is a binary relation from B to C, the *composition* of R with S is a binary relation from A to C defined by

$$R \circ S = \{(x, z) \in A \times C : \exists y \in B[(x, y) \in R, \ (y, z) \in S]\}.$$

(Caution: this notation can differ with the common notation for composition of functions—see comments following that definition.) Note that if $R \subseteq A \times B$, $S \subseteq B \times C$, and $T \subseteq C \times D$ are binary relations, then

$$(R \circ S) \circ T = R \circ (S \circ T);$$

so in a sense, composition of relations is associative.

If R is a binary relation from A to A (that is, $R \subset A \times A = A^2$) R is said to be a binary relation *on* A. For a binary relation on a set A, define R^1 to be simply R, and for each $n \geq 2$, define recursively another relation on A, called the *n*-th iterate of R, by

$$R^n = \{(x, z) : \text{ if there exists } y \in A \text{ so that } (x, y) \in R^{n-1} \text{ and } (y, z) \in R\}.$$

309

Under this definition of R^n, R^{n-1} is composed with R; the reader can check that
if one were to define it recursively as the composition of R with R^{n-1} instead, the
result is identical. For example, if $A = \{a, b, c, d\}$ and

$$R = \{(a, b), (a, c), (b, a), (b, c), (c, d)\},$$

then

$$R^2 = \{(a, a), (a, c), (a, d), (b, b), (b, c), (b, d)\}$$

and

$$R^3 = \{(a, b), (a, c), (a, d), (b, a), (b, c), (b, d)\}.$$

Recall that a binary relation R on A is *symmetric* iff $(a, b) \in R \Rightarrow (b, a) \in R$.

Exercise 591. *Let R be a symmetric relation on a set A. Prove that for each $n \geq 1$,
the relation R^n is also symmetric.*

18.2   Functions

A binary relation f from A to B is a *function* (or *mapping*) from A to B iff for every
$a \in A$, there exists precisely one $b \in B$ satisfying $(a, b) \in f$. If $(a, b) \in f$, one usually
writes $f(a) = b$ and say "f of a is equal to b". In this case, A is called the *domain*
of f, and $\{f(a) : a \in A\}$ is the *range* of f, sometimes abbreviated by "dom(f)" and
"ran(f)" respectively. The set B is sometimes called the *codomain* of f.

Exercise 592. *Let $R \subset A \times B$ be a binary relation from A to B so that for every
$a \in A$, there exists $b \in B$ with $(a, b) \in R$. (In other words, A is the domain for the
relation R.) Use Zorn's lemma to prove that there is a function $f \subset R$ that is a
function from A to B.*

(Compare Exercise 592 to Exercise 594.)

Exercise 593. *Find all functions $f : \mathbb{Z}^+ \to \mathbb{Z}^+$ that satisfy $f(2) = 2$, and for every
$n \in \mathbb{Z}^+$,*

$$f(n + 1) = 1 + 1 \cdot f(1) + 2 \cdot f(3) + \cdots n \cdot f(n).$$

The next exercises use the notation for composition for functions: if $f : A \to B$,
and $g : B \to C$, then the *composition* of f with g is $g \circ f : A \to C$ defined by
$(g \circ f)(x) = g(f(a))$ for each $a \in A$.

Note: In many texts, (for example [95]) the notation for composition of functions
is reversed so that in $(f \circ g)(x) = g(f(x))$; in this case, the function notation would
agree with the relation notation and ordered pairs, and to many students is the
most natural notation. The usage defined for this text, however, seems to be the
most common in American freshman texts; it also agrees with algebraic notation:
the order of action on x starts from the operator closest to x.

Recall that a function $f : A \to B$ is called *surjective* (or is called a *surjection*) (or *onto B*) if for every $b \in B$ there exists $a \in A$ so that $f(a) = b$. The notation I_B is also used to denote the identity function on B, that is, for every $b \in B$, $I_B(b) = b$. (Notation for the identity function on a set X varies widely, including ι_X and Id_X.)

Exercise 594. *Let A and B be sets and let $f : A \to B$ be a function. Prove, using the Axiom of Choice, that if f is surjective, there is a function $g : B \to A$ so that $f \circ g = I_B$.*

Exercise 595. *Let $O_{m,n}$ be the number of surjections (onto functions) from an m element set onto an n element set. Counting onto functions, establish the recursion*

$$O_{m+1,n} = n O_{m,n} + n O_{m,n-1},$$

and use it to prove by induction that for each $m, n \geq 0$,

$$O_{m,n} = \sum_{i=0}^{n} \binom{n}{i} (-1)^{n-i} i^m.$$

Prove also that this number is equal to $n!$ times $S_{m,n}$ (the Stirling number of the second kind, see Exercise 413). Then derive the formula for $O_{m,n}$ directly using the definition of $S_{m,n}$ (see Section 12.9).

Recall that a function $f : A \to B$ is *one-to-one* (or is *injective* or is an *injection*) if for any $a_1, a_2 \in A$, then $f(a_1) = f(a_2) \Rightarrow a_1 = a_2$. Alternatively, stating this in the contrapositive, f is one-to-one if $a_1 \neq a_2 \Rightarrow f(a_1) \neq f(a_2)$.

Exercise 596. *Let X and Y be fixed sets, each with n elements. Let m_n be the number of injections from some subset of X into Y. Prove the recursion*

$$m_n = n m_{n-1} + 1$$

and use this to show by induction that

$$m_n = n! \sum_{k=0}^{n} \frac{1}{k!}.$$

Can you prove this result by a direct argument? Remember that $0! = 1$.

If the domains and ranges of functions are appropriate, then composition of functions is associative, that is, for appropriate f, g, and h, $f \circ (g \circ h) = (f \circ g) \circ h$. In the special case where $f : A \to A$, define f^n to be f composed with itself n times (by associativity, it matters not in what order the composition is carried out). The function f^n is often called the n-th *iterate* of f. For example, if f is a function $f^n : \mathbb{R} \to \mathbb{R}$, defined recursively by $f^1 = f$ and for any $n \geq 2$ (and each $x \in \mathbb{R}$), $f^n(x) = f(f^{n-1}(x))$.

Exercise 597. *For fixed real numbers a and b, define $f : \mathbb{R} \to \mathbb{R}$ by $f(x) = a(x + b) - b$. Prove that for every $n \in \mathbb{Z}^+$, $f^n(x) = a^n(x + b) - b$.*

Exercise 598. *Prove that for any function $f : X \to X$ and all $m, n \in \mathbb{Z}^+$, $f^m \circ f^n = f^{m+n}$.*

Exercise 599. *Assume that a function $f : \mathbb{R} \to \mathbb{R}$ is so that for every $x, y \in \mathbb{R}$,*

$$f(x + y) = f(x) + f(y).$$

Show that $f(0) = 0$ and for every $n \in \mathbb{Z}^+$, $f(n) = nf(1)$.

In part of the solution to the next exercise, induction is used:

Exercise 600. *Find all functions $f : \mathbb{R}^+ \to \mathbb{R}^+$ so that for all $x, y \in \mathbb{R}^+$, (i) $f(x(f(y)) = yf(x)$, and (ii) $f(x) \to 0$ as $x \to \infty$.*

The next function has the properties of the derivative operator.

Exercise 601. *Assume that $f : \mathbb{R} \to \mathbb{R}$ is so that for every $x, y \in \mathbb{R}$,*

$$f(xy) = xf(y) + yf(x).$$

Show that $f(1) = 0$ and for every $n \in \mathbb{Z}^+$, $f(x^n) = nx^{n-1}f(x)$.

A function f is called *convex* on $[a, b]$ if for every $x, y \in [a, b]$, and for any $t \in (0, 1)$, $f(tx + (1-t)y) \le tf(x) + (1-t)f(y)$. (Often this definition is stated just for the case $t = 1/2$.)

Exercise 602. *Prove that if f is convex on some closed interval $[a, b]$, then for any $x_1, \ldots, x_n \in [a, b]$,*

$$f\left(\frac{\sum_{i=1}^n x_i}{n}\right) \le \frac{\sum_{i=1}^n f(x_i)}{n}.$$

One can extend Exercise 602 to the following:

Exercise 603 (Jensen's inequality). *Prove that for $p_1, p_2, \ldots, p_n \in \mathbb{R}^+$, if f is convex, then*

$$f\left(\frac{\sum_{i=1}^n p_i x_i}{\sum_{i=1}^n p_i}\right) \le \frac{\sum_{i=1}^n p_i f(x_i)}{\sum_{i=1}^n p_i}.$$

Exercise 604. *Fix some positive integer m. For any vector with positive integer entries $\mathbf{x} = (x_0, x_1, x_2, \ldots, x_\ell)$ (regardless of length), put $s(\mathbf{x}) = \sum_{i=0}^\ell ix_i$. Prove that over all vectors $\mathbf{x} = (x_0, x_1, x_2, \ldots, x_\ell)$ with positive integer entries that satisfy $x_0 + x_1 + x_2 + \cdots + x_\ell = m$, the function $s(\mathbf{x})$ is at most $\binom{m}{2}$, and reaches this maximum precisely for the vector $(1, 1, \ldots, 1)$ (of length m).*

The next problem uses $\mathbb{C}$, the complex numbers. If $z = a + bi$ (where $i^2 = -1$), the *complex conjugate* of z is $\bar{z} = a - bi$.

Exercise 605. *Prove that for any $z \in \mathbb{C}$ and any $n \in \mathbb{Z}$,*

$$(\bar{z})^n = \overline{z^n}.$$

Perhaps the above exercise can be used as a warm-up for the following:

Exercise 606. *Prove that for $z_1, z_2, \ldots, z_n \in \mathbb{C}$, if $f(z_1, z_2, \ldots, z_n)$ is obtained by a finite number of rational operations (i.e., addition, subtraction, multiplication, and division), then*

$$f(\bar{z_1}, \bar{z_2}, \ldots, \bar{z_n}) = \overline{f(z_1, z_2, \ldots, z_n)}.$$

The following few definitions are necessary to state a theorem that is central in functional analysis.

A *norm* on a vector space V over some field $\mathbb{F}$ is a function $\|\cdot\| : V \to \mathbb{R}$ that satisfies the following three conditions: (1) $\forall \alpha \in \mathbb{F}$ and $\forall \mathbf{v} \in V$, $\|\alpha \mathbf{v}\| = |\alpha| \|\mathbf{v}\|$; (2) $\|\mathbf{v}\| = 0$ if and only if $\mathbf{v} = \mathbf{0}$; (3) $\forall \mathbf{v}, \mathbf{w} \in V$, $\|\mathbf{v} + \mathbf{w}\| \le \|\mathbf{v}\| + \|\mathbf{w}\|$. If $\mathbb{F} = \mathbb{R}$, a *real linear functional* on V is a function $f : V \to \mathbb{R}$ so that for all $\mathbf{v}, \mathbf{w} \in V$ and all $\lambda \in \mathbb{R}$, both $f(\mathbf{v} + \mathbf{w}) = f(\mathbf{v}) + f(\mathbf{w})$ and $f(\lambda \mathbf{v}) = \lambda f(\mathbf{v})$. (So a linear functional on V is a linear transformation $f : V \to \mathbb{R}$.) The set of linear functionals on V is a vector space, called the *dual space*, denoted V^*. The norm on V^* is defined by

$$\|f\| = \sup\{|f(\mathbf{x})| : \mathbf{x} \in V, \|x\| \le 1\},$$

and a linear functional f on a real vector space is said to be bounded if there exists $c > 0$ so that for every $\mathbf{v} \in V$, $|f(\mathbf{v})| \le c \|\mathbf{v}\|$.

If $X \subseteq Y$ are sets and $f_0 : X \to \mathbb{R}$ is a function, then say that a function $f : Y \to \mathbb{R}$ *extends* f_0 if for every $x \in X$, $f(x) = f_0(x)$.

The next theorem is one of the so-called "three pillars of functional analysis". The theorem has a complex analogue, yet is given here only for reals.

Theorem 18.2.1 (Hahn–Banach Theorem). *Let X be a real normed linear space (i.e., a vector space over the field $\mathbb{R}$ with a norm) and $M \subset X$ be a subspace. If f is a bounded (real) linear functional on M, then f can be extended to a bounded linear functional F on X so that $\|F\| = \|f\|$.*

[Notes: M need not be closed, that is, M need not be finite dimensional. Also, norms are relative to their own domains. The Hahn–Banach theorem can be used to give a separating hyperplane defined by $F(x) = 0$, and $F(y)$ is positive or negative depending on which side y lies on.]

Exercise 607. *Prove the Hahn–Banach Theorem using Zorn's lemma (or Hausdorff's Maximality Principle) by extending f one dimension at a time.*

18.3   Calculus

There are a few other problems in this volume that use or mention calculus; for example, see Exercises 601 and 684.

18.3.1   Derivatives

Exercise 608. *For $m \geq n$, the n-th derivative of x^n with respect to x is*

$$\frac{d^n}{dx^n}[x^m] = \frac{m!}{(m-n)!}x^{m-n}.$$

Exercise 609. *If $u_1, u_2, \ldots, u_n$ are differentiable functions of x, prove that*

$$\frac{d}{dx}(u_1 + u_2 + \cdots + u_n) = \frac{du_1}{dx} + \frac{du_2}{dx} + \cdots + \frac{du_n}{dx}.$$

Exercise 610. *Let $f_1, f_2, \ldots, f_n$ be functions where for each i, $\lim_{x \to a} f_i(x)$ exists. Prove that*

$$\lim_{x \to a} \prod_{i=1}^{n} f_i(x) = \prod_{i=1}^{n} \lim_{x \to a} f_i(x),$$

and

$$\lim_{x \to a} \sum_{i=1}^{n} f_i(x) = \sum_{i=1}^{n} \lim_{x \to a} f_i(x).$$

Recall the basic product rule for derivatives: For differentiable real functions f and g of a single real variable, the derivative of their product is

$$(fg)' = f'g + fg'. \tag{18.1}$$

The result in the next exercise might be called "the general product rule for derivatives".

Exercise 611. *Assuming the product rule in (18.1), prove by induction on $n \geq 2$ that if $f_1, \ldots, f_n$ are differentiable functions then the derivative of their product is*

$$(f_1 f_2 \cdots f_n)' = f_1' f_2 f_3 \cdots f_n + f_1 f_2' f_3 \cdots f_n + \cdots + f_1 f_2 \cdots f_{n-1} f_n'. \tag{18.2}$$

The next exercise shows a different way to write the general product rule for derivatives. (See Exercise 611 and its solution.)

Exercise 612. *If $u_1, u_2, \ldots, u_n$ are differentiable functions of x and*

$$y = u_1 u_2 \cdots u_n,$$

then y is a differentiable function of x and

$$\frac{y'}{y} = \frac{u_1'}{u_1} + \frac{u_2'}{u_2} + \cdots + \frac{u_n'}{u_n}.$$

Exercise 613. *Prove that if $y = \ln x$, (the natural logarithm), then*

$$\frac{d^n y}{dx^n} = \frac{(-1)^{n-1}(n-1)!}{x^n}.$$

A similar exercise is found in Pólya's book *Mathematical discovery* [435, Prob 3.85, pp. 97–98, Vol I]:

Exercise 614. *Using induction, show that the nth derivative $f^{(n)}$ of the function $f(x) = \frac{\ln x}{x}$ is of the form*

$$f^{(n)}(x) = (-1)^n \frac{n! \ln x}{x^{n+1}} + (-1)^{n-1} \frac{c_n}{x^{n+1}},$$

where c_n is an integer depending only on n (and not on x); express c_n in terms of n.

Exercise 615 (Leibniz's theorem). *Prove that if u and v are differentiable functions of x, then (using the notation $\frac{d^0}{dx^0}[f(x)] = f(x)$)*

$$\frac{d^n}{dx^n}[uv] = \sum_{i=0}^{n} \binom{n}{i} \frac{d^i u}{dx^i} \cdot \frac{d^{n-i} v}{dx^{n-i}}.$$

18.3.2 Differential equations

The field of differential equations relies, in many ways, upon induction. The exposure given here is only very brief.

If y is an unknown function of x, but some information is known about $\frac{dy}{dx}$, the method of *undetermined coefficients* is often successful. Set $y = c_0 + c_1 x + c_2 x^2 + c_3 x^3 + \cdots$, and find the constants $c_0, c_1, c_2, c_3, \ldots$ by manipulating this power series, multiplying it, and most often, by differentiating (at least once).

The following example was given by Pólya (see Problem 3.81 in [435, Vol I, pp. 96–97]). If x and y satisfy $y = 1$ when $x = 0$ and

$$\frac{dy}{dx} = x^2 + y^2. \tag{18.3}$$

Express y in powers of x as follows: For unknown constants c_i, put $y = c_0 + c_1 x + c_2 x^2 + c_3 x^3 + \cdots$. Then equation (18.3) becomes

$$c_1 + 2c_2 x + 2c_3 x^2 + 4c_3 x^3 + \cdots = c_0^2 + 2c_0 c_1 x + (2c_0 c_2 + c_1^2 + 1)x^2 + \cdots . \tag{18.4}$$

Comparing coefficients of terms with the same power of x,

$$c_1 = c_0^2$$
$$2c_2 = 2c_0 c_1$$

$$3c_3 = 2c_0c_2 + c_1^2 + 1$$
$$4c_4 = 2c_0c_3 + 2c_1c_2$$

$$\vdots$$

Using the initial condition $y = 1$ when $x = 0$, $c_0 = 1$, and from the above system, it is not difficult to recursively compute the next few constants: $c_1 = 1$, $c_2 = 1$, $c_3 = 4/3$, and $c_4 = 7/6$. What is the general expression?

Exercise 616. *In the above example, prove by mathematical induction that for each $n \geq 3$, $c_n > 1$.*

Although the following is not exactly a question specifically requiring induction, solving it familiarizes one with the method of undetermined coefficients.

Exercise 617. *Show that the power series in x for the function y satisfying the differential equation*

$$\frac{d^2y}{dx^2} = -y,$$

and the initial conditions $y(0) = 1$ and $\frac{dy}{dx}\big|_{x=0} = 0$, has coefficients $c_{2n} = \frac{(-1)^n}{2n!}$ and $c_{2n-1} = 0$.

This next exercise was kindly written by Julien Arino for inclusion here, as my knowledge of differential equations forms nearly an empty set. He informs me that induction is a fundamental frequently used tool in differential equations. A standard reference he gave me was [46].

An initial value problem of the form

$$x' = f(t, x(t), x(t - 1))$$
$$x(t) = \phi_0(t), \quad t \in [-1, 0] \tag{18.5}$$

involves the *delay differential equation* $x' = f(t, x(t), x(t-1))$, with *delay* here equal to 1, and the *initial data* ϕ_0. The next exercise addresses the existence, uniqueness and regularity of the solutions to a delay differential equation.

Exercise 618. *Use the so-called* method of steps *to prove that if f is a C^1 (continuously differentiable) function, then the solution to (18.5) exists and is unique for all $t \geq 0$, given ϕ_0 a C^0 (continuous) function on $[-1, 0]$, and that if ϕ_0 is a function of class C^p on $[-1, 0]$, then the solution $x(t)$ to (18.5) is of class C^{p+k} on the interval $[k - 1, k]$, for all $k \in \mathbb{N} \setminus \{0\}$.*

18.3.3   Integration

Mathematical induction can sometimes be used when integrands have an integer exponent. Recall that integration by parts (for either definite or indefinite integrals) is applying the rule

$$\int u\, dv = uv - \int v\, du$$

for appropriate choices of u and dv that make the second integral "simpler", either simpler to solve or easier to approximate. For example, if n is a positive integer,

$$\int |\ln x|^n \, dx = |\ln x|^n x - \int xn|\ln x|^{n-1}\frac{1}{x}\, dx \quad \boxed{\begin{array}{ll} u = |\ln x|^n & dv = dx \\ du = n|\ln x|^{n-1}(\frac{1}{x})dx & v = x \end{array}},$$

whence one arrives at

$$\int |\ln x|^n \, dx = |\ln x|^n x - n \int |\ln x|^{n-1}\, dx \tag{18.6}$$

The equation (18.6) is called a *reduction formula* as it *reduces* an integral to an expression with another integral of the same form, however with a smaller exponent in the integrand. A more general form, derived in a similar manner is

$$\int x^m |\ln x|^n \, dx = \frac{1}{m+1}x^{m+1}|\ln x|^n - \frac{n}{m+1}\int x^m |\ln x|^{n-1}\, dx.$$

Reduction formulae can be found for some trigonometric integrals; for example, where $n \geq 2$,

$$\int \sin^n(x)\, dx \quad \boxed{\begin{array}{ll} u = \sin^{n-1}(x) & dv = \sin(x)\, dx \\ du = (n-1)\sin^{n-2}(x)\cos(x)dx & v = -\cos(x) \end{array}}$$

$$= \sin^{n-1}(x)(-\cos(x)) - \int (-\cos(x))(n-1)\sin^{n-2}(x)\cos(x)\, dx$$

$$= -\sin^{n-1}(x)\cos(x) + (n-1)\int \cos^2(x)\sin^{n-2}(x)\, dx$$

$$= -\sin^{n-1}(x)\cos(x) + (n-1)\int (1-\sin^2(x))\sin^{n-2}(x)\, dx$$

$$= -\sin^{n-1}(x)\cos(x) + (n-1)\int \sin^{n-2}(x)\, dx - \int \sin^n(x)\, dx,$$

and bringing the last term to the other side and dividing by n gives the reduction formula

$$\int \sin^n(x)\, dx = -\frac{1}{n}\sin^{n-1}(x) + \frac{n-1}{n}\int \sin^{n-2}(x)\, dx. \tag{18.7}$$

Application of (18.7) can be iterated until the most complicated integral remaining is either $\int \sin(x)\, dx$ (when n is odd) or $\int dx$ (when n is even).

Sometimes, integration by parts needs to be used twice to find a reduction formula. Other reduction formulae, also derived using integration by parts, include

$$\int \cos^n(x)\, dx = \frac{1}{n}\cos^{n-1}(x)\sin(x) + \frac{n-1}{n}\int \cos^{n-2}(x)\, dx;$$

$$\int \sec^n(x)\,dx = \frac{1}{n-1}\tan(x)\sec^{n-2}(x) + \frac{n-2}{n-1}\int \sec^{n-2}(x)\,dx \quad (n \neq 1);$$

$$\int x^n e^x\,dx = x^n e^x - n\int x^{n-1} e^x\,dx;$$

$$\int (x^2 + a^2)^n\,dx = \frac{x(x^2 + a^2)^n}{2n+1} + \frac{2na^2}{2n+1}\int (x^2 + a^2)^{n-1}\,dx \qquad (n \neq -\tfrac{1}{2});$$

$$\int \tan^n(x)\,dx = \frac{1}{n-1}\tan^{n-1}(x) - \int \tan^{n-2}(x)\,dx;$$

$$\int \cot^n(x)\,dx = -\frac{1}{n-1}\cot^{n-1}(x) - \int \cot^{n-2}\,dx \qquad (n \neq 1);$$

$$\int x^n \sin(x)\,dx = x^n \cos(x) + n\int x^{n-1}\cos(x)\,dx;$$

$$\int x^n \cos(x)\,dx = x^n \sin(x) - n\int x^{n-1}\sin(x)\,dx.$$

The last two equations are not precisely reduction formulae, but combining them produces actual reduction formulae:

$$\int x^n \sin(x)\,dx = -x^n \cos(x) + nx^{n-1}\sin(x) - n(n-1)\int x^{n-2}\sin(x)\,dx;$$

$$\int x^n \cos(x)\,dx = x^n \sin(x) + nx^{n-1}\cos(x) - n(n-1)\int x^{n-2}\sin(x)\,dx.$$

Successive applications of a reduction formula sometimes reveal a pattern that is provable by induction, however to spot such a pattern often takes a few applications and a lot of paper (try this for (18.6)). Once the pattern is observed, the reduction formula makes short work of the inductive step in proving the pattern. For example, the reduction formula

$$\int x^m e^{-x}\,dx = -x^n e^{-x} + n\int x^{n-1} e^x\,dx$$

is easily proved with one application of integration by parts with $u = x^n$ and $dv = e^{-x}dx$. For each $n \geq 0$, put

$$I_n = \int_0^1 x^n e^{-x}\,dx.$$

Then $I_0 = -e^{-1} + 1$, and $I_n = nI_{n-1} - e^{-1}$. Working out a few values suggests that

$$I_n = n! - \frac{n!}{e}\sum_{j=0}^{n}\frac{1}{j!}. \qquad (18.8)$$

Exercise 619. *Prove equation (18.8) by mathematical induction.*

The next exercise is a challenging application of induction (and is given without solution; see [534, p. 594]).

Exercise 620. *Use a reduction formula to prove that*

$$\int x^n \cos(x)\,dx = \sin(x)\sum_{i=0}^{\lfloor n/2 \rfloor}\frac{(-1)^i n!}{(n-2i)!}x^{n-2i} + \cos(x)\sum_{i=0}^{\lfloor (n-1)/2 \rfloor}\frac{(-1)^i n!}{(n-2i-1)!}x^{n-2i-1} + C.$$

The next two exercises occur in many calculus texts.

Exercise 621. *Prove that for any even integer $n \geq 2$,*

$$\int_0^{\pi/2} \sin^n(x)\,dx = \int_0^{\pi/2} \cos^n(x)\,dx = \frac{1\cdot 3\cdot 5\cdot 7 \cdots (n-1)}{2\cdot 4\cdot 6 \cdots n}\cdot\frac{\pi}{2}.$$

Exercise 622. *Prove that for any odd integer $n \geq 1$,*

$$\int_0^{\pi/2} \sin^n(x)\,dx = \int_0^{\pi/2} \cos^n(x)\,dx = \frac{2\cdot 4\cdot 6 \cdots (n-1)}{1\cdot 3\cdot 5 \cdots n}.$$

Applying Exercises 621 and 622 gives a strange formula for π: put

$$I_n = \int_0^{\pi/2} \sin^n(x)\,dx,$$

and discover that for each n, $I_n \geq I_{n+1}$. Also find that, for each k,

$$\frac{I_{2k+2}}{I_{2k}} = \frac{2k+1}{2k+2}.$$

It follows that

$$\frac{2k+1}{2k+2} \leq \frac{I_{2k+1}}{I_{2k}} \leq 1,$$

and so $\lim_{k\to\infty}\frac{I_{2k+1}}{I_{2k}} = 1$. Applying the Exercises, arrive at

$$\lim_{k\to\infty}\frac{2}{1}\cdot\frac{2}{3}\cdot\frac{4}{3}\cdot\frac{4}{5}\cdot\frac{6}{5}\cdot\frac{6}{7}\cdots\frac{2k}{2k-1}\cdot\frac{2k}{2k+1} = \frac{\pi}{2},$$

which is called the *Wallis formula* for π, which has many applications (see, *e.g.* [182]; this formula can also be derived using continued fractions).

Exercise 623. *For a positive integer n, derive the reduction formula*

$$\int_{-1}^1 (1-x^2)^n\,dx = \frac{2n}{2n+1}\int_{-1}^1 (1-x^2)^{n-1}\,dx,$$

and conclude by induction that

$$\int_{-1}^1 (1-x^2)^n\,dx = \frac{2^{2n+1}(n!)^2}{(2n+1)!}.$$

18.4   Polynomials

Many other exercises could appear in this section; for example, see Exercises 229, 292, and 684. See also Section 9.6.2 on Bernoulli numbers.

Exercise 624. *Prove that if polynomials p and q in any number of variables are of degrees m and n respectively, then the product pq is of degree m + n.*

The result in the next exercise may seem rather surprising. For example, consider the quadratic $q(x) = x^2 + 3x - 1$. Letting the roots of q be $\alpha = \frac{-3+\sqrt{13}}{2}$ and $\beta = \frac{-3-\sqrt{13}}{2}$, notice that $\alpha + \beta = -3$ (an integer) and $\alpha^2 + \beta^2 = 11$, another integer, and relatively prime to -3. Is there a pattern?

Exercise 625. *Let p be an odd positive integer and let α and β be roots of the equation $x^2 + px - 1 = 0$. For each $n \geq 0$, set $y_n = \alpha^n + \beta^n$. Prove that for each $n \geq 0$, y_n is an integer and y_n and y_{n+1} are relatively prime.*

Exercise 626. *Prove that, for any $n \geq 1$, a polynomial $p(x_1, x_2, \ldots, x_n)$ in n variables vanishes everywhere if and only if it vanishes in a neighborhood of a point.*

Exercise 627. *Prove that a polynomial in any number of variables vanishes everywhere if and only if all of its coefficients are zero.*

Related to the above problem is Exercise 229, where arithmetic is done modulo p. Also see Exercise 684.

Exercise 628. *Prove that for an n-th degree polynomial*

$$f(x) = a_n x^n + a_{n-1} x^{n-1} + \cdots + a_1 x_1 + a_0$$

there exist unique constants $\alpha_1, \alpha_2, \ldots, \alpha_n$ so that

$$f(x) = a_n(x - \alpha_1)(x - \alpha_2) \cdots (x - \alpha_n).$$

From [91], "To prove this theorem by complete induction one needs the fundamental theorem of algebra that there is at least one value of x for which such a polynomial $f(x)$ vanishes."

The following problem was recently posed by D. Marghidanu in *The College Mathematics Journal* "Problems and Solutions" section (number 879):

Problem: Suppose that a, b, c, and d are positive real numbers so that the polynomial $f(x) = x^4 - 4ax^3 + 6b^2x^2 - 4c^3x + d^4$ has four positive distinct roots. Show that $a > b > c > d$.

The solution given in the May 2009 issue of the same journal proves a generalization of the above problem. The statement in this generalization was also proved in 1729 by Maclaurin (see the book *Inequalities* [259, Thm. 52]), but the (inductive) proof supplied in [117, p.219] is due to, independently, R. Mosier and J. Nieto.

Exercise 629. *For $n \geq 2$, let $u_1, \ldots, u_n$ denote positive real numbers. Show that if the polynomial*

$$f(x) = x^n + \sum_{i=1}^{n}(-1)^i \binom{n}{i} u_i^i x^{n-i}$$

has n distinct positive roots, then $u_1 > u_2 > \cdots > c_n$.

Let $\mathbb{F}$ be a field and let $\mathbb{F}[x]$ be the ring of finite polynomials in the indeterminate x with coefficients chosen from $\mathbb{F}$. The following exercise might be viewed as a generalization of Exercise 229. See also Exercise 684.

Exercise 630. *Prove that if $p(x) \in \mathbb{F}[x]$ is of degree $n \geq 1$, then $p(x)$ has at most n roots in $\mathbb{F}$.*

Exercise 631. *Let $p(x, y)$ be a polynomial with x-degree m and y-degree n. Prove that $p(x, e^x)$ can have at most $mn + m + n$ real zeros.*

The next problem appeared in [91], however it is a standard question in many algebra texts. A function f of n variables is called *symmetric* iff for any permutation (see Section 19.2) $\sigma : \{1, \ldots, n\} \to \{1, \ldots, n\}$,

$$f(x_1, \ldots, x_n) = f(x_{\sigma(1)}, \ldots, x_{\sigma(n)}).$$

For example, if f is defined by

$$f(x, y, z) = 4x^2y^2z + 4x^2z^2y + 4xy^2z^2 - \frac{1}{(xyz)^3},$$

then $f(x, y, z) = f(y, z, x)$. In particular, a polynomial in n variables is symmetric iff permuting variables gives back the same polynomial. Among the symmetric polynomials in n variables, there are n polynomials called *elementary symmetric polynomials* (or functions) denoted by

$$
\begin{aligned}
s_1 &= x_1 + x_2 + \cdots + x_n \\
s_2 &= x_1x_2 + x_1x_3 + \cdots + x_{n-1}x_n \\
s_3 &= x_1x_2x_3 + x_1x_2x_4 + \cdots + x_{n-2}x_{n-1}x_n \\
&\vdots \\
s_n &= x_1x_2 \cdots x_n,
\end{aligned}
$$

where for each $k = 1, 2, \ldots, n$, the elementary symmetric polynomial s_k is formed by summing all possible products of k different variables. Observe that any polynomial in symmetric functions is again a symmetric function. For example, with $n = 2$, $s_1^2 - 2s_2 = (x + y)^2 - 2xy = x^2 + y^2$ is symmetric. The following is a standard theorem with proofs given in many advanced algebra books (*e.g.*, [23, p. 166]).

Exercise 632. *Prove that a symmetric polynomial in n variables can be expressed as a polynomial in elementary symmetric functions. Hint: Examine $p(x_1, \ldots, x_{n-1}, 0)$.*

In fact, the polynomial of elementary symmetric functions mentioned in Exercise 632 is unique, however this fact is not asked to be proved.

18.5 Primitive recursive functions

A certain class of (number theoretic) functions are defined from $\omega = \{0, 1, 2, \ldots\}$ to ω recursively. For example, the successor function defined by, for each $x \in \omega$, $s(x) = x + 1$, gives a recursive definition of positive integers. Let $x_1, x_2, \ldots$ denote non-negative integers, n be a positive integer; then for any (constant) $y \in \omega$,

$$f(x_1, \ldots, x_n) = y$$

is called a *constant* function. The function defined by

$$\pi_i(x_1, \ldots, x_n) = x_i$$

is called the i-th projection function.

Define a class P of functions recursively. First, let the successor function be in P. Put all constant functions (of a finite number of variables) in P. For each positive integer n and each $i = 1, \ldots, n$, put all projection functions $\pi_i : \omega^n \to \omega$ in P. If $f_1, \ldots, f_m \in P$ are functions of n variables, and $g \in P$ is a function in m variables, then the composition function $g(f_1, \ldots, f_m)$ is also a function (of n variables) in P. Finally, any function recursively defined by a finite number of compositions of functions in P is also in P.

The functions in P are called *primitive recursive*. With some work (all by induction, as in Chapter 2), a very large class of functions is proved to be primitive recursive, including all bounded sums and products (and so all finite polynomials in any finite number of variables) and even those with, say, 7 levels of exponents. Primitive recursive functions are, essentially, those that can be written in a finite form that does not depend on the input.

Examples of functions that are not primitive recursive arise out of attempts to solve very difficult computing or combinatorial problems where often double induction is used and bounds are of the form where a tower of exponents grows exponentially fast in height. See the comments surrounding the Hales–Jewett theorem in Chapter 21 on Ramsey theory (and further comments in [231]).

For a more formal and complete discussion of primitive recursive functions, see, *e.g.*, [118, pp. 8ff] or [316, pp. 43ff]. Also see [470, pp. 416ff] for a recent discussion of primitive recursive functions (and relations to Turing machines and DNA computers).

18.6 Ackermann's function

Wilhelm Ackermann (1896–1962) was a student of David Hilbert (1862–1943). The next exercises involve Ackermann's function (as described in, *e.g.*, [238] or [292]). The range of an Ackermann function is an example of a sequence that grows faster than can be described by any closed-form formula (it is not primitive recursive), and so it has been of interest to computer scientists, mathematicians, and logicians.

It has been used as a kind a measure of how fast certain algorithms run. There are many equivalent definitions of the Ackermann function; see [118] and [231] for forms other than the one given below.

For non-negative integers m, n define $A(m, n)$ recursively as follows: for $n \geq 0$, $A(0, n) = n + 1$, and for $m, n \geq 1$,

$$A(m, 0) = A(m - 1, 1), \quad \text{and}$$
$$A(m, n) = A(m - 1, A(m, n - 1)).$$

So, for example, $A(0, 1) = 2$ and

$$A(1, 1) = A(0, A(1, 0)) = A(0, A(0, 1)) = A(0, 2) = 3.$$

Exercise 633. *Prove by induction that for $n \geq 0$,*

$$A(1, n) = n + 2.$$

Exercise 634. *Prove by induction that for $n > 0$,*

$$A(2, n) = 2n + 3.$$

Exercise 635. *Prove by induction that for $n \geq 0$,*

$$A(3, n) = 2^{n+3} - 3.$$

Exercise 636. *Use double induction to prove that for all $m, n \geq 0$,*

$$A(m, n) > n.$$

Chapter 19

Linear and abstract algebra

If you don't like your analyst, see your local algebraist.

—Gert Almkvist,

founder and director of The Institute for Algebraic Meditation.

19.1 Matrices and linear equations

Unless otherwise mentioned, all matrices here are real-valued; however, many theorems and exercises given here apply to matrices over any field (including complex numbers).

The following exercise is proved in nearly every text on elementary linear algebra, and is shown in an article by Yuster [584].

Exercise 637. *Prove that a reduced row echelon form of a matrix is unique (justifying the expression "the reduced row echelon form of a matrix"). Hint: induct on the number of columns.*

The next exercise might be called "the marked matrix problem", and may be considered a classic old problem. [I do not know the original source.]

Exercise 638. *In an $m \times n$ matrix of real numbers, mark at least p of the largest numbers in each column ($p \leq m$), and at least q of the largest numbers in each row. Prove that at least pq of the numbers are marked at least twice. Hint: induct on $m + n$.*

Exercise 639. *Let A and B be matrices with $AB = BA$. Show that for each positive integer n, $(AB)^n = A^n B^n$.*

If $A = [a_{ij}]$ is an $n \times m$ matrix, the *transpose* of A is an $m \times n$ matrix B defined by $B = [b_{ij}]$ where $b_{ij} = a_{ji}$. The transpose of a matrix A is denoted A^T.

Exercise 640. *Show that for each positive integer n, if $A_1, A_2, \ldots, A_n$ are matrices of the same size, then*

$$(A_1 + A_2 + \cdots + A_n)^T = A_1^T + A_2^T + \cdots + A_n^T$$

Exercise 641. *Prove that if $A_1, A_2, \ldots, A_n$ are square matrices of the same size, then*

$$(A_1 A_2 \cdots A_n)^T = A_n^T \cdots A_2^T A_1^T.$$

Exercise 642. *Show that for each positive integer n, if $A_1, A_2, \ldots, A_n$ are matrices with sizes so that the product $A_1 A_2 \cdots A_n$ is defined, then*

$$(A_1 A_2 \cdots A_n)^T = A_n^T \cdots A_2^T A_1^T.$$

Exercise 643. *Show that if $A_1, A_2, \ldots, A_n$ are invertible matrices of the same size, then (letting A^{-1} denote the inverse of A)*

$$(A_1 A_2 \cdots A_n)^{-1} = A_n^{-1} \cdots A_2^{-1} A_1^{-1}.$$

For the next exercises, some definitions are needed (some of which probably come from linear programming). A non-invertible square matrix (with zero determinant) is called *singular* (so an invertible matrix is also called "non-singular"). A square matrix with integer entries is called *unimodular* if its determinant is 1 or -1. A *principal submatrix* of a matrix is one obtained by deleting rows and/or columns. An $m \times n$ integer matrix is *totally unimodular* iff every square non-singular principal submatrix of A is unimodular.

Exercise 644. *Let A be an $m \times n$ matrix with the following properties: The rows of A can be partitioned into disjoint sets R_1 and R_2 so that*
(i) Every column of A contains at most two non-zero entries.
(ii) Every entry in A is either -1, 0, or 1.
(iii) For $i \neq j$, if a_{ik} and a_{jk} are non-zero and have the same sign, then row i is in R_1 and row j is in R_2, or row i is in R_2 and row j is in R_1.
(iv) For $i \neq j$, if a_{ik} and a_{jk} are non-zero and have different signs, then row i and row j are both in R_1, or row i and row j are both in R_2.
Prove that A is totally unimodular.

Some of the following statements have direct proofs using the definition of determinant as the sum of signed elementary products, however, if one defines the determinant recursively by cofactor expansions (also called Laplace expansions, named after Pierre-Simon Laplace (1749–1827)), inductive proofs are natural. To quickly review this latter definition, here is some terminology. For an $n \times n$ matrix A, for each $i, j \in \{1, \ldots, n\}$ let A_{ij} denote the principal $(n-1) \times (n-1)$ submatrix of A formed by deleting the i-th row and j-th column. The determinant $\det(A_{ij})$ is called the (i,j)-minor of A, and the (i,j)-cofactor of A is $C_{ij} = (-1)^{i+j} \det(A_{ij})$. Letting

a_{ij} denote the (i, j)-entry of A, the *cofactor expansion* (or Laplace expansion) along row i is

$$a_{i1}C_{i1} + a_{i2}C_{i2} + \cdots + a_{in}C_{in},$$

and the cofactor expansion along column j is

$$a_{1j}C_{1j} + a_{2j}C_{2j} + \cdots + a_{nj}C_{nj}.$$

A main theorem in linear algebra is that the determinant of a matrix is its cofactor expansion along *any* row or column.

Exercise 645. *Prove by induction that if A is an $n \times n$ matrix containing a row of zeros, then $det(A) = 0$.*

Exercise 646. *Prove by induction that if A is an $n \times n$ matrix with two columns equal, then $det(A) = 0$.*

The following result is one key result in proving that the Laplace expansion along any row or column always yields the same number, the determinant.

Exercise 647. *Let A be an $n \times n$ matrix with columns $\mathbf{a}_1, \ldots, \mathbf{a}_n$. For a fixed k, let B and C be $n \times n$ matrices given by*

$$B = [\mathbf{a}_1 \mid \ldots \mid \mathbf{a}_{k-1} \mid \mathbf{b}_k \mid \mathbf{a}_{k+1} \mid \ldots \mid \mathbf{a}_n],$$

and

$$C = [\mathbf{a}_1 \mid \ldots \mid \mathbf{a}_{k-1} \mid \mathbf{c}_k \mid \mathbf{a}_{k+1} \mid \ldots \mid \mathbf{a}_n],$$

(where the $\mathbf{a}_i$'s, $\mathbf{b}_k$, $\mathbf{c}_k$ are all $n \times 1$ column vectors). Prove that if for some β and γ, $\mathbf{a}_k = \beta \mathbf{b}_k + \gamma \mathbf{c}_k$ then $det(A) = \beta det(B) + \gamma det(C)$.

Exercise 648. *Prove by induction that for every $n \geq 1$, if A is an $n \times n$ matrix, then $det(A^T) = det(A)$.*

A square matrix is called *upper triangular* if all entries below the main diagonal are zero, and *lower triangular* if all entries above the main diagonal are zero. A matrix is *triangular* if it is either upper or lower triangular.

Exercise 649. *Prove by induction that the determinant of a triangular matrix is the product of the entries on the main diagonal.*

A matrix A has an *LU-decomposition* if and only if there exists an invertible lower-triangular matrix L and an upper-triangular matrix U so that $A = LU$. [Note: in this definition, letters L and U have two different roles; in "LU-decomposition", they simply stand for "Lower" and "Upper", respectively, whereas in "$A = LU$", L and U denote specific matrices.] Not all matrices have an LU-decomposition; it turns out that if A can be row-reduced (to an upper triangular matrix) without using the row operation that switches rows, then A has an LU-decomposition (L is found by writing the row reduction as a product $R = E_k E_{k-1} \cdots E_1 A$). The next exercise determines precisely those matrices having an LU-decomposition.

Recall that a principal submatrix of a matrix A is a submatrix formed by deleting rows and or columns; a *leading principal submatrix* of A is a square principal submatrix formed by the first rows and columns; in notation, if $A = [a_{ij}]_{m \times n}$, then $A_k = [a_{ij}]_{k \times k}$ is the k-th leading principal submatrix.

Exercise 650. *Prove that a non-singular matrix has an LU-decomposition iff every one of its leading principal submatrices is invertible (non-singular).*

Recall that a square matrix A is *diagonalizable* iff there exists an invertible matrix Q so that $D = Q^{-1}AQ$ is diagonal; equivalently, A is diagonalizable iff there exists an invertible matrix A and a diagonal matrix D so that $A = QDQ^{-1}$. It follows that if A is diagonalizable, then Q is found with eigenvectors as columns. [Note: Some texts replace Q with Q^{-1}, so then $D = QAQ^{-1}$ is used.]

The next few exercises concern complex matrices. For positive integers m, n, let $M_{m \times n}(\mathbb{R})$ denote the set (or vector space) of all $m \times n$ matrices with real entries. This notation extends to any field, not just $\mathbb{R}$. So $M_{m \times n}(\mathbb{C})$ denotes the set of all $m \times n$ matrices with complex entries. For any $A \in M_{n \times n}(\mathbb{C})$, define the *complex conjugate transpose* $A^* = [a_{ij}^*]$ by $a_{ij}^* = \overline{a_{ji}}$.

A complex matrix P is called *unitary* iff P is invertible with inverse $P^{-1} = P^*$, its complex conjugate transpose, that is, P is unitary iff $P^*P = I$.

Exercise 651. *Prove Schur's decomposition theorem: For any $A \in M_{n \times n}(\mathbb{C})$, there exists a unitary matrix $U \in M_{n \times n}(\mathbb{C})$ and an upper triangular matrix $T \in M_{n \times n}(\mathbb{C})$ so that $T = U^*AU$. Note that an equivalent consequence is $A = UTU^*$, called the "Schur decomposition of A".*

As a corollary to the proof given for Exercise 651, by induction, the main diagonal of T (in Schur's decomposition theorem) consists of eigenvalues for A.

A matrix A is called *Hermitian* iff $A = A^*$. [Real Hermitian matrices are symmetric.] Corollaries of Schur's decomposition theorem include the following central results in linear algebra (see *e.g.* [561, 337–339] for more details):

Theorem 19.1.1 (Spectral theorem, Hermitian matrices). *Let $A \in M_{n \times n}(\mathbb{C})$ be Hermitian. Then there exists a unitary matrix $U \in M_{n \times n}(\mathbb{C})$ and diagonal matrix $D \in M_{n \times n}(\mathbb{C})$ so that $D = U^*AU$; furthermore, diagonal elements of D are eigenvalues of A, and corresponding columns of U are associated eigenvectors for A.*

Proof outline: By Schur's decomposition theorem, there exists a unitary U and triangular T so that $T = U^*AU$. Then $T^* = (U^*AU)^* = U^*A^*U^{**} = U^*A^*U = U^*AU$, and so T is Hermitian, hence diagonal. From the corollary to Schur's theorem mentioned above, diagonal entries of T are eigenvalues for A. To see that columns of U are eigenvectors, repeat the proof that a matrix A is diagonalizable iff there exist a basis for $\mathbf{C}^n$ consisting of eigenvectors of A. $\qquad\qquad\square$

Corollary 19.1.2. *Hermitian $n \times n$ matrices have real eigenvalues and an orthonormal set of n eigenvectors.*

A real unitary matrix is called *orthogonal*; in other words, a real matrix A is *orthogonal* iff $A^{-1} = A^T$. Two complex matrices A and B are *unitarily similar* iff there exists a unitary matrix P such that $A = PBP^*$. A complex matrix A is *unitarily diagonalizable* iff A is unitarily similar to some diagonal matrix D (so then $A = PDP^*$).

Which matrices are unitarily diagonalizable? Which real matrices are unitarily diagonalizable? Which real matrices are orthogonally diagonalizable?

Recall that a complex matrix M is called *normal* iff $M^*M = MM^*$.

Exercise 652. *Prove that a complex matrix A is unitarily diagonalizable iff A is normal.*

As a corollary to the result in Exercise 652, an $n \times n$ matrix A is normal iff A has a set of n orthonormal eigenvectors.

A matrix $A \in M_{n \times n}(\mathbb{C})$ is called *semisimple* or *non-defective* iff A has n linearly independent eigenvectors. As a consequence of the above exercises, a matrix is normal iff it is semisimple, and unitary matrices are semisimple. [In fact, even skew-Hermitian matrices, those satisfying $A^* = -A$, are also semisimple.] Following the above proofs also shows that real symmetric matrices are orthogonally diagonalizable, and they have a complete orthonormal set of real eigenvectors (and real eigenvalues); this fact is often called the "Spectral theorem for real symmetric matrices"..

Recall that a real symmetric $n \times n$ matrix A is *positive definite* iff for all non-zero $n \times 1$ column vectors $\mathbf{x}$, $\mathbf{x}^T A\mathbf{x} > 0$.

Exercise 653. *Let A be a real, symmetric and positive definite matrix. Prove that there exists a unique upper triangular R with positive diagonal entries such that $A = R^T R$ (called the Cholesky decomposition of A).*

Exercise 654. *For any constant c, prove that for each $n \geq 1$,*

$$\begin{bmatrix} 1 & c \\ 0 & 1 \end{bmatrix}^n = \begin{bmatrix} 1 & cn \\ 0 & 1 \end{bmatrix}.$$

Exercise 655. *For any fixed constants a, b, prove that for each $n \geq 1$,*

$$\begin{bmatrix} a & 0 \\ 0 & b \end{bmatrix}^n = \begin{bmatrix} a^n & 0 \\ 0 & b^n \end{bmatrix}.$$

Exercise 656. *Prove that for each $n \geq 2$,*

$$\begin{bmatrix} 4 & 2 \\ 2 & 1 \end{bmatrix}^n = 5^{n-1} \begin{bmatrix} 4 & 2 \\ 2 & 1 \end{bmatrix}.$$

The following exercise involves the standard rotation matrix for $\mathbb{R}^2$.

Exercise 657. *For any $\theta \in \mathbb{R}$ prove that for each $n \geq 1$,*

$$\begin{bmatrix} \cos(\theta) & -\sin(\theta) \\ \sin(\theta) & \cos(\theta) \end{bmatrix}^n = \begin{bmatrix} \cos(n\theta) & -\sin(n\theta) \\ \sin(n\theta) & \cos(n\theta) \end{bmatrix}.$$

For any $n \geq 1$, let I_n denote the $n \times n$ identity matrix and let J_n be the $n \times n$ matrix whose every entry is 1. For example,

$$I_3 = \begin{bmatrix} 1 & 0 & 0 \\ 0 & 1 & 0 \\ 0 & 0 & 1 \end{bmatrix}, \quad \text{and} \quad J_3 = \begin{bmatrix} 1 & 1 & 1 \\ 1 & 1 & 1 \\ 1 & 1 & 1 \end{bmatrix}.$$

Note that $J_n^2 = nJ_n$. Put $M_n = J_n - I_n$, the adjacency matrix for the complete graph K_n (see Exercise 490). By Exercise 490, if K_n has vertex set $v_1, v_2, \ldots, v_n$, the (i,j) entry of M_n^k is the number of walks from vertex v_i to v_j. For example,

$$\begin{aligned} M_n^2 &= (J_n - I_n)^2 \\ &= J_n^2 - 2J_n + I_n \\ &= nJ_n - 2J_n + I_n \\ &= (n-2)J_n + I_n \\ &= \begin{bmatrix} n-1 & n-2 & \cdots & n-2 \\ n-2 & n-1 & \cdots & n-2 \\ & & \ddots & \\ n-2 & n-2 & & n-1 \end{bmatrix}, \end{aligned}$$

and the number of walks of length 2 between distinct vertices is $n-2$ (for example, $v_1 v_3 v_2, v_1 v_4 v_2, \ldots, v_1 v_n v_2$ are the walks between v_1 and v_2), and the number of walks from v_i to v_i is $n-1$.

Exercise 658. *Let $n \geq 1$ be fixed. Prove by induction on k that for each $k \in \mathbb{Z}^+$,*

$$M_n^k = \left(\frac{(n-1)^k - (-1)^k}{n} \right) J_n + (-1)^k I_n.$$

A *Hadamard matrix* of order n is any $n \times n$ matrix H that satisfies $HH^T = nI$. For example,

$$H = \begin{bmatrix} +1 & +1 \\ +1 & -1 \end{bmatrix}$$

is a Hadamard matrix.

Exercise 659. *Prove that for any $k \geq 0$, there exists a Hadamard matrix of order 2^k.*

Exercise 660. *For each $n \geq 1$, let A_n be the $n \times n$ matrix with 0s on the main diagonal, 1's above the main diagonal, and -1's below. Prove by induction that if n is odd then $\det(A) = 0$, and if n is even then $\det(A) = 1$.*

The following exercise is about a kind of matrix called a Vandermonde matrix, named after Alexandre-Théophile Vandermonde (1735–1796), a significant contributor to the study of roots of equations and determinants. For each $n \geq 1$ and real numbers $c_0, \ldots, c_n$, a Vandermonde matrix is one of the form:

$$
M = \begin{bmatrix}
1 & c_0 & c_0^2 & \cdots & c_0^n \\
1 & c_1 & c_1^2 & \cdots & c_1^n \\
\vdots & \vdots & \vdots & & \vdots \\
1 & c_n & c_n^2 & \cdots & c_n^n
\end{bmatrix},
\tag{19.1}
$$

or its transpose.

Exercise 661 (Vandermonde determinant). *Prove that for a Vandermonde matrix of the form in (19.1),*

$$
\det(M) = \prod_{0 \leq i < j \leq n} (c_j - c_i).
$$

Exercise 662. *Let $x \neq 1$. Show that for each $n \geq 1$,*

$$
\begin{bmatrix} +1 & -1 \\ 0 & x \end{bmatrix}^n = \begin{bmatrix} +1 & \frac{x^n - 1}{1 - x} \\ 0 & x^n \end{bmatrix}.
$$

What if $x = 1$?

Exercise 663. *For each $n \geq 1$, prove that the determinant of the $n \times n$ matrix*

$$
A_n = \begin{bmatrix}
2 & -1 & 0 & \cdots & 0 & 0 \\
-1 & 2 & -1 & \cdots & 0 & 0 \\
0 & -1 & 2 & \cdots & 0 & 0 \\
0 & 0 & -1 & \cdots & 0 & 0 \\
\vdots & \vdots & \vdots & \ddots & \vdots \\
0 & 0 & 0 & \cdots & -1 & 2
\end{bmatrix}
$$

is $n + 1$.

Exercise 664. *For each $k \geq 1$, and scalars $a_0, \ldots, a_{k-1}$, prove that the $k \times k$ matrix of the form*

$$A = \begin{bmatrix} 0 & 0 & 0 & \ldots & 0 & -a_0 \\ 1 & 0 & 0 & \ldots & 0 & -a_1 \\ 0 & 1 & 0 & \ldots & 0 & . \\ 0 & 0 & 1 & \ldots & 0 & . \\ \vdots & \vdots & \vdots & \ddots & \vdots & \\ 0 & 0 & 0 & \ldots & 1 & -a_{k-1} \end{bmatrix}$$

has characteristic polynomial

$$det(A - xI) = c_A(x) = (-1)^n(a_0 + a_1 x + \cdots + a_{n-1}x^{n-1} + x^n).$$

(In this case, A, or its transpose, is called a companion matrix *for $c_A(x)$.)*

The following is an application in economics. In a paper [580] by Xie entitled "Mathematical induction applied on Leontief systems" the abstract reads [definitions appear after the quote]: "This note establishes a necessary and sufficient condition for a matrix to be positive. The recursive nature of the condition allows one to use mathematical induction to greatly simplify the proof of the existence and the solvability conditions of the Leontief system."

For notational convenience, if $\mathbf{x}$ is a vector (or column or row matrix) in $\mathbb{R}^n$, write $\mathbf{x} \geq \mathbf{0}$ iff every entry in $\mathbf{x}$ is non-negative. To state this next problem, a few definitions from Xie's work are needed. Define a square matrix $B = [b_{ij}]$ to be *Leontief* iff for every $i \neq j$, $b_{ij} \leq 0$, and define a matrix B to be *positive* iff B is Leontief and $B\mathbf{x} \geq \mathbf{0}$ implies $\mathbf{x} \geq \mathbf{0}$. [This definition does not appear to be universal; *e.g.*, compare [101].]

Here are two lemmas from Xie's paper that have relatively simple proofs (but omitted here).

Lemma 19.1.3. *The necessary and sufficient condition for a (square) matrix H of order $n + 1$ to be positive is:*

$$H = \begin{bmatrix} G & -A \\ -B & \gamma \end{bmatrix},$$

where G is Leontief of order n, A and B are non-negative $(n \times 1)$ and $(1 \times n)$ matrices respectively, and γ is a positive number. Furthermore, $G - \frac{1}{\gamma}AB$ is a positive matrix of order n.

Recall that a permutation matrix is a square 0-1 matrix with exactly one 1 in each row and in each column. Such a matrix obtains its name because multiplying one on the right by a column vector permutes the elements in the vector (or multiplication on the left by a row vector—giving another permutation). See Section 19.2 for definitions surrounding permutations.

Lemma 19.1.4. *A matrix B is positive iff for any (appropriately sized) permutation matrix P, $P^{-1}BP$ is positive.*

The following theorem [524, Thm 4.C.4]) was first proved without induction (and Xie reports that the early proof was more difficult than his inductive proof).

Theorem 19.1.5. *For any Leontief matrix H, the following conditions are equivalent.*

(1) There exists a vector $\mathbf{x} \geq 0$ so that $H\mathbf{x} > \mathbf{0}$.

(2) H is a positive matrix.

(3) H is nonsingular and all of the entries of H^{-1} are non-negative.

Exercise 665. *Using Lemmas 19.1.3 and 19.1.4, prove implications (1)$\Rightarrow$(2) and (2)$\Rightarrow$(3) of Theorem 19.1.5 by induction on the order of B. [That (3) implies (1) is trivial.]*

For the next exercise, recall that ω is the first infinite ordinal, and $\omega + \omega$ is an ordinal formed by following one copy of ω by another.

For positive integers m and n with $m \leq n$, an $m \times n$ *latin rectangle* is an $m \times n$ array (or matrix) $A = (a_{ij})$, where each entry $a_{ij} \in \{1, 2, \ldots, n\}$ and no element is repeated in any row or in any column. [In fact, one can use any n symbols to choose entries from; however, integers in $[1, n]$ are the most convenient.] A *latin square* is an $m \times n$ latin rectangle where $m = n$. In 1944, Marshall Hall [254] published the result in the next exercise. The solution uses a famous theorem in graph theory (covered in Chapter 15).

Exercise 666. *Use (Philip) Hall's theorem (Theorem 15.5.1) and induction to show that for $m < n$, any $m \times n$ latin rectangle can be completed to an $n \times n$ latin square by the addition of $n - m$ rows.*

See Exercise 667 below for an exercise showing the existence of an infinite latin square. For more on latin squares, see [129] or [130].

Exercise 667. *Show that it is possible to have an infinite matrix of size $(\omega + \omega) \times (\omega + \omega)$ so that every positive integer occurs in each row precisely once and in each column precisely once.*

Such an infinite matrix in Exercise 667 is called an *infinite latin square*. See also Section 19.4 for more exercises regarding eigenvalues and eigenvectors and linear independence.

The next exercise relies on something called a *Hankel* matrix and on the Catalan numbers (see Section 12.5). Given any sequence $a_1, a_2, a_3, \ldots$ of integers, and for any positive integer n, define the $n \times n$ Hankel matrix $H_n = [h_{i,j}]$ by $h_{i,j} = a_{i+j+1}$. Define the *shifted Hankel* matrix H'_n defined to be the Hankel matrix for the shifted sequence $a_2, a_3, a_4, \ldots$. This next non-trivial exercise has a solution by induction, which was donated kindly by Michael Doob for inclusion here.

Exercise 668. *Prove that the Catalan sequence sequence 1,1,2,5,14,... is the only sequence so that for every $n \geq 1$, both its Hankel matrix H_n and shifted Hankel matrix H_n' have determinant 1.*

Exercise 669. *Prove that for each $m, n \geq 1$, there exists an $m \times n$ matrix A so that all mn entries of A are different, each a perfect square, and all row sums and column sums are also perfect squares. Hint: induct on $m + n \geq 2$.*

19.2   Groups and permutations

19.2.1   Semigroups and groups

For more detailed discussion on the following topics, the reader might consult one of the standard references for algebra (*e.g.*, [152], [283]).

A closed *binary operation* on a set S is a function from $S \times S$ to S. Examples of binary operations are multiplication and addition of real numbers. The notation for the binary operation varies according to the situation; common notations are $a + b$, $a * b$, $a \cdot b$, ab, or (ab). A *semigroup* is a non-empty set together with a binary operation that is associative. The reader interested in semigroups might take a look at Seth Warner's article [559] to consider mathematical induction over an arbitrary commutative semigroup.

A *monoid* is a semigroup with a two-sided identity, and a *group* is a monoid with inverses. A group is called *abelian* iff the binary operation is commutative.

Exercise 670. *Let G be an abelian group, that is, G is a group and for every $x, y \in G$, $xy = yx$. Prove that for every positive integer n, $(ab)^n = a^n b^n$.*

Definition 19.2.1. For a prime p, a group G is a *p-group* if every element of G has order which is a power of p.

The following is a standard result in group theory:

Theorem 19.2.2 (Cauchy's theorem). *If G is a group and p is a prime dividing $|G|$, then G has an element of order p.*

As a consequence of Cauchy's theorem, if p divides $|G|$, then G contains a subgroup of order p (namely one generated by an element of order p) and hence a p-group. In fact, much more is true; namely, for any prime power dividing the order of a group, there is a subgroup whose order is that prime power:

Theorem 19.2.3 (Sylow's first theorem, 1872). *Let G be a finite group and let p be a prime. If p^k is the largest power of p that divides $|G|$, then for each $i = 1, \ldots, k$, there exists a subgroup H_i of order p^i.*

Sylow's first theorem is often cited as also saying that for each $i = 1, \ldots, k - 1$, H_i is a normal subgroup of H_{i+1}.

Exercise 671. *Prove part of Sylow's first theorem, namely that if p^k divides $|G|$, then for each $i = 1, 2, \ldots, k$, G contains a subgroup of order p^i.*

19.2.2 Permutations

In general, a *permutation* on a set X is a bijection from X to itself, but this definition can be stated in many ways. One can denote a permutation by functional notation; in this setting, say that a permutation on $\{1, 2, \ldots, n\}$ is a bijection

$$\sigma : \{1, 2, \ldots, n\} \to \{1, 2, \ldots, n\},$$

and represent σ by an ordered n-tuple $(\sigma(1), \sigma(2), \ldots, \sigma(n))$, where for $i \neq j$, $\sigma(i) \neq \sigma(j)$. In this fashion, a permutation of the symbols $1, 2, \ldots, n$ is seen as a linear (re)arrangement of these symbols.

By simple counting, there are $n!$ permutations on n symbols; however, this result also has an easy inductive proof.

Exercise 672. *Prove by induction that for each $n \geq 1$, there are $n!$ distinct permutations of the symbols $1, 2, \ldots, n$.*

An *inversion* in a permutation $\tau = (\tau(1), \ldots, \tau(n))$ is is any pair of the form $(\tau(i), \tau(j))$ where $i < j$ and $\tau(i) > \tau(j)$.

For example, the permutation $(2, 3, 4, 7, 5, 1, 6)$ has seven inversions: $(2, 1)$, $(3, 1)$, $(4, 1)$, $(7, 1)$, $(5, 1)$, $(7, 5)$, and $(7, 6)$. Define $I_{n,k}$ to be the number of permutations on $\{1, 2, \ldots, n\}$ with exactly k inversions.

Exercise 673. *Prove $I_{n,0} = 1$; $I_{n,1} = n - 1$; for $k \geq n$, $I_{n,k} = 0$; $I_{n,\binom{n}{2}-k} = I_{n,k}$; and for $k < n$, $I_{n,k} = I_{n,k-1} + I_{n-1,k}$.*

Permutations with an even [odd] number of permutations is called *even* [resp. *odd*]. Even and odd permutations are used in one way to define a determinant.

One of the most common groups is the *symmetric group* $\mathfrak{S}_n$. The symmetric group is commonly represented by permutations, $\mathfrak{S}_n$ being the set of all permutations on n given elements, with composition as the operation (see below for notation on composition). For example, when $n = 3$, there are six permutations on $\{1, 2, 3\}$; eliminating set brackets and commas, these can be denoted by 123, 132, 213, 231, 312, 321.

Notation for permutations varies. For example, the permutation denoted by 132 is the bijection

$$\sigma : \{1, 2, 3\} \longrightarrow \{1, 2, 3\}$$

given by $\sigma(1) = 1$, $\sigma(2) = 3$, and $\sigma(3) = 2$. Another standard way to describe a permutation on a set A is using two rows, the top row being an arbitrary (but fixed) ordering of the elements in A, and the second row showing the images:

$$\sigma = \begin{pmatrix} 1 & 2 & 3 & \cdots & n \\ \sigma(1) & \sigma(2) & \sigma(3) & \cdots & \sigma(n) \end{pmatrix}.$$

For example, the above permutation denoted by 132 would be written as

$$\begin{pmatrix} 1 & 2 & 3 \\ 1 & 3 & 2 \end{pmatrix}.$$

Yet another way to denote permutations is by *cycle notation*. Perhaps the easiest way to describe the notation is first by an example:

Example 19.2.4. *Let*

$$\tau = \begin{pmatrix} 1 & 2 & 3 & 4 & 5 & 6 & 7 & 8 & 9 \\ 2 & 5 & 4 & 9 & 7 & 1 & 6 & 8 & 3 \end{pmatrix}.$$

If one follows where elements are moved, cycles arise:

$$1 \to 2 \to 5 \to 7 \to 6 \to 1 \to 2 \cdots$$

and starting from the first unused element in the above cycle,

$$3 \to 4 \to 9 \to 3 \to 4 \ldots;$$

the element 8 remains fixed by σ. In cycle notation, one writes

$$\tau = (12576)(349)(8),$$

or simply $\tau = (12576)(349)$. Note that the cycle notation $(57612)(934)$ describes the same permutation.

If ρ and σ are both permutations on the same set, define the composition of ρ with σ by

$$\sigma \circ \rho(x) = \sigma(\rho(x))$$

for each element x in the set. Rather than write $\sigma \circ \rho$, it is common to use product notation $\sigma\rho$. Since permutations can be viewed as functions, composition of permutations inherits all properties of composition of functions, for example, associativity. In the above example, σ is the product (composition) of disjoint cycles of length 5, 3, (and 1) respectively. A cycle of length two is called a *transposition*, that is, a permutation σ on $\{1, 2, \ldots, n\}$ is a transposition if and only there exist j and k, $j \ne k$ so that $\sigma(j) = k$, $\sigma(k) = j$, and if $i \ne j$ and $i \ne k$ then $\sigma(i) = i$.

Exercise 674. *Prove that for each $r \ge 2$, every cycle of length r can be written as a product of transpositions, as in*

$$(x_1, \ldots, x_r) = (x_1, x_2) \circ (x_1, x_3) \circ \cdots \circ (x_1, x_r).$$

Exercise 675. *Prove that every permutation of a finite set that is not the identity permutation and is not itself a cycle, can be written as a product of disjoint cycles of length at least two. Show also that the decomposition into cycles is unique except for the order of the cycles.*

Exercise 676. *Show that for each positive integer n, if σ is a permutation on a finite set A, then so too is $\sigma^n = \underbrace{\sigma \circ \sigma \circ \ldots \circ \sigma}_{n \ times}$.*

A *derangement* is a permutation that fixes no elements, that is, a derangement of the (ordered) symbols $1, 2, \ldots, n$ is a permutation

$$(\sigma(1), \sigma(2), \ldots, \sigma(n)),$$

so that for every i, $\sigma(i) \neq i$. For example, $(3, 1, 4, 2)$ is a derangement of four elements. The number of derangements on n symbols is denoted by D_n. Counting derangements can be done by the inclusion-exclusion principle, but since that principle has an inductive proof (see Exercise 427), it might be no surprise that so does the following:

Exercise 677. *Prove that for $n \geq 1$,*

$$D_n = n! \sum_{i=0}^{n} \frac{(-1)^i}{i!}.$$

Exercise 678. *Denote the number of permutations of $\{1, 2, \ldots, n\}$ that fix precisely p points by $e_p(n)$. Prove that*

$$e_p(n) = \sum_{j=p}^{n} (-1)^{j-p} \binom{j}{p} \binom{n}{j} (n-j)!.$$

Exercise 679. *Prove that the number of permutations π of $\{1, 2, \ldots, n\}$ that satisfy (for each $k = 1, 2, \ldots, n$)*

$$|\pi(k) - k| \leq 1$$

is precisely F_n, the n-th Fibonacci number (where $F_0 = 0$, $F_1 = 1$, and for $n \geq 2$, $F_n = F_{n-2} + F_{n-1}$).

An extension of Exercise 679 is as follows.

Exercise 680. *Let S_n denote the set of permutations of $\{1, 2, \ldots, n\}$. For any $f, g \in S_n$, define $\delta(f, g) = \max_{1 \leq i \leq n} |f(i) - g(i)|$. (So δ is a metric on S_n.) For $n, r \in \mathbb{Z}^+$, and $f \in S_n$, define*

$$a_f(n, r) = |\{g \in S_n : \delta(f, g) \leq r\}|.$$

Prove that the numbers $a_f(n, r)$ are independent of f, and so $a(n, r) = a_f(n, r)$ is well defined. Show that for $n \geq 6$,

$$a(n, 2) = 2a(n - 1, 2) + 2a(n - 3, 2) - a(n - 5, 2),$$

and that $a(n, 1) = F_{n+1}$, the $(n + 1)$-st Fibonacci number (where $F_0 = 0$, $F_1 = 1$, and for $n \geq 2$, $F_n = F_{n-2} + F_{n-1}$).

For more on permutations, see also Section 12.7 on Eulerian numbers, where permutations are counted according to the number of ascents, descents, rises, and falls.

19.3 Rings

A ring $(R, +, \cdot)$ is a set R together with two binary operations addition $(+)$ and multiplication $(\cdot)$ so that $(R, +)$ is an abelian group, R is closed and associative under multiplication, and both distributive laws hold. A *ring with unity* is a ring containing a multiplicative identity 1_R. Multiplication is not necessarily commutative, nor need there be multiplicative inverses or cancellation laws. The set $\mathbb{Z}$ of integers is the most common example of a ring. Another common example is the ring $\mathbb{Z}[x]$ of finite polynomials in an indeterminate x with integer coefficients. More generally, if R is a ring, then $R[x]$ is also a ring, as is the set of all functions on R with ordinary addition and multiplication of functions.

Exercise 681. *Prove that if $f : (R, +, \cdot) \to (S, \oplus, \odot)$ is a ring homomorphism, then for every $n \geq 1$ and every $a \in R$, $f(a^n) = [f(a)]^n$.*

An *ideal* in a ring R is a subring $I \subseteq R$ such that $IR = RI = I$ (that is, for every $i \in I$ and $r \in R$, $ir \in I$ and $ri \in I$). An ideal I is *proper* if $I \neq R$. A *maximal ideal* in a ring is an ideal I with the property that if J is an ideal satisfying $I \subseteq J \subseteq R$, then $J = I$ or $J = R$.

Exercise 682. *Use Zorn's lemma to show that in a commutative ring R with unity, ever proper ideal is contained in a maximal ideal.*

An *algebraic integer* is a number $\alpha \in \mathbb{C}$ that is a root of a monic polynomial $f \neq 0$ in $\mathbb{Z}[x]$ (the ring of finite polynomials with integer coefficients). Let $\mathbb{Z}[\alpha]$ be the smallest subring containing α.

Exercise 683. *Prove that if $\alpha \in \mathbb{C}$ is an algebraic integer, then the additive group $\mathbb{Z}[\alpha]$ is finitely generated.*

19.4 Fields

A set $\mathbb{F}$, together with two binary operations, addition and multiplication, usually denoted by $+$ and $\cdot$, is called a *field* if it satisfies the following axioms:

A1 $\forall a, b \in \mathbb{F}, (a + b) \in \mathbb{F}$. (closure under addition)

A2 $\forall a, b, c \in \mathbb{F} \ (a + b) + c = a + (b + c)$. (associativity of addition)

A3 There exists an element called 0 so that $\forall a \in \mathbb{F}, a + 0 = a = 0 + a$. (In an additive group, this element is unique; 0 is called an additive identity)

A4 $\forall a \in \mathbb{F}, \exists b \in \mathbb{F}$ so that $a + b = 0 = b + a$. (In an additive group, such a b is unique and is denoted by $b = -a$, the additive inverse of a.)

A5 $\forall a, b \in \mathbb{F}, a + b = b + a$.

M1 $\forall a, b \in \mathbb{F}$, $(a \cdot b) \in \mathbb{F}$. (closure under multiplication)

M2 $\forall a, b, c \in \mathbb{F}$, $(a \cdot b) \cdot c = a \cdot (b \cdot c)$. (multiplication is associative)

M3 There exists an element called 1 so that $\forall a \in \mathbb{F}$, $a \cdot 1 = a = 1 \cdot a$. (1 is called a multiplicative identity)

M4 $\forall a \in \mathbb{F}$, $\exists b \in \mathbb{F}$ so that $a \cdot b = 1 = b \cdot a$. (In a group, such a b is unique and is denoted by $b = a^{-1}$, the multiplicative inverse of a.)

M5 $\forall a, b \in \mathbb{F}$, $a \cdot b = b \cdot a$. (multiplication is commutative)

LD $\forall a, b, c \in \mathbb{F}$, $a \cdot (b + c) = a \cdot b + a \cdot c$. (left distributivity)

RD $\forall a, b, c \in \mathbb{F}$, $(a + b) \cdot c = a \cdot c + b \cdot c$. (right distributivity)

II $0 \neq 1$.

The axiom [RD] follows from [LD] and [M5], so is sometimes not mentioned. The axioms [A1]–[A5] show that $(\mathbb{F}, +)$ is an abelian (commutative) group; setting $\mathbb{F}^* = \mathbb{F} \backslash \{0\}$, axioms [M1]–[M5] show that $(\mathbb{F}^*, \cdot)$ is an abelian group.

The number of elements in any finite field is a prime power (*i.e.*, the field has prime power order), and for each prime p and positive integer s, the field of order p^s is unique, called the Galois field of order p^s, denoted $GF(p^s)$. For any $z \in GF(p^s)$, $pz = 0$, where pz means $z + z + \cdots + z$ with z repeated p times. [$GF(p^s)$ has characteristic p.]

As in Section 18.4, if R is a ring, $R[x]$ denotes the ring of polynomials in indeterminate x with coefficients in R. In many situations, R is a field. For a field $\mathbb{F}$ and $f(x) \in \mathbb{F}[x]$, let $f^{(\ell)}(x)$ denote the ℓ-th derivative of $f(x)$ with respect to x.

Exercise 684. *Let p be a prime and let $\mathbb{F} = GF(p)$. For $L \leq p$, if $x_0 \in \mathbb{F}$ and $f(x) \in F[x]$ are so that for $0 \leq \ell < L$, $f^{(\ell)}(x_0) = 0$, then x_0 is a zero of f with multiplicity at least L.*

The next exercise relies on the division algorithm for polynomials, similar to Theorem 11.1.2 for integers: Let K be a field and let $f, g \in K[x]$, where $\deg(g) \leq \deg(f)$. Then there exist polynomials $q, r \in K[x]$ with $\deg(r) < \deg(g)$ so that $f = qg + r$. Furthermore, when $1_K \in K[x]$ is viewed as a constant polynomial, if f and g are relatively prime, then there exist polynomials $s, t \in K[x]$ so that $1 = sf + tg$ (similar to Bezout's Lemma 11.1.3). The following question might bring the method of partial fractions to mind.

Exercise 685. *Let K be a field and let $f, g \in K[x]$. Let $g = g_1 g_2 \cdots g_m$, where the g_i's are relatively prime. Prove that there exists $a_1, \ldots, a_m \in K[x]$ so that*

$$\frac{f(x)}{g(x)} = \sum_{i=1}^{m} \frac{a_i(x)}{g_i(x)}.$$

Exercise 686. *Let K be a field, $b(x) \in K[x]$ with $\deg(b) \geq 1$. Prove that each $f(x) \in K[x]$ has an expression*

$$f(x) = d_m(x)b(x)^m + d_{m-1}(x)b(x)^{m-1} + \cdots + d_0(x),$$

where for each j, $d_j(x) \in K[x]$ and either $d_j(x) = 0$ or $\deg(d_j) < \deg(b)$.

A field K can be *ordered* (or is *orderable*) iff there exists a subset $P \subset K$ (called a *domain of positivity*) satisfying (i) $a, b \in P \Rightarrow ab, a + b \in P$; (ii) $0_K \notin P$; (iii) $\{-p : p \in P\} \cup \{0\} \cup P = K$. An *ordered field* has a fixed domain of positivity. A field $\mathbb{F}$ is defined to be *formally real* iff -1 can not be represented as the sum of squares. An equivalent definition of a formally real field is one that satisfies

$$[a_1^2 + \cdots + a_n^2 = 0] \Rightarrow [a_1 = \cdots = a_n = 0].$$

The next exercise uses Zorn's lemma.

Exercise 687 (Artin–Schreier). *Every formally real field can be ordered.*

For the next exercise, some basics about field extensions are reviewed. For more details, see, *e.g.*, Stewart's *Galois Theory* [509, 512] or [261, p. 418]. If K is a field and a degree n polynomial $p(x) \in K[x]$ is irreducible, then for any zero a of $p(x)$, the extension $K(a)$ formed by adjoining a to K is a vector space over K with dimension n.

Lemma 19.4.1. *If a is a root of an irreducible polynomial over K, then every $b \in K(a)$ is a zero of an irreducible polynomial in $K[x]$ (in other words, b is algebraic over K).*

Proof of lemma: Since $K(a)$ is a ring, the elements $b^0 = 1, b^1 = b, b^2, \ldots, b^n$ are elements in $K(a)$, but since $K(a)$ has dimension n, these must be linearly dependent over K, and so for some constants $c_i \in K$ not all zero, $c_0 b^0 + c_1 b^1 + \cdots + b^n = 0_K$. $\square$

An extension K' of K is said to be *algebraic* (over K) if every element $b \in K'$ is algebraic over K. An extension E of K is called *finite* over K if E is a vector space of finite dimension over K. Then Lemma 19.4.1 implies:

Theorem 19.4.2. *If K is a field and E is a finite field extension of K, then E is algebraic over K.*

Exercise 688. *Let K be a field. Prove by induction that every extension of K formed by adding finitely many algebraic (over K) elements is finite and therefore algebraic over K.*

[Conversely, every finite extension of K can be generated by adding finitely many elements each algebraic over K.]

The result in the next exercise guarantees a "splitting field", and is due to Kronecker (1823–1891). The following fact is needed (and given without proof):

Fact: If K is a field and $p(x) \in K[x]$ is irreducible, then the quotient field $K[x]/\langle p(x) \rangle$ is a field containing (an isomorphic copy of) K *and* a root z of $p(x)$.

Exercise 689. *Let K be a field and $f(x) \in K[x]$. Prove that there exists a field E containing K as a subfield so that $f(x)$ is a product of linear polynomials in $E[x]$.*

19.5 Vector spaces

Given a field $\mathbb{F}$, a vector space over $\mathbb{F}$ is a set V of objects (called *vectors*), together with two operations, *addition* "+" and *scalar product* "$\cdot$" so that for every $\mathbf{u}, \mathbf{v}, \mathbf{w} \in V$ and $\alpha, \beta \in \mathbb{F}$,

1. $\mathbf{u} + \mathbf{v} \in V$. [closure of addition]

2. $\alpha \cdot \mathbf{v} \in V$. [closure of scalar product]

3. $\mathbf{u} + \mathbf{v} = \mathbf{v} + \mathbf{u}$. [addition is commutative]

4. $(\mathbf{u} + \mathbf{v}) + \mathbf{w} = \mathbf{u} + (\mathbf{v} + \mathbf{w})$. [addition is associative]

5. There is a vector $\mathbf{0} \in V$ so that for all $\mathbf{v} \in V$, $\mathbf{0} + \mathbf{v} = \mathbf{v}$. [additive identity exists]

6. If $1_\mathbb{F}$ is the identity in $\mathbb{F}$, then $1_\mathbb{F} \cdot \mathbf{v} = \mathbf{v}$. [scalar product identity]

7. $\alpha \cdot (\mathbf{u} + \mathbf{v}) = \alpha \cdot \mathbf{u} + \alpha \cdot \mathbf{v}$. [distributive property over vector addition]

8. $(\alpha + \beta)\mathbf{v} = \alpha \cdot \mathbf{v} + \beta \cdot \mathbf{v}$. [distributive property over field addition]

9. $(\alpha\beta) \cdot \mathbf{v} = \alpha \cdot (\beta \cdot \mathbf{v})$. [associativity of scalar product]

A vector space is sometimes called a *linear space*. The elements of $\mathbb{F}$ are called *scalars*. Vectors are often written in boldface, and scalars are in math italics. Notation for vectors varies: v, $\bar{v}$, $\overline{v}$, $\vec{v}$, and $\underset{\sim}{v}$ are all common notations for the vectors. (The squiggly line underneath a letter is a typesetter's notation for making something bold—this notation is often employed when writing math by hand.)

Perhaps the most common example of a vector space is, for a positive integer n, the Euclidean vector space $V = \mathbb{R}^n$, the set of n-tuples of real numbers, considered as a vector space over the field $\mathbb{R}$. For a set X, and a positive integer n, define $X^n = \{(x_1, x_2, \ldots, x_n) : x_1, x_2, \ldots, x_n \in X\}$, the set of all ordered n-tuples whose coordinates are in X. This notation is often used when X is one of the fields $\{0, 1\}$, $\mathbb{R}$, or $\mathbb{C}$. Elements of X^n can be called vectors, and are occasionally written as column matrices. The zero vector $\mathbf{0}$ is the vector all of whose coordinates are 0.

A *subspace* of a vector space V is a non-empty subset $W \subset V$ that is itself a vector space under the same operations as in V. .

For a vector space V over a field $\mathbb{F}$, a *linear combination* of vectors $\mathbf{v}_1, \mathbf{v}_2, \ldots, \mathbf{v}_k \in V$ is an expression of the form

$$c_1\mathbf{v}_1 + c_2\mathbf{v}_2 + \cdots + c_k\mathbf{v}_k,$$

where the c_i's are elements (scalars) of the field $\mathbb{F}$.

A collection of vectors $\mathbf{v}_1, \mathbf{v}_2, \ldots, \mathbf{v}_k \in \mathbb{C}^n$ is called *linearly independent* (over $\mathbb{C}$) iff the only complex numbers $c_1, \ldots, c_k$ for which the equation

$$c_1 \mathbf{v}_1 + c_2 \mathbf{v}_2 + \cdots + c_k \mathbf{v}_k = \mathbf{0}$$

has a solution are $c_1 = c_2 = \cdots = c_k = 0$. (The vectors are said to be linearly *dependent* if they fail to be linearly independent, that is, if there exists some non-trivial linear combination of these vectors that vanishes.)

The definitions for eigenvalues and eigenvectors from Section 19.1 apply in general; they are repeated here for convenience. Let A be an $n \times n$ matrix with entries from $\mathbb{C}$ (actually, any field will do). An element $\lambda \in \mathbb{C}$ is called an *eigenvalue* for A if there exists a non-zero column matrix $\mathbf{x}$ so that $A\mathbf{x} = \lambda\mathbf{x}$; such an $\mathbf{x}$ is called an *eigenvector* associated with λ.

Exercise 690. *Let A be an $n \times n$ matrix that has distinct eigenvalues $\lambda_1, \ldots, \lambda_r$. Prove that if $\mathbf{v}_1, \ldots, \mathbf{v}_r$ are eigenvectors each associated with a different eigenvalue, then they are linearly independent.*

Given two vector spaces V and W over a field $\mathbb{F}$, a function $T : V \to W$ is called a *linear transformation* iff for any $\mathbf{u}, \mathbf{v} \in V$ and any $k \in \mathbb{F}$,

$$T(k\mathbf{u} + \mathbf{v}) = kT(\mathbf{u}) + T(\mathbf{v}).$$

(Equivalently, T is a linear transformation iff for any $\mathbf{u}, \mathbf{v} \in V$ and any $k \in \mathbb{F}$, both $T(\mathbf{u} + \mathbf{v}) = T(\mathbf{u}) + T(\mathbf{v})$ and $T(k\mathbf{u}) = kT(\mathbf{u})$ hold.) A linear transformation $T : V \to V$ T is called a *linear operator* on V.

Let T be a linear operator on a vector space V over a field $\mathbb{F}$. For $\lambda \in \mathbb{F}$, if there exists a non-zero $\mathbf{v} \in V$ with $T(\mathbf{v}) = \lambda\mathbf{v}$, then λ is an *eigenvalue for T* and $\mathbf{v}$ is an *eigenvector* associated with λ. A subspace W of V is called *T-invariant* iff for any $\mathbf{w} \in W$, $T(\mathbf{w}) \in W$.

Exercise 691. *Let T be a linear operator on a finite dimensional vector space V and let W be a T-invariant subspace of V. Suppose that $\mathbf{v}_1, \ldots, \mathbf{v}_k$ are eigenvectors of T that correspond to distinct eigenvalues. Prove that if $\mathbf{v}_1 + \cdots + \mathbf{v}_k \in W$, then each $\mathbf{v}_i \in W$.*

Recall that if S is a subset of a vector space V over a field $\mathbb{F}$, the span of S is the set (denoted $\text{span}(S)$) of all finite linear combinations of vectors from S (where coefficients in any linear combination occur in $\mathbb{F}$). It is not hard to show that for any $S \subset V$, $\text{span}(S)$ is itself a vector space (and so is a subspace of V). If $\text{span}(S) = W$, the set S is said to "span W".

If V is a vector space, a subset $B \subset V$ of vectors is called a *basis* for V iff B is linearly independent and $\text{span}(B) = V$.

One standard application of Zorn's lemma is the following.

Theorem 19.5.1. *Every vector space has a basis.*

Proof idea: Let V be a vector space and let $\mathcal{B}$ be the set of all linearly independent subsets of V. Then show that the union of any chain of independent sets is again independent. $\qquad \square$

Exercise 692. *Fill in the details to the proof of Theorem 19.5.1.*

The dimension of a vector space is the number of elements in a basis for that space.

Exercise 693. *Let $A \in M_{n \times n}(\mathbb{F})$. Prove that $\dim(span(I_n, A, A^2, \ldots)) \leq n$.*

An *inner product* on a vector space V over a field $\mathbb{F}$ is a function $\langle \cdot, \cdot \rangle : V^2 \to \mathbb{F}$ that satisfies (for all $\mathbf{a}, \mathbf{b}, \mathbf{c} \in V$, and $k, \ell \in \mathbb{F}$)

(i) $\langle \mathbf{a}, \mathbf{b} \rangle = \overline{\langle \mathbf{b}, \mathbf{a} \rangle}$ (the complex conjugate, in the case that $\mathbb{F} = \mathbb{C}$).

(ii) $\langle k\mathbf{a} + \ell\mathbf{b}, \mathbf{c} \rangle = k\langle \mathbf{a}, \mathbf{c} \rangle + \ell\langle \mathbf{b}, \mathbf{c} \rangle$.

(iii) $\langle \mathbf{a}, \mathbf{a} \rangle \geq 0$ and $\langle \mathbf{a}, \mathbf{a} \rangle = 0$ if and only if $\mathbf{a} = \mathbf{0}$.

A vector space with an inner product is called an *inner product space*.

For any positive integer n, and vectors $\mathbf{a} = (a_1, a_2, \ldots, a_n) \in \mathbb{R}^n$ and $\mathbf{b} = (b_1, b_2, \ldots, b_n) \in \mathbb{R}^n$, define the *dot product* of $\mathbf{a}$ and $\mathbf{b}$ to be

$$\mathbf{a} \bullet \mathbf{b} = a_1 b_1 + a_2 b_2 + \cdots + a_n b_n.$$

One can easily verify that the dot product is indeed an inner product, and is sometimes called the *Euclidean inner product*.

In an inner product space V, for any $\mathbf{v} \in V$, define the norm of $\mathbf{v}$ to be

$$\|\mathbf{v}\| = \langle \mathbf{v}, \mathbf{v} \rangle^{1/2}.$$

When the inner product is the dot product, the norm is called the *Euclidean norm*; if $\mathbf{a} = (a_1, a_2, \ldots, a_n) \in \mathbb{R}^n$, then the Euclidean norm of $\mathbf{a}$ is

$$\|\mathbf{a}\| = \sqrt{\mathbf{a} \bullet \mathbf{a}} = \sqrt{a_1^2 + a_2^2 + \cdots + a_n^2}.$$

When $n = 1$, the norm is just the absolute value.

Consider the famous "Cauchy–Schwarz" inequality, named after Augustin Louis Cauchy and Karl Herman Amandus Schwarz. Its proof given here is not inductive, however it is a very useful inequality in most areas of mathematics, and it is needed later. The proof might also give one ideas as to how to prove some later results by induction.

Theorem 19.5.2 (Cauchy–Schwarz inequality for dot products). *For each positive integer n and real numbers $a_1, a_2, \ldots, a_n, b_1, b_2, \ldots, b_n$, then*

$$(a_1^2 + a_2^2 + \cdots + a_n^2) \cdot (b_1^2 + b_2^2 + \cdots + b_n^2) \geq (a_1 b_1 + a_2 b_2 + \cdots + a_n b_n)^2,$$

or equivalently,

$$\mathbf{a} \bullet \mathbf{b} \leq \|\mathbf{a}\| \cdot \|\mathbf{b}\|.$$

Proof: One common proof is to examine $\|\mathbf{a} + t\mathbf{b}\|^2 \geq 0$, and after some simplification, use known properties of quadratic polynomials. Such a proof is easily duplicated to prove the general inner product version of Cauchy–Schwarz: $|\langle \mathbf{a}, \mathbf{b} \rangle| \leq \|\mathbf{a}\| \cdot \|\mathbf{b}\|$. Here is another proof of Cauchy–Schwarz for the dot product: When $n = 1$, the theorem is true since it says $a^2 b^2 \leq (ab)^2$. When $n \geq 2$, relying on the fact that for any real numbers x and y, $0 \leq (x - y)^2 = x^2 + y^2 - 2xy$,

$$2xy \leq x^2 + y^2. \tag{19.2}$$

In particular, for $\mathbf{a} = (a_1, a_2, \ldots, a_n)$ and $\mathbf{b} = (b_1, b_2, \ldots, b_n)$, in the following sequence of inequalities, Equation (19.2) is used for each $i = 1, \ldots, n$, using $x = \frac{a_i}{\|\mathbf{a}\|}$ and $y = \frac{b_i}{\|\mathbf{b}\|}$:

$$
\begin{aligned}
\mathbf{a} \bullet \mathbf{b} &= \sum_{i=1}^{n} a_i b_i \\
&= \left(\sum_{i=1}^{n} \frac{a_i}{\|\mathbf{a}\|} \frac{b_i}{\|\mathbf{b}\|} \right) \|\mathbf{a}\| \cdot \|\mathbf{b}\| \\
&\leq \frac{1}{2} \sum_{i=1}^{n} \left[\left(\frac{a_i}{\|\mathbf{a}\|} \right)^2 + \left(\frac{a_i}{\|\mathbf{a}\|} \right)^2 \right] \|\mathbf{a}\| \cdot \|\mathbf{b}\| \\
&= \frac{1}{2} \left[\frac{1}{\|\mathbf{a}\|^2} \sum_{i=1}^{n} a_i^2 + \frac{1}{\|\mathbf{b}\|^2} \sum_{i=1}^{n} b_i^2 \right] \|\mathbf{a}\| \cdot \|\mathbf{b}\| \\
&= \frac{1}{2}(1 + 1)\|\mathbf{a}\| \cdot \|\mathbf{b}\|,
\end{aligned}
$$

and so $\mathbf{a} \bullet \mathbf{b} \leq \|\mathbf{a}\| \cdot \|\mathbf{b}\|$. □

An apparently stronger result also holds—for real numbers $a_1, a_2, \ldots, a_n$, and $b_1, b_2, \ldots, b_n$,

$$(a_1^2 + a_2^2 + \cdots + a_n^2) \cdot (b_1^2 + b_2^2 + \cdots + b_n^2) \geq (|a_1 b_1| + |a_2 b_2| + \cdots + |a_n b_n|)^2,$$

though one sees that this follows from the Cauchy–Schwarz inequality directly: for all $a_i b_i$ that are negative, replace a_i with $-a_i$ and apply Cauchy–Schwarz; the left side remains unchanged. On the other hand, this stronger-looking inequality implies Cauchy–Schwarz immediately because

$$(|a_1 b_1| + |a_2 b_2| + \cdots + |a_n b_n|)^2 \geq (a_1 b_1 + a_2 b_2 + \cdots + a_n b_n)^2$$

In the case that $\mathbf{b} = (1, 1, \ldots, 1)$, the Cauchy–Schwarz inequality yields

$$a_1 + a_2 + \cdots + a_n \leq \sqrt{a_1^2 + a_2^2 + \cdots + a_n^2} \, \sqrt{n},$$

and thus gives an upper bound for the arithmetic mean

$$\frac{a_1 + a_2 + \cdots + a_n}{n} \leq \sqrt{\frac{a_1^2 + a_2^2 + \cdots + a_n^2}{n}}.$$

In $\mathbb{R}^2$ and $\mathbb{R}^3$, if $\mathbf{u}$ and $\mathbf{v}$ are vectors with angle θ between them, then the dot product is often first defined by

$$\mathbf{u} \bullet \mathbf{v} = \|\mathbf{u}\| \|\mathbf{v}\| \cos \theta.$$

Thus in dimensions 2 and 3, Cauchy–Schwarz (for dot products) follows directly from this definition. (The cosine law is then used to show that this definition of dot product agrees with the more usual definition.) In higher dimensions, the dot product is defined first, then

$$\theta = \cos^{-1} \left(\frac{\mathbf{u} \bullet \mathbf{v}}{\|\mathbf{u}\| \|\mathbf{v}\|} \right)$$

is defined to be the angle between the vectors—so the Cauchy–Schwarz inequality guarantees that the argument of $\cos^{-1}$ is bounded between -1 and 1 and hence the angle is defined.

The next exercise generalizes the triangle inequality (see Exercise 193) to n dimensions; it is one form of what is known as "Minkowski's inequality"; Hermann Minkowski (1864–1909) was one of Einstein's mathematics teachers.

Exercise 694 (Minkowski's inequality). *Fix some non-negative integer n. Use the Cauchy–Schwarz inequality and induction on m to prove that for every $n \geq 1$, and every collection of vectors $\mathbf{v}_1, \mathbf{v}_2, \ldots, \mathbf{v}_m$ in $\mathbb{R}^n$,*

$$\|\mathbf{v}_1 + \mathbf{v}_2 + \cdots + \mathbf{v}_m\| \leq \|\mathbf{v}_1\| + \|\mathbf{v}_2\| + \cdots + \|\mathbf{v}_m\|.$$

Another form of Minkowski's inequality that generalizes the triangle inequality occurs when the notion of norm is generalized. If $\mathbf{a} = (a_1, a_2, \ldots, a_n) \in \mathbb{R}^n$ is a vector, and $p > 1$ is a real number, define

$$\|\mathbf{a}\|_p = (|a_1|^p + |a_2|^p + \cdots + |a_n|^p)^{1/p},$$

called the *p-norm* of $\mathbf{a}$. Just as the Cauchy–Schwarz inequality was required to prove Minkowski's inequality, something called Hölder's inequality (for vectors) is used to generalize Minkowski's inequality for p-norms. First a useful lemma (which appears in, for example, [74]) is given:

Lemma 19.5.3. *For real numbers* $x, y > 0$ *and* $0 < \alpha < 1$,

$$x^\alpha y^{1-\alpha} \leq \alpha x + (1 - \alpha)y.$$

Proof outline: For $u > 0$, consider the function $f(u) = u^\alpha - \alpha u - 1 + \alpha$. Simple calculus shows that on the interval $(0, \infty)$, f achieves a maximum of 0 at $u = 1$, that is, $f(u) \leq 0$ for all $u > 0$. Putting $u = \frac{x}{y}$ and a little algebra yields the desired inequality. $\square$

Theorem 19.5.4 (Hölder's inequality for vectors). *If* $p > 1$ *and* $q > 1$ *are real numbers with* $\frac{1}{p} + \frac{1}{q} = 1$, *then*

$$\left| \sum_{i=1}^{n} x_i y_i \right| \leq \left(\sum_{i=1}^{n} |x_i|^p \right)^{1/p} \left(\sum_{i=1}^{n} |y_i|^q \right)^{1/q},$$

that is, if $\mathbf{x}, \mathbf{y} \in \mathbb{R}^n$, *then* $|\mathbf{x} \bullet \mathbf{y}| \leq \|\mathbf{x}\|_p \|\mathbf{y}\|_q$.

Proof: If $\mathbf{x} = \mathbf{0}$ or $\mathbf{y} = \mathbf{0}$, then the result is trivial, so suppose that both $\|\mathbf{x}\|_p > 0$ and $\|\mathbf{y}\|_q > 0$. In the following sequence of inequalities, Lemma 19.5.3 is used n times, each time with $\alpha = \frac{1}{p}$, $x = \frac{|x_i|^p}{(\|\mathbf{x}\|_p)^p}$, and $y = \frac{|y_i|^q}{(\|\mathbf{y}\|_q)^q}$:

$$
\begin{aligned}
\frac{|\sum_{i=1}^{n} x_i y_i|}{\|\mathbf{x}\|_p \|\mathbf{y}\|_q}
&\leq \frac{\sum_{i=1}^{n} |x_i| \cdot |y_i|}{\|\mathbf{x}\|_p \|\mathbf{y}\|_q} \quad \text{(by the triangle inequality)} \\
&= \sum_{i=1}^{n} \left(\frac{|x_i|}{\|\mathbf{x}\|_p} \right) \left(\frac{|y_i|}{\|\mathbf{y}\|_q} \right) \\
&= \sum_{i=1}^{n} \left(\frac{|x_i|^p}{(\|\mathbf{x}\|_p)^p} \right)^{1/p} \left(\frac{|y_i|^q}{(\|\mathbf{y}\|_q)^q} \right)^{1/q} \\
&\leq \sum_{i=1}^{n} \left[\frac{1}{p} \left(\frac{|x_i|^p}{(\|\mathbf{x}\|_p)^p} \right) + \frac{1}{q} \left(\frac{|y_i|^q}{(\|\mathbf{y}\|_q)^q} \right) \right] \quad \text{(by Lemma 19.5.3)} \\
&= \frac{1}{p(\|\mathbf{x}\|_p)^p} \sum_{i=1}^{n} |x_i|^p + \frac{1}{q(\|\mathbf{y}\|_q)^q} \sum_{i=1}^{n} |y_i|^q \\
&= \frac{1}{p} + \frac{1}{q} \\
&= 1,
\end{aligned}
$$

and so

$$\left| \sum_{i=1}^{n} x_i y_i \right| \leq \|\mathbf{x}\|_p \|\mathbf{y}\|_q$$

as desired. In fact, it was proved that $\sum_{i=1}^{n} |x_i y_i| \leq \|\mathbf{x}\|_p \|\mathbf{y}\|_q$. $\square$

Exercise 695 (Minkowski's inequality for p-norm). *Use Hölder's inequality to prove that for any real $p > 1$, and any vectors $\mathbf{x}, \mathbf{y} \in \mathbb{R}^n$,*

$$\|\mathbf{x} + \mathbf{y}\|_p \leq \|\mathbf{x}\|_p + \|\mathbf{y}\|_p.$$

Then prove by induction that for any vectors $\mathbf{x}_1, \mathbf{x}_2, \ldots \mathbf{x}_n \in \mathbb{R}^n$,

$$\|\mathbf{x}_1 + \mathbf{x}_2 + \cdots + \mathbf{x}_n\|_p \leq \|\mathbf{x_1}\|_p + \|\mathbf{x_2}\|_p + \cdots + \|\mathbf{x_n}\|_p.$$

Two vectors $\mathbf{u}$ and $\mathbf{v}$ in an inner product space V are said to be *orthogonal* if and only if $\langle \mathbf{u}, \mathbf{v} \rangle = 0$. A set of vectors is called orthogonal iff any two vectors in the set are orthogonal.

Lemma 19.5.5. *Let V be an inner product space. Any set of non-zero orthogonal vectors in V is linearly independent.*

Proof: Suppose that $\{\mathbf{v}_1, \ldots, \mathbf{v}_r\}$ is a set of non-zero pairwise orthogonal vectors and let $c_1, \ldots, c_r$ be real constants so that

$$c_1 \mathbf{v}_1 + \cdots + c_r \mathbf{v}_r = \mathbf{0}.$$

For each $i \in \{1, \ldots, r\}$, taking the inner product of both sides with $\mathbf{v}_i$ yields $c_i \|\mathbf{v}_i\|^2 = 0$, and so each $c_i = 0$. $\square$

In the next theorem, the construction of the orthogonal set $\{\mathbf{v}_1, \ldots, \mathbf{v}_n\}$ is called the "Gram–Schmidt orthogonalization process".

Theorem 19.5.6 (Gram–Schmidt process). *Let V be an inner product space, and let $S = \{\mathbf{w}_1, \ldots, \mathbf{w}_n\} \subset V$. For each $i = 1, \ldots, n$ define $\mathbf{v}_i$ recursively as follows: Put $\mathbf{v}_1 = \mathbf{w}_1$, and for each $k = 2, \ldots, n$, put*

$$\mathbf{v}_k = \mathbf{w}_k - \sum_{j=1}^{k-1} \frac{\langle \mathbf{w}_k, \mathbf{v}_j \rangle}{\|\mathbf{v}_j\|^2} \mathbf{v}_j.$$

Then $T = \{\mathbf{v}_1, \ldots, \mathbf{v}_n\}$ is an orthogonal set of non-zero vectors with $span(T) = span(S)$.

So the Gram–Schmidt orthogonalization process produces an orthogonal basis for span(S). A a set W of vectors is called *orthonormal* if and only if W is orthogonal and each $\mathbf{v} \in W$ is a unit vector, that is, $\|\mathbf{v}\| = 1$. Any orthogonal basis T can be transformed to an orthonormal basis by simply taking multiples of vectors in T that are unit vectors (if $\mathbf{w} \in T$, use $\mathbf{v} = \frac{1}{\|\mathbf{w}\|}\mathbf{w}$).

Exercise 696. *Prove Theorem 19.5.6 by induction on n, and conclude that every finite dimensional vector space with an inner product has an orthonormal basis.*

Exercise 697. *Let V be a vector space, $T : V \to V$ be a linear operator on V, and let W be a T-invariant subspace of V. If $\mathbf{v}_1, \ldots, \mathbf{v}_k$ are eigenvectors of T corresponding to distinct eigenvalues, prove that if $v_1 + \cdots + v_k \in W$ then for each $i = 1, \ldots, k$, $\mathbf{v}_i \in W$. Hint: Induct on k.*

Chapter 20

Geometry

Geometry supplies sustenance and meaning to bare formulas... One can still believe Plato's statement that "Geometry draws the soul toward truth."

—Morris Kline

Exercise 698. *If a, b, c are sides of a right angle triangle with c being the hypotenuse, prove that for every natural number $n \geq 3$,*

$$a^n + b^n < c^n.$$

Exercise 699. *For every $n \geq 2$, given a line segment of length 1 and using a straightedge and compass only, inductively construct a line segment of length $\sqrt{n}$.*

One area of combinatorial geometry that Erdős seemed to love was the study of the frequency of the distances between points in the plane. For example, in 1946 [163], he conjectured that, in the real euclidean plane $\mathbb{R}^2$, with the usual distance metric the n vertices of any convex n-gon determine at least $\lfloor n/2 \rfloor$ distances. A regular n-gon shows that this number can not be increased. This conjecture was finally proved in 1963 by Altman [17]. In the 1946 paper, Erdős gave the solution to the following problem:

Exercise 700. *Show that for each $n \geq 1$, the greatest distance among a set of n points is realized by at most n different pairs of points.*

For many more problems of the above type, the reader is recommended to begin with [250, pp. 21–25, 47–50]. There are many more modern surveys. The list of mathematicians producing results in distance-realization problems is very impressive.

It is fairly easy to see that the maximum number k of points that can be chosen in the real interval $[-\sqrt{2}, \sqrt{2}]$ so any two are more than distance 2 apart is $k = 2$. At most $k = 3$ points can be chosen in a 2-D circle of radius $\sqrt{2}$ so that each of the k points has distance greater than 2 from all others. This pattern continues.

Exercise 701. *Prove by induction on d that if n points in $\mathbb{R}^d$ are contained in a ball of radius $\sqrt{2}$ so that every pair of points are at distance greater than 2, then $n \leq d + 1$.*

Exercise 702. *Mark n points around a circle and label them either red or blue. Prove that there are at most $\lfloor (3n+2)/2 \rfloor$ chords that join differently labelled points and that do not intersect inside the circle.*

Exercise 703. *Given $n \geq 2$ squares with respective side lengths $a_1 \leq a_2 \leq \cdots \leq a_n$, show that one can dissect the squares each into at most four pieces so that the pieces can be reassembled into a single square.*

20.1 Convexity

For information on convex sets not given here, see nearly any of the many books titled *Convex sets...*, (*e.g.*, [343]). Throughout this discussion, n is a positive integer and $\mathbb{R}^n$ is endowed with the usual metric (in which case, this space is often denoted by $\mathbb{E}^n$, the n-dimensional Euclidean space). If $\mathbf{x} = (x_1, \ldots, x_n) \in \mathbb{R}^n$, then $\|\mathbf{x}\| = \sqrt{x_1^2 + \cdots + x_n^2}$. The distance between points (or vectors) $\mathbf{x}$ and $\mathbf{y}$ is $\|\mathbf{x} - \mathbf{y}\|$.

Recall that for points $x_1, \ldots, x_k$ in $\mathbb{R}^n$, a linear combination of these points is any expression of the form

$$\lambda_1 \mathbf{x}_1 + \lambda_2 \mathbf{x}_2 + \cdots + \lambda_k \mathbf{x}_k, \tag{20.1}$$

where each $\lambda_i \in \mathbb{R}$. The linear combination (20.1) is called an *affine combination* if and only if $\sum_{i=1}^k \lambda_i = 1$, and is called a *convex combination* if and only if $\sum_{i=1}^k \lambda_i = 1$ and each $\lambda_i \geq 0$.

A set $C \subseteq \mathbb{R}^n$ is called *affine* if and only if for any two points $\mathbf{x}$ and $\mathbf{y}$ in C, for every $\lambda \in \mathbb{R}$, the affine combination $\lambda \mathbf{x} + (1 - \lambda)\mathbf{y}$ is also in C. Intuitively, this says that if two points are in an affine set, the entire (infinite) straight line containing these two points is also in the set. One can show that the intersection of any two affine sets is again affine, and so by induction, that the intersection of finitely many affine sets is again affine.

A set $C \subseteq \mathbb{R}^n$ is called *convex* if and only if for every $\lambda \in [0, 1]$, the convex combination $\lambda \mathbf{x} + (1 - \lambda)\mathbf{y}$ is also in C. Intuitively, this says that if two points are in a convex set, then the straight line *segment* containing these two points is also in the set. One can prove that if $C_1, \ldots, C_k \in \mathbb{R}^n$ are convex sets, then their intersection $\cap_{i=1}^k C_i$ is also convex. An affine set is, by definition (restricting the λ's) convex, however a convex set need not be affine (a straightline segment is convex, but is not affine). An affine set in $\mathbb{R}^n$ is sometimes called an affine *space* (for reasons made apparent below).

Recall that a linear subspace of $\mathbb{R}^n$ is closed under arbitrary finite linear combinations. The similar statements are true for affine and convex subspaces of $\mathbb{R}^n$.

Exercise 704. *Let $C \subseteq \mathbb{R}^n$ be a convex set. Prove by induction on $m \geq 1$, that if $\mathbf{x}_1, \ldots, \mathbf{x}_m \in C$, then for any $\alpha_1, \ldots, \alpha_m \in [0,1]$ satisfying $\sum_{i=1} \alpha_i = 1$, the convex combination*

$$\alpha_1 \mathbf{x}_1 + \cdots + \alpha_m \mathbf{x}_m$$

is also in C. Repeat this exercise for affine sets.

It is not difficult to check that every plane, line, or point in $\mathbb{R}^3$ is an affine space. In fact, these are the only affine spaces in $\mathbb{R}^3$, as is stated in the next theorem (whose simple proof is omitted—see [343, p. 14]).

Theorem 20.1.1. *Let $A \subset \mathbb{R}^n$ be an affine space. Then there exists a linear subspace (containing the origin) $W \subseteq \mathbb{R}^n$ and a vector (or point) $\mathbf{v} \in \mathbb{R}^n$ so that*

$$A = \mathbf{v} + W = \{\mathbf{v} + \mathbf{w} : \mathbf{a} \in W\}.$$

When W is a (linear) subspace and $\mathbf{v}$ is a vector, a set of the form $\mathbf{v} + W$ in Theorem 20.1.1 is called a *translate* of W, a *shifted linear space*, or a *flat*. The dimension of an affine space $A = \mathbf{v} + W$ is defined to be the dimension of W. A subspace $W \subseteq \mathbb{R}^n$ is called a *hyperplane* if W has dimension $n - 1$. The dimension of any set $S \subseteq \mathbb{R}^n$ is defined to be the dimension of the smallest flat containing S.

Recall from linear algebra, that vectors $\mathbf{v}_1, \ldots, \mathbf{v}_k$ in a vector space V (over the reals) are called *linearly independent* if and only if for any scalars $c_1, \ldots, c_k \in \mathbb{R}$, the equation

$$c_1 \mathbf{v}_1 + \ldots + c_k \mathbf{v}_k = \mathbf{0} \tag{20.2}$$

is satisfied only when $c_1 = \cdots = c_k = 0$. If there exist c_i's not all zero satisfying the above equation, the vectors are *linearly dependent*. If there exist $c_1, \ldots, c_k \in \mathbb{R}$ not all zero but with $\sum_{i=1}^k c_i = 0$ satisfying (20.2), then $\mathbf{v}_1, \ldots, \mathbf{v}_k$ are called *affinely dependent*, and are *affinely independent* if the only c_i's satisfying $\sum_{i=1}^k c_i = 0$ and (20.2) are all zeros. Recall from linear algebra that any $n + 1$ vectors in $\mathbb{R}^n$ are linearly dependent.

Lemma 20.1.2. *Any collection of $n + 2$ vectors in $\mathbb{R}^n$ are affinely dependent.*

Proof: Let $\mathbf{x}_1, \ldots, \mathbf{x}_{n+2} \in \mathbb{R}^n$. The $n + 1$ vectors $\mathbf{x}_2 - \mathbf{x}_1$, $\mathbf{x}_3 - \mathbf{x}_1$, $\ldots$, $\mathbf{x}_{n+2} - \mathbf{x}_1$ are linearly dependent, so let $\lambda_2, \ldots, \lambda_{n+2} \in \mathbb{R}$ satisfy

$$\lambda_2 (\mathbf{x}_2 - \mathbf{x}_1) + \lambda_3 (\mathbf{x}_3 - \mathbf{x}_1) + \cdots + \lambda_{n+2} (\mathbf{x}_{n+2} - \mathbf{x}_1) = \mathbf{0}.$$

Rewriting,

$$-\left(\sum_{i=2}^{n+2} \lambda_i \right) \mathbf{x}_1 + \lambda_2 \mathbf{x}_2 + \lambda_3 \mathbf{x}_3 + \cdots + \lambda_{n+2} \mathbf{x}_{n+2} = \mathbf{0}.$$

The coefficients in this expression sum to 0, so this expression shows the vectors $\mathbf{x}_1, \ldots, \mathbf{x}_{n+2}$ are affinely dependent. ∎

From the proof of Lemma 20.1.2, observe that by multiplication of an appropriate constant, any one of the $n + 2$ vectors can be written as an affine combination of the remaining ones.

For a set $S \subseteq \mathbb{R}^n$, define the *convex hull* of S, denoted conv(S), to be the intersection of all convex sets containing S. The convex hull of a set S is the smallest convex set containing all of S. If C is a convex set, then conv(C) = C. By Exercise 704, the convex hull of S is the set of all convex linear combinations of points in S; for later reference, this fact is identified as a lemma:

Lemma 20.1.3. *A set S is convex if and only if every convex combination of finitely many points in S is also in S.*

Theorem 20.1.4. *Let S be a set and let T consist of all (finite) convex linear combinations of points in S. Then conv(S) = T.*

Proof: By definition, $S \subseteq T$. By Lemma 20.1.3, applied to the convex set conv(S), $T \subseteq$ conv(S). It remains to show that conv(S) $\subseteq T$, and for this, it suffices to show that T is convex. Suppose that $\mathbf{x}$ and $\mathbf{y}$ are points in T, with

$$\mathbf{x} = \sum_{i=1}^{r} \alpha_i \mathbf{x}_i \quad \text{and} \quad \mathbf{y} = \sum_{j=1}^{s} \beta_i \mathbf{y}_j,$$

where all the $\mathbf{x}_i$'s and $\mathbf{y}_j$'s are in S. Then for any $\lambda \in [0, 1]$,

$$\lambda \mathbf{x} + (1 - \lambda)\mathbf{y} = \sum_{i=1}^{r} \lambda \alpha_i \mathbf{x}_i + \sum_{j=1}^{s} (1 - \lambda)\beta_i \mathbf{y}_j$$

is a linear combination of points in S, where the sum of the coefficients is

$$\sum_{i=1}^{r} \lambda \alpha_i + \sum_{j=1}^{s} (1 - \lambda)\beta_i = \lambda \sum_{i=1}^{r} \alpha_i + (1 - \lambda) \sum_{j=1}^{s} \beta_i = \lambda(1) + (1 - \lambda)(1) = 1.$$

Thus T is convex, and so conv(S) $\subseteq T$. ∎

The following theorem was proved by Caratheodory [96] in 1907.

Theorem 20.1.5 (Caratheodory). *If $S \subseteq \mathbb{R}^n$, then every $\mathbf{x} \in$ conv(S) can be expressed as a convex linear combination of at most $n + 1$ points from S.*

Proof outline: Let $\mathbf{x}$ be expressed as a convex combination of more than $n + 1$ points in S. By Lemma 20.1.2, there is an *affine* combination of these points that equals $\mathbf{0}$. Subtract suitable multiples of these two equations to eliminate one of the points, and then with scaling, make make this new sum a convex combination. If necessary, continue by induction until a new representation uses at most $n+1$ points.

Another proof (see [61, p. 88]) of Caratheodory's theorem uses the following theorem due to Radon, published in 1921.

Theorem 20.1.6 (Radon's theorem [445]). *Let $S = \{\mathbf{x}_1, \ldots, \mathbf{x}_r\}$ be set of points in $\mathbb{R}^n$. If $r \geq n+2$, then S can be partitioned into two disjoint sets $S = S_1 \cup S_2$ so that $conv(S_1) \cap conv(S_2) \neq \emptyset$.*

Proof: Suppose $r \geq n+2$. By Lemma 20.1.2, there exist $\lambda_1, \ldots, \lambda_r$, not all zero, with $\sum_{i=1}^r \lambda_i = 0$ and $\sum_{i=1}^r \lambda_i \mathbf{x}_i = \mathbf{0}$. Since the λ_i's sum to zero, some are positive, and some are negative; without loss of generality, let $k \in \{1, \ldots, r-1\}$ be so that $\lambda_1, \ldots, \lambda_k$ are non-negative and $\lambda_{k+1}, \ldots, \lambda_r$ are all negative. Again since their sum is zero,

$$\lambda_1 + \cdots + \lambda_k = -(\lambda_{k+1} + \cdots + \lambda_r).$$

Letting $s = \lambda_1 + \cdots + \lambda_k > 0$, for each $i = 1, \ldots, r$ put $\alpha_i = \frac{\lambda_i}{s}$; then $\sum_{i=1}^k \alpha_i = 1$, $\sum_{i=k+1}^r -\alpha_i = 1$, and

$$\sum_{i=1}^k \alpha_i \mathbf{x}_i = \sum_{i=k+1}^r -\alpha_i \mathbf{x}_i,$$

a vector expressed as a convex combination of $S_1 = \{\mathbf{x}_1, \ldots, \mathbf{x}_k\}$ and as a convex combination of $S_2 = \{\mathbf{x}_{k+1}, \ldots, \mathbf{x}_r\}$. □

Exercise 705. *Let $k \geq 2$ be an integer and $X = \{\mathbf{x}_1, \mathbf{x}_2, \ldots, \mathbf{x}_{k+1}\} \subset \mathbb{R}^{k-1}$. Prove that there exists a point $\mathbf{y} \in \mathbb{R}^{k-1}$ so that for any subcollection of k of the points in X, y is a convex linear combination of these k points.*

The following, perhaps surprising, theorem was published by Edward Helly (1884–1943) in 1923 [263], but appeared in a paper published by Radon in 1921. Helly actually discovered this theorem and told Radon of it in 1913, but was delayed in publishing by joining the Austrian army that year, getting wounded by Russians, taken prisoner to Siberia, and only finding his way back to Vienna two years after the war ended.

Theorem 20.1.7 (Helly's theorem). *For $n \geq 1$, if convex sets $C_1, C_2, \ldots, C_r$ in $\mathbb{R}^n$ have the property that any $n+1$ of them share a common point, then some point is contained in all of the sets.*

By a compactness argument (see, e.g. [250, p. 60]) the number of convex sets in Helly's theorem may also be infinite.

Using Exercise 705, one can show that Helly's theorem follows (see [58, Ex. 2, p. 86]). Using Radon's theorem is relatively straightforward:

Exercise 706. *Prove Helly's theorem by induction on r, using Radon's theorem.*

For more references and a survey of applications of Helly's theorem, see the article by Danzer, Grünbaum, and Klee [126].

20.2 Polygons

Occasionally, the term "n-gon" abbreviates "n-sided polygon". A polygon is called *convex* if every line segment joining two interior points lies entirely inside the polygon. See Figure 20.1.

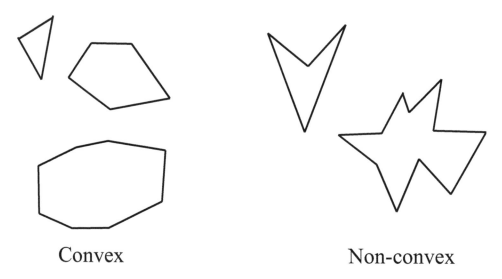

Convex   Non-convex

Figure 20.1: Convex and non-convex polygons

A polygon is *simple* if every vertex is incident with precisely two edges (so, for example, something resembling a figure eight is not simple).

Exercise 707. *Prove that for $n \geq 3$, the sum of interior angles of a simple convex n-gon is $(n-2)180$ degrees.*

Does the result in Exercise 707 generalize to non-convex polygons?

Exercise 708. *Prove that $n \geq 3$, the sum of the exterior angles of any polygon with n sides is $\pi(n+2)$ (in radians).*

A *diagonal* of a polygon is a line segment joining two non-consecutive vertices of the polygon. In a convex n-gon, each vertex is the endpoint of $n-3$ diagonals, so counting over all vertices, then dividing by 2 because each diagonal is counted twice shows that there are $n(n-3)/2$ diagonals in a convex n-gon. Another way to see the same result is to observe that the number of diagonals is the number of all segments minus those n used for the n-gon, giving

$$\binom{n}{2} - n = \frac{n(n-1)}{2} - n = \frac{n^2 - n - 2n}{2} = \frac{n(n-3)}{2}$$

diagonals. There is also a fairly simple inductive proof of the above result:

Exercise 709. *Prove, by induction, that for $n \geq 4$, a convex n-gon has $n(n-3)/2$ diagonals.*

The next exercise is really covered by Exercise 405, but is stated here without mention of Catalan numbers. A polygon is said to be *triangulated* if the polygon is divided into triangles, each of which has vertices that are vertices of the polygon. For convex n-gons, it is fairly easy to triangulate them, for one adds $n - 2$ diagonals that don't cross, say, all containing some fixed vertex, so induction is not really necessary in the next exercise.

Exercise 710. *Prove by induction that every convex polygon with three or more sides can be triangulated.*

Although induction was not necessary to solve Exercise 710, when generalizing from convex polygons to arbitrary simple polygons, the following is very useful in an inductive proof for the more general simple polygons.

Lemma 20.2.1. *Every simple polygon (convex or not) has at least one diagonal lying completely inside the polygon.*

Proof: In any polygon, there exists at least one triple of vertices v_1, v_2, v_3 (taken in counterclockwise order) so that the interior angle at v_2 is less than π. Fix such a triple. If segment $\overline{v_1 v_3}$ lies entirely inside P, then $\overline{v_1 v_3}$ is a diagonal as desired. So suppose that $\overline{v_1 v_3}$ is not entirely contained in P. Then $\triangle v_1 v_2 v_3$ contains additional points of P. Of these points, choose the one w so that $m \angle v_1 v_2 w$ is smallest. Since w is inside the triangle, w does not lie on the ray $\overrightarrow{v_1 v_2}$, so w is visible from v_2, and so $\overline{v_2 w}$ is the desired diagonal. $\qquad\square$

Exercise 711. *Using the result in Lemma 20.2.1, prove that any simple n-gon can be triangulated with diagonals that lie inside the n-gon, producing $n - 2$ triangles.*

Exercise 712. *Prove that in any triangulation of a simple polygon, there exists at least one triangle with two sides forming edges of the polygon. Hint: The solution is easier if one proves that there are always two such triangles!*

Exercise 713. *Finally, prove that the vertices of a triangulated n-gon can be colored with three colors so that no two vertices of the same color are connected by an edge.*

Exercise 714. *Suppose that $n \geq 1$ points are given in the interior of some square. Prove that the square can be divided into $2n + 2$ triangles with vertices chosen from the n given points and the four vertices of the square.*

Exercise 715. *Prove that if a polygon P is convex and contained in the polygon Q, the perimeter of P is shorter than the perimeter of Q.*

This next exercise asks to prove a famous problem in computational geometry; to find a solution without peeking might be challenging, however, the solution given is easy to read and may be entertaining.

Exercise 716 (The art gallery problem). *An art gallery has walls forming a polygon with n sides. Show that $\lfloor n/3 \rfloor$ guards can be placed so that all areas of the gallery are watched. For each $n \geq 3$, draw the floor plan of a gallery that requires $\lfloor n/3 \rfloor$ guards.*

For present purposes, a *lattice point* is a point $(x, y) \in \mathbb{R}^2$ in the real cartesian plane whose coordinates x, y are integers. In other words, a lattice point is an element of $\mathbb{Z}^2$.

To calculate the area of an arbitrary polygon might be very cumbersome, however if the polygon has vertices that are lattice points, then finding its area is nearly trivial by the spectacular 1899 result of Georg Alexander Pick (1859–1942) [430].

Theorem 20.2.2 (Pick's theorem). *Let P be a simple (non-intersecting) polygon whose vertices are lattice points, let $I(P)$ be the number of lattice points in the interior of P, and let $B(P)$ be the number of lattice points occurring on the boundary of P. Then the area of P is*

$$A(P) = I(P) + \frac{1}{2}B(P) - 1. \tag{20.3}$$

For example, in Figure 20.2, there are 4 interior points, and 9 boundary points, and the area is $4 + \frac{1}{2}9 - 1 = \frac{15}{2}$.

Pick's theorem has many proofs, but (at least) one is by induction:

Exercise 717. *Prove Pick's theorem in the following steps: (i) when P is a simple rectangle with sides parallel to the axes; (ii) when P is a right triangle with two legs parallel to the axes; (iii) when P is any triangle (by first surrounding P with a rectangle, then subtracting the area of the outside right triangles and rectangles thereby formed); (iv) when P is an arbitrary simple n-gon by induction on n (using Lemma 20.2.1, either by splitting the polygon into two pieces or by adjoining a triangle).*

The next few results show that the square is the only regular n-gon that can have all vertices as lattice points.

Lemma 20.2.3. *No three points in the integer lattice $\mathbb{Z}^2$ form an equilateral triangle.*

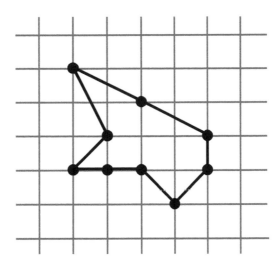

Figure 20.2: Pick's theorem: $I(P) = 4$, $B(P) = 9$; Area $= 7.5$.

Proof: Let T be an equilateral triangle with side length c, and suppose that the corners of T are lattice points. If two of these points have (integer) coordinates (x_1, y_1) and (x_2, y_2), then by the distance formula, $c^2 = (x_1 - x_2)^2 + (y_1 - y_2)^2$ is an integer. Hence the area $\frac{\sqrt{3}}{4}c^2$ is irrational. However, the area of any polygon with vertices on the integer lattice is rational (see Pick's theorem, if necessary).   □

For each $k \in \mathbb{Z}^+$, then no regular $3k$-gon can have all vertices as integer lattice points (because such a polygon has vertices that determine an equilateral triangle). For example, no regular hexagon can have all integer lattice points for vertices. Of course, it is easy to find a square with integer lattice points. The result in the next exercise might seem rather strong, but one proof is surprisingly simple.

Exercise 718. *Show that for each positive integer $n \geq 5$, no regular n-gon exists whose vertices are integer lattice points. Hint: Do not use induction on n, but instead use infinite descent.*

The next challenging exercise is answered several ways in [341], including by "Induction or Recursion".

Exercise 719. *Consider an equilateral triangle with side length n, drawn with a grid on it forming unit equilateral triangles whose sides are parallel to the large triangle (as in Figure 20.3). Prove that the number of triangles that can be counted in such a figure is*

$$\left\lfloor \frac{n(n+2)(2n+1)}{8} \right\rfloor.$$

(For example, when $n = 3$, there are 13 triangles.)

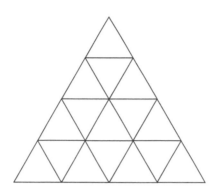

Figure 20.3: Count the triangles; $n = 4$

Exercise 720. *For $n \geq 1$, suppose that $3n$ points in a plane are given in general position (no three on a line). By induction on n, prove that these points form the vertices of n mutually disjoint triangles.*

20.3 Lines, planes, regions, and polyhedra

The first exercise in this section is an old classic; it asks to show that for any set of n non-collinear points, there are at least n different lines joining pairs of them. A very simple proof is available by induction if one first "notices" the following property:

Lemma 20.3.1. *If a finite set of points in the plane has the property that any line that passes through two of the points also contains a third, then all the points are on a line.*

This property was first posed by Sylvester [521] as a question in 1893. Erdős rediscovered the problem in 1933 (while reading the book *Geometry and the Imagination* by Hilbert and Cohn-Vossen), and Tibor Gallai found a proof the same year. (Gallai's name was previously Géza Grünwald.)

Some 10 years later, Erdős in posed the problem in the *Amer. Math. Monthly* [162], with the "ingenious proof" [167, p.208] due to Gallai appearing (together with a proof by Steinberg) a year later [172]. See [79], [167], [309], and [396] for more history and generalizations of this problem. The following simpler solution, attributed to L. M. Kelly (appearing in a 1948 paper, but by Coxeter [121]), might make one wonder how the problem went unsolved for so long!

Proof of Lemma 20.3.1: [Kelly] Suppose that a set S of points is given so that any line passing through two of these points also contains a third. In hope of a contradiction, suppose that this set is not collinear (and so some points do not lie on all lines).

Let $P \in S$ be a point not on a line ℓ (which contains three other points in S) so that the distance from P to ℓ is minimum (but not zero). Let X be the point (not necessarily in S) on ℓ that is closest to P. Since ℓ contains at least three points from S, either two points are on one side of F, or the middle of these three is F itself. Let A, B, F occur in order on ℓ (where $B = F$ is allowed). Let Y be the point on AP closest to B (see Figure 20.4). Since triangles $\triangle APX$ and $\triangle ABY$ are similar, the distance $\|BY\|$ from B to the line AP is less than $\|PX\|$, the supposed minimal distance. $\qquad\square$

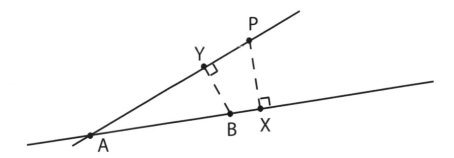

Figure 20.4: Kelly's proof using similar triangles

With Gallai's result in hand, the solution to the following exercise is nearly trivial by induction; this exercise is very popular, appearing in many modern works on problem solving (see, *e.g.*, [161, 8.35, p. 209]).

Exercise 721. *Use mathematical induction and Lemma 20.3.1 to prove that if $n \geq 3$ points do not all lie on a line, then at least n of the lines joining them are different.*

Among the next exercises, many are often referred to as *plane separation* problems (which is ambiguous, as sometimes it is the plane being separated, or it is a plane doing the separating).

Exercise 722. *Prove that if a convex region in the plane is crossed by ℓ lines with p interior points of intersection, then the number of disjoint regions created is $r = \ell + p + 1$. Hint: induct on ℓ.*

Using Exercise 722 helps to solve Exercise 29; this latter exercise is repeated next, but with a little more apparent conclusion. Lines in a geometry are called *concurrent* if they share a common point.

Exercise 723. *Place n points on a circle and draw in all possible chords joining these points. If no three chords are concurrent, show that the number of regions created is $\binom{n}{4} + \binom{n}{2} + 1$.*

Very much related to Exercise 722 is a problem (and solution) made famous by a young Lovász in a game show.

Lemma 20.3.2. *Chords determined by $n \geq 4$ points around a circle intersect in at most $\binom{n}{4}$ points.*

One proof of Lemma 20.3.2 is trivial: every four points determine two intersecting chords, and by perturbing the points slightly, all of these intersections can be made to be distinct.

Exercise 724. *Prove Lemma 20.3.2 by induction.*

A set of lines in the (Euclidean) plane are said to be in *general position* if no two are parallel and no three are concurrent.

Exercise 725. *For $n \geq 0$ prove that n lines in general position in the plane partition the plane into $1 + \binom{n+1}{2}$ regions.*

Although it is not inductive, Moore [392] gives a very elegant solution to Exercise 725 using Euler's formula for planar graphs (see Exercise 503): Draw a circle around all the points of intersection of the n lines in general position. Throw away the rays on the outside of this circle, and get a planar graph G. Then G has $\binom{n}{2}$ interior points, and since each line cuts the circle in two points, there are $2n$ exterior points. Since interior points have degree 4 and exterior points have degree 3, by the handshaking lemma, G has

$$\frac{1}{2}\left[4\binom{n}{2} + 3 \cdot 2n\right] = n(n-1) + 3n = n^2 + n$$

edges. The number of interior regions of G is the same as the number of regions in the plane determined, so by Euler's formula $v + f = e + 2$, the number of regions (disregarding the external face of G) is

$$f - 1 = e + 1 - v = n^2 + n + 1 - \left[\binom{n}{2} + 2n\right] = \frac{n^2 + n + 2}{2} = 1 + \binom{n+1}{2}.$$

The next result has a similar solution using Euler's formula.

Exercise 726. *Let n circles be in the plane so that any two circles intersect in two points, and no three intersect in a single point. Prove by induction on n that these circles divide the plane into $n^2 - n + 2$ regions.*

It might be interesting to note that Exercise 726 has an unexpected consequence: since four circles divide the plane into at most 14 regions, there can never be a Venn diagram using precisely 4 circles, no matter what their sizes, because a Venn diagram on four sets requires $2^4 = 16$ regions. [Note, if three circles are concurrent, one gets fewer regions.]

Rather than use circles or lines to partition the plane, the next two exercises use "bent lines" and "zig-zag lines", as in Figure 20.5.

Let two rays originating from the same point be called a *bent line*. A continuous line made from one segment and two rays (and not self-intersecting) is called a *zig-zag* line. Following notation from [230], let Z_n be the maximum number of regions that n bent lines can partition the plane into. It is not too difficult to check that $Z_1 = 2$ and $Z_2 = 7$.

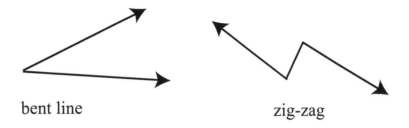

bent line   zig-zag

Figure 20.5: Partitioning the plane with other shapes

Exercise 727. *Prove that for each $n \geq 1$, $Z_n = 2n^2 - n + 1$.*

The similar problem for zig-zag lines is slightly more difficult.

Exercise 728. *Let ZZ_n be the maximum number regions determined by n zig-zags lines. For positive integers n, first prove the recursion $ZZ_n = ZZ_{n-1} + 9n - 8$, and conclude that*
$$ZZ_n = \frac{9n^2 - 7n + 2}{2}.$$

Recall that a point (x, y) in the real plane is called a *lattice point* iff both x and y are integers.

Exercise 729. *For each $n \geq 0$, consider the set S_n of lattice points bounded by $x \geq 0$, $y \geq 0$, and $x + y \leq n$. Prove that S_n can be covered by no fewer than $n + 1$ lines.*

Suppose the plane is divided into regions (by lines or curves). If one colors each region with one of k colors so that any two regions sharing a (non-trivial) common border receive two different colors, such a coloring is called a proper k-coloring.

Exercise 730. *For each $n \geq 0$ show that n lines in the plane in general position divide the plane into regions that can be properly 2-colored.*

The very same idea occurs in the next exercise:

Exercise 731. *Prove that for $n \geq 0$ circles in the plane, the regions thereby determined can be properly 2-colored.*

Exercise 732. *Let $n = r + k$ lines be given in the plane with no three concurrent and exactly k of the lines are parallel (but no others). If $f(r, k)$ is the number of regions the plane is partitioned into, prove that*
$$f(r, k) = \frac{r^2 + r + 2}{2} + k(r + 1).$$

Exercise 733. *Given $N \geq 1$ lines in a plane in general position, prove that it is possible to assign a non-zero integer of absolute value at most N to each region of the plane determined by these lines such that the sum of the integers on either side of any of the lines is 0.*

For use in the next exercise, a fairly easy lemma is provided (which does not need induction):

Lemma 20.3.3. *Suppose that a line (for example, the real line) is covered by a finite collection of rays (half-lines). Then some two of these rays cover the entire line.*

Proof: Let p be the rightmost endpoint of all rays pointing to the left, and let q be the leftmost endpoint of all rays pointing to the right. By assumption, all rays cover the line, and so p is not to the left of q. So the two rays starting at p and q respectively cover the line. $\square$

For the next exercise, a *half-plane* is a region of the plane on one side of a line. [It does not matter if one considers only open half-planes or only closed half-planes.]

Exercise 734. *Show by induction on n that if a plane is covered with $n \geq 2$ half-planes, then there exist two or three half-planes that cover all of the plane.*

Exercise 735. *Prove that n planes, passing through one point in a way that no three pass through the same line, divide space into $n(n-2)+2$ parts.*

Planes in three dimensional space are said to be in *general position* if no three planes share a common line and no two planes are parallel.

Exercise 736. *The maximum number of regions three dimensional space is divided into by n planes in general position is*

$$\binom{n}{0} + \binom{n}{1} + \binom{n}{2} + \binom{n}{3},$$

and the number of infinite (unbounded) regions is

$$2\binom{n}{0} + 2\binom{n}{2}.$$

Exercise 737. *Show that $n \geq 2$ spheres, any two of which intersect, partition 3-space into at most*

$$\frac{n(n^2 - 3n + 8)}{3}$$

regions.

The result in Exercise 734 has a three-dimensional version, with a solution that follows the same idea (not included here; see [220, Ex. 35, pp. 120–121]). A *half-space* is a region on one side of a plane.

Exercise 738. *Prove that if $n \geq 2$ half-spaces cover all of three-dimensional space, then there exist two, three, or four of them that cover the whole space.*

Exercise 739 (Euler's formula for polyhedra). *If v is the number of vertices, e the number of edges, and f the number of faces in a convex polyhedron, then*

$$v + f = e + 2.$$

In fact, this formula is true for any simple polyhedron, that is, one with no holes and all faces intersecting only in edges.

Exercise 740. *For $n \geq 0$, let $P_1, \ldots, P_{2n+1}$ be $2n+1$ points on a unit circle, all on the same side of some diameter. Using O to denote the center of the circle, prove that*

$$\|\overrightarrow{OP_1} + \cdots + \overrightarrow{OP_{2n+1}}\| \geq 1.$$

For an authoritative work on some of the fantastic applications of induction to (mostly) Euclidean geometry (inducting on points, lines, planes, spheres, dimension), the reader may be well rewarded with a glimpse at *Induction in Geometry* [220]. The more accessible book [161] contains a number of classic relations between induction and geometry. The textbooks *Concrete Mathematics* [230] and *Discrete Mathematics and its Applications*, 6th ed., [462] collectively cover many of the popular [and not so popular] theorems in geometry that are provable by induction. These are all highly recommended.

20.4 Finite geometries

In the study of finite projective planes (FPP's—briefly described below), there seem to be few inductive proofs, except for trivial ones like "if one line is finite, all are finite", after one shows that any two lines in a projective plane have the same number of points. There seem to be more inductive proofs in the theory of "block designs", a class of set systems that contain FPP's. (See [142] for surveys on design theory.) The only example presented here can also be thought of in terms of block designs, but comes from Lynn Batten's rich little book *Combinatorics of finite geometries* [40, pp. 15–16]. A few simple definitions are required. [I have changed the notation somewhat to be more consistent with general literature on designs and FPP's. The proof given in the solutions here seems far easier to read with this notation.]

For present purposes, define a *finite geometry* to be an ordered pair $(P, \mathcal{L})$, where P is a finite set of points and $\mathcal{L}$ is a set (so no repeats) of finite subsets of P, each subset $L \in \mathcal{L}$ called a *line*. A point $p \in P$ is said to be on a line $L \in \mathcal{L}$ (or *incident with L*) if and only if $p \in L$.

A finite geometry $\mathcal{S} = (P, \mathcal{L})$ is called a *near linear space* or *partial plane* if and only if

(i) Every line $L \in \mathcal{L}$ contains at least two points, and
(ii) For any distinct points $p, q \in P$, there exists at most one line incident with both.

In a near linear space, distinct lines intersect in at most one point (for if two lines intersect in two points, these two points violate (ii)).

A *linear space* is a near linear space where in addition, for every pair of points, there is a unique line containing them. One example of a linear space is a *finite projective plane*, a geometry that satisfies the following three axioms:

A1: For any pair of points, there exists a unique line containing them;
A2: For any pair of lines, there exists a unique point incident with both lines;
A3: There exist four points, no three of which are on a line.

So a linear space satisfies A1 and lines intersect in at most one point.

Theorem 20.4.1. *Let $\mathcal{S} = (P, \mathcal{L})$ be a near linear space with $|P| = v$ points, and lines labelled $L_1, L_2, \ldots, L_b$, where each L_i contains $|L_i| = k_i$ points. Then $\mathcal{S}$ is linear if and only if*

$$\sum_{i=1}^{b} \binom{k_i}{2} \geq \binom{v}{2}.$$

Proof: Only one direction is proved here, the remaining direction left as Exercise 741 below. Assume that $\mathcal{S}$ is linear. Counting all pairs of points give $\binom{v}{2}$, but since no pair of points is contained on different lines, and for every pair there is a line containing them, counting all pairs of points by just looking at the pairs on each line gives $\sum_{i=1}^{b} \binom{k_i}{2}$. [So the inequality is actually an equality.] $\square$

Exercise 741. *Prove the remaining direction of Theorem 20.4.1 by induction on v.*

Chapter 21

Ramsey theory

Mocking the air with colours idly spread.

—Shakespeare,

King John.

Roughly speaking, Ramsey theory is the study of partitions or colorings of a set and properties preserved (or not) in some partition class or color. It may seem like nearly every theorem in Ramsey theory is provable by induction, however there are still some "Ramsey-type" theorems proved by, say, the probabilistic method, for which no inductive proof has yet been found. The collection of problems addressed here is rather eclectic, including both simple and difficult results. For a far more comprehensive look at Ramsey theory and the many inductive proofs used in the field, see *e.g*, [231], [409], or [440]. Some of the material here is reproduced (with permission!) from [242].

An r-partition of a set X is a decomposition as a union of r sets,

$$X = X_1 \cup X_2 \cup \cdots \cup X_t,$$

where for $i \neq j$, $X_i \cap X_j = \emptyset$.

The above partition language can be couched in terms of colorings: An (ordered) r-partition of a set $X = X_1 \cup X_2 \cup \cdots \cup X_r$ can be viewed as a "coloring function" $c : X \to \{1, 2, \ldots, r\}$, where for each i, $c^{-1}(i) = X_i$. Any subset $Y \subseteq X$ is called *monochromatic* if for some i, $Y \subset X_i$. The elements $1, 2, \ldots, r$ are called colors, but any r-set can play the role of the colors; for example, when $r = 2$, often {red, blue} is used. Note that any r-coloring of a set defines an *ordered* partition of that set. The symbols χ or Δ are also common choices to denote a coloring function c as above. Often, the most common sets of r colors are (using ordinals) $r = \{0, 1, 2, \ldots, r-1\}$, or $[r] = \{1, 2, \ldots, r\}$.

For integers $a \leq b$, define the interval notation

$$[a, b] = \{x \in \mathbb{Z} : a \leq x \leq b\}.$$

For convenience write $[1, n] = [n]$.

A set X with $|X| = n$ is often called an n-set. For a set X and a positive integer k, denote the collection of all k-subsets of X by

$$[X]^k = \{Y \subseteq X : |Y| = k\}.$$

[A competing notation was once $\binom{X}{k}$.] The similar definition holds for any cardinal number k. The notation $[X]^\omega$ denotes the collection of all countably infinite subsets of X, and $[X]^{<\omega}$ denotes the family of *finite* subsets of X. Note that $[X]^k \neq X^k$, the expression X^k denoting a cartesian product. When denoting all k-subsets of integers from $[n]$, write $[n]^k$ rather than the more accurate notation $[[n]]^k$.

Exercise 742. *Prove that for $n \geq 1$ and any n-coloring of the 2^n subsets of $[n]$,*

$$\Delta : \mathcal{P}([n]) \to [n],$$

there exist subsets $A, B \in \mathcal{P}([n])$ with $A \subsetneq B$ and $\Delta(A) = \Delta(B)$.

Perhaps the simplest of all combinatorial results having a "Ramsey flavor" is the *pigeonhole principle*, also called "Dirichlet's Schubfach Prinzip" (named after P. G. Lejeune Dirichlet (1805–1859)), or the "box" or "drawer" principle. If $r + 1$ pigeons roost in r holes, then (at least) two pigeons have to share a hole. In general, if $rm + 1$ pigeons roost in r holes, then there is at least one hole with at least $m + 1$ pigeons in it. The easiest proof of this is by contradiction: if every hole has at most m pigeons, then the total is at most rm, contradicting that there are $rm + 1$ pigeons. An infinite version of the pigeonhole principle states that for any infinite set partitioned into finitely many parts, (at least) one part must be infinite.

The (finite) pigeonhole principle also has a fairly simple inductive proof:

Exercise 743 (Pigeonhole principle). *Prove by induction that for any $n \geq 0$ and $k \geq 1$, if X is a set with at least $nk + 1$ elements, then for any k-partition $X = X_1 \cup X_2 \cup \cdots \cup X_k$ there exists an i so that $|X_i| \geq n + 1$.*

21.1 The Ramsey arrow

The main results in the next sections use special notation. For any positive integers k, r, and positive integers $p_1, p_2, \ldots, p_r$, write

$$n \longrightarrow (p_1, p_2, \ldots, p_r)^k_r$$

iff for every r-partition $[n]^k = C_1 \cup \cdots \cup C_r$, there exists an i and a p_i-set $X \in [n]^{p_i}$ so that $[X]^k \subseteq C_i$. Any X so that $[X]^k$ is contained wholly in some C_i is said to be *monochromatic* (in color i). The above arrow notation is called a Ramsey arrow. In the Ramsey arrow notation, one often suppresses the subscript r when it is clear how many p_i's there are. When all p_i's are the same, use the shorthand

$$n \longrightarrow (p)_r^k$$

to denote

$$n \longrightarrow \underbrace{(p, p, \dots, p)}_{r}{}_r^k.$$

When $k = 1$, a Ramsey arrow notation boils down a pigeonhole statement. To be precise

$$(p_1 + \dots + p_r) - r + 1 \longrightarrow (p_1, p_2, \dots, p_r)_r^1, \tag{21.1}$$

and no smaller number arrows $(p_1, p_2, \dots, p_r)_r^1$ (for the partition into r parts each of size $p_i - 1$ fails the arrow). A simple proof of (21.1) is by contradiction; another is available by induction, either inducting on r or on some p_i (starting with all p_i's equal to 1).

21.2 Basic Ramsey theorems

Considering the Ramsey arrow when $k \geq 2$ is the starting point behind what is now called "Ramsey theory". The main theorem in the field is eponymous with Frank Plumpton Ramsey (1903–1930). It has two versions, the finite and the infinite; many authors prefer to first prove the infinite version (by induction) and derive the finite version from it by a compactness argument (using König's theorem which says that any infinite but locally finite rooted tree has an infinite branch—the vertices of the tree are restricted colorings, and vertices are connected iff one coloring extends another). The finite version also has an inductive proof, which is asked for as an exercise below after some initial observations.

The following theorem was proved in 1928, and published in 1930 [446].

Theorem 21.2.1 (Finite Ramsey theorem). *For any positive integers k, r, and positive integers $p_1, p_2, \dots, p_r$, all at least k, there exists a least $n = R_k(p_1, p_2, \dots, p_r)$ so that*

$$n \rightarrow (p_1, p_2, \dots, p_r)_r^k.$$

By the pigeonhole principle, when $k = 1$, the existence of such numbers is guaranteed.

When $k = 2$ and $r = 2$, the numbers $R_2(p_1, p_2)$ guaranteed above by Ramsey's theorem are called *Ramsey numbers*, and are abbreviated by simply $R(p_1, p_2)$, and are called *diagonal Ramsey numbers* when $p_1 = p_2$. One could, by definition, for any $m \geq 2$, define $R(1, m) = 1$. Observe that $R(a, b) = R(b, a)$, and for all $m \geq 2$, and $R(2, m) = R(m, 2) = m$. It is helpful to view any 2-coloring of the pairs $[n]^2$ as a 2-coloring of the edges of the complete graph K_n. For some a, b, and n, to show that $R(a,b) \leq n$, it suffices to show that under any red-blue coloring of edges of K_n, there exists either a copy of K_a with all edges red or a copy of K_b with all blue edges. To show that $R(a,b) > n - 1$, it suffices to show a red-blue coloring of $E(K_{n-1})$ for which there is no red K_a or blue K_b. Proving both inequalities shows $R(a,b) = n$.

A standard proof shows that $R(3,3) = 6$. First show that $R(3,3) \leq 6$ by the following: Consider any red-blue coloring of the edges in a copy of K_6 with vertices labelled A, B, C, D, E, F. Of the five edges from A, at least three are of the same color, say those joining B, C, D. If any of the three edges BC, CD, or BD are of the same color as edges AB, AC, AD, get a monochromatic triple including A; if they are all of the opposite color, BCD is a monochromatic triangle. To see that $R(3,3) > 5$, consider a pentagon with inner edges red and outer edges blue.

In 1935, Erdős and Szekeres [173] discovered Ramsey theory, not by looking at logic and Ramsey's proofs, but quite independently by looking at convex polygons. I think that their paper contained the first published proof of $R(3,3) = 6$. One of the main contributions from that original paper was a recursion that showed all Ramsey numbers $R(s,t)$ exist by specifying an upper bound; Ramsey's proof only proved the existence of these numbers with no attention devoted to calculating how large they are.

Theorem 21.2.2 (Erdős-Szekeres recursion). *For all $s,t \geq 2$,*

$$R(s,t) \leq R(s,t-1) + R(s-1,t).$$

For $k = 2$, and $r = 2$, Theorem 21.2.2 implies the Ramsey theorem:

Corollary 21.2.3. *For each $s,t \geq 2$, $R_2(s,t) = R(s,t)$ exists.*

The proof idea for Theorem 21.2.2 is to examine any 2-coloring of the edges of the complete graph on $R(s,t-1) + R(s-1,t)$ vertices, and fix some vertex x. Consider the two neighborhoods, those connected to x by one color, and those connected to x by the other color. By the pigeonhole principle, either the first neighborhood has at least $R(s,t-1)$ vertices or the second neighborhood has at least $R(s-1,t)$ vertices. The proof is complete by applying the appropriate Ramsey statement to each respective subgraph. [This proof is repeated below with more details.]

As an example,

$$R(3,4) \leq R(3,3) + R(2,4) = 6 + 4 = 10.$$

In fact, $R(3,4) = 9$, proved in 1955 [236]—the proof is not difficult.

Exercise 744. *Use the Erdős-Szekeres recursion to prove that for each $s,t \geq 2$,*

$$R(s,t) \leq \binom{s+t-2}{t-1}.$$

The Erdős-Szekeres recursion works analogously when partitioning subsets larger than just pairs:

Exercise 745. *By induction, prove the finite Ramsey theorem (Theorem 21.2.1) for $r = 2$ colors. Hint: Induct on k, and in the inductive step, look for a recursion formula similar to the Erdős-Szekeres recursion, then induct on $p_1 + p_2$.*

The technique used in the proof of Theorem 21.2.2 easily extends to give a many-color version of the Erdős-Szekeres recursion.

Theorem 21.2.4. *For $r \geq 2$, and $s_0, s_1, \ldots, s_{r-1}$, each $s_i > 2$, $R(s_0, s_1, \ldots, s_{r-1}) \leq R(s_0-1, s_1, \ldots, s_{r-1}) + R(s_0, s_1-1, \ldots, s_{r-1}) + \ldots + R(s_0, s_1, \ldots, s_{r-2}, s_{r-1}-1) - r + 2.$*

Proof: For each $i \in r$, put $n_i = R(s_0, s_1, \ldots, s_{i-1}, s_i - 1, s_{i+1} \ldots, s_{r-1})$, and set $n = (\sum_{i \in r} n_i) - r + 2$. Let $\Delta : E(K_n) \to r$ be a given coloring, and fix $x \in V(K_n)$. For each $i \in r$, set $X_i = \{y : \Delta(x, y) = i\}$. There exists an i so that $|X_i| \geq n_i$, since if not, $n = 1 + \sum_{i \in r} |X_i| = 1 + \sum_{i \in r}(n_i - 1) = \sum_{i \in r} n_i - r + 1$, a contradiction. Fix such an i with $|X_i| \geq n_i$. If for every $j \neq i$, X_i does not induce a j-monochromatic copy of K_{s_j}, then by the choice of n_i, there exists $Y_i \subset X_i$ which induces an i-monochromatic copy of K_{s_i-1}, and, in this case, $\{x\} \cup Y_i$ induces an i-monochromatic copy of K_{s_i}. $\square$

All elements of the inductive proof for the finite Ramsey theorem are now in place:

Exercise 746. *By induction, prove the finite Ramsey theorem (Theorem 21.2.1) for any number of $r \geq 2$ colors. Hint: induct on k, and in the inductive step, then induct on the sum of the p_i's.*

Exercise 747. *Prove by induction on $r \geq 2$ that*

$$R_2(\underbrace{3, 3, \ldots, 3}_{r}) \leq \lfloor er! \rfloor + 1. \tag{21.2}$$

Hint (as in [59]): $\lfloor er! \rfloor + 1 = \lfloor e(r-1)! \rfloor r + 1$.

An infinite form of Ramsey's theorem has an inductive proof:

Theorem 21.2.5 (Ramsey's theorem, infinite, 2 colors). *For any infinite set X and any 2-coloring $\chi : [X]^2 \to \{1, 2\}$ there exists an infinite subset $Y \subseteq X$ so that $[Y]^2$ is monochromatic.*

Proof: Fix a 2-coloring $\chi : [X]^2 \to \{1, 2\}$. Pick an arbitrary $x_0 \in X$. Then χ induces a coloring $\chi_0 : X \setminus \{x_0\} \longrightarrow \{1, 2\}$ by $\chi_0(x) = \chi(\{x, x_0\})$. By the infinite pigeonhole principle, one of the two color classes $\chi_0^{-1}(1), \chi_0^{-1}(2)$ is infinite; call this infinite set A_0. Hence χ is constant on

$$\{x_0\} \times A_0 = \{\{x_0, y\} : y \in A_0\} \subset [X]^2,$$

say $\chi(\{x_0\} \times A_0) = r_0 \in \{1,2\}$.

Select any element $x_1 \in A_0$. Repeating the same argument as above, there exists an infinite set $A_1 \subset A_0$ so that χ is constant on $\{x_1\} \times A_1$, say $\chi(\{x_1\} \times A_1) = r_1 \in \{1,2\}$. [Note: r_0 and r_1 may differ, while still $\chi(\{x_0\} \times A_1) = r_0$.]

Continue in this manner (inductively), and get a set $X^* = \{x_i : i = 0, 1, 2, \ldots\} \subset X$ so that for any E and E' in $[X^*]^2$, $\chi(E) = \chi(E')$ whenever $\min(E) = \min(E')$. [If $i = \min\{j : x_j \in E\}$, write $\min(E) = x_i$.] This induces a 2-coloring χ^* of X^* by $\chi^*(x_i) = \chi(E)$ for any $E \in [X^*]^2$ satisfying $\min(E) = x_i$.

By the pigeonhole principle, there is an infinite set $Y \subset X^*$ so that χ^* is constant on Y, and since

$$[Y]^2 \subseteq \{E \in [X^*]^2 : \min(E) \in Y\},$$

the coloring χ is constant on $[Y]^2$. $\qquad\square$

Exercise 748. *By induction on r, prove the following r-coloring version of the infinite form of Ramsey's theorem: Let T be an infinite set and let $r \geq 2$ be an integer. For any r-coloring of $[T]^2$, there exists an infinite set $S \subseteq T$ so that $[S]^2$ is monochromatic.*

Before looking at other classic Ramsey-type theorems, the Erdős-Szekeres result for polygons is given, and for this, a definition is needed.

Definition 21.2.6. Points $p_1, \ldots, p_m$ in $\mathbb{R}^2$ with strictly increasing x-coordinates, (so no two have the same x-coordinate, and they are given in order from left to right) form an *m-cup* iff for $i = 1, 2, \ldots, m-1$, the slopes of the line segments $p_i p_{i+1}$ are increasing; these points form an *m-cap* if the slopes are decreasing.

So an m-cup is a collection of points on a graph of a (strictly) convex function.

Theorem 21.2.7 (Erdős-Szekeres [173]). *For each $k, \ell \geq 2$, if $\binom{k+\ell-4}{k-2} + 1$ points in the plane have increasing x-coordinates, either some k of these points forms a k-cup, or some ℓ of these points form an ℓ-cap.*

Exercise 749. *Prove Theorem 21.2.7 by induction on $k + \ell$.*

Twenty-five years later, the same two authors finally published the proof that Theorem 21.2.7 was optimal.

Theorem 21.2.8 ([174]). *For each $k, \ell \geq 2$, there exists a configuration of $\binom{k+\ell-4}{k-2}$ points that contains no k-cup, nor any ℓ-cap.*

Exercise 750. *Study the proof in Exercise 749 and recover the inductive construction that proves Theorem 21.2.8.*

Since any k-cap or k-cup forms the vertices of a convex k-gon, it follows that Erdős and Szekeres showed a Ramsey-type theorem for strictly convex (convex, and no three points in a straight line) k-gons.

Theorem 21.2.9. *For any given $k \geq 3$, there exists a least number $f(k)$ so that if $f(k)$ points in $\mathbb{R}^2$ occur in general position (no three in a line), then this set of points contains k points that are the vertices of a strictly convex k-gon.*

It is conjectured that $f(k) = 2^{k-2} + 1$, and if this is true, it is known to be best possible by an example on 2^{k-1} points containing no convex k-gon. [This example may be constructed by induction. For other proofs of this result by Ramsey's theorem, see [231].]

Definition 21.2.10. For $d + 1$ positive integers $x_0, x_1, \ldots, x_d$, the collection

$$H(x_0, x_1, \ldots, x_d) = \left\{ x_0 + \sum_{i \in I} x_i : I \subseteq [d] \right\}$$

is called a d-dimensional *affine cube*, or simply, an affine d-cube.

In 1892, Hilbert [268] showed that for all positive integers d and r, there exists a least number $h(d; r)$ so that for every r-coloring $\chi : [1, h(d; r)] \to [1, r]$, there exists an affine d-cube monochromatic under χ. At first, proving Hilbert's "affine cube lemma" might appear daunting, but a proof by induction is surprisingly simple.

Exercise 751. *Prove Hilbert's affine cube lemma by induction on d. Hint: show that $h(d + 1; r) \leq r^{h(d;r)} + h(d; r)$.*

Historically, the next major result in Ramsey theory is known as "Schur's Theorem" [482], a slight strengthening of the original. [Other results are also called "Schur's theorem"; for example, see Exercise 651.]

Theorem 21.2.11 (Schur's theorem). *For any positive integer r, there exists a least positive integer $n = S(r)$ so that for any coloring $\Delta : [1, n] \longrightarrow [1, r]$, there exist positive integers $x, y \in [1, n]$ so that*

$$\Delta(x) = \Delta(y) = \Delta(x + y).$$

Schur's theorem remains true when the condition $x \neq y$ is added. For a standard inductive proof of Schur's theorem that shows

$$S(r) \leq r! \left(1 + \frac{1}{1!} + \frac{1}{2!} + \cdots + \frac{1}{r!} \right) = \lfloor e \cdot r! \rfloor,$$

see, *e.g.*, [266, pp. 177–178]. (For a simple modern proof of Schur's theorem using Ramsey's theorem, see [231].) The interested reader might be impressed as to how many more Ramsey-type theorems concern sums and "sum-sets"; only a tiny sample is given here but the books *Additive Combinatorics* by Tao and Vu [527] and *Additive Number Theory: Inverse Problems and the Geometry of Sumsets* [401] by Nathanson might be excellent references to begin further research in this area.

For sets A and B and a binary operation $+$, define $A + B = \{a + b : a \in A, b \in B\}$. The notation $a + B$ is also used to denote $\{a\} + B$. Martin Kneser [320] proved the following in 1953.

Theorem 21.2.12 (Kneser's theorem). *Let G be a non-trivial abelian group, and let A and B be non-empty finite subsets of G. If $|A| + |B| \leq |G|$, then there exists a proper subgroup H of G with*

$$|A + B| \geq |A| + |B| - |H|.$$

Exercise 752. *Prove Kneser's theorem by induction on $|B|$.*

One kind of affine cube is an arithmetic progression; in particular,

$$H(a, \underbrace{d, d, \ldots, d}_{k-1}) = \{a, a + d, a + 2d, \ldots, a + (k-1)d\}$$

is called an *arithmetic progression of k terms with difference d*. One often abbreviates "k-term arithmetic progression" by "AP_k". Hilbert's affine cube lemma does not prove the existence of monochromatic progressions.

B. L. van der Waerden attended a lecture given by Baudet in which he learned of a conjecture by Schur (for his own account of the story see [547]). He managed to give a proof [546] for the so-called "Baudet's conjecture", now becoming eponymous with van der Waerden. Here is one form of van der Waerden's theorem:

Theorem 21.2.13 (van der Waerden's theorem). *For any positive integers r and k, there exists a smallest integer $n = W(k; r)$ so that for any coloring $\Delta : [n] \longrightarrow r$, there exists a monochromatic AP_k.*

The original proof is inductive, and is now called "the block proof", or the "fan proof" (see [525] for a proof using fans). This proof is also given in [231]; another approach to the same proof is in [337] and variants of this proof for $r = 3$ are detailed in [243]. For the student of induction, this proof is a gem (but takes a few pages to describe well). Van der Waerden's theorem also follows from the more general Hales-Jewett theorem (Theorem 21.3.1 below). Since the block proof is a double induction, no useful bounds for $W(k; r)$ arise. Only in 1988 was it shown by Shelah [488] that the van der Waerden function $W(k; r)$ is primitive recursive (see Section 18.5 for definition) (and the proof is inductive—unfortunately, not included here). More recently, Gowers showed that $W(k; r)$ can be bounded above by a tower function (see, for one of many sources, [227]). For a survey of theorems relating to van der Waerden's theorem, see [291].

See also Exercise 524 for a related result (for arithmetic progressions of length 3).

In 1928, very soon after van der Waerden's original 1927 proof, Brauer published a paper proving that for each k, and sufficiently large primes p, there are k consecutive integers that are quadratic residues modulo p, and there are k consecutive integers that are quadratic nonresidues modulo p. In his proof, Brauer uses a result whom he attributes to Schur; this result generalizes both van der Waerden's theorem and Schur's theorem, and has a fairly straightforward inductive proof (based on van der Waerden's theorem):

Theorem 21.2.14. *Let k and r be positive integers. There exists a least integer $SB(k;r)$ so that for any $n \geq SB(k;r)$ and any r-coloring of $[1,n]$, there exists a, d so that $\{d, a, a+d, a+2d, \ldots, a(k-1)d\}$ is monochromatic.*

Exercise 753. *Using van der Waerden's theorem, prove Theorem 21.2.14 by induction on r.*

Virtually the same proof as that for Theorem 21.2.14 works to prove a slightly stronger theorem:

Theorem 21.2.15. *For positive integers k, r, s, there exists a least $n = n(k, r, s)$ so that for any coloring $\Delta : [1, n] \to [1, r]$, there exist positive integers a and d so that the set*

$$\{a, a+d, a+2d, \ldots, a+(k-1)d\} \cup \{sd\}$$

is monochromatic.

Exercise 754. *Prove Theorem 21.2.15 by induction on r.*

Proving lower bounds for Ramsey-type functions often consists in finding (or showing the existence of) examples of large structures that are somehow "balanced". This notion of balance is a central aspect of the next problem, although apparently it has not yet found application in Ramsey theory (if it has, this fact does not seem to be well-known).

The problem is called the *Tarry-Escott* problem, and has a long history (see [85] for further references) going back to at least 1851, a generalization called "Prouhet's problem". The goal is to partition numbers of the form $0, 1, 2, 3, \ldots, 2^{n+1} - 1$ into two equal classes so that not only the sum of numbers in each class is the same, but the sums of squares, cubes, and so on up until nth powers, are also (respectively) equal. For example, with $n = 2$, let 1,2,4,7 be in one class, 0,3,5,6 in the other. Then the sum in each class is 14, and the sum of the squares in each class is 70. In the following statement of the problem, the convention that $0^0 = 1$ is used.

Exercise 755. *For each non-negative integer n, prove that there is a partition of $\{0, 1, 2, \ldots, 2^{n+1} - 1\}$ into two classes A_n and B_n so that for each $j = 0, 1, \ldots, n$,*

$$\sum_{a \in A_n} a^j = \sum_{b \in B_n} b^j.$$

21.3 Parameter words and combinatorial spaces

The triple 200, 210, 220 may be interpreted as words over the alphabet $\{0, 1, 2\}$, or as a horizontal line in some geometric space, or, if viewed as base 3 numbers, the arithmetic progression 18, 21, 24. This triple of words may be abbreviated by the "parameter word" $2\lambda 0$, where the parameter λ varies in $\{0, 1, 2\}$. One Ramsey-type result for words, called the Hales–Jewett theorem (Theorem 21.3.1), not only generalizes van der Waerden's theorem (Theorem 21.2.13), but perhaps surprisingly, one proof of the Hales–Jewett theorem gives better bounds on the van der Waerden function than could be previously obtained without the generalization. For the statement and proof the Hales–Jewett theorem, some notation is convenient.

Recall that for $n \in \mathbb{Z}^+$, $[n] = \{1, \ldots, n\}$, or in ordinal notation, $n = \{0, 1, \ldots, n-1\}$ (so $i \in n$ means $0 \leq i \leq n-1$). An *alphabet* is a set of symbols; throughout this section, A denotes an alphabet with a finite number of symbols. A *word* over A is a sequence, usually written without commas, of symbols from A. For each $n \in \mathbb{Z}^+$, a word $\ell_1\ell_2\cdots\ell_n$ of length n over A can be viewed as a function $f : [n] \longrightarrow A$, where for each $i \in [n]$, $f(i) = \ell_i$. As usual, $A^n = \{f : [n] \longrightarrow A\}$. Let $\lambda_0, \lambda_1, \ldots, \lambda_{m-1}$ be symbols not in A, called *parameters*. For $m \leq n$ define the set of *m-parameter words* of length n over A by

$$[A]\binom{n}{m} = \{f : [n] \longrightarrow (A \cup \{\lambda_0, \lambda_1, \ldots, \lambda_{m-1}\}) :$$

$$\forall j \in m, \ f^{-1}(\lambda_j) \neq \emptyset \text{ and for } i < j, \ \min f^{-1}(\lambda_i) < \min f^{-1}(\lambda_j)\}.$$

The last condition (that the first occurrences of each parameter are in order) is not really required, although it helps to fix ideas. So $[A]\binom{n}{m}$ can be viewed as a set of ordered n-tuples containing each λ_i at least once, and whenever $i < j$, the first occurrence of λ_i must precede the first occurrence of λ_j. For convenience, these ordered n-tuples are often written as "strings" or "words" by omitting the parentheses and commas. For example, if $A = \{a, b, c\}$, $ab\lambda_1\lambda_2$ and $\lambda_1 c\lambda_2\lambda_1$ are in $[A]\binom{4}{2}$ but $a\lambda_2\lambda_1\lambda_2$ and $a\lambda_1\lambda_1 b$ are not. Note that $A^n = [A]\binom{n}{0}$.

For $f \in [A]\binom{n}{m}$ and $g \in [A]\binom{m}{k}$ define the composition $f \circ g \in [A]\binom{n}{k}$ by

$$f \circ g(i) = \begin{cases} f(i) & \text{if } f(i) \in A, \\ g(j) & \text{if } f(i) = \lambda_j. \end{cases}$$

For example if $f = a\lambda_1\lambda_2\lambda_1$ and $g = b\lambda_1$, then $f \circ g = ab\lambda_1 b$.

It is not difficult to prove that the composition of parameter words is associative. For $f \in [A]\binom{n}{m}$, define the *space* of f,

$$\mathrm{sp}(f) = \left\{ f \circ g : g \in [A]\binom{m}{0} \right\},$$

sometimes denoted $f \circ [A]\binom{m}{0}$, to be the set of words from $[A]\binom{n}{0}$ that are formed by faithfully replacing parameters in f with elements from A. Define an *m-dimensional combinatorial subspace* of A^n to be the space of some word in $[A]\binom{n}{m}$. If $f \in [A]\binom{n}{1}$ then $\mathrm{sp}(f)$ is called a *combinatorial line* in A^n, or simply, a *line*. For example, if $f = a\lambda_1\lambda_2\lambda_1$, then $\mathrm{sp}(f) = \{ax_1x_2x_1 : x_1, x_2 \in A\} \subset A^4$. In this example $\mathrm{sp}(f)$ can be seen as a 2-dimensional subspace of A^4 where $f \in [A]\binom{4}{2}$.

The set $A^n = \{f : [n] \to A\}$ is often called an *n-dimensional cube*. Elements of this cube can be viewed as words. Note that if $A = \{0, 1, \ldots, t-1\}$ and if

A^n is viewed as a discrete "geometric" n-cube, then not all "geometric lines" are combinatorial lines. For example, $(\{2,0,0\}, \{1,1,0\}, \{0,2,0\})$ is a geometric line with equations $x + y = 2$, $z = 0$ in the three dimensional cube over $\{0,1,2\}$ but 200, 110, 020 is not a combinatorial line in $[\{0,1,2\}]\binom{3}{0}$.

A type of *concatenation* of parameter words distinguishes between parameters from respective words. If $f \in [A]\binom{n}{m}$ and $g \in [A]\binom{k}{\ell}$ define $f \,\hat{}\, g \in [A]\binom{n+k}{m+\ell}$ as follows:

$$f \,\hat{}\, g(i) = \begin{cases} f(i) & \text{if } i \in [n], \\ g(i - n) & \text{if } i > n \text{ and } g(i - n) \in A, \\ \lambda_{m+j} & \text{if } i > n \text{ and } g(i - n) = \lambda_j. \end{cases}$$

For example, $a\lambda_1 b\lambda_2 \,\hat{}\, c\lambda_1\lambda_2\lambda_1 a = a\lambda_1 b\lambda_2 c\lambda_3\lambda_4\lambda_3 a$. Note that

$$\mathrm{sp}(f \,\hat{}\, g) = \{f' \,\hat{}\, g' : f' \in \mathrm{sp}(f), g' \in \mathrm{sp}(g)\}.$$

Another way to talk about combinatorial subspaces avoids the $[A]\binom{n}{m}$ notation and composition of functions. For any $m \leq n$, an m-dimensional combinatorial subspace of A^n can be defined to be any subset $S \subset A^n$ obtained in the following manner:

Fix a partition $[n] = F \cup M_1 \cup \cdots \cup M_m$ where each $M_i \neq \emptyset$ ($F = \emptyset$ may occur) and fix some $f_0 \in F^A$, that is, for each $j \in F$, fix $f_0(j) \in A$. Now let S be the set of all $g \in A^n$ satisfying
   (1) for each $j \in F$, $g(j) = f_0(j)$, and
   (2) for each M_i, if $j, k \in M_i$, then $g(j) = g(k)$.

In the above definition, F is called the set of *fixed coordinates* and each M_i is called a set of *moving coordinates*. For example, when $A = \{0,1\}$, $n = 5$, $m = 2$, $F = \{3,4\}$, $M_1 = \{1\}$, and $M_2 = \{2,5\}$, $f_0 = 10$, the 2-dimensional combinatorial subspace determined is the set of all words of the form $xy10y$, that is, the set $\{00100, 01101, 10100, 11101\}$.

A 1-dimensional combinatorial subspace is called a *combinatorial line*, and an m-dimensional combinatorial subspace is called a combinatorial m-space. A 0-dimensional combinatorial subspace of A^n is a single word of length n over the alphabet A. If one needs to identify S by the partition and the constant portion, one might write $S = S(F, M_1, \ldots, M_m, f_0)$.

The following result is one of the main tools in Ramsey theory. The proof is too complicated to be given as an exercise; however, the result is so central in Ramsey theory that an inductive proof is given below.

Theorem 21.3.1 (Hales-Jewett Theorem [253]). *Let $m, r \in \mathbb{Z}^+$ and a finite alphabet A be given. There exists a smallest number $n = HJ(|A|, m, r) \in \mathbb{Z}^+$ so that for every coloring $\Delta : A^n \longrightarrow [r]$ there is an $f \in [A]\binom{n}{m}$ for which $\mathrm{sp}(f)$ is monochromatic.*

Proof: The proof given here is by mathematical induction; this proof uses an outer loop inducting on t, and for each fixed t, induction on both m and r is used.

For this proof, the induction is driven by the following two inequalities, whose proofs are delayed until after the induction is explained:

$$\mathrm{HJ}(t, m+1, r) \;\leq\; \mathrm{HJ}(t, 1, r) + \mathrm{HJ}(t, m, r^{t^{\mathrm{HJ}(t,1,r)}}), \tag{21.3}$$

$$\mathrm{HJ}(t+1, 1, r+1) \;\leq\; \mathrm{HJ}(t, 1 + \mathrm{HJ}(t+1, 1, r), r+1). \tag{21.4}$$

For each $t, m, r \in \mathbb{Z}^+$, let $P(t, m, r)$ be the proposition that $\mathrm{HJ}(t, m, r)$ exists.

BASE STEP ($t = 1$): For all m and r, $\mathrm{HJ}(1, m, r) = m$ since for any $n \geq m$ the space of any word in $[\{a\}]\binom{n}{m}$ is the only word in the trivial space $\{a\}^n$. Hence, for any m and r, $P(1, m, r)$ holds true.

INDUCTIVE STEP: Fix $t_0 \geq 1$ and suppose that for all m and r, $P(t_0, m, r)$ holds. The first step is to show that for every r, $P(t_0 + 1, 1, r)$ holds, and this is shown by induction on r. Since $P(t_0 + 1, 1, 1)$ holds trivially, the base step holds. For the inductive step, suppose that $r_0 \geq 1$ is so that $P(t_0 + 1, 1, r_0)$ is true. Then $HJ(t_0 + 1, 1, r_0)$ exists, and using $m_0 = 1 + HJ(t_0 + 1, 1, r_0)$ and $P(t_0, m_0, r_0 + 1)$ (which is part of the inductive hypothesis with $m = m_0$ and $r = r_0 + 1$), equation (21.4) implies $P(t_0 + 1, 1, r_0 + 1)$, completing the inductive step for r. By mathematical induction on r, for every r, $P(t_0 + 1, 1, r)$ holds.

For a fixed t, a brief induction on m using (21.3) implies

$$[\forall r, P(t, 1, r)] \Rightarrow [\forall m, \forall r, P(t, m, r)]. \tag{21.5}$$

So with $t = t_0 + 1$ and for every r, $P(t_0 + 1, 1, r)$, it follows from (21.5) that for every m and r, $P(t_0 + 1, m, r)$ is true, completing the inductive step for t.

By mathematical induction on t, equations (21.3) and (21.4) imply that for all t, m and r, $P(t, m, r)$ holds.

Thus it remains to prove (21.3) and (21.4). Throughout the remainder of the proof, fix t, m, and r.

Proof of (21.3): Set $M = \mathrm{HJ}(t, 1, r)$ and $N = \mathrm{HJ}(t, m, r^{t^M})$. Fix a coloring

$$\Delta : A^{M+N} \longrightarrow [r]$$

and define $\Delta_N : A^N \longrightarrow [r^{t^M}]$ by

$$\Delta_N(f) = \langle \Delta(g \,\hat{}\, f) : g \in A^M \rangle,$$

i.e., each f is colored with a sequence induced by Δ. Let $f_N \in [A]\binom{N}{m}$, guaranteed by the choice of N, be so that $\mathrm{sp}(f_N)$ is monochromatic with respect to Δ_N. For any $h \in A^m$, define $\Delta_M : A^M \longrightarrow [r]$ by

$$\Delta_M(g) = \Delta(g \,\hat{}\, (f_N \circ h)).$$

By the choice of f_N, $\Delta_M(g)$ does not depend on h. So there exists $f_M \in [A]\binom{M}{1}$ so that $\mathrm{sp}(f_M)$ is monochromatic with respect to Δ_M.

Setting

$$f = f_M \,\hat{}\, f_N \in [A]\binom{M+N}{1+m},$$

$\mathrm{sp}(f)$ is monochromatic with respect to Δ by the following: For any $l \in [A]\binom{1}{0}$ and $h \in [A]\binom{m}{0}$,

$$f \circ (l \,\hat{}\, h) = (f_M \circ l) \,\hat{}\, (f_N \circ h)$$

and so $\Delta(f \circ (l \,\hat{}\, h)) = \Delta_M(f_M \circ l)$ is constant. Hence (21.3) holds.

Proof of (21.4): In this part, put $M = \mathrm{HJ}(t+1, 1, r)$ and $N = \mathrm{HJ}(l, 1+M, r+1)$. With $|A| = t$, choose a symbol $b \notin A$ and put $B = A \cup \{b\}$. Fix a coloring

$$\Delta : B^N \longrightarrow [r+1].$$

It remains to show there exists a monochromatic line in B^N, i.e., an $h \in [B]\binom{N}{1}$ so that $\mathrm{sp}(h) = h \circ [B]\binom{1}{0}$ is monochromatic with respect to Δ.

Define

$$\Delta_A : A^N \longrightarrow [r+1],$$

the restriction of Δ to A, in the natural way (i.e., for any $f \in A^N$, $\Delta_A(f) = \Delta(f)$). By the choice of N, there is $f_A \in [A]\binom{N}{1+M}$ that has $f_A \circ A^{1+M}$ monochromatic with respect to Δ_A. Without loss, for every $f \in f_A \circ A^{1+M}$, let $\Delta_A(f) = r$.

If there is a $g \in B^M$ so that also $\Delta(f_A \circ (\langle b \rangle \,\hat{}\, g)) = r$ (i.e., if there is a word in B^N containing b's that occur in the same positions as λ_1 occurs in f_A and is colored the same as f_A), construct h by replacing the occurrence of each b in $f_A \circ (\langle b \rangle \,\hat{}\, g)$ with λ_1. Then $h \in [B]\binom{N}{1}$ and whenever $x \in A$,

$$h \circ \langle x \rangle \in f_A \circ A^{1+M},$$

and so in this case $\Delta(h \circ \langle x \rangle) = r$. Also

$$h \circ \langle b \rangle = f_A \circ (\langle b \rangle \,\hat{}\, g)$$

and so $\Delta(h \circ \langle b \rangle) = r$. So in this case $h \circ [B]\binom{1}{0}$ is monochromatic with respect to Δ.

So suppose there is no such $g \in B^M$ satisfying $\Delta(f_A \circ (\langle b \rangle \,\hat{}\, g)) = r$. Define $\Delta_M : [B]\binom{M}{0} \longrightarrow [r]$ by

$$\Delta_M(g) = \Delta(f_A \circ (\langle b \rangle \,\hat{}\, g)).$$

By the theorem, there is $f_M \in [B]\binom{M}{1}$ with $f_M \circ [B]\binom{1}{0}$ monochromatic with respect to Δ_M. For

$$h = f_A \circ (\langle b \rangle \,\hat{}\, f_M) \in [B]\binom{N}{1},$$

$h \circ [B]\binom{M}{0}$ is monochromatic with respect to Δ, finishing the proof of (21.4). $\quad\square$

21.4 Shelah bound

The inequalities (21.3) and (21.4) do not yield primitive recursive (see Section 18.5 for definition); bounds for $\mathrm{HJ}(t, m, r)$. In fact, it was not until 1988 before it was shown that the function HJ is primitive recursive; this was accomplished by Shelah [488] (given here as Theorem 21.4.5, below) with an elementary inductive proof of the Hales–Jewett theorem that avoided the double induction used in the above proof, and yielded bounds that were expressible in terms of "simple" functions. Shelah's proof is a discovery that may already be deemed to be a classic example of an inductive proof that overcomes the weakness of a double induction and yields a much stronger result; the style in which induction is applied may also be very instructive. Also, since the Hales–Jewett theorem is a central theorem in Ramsey theory, Shelah's proof seemed worthy of inclusion here, even though it takes a non-trivial effort to work through the details.

The innovative tool that Shelah used in his proof of the Hales–Jewett theorem (Theorem 21.3.1) is now often called "Shelah's cube lemma"; to avoid confusion with other "cube lemmas", here it is called "Shelah's string lemma" (Lemma 21.4.1, below), a name that may be slightly more apt. The proof of the string lemma is elementary (although the proof requires special notation that makes it look unwieldy) and is by induction. Using this string lemma, a short inductive proof shows that $\mathrm{HJ}(t, 1, r)$ exists and is primitive recursive. The more general result for $\mathrm{HJ}(t, m, r)$ can be proven similarly, but also follows from the 1-dimensional case with a simple observation.

Even for the case $m = 1$, the bounds obtained for $\mathrm{HJ}(t, 1, r)$ are extremely large; one can see a more complete discussion in, *e.g.*, [231, pp. 60ff], together with many other Ramsey-type results related to the Hales–Jewett theorem. For a brief version of the Shelah proof, see [414], or for a variant of Shelah's proof, see [368]. (See also comments surrounding Theorem 21.2.13 concerning bounds on the van der Waerden function.) In the literature and on the web, Shelah's proof is explained in many different ways; the presentation here arose from lectures by Norbert Sauer [476] (who learned this proof from Deuber while in Germany in in 1988; also see [134] regarding a lecture in Norwich, 1989) for the case $m = 1$, and uses standard parameter-set notation.

For a positive integer n, recall the notations $[n] = [1, n] = \{1, 2, \ldots, n\}$ and $[n]^2 = \{\{x, y\} : x, y \in [n], x \neq y\}$. The main objects of consideration in Shelah's string lemma are "strings" of the form (where $m, n \in \mathbb{Z}^+$ and $i \in [m]$)

$$(\{x_1, y_1\}, \{x_2, y_2\}, \ldots, \{x_{i-1}, y_{i-1}\}, \{x_i\}, \{x_{i+1}, y_{i+1}\}, \ldots, \{x_m, y_m\})$$

chosen from

$$[n]^2 \times [n]^2 \times \cdots \times [n]^2 \times \underbrace{[n]}_{i\text{-th position}} \times [n]^2 \times \cdots \times \underbrace{[n]^2}_{m\text{-th position}} \, .$$

Slightly abusing notation, subscripts are added to indicate position of each $[n]^2$ and $[n]$; so the set of such strings with a singleton in the i-th position (call them "type i" strings) is indicated by

$$[n]_1^2 \times \cdots \times [n]_{i-1}^2 \times [n]_i \times [n]_{i+1}^2 \times \cdots \times [n]_m^2.$$

Lemma 21.4.1 (Shelah's string lemma [488]). *For each $m, r \in \mathbb{Z}^+$, there exists a smallest $n = Sh(m, r)$ so that for any family of m r-colorings,*

$$\Delta_1 : [n]_1 \times [n]_2^2 \times [n]_3^2 \times \cdots \times \cdots \times [n]_m^2 \longrightarrow [r]$$

$$\vdots$$

$$\Delta_i : [n]_1^2 \times \cdots \times [n]_{i-1}^2 \times [n]_i \times [n]_{i+1}^2 \times \cdots \times [n]_m^2 \longrightarrow [r]$$

$$\vdots$$

$$\Delta_m : [n]_1^2 \times [n]_2^2 \times \cdots \times [n]_{m-1}^2 \times [n]_m \longrightarrow [r].$$

there exist $\{x_1, y_1\}, \ldots, \{x_m, y_m\} \in [n]^2$ so thal for each $i = 1, \ldots, m$, the two type i words have the same color:

$$\Delta_i(\{x_1, y_1\}, \ldots, \{x_{i-1}, y_{i-1}\}, \{x_i\}, \{x_{i+1}, y_{i+1}\}, \ldots, \{x_m, y_m\}) =$$
$$\Delta_i(\{x_1, y_1\}, \ldots, \{x_{i-1}, y_{i-1}\}, \{y_i\}, \{x_{i+1}, y_{i+1}\}, \ldots, \{x_m, y_m\}).$$

(Each Δ_i is "insensitive to a switch" in the i-th position of a string.)

Proof: Fix $r \in \mathbb{Z}^+$ throughout the proof. The proof is by induction on m. For each positive integer m, let $C(m)$ denote that $\mathrm{Sh}(m, r)$ exists.

BASE STEP: When $m = 1$, the theorem says only that there exists n so that for any coloring $\Delta_1 : [n] \to [r]$, there exists $\{x, y\} \in [n]^2$ so that $\Delta_1(x) = \Delta_1(y)$, which, by the pigeonhole principle, is true iff $n = r + 1$; so, $\mathrm{Sh}(1, r) = r + 1$, and so $C(1)$ is true.

INDUCTION STEP: Fix $\ell \geq 1$ and assume that $C(\ell)$ holds, *i.e.*, $\mathrm{Sh}(\ell, r)$ exists. The statement $C(\ell + 1)$ follows by proving the following claim:

$$\mathrm{Sh}(\ell + 1, r) \leq 1 + r^{\binom{\mathrm{Sh}(\ell, r)}{2}^{\ell}}. \tag{21.6}$$

To prove (21.6), set $n = 1 + r^{\binom{\mathrm{Sh}(\ell, r)}{2}^{\ell}}$, and consider any $\ell + 1$ colorings

$$\Delta_1 : [n]_1 \times [n]_2^2 \times [n]_3^2 \times \cdots \times \cdots \times [n]_{\ell+1}^2 \longrightarrow [r]$$

$$\vdots$$

$$\Delta_i : [n]_1^2 \times \cdots \times [n]_{i-1}^2 \times [n]_i \times [n]_{i+1}^2 \times \cdots \times [n]_{\ell+1}^2 \longrightarrow [r]$$

$$\vdots$$

$$\Delta_{\ell+1} : [n]_1^2 \times [n]_2^2 \times \cdots \times [n]_\ell^2 \times [n]_{\ell+1} \longrightarrow [r].$$

Consider $\Delta_{\ell+1}$ restricted to the set

$$[\mathrm{Sh}(\ell,r)]_1^2 \times \cdots \times [\mathrm{Sh}(\ell,r)]_\ell^2 \times [n]_{\ell+1}. \tag{21.7}$$

For each $z \in [n]_{\ell+1}$, there are $\binom{Sh(\ell,r)}{2}^\ell$ strings in

$$T(z) = [\mathrm{Sh}(\ell,r)]_1^2 \times \cdots \times [\mathrm{Sh}(\ell,r)]_\ell^2 \times \{z\} \tag{21.8}$$

and so for each z, the coloring $\Delta_{\ell+1}$ restricted to $T(z)$ can be represented by a "color-pattern" vector of length $|T(z)| = \binom{Sh(\ell,r)}{2}^\ell$ with entries from $[r]$ (where the strings in $T(z)$ are ordered in some way). There are only there are only $r^{\binom{\mathrm{Sh}(\ell,r)}{2}^\ell}$ such color-pattern vectors. Since there are $n > r^{\binom{\mathrm{Sh}(\ell,r)}{2}^\ell}$ choices for z, by the pigeonhole principle, there are $z_1, z_2 \in [n]_{\ell+1}$, $z_1 \neq z_2$ such that the color-pattern vector of $\Delta_{\ell+1}$ restricted to $T(z_1)$ equals the color-pattern vector of $\Delta_{\ell+1}$ restricted to $T(z_1)$; that is, for each $\{x_1, y_1\} \in [\mathrm{Sh}(\ell,r)]^2$, $\ldots$, $\{x_\ell, y_\ell\} \in [\mathrm{Sh}(\ell,r)]^2$,

$$\Delta_{\ell+1}((\{x_1,y_1\},\ldots,\{x_\ell,y_\ell\},\{z_1\})) = \Delta_{\ell+1}((\{x_1,y_1\},\ldots,\{x_\ell,y_\ell\},\{z_2\})).$$

Finally, consider the remaining ℓ colorings $\Delta_1, \ldots, \Delta_\ell$, and for each $i = 1, \ldots, \ell$, the restriction of Δ_i to

$$\Delta_i^* : [\mathrm{Sh}(\ell,r)]^2 \times \cdots \times [\mathrm{Sh}(\ell,r)]_i \times \cdots \times [\mathrm{Sh}(\ell,r)]_\ell^2 \times \{z_1, z_2\} \longrightarrow [r].$$

By the definition of $\mathrm{Sh}(\ell,r)$ for each $i = 1, \ldots, \ell$ select a pair $\{x_i, y_i\} \in [\mathrm{Sh}(\ell,r)]^2$ so that Δ_i^* (and so also Δ_i) is insensitive to a switch in position i. Thus the pairs $\{x_1,y_1\},\ldots,\{x_\ell,y_\ell\},\{z_1,z_2\}$ witness the conclusion of $C(\ell+1)$.

Thus (21.6) holds, completing the inductive step.

For a fixed r, by mathematical induction, for every $m \in \mathbb{Z}^+$, $\mathrm{Sh}(m,r)$ exists. The theorem follows since r was arbitrary.   □

Lemma 21.4.2. *The function $Sh(m,r)$ is primitive recursive.*

Proof: Replacing m by $m-1$ in the recursion (21.6),

$$\mathrm{Sh}(m,r) < 2r^{\mathrm{Sh}(m-1,r)^{2(m-1)}}. \tag{21.9}$$

Applying (21.9) twice more gives

$$2r^{2r^{2r^{\mathrm{Sh}(m-3,r)^{2^3}(m-3)(m-2)(m-1)}}}.$$

An easy proof by induction then shows (again, with very loose bounds),

$$\left. \mathrm{Sh}(m,r) \le 2r^{2r^{\cdot^{\cdot^{\cdot^{2r^{2r(r+1)2^m m!}}}}}} \right\} \text{ a tower of height } m+2.$$

The next result by Shelah not only shows (by induction) that $HJ(t,1,r)$ exists, but it also leads to showing that $HJ(t,1,r)$ is primitive recursive.

Theorem 21.4.3 (Shelah [488]). *For any positive integers t, r,*

$$HJ(t+1,1,r) \le HJ(t,1,r) \cdot \mathrm{Sh}\left(HJ(t,1,r^{(t+1)HJ(t,1,r)^{m-1}}) \right). \tag{21.10}$$

Proof: Let $m = HJ(t,1,r)$ and $n = \mathrm{Sh}\left(m, r^{(t+1)^{m-1}}\right)$. Fix an alphabet A with $|A| = t$ and let $a \in A$ and $b \notin A$ be given. Set $B = A \cup \{b\}$. Let $\Delta : B^{mn} \longrightarrow [r]$ be given. To prove the theorem, one must show the existence of a monochromatic line in B^{mn}. This is accomplished by first looking only at special words with $m-1$ parameters $\lambda_1, \ldots, \lambda_{m-1}$ so that for each i, all copies of λ_i occur in a single block. These special words are constructed by concatenating words of only two kinds: For each pair of positive integers $x < y \le n$, define the words

$$u_{xy} = \underbrace{aa \cdots a}_{x} \underbrace{\lambda\lambda \cdots \lambda}_{y-x} \underbrace{bb \cdots b}_{n-y} \in [B]\binom{n}{1},$$
$$v_x = \underbrace{aa \cdots a}_{x} \underbrace{bb \cdots bbb \cdots b}_{n-x} \in B^n.$$

For the moment, fix $i \in [m]$. For each positive integer sequence $c_1 < d_1 \le n$, $c_2 < d_2 \le n, \ldots, c_m < d_m \le n$, define two words in $[B]\binom{mn}{m-1}$:

$$f_1 = u_{c_1 d_1}{}^{\wedge} \cdots {}^{\wedge} u_{c_{i-1} d_{i-1}}{}^{\wedge} v_{c_i}{}^{\wedge} u_{c_{i+1} d_{i+1}}{}^{\wedge} \cdots {}^{\wedge} u_{c_m, u_m},$$
$$f_2 = u_{c_1 d_1}{}^{\wedge} \cdots {}^{\wedge} u_{c_{i-1} d_{i-1}}{}^{\wedge} v_{d_i}{}^{\wedge} u_{c_{i+1} d_{i+1}}{}^{\wedge} \cdots {}^{\wedge} u_{c_m, u_m}.$$

Say that these words are of "type i". Note that in the concatenation process, the λ's in the subword u_{c_j, d_j} are replaced by λ_j's in either f_1 or f_2. A word in $\mathrm{sp}(f_1)$ is of the form $f_1 \circ w$ for some $w \in B^{m-1}$. List $B^{m-1} = \{w_1, \ldots, w_k\}$, where $k = (t+1)^{m-1}$.

For any "type i" sequence (or Shelah string) of the form

$$s = (\{c_1, d_1\}, \ldots, \{c_{i-1}, d_{i-1}\}, \{c_i\}, \{c_{i+1}, d_{i+1}\}, \ldots, \{c_m, d_m\})$$
$$\in [n]^2 \times \cdots [n]_i \times \cdots \times [n]^2,$$

where each $1 \le c_j < d_j \le n$, and f_1 as defined above, define the vector

$$\Delta_i(s) = (\Delta(f_1 \circ w_1)), \Delta(f_1 \circ w_2), \ldots, \Delta(f_1 \circ w_k)).$$

So Δ_i is an $r^k = r^{(t+1)^{m-1}}$-coloring of all type i Shelah strings.

So by the choice of n and Lemma 21.4.1, there exists a fixed choice of m pairs $\{x_1, y_1\} \in [n]_1^2, \ldots, \{x_m, y_m\} \in [n]_m^2$ so that for each $i \in [m]$,

$$\Delta_i(u_{x_1 y_1} \hat{} \cdots \hat{} u_{x_{i-1} y_{i-1}} \hat{} v_{x_i} \hat{} u_{x_{i+1} y_{i+1}} \hat{} \cdots \hat{} u_{x_m y_m})$$
$$= \Delta_i(u_{x_1 y_1} \hat{} \cdots \hat{} u_{x_{i-1} y_{i-1}} \hat{} v_{y_i} \hat{} u_{x_{i+1} y_{i+1}} \hat{} \cdots \hat{} u_{x_m y_m}).$$

Since these two color vectors agree, by the definition of Δ_i, for each $w \in B^{m-1}$,

$$\Delta(f_1 \circ w) = \Delta(f_2 \circ w). \tag{21.11}$$

Consider the m-parameter word

$$u = u_{x_1 y_1} \hat{} \cdots \hat{} u_{x_m y_m} \in [B]\binom{mn}{m}.$$

Then Δ induces an r-coloring $\Delta_u : A^m \longrightarrow [r]$ defined by $\Delta_u(g) = \Delta(u \circ g)$. [Recall, $b \notin A$.] By the choice of $m = \mathrm{HJ}(t, 1, r)$, there exists a Δ_u-monochromatic line $h \in [A]\binom{m}{1}$. In other words, for any $z_1, z_2 \in A$,

$$\Delta_u(h \circ z_1) = \Delta_u(h \circ z_2) \Rightarrow \Delta(u \circ h \circ z_1) = \Delta(u \circ h \circ z_2).$$

To show that $u \circ h$ is the required line in $[B]\binom{mn}{1}$ to finish the proof that $\mathrm{HJ}(t + 1, 1, r) \leq mn$, it remains only to show for some $z \in A$ that $\Delta(u \circ h \circ b) = \Delta(u \circ h \circ z)$. The natural choice for z is a, so to conclude the proof of (21.10), it remains to show that $\Delta(u \circ h \circ b) = \Delta(u \circ h \circ a)$.

Observe that for each $i \in [m]$, $u_{x_i, y_i} \circ a = v_{y_i}$ and $u_{x_i, y_i} \circ b = v_{x_i}$. Letting f_1 and f_2 denote the type i words as defined at the beginning of this proof but instead using the pairs $\{x_1, y_1\}, \ldots, \{x_m, y_m\}$, equation (21.11) says that for any $w \in B^{m-1}$, $\Delta(f_1 \circ w) = \Delta(f_2 \circ w)$. Write $h = (h(1), \ldots, h(m))$, and suppose that $h(i) = \lambda$ (there may be more λ's). Then $h \circ b$ is of the form $\alpha_1 \cdots \alpha_{i-1} b \alpha_{i+1} \cdots \alpha_m$, where each $\alpha_j \in B$. The w to be used when applying (21.11) below is $w = \alpha_1 \cdots \alpha_{i-1} \alpha_{i+1} \cdots \alpha_m$.

$$
\begin{aligned}
\Delta((u \circ h) \circ b) &= \Delta(u \circ (h \circ b)) \\
&= \Delta(u_{x_1, y_1} \circ \alpha_1 \hat{} \cdots \hat{} u_{x_i, y_i} \circ b \hat{} \cdots u_{x_m, y_m} \circ \alpha_m) \\
&= \Delta(u_{x_1, y_1} \circ \alpha_1 \hat{} \cdots \hat{} v_{x_i} \hat{} \cdots u_{x_m, y_m} \circ \alpha_m) \\
&= \Delta(u_{x_1, y_1} \circ \alpha_1 \hat{} \cdots \hat{} v_{y_i} \hat{} \cdots u_{x_m, y_m} \circ \alpha_m) \qquad \text{(by (21.11))} \\
&= \Delta(u_{x_1, y_1} \circ \alpha_1 \hat{} \cdots \hat{} u_{x_i, y_i} \circ a \hat{} \cdots u_{x_m, y_m} \circ \alpha_m).
\end{aligned}
$$

If there h has another λ in, say, position j, repeat the process above calculations as applied to words of j-type; continue so that every occurrence of b in $h \circ b$ is changed to an a in the above calculation, arriving at $\Delta((u \circ h) \circ b) = \Delta((u \circ h) \circ a)$ as desired. Hence (21.10) holds. $\qquad \square$

In the proof of Theorem 21.4.3, nowhere was it (essentially) relied upon that $m = 1$, except in some details near the end. The interested reader can verify that the proof works for arbitrary m, but if this assertion is not convincing, the following Lemma 21.4.4 can be applied. This alternative method (see *e.g.* [231, p. 40]) relates the Hales-Jewett number for m-spaces to that for lines. The simple idea used is that if B is an alphabet and $A = B^m$, then a combinatorial line in A^n is an m-space in B^{nm}.

Lemma 21.4.4. *For any $t, m, r \in \mathbb{Z}^+$,*

$$HJ(t, m, r) \leq m \cdot HJ(t^m, 1, r). \tag{21.12}$$

Proof: Fix t, m, r. Let B be an alphabet with $|B| = t$ and set $A = B^m = \{f : m \to B\}$. Let $n = HJ(t^m, 1, r)$. If $g \in B^{nm}$, then for each $i = 0, 1, \ldots, n-1$, set $g_i = g(im+1)g(im+2)\ldots g((i+1)m)$, where

$$g = g_0 \hat{\ } g_1 \hat{\ } \cdots \hat{\ } g_{n-1} \in A^n,$$

showing $B^{nm} \subseteq A^n$. In fact, B^{nm} can be viewed as precisely A^n. Fix a coloring $\Delta : B^{nm} \longrightarrow [r]$ (which is also a coloring of A^n). By the choice of n, there exists $g \in [A]\binom{n}{1}$ so that Δ is constant on $\mathrm{sp}(g)$. But since the parameter of g can be replaced by any $f : m \to B$, $g \in [B]\binom{nm}{m}$ and Δ is constant on $\mathrm{sp}(g)$, (21.12) holds. $\qquad\square$

Beginning with $HJ(1, 1, r) = 1$, applying Theorem 21.4.3 $t - 1$ times (and then Lemma 21.4.4), another inductive proof of the Hales–Jewett theorem is achieved. This induction reveals an explicit (but very large) upper bound for $HJ(t, m, r)$; see [231, pp. 60–68] for discussion of just how large this bound is; it falls into a class of functions called "wowzer" functions (towers of towers). This outlines the proof for the theorem mentioned at the beginning of this section:

Theorem 21.4.5 (Shelah [488]). *The function $HJ(t, m, r)$ is primitive recursive.*

21.5 High chromatic number and large girth

The next result concerns the existence of hypergraphs with high chromatic number *and* containing no small cycles. [See Chapter 15 for notation and definitions.] Naively, one might think that since short cycles force the chromatic number up, short cycles are necessary to do so. In fact, quite the opposite is true.

A simple construction produces triangle-free graphs with arbitrarily high chromatic number. The following construction is due to Mycielski [399]: given a graph G on vertices $V = \{v_1, \ldots, v_n\}$, construct a graph G^* as follows: let $U = \{u_1, \ldots, u_n\} \cap V = \emptyset$, and let x be a vertex not in $U \cup V$. Define G^* on vertex set $U \cup V \cup \{x\}$ by including all edges of G, and for each $i = 1, \ldots, n$, add edges joining u_i to each of the *neighbors* of v_i, and then finally join x to each vertex in U. It is not difficult to verify that $K_2^* = C_5$, and C_5^* is the graph in Figure 21.1, also called the Grötzsch graph (independently, Grötzsch found this graph just a few years later).

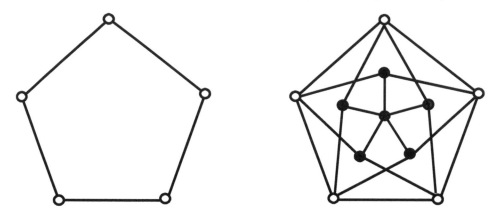

Figure 21.1: The Grötzsch graph and the Mycielski construction

Exercise 756. *Using the Mycielski construction described above, let $G_2 = K_2$, and recursively, for each $i \geq 3$, put $G_i = G_{i-1}^*$. Prove that for each $i \geq 2$, G_i is triangle-free and $\chi(G_n) \geq n$.*

There are graphs with both arbitrarily large girth *and* arbitrarily large chromatic number. According to [406], the study of such questions began with Tutte and Zykov (see [132]) in the 1940s. The question was finally answered by Erdős [164] and Erdős and Hajnal [170] in the 1960s using *probabilistic* methods. The first constructive proof was given by Lovász [352] in 1968.

A later inductive construction (given below) is due to Nešetřil and Rödl [406] and is apparently an extension of an idea of Tutte [132]. This construction is a simple example of a more general technique called *partite amalgamation*, a technique developed to prove theorems in Ramsey theory. One may think of partite amalgamation as inductively *gluing* partite graphs together but only along certain partite sets or "coordinates"; this particular application consists of gluing only at a single partite set (part). Because of the difficulty of the general process of partite amalgamation, the proof given here of the existence of *sparse* (large girth) highly chromatic hypergraphs serves as an introduction to the process. [For more information on the powerful applications of partite amalgamation, see, *e.g.*, [407], [408] or [409].]

As usual, let $\chi(G)$ denote the chromatic number of G. The *girth* of a hypergraph is the length of a smallest cycle contained in G.

Theorem 21.5.1. *For positive integers $k \geq 2$, n, p there exists a k-uniform hypergraph G so that girth$(G) > p$ and $\chi(G) > n$.*

Before seeing an inductive proof of Theorem 21.5.1, one might wonder why this result is in a section on Ramsey theory. The proof below follows that of a Ramsey theorem (see above references) for edge partitions. This is not surprising since one can interpret Theorem 21.5.1 as a Ramsey result. For a k-uniform graph G, let a copy of an edge of G (all of which are isomorphic) be denoted by E_G. Also, let C_i be a k-uniform hypergraph that is a cycle of length i (cycles are not unique in a hypergraph setting, so further suppose that C_i contains no smaller cycles, and so hyperedges intersect in at most one vertex, making C_i unique). Then Theorem 21.5.1 can be restated as follows.

Theorem 21.5.2. *For positive integers $k \geq 2$, r, and p, there exists a k-uniform hypergraph G with $girth(G) > p$ so that $G \longrightarrow (E_G)_r^1$, that is, under any r-coloring of the vertices of G, at least one edge is monochromatic.*

For a short probabilistic proof of Theorem 21.5.1, see, *e.g.*, [13].

Proof of Theorem 21.5.1: For subscripts, ordinal notation ($a = \{0, 1, \ldots, a-1\}$ is convenient. For an a-partite k-uniform hypergraph $G = ((V_i)_{i \in a}, E)$ and some $r \in a$ with $|V_r| = \ell$, and an ℓ-uniform hypergraph $H = (X, D)$, define the a-partite k-uniform hypergraph $H_r^* G = ((V_i')_{i \subset a}, E')$ as follows:

For $i \neq r$, set $V_i' = V_i \times D$ and set $V_r' = X$. For each $d \in D$ fix an injection $\psi_d : \cup_{i \in a} V_i \to \cup_{i \in a} V_i'$ taking V_r to $d \subset V_r'$ and for $i \neq r$, $\psi_d(V_i) = \{(v, d) : v \in V_i\}$. Define

$$E' = \{\{\psi_d(v_1), \ldots, \psi_d(v_k)\} : \{v_1, \ldots, v_k\} \in E, d \in D\}.$$

An edge $e \in E'$ will be denoted $\psi_d(e)$ for some $e \in E$, and $d \in D$. So $H_r^* G$ is formed by taking $|D|$ copies of G and identifying the copies of V_r with edges of H. Copies of G have been *amalgamated* along the r-th part, using H as a template for the new r-th part.

The theorem is proved by induction on p.

BASE STEP: For $p = 1$, observe that any loopless hypergraph G satisfies $girth(G) \geq 2$, and for each k, trivial examples of k-graphs exist with $\chi(G) \geq n$.

INDUCTIVE STEP: Fix $p > 1$ and suppose that the theorem holds for every q satisfying $1 \leq q < p$ and for all edge sizes k'. Put $a = (k-1)n + 1$ and let

$$G^0 = ((V_i^0)_{i \in a}, E^0)$$

be an a-partite k-uniform hypergraph so that for every set $A \in [a]^k$ and for every $i \in A$ there is an edge $e \in E^0$ with $e \cap V_i' \neq \emptyset$ and has $girth(G^0) > p$. (One could take G^0 to be a collection of disjoint k-edges.)

For each $j = 1, \ldots, a$, inductively define a-partite k-graphs $G^j = ((V_i^j)_{i \in a}, E^j)$, as follows. Having defined $G^m = ((V_i^m)_{i \in a}, E^m)$, ($m < a$), put $|V_m^m| = \ell_m$ and let $H^m = (X^m, D^m)$ be an ℓ_m-graph which satisfies $girth(H^m) \geq p$ and $\chi(H^m) > n$. (Such a hypergraph exists by induction hypothesis using a different value for k.) Put

$$G^{m+1} = H^m *_m G^m = ((V_i^{m+1})_{i \in a}, E^{m+1}).$$

Claim: The graph $G^a = ((V_i^a)_{i \in a}, E^a)$ satisfies the theorem.

To see that for each $j \leq a$, girth$(G^j) > p$, induct on j: suppose that for some fixed j, girth$(G^j) > p$. In E^{j+1}, pick a sequence of vertices

$$C = \{\psi_{d_0}(v_0), \ldots, \psi_{d_{q-1}}(v_{q-1})\}$$

of minimal length q that determine a cycle. If all the d_i, $i \in q$ are equal, by the induction hypothesis there are no small cycles in a copy of G^j and so $q > p$ holds as desired. So suppose that not all the d_i are equal. In this case, the only way that C can be a cycle is if C uses vertices from V_j^{j+1}, the j-th part of G^{j+1}. Now use the fact that, by induction hypothesis, girth$(H^j) \geq p$ and conclude that $q > p$ (in fact, q, in this case, is at least $2p$) as desired.

To see the proof of $\chi(G^a) > n$, fix a coloring $\Delta : V(G^a) \longrightarrow n$. The restriction of Δ to X_{a-1}^a, the last part of G^a, imposes a coloring on X^{a-1}, the vertices of H^{a-1}. By the inductive hypothesis $\chi(H_{a-1}) > n$, and so there exists a monochromatic edge $d_{a-1} \in D_{a-1}$. Setting $Z_{a-1} = d_{a-1} \in [X_{a-1}^a]^{\ell_{a-1}}$ to be the last part of a new graph $F^{a-1} \preceq G^{a-1}$, that is,

$$F^{a-1} = ((X_i^{a-1})_{i \in a-1}, Z_{a-1}, E^{a-1} \cap [\cup_{i \in a-1} X_i^{a-1} \cup Z_{a-1}]^k),$$

look only at how Δ colors $V(F^{a-1})$. Certainly Δ is constant on Z_{a-1} by design. Repeat in this manner, using the vertices of a monochromatic edge of H^{a-2} as a new part, a subset of X_{a-2}^{a-1}, create F^{a-2}, Δ being constant on the second last part thereof. Continuing inductively in this manner, get $F^0 = ((Z_i^0)_{i \in a}, E(F^0))$, a copy of G^0, the vertices of which have colors depending only on the part whence they came. By the choice of a, there exist k parts all colored the same. By the design of G^0, there exists a k-edge determined by those parts, guaranteed now to be monochromatic.

By MI, for any $p \geq 1$, the construction works. $\qquad\square$

Ramsey theory for graphs is a very popular area of research with hundreds of amazing [to me, at least] and deep theorems proved by induction. See [65] for an extensive survey of graph Ramsey theory available on-line. Other surveys occur in references cited above, but I neglected to mention some other excellent surveys and other areas in Ramsey theory (like Euclidean Ramsey theory, where induction is applied quite handily); other recommended surveys include those by Graham [228] and Nešetřil [405].

Chapter 22

Probability and statistics

> *It is remarkable that a science [probabilities] which began with the consideration of games of chance, should have become the most important object of human knowledge,*
>
> — Pierre-Simon Laplace (1749–1827),
>
> *Théorie Analytique des Probabilités.*

Here is an easy exercise to start.

Exercise 757. *Show that if $x_1, x_2, \ldots, x_n$ are real numbers each satisfying $a \leq x_i \leq b$, then the mean of the x_i's is also in the closed interval $[a, b]$.*

The next exercise computes the mean of the binomial distribution.

Exercise 758. *For non-negative p and q with $p + q = 1$, prove that for each $n \geq 1$,*

$$\sum_{j=1}^{n} j \binom{n}{j} p^j q^{n-j} = np.$$

Exercise 759. *Let $a_1, a_2, \ldots, a_k$ be positive integers, and set*

$$S = \{x \in \mathbb{Z}^+ : \forall i, a_i \text{ does not divide } x\}.$$

Prove that the density of S in $\mathbb{Z}^+$ is at least

$$\left(1 - \frac{1}{a_1}\right)\left(1 - \frac{1}{a_2}\right) \cdots \left(1 - \frac{1}{a_k}\right).$$

22.1 Probability basics

This section is a collection of basic definitions and statements used in probability theory (and implicitly also in statistics). This section contains no exercises.

22.1.1 Probability spaces

For most purposes in combinatorics and computing science, a *probability space* is a pair (Ω, P) where Ω is a (usually finite) set (called a *sample space*) and $P : \mathcal{P}(\Omega) \to [0,1]$ is a function satisfying

$$\sum_{G \in \Omega} P(G) = P(\Omega) = 1.$$

The function P is called a *probability function* or *probability measure*. There are many notations for a probability function, including $\mathrm{Prob}[G]$ or $\mathrm{Pr}(G)$.

If (Ω, P) is a probability space where for each $\omega \in \Omega$, $P(\omega) = \frac{1}{|\Omega|}$, then the space is called a *uniform probability space*. [Infinite uniform probability spaces take a little more to define.]

Definition 22.1.1. Let (Ω, P) be a probability space. An *event* is some subset $\mathcal{A} \subseteq \Omega$, and the *probability of event* $\mathcal{A}$ is $P(\mathcal{A}) = \sum_{G \in \mathcal{A}} P(G)$.

One is often able to equate events (that is, certain subsets $\mathcal{A} \subseteq \Omega$) with qualities, as is shown in Section 22.1.2 and others below.

In a finite uniform probability space, calculating probabilities is tantamount to counting, that is, $P(\mathcal{A}) = \frac{|\mathcal{A}|}{|\Omega|}$. In an infinite "uniform probability space" (and many other infinite probability spaces), one must be a bit more careful about counting. For example, let $\Omega = \{G_1, G_2, G_3, \ldots\}$ be an infinite (but countable) sequence of graphs and put $\mathcal{A} = \{G_2, G_3, G_4, \ldots\}$. Then $P(\mathcal{A}) = 1 = P(\Omega)$ but $\mathcal{A} \neq \Omega$. In other words, one says that a certain property of graphs in Ω holds with probability 1, yet there are graphs (namely G_1) that do not satisfy that property. In such cases, one says that a property holds *almost surely* or *almost always*, or *almost all* graphs have the property.

Lemma 22.1.2. *For events $\mathcal{A}$ and $\mathcal{B}$ in a probability space,*

$$P(\mathcal{A} \backslash \mathcal{B}) = P(\mathcal{A}) - P(\mathcal{A} \cap \mathcal{B}).$$

Lemma 22.1.3. *For any events $\mathcal{A}$ and $\mathcal{B}$ in a probability space,*

$$P(\mathcal{A} \cup \mathcal{B}) = P(\mathcal{A}) + P(\mathcal{B}) - P(\mathcal{A} \cap \mathcal{B}).$$

Proof idea?: For any sets X and Y, by the inclusion-exclusion principle, $|X \cup Y| = |X| + |Y| - |X \cap Y|$.

It follows that $P(\cup_{i \in I} \mathcal{A}_i) \leq \sum_{i \in I} P(\mathcal{A}_i)$.

Let $\Omega \backslash \mathcal{A} = \overline{\mathcal{A}}$ denote the complementary event to $\mathcal{A}$. Since events correspond to sets, often the script notation is dispensed with and an event is some $A \subseteq \Omega$. Also, events can be thought of "occurring", so the event $A \cap B$ can be taken as the event "A and B", written $A \wedge B$, and the event $\bar{A}$ can be written $\neg A$.

22.1.2 Independence and random variables

Two events $\mathcal{A}$ and $\mathcal{B}$ in a probability space are said to be *independent* iff $P(\mathcal{A} \cap \mathcal{B}) = P(\mathcal{A}) \cdot P(\mathcal{B})$.

Lemma 22.1.4. *Let $\mathcal{A}$ and $\mathcal{B}$ be independent events. Then the events $\bar{\mathcal{A}} = \Omega \backslash \mathcal{A}$ and $\bar{\mathcal{B}}$ are also independent.*

Proof: Let $P(\mathcal{A}) = p_1$ and $P(\mathcal{B}) = p_2$, and so $P(\bar{\mathcal{A}}) = 1 - p_1$ and $P(\bar{\mathcal{B}}) = 1 - p_2$. Since $\mathcal{A}$ and $\mathcal{B}$ are independent, $P(\mathcal{A} \cap \mathcal{B}) = p_1 p_2$. Then

$$
\begin{aligned}
P(\bar{\mathcal{A}} \cap \bar{\mathcal{B}}) &= P((\Omega \backslash \mathcal{A}) \cap (\Omega \backslash \mathcal{B})) \\
&= P(\Omega \backslash (\mathcal{A} \cup \mathcal{B})) \\
&= 1 - P(\mathcal{A} \cup \mathcal{B}) \\
&= 1 - (P(\mathcal{A}) + P(\mathcal{B}) - P(\mathcal{A} \cap \mathcal{B})) \quad \text{(by Lemma 22.1.3)} \\
&= 1 - (p_1 + p_2 - p_1 p_2) \\
&= (1 - p_1)(1 - p_2) \\
&= P(\bar{\mathcal{A}}) \cdot P(\bar{\mathcal{B}}),
\end{aligned}
$$

which shows that $(\bar{\mathcal{A}})$ and $(\bar{\mathcal{B}})$ are independent. $\qquad \square$

Definition 22.1.5. An event A is *mutually independent* of events $B_1, \ldots, B_d$ iff A is independent of any boolean combination of the B_i's.

As mentioned by Jukna [296, p. 222], an event A can be independent of each of B_1 and B_2, but not mutually independent of B_1 and B_2. For example, flip a fair coin twice; let B_1 be the event that the first flip is heads, B_2 be the event that the second flip is heads, and let A be the event that both flips are the same. Then $P(A) = \frac{1}{2}$, $P(B_1) = P(B_2) = \frac{1}{2}$, $P(A \cap B_1) = P(A \cap B_2) = \frac{1}{4}$, and so A is independent of each B_1 and B_2, but

$$
\frac{1}{4} = P(A \cap B_1 \cap B_2) \neq P(A)P(B_1 \cap B_2) = \frac{1}{8},
$$

so A is not independent of $B_1 \cap B_2$.

Definition 22.1.6. A (real valued) *random variable* is a function $X : \Omega \to \mathbb{R}$.

If a random variable X takes on exactly two values, then X is called a *Bernoulli random variable*. For an event $\mathcal{A} \subset \Omega$, an *indicator random variable* $X = X_{\mathcal{A}}$ of the event $\mathcal{A}$ is the 0-1 Bernoulli random variable X defined by

$$
X(G) = \begin{cases} 1 \text{ if } G \in \mathcal{A} \\ 0 \text{ if } G \notin \mathcal{A}. \end{cases}
$$

Indicator random variables are often used to count the number of sets with a given property Q. For example, suppose that Ω is a set of graphs $G_1, \ldots, G_m$, and let $\mathcal{A}$ be the set of all graphs with some property Q. Letting each of $X_1, \ldots, X_m$ be indicator random variables defined by $X_i = X_i(G_i) = 1$ iff G_i has property Q, then $X = X_1 + \cdots + X_m$ counts the number of graphs with property Q, that is, $|\mathcal{A}|$.

In general, if the image of X is a finite (or countable) set, X is called a *discrete random variable*. If for some $k \in \mathbb{R}$, some X,

$$\mathcal{A} = \{G \in \Omega : X(G) \geq k\},$$

then $P(\mathcal{A})$ can be written $P(X \geq k)$.

Given a real valued random variable X, define the function $F_X : \mathbb{R} \longrightarrow [0,1]$ by $F_X(x) = P(X < x)$; then F_X is called the *distribution function* for X. For example, it might be that for some $f : \mathbb{R} \to \mathbb{R}$, that $F_X(x) = \int_{-\infty}^{x} f(z)dz$. In such a case, f is called the *density function* for X. Often, when a random variable X arises from a probability experiment, f is called the probability density function for X. A sequence of random variables, $X_1, X_2, \ldots$ is said to *converge in distribution* to that of X if for each $t \in \mathbb{R}$, $\lim_{n \to \infty} P(X_n < t) = P(X < t) = F_X(t)$.

22.1.3   Expected value and conditional probability

The *expected value* (or *mean value*) of a random variable X in a probability space (Ω, P) is

$$E[X] = \sum_{G \in \Omega} X(G)P(G).$$

The expected value $E[X]$ is often denoted by μ. Expectation is linear, that is, for any random variables X, Y, and $c \in \mathbb{R}$,

$$E[cX + Y] = cE[X] + E[Y].$$

For an event $\mathcal{B}$ with $P(\mathcal{B}) > 0$, the *conditional probability* of an event $\mathcal{A}$ given that event $\mathcal{B}$ has occurred (or relative to $\mathcal{B}$) is defined to be

$$P(\mathcal{A} \mid \mathcal{B}) = \frac{P(\mathcal{A} \cap \mathcal{B})}{P(\mathcal{B})}. \tag{22.1}$$

Equation (22.1) is sometimes called "Bayes' formula", and Bayes' theorem is an extension of this (see Lemma 22.6.1). A common application of conditional probability is the following:

Lemma 22.1.7. *If $P(\mathcal{B}) > 0$, then $\mathcal{A}$ and $\mathcal{B}$ are independent if and only if $P(\mathcal{A} \mid \mathcal{B}) = P(\mathcal{A})$.*

22.1.4 Conditional expectation

Just as expectation is defined in terms of probability, one can define "conditional expectation" in terms of conditional probability.

Definition 22.1.8. Given an event $\mathcal{E}$ in a probability space and a discrete random variable X, define the conditional expectation of X conditioned on the event $\mathcal{E}$ by

$$E[X \mid \mathcal{E}] = \sum_{G \in \omega} X(G) P(G \mid \mathcal{E}).$$

Note that depending on the event $\mathcal{E}$, many of the summands in the above definition might be zero. Written another way,

$$E[X \mid \mathcal{E}] = \sum_x x P(X = x \mid \mathcal{E}).$$

Lemma 22.1.9. *For any random variables X and Y,*

$$E[Y] = \sum_x P(X = x) E[Y \mid X = x], \tag{22.2}$$

where the sum is taken over $x \in \{X(G) : G \in \Omega\}$ and all of the expectations exist.

Proof: Beginning with the right-hand side of equation (22.2),

$$\sum_x P(X = x) E[Y \mid X = x] = \sum_x P(X = x) \sum_y y P(Y = y \mid X = x)$$

$$= \sum_y \sum_x y P(Y = y \mid X = x) P(X = x)$$

$$= \sum_y \sum_x P((Y = y) \wedge (X = x)) \qquad \text{(by Bayes')}$$

$$= \sum_y P(Y = y)$$

$$= E[Y].$$

$\square$

It follows from the above that conditional expectation is also linear, and so for any random variables $X_1, X_2, \ldots, X_n$ and Y where each $E[X_i] < \infty$,

$$E\left[\sum_{i=1}^n X_i \mid Y = y\right] = \sum_{i=1}^n E[X_i \mid Y = y].$$

For a fixed X, $E[X \mid Y]$ is a function of the random variable Y, and hence is itself a random variable.

Lemma 22.1.10. *For random variables X and Y,*

$$E[X] = E[E[X \mid Y]].$$

22.2 Basic probability exercises

No discussion of probability might seem complete without at least one "urn problem".

Exercise 760. *Suppose that an urn initially contains one red and one black marble and that, at each time $n = 1, 2, \ldots$, you randomly select a ball from the urn and replace it with two balls of the same color. Let X_n denote the number of red balls in the urn at time n (note $X_0 = 1$).*

a) *What is $E(X_1)$?*

b) *What is $E(X_2)$?*

c) *What is $E(X_3)$?*

d) *Conjecture a formula for $E(X_n)$.*

e) *Using mathematical induction, prove the conjectured formula.*

f) *Suppose that instead of replacing the ball with an additional ball of the same color, you replace it with an additional ball of the opposite color. Conjecture and prove.*

The following simple lemma sets the stage for a famous problem.

Lemma 22.2.1. *Let $X_1, X_2, \ldots$ be distinct real numbers chosen at random from the interval [0,1] and define $N = \min\{n \geq 2 : X_{n-1} < X_n\}$. Then $E[X] = e$.*

Proof: For each $n \geq 2$, there are $n!$ possible orderings of $X_1, \ldots, X_n$, and only the ordering $X_1 > X_2 > \cdots > X_n$ has no increasing adjacent pair, so $\text{Prob}(N > n) = \frac{1}{n!}$. Thus

$$E[N] = \sum_{n=0}^{\infty} \text{Prob}(N > n) = \sum_{n=0}^{\infty} \frac{1}{n!} = e,$$

which concludes the proof. $\qquad\qquad\square$

The following exercise is quite popular, appearing in many settings. For example, the case $x = 1$ appeared as a Putnam problem I.3 from the 1958 contest (see [90]) and a restricted version appears in [491]. This next exercise also appears in [467, 120–121], where it is combined with Lemma 22.2.1, and stated with $a = 1$, but proved more generally. The solution given here uses integrals.

Exercise 761. *Let $X_1, X_2, \ldots$ be distinct and chosen at random from the interval [0,1], and let $a \in (0, 1]$ be fixed. Define $M_a = \min\{n \geq 2 : X_1 + \cdots + X_n > a\}$. Prove by induction that for any positive integer n,*

$$Prob(M_a > n) = \frac{a^n}{n!}.$$

Setting $a = 1$ and $M_1 = M$ in Exercise 761, by the same reasoning in Lemma 22.2.1, conclude that $E[M] = e$. It may be somewhat surprising to have N and M with precisely the same distribution.

22.3 Branching processes

Exercise 762. *A microbe either splits into two perfect copies of itself or it disintegrates. If the probability of splitting is $p > \frac{1}{2}$, prove that the probability one microbe will produce an everlasting colony is at least $2 - \frac{1}{p}$.*

This next example appears in *Stochastic processes* [466] by Sheldon Ross. Consider organisms that don't split, but yet have finitely many offspring (of the same kind). Such reproduction is considered to be asexual, so the number of offspring from a particular organism in no way depends on the number of offspring of another. To each individual, there is the same probability, P_j, that it will produce $j \geq 0$ descendants.

Let G_0 be an initial population (the "0-generation) of organisms, and for each $j \in \mathbb{Z}^+$, let G_j be the generation (called the j-th generation) of offspring collectively produced by all of those in the previous generation G_{j-1}.

For each $i = 0, 1, 2, \ldots$, let $X_i = |G_i|$, the random variable denoting the population size after i generations. The Markov chain $X_0, X_1, X_2, \ldots$ is called a *branching process*. One can calculate the expected value of each X_n as follows. For any individual x, let $d(x)$ be the number of descendants of x. Then for each $n \geq 1$,

$$X_n = \sum_{x \in G_{n-1}} d(x).$$

For simplicity, consider only the case $X_0 = 1$, where a population starts from a single individual. Let μ be the expected value of $d(x)$. Then conditioning on X_{n-1},

$$\begin{aligned}
E[X_n] &= E[E[X_n | X_n - 1]] \\
&= \mu E[X_{n-1}] \\
&= \mu^2 E[X_{n-2}] \\
&\ \ \vdots \qquad\qquad\qquad\qquad \text{(by induction?)} \\
&= \mu^{n-1} E[X_1] \\
&= \mu^n.
\end{aligned}$$

Let π be the probability that (with $X_0 = 1$) the population dies out. Then

$$\pi = \sum_{j=0}^{\infty} \text{Prob}(\text{population dies out} \mid X_1 = j) P_j.$$

For the population to die out, given $X_1 = j$, all the j members of G_1 must spawn families that also die out. The $d(x)$'s are independent, and since the probability that the family started by x dies out is π (applying the model to that sub-family),

$$\pi = \sum_{j=0}^{\infty} \mu^j P_j.$$

Theorem 22.3.1. *[466, p. 192] Let $P_0 > 0$, and $P_0 + P_1 < 1$. Then π is the smallest number satisfying*

$$\pi = \sum_{j=0}^{\infty} \pi^j P_j.$$

Exercise 763. *Prove Theorem 22.3.1 by induction.*

22.4   The ballot problem and the hitting game

Consider an election for one of two candidates, A and B, where A wins. How likely is it that after each vote, A has more votes than B? The result in the following exercise is often called the "ballot problem", or "ballot theorem", sometimes attributed to Joseph Bertrand [50] in a half-page article published in 1887, with generalizations due to many authors. See *e.g.*, the 2007 article by Renault [450] for references, proofs, and the connection to Catalan numbers.

Exercise 764 (Ballot problem). *Let a and b be non-negative integers with $a > b$. In an election suppose that $a+b$ ballots are cast, a votes for candidate A and b votes for candidate B. Let $N(a, b)$ be the number of ways $a + b$ votes can be ordered so that after each vote, candidate A is winning. Prove that for $a + b \geq 1$,*

$$N(a, b) = \frac{a - b}{a + b} \binom{a + b}{a}.$$

Considering all $(a + b)!$ possible orderings of votes, the ballot problem result shows that the probability A stayed ahead during the entire vote is $N(a, b)/(a+b)!$. Variants of the ballot problem occur in the study of branching processes, random walks, Markov chains, and martingales, but little discussion of these topics is given here. Very briefly, if an individual is taking a walk on the integers, a vote for A can be interpreted as a step to the right, and a vote for B as a step to the left. Then the position in the walk measures the advantage of A over B; analysis similar to that used for the ballot problem can be used to find the probability (given a certain probability distribution of right-vs.-left steps) that a walk starting at 0 stays to the right of 0, even if the walk is infinite. Related questions might ask what is the expected time for the individual to first return to 0 (see the hitting time problem in Exercise 767 below) or what is the probability that a random walk ever extends past some threshold k (at which time, the walk may terminate, like falling off a cliff).

The following exercise can be given in many disguises (see Exercise 404, for example), and establishes a connection to the Catalan numbers. A solution to the next exercise is nearly direct from Exercise 764; however, an inductive proof is available by nearly duplicating the solution to Exercise 764. [The simple proof follows the exercise.]

Exercise 765 (Ballot problem weakened). *Let a and b be non-negative integers with $a > b$. In an election suppose that $a + b$ ballots are cast, a votes for candidate A and b votes for candidate B. Let $N^*(a,b)$ be the number of ways $a + b$ votes can be ordered so that after each vote, candidate A is not behind (so is either tied or is ahead). Prove by induction that for $a + b \geq 1$,*

$$N * (a,b) = \frac{a - b + 1}{a + 1} \binom{a + b}{a}.$$

Note that $N^*(a,a) = \frac{1}{2a+1}\binom{2a}{a}$, the Catalan number C_a. The weakened ballot problem follows from the ballot problem fairly easily: If an ordering of votes keeps A at least even, then by adding one more vote for A at the beginning, get an ordering for $a + 1 + b$ votes where A (receiving $a + 1$ votes) is always ahead. So

$$N^*(a,b) = N(a+1,b) = \frac{a+1-b}{a+1+b}\binom{a+1+b}{a+1} = \frac{a-b+1}{a+1}\binom{a+b}{a}.$$

This tidy little proof uses a technique quite common to these problems. Another is called the "reflection principle", or the "reversing technique", which essentially looks at the votes occurring in precisely the opposite order; see [450] for more details (or the solution to Exercise 668 here).

Essentially repeating the solution to Exercise 764 also solves a stronger version of the ballot problem, first mentioned by Barbier [37] in the same year (and journal) as the Bertrand note.

Exercise 766 (Ballot problem, generalized). *Let a and b be non-negative integers and let $m \in \mathbb{Z}^+$, where $a > mb$. In an election suppose that $a + b$ ballots are cast, a votes for candidate A and b votes for candidate B. Let $N_m(a,b)$ be the number of ways $a + b$ votes can be ordered so that after each vote, candidate A is winning with a margin of more than m times the number of votes for B. Prove that for $a + b \geq 1$,*

$$N_m(a,b) = \frac{a - mb}{a + b}\binom{a + b}{a}.$$

See also the solution to Exercise 668, where the ballot problem terminology reveals a useful identity for a matrix theory problem involving Catalan numbers.

The result in the next exercise is sometimes called the "hitting time theorem" or the "first return to 0" theorem. For a recent paper including many references (as well as relationships with the ballot problem and branching process) and applications, see [545]. The hitting time theorem is equivalent to the ballot problem result (each can be easily derived from the other). The "elementary" solution in [545] to this exercise is not quite so elementary for the novice, and so is not included here; their inductive solution does, however, show that the distributions for different starting points need not be the same.

Exercise 767. *Let $Y_1, Y_2, Y_3, \ldots$ be integers chosen independently and at random from $\{-1, 1, 2, 3, \ldots\} = [-1, \infty)$, and let k be a positive integer. Denote the position in a random walk starting at k after n steps by $S_n = k + Y_1 + \cdots + Y_n$. Given that such a walk eventually lands on 0, prove that the conditional probability that the walk hits 0 at time n is $\frac{k}{n}$.*

As an easy example of the hitting time theorem, let $k = 1$ and $n = 3$. The only three sequences of steps (Y_1, Y_2, Y_3) that return to 0 are $(-1, -1, 1)$, $(-1, 1, -1)$, and $(1, -1, -1)$, and all are equally likely, so the conditional probability desired is $1/3$.

22.5 Pascal's game

The next result is due to Pascal himself and regards a gambling problem. Some of the details presented here come from [91]. For an easy to read account of Pascal's dealings with Chevalier de Méré and probability problems arising from casino games, see [4, pp. 209–212]. It seems that Pascal might have had help from Fermat in solving such problems. You might have heard of this problem under the name *problem of the points*. In fact, this problem goes back to Luca Pacioli (1445–ca. 1509), who wrote about it in his *Sūma* [422] of 1494. According to [180], Girolamo Cardano (1501–1576) and Nicola Fontana (*ca.* 1499–1557), also known as Tartaglia, discussed the mathematics of this problem as well.

Exercise 768 (Pascal's game). *Two players A and B of equal skill are playing for a stake of P, and wish to leave the game table before finishing their game. Their scores and the number of points that constitute the game are given indirectly: Player A lacks α points of winning and player B needs β points. If $\alpha + \beta = n$, Pascal says that the stakes should be divided between B and A in the ratio*

$$\sum_{i=0}^{\alpha-1} \binom{n-1}{i} \quad : \quad \sum_{j=\alpha}^{n-1} \binom{n-1}{j}.$$

Since $\sum_{i=0}^{n-1} \binom{n-1}{i} = 2^{n-1}$, this is the same as saying that A's share is

$$\frac{P}{2^{n-1}} \sum_{i=\alpha}^{n-1} \binom{n-1}{j}$$

and B's share is

$$\frac{P}{2^{n-1}} \sum_{i=0}^{\alpha-1} \binom{n-1}{i}.$$

Use Pascal's identity (see Exercise 90) to prove this result by induction (on n) for all $n \geq 2$.

22.6 Local Lemma

The main result in this section, called the "Lovász Local Lemma" is a tool developed for some questions in Ramsey theory. Its proof is by induction and relies heavily on an extension of Bayes' formula. For more detailed examination of this topic, see either [13] or [391]; most of the material here is taken from these two classics.

Recall that Bayes' formula for conditional probability says that for a probability P, $P(A|B) = P(A \cap B)/P(B)$. Bayes' theorem can take on many forms; one form is useful here, where I now use the notation "Prob" instead of "P" for clarity later:

Lemma 22.6.1.
$$Prob(A \mid B \wedge C) = \frac{Prob(A \wedge B \mid C)}{Prob(B \mid C)}$$

Proof outline: Expand the left side by Bayes' formula, then divide both numerator and denominator by $Prob(C)$. □

Let $A_1, \ldots, A_n$ be events in a probability space Ω. Let $G = (V, E)$ be a graph on $V = [n] = \{1, 2, \ldots, n\}$. If for each $i \in V$, A_i is mutually independent of all $A_j : \{j, i\} \notin E$, then G is a *dependency graph* for the events $A_1, \ldots, A_n$. [Note: A digraph could have been used here.]

The following appeared in [171], a paper co-authored with Erdős, however, is now eponymous with its second author:

Theorem 22.6.2 (Lovász Local Lemma). *Let $A_1, \ldots, A_n$ be events in some probability space (Ω, P), and let G be a dependency graph for these events. If there exist reals $x_1, \ldots, x_n \in [0, 1)$ such that for each $i = 1, \ldots, n$,*

$$P(A_i) \leq x_i \prod_{\{i,j\} \in E} (1 - x_j),$$

then

$$P\left(\bigwedge_{i=1}^{n} \overline{A_i} \right) \geq \prod_{i=1}^{n} (1 - x_i).$$

In particular, the probability that no event A_i occurs is positive.

Proof: The following claim is central:

Claim: For any $S \subseteq [n]$, $S \neq [n]$, and any $i \notin S$,

$$P\left(A_i \mid \bigwedge_{j \in S} \overline{A_j} \right) \leq x_i.$$

Exercise 769. *Prove the above claim by induction on $|S|$. Hint: Use Lemma 22.6.1.*

Using $P(A \wedge B) = P(A)P(B \mid A)$ and $P(A \wedge B \mid C) = P(B \mid C) \cdot P(A \mid B \wedge C)$ repeatedly,

$$
P\left(\bigwedge_{i=1}^{n} \overline{A_i}\right) = P\left(\overline{A_1}\right) \cdot P\left(\bigwedge_{i=2}^{n} \overline{A_i} \mid \overline{A_1}\right)
$$

$$
= P\left(\overline{A_1}\right) \cdot P\left(\overline{A_2} \mid \overline{A_1}\right) P\left(\bigwedge_{i=3}^{n} \overline{A_i} \mid \overline{A_1} \wedge \overline{A_2}\right)
$$

$$
\vdots
$$

$$
= P(\overline{A_1}) \cdot P\left(\overline{A_2} \mid \overline{A_1}\right) \cdots P\left(\overline{A_n} \mid \bigwedge_{i=1}^{n-1} \overline{A_i}\right)
$$

$$
= (1 - P(A_1))(1 - P(A_2 \mid \overline{A_1})) \cdots \left(1 - P\left(A_n \mid \bigwedge_{i=1}^{n-1} \overline{A_i}\right)\right)
$$

$$
\geq (1 - x_1)(1 - x_2) \cdots (1 - x_n) \quad \text{(by claim with } S = \{1, \ldots, i-1\}\text{)}.
$$

$\square$

In the following well-known inequality, e is the base of the natural logarithm: for any positive real d,

$$
\left(1 - \frac{1}{d+1}\right)^d > \frac{1}{e} \tag{22.3}
$$

Here is one way to see equation (22.3): The Maclaurin expansion $e^x = 1 + x + x^2/2 + x^3/3! + \ldots$ shows that $e^x \geq 1 + x$ for $x > 0$. Set $x = \frac{1}{d}$; then $e^{1/d} > 1 + \frac{1}{d} = \frac{d+1}{d}$; this implies $e > \left(\frac{d+1}{d}\right)^d$, and thus

$$
\frac{1}{e} < \left(\frac{d}{d+1}\right)^d = \left(1 - \frac{1}{d+1}\right)^d,
$$

as desired. Another way to see this inequality is to observe that equation (22.3) is equivalent to $\ln(1 + 1/d) < 1/d$, and using the series expansion of $\ln(x)$ confirms this last inequality.

Corollary 22.6.3. *Suppose that G is a dependency graph for events $A_1, \ldots, A_n$, $0 < \Delta(G) \leq d$, and for every i, $P(A_i) \leq p$. If $ep(d+1) \leq 1$, then $P(\bigwedge_{i=1}^{n} \overline{A_i}) > 0$.*

Proof: Let $d > 0$, where $|\{j : \{i, j\} \in E(G)\}| \leq d$. For each i, let $x_i = \frac{1}{d+1}$. Then for every i, $P(A_i) \leq p$, but

$$
x_i \prod_{\{i,j\} \in E(G)} (1 - x_j) = \frac{1}{d+1}\left(1 - \frac{1}{d+1}\right)^d > \frac{1}{e} \cdot \frac{1}{d+1}.
$$

So if $\frac{1}{e(d+1)} \geq p$, LLL applies, that is, if $ep(d+1) \leq 1$. In this case, LLL yields
$$P(\bigwedge_{i=1}^{n} \overline{A_i}) \geq \left(1 - \frac{1}{d+1}\right)^n > 0. \qquad \square$$

Note: The inequality $ep(d+1) < 1$ is often relaxed to $4pd \leq 1$. Shearer [486] proved that the e cannot be replaced by a smaller constant.

Part III

Solutions and hints to exercises

Everything should be made as simple as possible, but not simpler.

—Albert Einstein

It may be that most mathematical exercises have more than one solution; the ones given here might serve as only a beginning to any investigations. As Erdős might have said, not all solutions here are necessarily proofs from "The Book" (a book in the "heavens" containing the most elegant proofs, see [7] for more about The Book).

Chapter 23

Solutions: Foundations

To think the thinkable—that is the mathematician's aim.

—C. J. Keyser,

The universe and beyond; Hibbert Journal.

23.1 Solutions: Properties of natural numbers

Exercise 1: (Multiplication is well defined) This exercise asks to show that there exists a unique function $g : \mathbb{N} \times \mathbb{N} \longrightarrow \mathbb{N}$ so that for all $x, y \in \mathbb{N}$

  (e) $g(x, 1) = x$;
  (f) $g(x, y') = x + g(x, y)$.

The proof parallels that given for Theorem 2.5.4. There are two parts to this proof, the first showing existence and the second showing uniqueness.

(Existence) Define $B \subseteq \mathbb{N}$ to be the set of those x for which there exists a set $\{g(x, y) : y \in \mathbb{N}\}$ such that for all $y \in \mathbb{N}$ (and the fixed x) both (e) and (f) hold. It suffices to show that $B = \mathbb{N}$; this is done by induction on x.

BASE STEP $(x = 1)$: First, for every $y \in \mathbb{N}$, define $g(1, y) = y$. Then by definition, $g(1, 1) = 1$, and so (e) holds for $x = 1$. Also, by definition, $g(1, y') = y' = 1 + y = 1 + g(1, y)$, and so (f) holds for $x = 1$. Hence, $1 \in B$.

INDUCTIVE STEP: Suppose that $x \in B$, that is, for all $y \in \mathbb{N}$, $g(x, y)$ is defined so that (e) and (f) hold for this fixed x. For all $y \in \mathbb{N}$, define $g(x', y) = y + g(x, y)$. Then $g(x', 1) = 1 + g(x, 1) = g(x, 1)'$ and so (e) holds for x'. Also, for all $y \in \mathbb{N}$, for all $y \in \mathbb{N}$,

$$
\begin{aligned}
g(x', y') &= y' + g(x, y') && \text{(by definition)} \\
&= y' + (x + g(x, y)) && \text{(by (f)), since } x \in B
\end{aligned}
$$

405

$$
\begin{aligned}
&= (y' + x) + g(x, y) && \text{(addition is associative)} \\
&= (x + y') + g(x, y) && \text{(addition is commutative)} \\
&= (x' + y) + g(x, y) && \text{(by (b') and (d'))} \\
&= x' + (y + g(x, y)) && \text{(addition is associative)} \\
&= x' + g(x', y) && \text{(by definition)},
\end{aligned}
$$

showing that (f) holds for x'. Hence $x' \in B$, completing the inductive step.

Hence, by P5, $B = \mathbb{N}$. This completes the existence proof.

(Uniqueness) Suppose that g is defined so that for all $x, y \in \mathbb{N}$ (e) and (f) hold, and further suppose that h is a function satisfying the corresponding equalities (for all $x, y \in \mathbb{N}$):

(e') $h(x, 1) = x$, and
(f') $h(x, y') = x + g(x, y)$.

Let $x \in \mathbb{N}$ be fixed and put $A_x = \{y \in \mathbb{N} : g(x, y) = h(x, y)\}$. Induction is now used to show that $A_x = \mathbb{N}$:

BASE STEP: By assumptions (e) and (e'),

$$
g(x, 1) = x = h(x, 1),
$$

and so $1 \in A_x$, proving the base case.

INDUCTIVE STEP: For some $y \in \mathbb{N}$, assume that $y \in A_x$, that is, $g(x, y) = h(x, y)$. By (f) and (f'),

$$
\begin{aligned}
g(x, y') &= x + g(x + y) && \text{(by (f))} \\
&= x + h(x + y) && \text{(by induction hypothesis)} \\
&= h(x, y') && \text{(by (f'))},
\end{aligned}
$$

and so $y' \in A_x$, completing the inductive step.

Thus, by P5, $A_x = \mathbb{N}$, and since x was arbitrary, this completes the uniqueness part of the proof, and hence the whole proof. $\qquad\square$

Note: It was probably clear to the reader that the notation "$g(x, y)$" really denoted the product of x and y. Now that it has been established that multiplication is well defined (for natural numbers), one can replace the notation "$g(x, y)$" with any of the more familiar notations, "$x \times y$", "$x \cdot y$" or simply "xy". Also, the above proof shows that for every natural number x, $1 \cdot x = x = x \cdot 1$, so 1 is truly a multiplicative identity.

Exercise 2: (Distributivity in $\mathbb{N}$) The proof is by induction on x that for any $x, y, z \in \mathbb{N}$,

$$
x(y + z) = xy + xz.
$$

Fix natural numbers y and z. Let

$$A = \{x \in \mathbb{N} : x(y + z) = xy + xz\}.$$

(Below, it is convenient to use notation "g" from Exercise 1.)

BASE STEP: Since $1(x + y) = g(1, x + y) = x + y$, it follows that $1 \in A$.

INDUCTIVE STEP: For some $x \in \mathbb{N}$, assume that $x \in A$. Then

$$
\begin{aligned}
x'(z + y) &= g(x', y + z) \\
&= y + z + g(x, y + z) \\
&= y + z + x(y + z) \\
&= y + z + xy + xz && \text{(by induction hypothesis)} \\
&= y + g(x, y) + z + g(x, z) \\
&= g(x', y) + g(x', z) \\
&= x'y + x'z,
\end{aligned}
$$

so $x' \in A$, completing the inductive step.

By P5, $A = \mathbb{N}$, finishing the proof of $x(y + z) = xy + xz$. The proof of the other equality is similar and is left to the reader. □

Exercise 3: (General distributivity in $\mathbb{N}$) For $n \geq 1$, let $S(n)$ be the statement that for any $x_1, x_2, \ldots, x_n, c \in \mathbb{N}$,

$$c\left(\sum_{i=1}^{n} x_i\right) = \sum_{i=1}^{n} cx_i.$$

The one inductive proof given here is to induct upon n and apply the recursive definition of the sum. To be consistent with previous notation, the proof here shows

$$\{n : S(n) \text{ is true}\} = \mathbb{N},$$

without naming the set on the left as A. (For another proof, one could induct on, say, x_1, as in the solution to Exercise 2, where each step is for all n.)

BASE STEP: When $n = 1$, $S(1)$ says that $cx_1 = cx_1$, which is true, so $S(1)$ holds. Also, by Exercise 2, $S(2)$ is also true.

INDUCTIVE STEP: Fix $k \geq 1$ and assume that $S(k)$ is true. Let $x_1, x_2, \ldots, x_k, x_{k+1}$ and c be natural numbers. To see that

$$S(k + 1): \quad c\left(\sum_{i=1}^{k+1} x_i\right) = \sum_{i=1}^{k+1} cx_i$$

holds, beginning with the left side,

$$c\left(\sum_{i=1}^{k+1} x_i\right) = c\left(\left(\sum_{i=1}^{k} x_i\right) + x_{k+1}\right) \qquad \text{(by def'n of sum)}$$

$$= c\left(\sum_{i=1}^{k} x_i\right) + cx_{k+1} \qquad \text{(by } S(2)\text{)}$$

$$= \left(\sum_{i=1}^{k} cx_i\right) + cx_{k+1} \qquad \text{(by } S(k)\text{)}$$

$$= \left(\sum_{i=1}^{k+1} cx_i\right),$$

the right side of $S(k+1)$ is arrived at. Thus $S(k+1)$ is also true, completing the inductive step.

Therefore, by P5, the set of all positive integers n for which $S(n)$ is true is indeed all of $\mathbb{N}$. □

Exercise 4: (Multiplication in $\mathbb{N}$ is associative) The exercise asks to prove that for any $x, y, z \in \mathbb{N}$,
$$(xy)z = x(yz).$$

Let x and y be fixed natural numbers and put
$$A = \{z \in \mathbb{N} : (xy)z = x(yz)\}.$$

Since x and y are arbitrary, it suffices to show that $A = \mathbb{N}$; this is done by induction.

BASE STEP: Using the notation from Exercise 1,

$$(xy) \cdot 1 = g(xy, 1)$$
$$= xy \qquad \text{(by (e))}$$
$$= x \cdot g(y, 1) \qquad \text{(by (e))}$$
$$= x(y \cdot 1),$$

and so $1 \in A$.

INDUCTIVE STEP: Suppose that $z \in A$. Then

$$(xy)z' = g(xy, z')$$
$$= xy + g(xy, z) \qquad \text{(by (f))}$$
$$= xy + (xy)z$$
$$= xy + x(yz) \qquad \text{(because } z \in A\text{)}$$
$$= x(y + yz) \qquad \text{(by distributivity)}$$

$$= x(y + g(y, z)) \qquad \text{(by (f))}$$
$$= x \cdot g(y, z')$$
$$= x(yz'),$$

showing that $z' \in A$, completing the inductive step.

Thus, by P5, $A = \mathbb{N}$; this completes the proof that multiplication is associative.

$\square$

23.2 Solutions: Well-ordered sets

Exercise 5 (Law of trichotomy): To be shown is that for any $x, y \in \mathbb{N}$, exactly one of $x < y$, $x = y$, or $y < x$ holds. (There may be a more elegant solution than the one given here.)

Let $P(x)$ be the property that for each $y \in \mathbb{N}$, at least one of $x < y$, $x = y$, or $y < x$ holds. Let $A = \{x \in \mathbb{N} : P(x) \text{ is true}\}$. The first goal is to show that $A = \mathbb{N}$.

BASE STEP ($1 \in A$: To show $1 \in A$ one must show that for each $y \in \mathbb{N}$, exactly one of $1 < y$, $1 = y$, or $y < 1$ holds.

Begin by showing that $y < 1$ never holds: Suppose, in hopes of a contradiction, that $y < 1$; then for some $n \in \mathbb{N}$, $y + n = 1$. If $n = 1$, then 1 has a predecessor, contrary to P3. If $n \neq 1$ then, by Theorem 2.5.3, n has a unique predecessor, z so that $z' = n$, that is, $z + 1 = n$; in this case, $y + (z + 1) = 1$ implies $(y + z) + 1 = 1$, again violating P3. So $y < 1$ can never occur.

Fix some $y \in \mathbb{N}$. It remains to show that either $y = 1$ or $1 < y$. If $y \neq 1$, then y has a predecessor $n \in \mathbb{N}$, in which case $n + 1 = y$, or equivalently (by commutativity), $1 + n = y$; but this satisfies the definition for $1 < y$. This completes the base step showing $1 \in A$.

INDUCTIVE STEP: For some fixed $x \in \mathbb{N}$, assume that $x \in A$, that is, for any $y \in \mathbb{N}$, one of $x < y$, $x = y$, or $y < x$ holds. To be shown is that $x' \in A$. Fix some $y \in \mathbb{N}$. If $x = y$, then by the observation just before Lemma 2.5.1, $x' = y'$, and so $x' = y + 1$; hence, by definition of "$<$", $y < x'$.

Consider the situation where $x \neq y$. Since $x \in A$, one of $y < x$ or $x < y$. Suppose that $y < x$, that is, there is an $n \in \mathbb{N}$ so that $y + n = x$. Then $(y + n)' = x'$, and so $y + (n + 1) = x'$, showing that $y < x'$.

Suppose that $x < y$. There are two cases: $x' = y$ and $x' \neq y$. If $x' = y$, then $x' \in A$; so assume that $x' \neq y$ (and $x < y$). Since $x < y$, there is some $m \in \mathbb{N}$ so that $x + m = y$. However, $x' \neq y$, and so $m \neq 1$. Then m has a unique predecessor (again, by Theorem 2.5.3), say ℓ, where $\ell + 1 = m$. Then $x + \ell + 1 = y$ implies $x' + \ell = y$, giving $x' < y$. This finishes the proof of $x' \in A$, and hence the inductive step.

Therefore, by P5, $A = \mathbb{N}$.

It remains to show that for any $x, y \in \mathbb{N}$, exactly one of $x < y$, $x = y$, $y < x$ holds. Suppose that $x, y \in \mathbb{N}$ and $x \neq y$. It suffices to show that at most one of $x < y$ and $y < x$ hold. Begin by assuming (in hopes of a contradiction) that both $x < y$ and $y < x$ hold, and that $m, n \in \mathbb{N}$ are so that

$$x + m = y \quad \text{and} \quad y + n = x.$$

Replacing the y with $x+m$ from the first equation into the second gives $x+m+n = x$. By Theorem 2.5.8 (using $m + n$ instead of y in the statement of that theorem), this is impossible. Therefore, abandon the assumption that both $x < y$ and $x > y$ hold. $\square$

Exercise 6 (Addition preserves order): There are two directions to prove.

($\rightarrow$): Let x and y be fixed natural numbers with $x < y$. Let $A_{x,y} = \{p \in \mathbb{N} : x + p < y + p\}$. Since $x < y$, there is some $m \in \mathbb{N}$ so that $x + m = y$. Then $(x+m)' = y'$, and by property (d'), $x'+m = y'$; hence $x' < y'$, that is, $x+1 < y+1$. This says that $1 \in A_{x,y}$. Assume that $p \in A$, that is, $x+p < y+p$. Then there exists an n so that $x+p+n = y+p$. Using properties (b'), (d'), $(x+p+n)' = x+p'+n$ and $(y + p)' = y + p'$, so $x + p' + n = y + p'$ and therefore $x + p' < y + p'$. Hence $p' \in A_{x,y}$. By P5, $A_{x,y} = \mathbb{N}$. Since x and y were arbitrary, this finishes the proof of the forward direction.

($\leftarrow$): Suppose that $x + p < y + p$. If $x = y$, then since the addition function f produces a unique number, $x+p = y+p$, contradicting (by the Law of trichotomy) that $x + p < y + p$. If $y < x$, there is some n so that $y + n = x$; then $x + p = y + n + p = y + p + n$ shows that $y + p < x + p$, contradicting $x + p < y + p$. Hence $x = y$ and $y < x$ do not occur and so by the Law of trichotomy, $x < y$. $\square$

Exercise 7: This proof appears in [95, Lemma 2.1, p. 41].

Let $(X, \leq)$ be a well-ordered set and let $Y \subseteq X$. Suppose $f : X \longrightarrow Y$ is an isomorphism (an order preserving bijection).

Let $W = \{x \in X : f(x) < x\}$. If $W \neq \emptyset$, then W has a least element, call it w_0. Since f is order preserving, $f(w_0) < w_0$ implies that $f(f(w_0)) < f(w_0)$, and so $f(w_0) \in W$. But $f(w_0) < w_0$ contradicts that w_0 is the least element of W. Hence, $W = \emptyset$. $\square$

23.3   Solutions: Fermat's method of infinite descent

Exercise 8: This exercise occurred in [437, 20-4, Challenge 2, pp. 61, 230], however the solution there is more direct, not using infinite descent.

Suppose, in hopes of a contradiction, that n is so that $\sqrt{4n - 1}$ is rational, say,

$$\sqrt{4n - 1} = \frac{p}{q}.$$

Squaring each side, $4n - 1 = \frac{p^2}{q^2}$, and so $(4n-1)q^2 = p^2$.

If q is even, say $q = 2\ell$, then $(4n-1)4\ell^2 = p^2$ implies that p is even, say $p = 2k$. So, if q is even, $\frac{k}{\ell} = \frac{p}{q}$ with $k < p$ and $\ell < p$.

Examine the case when q is odd, say $q = 2b+1$. Then q^2 is odd, and since $4n-1$ is odd, then so too is $(4n-1)q^2 = p^2$, and thus also p is odd, say $p = 2a+1$. Then

$$(4n-1)(4b^2 + 4b + 1) = 4a^2 + 4a + 1,$$

and multiplying out,

$$4n(4b^2 + 4b + 1) - 4b^2 - 4b - 1 = 4a^2 + 4a + 1.$$

The left side has remainder 3 (or -1) when divided by 4, and the right side has remainder 1 when divided by 4. Hence, q can not be odd.

So, return to the only possibility, namely that q is even. In this case, it has been shown that one can find $\frac{k}{\ell} = \frac{p}{q}$ with $k < p$ and $\ell < q$. If ℓ is odd, then a contradiction is arrived at by the second part above. If ℓ is even, then continue to find yet another representation for $\frac{p}{q}$ with even smaller numerator and denominator. Applying the same reasoning, this process can continue *ad infinitum* (if every new denominator produced is even). Since the natural numbers are well-ordered, such a process must stop, the final desired contradiction. $\qquad\square$

Chapter 24

Solutions: Inductive techniques applied to the infinite

24.1   Solutions: More on well-ordered sets

Exercise 9: Let $(W, <)$ be a well-ordered set, and let $X \subseteq W$. Put

$$S = \cup_{x \in X} \mathrm{seg}_W(x).$$

Since each initial segment in W is a subset of W, $S \subseteq W$. Suppose that $S \neq W$, and put $Y = W \setminus S$. Since $\emptyset \neq Y \subset W$ and W is well-ordered, Y has a least element y. Then $S = \mathrm{seg}(y)$. The proof for closed segments is similar. $\square$

24.2   Solutions: Axiom of choice and equivalent forms

Exercise 10: Assume that version 1 of Zorn's lemma holds, and suppose that $\mathcal{F}$ is a family of subsets of some set X has the property that for every chain $\mathcal{C} \subseteq \mathcal{F}$, $\cup \mathcal{C} \in \mathcal{F}$. Order $\mathcal{F}$ by inclusion; then every chain in $\mathcal{C} \subseteq \mathcal{F}$ is a totally ordered set, and $\cup \mathcal{C}$ is an upper bound for $\mathcal{C}$. The proof is complete by applying version 1 of Zorn's lemma. $\square$

Chapter 25

Solutions: Paradoxes and sophisms

25.1 Solutions: Trouble with the language?

Exercise 11: To think about: Can one prove a statement about a particular language using that same language? Richard's paradox is discussed at length in [400, pp. 1686–1688]; I recommend the reader to have a look at that article, from the beginning, to become familiar with the kinds of questions that arise surrounding this paradox. Richard's paradox is also treated in Kleene's *Introduction to Metamathematics* [316, p. 38]. Also see [274] and [275], popular readings for "self reference" in logic and logic in general. You might be drawn into learning things like Gödel's incompleteness theorem, a result which says, roughly, if a language is rich enough so as to be able to decide the truth (in a finite mechanistic way) of any statement in the language, then contradictions arise; it also says that, essentially, any consistent system is not complete enough so as to be able to prove (from within the system) all the truths expressible in that system. For example, second order logic (where quantification is allowed over subsets, rather than just individual elements) is not complete, and so there are statements in mathematics whose truth is never decidable. See, for example, Section 6.12 for the discussion of such a truth.

The amount of literature on these phenomena is incredible (phenomenal?), so to gain a bit more understanding of these issues, see almost any text in formal logic, or, say, [95], for a fairly complete, yet concise and readable introduction.

Exercise 12: (The unexpected exam) For an expository article on this paradox, see [185, pp. 161–166]. This paradox was first published in the British logic journal *Mind* (1948). A Swedish professor of mathematics, Lennart Ekbom, apparently first noticed the paradoxical aspect of a similar claim made on Swedish radio regarding a civil defense exercise.

It seems that no satisfactory resolution to this paradox has been found; however, there might be some recent references shedding more light on this paradox.

25.2 Solutions: Missed a case?

Exercise 13: The statement $S(1)$ does not follow from $S(0)$. □

Exercise 14: This example is apparently due to Pólya. The "proof" does not prove the case that two horses are the same color, that is, one can not get from the base case (one horse is the same color) to the first non-trivial case, since the induction hypothesis used implicitly requires two. The faulty reasoning is similar to that in Exercises 13 and 15. □

Exercise 15: The inductive step does not work when $k = 2$; a fourth line is necessary in the "proof" of the inductive step. □

25.3 Solutions: More deceit?

Exercise 16: This example appears in [462, 48, p.281]. In fact, the base case does not hold. The left side of $A(1)$ is $\sum_{i=1}^{1} i = 1$. The right side of $A(1)$ is $\left(1 + \frac{1}{2}\right)^2 / 2 = 9/8$. □

Exercise 18: Since the number of weighings (in this case, four) does not seem to be used in the proof, the same "proof" seems to reveal the stronger statement:

$S(m)$: For any m coins, at most one weighing is required to identify the lighter counterfeit coin.

When $m = 2$, exactly one weighing determines the ersatz coin, so $S(2)$ is true. Assume that for some $k \geq 2$, $S(k)$ holds. Consider $k + 1$ coins, and set one coin aside. There are two cases:

Case 1: The coin set aside is genuine.

Case 2: The coin set aside is counterfeit.

In Case 1, the remaining k coins contain one counterfeit, and so $S(k)$ applies to this set, and the counterfeit coin is found in one weighing. In Case 2, the remaining k coins are identical, so $S(k)$ can not be applied. One can verify directly that $S(3)$ holds by any weighing with one coin in each pan. However, with four coins, if the coin set aside is the counterfeit coin, the truth of $S(3)$ can not be applied to the remaining three genuine coins.

Perhaps a stronger statement could be proved:

$S'(m)$: For any m coins, at most one weighing is required to determine if there is a lighter counterfeit coin among these, and if so, to identify it.

Since one weighing of two coins determines whether or not they are the same or which is lighter, $S'(2)$ remains true. Consider the case with three coins; if any two coins are put on the scale and the scale balances, the third coin set aside is not identifiable as either genuine or counterfeit. So $S'(3)$ fails, and thus the stronger statement is not true either.

In general, by Exercise 585, 3^n is the most number of coins so that n weighings identify the lighter counterfeit coin; so the statement in this exercise (with four weighings) is false for 82 coins. $\square$

Chapter 26

Solutions: Empirical induction

26.1 Solutions: Introduction

Exercise 19: There are many obvious choices for such a statement. Here is one (albeit not terribly inventive): $S(n): \quad n \leq 1,000,000$. $\qquad \square$

26.2 Solutions: A sequence of integers?

Exercise 20: This problem was given in a lecture by (the late) Dr. Eric Milner, 9 October 1987; I do not know the original source. The solution presented here might not be the intended one. It turns out that this sequence is well-known, called a *Göbel sequence* and is related to a family of sequences called *Somos sequences*. For reference, see E15 of Guy's book [248] *Unsolved Problems in Number Theory* (pp. 214–5). [Also, an article by Fritz and Lenstra appeared in, I think, *Crux Mathematicorum* perhaps volume **15** (1989), however, I can not now find this reference]. No, it is not always an integer; the hint given in the lecture was to work modulo 43. My guess is that what was intended was the following idea: if each of $s_0, s_1, \ldots, s_{42}$ is indeed an integer, calculate them modulo 43, then find the numerator of s_{43} is not divisible by 43 (the denominator of s_{43} is 43). [For those not familiar with modulo notation, see Section 11.2 (Congruences).]

Define $t_n = 1 + s_0^2 + \cdots + s_{n-1}^2$, the numerator of s_n. Then $t_{n+1} = t_n + s_n^2$. The following table will help to verify the technique given later:

n	s_n	$s_n \bmod 43$	$s_n^2 \bmod 43$		t_n	$t_n \bmod 43$
0	1	1	1			
1	2	2	4		2	2
2	3	3	9		6	6
3	5	5	25		15	15
4	10	10	14		40	40
5	28	28	10		140	11
6	154	25	23		924	21
7	3520	37	36		24640	1
8		10	14	12415040		37
9		20	13			8
10		15	10			21

Since 43 is prime, every number $n \in \{1, 2, 3, \ldots, 42\}$ has a unique inverse n^{-1} modulo 43, that is, the number m so that $mn \equiv 1 \pmod{43}$. These inverses can found by the Euclidean algorithm. For example, since 5 is relatively prime to 43, $\gcd(5, 43) = 1$ and so, by the Euclidean division algorithm, (see Exercise 211) there are integers x and y so that $1 = 5x + 43y$. (The fact that if $\gcd(a, b) = d$, there exists x and y so that $d = ax + by$ is a standard consequence of the Euclidean division algorithm, and is called *Bézout's Lemma*—see Lemma 11.1.3.)

To find these x and y, the Euclidean division algorithm yields

$$
\begin{aligned}
43 &= 8(5) + 3 \\
5 &= 1(3) + 2 \\
3 &= 1(2) + 1,
\end{aligned}
$$

and undoing this sequence, beginning at the bottom,

$$
\begin{aligned}
1 &= 3 - 2 \\
&= 3 - (5 - 3) \\
&= 2(3) - 5 \\
&= 2(43 - 8(5)) - 5 \\
&= 2(43) - 17(5).
\end{aligned}
$$

Hence, $1 \equiv 2(43) - 17(5) \equiv -17(5) \equiv 26(5) \pmod{43}$ shows that $5^{-1} = 26$. This establishes the existence of inverses modulo 43, however calculating them is a different story. In practice, one might write a computer program to do this, especially when one requires all inverses modulo some large number. (Actually, one can cut the calculations in half, since $(-x)^{-1} = -x^{-1}$.)

Since

$$
s_{n+1} = \frac{t_n + s_n^2}{n + 1}
$$

write $s_{n+1} = (t_n + s_n^2)(n+1)^{-1}$. Then calculating modulo 43,

$$s_n = ([t_{n-1} + s_{n-1}^2)n^{-1} = t_n n^{-1}$$

helps to extend the table above without ever actually having to find s_n for larger n. Remember, the values are calculated as if each previous s_n is an integer; they would be meaningless if, say, s_{30} was not an integer (in which case this would answer the question anyway). The results are tabulated in the next chart; these should be checked by a computer program, since calculations were performed individually (checking by hand took a few hours!). All values (except n) are calculated modulo 43:

n	s_{n-1}	s_{n-1}^2	t_n	n^{-1}	s_n	n	s_{n-1}	s_{n-1}^2	t_n	n^{-1}	s_n
0					1	22	2	4	3	2	6
1	1	1	2	1	2	23	6	36	39	15	26
2	2	4	6	22	3	24	26	31	27	9	28
3	3	9	15	29	5	25	28	10	37	31	29
4	5	25	40	11	10	26	29	24	18	5	4
5	10	14	11	26	28	27	4	16	34	8	14
6	28	10	21	36	25	28	14	24	15	20	42
7	25	23	1	37	37	29	43	1	16	3	15
8	37	36	37	27	10	30	15	10	26	33	41
9	10	14	8	24	20	31	37	36	18	25	20
10	20	13	21	13	15	32	20	13	31	39	5
11	15	10	31	4	38	33	5	25	13	30	3
12	38	25	13	18	19	34	3	9	22	19	31
13	19	17	30	10	42	35	31	15	37	16	33
14	42	1	31	40	36	36	33	14	8	6	5
15	36	6	37	23	34	37	5	25	33	7	16
16	34	38	32	35	2	38	16	41	31	17	11
17	2	4	36	38	35	39	11	35	22	32	16
18	35	21	14	12	39	40	16	41	19	14	8
19	39	16	30	34	31	41	8	23	42	21	22
20	31	15	2	28	13	42	22	11	10	42	33
21	13	40	42	41	2	43	33	14	24		

So, if each of $s_0, s_1, \ldots, s_{42}$ is an integer, (and assuming that calculations are correct) then the numerator of s_{43} is congruent to 24 modulo 43, but the denominator of s_{43} is 43, hence s_{43} is not an integer. □

26.3 Solutions: Sequences with only primes?

Exercise 21: (Last digit in Fermat numbers) Let $S(t)$ be the statement "The last digit of F_t is a 7."

BASE STEP $(t = 2)$: $F_2 = 2^{2^2} + 1 = 2^4 + 1 = 16 + 1 = 17$, which shows $F(2)$ is true.

INDUCTIVE STEP: Suppose that for some $k \geq 2$, $S(k)$ is true, that is, the last digit of F_k is indeed 7, say $F_k = 10n + 7$ for some $n \geq 1$. Then

$$
\begin{aligned}
F_{k+1} &= 2^{2k+1} + 1 \\
&= (2^{2^k})^2 + 1 \\
&= (F_k - 1)^2 + 1 \\
&= (10n + 7 - 1)^2 + 1 \qquad \text{(by induction hypothesis)} \\
&= 100n^2 + 120n + 36 + 1,
\end{aligned}
$$

a number ending in 7. This shows that $S(k+1)$ is true and therefore completes the inductive step $S(k) \to S(k+1)$.

Hence, by mathematical induction, for all $t \geq 2$, the statement $S(t)$ is true. $\square$

Exercise 22: This proof is also found in [7]. For $n = 0, 1, 2, \ldots$, let $A(n)$ be the assertion that

$$
F_n = \left(\prod_{i=0}^{n-1} F_i \right) + 2.
$$

BASE STEP: Considering an empty product to be 1, $F_0 = 3 = 1 + 2$, and so $A(0)$ is true. To be sure, however, also check that since $F_1 = 2^2 + 1 = 5 = 3 + 2 = F_0 + 2$, $A(1)$ also holds.

INDUCTION STEP: Fix $k \geq 1$, and suppose that $A(k)$ holds. It remains to show

$$
A(k+1): \qquad F_{k+1} = \left(\prod_{i=0}^{k} F_i \right) + 2.
$$

Beginning with the left side of $A(k+1)$,

$$
\begin{aligned}
F_{k+1} &= 2^{2^{k+1}} + 1 \\
&= 2^{2^k} \cdot 2^{2^k} + 1 \\
&= (F_k - 1)(F_k - 1) + 1 \\
&= F_k(F_k - 1) - F_k + 2 \\
&= F_k\left(\prod_{i=0}^{k-1} F_i + 2 - 1 \right) - F_k + 2 \qquad \text{(by $A(k)$)}
\end{aligned}
$$

$$= \prod_{i=0}^{k} F_i + F_k - F_k + 2$$

$$= \prod_{i=0}^{k} F_i + 2,$$

which finishes the proof of $A(k+1)$ and hence the inductive step.

By MI, for each $n \geq 0$, the expression $A(n)$ holds. $\qquad\square$

Exercise 23: It turns out that the first seven terms are prime numbers, however the eighth is 17×19607843. By empirical induction it might be easy to conclude that all such numbers are prime. $\qquad\square$

Exercise 24: Observe that $f(40) = 1641 = 3 \cdot 547$. Perhaps even easier is to examine $f(41)$; in fact, for any n that is a multiple of 41, $n^2 + n + 41$ is easily seen to be not prime. $\qquad\square$

The reader might check that $f(-40), f(-39), \ldots, f(39)$ are all prime. According to [150], it is not known whether or not $n^2 + n + 41$ is prime for infinitely many values of n.

Two other popular "prime producing polynomials" are $p(n) = n^2 + n + 17$, which gives primes for $n = 0, 1, \ldots, 16$, and $q(n) = 2n^2 + 29$, which gives primes up to $n = 28$. Another, found in [205, p. 38] is $n^2 - 79 + 1601$, which gives primes with n all the way up to 79 but fails at 80.

26.4   Solutions: Divisibility

Exercise 25: For the prime $p = 1093$, $(1093)^2$ divides $2^{1092} - 1$. $\qquad\square$

26.5   Solutions: Never a square?

Exercise 26: According to [519, p.69], the first time that $f(n)$ is a perfect square is when
$$n = 12055735790331359447442538767.$$

26.6   Solutions: Goldbach's conjecture

Exercise 27: The first number for which this is not true is 127. $\qquad\square$

26.7   Solutions: Cutting the cake

Exercise 28: The next number in the sequence is 31.   □

Exercise 29: This appears as Problem 36 in [247] and Example 5 of [245], and also as sequence # 427 in [497]. First compare Exercise 725. It might be helpful to observe that

$$\frac{n^4 - 6n^3 + 24n^2 - 18n + 24}{24} =$$
$$\binom{n-1}{0} + \binom{n-1}{1} + \binom{n-1}{2} + \binom{n-1}{3} + \binom{n-1}{4}.$$

26.8   Solutions: Sums of hex numbers

Exercise 30: At each step, one adds a surrounding ring of pennies; the number of pennies in each ring increases by 6 each time (one extra for each "corner" of the hexagon) and so the n-th hex number is

$$h_n = 1 + 6 + 12 + 18 + \cdots + 6(n-1).$$

This formula can be easily be proved by induction if one is not convinced by the above discussion.

Factoring out the 6 and using the formula for the sum of the first $n-1$ positive integers gives,

$$
\begin{aligned}
h_n &= 1 + 6 + 12 + 18 + \cdots + 6(n-1) \\
&= 1 + 6(1 + 2 + 3 + \cdots + (n-1)) \\
&= 1 + 6\left(\frac{(n-1)n}{2}\right) \quad \text{(by Thm. 1.6.1),} \\
&= 1 + 3(n-1)n \\
&= 3n^2 - 3n + 1 \\
&= n^3 - [n^3 - 3n^2 + 3n - 1] \\
&= n^3 - (n-1)^3,
\end{aligned}
$$

a difference of adjacent cubes. Knowing this formula for h_n, one can prove inductively that $h_1 + h_2 + \cdots + h_n = n^3$ or simply observe that the sum telescopes (neighboring terms cancel; see Exercise 137):

$$h_1 + h_2 + \cdots + h_n = [1^3 - 0^3] + [2^3 - 1^3] + \ldots + [n^3 - (n-1)^3] = n^3.$$

□

Chapter 27

Solutions: Identities

27.1 Solutions: Arithmetic progressions

Exercise 32: Let $S(n)$ be the statement "$O_n + 2 = O_{n+1}$." This statement was called *Proposition IV* by Maurolycus; in fact, it read "The odd numbers are obtained from unity by successive additions of 2."

Define an *odd number* as a natural number which, upon division by 2 has a remainder 1. To prove $S(n)$ for any specific $n \geq 1$, induction is not necessary: If O_{n+1} is merely defined by being the next number after O_n that is odd, then since O_n is the n-th odd number, it has remainder 1 upon division by 2. The next natural number is $O_n + 1$, which will have remainder 0 upon division by 2, so it is not the next odd number. The next number is $O_n + 1 + 1$, which will have remainder 1 upon division by 2, so it is odd, indeed the next odd number after O_n. Hence $O_n + 2 = O_{n+1}$, and so $S(n)$ is true.

If one were to try and transform the above idea into an inductive proof, the truth of $S(k)$ is not needed for the proof of $S(k+1)$, in fact, it seems difficult to find a way to prove $S(k+1)$ based on the truth of $S(k)$. So let us try and interpret this exercise in another way.

Starting with the first odd number $1 = O_1$, $S(1)$ says that the way to find the next odd number is to add 2. To find the next odd number, $S(2)$ says to add 2 more. In general, $S(n)$ says that the $(n+1)$ odd number is found by beginning with 1 and adding n 2's. So one could translate $S(n)$ into the statement

$$S^*(n): \quad O_{n+1} = 1 + 2n,$$

a statement that seems to be true for $n \geq 0$. Proving $S^*(n)$ is easy by induction:

BASE STEP $(n = 0)$: The first odd number is $O_1 = 1$ which is equal to $1 + 2(0)$, proving $S(0)$.

INDUCTIVE STEP: Assume that for some $k \geq 0$, $S^*(k)$ is true, that is, $S(k+1) =$

$1 + 2k$. To be proved is

$$S^*(k+1): \quad O_{k+2} = 1 + 2(k+1).$$

The number O_{k+1} is odd (by definition, it is the $(k+1)$-th odd number) and so it has remainder 1 upon division by 2. The next natural number is $O_{k+1} + 1$, which will have remainder 0 upon division by 2, so it is not the next odd number. The next number is $O_{k+1} + 1 + 1$, which will have remainder 1 upon division by 2, so it is odd, indeed the next odd number after O_{k+1}. Hence $O_{k+1} + 2 = O_{k+2}$, and so by induction hypothesis,

$$O_{k+1} + 2 = (1 + 2k) + 2 = 1 + 2(k+1),$$

and so $S^*(k+1)$ is true.

Therefore, by mathematical induction, for all $n \geq 0$, the statement $S^*(n)$ is true. $\qquad\square$

Note: Above, $S^*(n)$ and $S^*(n+1)$ together imply the truth of $S(n)$, so one can now freely use $S(n)$ in any subsequent proofs.

Exercise 33: For interest's sake, here is the translation of Maurolycus's proof as reproduced in [91]. [Note: in this quotation, "collateral" numbers are numbers in the same row of some table whose columns were the natural numbers n, the even numbers $2(n-1)$, the odd numbers $2n-1$, the triangular numbers $\frac{n(n-1)}{2}$ (see Section 1.6), the squares n^2, and *numerus parte altera longior* numbers $n(n-1)$.]

> The integer 2 added to unity makes the integer 3 but when added to 3 it makes an amount greater by 2 and this (by virtue of *Proposition IV*) is the next odd integer, namely 5. Again since the integer 3 added to 2 makes 5, which is the collateral odd integer, when it is added to 4 the result will be greater by 2, that is (by virtue of *Proposition IV*) it will be the next odd integer which is 7. And in like manner to infinity as the proposition states.

In modern parlance, one might give the solution of this exercise as follows: For each $n \geq 1$, let $Q(n)$ be the statement "$n + (n-1) = O_n$."

BASE STEP: $1 + (1-1) = 1$, which is O_1, so $Q(1)$ holds.

INDUCTIVE STEP: For some fixed $k \geq 1$, assume that $Q(k) : k + (k-1) = O_k$ is true; to show is

$$Q(k+1): \quad (k+1) + k = O_{k+1}.$$

Starting with the left side,

$$(k+1) + k = k + (k-1) + 2$$

$$= O_k + 2 \qquad \qquad \text{(by } Q(k)\text{)}$$
$$= O_{k+1} \qquad \qquad \text{(by Exercise 32)},$$

which is the right side of $Q(k+1)$. This completes the inductive step.

Therefore, conclude by the principle of mathematical induction that for all $n \geq 1$, the statement $Q(n)$ is true. $\qquad\square$

Exercise 34: Duplicate the proof of Exercise 32.

Exercise 35: This was already proved as Theorem 1.6.1.

Exercise 36: Induction is not really required, since it is also provable by direct application of Theorem 1.6.1:

$$\sum_{i=m+1}^{n} i = \sum_{i=1}^{n} i - \sum_{i=1}^{m} i$$
$$= \frac{n(n+1)}{2} - \frac{m(m+1)}{2} \qquad \text{(by Thm 1.6.1)}$$
$$= \frac{n^2 + n - m^2 - m}{2}$$
$$= \frac{(n-m)(n+m+1)}{2}.$$

The reader is invited to provide a strictly inductive proof. It may be interesting to note that this exercise implies that any sum of consecutive integers can never be a power of 2 (see [583]).

Exercise 37: By Exercise 36 with m replaced by $n^2 - 1$ and n replaced with $n^2 + n$,

$$n^2 + (n^2 + 1) + (n^2 + 2) + \cdots + (n^2 + n) = \frac{(n^2 + n - (n^2 - 1))(n^2 + n + n^2)}{2}$$
$$= \frac{(n+1)(2n^2 + n)}{2}$$
$$= (2n+1)T_n.$$

The equality

$$n^2 + (n^2 + 1) + (n^2 + 2) + \cdots + (n^2 + n) = (n^2 + n + 1) + \cdots + (n^2 + 2n)$$

is simple algebra. $\qquad\square$

Both equalities also have fairly simple inductive proofs, which are left to the reader.

Exercise 38: For $n \geq 1$, denote the statement

$$S(n): \quad 1 + 3 + 5 + \cdots + (2n - 1) = n^2.$$

BASE STEP: $S(1)$ says $1 = 1^2$, clearly true.

INDUCTION STEP: Let some $k \geq 1$ be fixed, and suppose that

$$S(k): \quad 1 + 3 + 5 + \ldots + (2k - 1) = k^2$$

holds. It remains to show that

$$S(k + 1): \quad 1 + 3 + 5 + \ldots + (2k - 1) + (2k + 1) = (k + 1)^2$$

follows. Beginning with the left-hand side of $S(k + 1)$,

$$1 + 3 + 5 + \cdots + (2k - 1) + (2k + 1) = k^2 + (2k + 1) \qquad \text{(by } S(k)\text{)},$$
$$= (k + 1)^2,$$

the right-hand side of $S(k+1)$. This completes the inductive step $S(k) \to S(k+1)$.

Hence, by mathematical induction, $S(n)$ holds for all $n \geq 1$. $\qquad\square$

Comment: The above proof was due to Maurocylus (*Proposition XV*). He first proved a lemma (called *Proposition XIII*) that said "every square number plus the following odd number equals the following square number"—in other words, $n^2 + (2n + 1) = (n + 1)^2$. According to Bussey [91], the translated version was[1]:

> By a previous proposition the first square number (unity) added to the following odd number (3) makes the following square number (4); and this second square number (4) added to the 3d odd number (5) makes the 3d square number (9); and likewise the 3d square number (9) added to the 4th odd number (7) makes the 4th square number; and so successively to infinity the proposition is demonstrated by the repeated application of *Proposition XIII*.

Bussey says "This is a clear case of complete induction proof" and seems to indicate that this might be the first real example of an inductive proof (given in the year 1575). Bussey then goes on to say that "In modern symbols the proof would be this:"

> *1st.* The theorem is true when $n = 1$. *2d.* Assume that it is true when $n = k$, *i. e.*, assume $O_1 + O_2 + \cdots + O_k = S_k$; add O_{k+1} to both sides of this equation and get $O_1 + O_2 + \cdots + O_{k+1} = S_k + O_{k+1}$ which equals S_{k+1} by *Proposition XIII*.

Bussey's interpretation of the proof relies on adding something to each side of the statement $S(k)$; notice that in the proof given above, this was not necessary. In general, it may be poor practice to add something to each side of an equation to obtain a desired equation when the desired result can be obtained by a direct sequence of equalities just as easily. See Section 7.1 for more comments on this.

Exercise 39: For $n \geq 1$, let $S(n)$ be the statement

$$S(n): \quad 2 + 4 + 6 + \cdots + 2n = n(n+1).$$

BASE STEP $(n = 1)$: The statement $S(1)$ says $2 = 1 \cdot (1+1)$, clearly a true statement.

INDUCTIVE STEP: Fix some $k \geq 1$ and suppose that $S(k)$ holds; that is, the inductive hypothesis is

$$S(k): \quad 2 + 4 + 6 + \cdots + 2k = k(k+1).$$

To be shown is

$$S(k+1): \quad 2 + 4 + 6 + \cdots + 2k + 2(k+1) = (k+1)(k+2)$$

holds. Starting with the left-hand side of $S(k+1)$,

$$
\begin{aligned}
2 + 4 + 6 + \cdots + 2k + 2(k+1) &= k(k+1) + 2(k+1) \quad \text{(by ind. hyp.)}, \\
&= (k+1)(k+2),
\end{aligned}
$$

the right-hand side of $S(k+1)$, completing the inductive step $S(k) \rightarrow S(k+1)$.

By the principle of mathematical induction, conclude that for all $n \geq 1$, the statement $S(n)$ holds. $\qquad \square$

Exercise 40: For $n \geq 1$, let $S(n)$ be the statement

$$S(n): \quad 2 + 5 + 8 + \cdots + (3n - 1) = \frac{n(3n+1)}{2}.$$

BASE STEP $(n = 1)$: $S(1)$ says $2 = \frac{1(3 \cdot 1 + 1)}{2}$, a true statement.

INDUCTION STEP $(S(k) \rightarrow S(k+1))$: Fix some $k \geq 1$ and let the induction hypothesis

$$S(k): \quad 2 + 5 + 8 + \cdots + (3k - 1) = \frac{k(3k+1)}{2}$$

be assumed to be true. It is yet to be proved that

$$S(k+1): \quad 2 + 5 + 8 + \cdots + (3k - 1) + (3(k+1) - 1) = \frac{(k+1)(3(k+1) + 1)}{2}$$

follows. It might help to simplify $S(k+1)$ a bit first so that it is easier to see what is needed to prove:

$$S(k+1): \quad 2+5+8+\cdots+(3k-1)+(3k+2) = \frac{(k+1)(3k+4)}{2}.$$

Starting with the left side of $S(k+1)$,

$$
\begin{aligned}
2+5+\cdots+(3k-1)+(3k+2) &= \frac{k(3k+1)}{2} + 3k + 2 \qquad \text{(by ind. hyp.)} \\
&= \frac{3k^2+k}{2} + \frac{6k+4}{2} \\
&= \frac{3k^2+7k+4}{2} \\
&= \frac{(k+1)(3k+4)}{2},
\end{aligned}
$$

which agrees with the right side of $S(k+1)$. This completes the inductive step.

Consequently, by the principle of mathematical induction, for all $n \geq 1$, the statement $S(n)$ is true. $\square$

Note: Observe how much easier the proof was made by cleaning up the expression in advance that needed to be derived at the end of the inductive step. Without this preliminary calculation, one must transform $3k^2+7k+4$ into $(k+1)(3(k+1)+1)$, a slightly clumsy calculation.

Exercise 41: For $n \geq 1$, let $S(n)$ be the statement

$$S(n): \quad 3+11+19+\cdots+(8n-5) = 4n^2 - n.$$

BASE STEP ($n=1$): $S(1)$ says $3 = 4 \cdot 1^2 - 1$, which is true.

INDUCTIVE STEP ($S(k) \rightarrow S(k+1)$): Let some $k \geq 1$ be fixed, and suppose that

$$S(k): \quad 3+11+19+\cdots+(8k-5) = 4k^2 - k$$

is true. The next identity to be shown is

$$S(k+1): \quad 3+11+19+\cdots+(8k-5)+(8(k+1)-5) = 4(k+1)^2 - (k+1).$$

Beginning with the left side of $S(k+1)$,

$$
\begin{aligned}
3+11+\cdots &+(8k-5)+(8(k+1)-5) \\
&= 4k^2 - k + (8(k+1)-5) \qquad \text{(by } S(k)) \\
&= 4k^2 - k + 8k + 3 \\
&= 4k^2 + 8k + 4 - k - 1
\end{aligned}
$$

$$= \quad 4(k+1)^2 - (k+1),$$

which is the right side of $S(k+1)$. [*Note:* The second last expression above was determined by looking at the last line and regrouping accordingly—it was not a natural step that one should see without "looking ahead".] This completes the inductive step $S(k) \to S(k+1)$.

By the principle of mathematical induction, for all $n \geq 1$, the statement $S(n)$ is true. $\qquad\square$

Exercise 42: For $n \geq 1$, let $S(n)$ be the statement

$$S(n): \quad \sum_{i=1}^{n}(3i-1) = \left(\sum_{i=1}^{n} i\right) + n^2.$$

BASE STEP ($n = 1$): $S(1)$ says $3 \cdot 1 - 1 = 1 + 1^2$, which is true.

INDUCTIVE STEP ($S(k) \to S(k+1)$): Fix some $k \geq 1$ and suppose that

$$S(k): \quad \sum_{i=1}^{k}(3i-1) = \left(\sum_{i=1}^{k} i\right) + k^2$$

is true. Yet to be proved is

$$S(k+1): \quad \sum_{i=1}^{k+1}(3i-1) = \left(\sum_{i=1}^{k+1} i\right) + (k+1)^2.$$

Beginning with the left side of $S(k+1)$,

$$
\begin{aligned}
\sum_{i=1}^{k+1}(3i-1) &= \sum_{i=1}^{k}(3i-1) + 3(k+1) - 1 \\[2mm]
&= \left(\sum_{i=1}^{k} i\right) + k^2 + 3(k+1) - 1 \quad \text{(by } S(k)) \\[2mm]
&= \left(\sum_{i=1}^{k} i\right) + k^2 + 3k + 2 \\[2mm]
&= \left(\sum_{i=1}^{k} i\right) + (k+1) + k^2 + 2k + 1 \\[2mm]
&= \left(\sum_{i=1}^{k+1} i\right) + k^2 + 2k + 1 \\[2mm]
&= \left(\sum_{i=1}^{k+1} i\right) + (k+1)^2,
\end{aligned}
$$

the right-hand side of $S(k+1)$. This completes the inductive step.

Thus, by the principle of mathematical induction, for all $n \geq 1$, $S(n)$ holds. $\square$

Exercise 43: For $n \geq 1$, let $S(n)$ be the statement

$$5 + 9 + 13 + \cdots + 4n + 1 = n(2n + 3).$$

BASE STEP $(n = 1)$: $S(1)$ says $5 = 1(1 \cdot 2 + 3)$, which is true.

INDUCTIVE STEP $(S(k) \rightarrow S(k+1))$: Fix some $k \geq 1$, and assume that

$$S(k): \quad 5 + 9 + 13 + \cdots + (4k + 1) = k(2k + 3)$$

is true. To complete the inductive step, it suffices to prove

$$S(k+1): \quad 5 + 9 + 13 + \cdots + (4k + 1) + 4(k + 1) + 1 = (k+1)(2(k+1) + 3).$$

Beginning with the left-hand side of $S(k+1)$,

$$
\begin{aligned}
5 + 9 + \cdots + 4k + 1 + 4(k+1) + 1 &= k(2k+3) + 4(k+1) + 1 \text{ (by } S(k)) \\
&= 2k^2 + 3k + 4k + 5 \\
&= 2k^2 + 7k + 5 \\
&= (k+1)(2k+5),
\end{aligned}
$$

which agrees with the right-hand side of $S(k+1)$, completing the inductive step.

Hence, by the principle of mathematical induction, for all $n \geq 1$, $S(n)$ holds. $\square$

Exercise 44: For $n \geq 1$, let $S(n)$ be the statement

$$S(n): \quad (2n + 1) + (2n + 3) + (2n + 5) + \ldots + (4n - 1) = 3n^2.$$

It is easier to see what is going on $S(n)$ is rewritten as

$$S(n): \quad (2n + 1) + (2n + 3) + (2n + 5) + \ldots + (2n + 2n - 3) + (2n + 2n - 1) = 3n^2.$$

BASE STEP $(n = 1)$: The statement $S(1)$ says $2 \cdot 1 + 1 = 4 \cdot 1 - 3$, which is true.

INDUCTION STEP: Fix some $k \geq 1$, and assume that

$$S(k): \quad (2k + 1) + (2k + 3) + (2k + 5) + \cdots + (2k + 2k - 3) + (2k + 2k - 1) = 3k^2$$

is true. To be shown is

$$S(k+1): \quad (2(k+1) + 1) + (2(k+1) + 3) + \cdots + (2(k+1) + 2(k+1) - 1) = 3(k+1)^2.$$

Beginning with the left-hand side of $S(k+1)$,

$$(2(k+1) + 1) + \cdots + (2(k+1) + 2(k+1) - 3) + (2(k+1) + 2(k+1) - 1)$$

$$
\begin{aligned}
&= (2k+1) + 2 + \cdots + (2k + 2k - 1) + 2 + (2k + 2k + 1) + 2 \\
&= (2k+1) + (2k+3) + \cdots + (2k + 2k - 1) + 2k + (4k+3) \\
&= 3k^2 + 6k + 3 \quad \text{(by } S(k)\text{)} \\
&= 3(k^2 + 2k + 1)
\end{aligned}
$$

which is the right side of $S(k+1)$, completing the inductive step.

Hence, by mathematical induction, for all $n \geq 1$, $S(n)$ is true.   □

Exercise 45: Let a and d be fixed real numbers. For any natural number $n \geq 1$, let $S(n)$ be the statement

$$
S(n): \quad a + (a + d) + (a + 2d) + \cdots + (a + (n-1)d) = \frac{n}{2}[2a + (n-1)d].
$$

BASE STEP: The case $n = 1$ says $a = \frac{1}{2}[2a + (1-1)d]$, which holds.

INDUCTION STEP: Let $k \geq 1$ be fixed and suppose that

$$
S(k): \quad a + (a + d) + (a + 2d) + \cdots + (a + (k-1)d) = \frac{k}{2}[2a + (k-1)d]
$$

holds. To show that

$$
S(k+1): \quad a + (a+d) + (a+2d) + \cdots + (a+kd) = \frac{k+1}{2}[2a + kd]
$$

follows, starting with the left side of $S(k+1)$,

$$
\begin{aligned}
&a + (a+d) + (a+2d) + \cdots + (a + (k-1)d) + (a + kd) \\
&= \frac{k}{2}[2a + (k-1)d] + (a + kd) \quad \text{(by } S(k)\text{)} \\
&= \frac{1}{2}[k(2a) + k^2 d - kd] + \frac{2a + 2kd}{2} \\
&= \frac{1}{2}[k(2a) + 2a + k^2 d + kd] \\
&= \frac{1}{2}[(k+1)2a + (k+1)kd] \\
&= \frac{k+1}{2}[2a + kd],
\end{aligned}
$$

precisely the right-hand side of $S(k+1)$, completing the inductive step.

Thus, by mathematical induction, for each $n \geq 1$, $S(n)$ holds.   □

Note: A direct proof is also available by rearranging terms, factoring, and using Theorem 1.6.1 as follows:

$$
a + (a + d) + \cdots + (a + (n-1)d) \quad = \quad na + [1 + 2 + \cdots + (n-1)]d
$$

$$= na + \frac{(n-1)n}{2}d \quad \text{(by Thm. 1.6.1)}$$

$$= \frac{n}{2}[a + (n-1)d].$$

□

Exercise 46: The result in this exercise is obtained by simple algebra and Theorem 1.6.1:

$$T_n^2 - T_{n-1}^2 = \left[\frac{n(n+1)}{2}\right]^2 - \left[\frac{(n-1)n}{2}\right]^2$$

$$= \frac{n^2(n+1)^2 - (n-1)^2 n^2}{4}$$

$$= \frac{n^2[n^2 + n + 1 - (n^2 - 2n + 1)]}{4}$$

$$= \frac{n^2(4n)}{4},$$

completes the proof. An inductive proof is also available: For each $n \geq 2$, let $S(n)$ denote the statement

$$T_n^2 - T_{n-1}^2 = n^3.$$

BASE CASE: $S(2)$ says $T_2^2 - T_1^2 = 9 - 1 = 2^3$, so $S(2)$ holds.

INDUCTIVE STEP: Let $k \geq 2$ and suppose that

$$S(k): \quad T_k^2 - T_{k-1}^2 = k^3$$

holds. It remains to show that

$$S(k+1): \quad T_{k+1}^2 - T_k^2 = (k+1)^3$$

follows. Starting with the LHS of $S(k+1)$,

$$T_{k+1}^2 - T_k^2 = (T_k + (k+1))^2 - (T_{k-1} + k)^2$$
$$= T_k^2 + 2T_k(k+1) + (k+1)^2 - T_{k-1}^2 - 2T_{k-1}k + k^2$$
$$= T_k^2 - T_{k-1}^2 + 2T_k(k+1) - 2T_{k-1}k + 2k + 1$$
$$= k^3 + 2T_k(k+1) - 2T_{k-1}k + 2k + 1 \qquad \text{(by } S(k))$$
$$= k^3 + k(k+1)(k+1) - (k-1)k \cdot k + 2k + 1 \qquad \text{(by Thm 1.6.1)}$$
$$= k^3 + k[k^2 + 2k + 1 - k^2 + k] + 2k + 1$$
$$= k^3 + 3k^2 + 3k + 1,$$

which is $(k+1)^3$, the RHS of $S(k+1)$, completing the inductive step.

By MI, for each $n \geq 2$, $S(n)$ is true.

□

Exercise 47: See the solution to Exercise 49 when $r = 2$. $\square$

Exercise 48: Again, this is the special case $r = 3$ in Exercise 49. $\square$

Exercise 49: Fix $r \in \mathbb{R}$, $r \neq 1$, and for $n \geq 1$, let $S(n)$ be the statement

$$S(n): \quad 1 + r + r^2 + \cdots + r^n = \frac{r^{n+1} - 1}{r - 1}.$$

To solve the problem, it suffices to prove that for each $n \geq 1$, $S(n)$ holds, that is, it suffices to consider only $a = 1$. For $r = 0$, the statement $S(n)$ becomes $1 = 1$, so assume without loss of generality that $r \neq 0$. (This assumption might not be needed, but sometimes simple initial observations save headaches later.)

BASE STEP ($n = 1$): The statement $S(1)$ says $1 + r = \frac{r^2-1}{r-1}$ which is true since $r^2 - 1 = (r+1)(r-1)$ and $r \neq 1$. [Note, one could actually begin the induction at $n = 0$ since $S(0)$ says $1 = \frac{r-1}{r-1}$.]

INDUCTIVE STEP ($S(k) \to S(k+1)$): Fix some $k \geq 1$ and suppose that $S(k)$ is true, that is, assume

$$S(k): \quad 1 + r + r^2 + \cdots + r^n = \frac{r^{k+1} - 1}{r - 1} \quad \text{(inductive hypothesis)}$$

is true. To complete the inductive step, one needs to show that

$$S(k+1): \quad 1 + r + r^2 + \cdots + r^k + r^{k+1} = \frac{r^{k+2} - 1}{r - 1}$$

is also true. Beginning with the left side of $S(k+1)$,

$$\underbrace{1 + r + r^2 + \cdots + r^k}_{} + r^{k+1} = \underbrace{\frac{r^{k+1} - 1}{r - 1}}_{} + r^{k+1} \quad \text{(by ind. hyp.)}$$

$$= \frac{r^{k+1} - 1}{r - 1} + \frac{r^{k+1}(r - 1)}{r - 1}$$

$$= \frac{r^{k+1} - 1 + r^{k+1}(r - 1)}{r - 1}$$

$$= \frac{r^{k+1} - 1 + r^{k+2} - r^{k+1}}{r - 1}$$

$$= \frac{r^{k+2} - 1}{r - 1},$$

which is the righthand side of $S(k+1)$. So $S(k) \to S(k+1)$, completing the inductive step.

Hence, by the principle of MI, for all $n \geq 1$, the statement $S(n)$ holds. $\square$

Exercise 50: For $n \geq 1$, let $S(n)$ be the statement

$$1 \cdot 2^1 + 2 \cdot 2^2 + 3 \cdot 2^3 + \cdots + n \cdot 2^n = 2 + (n-1)2^{n+1}.$$

BASE STEP ($n = 1$): $S(1)$ says $1 \cdot 2^1 = 2 + (1-1)2^2$, which is true.

INDUCTIVE STEP ($S(k) \to S(k+1)$): Fix some $k \geq 1$ and suppose that

$$S(k): \quad 1 \cdot 2^1 + 2 \cdot 2^2 + \cdots + k \cdot 2^k = 2 + (k-1)2^{k+1}$$

is true. To be shown is

$$S(k+1): \quad 1 \cdot 2^1 + 2 \cdot 2^2 + \cdots + k \cdot 2^k + (k+1)2^{k+1} = 2 + (k)2^{k+2}.$$

The left side of $S(k+1)$ is

$$
\begin{aligned}
1 \cdot 2^1 + 2 \cdot 2^2 &+ \cdots + k \cdot 2^k + (k+1)2^{k+1} \\
&= 2 + (k-1)2^{k+1} + (k+1)2^{k+1} \quad \text{(by } S(k)) \\
&= 2 + (k-1+k+1)2^{k+1} \\
&= 2 + (2k)2^{k+1} \\
&= 2 + (k)2^{k+2},
\end{aligned}
$$

which is the right side of $S(k+1)$, so $S(k+1)$ is also true. This completes the inductive step $S(k) \to S(k+1)$.

By mathematical induction, for all $n \geq 1$, $S(n)$ is true. $\square$

Exercise 51: For each $n \geq 1$, let $S(n)$ denote the statement

$$1 + 2 \cdot 2 + 3 \cdot 2^2 + \cdots + n2^{n-1} = (n-1)2^n + 1.$$

BASE STEP ($n = 1$): $1 = (1-1)2^0 + 1$, so $S(1)$ holds.

INDUCTIVE STEP: For some fixed $k \geq 1$, assume the inductive hypothesis $S(k)$

$$1 + 2 \cdot 2 + 3 \cdot 2^2 + \cdots + k2^{k-1} = (k-1)2^k + 1.$$

to be true. To be shown is that $S(k+1)$

$$1 + 2 \cdot 2 + 3 \cdot 2^2 + \cdots + k2^{k-1} + (k+1)2^k = (k)2^{k+1} + 1.$$

follows. Starting with the left side of $S(k+1)$,

$$
\begin{aligned}
1 + 2 \cdot 2 + 3 \cdot 2^2 &+ \cdots + k2^{k-1} + (k+1)2^k \\
&= (k-1)2^k + 1 + (k+1)2^k \quad \text{(by } S(k))
\end{aligned}
$$

$$
\begin{aligned}
&= (k - 1 + k + 1)2^k + 1 \\
&= 2k2^k + 1 \\
&= k2^{k+1} + 1,
\end{aligned}
$$

agreeing with the right side of $S(k + 1)$. This completes the proof of the inductive step $S(k) \to S(k + 1)$.

Therefore, by the principle of mathematical induction, for each $n \geq 1$, $S(n)$ is indeed true. $\qquad\square$

Exercise 54: For $n \geq 1$, let $S(n)$ be the statement

$$
S(n): \quad 1^2 + 2^2 + 3^2 + \cdots + n^2 = \frac{n(n + 1)(2n + 1)}{6}.
$$

BASE STEP $(n = 1)$: The statement $S(1)$ says $1^2 = 1(2)(3)/6$ which is okay.

INDUCTIVE STEP $(S(k) \to S(k + 1))$: Fix some $k \geq 1$ and suppose that

$$
S(k): \quad 1^2 + 2^2 + 3^2 + \cdots + k^2 = \frac{k(k + 1)(2k + 1)}{6}
$$

holds. Needed to be shown is that

$$
S(k + 1): \quad 1^2 + 2^2 + 3^2 + \cdots k^2 + (k + 1)^2 = \frac{(k + 1)(k + 2)(2(k + 1) + 1)}{6}
$$

follows. Starting with the left-hand side of $S(k + 1)$,

$$
\begin{aligned}
1^2 + 2^2 + 3^2 &+ \cdots + k^2 + (k + 1)^2 \\
&= \frac{k(k + 1)(2k + 1)}{6} + (k + 1)^2 \quad \text{(by ind. hyp.)}, \\
&= (k + 1) \left[\frac{k(2k + 1)}{6} + (k + 1) \right] \\
&= (k + 1) \frac{k(2k + 1) + 6(k + 1)}{6} \\
&= (k + 1) \frac{2k^2 + k + 6k + 6)}{6} \\
&= (k + 1) \frac{2k^2 + 7k + 6)}{6} \\
&= (k + 1) \frac{(k + 2)(2k + 3)}{6},
\end{aligned}
$$

which equals the right-hand side of $S(k + 1)$. This completes the inductive step.

By MI, for every $n \geq 1$, $S(n)$ is true. $\square$

Exercise 56: For any $n \geq 1$, let $S(n)$ be the statement

$$S(n): \quad 1^3 + 2^3 + 3^3 + \cdots + n^3 = \left[\frac{n(n+1)}{2}\right]^2.$$

BASE STEP $(n = 1)$: $S(1)$ says $1^3 = [\frac{(1)(2)}{2}]^2 = 1$, which holds.

INDUCTIVE STEP $(S(k) \to S(k+1))$: Fix some $k \geq 1$, and assume that

$$S(k): \quad 1^3 + 2^3 + 3^3 + \cdots + k^3 = \left[\frac{k(k+1)}{2}\right]^2$$

holds (the inductive hypothesis). To be shown is

$$S(k+1): \quad 1^3 + 2^3 + 3^3 + \cdots + k^3 + (k+1)^3 = \left[\frac{(k+1)(k+2)}{2}\right]^2.$$

Beginning with the left side of $S(k+1)$,

$$
\begin{aligned}
1^3 + 2^3 + \cdots + k^3 + (k+1)^3 &= \left[\frac{k(k+1)}{2}\right]^2 + (k+1)^3 \quad \text{(by ind. hyp.)}, \\
&= (k+1)^2 \left(\frac{k^2}{4} + (k+1)\right) \\
&= (k+1)^2 \frac{k^2 + 4(k+1)}{4} \\
&= (k+1)^2 \frac{(k+2)^2}{4},
\end{aligned}
$$

which is precisely the right-hand side of $S(k+1)$, completing the inductive step.

    Hence, by the principle of mathematical induction, for each $n \geq 1$, the statement $S(n)$ holds. $\square$

Exercise 57: For $n \geq 1$, denote the proposition in the exercise by

$$P(n): \quad 1^2 + 3^2 + 5^2 + \cdots + (2n-1)^2 = \frac{n(2n-1)(2n+1)}{3}.$$

BASE STEP $(n = 1)$: Since $1^2 = \frac{1 \cdot 1 \cdot 3}{3}$, the statement $P(1)$ holds.

INDUCTIVE STEP: For some fixed $k \geq 1$, assume the inductive hypothesis

$$P(k): \quad 1^2 + 3^2 + \cdots + (2k-1)^2 = \frac{k(2k-1)(2k+1)}{3}.$$

to be true. To show that

$$P(k+1): \quad 1^2 + 3^2 + \cdots + (2(k+1) - 1)^2 = \frac{(k+1)(2(k+1) - 1)(2(k+1) + 1)}{3}.$$

follows, begin with the left side of $P(k+1)$ (adding the second last term so that it is easy to see how to apply $P(k)$):

$$1^2 + 3^2 + 5^2 + \cdots + (2k - 1)^2 + (2k + 1)^2$$

$$= \frac{k(2k - 1)(2k + 1)}{3} + (2k + 1)^2 \quad \text{(by } P(k)\text{)}$$

$$= (2k + 1)\left[\frac{k(2k - 1)}{3} + (2k + 1)\right]$$

$$= (2k + 1)\frac{k(2k - 1) + 3(2k + 1)}{3}$$

$$= (2k + 1)\frac{2k^2 - k + 6k + 3}{3}$$

$$= (2k + 1)\frac{2k^2 + 5k + 3}{3}$$

$$= (2k + 1)\frac{(k + 1)(2k + 3)}{3}$$

$$= \frac{(k + 1)(2k + 1)(2k + 3)}{3},$$

which is the same as the right side of $P(k+1)$. This concludes the inductive step $P(k) \rightarrow P(k+1)$.

By the principle of mathematical induction, for each $n \geq 1$, $P(n)$ is true. $\qquad \square$

Exercise 58: For $n \geq 1$, denote the statement in the exercise by

$$S(n): \quad 2^2 + 4^2 + 6^2 + \cdots + (2n)^2 = \frac{2n(n + 1)(2n + 1)}{3}.$$

BASE STEP ($n = 1$): Since $2^2 = 4 = \frac{2(1+1)(2+1)}{3}$, the statement $S(1)$ holds.

INDUCTIVE STEP: For some fixed $k \geq 1$, assume the inductive hypothesis

$$S(k): \quad 2^2 + 4^2 + 6^2 + \cdots + (2k)^2 = \frac{2k(k + 1)(2k + 1)}{3}.$$

to be true. It remains to show that

$$S(k+1): \quad 2^2 + 4^2 + 6^2 + \cdots + (2(k + 1))^2 = \frac{2(k + 1)(k + 2)(2(k + 1) + 1)}{3}$$

follows from $S(k)$. Starting with the left side of $S(k+1)$, derive the right side:

$$2^2 + 4^2 + 6^2 + \cdots + (2k)^2 + (2(k+1))^2$$

$$= \frac{2k(k+1)(2k+1)}{3} + (2(k+1))^2 \quad \text{(by ind. hyp. } S(k))$$

$$= (k+1)\frac{2k(2k+1)}{3} + \frac{(3(4(k+1)^2)}{3}$$

$$= 2(k+1)\left[\frac{k(2k+1)}{3} + \frac{(3(2(k+1))}{3}\right]$$

$$= 2(k+1)\frac{2k^2 + k + 6k + 6}{3}$$

$$= 2(k+1)\frac{2k^2 + 7k + 6}{3}$$

$$= 2(k+1)\frac{(k+2)(2k+3)}{3},$$

which agrees with the right side of $S(k+1)$. This completes the inductive step $S(k) \to S(k+1)$.

Therefore, by the principle of mathematical induction, $S(n)$ is true for all $n \geq 1$. $\qquad\qquad\Box$

Exercise 59: For each $n \geq 1$, denote the statement in the exercise by

$$S(n): \quad 1 - 4 + 9 - 16 + \cdots + (-1)^{n+1}n^2 = (-1)^{n+1}(1 + 2 + 3 + \ldots + n).$$

BASE STEP ($n = 1$): The statement $S(1)$ says $1 = (-1)^{1+1}(1)$, which is true.

INDUCTIVE STEP: For some fixed $k \geq 1$, assume the inductive hypothesis $S(k)$ to be true. To show that $S(k+1)$ follows from $S(k)$, use two applications of Theorem 1.6.1, one with $n = k$ and another with $n = k+1$. The left side of $S(k+1)$ is

$$1 - 4 + 9 - 16 + \cdots + (-1)^{k+1}k^2 + (-1)^{k+2}(k+1)^2$$

$$= (-1)^{k+1}(1 + 2 + 3 + \ldots + k) + (-1)^{k+2}(k+1)^2 \qquad \text{(by } S(k))$$

$$= (-1)^{k+1}[(1 + 2 + 3 + \ldots + k) - (k+1)^2]$$

$$= (-1)^{k+1}\left[\frac{k(k+1)}{2} - (k+1)^2\right] \qquad\qquad \text{(by Thm 1.6.1)}$$

$$= (-1)^{k+1}\frac{k(k+1) - 2(k+1)^2}{2}$$

$$= (-1)^{k+1} \frac{k^2 + k - 2k^2 - 4k - 2}{2}$$

$$= (-1)^{k+1} \frac{-k^2 - 3k - 2}{2}$$

$$= (-1)^{k+2} \frac{k^2 + 3k + 2}{2}$$

$$= (-1)^{k+2} \frac{(k+1)(k+2)}{2}$$

$$= (-1)^{k+2}(1 + 2 + \cdots + k + (k+1)) \qquad \text{(by Thm 1.6.1),}$$

which is the right side of $S(k+1)$, completing the inductive step.

Therefore, by the principle of mathematical induction, for every $n \geq 1$, $S(n)$ is true. $\qquad\square$

Exercise 60: For each $n \geq 1$, let $S(n)$ denote the statement

$$n^2 - (n-1)^2 + (n-2)^2 + \cdots + (-1)^{n-1}(1)^2 = \frac{n(n+1)}{2}.$$

BASE STEP $(n = 1)$: $S(1)$ says $1^2 = \frac{1(1+1)}{2}$ which is true.

INDUCTIVE STEP: For some fixed $k \geq 1$, assume the inductive hypothesis

$$S(k): \quad k^2 - (k-1)^2 + (k-2)^2 + \cdots + (-1)^{k-1}(1)^2 = \frac{k(k+1)}{2}$$

to be true. To see that

$$S(k+1): \quad (k+1)^2 - k^2 + (k-1)^2 + \cdots + (-1)^k(1)^2 = \frac{(k+1)(k+2)}{2}$$

follows, start with the left-hand side of $S(k+1)$:

$$(k+1)^2 - k^2 + (k-1)^2 + \cdots + (-1)^k(1)^2$$
$$= (k+1)^2 - [k^2 - (k-1)^2 + (k-2)^2 + \cdots + (-1)^{k-1}(1)^2]$$
$$= (k+1)^2 - \frac{k(k+1)}{2} \qquad \text{(by } S(k))$$
$$= (k+1)\left[(k+1) - \frac{k}{2}\right]$$
$$= (k+1)\frac{2k+2-k}{2}$$
$$= (k+1)\frac{k+2}{2},$$

which is the right-hand side of $S(k + 1)$. This completes the inductive step.

Therefore, by the principle of mathematical induction, for all $n \geq 1$, the statement $S(n)$ is true. $\qquad\square$

Exercise 61: For each $n \geq 1$, denote the equality in the exercise by

$$E(n): \quad 1^3 + 3^3 + 5^3 + \cdots + (2n - 1)^3 = n^2(2n^2 - 1).$$

BASE STEP $(n = 1)$: Since $1^3 = (1)^2(2(1)^2 - 1)$, $E(1)$ is true.

INDUCTIVE STEP: For some fixed $k \geq 1$, assume the inductive hypothesis $E(k)$ to be true. To see that $E(k + 1)$ follows,

$$
\begin{aligned}
1^3 &+ 3^3 + 5^3 + \cdots + (2k - 1)^3 + (2(k + 1) - 1)^3 \\
&= \quad 1^3 + 3^3 + 5^3 + \cdots + (2k - 1)^3 + (2k + 1)^3 \\
&= \quad k^2(2k^2 - 1) + (2k + 1)^3 \quad \text{(by } E(k)) \\
&= \quad 2k^4 - k^2 + 8k^3 + 12k^2 + 6k + 1 \\
&= \quad 2k^4 + 4k^3 + k^2 + 4k^3 + 8k^2 + 2k + 2k^2 + 4k + 1 \\
&= \quad k^2(2k^2 + 4k + 1) + 2k(2k^2 + 4k + 1) + 2k^2 + 4k + 1 \\
&= \quad (k^2 + 2k + 1)(2k^2 + 4k + 1) \\
&= \quad (k + 1)^2(2(k + 1)^2 - 1),
\end{aligned}
$$

and so $1^3 + 3^3 + 5^3 + \cdots + (2k - 1)^3 + (2(k + 1) - 1)^3 = (k + 1)^2(2(k + 1)^2 - 1)$, which is $E(k + 1)$. This concludes the inductive step $E(k) \rightarrow E(k + 1)$.

By the principle of mathematical induction, for each $n \geq 1$, $E(n)$ is true. $\qquad\square$

Exercise 63: For each $n \geq 1$, define the statement

$$S(n): \quad 1 \cdot 2 + 2 \cdot 3 + \cdots + n(n + 1) = \frac{n(n + 1)(n + 2)}{3}.$$

BASE STEP: The statement $S(1)$ says $1 \cdot 2 = \frac{1(2)(3)}{3}$, which is true.

INDUCTION STEP: Let $k \geq 1$, and suppose that the inductive hypothesis

$$S(k): \quad 1 \cdot 2 + 2 \cdot 3 + \cdots + k(k + 1) = \frac{k(k + 1)(k + 2)}{3}$$

is true. Yet to be proved is

$$S(k + 1): \quad 1 \cdot 2 + 2 \cdot 3 + \cdots + k(k + 1) + (k + 1)(k + 2) = \frac{(k + 1)(k + 2)(k + 3)}{3}.$$

Beginning with the left side of $S(k + 1)$,

$$1 \cdot 2 + 2 \cdot 3 + \cdots + k(k + 1) + (k + 1)(k + 2)$$

$$= \frac{k(k+1)(k+2)}{3} + (k+1)(k+2) \quad \text{(by } S(k))$$

$$= (k+1)(k+2)\left(\frac{k}{3}+1\right)$$

$$= (k+1)(k+2)\frac{k+3}{3},$$

which is the right side of $S(k+1)$, completing the inductive step.

Hence, by the principle of mathematical induction, for every $n \geq 1$, $S(n)$ holds.

□

Exercise 64: For each $n \geq 1$, let $A(n)$ be the assertion

$$A(n): \quad 1\cdot3 + 2\cdot4 + \cdots + (n-1)(n+1) = \frac{(n-1)(n)(2n+5)}{6}.$$

BASE STEP $(n=2)$: $A(2)$ says $1\cdot3 = \frac{(1)(2)(9)}{6}$ which is true.

INDUCTIVE STEP: For some $k \geq 2$ assume that

$$A(k): \quad 1\cdot3 + 2\cdot4 + \cdots + (k-1)(k+1) = \frac{(k-1)(k)(2k+5)}{6}$$

holds. To see that

$$A(k+1): \quad 1\cdot3 + 2\cdot4 + \cdots + (k-1)(k+1) + k(k+2) = \frac{(k)(k+1)(2k+7)}{6}$$

holds, start with the left side:

$$1\cdot3 + 2\cdot4 + \cdots + (k-1)(k+1) + k(k+2)$$

$$= \frac{(k-1)(k)(2k+5)}{6} + k(k+2) \quad \text{(by ind. hyp.)}$$

$$= k\left[\frac{(k-1)(2k+5)}{6} + (k+2)\right]$$

$$= k\left[\frac{2k^2 + 3k - 5}{6} + \frac{6k+12}{6}\right]$$

$$= k\left[\frac{2k^2 + 9k + 7}{6}\right]$$

$$= \frac{k(k+1)(2k+7)}{6}$$

which is the right-hand side of $A(k+1)$. So $A(k+1)$ follows from $A(k)$, completing the inductive step.

Thus by mathematical induction, for each $n \geq 2$, $A(n)$ holds.   □

Exercise 65: For each $n \geq 1$, let $S(n)$ denote the statement

$$1 \cdot 2 \cdot 3 + 2 \cdot 3 \cdot 4 + 3 \cdot 4 \cdot 5 + \cdots + n(n+1)(n+2) = \frac{1}{4}n(n+1)(n+2)(n+3).$$

BASE STEP ($n = 1$): Since $1 \cdot 2 \cdot 3 = 6 = \frac{1}{4}1 \cdot 2 \cdot 3 \cdot 4$, the base case $S(1)$ is true.

INDUCTIVE STEP: For some fixed $k \geq 1$, assume the inductive hypothesis

$$S(k): \quad 1 \cdot 2 \cdot 3 + 2 \cdot 3 \cdot 4 + \cdots + k(k+1)(k+2) = \frac{1}{4}k(k+1)(k+2)(k+3).$$

to be true. To be shown is that

$$S(k+1): \quad 1 \cdot 2 \cdot 3 + 2 \cdot 3 \cdot 4 + \cdots + (k+1)(k+2)(k+3) = \frac{1}{4}(k+1)(k+2)(k+3)(k+4).$$

follows. Beginning with the left side of $S(k + 1)$, (rewritten with the penultimate term inserted for clarity)

$$
\begin{aligned}
&1 \cdot 2 \cdot 3 + 2 \cdot 3 \cdot 4 + \cdots + k(k+1)(k+2) + (k+1)(k+2)(k+3) \\
&= \frac{1}{4}k(k+1)(k+2)(k+3) + (k+1)(k+2)(k+3) \quad \text{(by ind. hyp.)} \\
&= (k+1)(k+2)(k+3)\left[\frac{1}{4}k + 1\right] \\
&= (k+1)(k+2)(k+3)\left[\frac{1}{4}(k+4)\right],
\end{aligned}
$$

which is indeed the right side of $S(k + 1)$, concluding the inductive step $S(k) \rightarrow S(k + 1)$.

Therefore, by the principle of mathematical induction, for each $n \geq 1$, the statement $S(n)$ is true.   □

Exercise 66: For each $n \geq 1$, let $S(n)$ denote the statement

$$S(n): \quad \sum_{j=1}^{n} j(j+1)(j+2)(j+3) = \frac{n(n+1)(n+2)(n+3)(n+4)}{5}.$$

BASE STEP ($n = 1$): The left side of $S(1)$ is $1 \cdot 2 \cdot 3 \cdot 4 = 24$, and the right side of $S(1)$ is $\frac{1 \cdot 2 \cdot 3 \cdot 4 \cdot 5}{5} = 24$ as well, proving the base case.

INDUCTIVE STEP: For some fixed $k \geq 1$, assume the inductive hypothesis

$$S(k): \sum_{j=1}^{k} j(j+1)(j+2)(j+3) = \frac{k(k+1)(k+2)(k+3)(k+4)}{5}$$

to be true. It remains to show that

$$S(k+1): \sum_{j=1}^{k+1} j(j+1)(j+2)(j+3) = \frac{(k+1)(k+2)(k+3)(k+4)(k+5)}{5}$$

follows. Starting with the left side of $S(k+1)$ and separating the last term,

$$\sum_{j=1}^{k+1} j(j+1)(j+2)(j+3)$$

$$= \left(\sum_{j=1}^{k} j(j+1)(j+2)(j+3) \right)$$
$$+ (k+1)(k+2)(k+3)(k+4)$$

$$= \frac{k(k+1)(k+2)(k+3)(k+4)}{5}$$
$$+ (k+1)(k+2)(k+3)(k+4) \quad \text{(by } S(k)\text{)}$$

$$= (k+1)(k+2)(k+3)(k+4) \left[\frac{k}{5} + 1 \right]$$

$$= (k+1)(k+2)(k+3)(k+4) \frac{k+5}{5},$$

which agrees with the right side of $S(k+1)$, completing the inductive step $S(k) \rightarrow S(k+1)$.

Therefore, by the principle of mathematical induction, for all $n \geq 1$, $S(n)$ is true. $\qquad \square$

Exercise 67: Fix some $k \in \mathbb{Z}^+$, and for each $n \geq 1$, denote the statement

$$S(n): \sum_{j=1}^{n} j(j+1) \cdots (j+k-1) = \frac{(k+n)!}{(k+1) \cdot (n-1)!}.$$

BASE STEP ($n = 1$): The statement $S(1)$ says (where $j = 1$ produces the only summand on the left)

$$1 \cdot 2 \cdots k = \frac{(k+1)!}{k+1 \cdot (0!)};$$

since $(k+1)! = (k+1)(k!)$ and $0! = 1$, the two sides agree, completing the proof of the base case $S(1)$.

INDUCTIVE STEP: For some fixed $m \geq 1$, assume the inductive hypothesis

$$S(m): \quad \sum_{j=1}^{m} j(j+1)\cdots(j+k-1) = \frac{(k+m)!}{(k+1)\cdot(m-1)!}.$$

to be true. [*Note*: One shouldn't use $S(k)$ here, because the statement needed to prove already has a k in it, so employ a new variable m in the inductive step.] It remains to show that

$$S(m+1): \quad \sum_{j=1}^{m+1} j(j+1)\cdots(j+k-1) = \frac{(k+m+1)!}{(k+1)\cdot m!}$$

follows. Beginning with the left side of $S(m+1)$ and separating the last summand,

$$\sum_{j=1}^{m+1} j(j+1)\cdots(j+k-1)$$

$$= \left(\sum_{j=1}^{m} j(j+1)\cdots(j+k-1) \right) + (m+1)(m+2)\cdots(m+k)$$

$$= \frac{(k+m)!}{(k+1)\cdot(m-1)!} + (m+1)(m+2)\cdots(m+k) \quad \text{(by } S(m))$$

$$= \frac{(k+m)!}{(k+1)\cdot(m-1)!} + \frac{(m+k)!}{m!}$$

$$= \frac{(k+m)!\,m}{(k+1)\cdot(m)!} + \frac{(k+m)!}{m!}$$

$$= \frac{(k+m)!}{m!} \left[\frac{m}{k+1} + 1 \right]$$

$$= \frac{(k+m)!}{m!} \cdot \frac{m+k+1}{k+1},$$

which reduces to the right side of $S(m+1)$ as desired. This concludes the inductive step $S(m) \to S(m+1)$.

Therefore, by the principle of mathematical induction, for all $n \geq 1$, $S(n)$ is true. $\qquad \square$

Exercise 68: This identity was mentioned by Peter Ross in [465], a media review of the article *A LISP prover for induction formulae*, where he says that an induction proof of this identity "is tedious, although straightforward."

For each $n \geq 1$, denote the statement in the exercise by

$$S(n): \quad \sum_{k=1}^{n}(2k-1)(2k+1)(2k+3) = n(2n^3 + 8n^2 + 7n - 2).$$

BASE STEP ($n = 1$): The statement $S(1)$ says (where $k = 1$ gives the only summand on the left)

$$(2 \cdot 1 - 1)(2 \cdot 1 + 1)(2 \cdot 1 + 3) = 1(2(1^3) + 8(2^2) + 7(1) - 2).$$

This simplifies to $1 \cdot 3 \cdot 5 = 2 + 8 + 7 - 2$, or $15 = 15$, showing that indeed $S(1)$ holds.

INDUCTIVE STEP: For some fixed $t \geq 1$, assume the inductive hypothesis

$$S(t): \quad \sum_{k=1}^{t}(2k-1)(2k+1)(2k+3) = t(2t^3 + 8t^2 + 7t - 2)$$

to be true. It remains to show that

$$S(t+1): \quad \sum_{k=1}^{t+1}(2k-1)(2k+1)(2k+3) = (t+1)(2(t+1)^3 + 8(t+1)^2 + 7(t+1) - 2)$$

follows. Beginning with the left side of $S(t+1)$,

$$\sum_{k=1}^{t+1}(2k-1)(2k+1)(2k+3)$$

$$= \left[\sum_{k=1}^{t}(2k-1)(2k+1)(2k+3)\right] + (2t+1)(2t+3)(2t+5)$$

$$= t(2t^3 + 8t^2 + 7t - 2) + (2t+1)(2t+3)(2t+5) \quad \text{(by } S(t))$$

$$= 2t^4 + 8t^3 + 7t^2 - 2t + 8t^3 + 36t^2 + 46t + 15$$

$$= 2t^4 + 16t^3 + 43t^2 + 44t + 15$$

$$= (t+1)(2t^3 + 14t^2 + 25t + 15) \quad \text{(by polynomial division)}$$

$$= (t+1)(2(t+1)^3 + 8(t+1)^2 + 7(t+1) - 2),$$

where the last equality is easiest to see by multiplying out the second expression in the last line. This completes the derivation of $S(t+1)$ using $S(t)$, concluding the inductive step.

Therefore, by the principle of mathematical induction, for all $n \geq 1$, $S(n)$ is true. $\qquad \square$

Exercise 70: For each $n \geq 1$, let $S(n)$ denote the statement

$$0 \cdot 0! + 1 \cdot 1! + 2 \cdot 2! + 3 \cdot 3! + \cdots + n \cdot n! = (n+1)! - 1.$$

BASE STEP ($n = 1$): $S(1)$ says $0 \cdot 0! + 1 \cdot 1! = 2! - 1$, which is correct, since both sides equal 1.

INDUCTIVE STEP: For some fixed $k \geq 1$, assume the inductive hypothesis $S(k)$ to be true. To see that $S(k+1)$ follows, the left side of $S(k+1)$ is

$$
\begin{aligned}
0 \cdot 0! &+ 1 \cdot 1! + 2 \cdot 2! + \cdots + k \cdot k! + (k+1) \cdot (k+1)! \\
&= (k+1)! - 1 + (k+1) \cdot (k+1)! \quad \text{(by } S(k)) \\
&= (1 + (k+1)) \cdot (k+1)! - 1 \\
&= (k+2) \cdot (k+1)! - 1 \\
&= (k+2)! - 1,
\end{aligned}
$$

which equals the right side of $S(k+1)$. This completes the inductive step $S(k) \rightarrow S(k+1)$.

Therefore, by the principle of mathematical induction, for all $n \geq 1$, $S(n)$ is true. $\qquad \square$

Exercise 71: For each $n \geq 1$, let $S(n)$ denote the statement

$$\frac{1}{1 \cdot 2} + \frac{1}{2 \cdot 3} + \frac{1}{3 \cdot 4} + \cdots + \frac{1}{n(n+1)} = \frac{n}{n+1}.$$

BASE STEP ($n = 1$): $S(1)$ merely says $\frac{1}{1 \cdot 2} = \frac{1}{1+1}$, a true statement.

INDUCTIVE STEP: For some fixed $k \geq 1$, assume the inductive hypothesis $S(k)$ to be true. To be shown is that $S(k+1)$ follows:

$$
\begin{aligned}
\frac{1}{1 \cdot 2} + \frac{1}{2 \cdot 3} &+ \cdots + \frac{1}{k(k+1)} + \frac{1}{(k+1)(k+2)} \\
&= \frac{k}{k+1} + \frac{1}{(k+1)(k+2)} \qquad \text{(by } S(k)) \\
&= \frac{k(k+2) + 1}{(k+1)(k+2)} \\
&= \frac{k^2 + 2k + 1}{(k+1)(k+2)} \\
&= \frac{(k+1)^2}{(k+1)(k+2)}
\end{aligned}
$$

$$= \frac{k+1}{k+2},$$

which proves $S(k+1)$ from the truth of $S(k)$, thereby completing the inductive step.

Therefore, by the principle of mathematical induction, for all $n \geq 1$, $S(n)$ is true. □

Another solution to Exercise 71: Proving the equality in the exercise can be done directly with an old trick: the partial fraction identity $\frac{1}{n(n+1)} = \frac{1}{n} - \frac{1}{n+1}$. The sum can then be seen to telescope, a phenomenon that itself can be proved by induction (see Exercise 137).

$$\frac{1}{1 \cdot 2} + \frac{1}{2 \cdot 3} + \frac{1}{3 \cdot 4} + \cdots + \frac{1}{n(n+1)}$$

$$= \left(\frac{1}{1} - \frac{1}{2}\right) + \left(\frac{1}{2} - \frac{1}{3}\right) + \left(\frac{1}{3} - \frac{1}{4}\right) + \cdots + \left(\frac{1}{n} - \frac{1}{n+1}\right)$$

$$= 1 - \frac{1}{n+1}$$

$$= \frac{n}{n+1}.$$

This completes the second proof. □

Exercise 72: For each $n \geq 1$, let $S(n)$ denote the statement

$$\sum_{i=0}^{n-1} \frac{1}{(n+i)(n+i+1)} = \frac{1}{2n}.$$

The truth of $S(n)$ for all $n \geq 1$ follows from Exercise 71, since the sum in $S(n)$ is the sum of roughly the last half of a much longer series beginning at $\frac{1}{1 \cdot 2}$ rather than at $\frac{1}{n(n+1)}$. These calculations are given first, followed by an inductive proof of $S(n)$.
Direct proof of Exercise 72:

$$\sum_{i=0}^{n-1} \frac{1}{(n+i)(n+i+1)} = \sum_{i=0}^{2n-2} \frac{1}{(i+1)(i+2)} - \sum_{i=0}^{n-2} \frac{1}{(i+1)(i+2)}$$

$$= \frac{2n-1}{2n} - \frac{n-1}{2n} \qquad \text{(by Exercise 71)}$$

$$= \frac{2n-1-2n+2}{2n}$$

$$= \frac{1}{2n}.$$

□

Inductive proof of Exercise 72:

BASE STEP ($n = 1$): In $S(1)$, the sum on the left is the trivial sum $\frac{1}{(1+0)(1+0+1)}$, whereas the right-hand side is $\frac{1}{2}$; these two are equal, so $S(1)$ is true.

INDUCTIVE STEP: For some fixed $k \geq 1$, assume the inductive hypothesis

$$S(k): \quad \sum_{i=0}^{k-1} \frac{1}{(k+i)(k+i+1)} = \frac{1}{2k}$$

to be true. To see that

$$S(k+1): \quad \sum_{i=0}^{k} \frac{1}{(k+1+i)(k+i+2)} = \frac{1}{2(k+1)}$$

follows, starting with the left-hand side,

$$\sum_{i=0}^{k} \frac{1}{(k+1+i)(k+2+i)}$$

$$= \sum_{j=1}^{k+1} \frac{1}{(k+j)(k+j+1)}$$

$$= \sum_{j=0}^{k-1} \frac{1}{(k+j)(k+j+1)} - \frac{1}{k(k+1)} + \frac{1}{2k(2k+1)} + \frac{1}{(2k+1)(2k+2)}$$

$$= \frac{1}{2k} - \frac{1}{k(k+1)} + \frac{1}{2k(2k+1)} + \frac{1}{(2k+1)(2k+2)} \qquad \text{(by } S(k)\text{)}$$

$$= \frac{1}{2k} - \frac{1}{k(k+1)} + \frac{2k+2+2k}{2k(2k+1)(2k+2)}$$

$$= \frac{1}{2k} - \frac{1}{k(k+1)} + \frac{1}{2k(k+1)}$$

$$= \frac{k+1-2+1}{2k(k+1)}$$

$$= \frac{1}{2(k+1)},$$

completing the proof of $S(k+1)$ and hence the inductive step.

Therefore, by the principle of mathematical induction, for all $n \geq 1$, $S(n)$ is true. $\square$

Exercise 73: For each $n \geq 1$, let $S(n)$ denote the statement

$$\frac{1}{1\cdot 2\cdot 3} + \frac{1}{2\cdot 3\cdot 4} + \frac{1}{3\cdot 4\cdot 5} + \cdots + \frac{1}{n(n+1)(n+2)} = \frac{n(n+3)}{4(n+1)(n+2)}.$$

BASE STEP: $S(1)$ says $\frac{1}{1 \cdot 2 \cdot 3} = \frac{1 \cdot 4}{4 \cdot 2 \cdot 3}$ which is correct, since both sides are equal to $\frac{1}{6}$.

INDUCTIVE STEP: For some fixed $k \geq 1$, assume the inductive hypothesis

$$S(k): \quad \sum_{i=1}^{k} \frac{1}{i(i+1)(i+2)} = \frac{k(k+3)}{4(k+1)(k+2)}$$

to be true. It remains to show that

$$S(k+1): \quad \sum_{i=1}^{k+1} \frac{1}{i(i+1)(i+2)} = \frac{(k+1)(k+4)}{4(k+2)(k+3)}$$

follows. Starting with the left-hand side of $S(k+1)$,

$$\sum_{i=1}^{k+1} \frac{1}{i(i+1)(i+2)} = \sum_{i=1}^{k} \frac{1}{i(i+1)(i+2)} + \frac{1}{(k+1)(k+2)(k+3)}$$

$$= \frac{k(k+3)}{4(k+1)(k+2)} + \frac{4}{4(k+1)(k+2)(k+3)} \qquad \text{(by } S(k)\text{)}$$

$$= \frac{k(k+3)(k+3)}{4(k+1)(k+2)(k+3)} + \frac{4}{4(k+1)(k+2)(k+3)}$$

$$= \frac{k^3 + 6k^2 + 9k + 4}{4(k+1)(k+2)(k+3)}$$

$$= \frac{(k+1)^2(k+4)}{4(k+1)(k+2)(k+3)}$$

$$= \frac{(k+1)(k+4)}{4(k+2)(k+3)},$$

one arrives at the right-hand side of $S(k+1)$. This completes the inductive step.

Therefore, by the principle of mathematical induction, for each $n \geq 1$, $S(n)$ is true. $\qquad \square$

Exercise 74: For each $n \geq 1$, let $S(n)$ denote the statement

$$1 - \frac{1}{2} + \frac{1}{3} - \frac{1}{4} + \cdots + \frac{1}{2n-1} - \frac{1}{2n} = \frac{1}{n+1} + \frac{1}{n+2} + \cdots + \frac{1}{2n}.$$

The expression on the left constitutes the first $2n - 1$ terms of what is called the "alternating harmonic series".

BASE STEP ($n = 1$): Note that the left side of $S(n)$ has denominators which range from 1 to $2n$, whereas the denominators on the right range from $n + 1$ to $2n$. So, for $n = 1$, the denominators on the left range from 1 to 2, whereas on the right, they range from $1 + 1 = 2$ to $2 \cdot 1 = 2$, that is, on the right, there is only one term. Hence, $S(1)$ says $1 - \frac{1}{2} = \frac{1}{2}$, which is true.

INDUCTIVE STEP: For some fixed $k \geq 1$, assume the inductive hypothesis $S(k)$:

$$1 - \frac{1}{2} + \frac{1}{3} - \frac{1}{4} + \cdots + \frac{1}{2k - 1} - \frac{1}{2k} = \frac{1}{k + 1} + \frac{1}{k + 2} + \cdots + \frac{1}{2k}.$$

to be true. It remains to prove that $S(k + 1)$:

$$1 - \frac{1}{2} + \frac{1}{3} - \frac{1}{4} + \cdots + \frac{1}{2k + 1} - \frac{1}{2k + 2} = \frac{1}{k + 2} + \frac{1}{k + 3} + \cdots + \frac{1}{2k + 2}$$

follows. Beginning with the left side of $S(k + 1)$ (and filling two more penultimate terms)

$$1 - \frac{1}{2} + \frac{1}{3} - \frac{1}{4} + \cdots + \frac{1}{2k - 1} - \frac{1}{2k} + \frac{1}{2k + 1} - \frac{1}{2k + 2}$$

$$= \frac{1}{k + 1} + \frac{1}{k + 2} + \cdots + \frac{1}{2k} + \frac{1}{2k + 1} - \frac{1}{2k + 2} \qquad \text{(by } S(k))$$

$$= \frac{1}{k + 2} + \cdots + \frac{1}{2k} + \frac{1}{2k + 1} + \frac{1}{k + 1} - \frac{1}{2k + 2}$$

$$= \frac{1}{k + 2} + \cdots + \frac{1}{2k} + \frac{1}{2k + 1} + \frac{2}{2k + 2} - \frac{1}{2k + 2}$$

$$= \frac{1}{k + 2} + \cdots + \frac{1}{2k} + \frac{1}{2k + 1} + \frac{1}{2k + 2},$$

agreeing with the right side of $S(k+1)$. (This sequence of equalities turned out to be far easier than one might have thought at the onset!) This completes the inductive step $S(k) \to S(k + 1)$.

Therefore, by the principle of mathematical induction, for all $n \geq 1$, $S(n)$ is true. $\qquad \square$

Exercise 75: (This exercise is the subject of an exposition in [433, Vol I, p. 112–114].)

For each $n \geq 1$, let $P(n)$ be the proposition

$$\frac{1}{1 \cdot 3} + \frac{1}{3 \cdot 5} + \frac{1}{5 \cdot 7} + \cdots + \frac{1}{(2n - 1)(2n + 1)} = \frac{n}{2n + 1}.$$

BASE STEP: $P(1)$ says $\frac{1}{1 \cdot 3} = \frac{1}{2(1) + 1}$, which is true.

INDUCTIVE STEP: For some fixed $k \geq 1$, assume the inductive hypothesis $P(k)$:

$$\frac{1}{1 \cdot 3} + \frac{1}{3 \cdot 5} + \frac{1}{5 \cdot 7} + \cdots + \frac{1}{(2k-1)(2k+1)} = \frac{k}{2k+1}$$

to be true. The consequence to be proved is $P(k+1)$:

$$\frac{1}{1 \cdot 3} + \frac{1}{3 \cdot 5} + \frac{1}{5 \cdot 7} + \cdots + \frac{1}{(2k+1)(2k+3)} = \frac{k+1}{2k+3}.$$

The expression to the left of the equal sign in $P(k+1)$ (written with the second last summand explicit) is equal to

$$\frac{1}{1 \cdot 3} + \frac{1}{3 \cdot 5} + \frac{1}{5 \cdot 7} + \cdots + \frac{1}{(2k-1)(2k+1)} + \frac{1}{(2k+1)(2k+3)}$$

$$= \frac{k}{2k+1} + \frac{1}{(2k+1)(2k+3)} \qquad \text{(by } S(k)\text{)}$$

$$= \frac{k(2k+3)+1}{(2k+1)(2k+3)}$$

$$= \frac{2k^2 + 3k + 1}{(2k+1)(2k+3)}$$

$$= \frac{(2k+1)(k+1)}{(2k+1)(2k+3)}$$

$$= \frac{k+1}{2k+3},$$

which equals the expression on the right side of $P(k+1)$. This completes the proof of $P(k+1)$, and hence the inductive step $P(k) \to P(k+1)$.

By mathematical induction, one concludes that for all $n \geq 1$, $P(n)$ is true. $\square$

Exercise 77: For $n \geq 1$, denote the statement in the exercise by

$$S(n): \quad \frac{1}{1 \cdot 3} + \frac{1}{2 \cdot 4} + \cdots + \frac{1}{(n)(n+2)} = \frac{n(3n+5)}{4(n+1)(n+2)}.$$

BASE STEP: Since $\frac{1}{1 \cdot 3} = \frac{3+5}{4 \cdot 2 \cdot 3}$, $S(1)$ is true.

INDUCTIVE STEP: For some fixed $m \geq 1$, assume that

$$S(m): \quad \frac{1}{1 \cdot 3} + \frac{1}{2 \cdot 4} + \cdots + \frac{1}{(m)(m+2)} = \frac{m(3m+5)}{4(m+1)(m+2)}.$$

is true. It remains to show that

$$S(m+1): \quad \frac{1}{1\cdot 3} + \frac{1}{2\cdot 4} + \cdots + \frac{1}{(m+1)(m+3)} = \frac{(m+1)(3m+8)}{4(m+2)(m+3)}$$

follows. Starting with the left-hand side of $S(m+1)$,

$$\frac{1}{1\cdot 3} + \frac{1}{2\cdot 4} + \cdots + \frac{1}{m(m+2)} + \frac{1}{(m+1)(m+3)}$$

$$= \frac{m(3m+5)}{4(m+1)(m+2)} + \frac{1}{(m+1)(m+3)} \qquad \text{(by } S(m))$$

$$= \frac{m(3m+5)(m+3) + 4(m+2)}{4(m+1)(m+2)(m+3)}$$

$$= \frac{m^3 + 14m^2 + 19m + 8}{4(m+1)(m+2)(m+3)}$$

$$= \frac{(3m^2 + 11m + 8)(m+1)}{4(m+1)(m+2)(m+3)}$$

$$= \frac{3m^2 + 11m + 8}{4(m+2)(m+3)} \qquad \text{(since } m+1 \neq 0)$$

$$= \frac{(m+1)(3m+8)}{4(m+2)(m+3)},$$

proving $S(m+1)$ as desired, thereby concluding the inductive step.

By MI, for each $n \geq 1$, $S(n)$ is true. $\qquad \square$

Exercise 78: For $n \geq 1$, denote the equality in the exercise by

$$Q(n): \quad \frac{1}{1\cdot 5} + \frac{1}{5\cdot 9} + \frac{1}{9\cdot 13} + \cdots + \frac{1}{(4n-3)(4n+1)} = \frac{n}{4n+1}.$$

BASE STEP: It is trivial that $\frac{1}{1\cdot 5} = \frac{1}{4\cdot 1+1}$, and so $Q(1)$ is true.
INDUCTIVE STEP: Suppose that for some $k \geq 1$, $Q(k)$ holds. It suffices to show that

$$Q(k+1): \quad \frac{1}{1\cdot 5} + \frac{1}{5\cdot 9} + \frac{1}{9\cdot 13} + \cdots + \frac{1}{(4k+1)(4k+5)} = \frac{k+1}{4k+5}$$

follows. Beginning with the expression on the left side of the equality in $Q(k+1)$,

$$\frac{1}{1\cdot 5} + \frac{1}{5\cdot 9} + \frac{1}{9\cdot 13} + \cdots + \frac{1}{(4k-3)(4k+1)} + \frac{1}{(4k+1)(4k+5)}$$

$$= \frac{k}{4k+1} + \frac{1}{(4k+1)(4k+5)} \qquad \text{(by } Q(k))$$

$$= \frac{k(4k+5)+1}{(4k+1)(4k+5)}$$

$$= \frac{4k^2+5k+1}{(4k+1)(4k+5)}$$

$$= \frac{(4k+1)(k+1)}{(4k+1)(4k+5)}$$

$$= \frac{k+1}{4k+5},$$

proving $Q(k+1)$, and hence completing the inductive step $Q(k) \to Q(k+1)$.

Therefore, by the principle of mathematical induction, for all $n \geq 1$, $Q(n)$ is true. $\qquad\square$

Exercise 79: This appeared in [499, Prob. 14], for example.

For $n \geq 1$, let $P(n)$ be the proposition

$$\frac{1^2}{1\cdot 3} + \frac{2^2}{3\cdot 5} + \frac{3^2}{5\cdot 7} + \cdots + \frac{n^2}{(2n-1)(2n+1)} = \frac{n(n+1)}{2(2n+1)}.$$

BASE STEP: The left side of $P(1)$ is $\frac{1}{3}$, and the right side of $P(1)$ is $\frac{1\cdot 2}{2\cdot 3}$, which is also $\frac{1}{3}$, so $P(1)$ is true.

INDUCTIVE STEP: Let $k \geq 1$ be fixed and suppose that

$$P(k): \quad \sum_{i=1}^{k} \frac{i^2}{(2i-1)(2i+1)} = \frac{k(k+1)}{2(2k+1)}$$

is true. To be proved is that

$$P(k+1): \quad \sum_{i=1}^{k+1} \frac{i^2}{(2i-1)(2i+1)} = \frac{(k+1)(k+2)}{2(2k+3)}$$

follows; this is accomplished by:

$$\sum_{i=1}^{k+1} \frac{i^2}{(2i-1)(2i+1)} = \left[\sum_{i=1}^{k} \frac{i^2}{(2i-1)(2i+1)} \right] + \frac{(k+1)^2}{(2(k+1)-1)(2(k+1)+1)}$$

$$= \frac{k(k+1)}{2(2k+1)} + \frac{(k+1)^2}{(2k+1)(2k+3)} \qquad \text{(by } P(k))$$

$$= \frac{k(k+1)(2k+3) + 2(k+1)^2}{2(2k+1)(2k+3)}$$

$$= \frac{(k+1)[k(2k+3) + 2(k+1)]}{2(2k+1)(2k+3)}$$

$$= \frac{(k+1)[2k^2 + 5k + 2]}{2(2k+1)(2k+3)}$$

$$= \frac{(k+1)(2k+1)(k+2)}{2(2k+1)(2k+3)}$$

$$= \frac{(k+1)(k+2)}{2(2k+3)},$$

proves $P(k+1)$. This completes the inductive step.

By MI, for all $n \geq 1$, that $P(n)$ is true.   □

Exercise 80: For each $n \geq 1$, let $Q(n)$ denote the statement

$$\sum_{k=1}^{n} \frac{1}{k^2 + 3k + 2} = \frac{n}{2(n+2)}.$$

Proof using Exercise 71: Notice that the denominators in each summand factors:

$$\sum_{k=1}^{n} \frac{1}{k^2 + 3k + 2} = \sum_{k=1}^{n} \frac{1}{(k+1)(k+2)}$$

$$= \sum_{j=2}^{n+1} \frac{1}{j(j+1)}$$

$$= \sum_{j=1}^{n+1} \frac{1}{j(j+1)} - \frac{1}{2}$$

$$= \frac{n+1}{n+2} - \frac{1}{2} \qquad \text{(by Exercise 71)}$$

$$= \frac{2n + 2 - (n+2)}{2(n+2)}$$

$$= \frac{n}{2(n+2)},$$

as desired.   □

   Inductive proof of $Q(n)$:
BASE CASE: $Q(1)$ says $\frac{1}{(1)^2 + 3(1) + 2} = \frac{1}{2(1+2)}$, which reduces to $\frac{1}{6} = \frac{1}{6}$, and so is true.

INDUCTIVE STEP: For some $m \geq 1$, suppose that

$$Q(m) : \sum_{k=1}^{m} \frac{1}{k^2 + 3k + 2} = \frac{m}{2(m+2)}$$

is true. Prove the statement

$$Q(m+1) : \sum_{k=1}^{m+1} \frac{1}{k^2 + 3k + 2} = \frac{m+1}{2(m+3)}$$

as follows:

$$\sum_{k=1}^{m+1} \frac{1}{k^2 + 3k + 2} = \sum_{k=1}^{m} \frac{1}{k^2 + 3k + 2} + \frac{1}{(m+1)^2 + 3(m+1) + 2}$$

$$= \frac{m}{2(m+2)} + \frac{1}{m^2 + 5m + 6} \qquad \text{(by } Q(k))$$

$$= \frac{m}{2(m+2)} + \frac{1}{(m+2)(m+3)}$$

$$= \frac{m(m+3) + 2}{2(m+2)(m+3)}$$

$$= \frac{m^2 + 3m + 2}{2(m+2)(m+3)}$$

$$= \frac{(m+1)(m+2)}{2(m+2)(m+3)}$$

$$= \frac{m+1}{2(m+3)}.$$

This completes the inductive step $Q(m) \to Q(m+1)$.

By mathematical induction, for any $n \geq 2$, $Q(n)$ is true. $\qquad\square$

Exercise 81: This problem occurs in an old Canadian grade school text [385] without solution. For $n \geq 1$, denote the statement of the exercise by

$$S(n) : \sum_{i=1}^{n} \frac{i}{1 + i^2 + i^4} = \frac{n(n+1)}{2(n^2 + n + 1)}.$$

BASE STEP: $S(1)$ says $\frac{1}{1+1+1} = \frac{1 \cdot 2}{2(1+1+1)}$, which is true.

INDUCTIVE STEP: Fix $m \geq 1$ and suppose that $S(m)$ is true. To prepare for $S(m+1)$, observe that

$$1 + (m+1)^2 + (m+1)^4 = (m^2 + m + 1)(m^2 + 3m + 3).$$

To prove $S(m+1)$, start with the LHS of $S(m+1)$:

$$\sum_{i=1}^{m+1} \frac{i}{1+i^2+i^4} = \sum_{i=1}^{m} \frac{i}{1+i^2+i^4} + \frac{m+1}{1+(m+1)^2+(m+1)^4}$$

$$= \frac{m(m+1)}{2(m^2+m+1)} + \frac{m+1}{1+(m+1)^2+(m+1)^4} \qquad \text{(by } S(m)\text{)}$$

$$= \frac{m(m+1)}{2(m^2+m+1)} + \frac{m+1}{(m^2+m+1)(m^2+3m+3)}$$

$$= \frac{m+1}{m^2+m+1}\left[\frac{m}{2} + \frac{1}{m^2+3m+3}\right]$$

$$= \frac{m+1}{m^2+m+1} \cdot \frac{m(m^2+3m+3)+2}{2(m^2+3m+3)}$$

$$= \frac{m+1}{m^2+m+1} \cdot \frac{m^3+3m^2+3m+2}{2((m+1)^2+(m+1)+1)}$$

$$= \frac{m+1}{m^2+m+1} \cdot \frac{(m^2+m+1)(m+2)}{2((m+1)^2+(m+1)+1)}$$

$$= \frac{(m+1)(m+2)}{2((m+1)^2+(m+1)+1)},$$

which is the right side of $S(m+1)$, concluding the inductive step.

By the principle of mathematical induction, for each $n \geq 1$, $S(n)$ is true. [In fact, $S(0)$ is also true.] □

Exercise 82: For each $n \geq 1$, let $S(n)$ be the statement

$$\frac{1}{1\cdot 4} + \frac{1}{4\cdot 7} + \frac{1}{7\cdot 10} + \cdots + \frac{1}{(3n-2)(3n+1)} = \frac{n}{3n+1},$$

or written in sigma notation,

$$\sum_{i=1}^{n} \frac{1}{(3i-2)(3i+1)} = \frac{n}{3n+1}.$$

BASE STEP: When $n=1$, $S(1)$ says $\frac{1}{(1)(4)} = \frac{1}{4}$, a true statement.

INDUCTIVE STEP: For some fixed $k \geq 1$, suppose that $S(k)$ is true. To see that $S(k+1)$ is true,

$$\sum_{i=1}^{k+1} \frac{1}{(3i-2)(3i+1)} = \left[\sum_{i=1}^{k} \frac{1}{(3i-2)(3i+1)}\right] + \frac{1}{(3(k+1)-2)(3(k+1)+1)}$$

$$= \left[\frac{k}{3k+1}\right] + \frac{1}{(3k+1)(3k+4)} \qquad \text{(by } S(k)\text{)}$$

$$= \frac{k(3k+4)+1}{(3k+1)(3k+4)}$$

$$= \frac{3k^2+4k+1}{(3k+1)(3k+4)}$$

$$= \frac{(k+1)(3k+1)}{(3k+1)(3k+4)}$$

$$= \frac{k+1}{3k+4}$$

$$= \frac{k+1}{3(k+1)+1},$$

which proves $S(k+1)$. This completes the inductive step.

By the principle of mathematical induction, for all $n \geq 2$, the statement $S(n)$ is true. $\qquad \square$

Exercise 84: For each $n \geq 2$, let $S(n)$ be the statement

$$\left(1-\frac{1}{2}\right)\left(1-\frac{1}{3}\right)\left(1-\frac{1}{4}\right)\cdots\left(1-\frac{1}{n}\right) = \frac{1}{n},$$

or using product notation,

$$\prod_{i=2}^{n}\left(1-\frac{1}{i}\right) = \frac{1}{n}.$$

BASE STEP: When $n = 2$, the equality says $1 - \frac{1}{2} = \frac{1}{2}$, which is true.

INDUCTIVE STEP: For some $k \geq 2$, suppose that $S(k)$ is true. Then

$$\prod_{i=2}^{k+1}\left(1-\frac{1}{i}\right) = \left[\prod_{i=2}^{k}\left(1-\frac{1}{i}\right)\right]\left(1-\frac{1}{k+1}\right)$$

$$= \frac{1}{k}\left(1-\frac{1}{k+1}\right) \qquad \text{(by } S(k)\text{)}$$

$$= \frac{1}{k}\cdot\frac{k}{k+1}$$

$$= \frac{1}{k+1},$$

proves that $S(k+1)$ follows. This completes the inductive step.

By mathematical induction, for all $n \geq 2$, $S(n)$ is true. $\qquad \square$

Exercise 85: For each $n \geq 2$, let $S(n)$ be the statement

$$\left(1-\frac{1}{4}\right)\left(1-\frac{1}{9}\right)\left(1-\frac{1}{16}\right)\cdots\left(1-\frac{1}{n^2}\right) = \frac{n+1}{2n},$$

or in product notation,

$$\prod_{i=2}^{n}\left(1-\frac{1}{i^2}\right)=\frac{n+1}{2n}.$$

BASE STEP $(n=2)$: $S(2)$ says $1-\frac{1}{4}=\frac{2+1}{2\cdot 2}$, which is correct, both sides being $\frac{3}{4}$.

INDUCTION STEP: For some fixed $k \geq 2$, suppose that $S(k)$ is true. Then

$$\prod_{i=2}^{k+1}\left(1-\frac{1}{i^2}\right)=\left[\prod_{i=2}^{k}\left(1-\frac{1}{i^2}\right)\right]\left(1-\frac{1}{(k+1)^2}\right)$$

$$=\frac{k+1}{2k}\left(1-\frac{1}{(k+1)^2}\right) \qquad\qquad \text{(by } S(k))$$

$$=\frac{k+1}{2k}\cdot\frac{k^2+2k}{(k+1)^2}$$

$$=\frac{k+2}{2(k+1)},$$

shows that $S(k+1)$ follows, completing the inductive step.

Therefore, by mathematical induction, for all $n \geq 2$, the statement $S(n)$ holds true. $\qquad\qquad\qquad\qquad\qquad\qquad\qquad\qquad\qquad\qquad\qquad\qquad\qquad\square$

Exercise 86: See [266, pp. 125–6] for solution.

27.2   Solutions: Sums with binomial coefficients

Exercise 88: For each $n \geq 1$, let $S(n)$ be the statement

$$\binom{2}{2}+\binom{3}{2}+\binom{4}{2}+\cdots+\binom{n+1}{2}=\frac{n(n+1)(n+2)}{6}.$$

BASE STEP $(n=1)$: The statement $S(1)$ says $\binom{2}{2}=\frac{(1)(2)(3)}{6}$, so $S(1)$ holds.

INDUCTION STEP $(S(k) \rightarrow S(k+1))$: Fix some $k \geq 1$ and suppose that the induction hypothesis

$$S(k): \quad \binom{2}{2}+\binom{3}{2}+\binom{4}{2}+\cdots+\binom{k+1}{2}=\frac{k(k+1)(k+2)}{6}$$

is true. Next, show that

$$S(k+1): \quad \binom{2}{2}+\binom{3}{2}+\binom{4}{2}+\cdots+\binom{k+1}{2}=\frac{(k+1)(k+2)(k+3)}{6}$$

follows. Starting with the left side of $S(k+1)$,

$$\binom{2}{2} + \cdots + \binom{k+1}{2} + \binom{k+2}{2}$$
$$= \frac{k(k+1)(k+2)}{6} + \binom{k+2}{2} \qquad \text{(by } S(k))$$
$$= \frac{k(k+1)(k+2)}{6} + \frac{(k+1)(k+2)}{2}$$
$$= \frac{k(k+1)(k+2)}{6} + \frac{3(k+1)(k+2)}{6}$$
$$= \frac{k(k+1)(k+2) + 3(k+1)(k+2)}{6},$$

which yields the right side of $S(k+1)$. the inductive step $S(k) \to S(k+1)$ is completed.

By mathematical induction, for all $n \geq 1$, $S(n)$ holds. $\qquad \square$

Exercise 89: For $n \geq 1$, denote the first proposition in the statement of the exercise by

$$P(n): \quad \sum_{i=1}^{n} \frac{1}{\binom{i+1}{2}} = 2 - \frac{2}{n+1}.$$

BASE STEP: Since

$$\frac{1}{\binom{2}{2}} = 1 = 2 - \frac{2}{2}$$

$P(1)$ is true.

INDUCTIVE STEP: For some fixed $k \geq 1$, assume that $P(k)$ holds. Then

$$\sum_{i=1}^{k+1} \frac{1}{\binom{i+1}{2}} = \left[\sum_{i=1}^{k} \frac{1}{\binom{i+1}{2}} \right] + \frac{1}{\binom{k+2}{2}}$$
$$= 2 - \frac{2}{k+1} + \frac{1}{\binom{k+2}{2}} \qquad \text{(by } P(k))$$
$$= 2 - \frac{2}{k+1} + \frac{2}{(k+2)(k+1)}$$
$$= 2 - \frac{2}{k+1} \left[1 - \frac{1}{k+2} \right]$$
$$= 2 - \frac{2}{k+1} \left[\frac{k+1}{k+2} \right]$$
$$= 2 - \frac{2}{k+2},$$

proving $P(k+1)$, and completing the inductive step.

By mathematical induction, for all $n \geq 1$, $P(n)$ is true.

An infinite series converges iff the limit of the partial sums exists. In this case, the limit exists because the partial sums are increasing and each is less than 2. Hence, conclude that this limit exists (and is at most 2). Indeed, one can calculate the limit directly:

$$\sum_{i=1}^{\infty} \frac{1}{\binom{i+1}{2}} \leq 2 = \lim_{n \to \infty} \sum_{i=1}^{n} \frac{1}{\binom{i+1}{2}} = \lim_{n \to \infty} 2 - \frac{2}{n+1} = 2.$$

$\square$

See [402] for a related diagram.

It turns out that all of this machinery was unnecessary, since $\frac{1}{\binom{i}{2}} = \frac{2}{i(i+1)} = \frac{2}{i} - \frac{2}{i+1}$, and so the sum $\sum_{i=1}^{n} \frac{1}{\binom{i+1}{2}}$ telescopes to $2 - \frac{2}{n+1}$.

Exercise 90 (Pascal's identity): Two proofs are presented; the first is by induction and is rather cumbersome, whereas the second is direct and very simple. This demonstrates that induction is not always the preferred proof, but the inductive method is presented here anyway, if only to show its utility.

Proof by induction: Fix r, and for all $n \geq r$, let $P(n)$ be the proposition

$$P(n): \quad \binom{n+1}{r} = \binom{n}{r} + \binom{n}{r-1}.$$

BASE STEP ($n = r$): The statement $P(r)$ says $\binom{r+1}{r} = \binom{r}{r} + \binom{r}{r-1}$, that is, $r+1 = 1 + r$, and so $P(r)$ is true.

INDUCTION STEP: Fix some $k \geq r$ and suppose that

$$P(k): \quad \binom{k+1}{r} = \binom{k}{r} + \binom{k}{r-1}$$

is true. Consider the statement

$$P(k+1): \quad \binom{k+2}{r} = \binom{k+1}{r} + \binom{k+1}{r-1}.$$

Beginning with the left side of $P(k+1)$, apply Lemma 9.6.1 four times, in the first, third and sixth lines below (in the first line, use $m = k+2$ and $s = r$; in the third line, use $m = k+1$ and $s = r$ and $s = r-1$ respectively; in the sixth line, use $m = k+1$ and $s = r$):

$$\binom{k+2}{r}$$

$$= \frac{k+2}{k+2-r} \binom{k+1}{r}$$

$$= \frac{k+2}{k+2-r} \left[\binom{k}{r} + \binom{k}{r-1} \right] \quad \text{(by } P(k))$$

$$= \frac{k+2}{k+2-r} \left[\frac{k+1-r}{k+1} \binom{k+1}{r} + \frac{k+1-(r-1)}{k+1} \binom{k+1}{r-1} \right]$$

$$= \frac{(k+2)(k+1-r)}{(k+2-r)(k+1)} \binom{k+1}{r} + \frac{k+2}{k+1} \binom{k+1}{r-1}$$

$$= \frac{(k+2)(k+1-r)}{(k+2-r)(k+1)} \binom{k+1}{r} + \frac{1}{k+1} \binom{k+1}{r-1} + \binom{k+1}{r-1}$$

$$= \frac{(k+2)(k+1-r)}{(k+2-r)(k+1)} \binom{k+1}{r} + \frac{r}{(k+1)(k-r+2)} \binom{k+1}{r} + \binom{k+1}{r-1}$$

$$= \frac{(k+2)(k+1-r)+r}{(k+2-r)(k+1)} \binom{k+1}{r} + \binom{k+1}{r-1}$$

$$= \binom{k+1}{r} + \binom{k+1}{r-1}$$

which is the right-hand side of $P(k+1)$, completing the inductive step.

Consequently, by mathematical induction, for all $n \geq 0$, $P(n)$ holds. $\qquad\square$

Direct proof: Let S be a set with $n+1$ elements, and consider some fixed $x \in S$. There are $\binom{n+1}{r}$ r-subsets of S—count them according to whether or not they contain x: there are $\binom{n}{r}$ not containing x, (each formed by choosing r of the remaining n elements in $S \backslash \{x\}$), and there are $\binom{n}{r-1}$ r-sets containing x, (each formed by selecting an additional $r-1$ elements in $S \backslash \{x\}$). $\qquad\square$

Exercise 92 [Pascal]: For each $n \geq 2$ let $S(n)$ be the assertion that for all k satisfying $1 \leq k \leq n-1$,

$$\frac{\binom{n}{k}}{\binom{n}{k+1}} = \frac{k+1}{n-k}.$$

The proof is by induction on n, and in the inductive step, Lemma 9.6.1 is applied.

BASE STEP ($n = 2$): When $n = 2$, the only choice for k is $k = 1$. In this case,

$$\frac{\binom{2}{1}}{\binom{2}{1+1}} = 2 = \frac{1+1}{2-1},$$

and so $S(2)$ is true.

INDUCTIVE STEP $(S(m) \to S(m+1))$: Fix $m \geq 2$, and suppose that $S(m)$ is true, that is, for any k satisfying $1 \leq k \leq m-1$,

$$S_k(m): \quad \frac{\binom{m}{k}}{\binom{m}{k+1}} = \frac{k+1}{m-k}$$

is true. To be shown is that for any $1 \leq k \leq m$,

$$\frac{\binom{m+1}{k}}{\binom{m+1}{k+1}} = \frac{k+1}{m+1-k}.$$

For $m = k$, the desired equality is true, so assume that $k < m$. Then

$$\frac{\binom{m+1}{k}}{\binom{m+1}{k+1}} = \frac{\frac{m+1}{m+1-k}\binom{m}{k}}{\frac{m+1}{m-k}\binom{m}{k+1}} \qquad \text{(by Lemma 9.6.1, twice)}$$

$$= \frac{m-k}{m+1-k} \cdot \frac{\binom{m}{k}}{\binom{m}{k+1}}$$

$$= \frac{m-k}{m+1-k} \cdot \frac{k+1}{m-k} \qquad \text{(by } S_k(m)\text{)}$$

$$= \frac{k+1}{m+1-k},$$

as desired. This completes the inductive step.

Hence, by mathematical induction, for each $n \geq 2$, the statement $S(n)$ holds true. $\qquad \square$

Exercise 93 [Pascal]: The proof provided here is far less intuitive than that in Exercise 92, and was arrived at by working backwards from what was required.

For each $n \geq 2$ let $S(n)$ be the claim that for all k satisfying $1 \leq k \leq n-1$,

$$\frac{\binom{n}{k}}{\binom{n}{k+1}} = \frac{k+1}{n-k}.$$

Proceed by induction on n, and use Pascal's identity:

BASE STEP $(n = 2)$: When $n = 2$, the only choice for k is $k = 1$. In this case,

$$\frac{\binom{2}{1}}{\binom{2}{1+1}} = 2 = \frac{1+1}{2-1},$$

and so $S(2)$ is true.

INDUCTIVE STEP $(S(m) \to S(m+1))$: Fix $m \geq 2$, and suppose that $S(m)$ is true, that is, for any k satisfying $1 \leq k \leq m-1$,

$$S_k(m): \quad \frac{\binom{m}{k}}{\binom{m}{k+1}} = \frac{k+1}{m-k}$$

is true. It remains to prove that for any $1 \leq k \leq m$,

$$\frac{\binom{m+1}{k}}{\binom{m+1}{k+1}} = \frac{k+1}{m+1-k}.$$

For $m = k$, the desired equality is true, so assume that $k < m$. By the inductive hypothesis,

$$\frac{k+1}{m-k} = \frac{\binom{m}{k}}{\binom{m}{k+1}},$$

and simple algebra shows

$$\frac{1 + \frac{k}{m-k+1}}{1 + \frac{k+1}{m-k}} \cdot \frac{k+1}{m+1-k} = \frac{k+1}{m-k}.$$

Thus,

$$\frac{\binom{m}{k}}{\binom{m}{k+1}} = \frac{1 + \frac{k}{m-k+1}}{1 + \frac{k+1}{m-k}} \cdot \frac{k+1}{m+1-k}.$$

By inductive hypotheses $S_{k-1}(m)$ and $S_k(m)$,

$$\frac{\binom{m}{k}}{\binom{m}{k+1}} = \frac{1 + \frac{\binom{m}{k-1}}{\binom{m}{k}}}{1 + \frac{\binom{m}{k}}{\binom{m}{k+1}}} \cdot \frac{k+1}{m+1-k}.$$

Multiplying out, one arrives at

$$(m+1-k)\left[\binom{m}{k} + \binom{m}{k-1}\right] = (k+1)\left[\binom{m}{k+1} + \binom{m}{k}\right].$$

Applying Pascal's identity,

$$(m+1-k)\binom{m+1}{k} = (k+1)\binom{m+1}{k+1},$$

whence the desired equality follows, completing the inductive step.

So, by mathematical induction, for all $n \geq 2$, $S(n)$ is true.   □

Exercise 95 [Euler]: For $t \geq 0$, let $S(t)$ be the statement that for any non-negative integers m and n with $m + n = t$, and any $p \geq 0$,

$$\binom{m+n}{p} = \sum_{i=0}^{p} \binom{m}{i}\binom{n}{p-i}.$$

BASE STEP $(t = 0)$: When $t = 0$, only $m = n = 0$ is possible. In this case, $S(0)$ says

$$\binom{0}{p} = \sum_{i=0}^{p} \binom{0}{i}\binom{0}{p-i}.$$

If $p = 0$, the left side of $S(0)$ is equal to 1, and the right side has only one summand, namely $\binom{0}{0}\binom{0}{0-0}$, also equal to 1. If $p \geq 1$, the left side is equal to 0, and every summand on the right-hand side will have a factor of the form $\binom{0}{i}$ where $i > 0$, and so every summand on the right is also 0. Thus, $S(0)$ holds.

Suppose that for some $k \geq 0$, $S(k)$ holds, that is, for every $m \geq 0$ and $n \geq 0$ with $m + n = k$, and any $p \geq 0$,

$$\binom{m+n}{p} = \sum_{i=0}^{p} \binom{m}{i}\binom{n}{p-i}$$

holds. To show that $S(k + 1)$ holds, show that for any $m \geq 0$ and $n \geq 0$ with $m + n = k$, and any $p \geq 0$,

$$\binom{m+n+1}{p} = \sum_{i=0}^{p} \binom{m}{i}\binom{n+1}{p-i}.$$

If $p = 0$, then both sides are equal to 1, so assume that $p \geq 1$. Beginning with the left side of the above equality,

$$\binom{m+n+1}{p} = \binom{m+n}{p} + \binom{m+n}{p-1} \qquad \text{(Pascal's id.)}$$

$$= \sum_{1-0}^{p} \binom{m}{i}\binom{n}{p-i} + \sum_{1-0}^{p-1} \binom{m}{i}\binom{n}{p-1-i} \qquad \text{(ind. hyp.)}$$

$$= \sum_{i=0}^{p-1} \binom{m}{i}\left[\binom{n}{p-i} + \binom{n}{p-1-i}\right] + \binom{m}{p}$$

$$= \left[\sum_{i=0}^{p-1} \binom{m}{i}\binom{n+1}{p-i}\right] + \binom{m}{p}\binom{n+1}{0} \qquad \text{(Pascal's id.)}$$

$$= \sum_{i=0}^{p} \binom{m}{i}\binom{n+1}{p-i},$$

finishing the proof of the desired equality, and hence the proof of $S(k+1)$. This completes the inductive step.

Therefore, by mathematical induction, for all $t \geq 0$, $S(t)$ holds. $\qquad\square$

Exercise 96 [Lagrange]: This is a very interesting equality, because proving it directly by induction seems far more difficult than proving the more general statement of Theorem 9.6.2. One attempt to prove this by induction was given in [350, Ex. 55], however the careful reader will spot that this proof is not "purely inductive". In fact, in the inductive step of this proof, another special case of Theorem 9.6.2 is invoked; this seems rather pointless, since the Corollary 9.6.3 follows directly in one step from another special case of Theorem 9.6.2! Nevertheless, there is value in examining such an attempt. The presentation given here differs slightly from that in [350], mostly in the order of equalities used; the main idea is the same.

Proof of Corollary 9.6.3: For each integer $n > 0$, let $S(n)$ denote the statement

$$S(n) : \binom{2n}{n} = \sum_{i=0}^{n} \binom{n}{i}^2.$$

BASE STEP $(n = 0, 1)$: The statement $S(0)$ says $\binom{0}{0} = \binom{0}{0}^2$, which is true since both sides are equal to 1. To start the induction step for $k \geq 1$, also check the case $n = 1$: the statement $S(1)$ says $\binom{2}{1} = \binom{0}{0}^2 + \binom{1}{0}^2$, and upon evaluating, says $2 = 1 + 1$, which is true.

INDUCTIVE STEP $(S(k) \rightarrow S(k+1))$: Fix some $k \geq 1$ and assume that the inductive hypothesis

$$S(k) : \binom{2k}{k} = \sum_{i=0}^{k} \binom{k}{i}^2$$

is true. It remains to prove

$$S(k+1) : \binom{2k+2}{k+1} = \sum_{i=0}^{k+1} \binom{k+1}{i}^2.$$

Starting with the left side of $S(k+1)$ and applying Pascal's identity twice,

$$\binom{2k+2}{k+1} = \binom{2k+1}{k+1} + \binom{2k+1}{k}$$

$$= \binom{2k}{k+1} + \binom{2k}{k} + \binom{2k}{k} + \binom{2k}{k-1}$$

$$= \binom{2k}{k} + 2\binom{2k}{k+1} + \binom{2k}{k},$$

where the last line used $\binom{2k}{k+1} = \binom{2k}{k-1}$. In applying the inductive hypothesis to the first summand in the bracket in the last line, how is the last term handled? Notice that by Theorem 9.6.2,

$$\binom{2k}{k-1} = \sum_{i=0}^{k+1} \binom{k}{i}\binom{k}{k+1-i},$$

and so, by also applying the inductive hypothesis,

$$\binom{2k+2}{k+1} = \sum_{i=0}^{k} \binom{k}{i}^2 + 2\sum_{i=0}^{k+1} \binom{k}{i}\binom{k}{k+1-i} + \sum_{j=0}^{k} \binom{k}{j}^2$$

$$= \sum_{i=0}^{k} \binom{k}{i}^2 + 2\sum_{i=1}^{k} \binom{k}{i}\binom{k}{k+1-i} + \sum_{j=0}^{k} \binom{k}{j}^2$$

$$= \sum_{i=0}^{k} \binom{k}{i}^2 + 2\sum_{i=1}^{k} \binom{k}{i}\binom{k}{i-1} + \sum_{i=1}^{k+1} \binom{k}{i-1}^2$$

$$= \binom{k}{0}^2 + \sum_{i=1}^{k} \left(\binom{k}{i} + \binom{k}{i-1}\right)^2 + \binom{k}{k}^2$$

$$= \binom{k+1}{0}^2 + \sum_{i=1}^{k} \binom{k+1}{i}^2 + \binom{k+1}{k+1}^2 \qquad \text{(Pascal's id.)}$$

$$= \sum_{i=0}^{k+1} \binom{k+1}{i}^2.$$

This completes the inductive step.

Therefore, by induction, for each $n \geq 0$, the statement $S(n)$ is true. $\qquad \square$

Note: The technique used in [350] to begin the sequence of equalities above was to use the identity $\binom{m}{s} = \frac{m}{s}\binom{m-1}{s-1}$ (which follows by direct computation or by Lemma 9.6.1 twice) as follows:

$$\binom{2k+2}{k+1} = \frac{2k+2}{k+1}\binom{2k+1}{k} = 2\binom{2k+1}{k},$$

and then apply Pascal's identity. This technique does not seem as natural as the one employed above.

It seems that Lagrange's identity is more difficult to prove than Euler's equality because upon using Pascal's identity, certain variables are reduced only by one, and then the inductive hypothesis is of no help because there are binomial coefficients which are not of the form $\binom{2k}{k}$. It is entirely possible, however, that a purely inductive proof exists, though it seems as if it would be either messy or intricate.

Exercise 97: This exercise occurs in [350, Ex. 59], however there strong induction is used, although not needed. The proof is rather simple.

For each $n \geq 0$, let $S(n)$ be the statement that for any $m \geq 0$,

$$\sum_{i=0}^{n} \binom{m+i}{m} = \binom{m+n+1}{m+1}.$$

BASE STEP: When $n = 0$, $S(0)$ says $\sum_{i=0}^{0} \binom{m+i}{m} = \binom{m+1}{m+1}$, which is true since both sides are equal to 1.

INDUCTION STEP: Suppose that for some $k \geq 0$, $S(k)$ is true. To be shown is that $S(k+1)$ is true, that is, for any $m \geq 0$,

$$\sum_{i=0}^{k+1} \binom{m+i}{m} = \binom{m+k+2}{m+1}.$$

Starting with the left side of this equation,

$$\sum_{i=0}^{k+1} \binom{m+i}{m} = \sum_{i=0}^{k} \binom{m+i}{m} + \binom{m+k+1}{m}$$

$$= \binom{m+k+1}{m+1} + \binom{m+k+1}{m} \qquad \text{(by } S(k)\text{)}$$

$$= \binom{m+k+2}{m+1} \qquad \text{(by Pascal's id.)},$$

finishing the proof of $S(k+1)$ and hence the inductive step.

Therefore, by the principle of mathematical induction, for all $n \geq 0$, the statement $S(n)$ is true. $\qquad\square$

Exercise 98: This exercise occurs in [350, Ex. 60], where strong induction is used, however unnecessarily so. The equality in this exercise is the very same as that in Exercise 97, because binomial coefficients are symmetric, that is, for $M \leq N$, $\binom{n}{M} = \binom{n}{N-M}$; so in this case, $\binom{m+i}{i} = \binom{m+i}{m}$ and $\binom{m+n+1}{m+1} = \binom{m+n+1}{n}$ yield precisely the same problem! However, pretending not to notice this, here is a proof of the result by induction nevertheless.

For each $n \geq 0$, let $S(n)$ be the declaration that for every $m \geq 0$,

$$\sum_{i=0}^{n} \binom{m+i}{i} = \binom{m+n+1}{n}.$$

BASE STEP: $S(0)$ says $\sum_{i=0}^{0} \binom{m+i}{i} = \binom{m+1}{0}$, which is true since both sides equal 1.

INDUCTION STEP: For some $k \geq 0$, assume that $S(k)$ is true. To be shown is that $S(k+1)$ is true, that is, for any $m \geq 0$,

$$\sum_{i=0}^{k+1} \binom{m+i}{i} = \binom{m+k+2}{k+1}.$$

Beginning with the left side of this equation,

$$
\begin{aligned}
\sum_{i=0}^{k+1} \binom{m+i}{i} &= \sum_{i=0}^{k} \binom{m+i}{i} + \binom{m+k+1}{k+1} \\
&= \binom{m+k+1}{k} + \binom{m+k+1}{k+2} && \text{(by } S(k)\text{)} \\
&= \binom{m+k+2}{k+1} && \text{(by Pascal's id.)}
\end{aligned}
$$

completing the proof of $S(k+1)$, and hence the inductive step.

Therefore, by the principle of MI, for all $n \geq 0$, $S(n)$ is true. $\qquad\square$

Exercise 99: There are two choices for the variable to induct on. The most natural choice is to fix m and induct on $n \geq m$. There is a small problem here: while inducting on n, say, from $n = k$ to $n = k+1$, the truth of the statement is needed not just for k and m, but for k and $m - 1$ as well. Hence, fixing m in advance might be troublesome. Instead, induct on n, but for each n, "do" all $m \leq n$. In the induction step then, with $n = k+1$, the case $m = k+1$ is unavailable, and so this is handled separately.

Let $S(m,n)$ be the statement,

$$\sum_{i=0}^{n-m} (-1)^i \binom{n}{m+i} = \binom{n-1}{m-1}.$$

Let $T(n)$ be the statement that for all $m \leq n$, $S(m,n)$ holds.

BASE STEP($n = 1$): When $n = 1$, the only choice for m is $m = 1$, in which case $T(1) = S(1,1)$ says $\sum_{i=0}^{1-1} (-1)^i \binom{1}{1+i} = \binom{1-1}{1-1}$; this sum has only one term (when $i = 0$), which is $\binom{1}{1} = 1$, and since $\binom{1-1}{1-1} = 1$ as well, $T(1)$ is true.

INDUCTION STEP: For some $k \geq 1$, assume that $T(k)$ is true, that is for all $m = 1, \ldots, k$, $S(m, k)$, is true. That is, assume that for all $m = 1, \ldots, k$,

$$\sum_{i=0}^{k-m} (-1)^i \binom{k}{m+i} = \binom{k-1}{m-1}.$$

It remains to show $T(k+1)$, that is, to show that for all $m = 1, \ldots, k, k+1$,

$$\sum_{i=0}^{k+1-m} (-1)^i \binom{k+1}{m+i} = \binom{k}{m-1}.$$

Examine two cases, $m \leq k$, and $m = k+1$.

First, let $m \leq k$. Then

$$\sum_{i=0}^{k+1-m} (-1)^i \binom{k+1}{m+i}$$

$$= \sum_{i=0}^{k+1-m} (-1)^i \left[\binom{k}{m+i} + \binom{k}{m+i-1} \right]$$

$$= \sum_{i=0}^{k-m} (-1)^i \binom{k}{m+i} + (-1)^{k+1-m} \binom{k}{k+1} + \sum_{i=0}^{k+1-m} (-1)^i \binom{k}{m+i-1}$$

$$= \sum_{i=0}^{k-m} (-1)^i \binom{k}{m+i} + 0 + \sum_{i=0}^{k-(m-1)} (-1)^i \binom{k}{m-1+i}$$

$$= \binom{k-1}{m-1} + \binom{k-1}{m-2} \quad \text{(by ind. hyp. } S(m, k) \text{ and } S(m-1, k))$$

$$= \binom{k+1}{m-1},$$

finishing the inductive step for the case $m \leq k$.

Let $m = k+1$. Then

$$\sum_{i=0}^{k+1-(k+1)} (-1)^i \binom{k+1}{k+1+i} = \sum_{i=0}^{0} (-1)^i \binom{k+1}{k+1+i} = \binom{k+1}{k+1} = 1;$$

on the other hand, $\binom{k+1-1}{m-1} = \binom{k}{k} = 1$ as well. Even though the inductive hypothesis was not required here, this completes the inductive step when $m = k+1$. Hence $T(k) \rightarrow T(k+1)$, completing the inductive step.

Therefore, by mathematical induction, for each $n \geq 1$, $T(n)$ holds. $\square$

Exercise 100: For each $n \geq 1$, let $A(n)$ be the assertion that

$$\sum_{i=0}^{n}(-1)^i \binom{n}{i} = 0.$$

BASE STEP: $A(1)$ says $(-1)^0 \binom{1}{0} + (-1)^1 \binom{1}{1} = 0$, which is true.

INDUCTIVE STEP: Fix some $k \geq 1$ and assume that $A(k)$ is true. Then

$$\sum_{i=0}^{k+1}(-1)^i \binom{k+1}{i}$$

$$= \binom{k+1}{0} + (-1)^{k+1}\binom{k+1}{k+1} + \sum_{i=1}^{k}(-1)^i \binom{k+1}{i}$$

$$= 1 + (-1)^{k+1} + \sum_{i=1}^{k}(-1)^i \left[\binom{k}{i} + \binom{k}{i-1}\right]$$

$$= 1 + \left[\sum_{i=1}^{k}(-1)^i \binom{k}{i}\right] + \left[\sum_{i=1}^{k}(-1)^i \binom{k}{i-1}\right] + (-1)^{k+1}$$

$$= \binom{k}{0} + \left[\sum_{i=1}^{k}(-1)^i \binom{k}{i}\right] - \left[\sum_{j=0}^{k-1}(-1)^j \binom{k}{j}\right] - (-1)^k \binom{k}{k}$$

$$= \binom{k}{0} + \left[\sum_{i=1}^{k}(-1)^i \binom{k}{i}\right] - \left[\sum_{j=0}^{k}(-1)^j \binom{k}{j}\right]$$

$$= \binom{k}{0} + \sum_{i=1}^{k}(-1)^i \binom{k}{i} - 0 \quad \text{(by } A(k)\text{)}$$

$$= \sum_{i=0}^{k}(-1)^i \binom{k}{i}$$

$$= 0 \quad \text{(again by } A(k)\text{)}$$

shows that $A(k+1)$ is also true, completing the inductive step.

By the principle of mathematical induction, for all $n \geq 1$, $A(n)$ holds. $\square$

Exercise 101: This "well-known" identity is mentioned by Klee in [315], complete with proof, where it appears in discussion regarding the Euler characteristic. By convention, $\binom{-1}{0} = 1 = \binom{n}{0}$ and for $i > n$, $\binom{n}{i} = 0$. First, the exercise is done for $0 \leq m < n$. (See the technique used in Exercise 99.)

For each $n \geq 1$, let $Q(n)$ be the proposition that for every m satisfying $0 \leq m < n$,

$$P(m, n): \quad \sum_{i=0}^{m} (-1)^i \binom{n}{i} = (-1)^m \binom{n-1}{m}.$$

BASE STEP: When $n = 1$, the only possible m is $m = 0$. In this case, the above equation reads $(-1)^0 \binom{n}{0} = \binom{0}{0}$, and since both sides equal 1, $Q(1)$ is true.

INDUCTION STEP: For some fixed $k \geq 1$, assume that $Q(k)$ is true. It remains show that $Q(k+1)$ is also true, namely, that for every $m = 0, 1, \ldots, k$,

$$\sum_{i=0}^{m} (-1)^i \binom{k+1}{i} = (-1)^m \binom{k}{m}.$$

Divide this inductive step into two cases.

If $m < k$, then

$$\sum_{i=0}^{m} (-1)^i \binom{k+1}{i} = \binom{k+1}{0} + \sum_{i=1}^{m} (-1)^i \binom{k+1}{i}$$

$$= 1 + \sum_{i=1}^{m} \left[\binom{k}{i} + \binom{k}{i-1} \right]$$

$$= \sum_{i=0}^{m} (-1)^i \binom{k}{i} + \sum_{j=0}^{m-1} (-1)^{j+1} \binom{k}{j}$$

$$= (-1)^m \binom{k-1}{m} - \sum_{j=0}^{m-1} (-1)^j \binom{k}{j} \quad \text{(by } P(m, k))$$

$$= (-1)^m \binom{k-1}{m} - (-1)^{m-1} \binom{k-1}{m-1} \quad \text{(by } P(m-1, k))$$

$$= (-1)^m \binom{k-1}{m} + (-1)^m \binom{k-1}{m-1}$$

$$= (-1)^m \left[\binom{k-1}{m} + \binom{k-1}{m-1} \right]$$

$$= (-1)^m \binom{k}{m},$$

which completes the inductive step in this case, showing $P(m, k+1)$ where $m < k$.

For $m = k$, to be shown is $P(k, k+1) : \sum_{i=0}^{k}(-1)^i \binom{k+1}{i} = (-1)^k \binom{k}{k}$. This follows from equation (9.5) with $n = k+1$ by examining all but the last summand in the sum $\sum_{i=0}^{k+1}(-1)^k \binom{k+1}{i}$, or the binomial theorem and using $\binom{k+1}{k+1} = \binom{k}{k}$. So, in any case, $P(m, k+1)$ holds and so $Q(k+1)$ is true; so the inductive step holds.

Therefore, by mathematical induction, the result holds for all $0 \le m < n$.

To be shown is that the desired equality holds for any choice of non-negative integers m and n (not just $0 \le m < n$). If $m = n$, then the equality

$$\sum_{i=0}^{m}(-1)^i \binom{n}{i} = (-1)^m \binom{n-1}{m}$$

becomes

$$\sum_{i=0}^{n}(-1)^i \binom{n}{i} = 0,$$

precisely the statement in equation (9.5). If $n < m$, the right-hand side remains 0, and the additional summands on the left are all zero, so the result still holds. $\square$

Exercise 102: This problem appeared in [350, Ex. 62].

For each $n \ge 0$, let $S(n)$ denote the expression

$$\sum_{i=0}^{n} \binom{i}{2}\binom{i}{5} = \binom{n+1}{2}\binom{n+1}{6} + \binom{n+2}{8} - n\binom{n+2}{7}.$$

BASE STEP: When $n = 0$, both sides are 0; similarly, one can verify that both sides are 0 for $n = 1, 2, 3, 4$. When $n = 5$, $S(n)$ reduces to

$$\binom{5}{2}\binom{5}{5} = \binom{6}{2}\binom{6}{6} + \binom{7}{8} - 5\binom{7}{7},$$

or $10 = 15 + 0 - 5$ which is correct, so $S(5)$ is true. The case $n = 5$ can be the base case for the induction (although $n = 0$ works, too, but with $n = 0$, there are a lot of trivial binomial coefficients floating around, and applying certain tricks, like Pascal's identity, requires more checking to make sure things make sense with zeros everywhere).

INDUCTIVE STEP: Fix some $k \ge 5$ and suppose that

$$S(k) : \sum_{i=0}^{k} \binom{i}{2}\binom{i}{5} = \binom{k+1}{2}\binom{k+1}{6} + \binom{k+2}{8} - k\binom{k+2}{7}$$

is true. To be shown is that

$$S(k+1): \quad \sum_{i=0}^{k+1}\binom{i}{2}\binom{i}{5} = \binom{k+2}{2}\binom{k+2}{6} + \binom{k+3}{8} - (k+1)\binom{k+3}{7}$$

follows. (The sequence of equalities that works is a bit tricky, applying Pascal's identity four times, in two different ways.) Beginning with the left side of $S(k+1)$,

$$\sum_{i=0}^{k+1}\binom{i}{2}\binom{i}{5}$$

$$= \sum_{i=0}^{k}\binom{i}{2}\binom{i}{5} + \binom{k+1}{2}\binom{k+1}{5}$$

$$= \binom{k+1}{2}\binom{k+1}{6} + \binom{k+2}{8} - k\binom{k+2}{7} + \binom{k+1}{2}\binom{k+1}{5} \quad \text{(by } S(k)\text{)}$$

$$= \binom{k+1}{2}\left[\binom{k+1}{6} + \binom{k+1}{5}\right] + \binom{k+2}{8} - k\binom{k+2}{7}$$

$$= \binom{k+1}{2}\binom{k+2}{6} + \binom{k+2}{8} - k\binom{k+2}{7}$$

$$= \left[\binom{k+2}{2} - \binom{k+1}{1}\right]\binom{k+2}{6} + \left[\binom{k+3}{8} - \binom{k+2}{7}\right] - k\binom{k+2}{7}$$

$$= \binom{k+2}{2}\binom{k+2}{6} + \binom{k+3}{8} - (k+1)\binom{k+2}{7} - (k+1)\binom{k+2}{6}$$

$$= \binom{k+2}{2}\binom{k+2}{6} + \binom{k+3}{8} - (k+1)\left[\binom{k+2}{7} + \binom{k+2}{6}\right]$$

$$= \binom{k+2}{2}\binom{k+2}{6} + \binom{k+3}{8} - (k+1)\binom{k+3}{7},$$

which is equal to the right side of $S(k+1)$. This finishes the inductive step $S(k) \to S(k+1)$.

By the principle of mathematical induction, for all $n \geq 0$, the statement $S(n)$ is true. $\square$

Exercise 103: For each $n \geq 1$, let $S(n)$ be the statement

$$(1+x)^n = \binom{n}{0} + \binom{n}{1}x + \binom{n}{2}x^2 + \cdots + \binom{n}{n}x^n.$$

BASE STEP: $S(1)$ says $(1+x)^1 = \binom{1}{0} + \binom{1}{1}x$ which is valid.

INDUCTIVE STEP: Suppose that for some $k \geq 1$, $S(k)$ is true. Then

$$
\begin{aligned}
(1+x)^{k+1} &= (1+x)^k(1+x) \\
&= \left[\binom{k}{0} + \binom{k}{1}x + \cdots + \binom{k}{k}x^k \right](1+x) \quad \text{(by } S(k)\text{)} \\
&= \binom{k}{0} + \binom{k}{1}x + \cdots + \binom{k}{k}x^k \\
&\quad + \binom{k}{0}x + \binom{k}{1}x^2 + \cdots + \binom{k}{k}x^{k+1} \\
&= \binom{k}{0} + \sum_{i=1}^{k}\left[\binom{k}{i} + \binom{k}{i-1} \right]x^i + \binom{k}{k}x^{k+1} \\
&= \binom{k+1}{0} + \sum_{i=1}^{k}\binom{k+1}{i}x^i + \binom{k+1}{k+1}x^{k+1} \quad \text{(Pascal's id.)} \\
&= \binom{k+1}{0} + \binom{k+1}{1}x^1 + \binom{k+1}{2}x^2 + \cdots + \binom{k+1}{k+1}x^{k+1},
\end{aligned}
$$

finishing the proof of $S(k+1)$, and hence the inductive step.

Therefore, by MI, $S(n)$ holds for all $n \geq 1$. $\qquad\square$

Exercise 104: The letters x and y are variables indicating non-zero values for which $x^0 = 1$ and $y^0 = 1$ make sense—these values can be taken from any number field, for example. For each $n \geq 1$, let $S(n)$ be the statement

$$
(x+y)^n = \sum_{j=0}^{n}\binom{n}{j}x^{n-j}y^j.
$$

BASE STEP: $S(1)$ says $(x+y)^1 = \binom{1}{0}x + \binom{1}{1}y$ which is valid.

INDUCTIVE STEP: For some $k \geq 1$, suppose that $S(k)$ is true. To complete the inductive step, it suffices to prove that

$$
S(k+1): \quad (x+y)^{k+1} = \sum_{j=0}^{k+1}\binom{k+1}{j}x^{k+1-j}y^j
$$

holds. To prove $S(k+1)$, the same trick is used as in the inductive step for Exercise 103, namely break off one factor, apply the inductive hypothesis, use distributivity, combine like terms (that is, ones with the same $x^a y^b$'s), and use Pascal's equality. Just how to collect like terms is most easily seen by writing out a few terms in each sum, however due to space constraints, sigma notation is used throughout. [If the notation is at all confusing, the reader might write out the sigma expressions over two lines, lining up like terms.] Beginning with the left side of $S(k+1)$,

$$(x+y)^{k+1}$$

$$= (x+y)^k(x+y)$$

$$= \left[\sum_{j=0}^{k}\binom{k}{j}x^{k-j}y^j\right](x+y) \quad (\text{by } S(k))$$

$$= \left[\sum_{j=0}^{k}\binom{k}{j}x^{k-j}y^j\right]x + \left[\sum_{j=0}^{k}\binom{k}{j}x^{k-j}y^j\right]y$$

$$= \sum_{j=0}^{k}\binom{k}{j}x^{k+1-j}y^j + \sum_{j=0}^{k}\binom{k}{j}x^{k-j}y^{j+1}$$

$$= \binom{k}{0}x^{k+1}y^0 + \sum_{j=1}^{k}\binom{k}{j}x^{k+1-j}y^j + \sum_{j=0}^{k-1}\binom{k}{j}x^{k-j}y^{j+1} + \binom{k}{k}y^{k+1}$$

$$= x^{k+1}y^0 + \sum_{j=1}^{k}\binom{k}{j}x^{k+1-j}y^j + \sum_{j=1}^{k}\binom{k}{j-1}x^{k+1-j}y^j + y^{k+1}$$

$$= \binom{k+1}{0}x^{k+1}y^0 + \sum_{j=1}^{k}\left[\binom{k}{j}+\binom{k}{j-1}\right]x^{k+1-j}y^j + \binom{k+1}{k+1}y^{k+1}$$

$$= \binom{k+1}{0}x^{k+1}y^0 + \sum_{j=1}^{k}\binom{k+1}{j}x^{k+1-j}y^j + \binom{k+1}{k+1}y^{k+1} \quad (\text{Pasc. id.})$$

$$= \sum_{j=0}^{k+1}\binom{k+1}{j}x^{k+1-j}y^j,$$

finishing the proof of $S(k+1)$, and hence the inductive step.

Therefore, by MI, $S(n)$ holds for all $n \geq 1$. □

Exercise 106: Let $S(n)$ be the statement

$$x^n - y^n = (x-y)\left(\sum_{\nu=1}^{n}x^{n-\nu}y^{\nu-1}\right).$$

BASE STEP: When $n = 1$,

$$x^1 - y^1 = (x - y)x^0 y^0 = (x - y)\left(\sum_{\nu=1}^{1} x^{1-\nu} y^{\nu-1}\right),$$

and so $S(1)$ is true.

INDUCTIVE STEP: For some fixed $k \geq 1$, assume that $S(k)$ is true, that is,

$$x^k - y^k = (x - y)\left(\sum_{\nu=1}^{k} x^{k-\nu} y^{\nu-1}\right).$$

Next to show is that $S(k+1)$

$$x^{k+1} - y^{k+1} = (x - y)\left(\sum_{\nu=1}^{k+1} x^{k+1-\nu} y^{\nu-1}\right)$$

is true. To accomplish this, perform a trick of adding and subtracting an extra term:

$$\begin{aligned}
x^{k+1} - y^{k+1} &= x^{k+1} - xy^k + xy^k - y^{k+1} \\
&= x(x^k - y^k) + (x - y)y^k \\
&= x(x - y)\left(\sum_{\nu=1}^{k} x^{k-\nu} y^{\nu-1}\right) + (x - y)y^k \quad \text{(by } S(k)\text{)} \\
&= (x - y)\left[x\left(\sum_{\nu=1}^{k} x^{k-\nu} y^{\nu-1}\right) + y^k\right] \\
&= (x - y)\left[\left(\sum_{\nu=1}^{k} x^{k+1-\nu} y^{\nu-1}\right) + y^k\right] \\
&= (x - y)\left[\sum_{\nu=1}^{k+1} x^{k+1-\nu} y^{\nu-1}\right],
\end{aligned}$$

thereby proving $S(k+1)$.

Mathematical induction proves that for all $n \geq 1$, the statement $S(n)$ is true. $\quad\square$

Exercise 107: For each $n \geq 2$ and $0 < j < n$, let $P(n, j)$ be the statement that $\sum_{k=0}^{n} \binom{n}{k}(-1)^{n-k} k^j = 0$. As suggested in the exercise, one proof is by induction on j. Note that when $j = 0$, if one interprets 0^0 as 1, then $P(n, 0)$ is true by equation (9.5).

BASE STEP $(P(n,1))$: When $n \geq 2$, $P(n,1)$ is simply equation (9.7).

INDUCTIVE STEP: $([P(n,i-1) \wedge P(n-1,i-1)] \to P(n,i))$: Fix $i \geq 2$ and let $n > i$. Suppose that both $P(n,i-1)$ and $P(n-1,i-1)$ are true. Then

$$\sum_{k=0}^{n} \binom{n}{k}(-1)^{n-k}k^i$$

$$= \left(\sum_{k=1}^{n-1} \binom{n}{k}(-1)^{n-k}k^i\right) + n^i$$

$$= \left(\sum_{k=1}^{n-1} \frac{n}{k}\binom{n-1}{k-1}(-1)^{n-k}k^i\right) + n^i \qquad \text{(by Lemma 9.6.1)}$$

$$= n\left(\sum_{k=1}^{n-1} \binom{n-1}{k-1}(-1)^{n-k}k^{i-1}\right) + n^i$$

$$= n\left(\sum_{k=1}^{n-1} \left[\binom{n}{k} - \binom{n-1}{k}\right](-1)^{n-k}k^{i-1}\right) + n^i \qquad \text{(by Pascal's id.)}$$

$$= n\left(\sum_{k=0}^{n-1} \binom{n}{k}(-1)^{n-k}k^{i-1}\right) + (-1)^0\binom{n}{n}n^i - n\sum_{k=0}^{n-1}\binom{n-1}{k}(-1)^{n-k}k^{i-1}$$

$$= n\sum_{k=0}^{n} \binom{n}{k}(-1)^{n-k}k^{i-1} - n\sum_{k=0}^{n-1}\binom{n-1}{k}(-1)^{n-k}k^{i-1}$$

$$= n\cdot 0 - n\cdot 0 \qquad \text{(by } P(n,i-1) \text{ and } P(n-1,i-1)),$$

shows that $P(n,i)$ holds, completing the inductive step.

By mathematical induction, for all $j \geq 1$, and any $n > j$, $P(n,j)$ is true. $\qquad\square$

Exercise 108 [Abel identity 1]: Let $a \in \mathbb{R}$ and for each $n \geq 1$, let $S(n)$ denote the statement

$$\sum_{k=0}^{n} \binom{n}{k}x(x-ka)^{k-1}(y+ka)^{n-k} = (x+y)^n.$$

It seems quite difficult to prove $S(n)$ by a standard inductive argument, so a trick contained in the following lemma (which needs a bit of calculus for its proof) is used.

Lemma 27.2.1. *Polynomials $p(x,y)$ and $q(x,y)$ agree if and only if both $p(x,-x) = q(x,-x)$, and*

$$\frac{\partial p}{\partial y} = \frac{\partial q}{\partial y}.$$

Proof outline: If p and q are differentiable functions with domain $\mathbb{R}^2$, the conditions imply that for every fixed x_0, the restriction of p and q to single variable functions $f(y) = p(x_0, y)$ and $g(y) = q(x_0, y)$ is identical because they agree at the point $y = -x_0$ and their derivatives are the same. (This is a standard result in first-year calculus, following easily from the Mean Value Theorem.) Then use the fact that if two polynomials agree everywhere, they must be the same polynomial. $\square$

Let $p_n(x, y)$ be the left-hand side of $S(n)$ and let $q_n(x, y)$ be the right side. The conditions in Lemma 27.2.1 are proved separately; denote these two statements by

$$T(n): \quad p_n(x, -x) = q_n(x, -x),$$

and

$$U(n): \quad \frac{\partial p_n}{\partial y} = \frac{\partial q_n}{\partial y}.$$

Statement $T(n)$ is proved directly, and $U(n)$ is proved by induction.

Proof of T(n): First rewrite

$$T(n): \quad x \sum_{k=0}^{n} \binom{n}{k} (-1)^{n-k} (x - ka)^{n-1} = 0.$$

Indeed, by expanding the term $(x - ka)^{n-1}$ and reinterpreting the order of summation, find an expression equal to zero inside:

$$x \sum_{k=0}^{n} \binom{n}{k} (-1)^{n-k} (x - ka)^{n-1}$$

$$= x \sum_{k=0}^{n} \binom{n}{k} (-1)^{n-k} \sum_{j=0}^{n-1} (-ka)^j x^{n-j} \quad \text{(by binomial thm)}$$

$$= x \sum_{k=0}^{n} \binom{n}{k} (-1)^{n-k} \sum_{j=0}^{n-1} \binom{n-1}{j} (-ka)^j x^{n-j}$$

$$= x \sum_{j=0}^{n-1} a^j x^{n-j} \binom{n-1}{j} (-1)^j \sum_{k=0}^{n} \binom{n}{k} (-1)^{n-k} k^j$$

$$= x \sum_{j=0}^{n-1} a^j x^{n-j} \binom{n-1}{j} (-1)^j \cdot 0 \quad \text{(by Exercise 107)}$$

$$= 0$$

as desired.

Proof of U(n):

BASE STEP: For $n = 1$, it is straightforward to check that both partials equal 1.

INDUCTIVE STEP: For some fixed $m \geq 1$, assume that $U(m)$ holds. Then by $T(m)$, $S(m)$ holds. To prove that $U(m+1)$ follows from $S(m)$,

$$\frac{\partial p_{m+1}}{\partial y} = \frac{\partial}{\partial y}[p_{m+1}(x, y)]$$

$$= \frac{\partial}{\partial y}\left[\sum_{k=0}^{m+1} \binom{m+1}{k} x(x - ka)^{k-1}(y + ka)^{m+1-k}\right]$$

$$= \frac{\partial}{\partial y}\left[x(x - (m+1)a)^m + \sum_{k=0}^{m} \binom{m+1}{k} x(x - ka)^{k-1}(y + ka)^{m+1-k}\right]$$

$$= \frac{\partial}{\partial y}\left[\sum_{k=0}^{m} \binom{m+1}{k} x(x - ka)^{k-1}(y + ka)^{m+1-k}\right]$$

$$= \sum_{k=0}^{m} \binom{m+1}{k}(m+1-k)x(x - ka)^{k-1}(y + ka)^{m-k}$$

$$= \sum_{k=0}^{m}(m+1)\binom{m}{k} x(x - ka)^{k-1}(y + ka)^{m-k}$$

$$= (m+1)\sum_{k=0}^{m} \binom{m}{k} x(x - ka)^{k-1}(y + ka)^{m-k}$$

$$= (m+1)(x + y)^m \qquad \text{(by } S(m))$$

$$= \frac{\partial}{\partial y}[(x + y)^{m+1}],$$

which is the partial with respect to y of $q_{m+1}(x, y)$ as desired. This completes the inductive step $U(m) \rightarrow U(m+1)$.

Hence, by MI, for every $n \geq 1$, $U(n)$ is true, and so it follows that for every $n \geq 1$, $S(n)$ is true.

Exercise 109 [Abel identity 2]: A proof appeared in [354, Prob. 1.44(a)]. Here is one outline of a proof. For each $n \geq 1$, let $S(n)$ denote the statement

$$\sum_{k=0}^{n} \binom{n}{k} x(x + k)^{k-1}(y + n - k)^{n-k} = (x + y + n)^n.$$

A direct proof of $S(n)$ is available by Exercise 108 with $a = -1$ and replacing y with $y + n$. The inductive proof can be found by imitating the solution of Exercise 108 using these replacements. □

Exercise 110: For $m \geq 0$, let $A(m)$ be the assertion that $S_m(n)$ is a polynomial in n of degree $m + 1$ with constant term 0.

BASE STEP: $S_0(n) = 1^0 + 2^0 + \cdots + n^0 = n$, which is a polynomial of degree 1.

INDUCTIVE STEP: Fix $p \geq 0$, and assume that for $0 \leq j < p$ that $A(j)$ holds, that is, each of $S_0(n), \ldots, S_p(n)$ is a polynomial of appropriate degree with constant term 0. Using $m = p$ in (9.9),

$$\binom{p+1}{p} S_p(n) = (n+1)^{p+1} - \sum_{j=0}^{p-1} \binom{p+1}{j} S_j(n). \tag{27.1}$$

By induction hypothesis, for each $j < p$, $S_j(n)$ is a polynomial in n of degree $j + 1$ with constant term 0, and so the right-hand side of (27.1) is a polynomial in n of degree $p + 1$ with constant term 0. Thus $S_p(n)$ is a polynomial in n of degree $p + 1$ with constant term 0, proving $A(p)$, completing the inductive step.

By mathematical induction, for each $m \geq 0$, $S_m(n)$ is a polynomial in n of degree $m + 1$ and with constant term 0. □

27.3 Solutions: Trigonometry

Exercise 113: This exercise appeared, *e.g.*, in [582, Prob. 39]. There are many proofs; one simple proof is given here with two base cases and an inductive step that jumps by two (similar to the proof of Theorem 3.4.1, where there are three base cases).

Let $S(n)$ denote the proposition $\cos(n\pi) = (-1)^n$.

BASE STEPS ($n = 1, 2$): Since $\cos(\pi) = -1$, $S(1)$ holds. Since $\cos(2\pi) = 1 = (-1)^2$, and so $S(2)$ also holds.

INDUCTIVE STEP ($S(m) \rightarrow S(m + 2)$): Suppose, for some $m \geq 1$, that $S(m)$ holds. Recall the identity $\cos(\theta + 2\pi) = \cos(\theta)$. Starting with the left side of $S(m + 2)$,

$$\begin{aligned}
\cos((m+2)\pi) &= \cos(m\pi + 2\pi) \\
&= \cos(m\pi) \text{ (by above identity with } \theta = m\pi) \\
&= (-1)^m \\
&= (-1)^{m+2},
\end{aligned}$$

which is the right side of $S(m + 2)$. This completes the inductive step $S(m) \rightarrow S(m + 2)$.

By the principle of mathematical induction, (or as referred to in Section 3.4, "alternative mathematical induction") for all $n \geq 1$, $S(n)$ holds. □

Exercise 114: Let $S(n)$ be the statement

$$S(n): \quad \text{for any real } x, \ |\sin(nx)| \leq n|\sin(x)|.$$

BASE STEP: The statement $S(1)$ says $|\sin(x)| \leq |\sin(x)|$, which is trivially true.

INDUCTIVE STEP: Fix some $k \geq 1$ and assume that

$$S(k): \quad |\sin(kx)| \leq k|\sin(x)|$$

holds. To be proved is that

$$S(k+1): \quad |\sin((k+1)x)| \leq (k+1)|\sin(x)|$$

holds. Beginning with the left-hand side of $S(k+1)$,

$$
\begin{aligned}
|\sin((k+1)x)| &= |\sin(kx + x)| \\
&= |\sin(kx)\cos(x) + \cos(kx)\sin(x)| \\
&\leq |\sin(kx)\cos(x)| + |\cos(kx)\sin(x)| \quad \text{(triangle ineq.)} \\
&= |\sin(kx)| \cdot |\cos(x)| + |\cos(kx)| \cdot |\sin(x)| \\
&\leq |\sin(kx)| + |\sin(x)| \quad \text{(because } |\cos(\theta)| \leq 1) \\
&\leq k|\sin(x)| + |\sin(x)| \quad \text{(by ind. hyp.)} \\
&\leq (k+1)|\sin(x)|,
\end{aligned}
$$

the right-hand side of $S(k+1)$, completing the inductive step.

By mathematical induction, for all $n \geq 1$, the statement $S(n)$ is true. □

Exercise 115 (De Moivre's Theorem): Let $D(n)$ be the statement

$$D(n): \quad [\cos(\theta) + i\sin(\theta)]^n = \cos(n\theta) + i\sin(n\theta).$$

BASE STEP: $D(1)$ is trivially true because it says only $[\cos(\theta) + i\sin(\theta)]^1 = \cos(\theta) + i\sin(\theta)$.

INDUCTIVE STEP: Let $k \geq 1$ be fixed and assume that $D(k)$ is true. Then

$$
\begin{aligned}
[\cos(\theta) &+ i\sin(\theta)]^{k+1} \\
&= (\cos(\theta) + i\sin(\theta))[\cos(\theta) + i\sin(\theta)]^k \\
&= (\cos(\theta) + i\sin(\theta))(\cos(k\theta) + i\sin(k\theta)) \quad \text{(by } D(k))
\end{aligned}
$$

$$= \cos(\theta)\cos(k\theta) - \sin(\theta)\sin(k\theta) + i(\sin(\theta)\cos(k\theta) + \cos(\theta)\sin(k\theta))$$
$$= \cos(\theta + k\theta) + i\sin(\theta + k\theta) \quad \text{(by eq'ns (9.12) and (9.11))}$$
$$= \cos((k+1)\theta) + i\sin((k+1)\theta)$$

proves $D(k+1)$, completing the inductive step.

By MI, for each $n \geq 1$, $D(n)$ is true, completing the inductive proof of De Moivre's theorem (also called De Moivre's formula). □

Exercise 116: For any $n \geq 1$ let

$$S(n): \quad \sin(\theta + n\pi) = (-1)^n \sin(\theta)$$

denote the statement in the exercise. In fact, $S(n)$ is defined for any integer n—see remark following the proof. The identity $\sin(\alpha + \pi) = -\sin(\alpha)$ is relied on, which can be proved either by noticing that the angles α and $\alpha + \pi$ correspond to antipodal points on the unit circle, or directly by equation (9.11) as follows:

$$\sin(\alpha + \pi) = \sin(\alpha)\cos(\pi) + \cos(\alpha)\sin(\pi) = \sin(\alpha)(-1) + \cos(\alpha) \cdot 0 = -\sin(\alpha).$$

BASE STEP: Since $\sin(\theta + \pi) = -\sin(\theta)$, $S(n)$ is true. [Notice that $S(0)$ is true as well, so the base case could have been $n = 0$.]
INDUCTIVE STEP: For some fixed $k \geq 1$, assume that $S(k)$ is true. Then

$$\begin{aligned}
\sin(\theta + (k+1)\pi) &= \sin((\theta + k\pi) + \pi) \\
&= -\sin(\theta + k\pi) \\
&= -(-1)^k \sin(\theta) \quad \text{(by } S(k)) \\
&= (-1)^{k+1} \sin(\theta)
\end{aligned}$$

proves $S(k+1)$, completing the inductive step $S(k) \to S(k+1)$.

Therefore, by MI, for all $n \geq 1$ the statement $S(n)$ is true. □

Remark: Notice that since $S(0)$ is also true, one could have concluded that $S(n)$ is true for all $n \geq 0$. In fact, $S(n)$ is true for all $n \in \mathbb{Z}$; this can be shown in a number of ways. One could prove this by induction for the negative integers using $\sin(\alpha - \pi) = -\sin(\alpha)$, and imitating the proof above, showing $\sin(\theta - n\pi) = (-1)^n \sin(\theta)$ for $n \geq 0$. One can also see that $\sin(\theta - n\pi) = (-1)^n \sin(\theta)$ follows directly from $S(n)$ by

$$\sin(\theta - n\pi) = \sin(\theta + n\pi - n2\pi) = \sin(\theta + n\pi).$$

Lastly, $\sin(\theta - n\pi) = (-1)^n \sin(\theta)$ follows from $S(n)$ using equation (9.11) and Exercise 113:

$$\sin(\theta - n\pi) = \sin(\theta)\cos(-n\pi) + \cos(\theta)\sin(-n\pi)$$

$$
\begin{aligned}
&= \sin(\theta)\cos(n\pi) + \cos(\theta) \cdot 0 \\
&= \sin(\theta)(-1)^n.
\end{aligned}
$$

Exercise 117: This exercise appeared in, for example, [550].
For each $n \geq 1$, let

$$
E(n): \quad \cos(\theta + n\pi) = (-1)^n \cos(\theta)
$$

denote the equality in the exercise. Use the identity $\cos(\alpha + \pi) = -\cos(\alpha)$ which follows because the angles $\alpha + \pi$ and α correspond to antipodal points on the unit circle; one can also derive this identity by applying equation (9.12) as follows: $\cos(\alpha + \pi) = \cos(\alpha)\cos(\pi) - \sin(\alpha)\sin(\pi) = \cos(\alpha)(-1) - \sin(\alpha) \cdot 0 = -\cos(\alpha)$.

BASE STEP: By the identity mentioned above, $\cos(\theta + \pi) = -\cos(\theta)$, and so $E(1)$ is true. (Notice that $E(0)$ is also true, so this could have been the base case.)
INDUCTIVE STEP: For some fixed $k \geq 1$, assume that $E(k)$ is true. Then

$$
\begin{aligned}
\cos(\theta + (k+1)\pi) &= \cos((\theta + k\pi) + \pi) \\
&= -\cos(\theta + k\pi) \\
&= -(-1)^k \cos(\theta) \quad \text{(by } E(k)) \\
&= (-1)^{k+1} \cos(\theta)
\end{aligned}
$$

proves $E(k+1)$, completing the inductive step $E(k) \to E(k+1)$.

Therefore, by MI, for all $n \geq 1$ the statement $E(n)$ is true. $\qquad \square$

Remark: Notice that since $E(0)$ is also true, one could conclude that $E(n)$ is true for all $n \geq 0$. Furthermore, just as in Exercise 116, one can show that $E(n)$ is true for all $n \in \mathbb{Z}$.

Exercise 118: This exercise appeared in many places, for example, [550] and [499, Prob. 33].
For each $n \geq 1$, denote the statement in the exercise by

$$
S(n): \quad \sum_{j=1}^{n} \sin j\theta = \frac{\sin(\frac{n+1}{2}\theta)\sin(\frac{n\theta}{2})}{\sin(\theta/2)}.
$$

BASE STEP: The statement $S(1)$ says

$$
\sin(\theta) = \frac{\sin(\frac{1+1}{2}\theta)\sin(\frac{\theta}{2})}{\sin(\theta/2)},
$$

which is true.

INDUCTIVE STEP: For some fixed $k \geq 1$, suppose that

$$S(k) : \quad \sum_{j=1}^{k} \sin j\theta = \frac{\sin(\frac{k+1}{2}\theta) \sin(\frac{k\theta}{2})}{\sin(\theta/2)}$$

is true. It remains to prove that

$$S(k+1) : \quad \sum_{j=1}^{k+1} \sin j\theta = \frac{\sin(\frac{k+2}{2}\theta) \sin(\frac{(k+1)\theta}{2})}{\sin(\theta/2)}$$

is true.

To streamline the proof of $S(k+1)$, use the identity

$$2 \cos \left(\frac{k+1}{2}\theta \right) \sin(\theta/2) = \sin \left(\frac{k+2}{2}\theta \right) - \sin(k\theta/2), \qquad (27.2)$$

which follows from using $A = \frac{k+1}{2}\theta$ and $B = \frac{1}{2}\theta$ in the following identity

$$\begin{aligned}
&\sin(A+B) - \sin(A-B) \\
&= \sin A \cos B + \cos A \sin B - (\sin A \cos(-B) + \cos A \sin(-B)) \\
&= \sin A \cos B + \cos A \sin B - (\sin A \cos B - \cos A \sin B) \\
&= 2 \cos A \sin B.
\end{aligned}$$

Beginning with the left side of $S(k+1)$,

$$\sum_{j=1}^{k+1} \sin(j\theta)$$

$$= \left[\sum_{j=1}^{k} \sin(j\theta) \right] + \sin((k+1)\theta)$$

$$= \frac{\sin(\frac{k+1}{2}\theta) \sin(\frac{k\theta}{2})}{\sin(\theta/2)} + \sin((k+1)\theta) \quad \text{(by } S(k))$$

$$= \frac{\sin \left(\frac{k+1}{2}\theta \right) \sin \left(\frac{k\theta}{2} \right)}{\sin(\theta/2)} + \sin \left(\frac{k+1}{2}\theta + \frac{k+1}{2}\theta \right)$$

$$= \frac{\sin(\frac{k+1}{2}\theta) \sin(\frac{k\theta}{2})}{\sin(\theta/2)} + 2 \sin \left(\frac{k+2}{2}\theta \right) \cos \left(\frac{k+2}{2}\theta \right) \quad \text{(by eqn (9.11))}$$

$$= \sin \left(\frac{k+1}{2}\theta \right) \frac{\sin(k\theta/2) + 2 \cos \left(\frac{k+1}{2}\theta \right) \sin(\theta/2)}{\sin(\theta/2)}$$

$$= \sin \left(\frac{k+1}{2}\theta \right) \frac{\sin(k\theta/2) + \sin \left(\frac{k+2}{2}\theta \right) - \sin(k\theta/2)}{\sin(\theta/2)} \quad \text{(by eqn (27.2))}$$

$$= \sin \left(\frac{k+1}{2}\theta \right) \frac{\sin \left(\frac{k+2}{2}\theta \right)}{\sin(\theta/2)},$$

proves that $S(k+1)$ follows, completing the inductive step.

By MI, for every $n \geq 1$, $S(n)$ holds.   □

Exercise 119: This problem appeared in, for example, [550].

For each integer $n \geq 1$, let the statement $S(n)$ denote the statement

$$\cos\theta + \cos(2\theta) + \cdots + \cos(n\theta) = \frac{\cos(\frac{n+1}{2}\theta)\sin(\frac{n\theta}{2})}{\sin(\theta/2)}.$$

BASE STEP: Since

$$\frac{\cos(\frac{1+1}{2}\theta)\sin(\frac{\theta}{2})}{\sin(\theta/2)} = \cos(\theta),$$

$S(1)$ is true.

In a naive attempt to prove the inductive step, one sees that a rather strange identity is necessary; one which boils down to the following:

$$\cos(A+B)\sin(A) + \cos(2A+2B)\sin(B) = \cos(A+2B)\sin(A+B). \qquad (27.3)$$

The following sequence of identities shows one way to derive equation (27.3), though there might be other simpler ways: Beginning with

$$\sin A(1 - \sin^2 B) = \sin A \cos^2 B,$$

and adding $\cos A \cos B \sin B$ to each side gives

$$\sin A + \cos A \cos B \sin B - \sin A \sin B \sin B = \sin A \cos B \cos B + \cos A \sin B \cos B.$$

First use the identities (9.12) and (9.11) and then multiply each side by $\cos(A+B)$, giving

$$\cos(A+B)[\sin A + \cos(A+B)\sin B] = \cos(A+B)\sin(A+B)\cos B.$$

Multiplying out the left side,

$$\cos(A+B)\sin A + \cos^2(A+B)\sin B = \cos(A+B)\sin(A+B)\cos B.$$

Subtracting $\sin^2(A+B)\sin B$ from each side gives

$$\cos(A+B)\sin A + [\cos^2(A+B) - \sin^2(A+B)]\sin B$$
$$= [\cos(A+B)\cos B - \sin(A+B)\sin B]\sin(A+B).$$

Finally, applying identity (9.12) twice, on the left with $\alpha = A+B = \beta$, and on the right with $\alpha = A+B$ and $\beta = B$, arrive at

$$\cos(A+B)\sin(A) + \cos(2A+2B)\sin(B) = \cos(A+2B)\sin(A+B),$$

finishing the proof of equation (27.3)

INDUCTIVE STEP: For some fixed $k \geq 1$, assume that

$$S(k): \quad \sum_{j=1}^{k} \cos j\theta = \frac{\cos(\frac{k+1}{2}\theta)\sin(\frac{k\theta}{2})}{\sin(\theta/2)}$$

is true. It remains to prove that

$$S(k+1): \quad \sum_{j=1}^{k+1} \cos j\theta = \frac{\cos(\frac{k+2}{2}\theta)\sin(\frac{(k+1)\theta}{2})}{\sin(\theta/2)}$$

follows. Beginning with the left side of $S(k+1)$,

$$
\begin{aligned}
\sum_{j=1}^{k+1} \cos(j\theta) &= \left[\sum_{j=1}^{k} \cos(j\theta)\right] + \cos((k+1)\theta) \\
&= \frac{\cos(\frac{k+1}{2}\theta)\sin(\frac{k\theta}{2})}{\sin(\theta/2)} + \cos((k+1)\theta) \quad \text{(by } S(k)) \\
&= \frac{\cos(\frac{k+1}{2}\theta)\sin(\frac{k\theta}{2}) + \cos((k+1)\theta)\sin(\frac{\theta}{2})}{\sin(\theta/2)} \\
&= \frac{\cos(\frac{k+2}{2}\theta)\sin(\frac{(k+1)\theta}{2})}{\sin(\theta/2)} \quad \text{(by eq'n 27.3)}
\end{aligned}
$$

where the last equality follows from equation (27.3) with $A = k\theta/2$ and $B = \theta/2$. This completes the proof of $S(k+1)$ and hence the inductive step.

By MI, for all $n \geq 1$ the statement $S(n)$ is true. $\square$

Exercise 120: This problem appeared in, for example, [550].

For $n \geq 1$, let the statement in the exercise be denoted by

$$P(n): \quad \sin(\theta) + \sin(3\theta) + \cdots + \sin((2n-1)\theta) = \frac{\sin^2(n\theta)}{\sin(\theta)}.$$

BASE STEP: Since $\sin(\theta) = \frac{\sin^2(\theta)}{\sin(\theta)}$, $P(1)$ is clearly true.

In the inductive step, the following identity is used:

$$\sin(A+B)\sin(A-B) + \sin^2(B) = \sin^2(A). \tag{27.4}$$

The proof of equation (27.4) is fairly straightforward:

$$
\begin{aligned}
&\sin(A+B)\sin(A-B) + \sin^2(B) \\
&= (\sin A \cos B + \cos A \sin B)(\sin A \cos(-B) + \cos A \sin(-B)) + \sin^2 B \\
&= (\sin A \cos B + \cos A \sin B)(\sin A \cos B - \cos A \sin B) + \sin^2 B
\end{aligned}
$$

$$
\begin{aligned}
&= \ \sin^2 A \cos^2 B - \cos^2 A \sin^2 B + \sin^2 B \\
&= \ \sin^2 A \cos^2 B + (1 - \cos^2 A) \sin^2 B \\
&= \ \sin^2 A \cos^2 B + \sin^2 A \sin^2 B \\
&= \ \sin^2 A(\cos^2 B + \sin^2 B) \\
&= \ \sin^2 A.
\end{aligned}
$$

INDUCTIVE STEP: For some fixed $k \geq 1$, suppose that

$$
P(k): \ \sum_{j=1}^{k} \sin((2j - 1)\theta) = \frac{\sin^2(k\theta)}{\sin(\theta)}
$$

is true. It remains to show that

$$
P(k+1): \ \sum_{j=1}^{k+1} \sin((2j - 1)\theta) = \frac{\sin^2((k + 1)\theta)}{\sin(\theta)}
$$

follows. Indeed,

$$
\begin{aligned}
\sum_{j=1}^{k+1} \sin((2j - 1)\theta) &= \ \left[\sum_{j=1}^{k} \sin((2j - 1)\theta) \right] + \sin((2k + 1)\theta) \\
&= \ \frac{\sin^2(k\theta)}{\sin(\theta)} + \sin((2k + 1)\theta) \quad \text{(by } P(k)\text{)} \\
&= \ \frac{\sin^2(k\theta) + \sin((2k + 1)\theta) \sin(\theta)}{\sin(\theta)} \\
&= \ = \frac{\sin^2((k + 1)\theta)}{\sin(\theta)},
\end{aligned}
$$

where the last equality follows from equation (27.4) using $A = (k+1)\theta$ and $B = k\theta$. This finishes the proof of $P(k + 1)$, and hence the inductive step.

Therefore, by MI, for all $n \geq 1$, $P(n)$ is true. $\qquad\square$

Exercise 121: This exercise appeared, for example, in [550].

Fix some angle θ which is not a multiple of π. For $n \geq 1$, denote the statement in the exercise by

$$
S(n): \ \sum_{j=1}^{n} \cos (2j - 1)\theta = \frac{\sin(2n\theta)}{2\sin(\theta)}.
$$

BASE STEP: $S(1)$ says

$$
\cos(\theta) = \frac{\sin(2\theta)}{2\sin(\theta)},
$$

which is true because $\sin(2\theta) = 2\cos(\theta)\sin(\theta)$.

INDUCTIVE STEP: For some fixed $k \geq 1$, assume that

$$S(k) : \quad \sum_{j=1}^{k} \cos((2j-1)\theta) = \frac{\sin(2k\theta)}{2\sin(\theta)}$$

is true. To be proved is

$$S(k+1) : \quad \sum_{j=1}^{k+1} \cos((2j-1)\theta) = \frac{\sin(2(k+1)\theta)}{2\sin(\theta)}.$$

To this end, use the following identity:

$$\sin(2k\theta) + 2\cos((2k+1)\theta)\sin(\theta) = \sin(2(k+1)\theta). \qquad (27.5)$$

A proof of equation (27.5) might go as follows:

$$\begin{aligned}
&\sin(2k\theta) + 2\cos((2k+1)\theta)\sin(\theta) \\
={}& \sin(2k\theta) + \cos((2k+1)\theta)\sin(\theta) + \cos((2k+1)\theta)\sin(\theta) \\
={}& \sin(2k\theta) + \cos(2k\theta + \theta)\sin(\theta) + \cos((2k+1)\theta)\sin(\theta) \\
={}& \sin(2k\theta) + [\cos(2k\theta)\cos(\theta) - \sin(2k\theta)\sin(\theta)]\sin(\theta) + \cos((2k+1)\theta)\sin(\theta) \\
={}& \sin(2k\theta) - \sin(2k\theta)\sin^2(\theta) + \cos(2k\theta)\sin(\theta)\cos(\theta) + \cos((2k+1)\theta)\sin(\theta) \\
={}& \sin(2k\theta)\cos^2(\theta) + \cos(2k\theta)\sin(\theta)\cos(\theta) + \cos((2k+1)\theta)\sin(\theta) \\
={}& [\sin(2k\theta)\cos(\theta) + \cos(2k\theta)\sin(\theta)]\cos(\theta) + \cos((2k+1)\theta)\sin(\theta) \\
={}& [\sin(2k\theta + \theta)]\cos(\theta) + \cos((2k+1)\theta)\sin(\theta) \\
={}& [\sin((2k+1)\theta)]\cos(\theta) + \cos((2k+1)\theta)\sin(\theta) \\
={}& \sin((2k+1)\theta + \theta) \\
={}& \sin((2k+2)\theta).
\end{aligned}$$

Now that the tools are assembled, here is the proof of $S(k+1)$:

$$\begin{aligned}
\sum_{j=1}^{k+1} \cos(2j-1)\theta &= \left[\sum_{j=1}^{k} \cos((2j-1)\theta)\right] + \cos((2k+1)\theta) \\
&= \frac{\sin(2k\theta)}{2\sin(\theta)} + \cos((2k+1)\theta) \quad \text{(by } S(k)) \\
&= \frac{\sin(2k\theta) + 2\cos((2k+1)\theta)\sin(\theta)}{2\sin(\theta)} \\
&= \frac{\sin(2(k+1)\theta)}{2\sin(\theta)} \quad \text{(by eq'n (27.5)).}
\end{aligned}$$

This proves $S(k+1)$, and hence finishes the inductive step.

By MI, for every $n \geq 1$, $S(n)$ is true.   □

Exercise 122: This exercise appeared in *e.g.*, [550].
It is given that $s_0 = 0$, $s_1 = 1$, and for $n \geq 2$,

$$s_n = 2\cos(\theta)s_{n-1} - s_{n-2}.$$

To be proved is that for each $n \geq 0$ the statement

$$S(n): \quad s_n = \frac{\sin(n\theta)}{\sin(\theta)},$$

and for each $n \geq 1$,

$$C(n): \quad \cos(n\theta) = \cos(\theta)s_n - s_{n-1}$$

are true. Both $S(n)$ and $C(n)$ are proved by simple induction, inducting on each at the same time.

BASE STEP $S(0)$, $S(1)$, AND $C(1)$: Since s_n is defined by two previous values, one would think that there should be two base cases for $S(n)$. In fact, the only reason that two base cases are needed for $S(n)$ is to have $S(1)$, since the induction is started on $C(n)$ at $n = 1$ as well; the actual induction is simple (as opposed to strong).

$S(0)$ says that $s_0 = \frac{sin(0 \cdot \theta)}{sin(\theta)}$ which is true since both sides are 0. $S(1)$ says $s_1 = \frac{\sin(1 \cdot \theta)}{\sin(\theta)}$, also true since each side equals 1. Finally, $C(1)$ says $\cos(1 \cdot \theta) = \cos(\theta)s_1 - s_0$, which is true since $s_1 = 1$ and $s_0 = 0$.

INDUCTIVE STEP: For some fixed $k \geq 1$, assume that both $S(k)$ and $C(k)$ are true.

The inductive step is accomplished in two stages, one for $S(k+1)$, and the other for $C(k+1)$. First show

$$S(k+1): \quad s_{k+1} = \frac{\sin((k+1)\theta)}{\sin(\theta)}.$$

Here is the derivation:

$$
\begin{aligned}
s_{k+1} &= 2\cos(\theta)s_k - s_{k-1} \quad \text{(def'n of } s_n\text{)} \\
&= \cos(\theta)s_k + \cos(\theta)s_k - s_{k-1} \\
&= \cos(\theta)s_k + \cos(k\theta) \quad \text{(by } C(k)\text{)} \\
&= \cos(\theta)\frac{\sin(k\theta)}{\sin(\theta)} + \cos(k\theta) \quad \text{(by } S(k)\text{)} \\
&= \frac{\cos(\theta)\sin(k\theta) + \cos(k\theta)\sin(\theta)}{\sin(\theta)} \\
&= \frac{\sin(k\theta + \theta)}{\sin(\theta)} \\
&= \frac{\sin((k+1)\theta)}{\sin(\theta)},
\end{aligned}
$$

completing the proof of $S(k+1)$.

To finish the inductive step, one needs to prove

$$C(k+1): \cos((k+1)\theta) = \cos(\theta)s_{k+1} - s_k.$$

Having both $S(k)$ and $S(k+1)$ in hand,

$$
\begin{aligned}
\cos((k+1)\theta) &= \cos(\theta + k\theta) \\
&= \cos(\theta)\cos(k\theta) - \sin(\theta)\sin(k\theta) \\
&= \frac{\cos(\theta)\cos(k\theta)\sin(\theta) - \sin^2(\theta)\sin(k\theta)}{\sin(\theta)} \\
&= \frac{\cos(\theta)\cos(k\theta)\sin(\theta) - (1 - \cos^2(\theta))\sin(k\theta)}{\sin(\theta)} \\
&= \frac{\cos(\theta)\cos(k\theta)\sin(\theta) + \cos^2(\theta)\sin(k\theta) - \sin(k\theta)}{\sin(\theta)} \\
&= \frac{\cos(\theta)[\cos(k\theta)\sin(\theta) + \cos(\theta)\sin(k\theta)] - \sin(k\theta)}{\sin(\theta)} \\
&= \frac{\cos(\theta)\sin((k+1)\theta) - \sin(k\theta)}{\sin(\theta)} \\
&= \frac{\cos(\theta)\sin((k+1)\theta)}{\sin(\theta)} - \frac{\sin(k\theta)}{\sin(\theta)} \\
&= \cos(\theta)s_{k+1} - s_k \quad \text{(by } S(k+1) \text{ and } S(k)),
\end{aligned}
$$

completing the proof of $C(k+1)$. Since the implications $S(k) \wedge C(k) \to S(k+1)$ and $S(k) \wedge S(k+1) \to C(k+1)$, are proved, the implication

$$[S(k) \wedge C(k)] \to [S(k+1) \wedge C(k+1)]$$

is proved, completing the inductive step.

By MI, for all $n \geq 1$, $S(n)$ and $C(n)$ hold. Together with the base case $S(0)$, all that was required is proved. $\qquad\square$

Exercise 123: This exercise appears in [499, Prob. 32], together with a sketch of the solution.

Write $s_1 = \cos(\theta)$, $s_2 = \cos(2\theta)$ and for $n > 2$,

$$s_n = 2\cos(\theta)s_{n-1} - s_{n-2}.$$

For every $n \geq 1$, denote the assertion in the exercise by

$$A(n): \quad s_n = \cos(n\theta).$$

BASE STEP: Both $A(1)$ and $A(2)$ are true by definition of s_1 and s_2.

INDUCTIVE STEP: For some fixed $k \geq 2$, assume that both $A(k-1)$ and $A(k)$ are true. To be shown is that

$$A(k+1): \quad s_{k+1} = \cos((k+1)\theta)$$

follows. Indeed,

$$
\begin{aligned}
s_{k+1} &= 2\cos(\theta)\cos(k\theta) - \cos((k-1)\theta) \quad \text{(by } A(k) \text{ and } A(k-1)) \\
&= 2\cos(\theta)\cos(k\theta) - \cos(k\theta - \theta) \\
&= 2\cos(\theta)\cos(k\theta) - [\cos(k\theta)\cos(-\theta) - \sin(k\theta)\sin(-\theta)] \\
&= 2\cos(\theta)\cos(k\theta) - [\cos(k\theta)\cos(\theta) + \sin(k\theta)\sin(\theta)] \\
&= \cos(\theta)\cos(k\theta) - \sin(k\theta)\sin(\theta) \\
&= \cos(k\theta + \theta) \quad \text{(by identity (9.12))} \\
&= \cos((k+1)\theta),
\end{aligned}
$$

as desired. This completes the inductive step.

By MI, for all $n \geq 1$, $A(n)$ is true. $\qquad \square$

Exercise 124: Use Exercise 194 with x replaced by $\cos^2(x)$ and y replaced by $\sin^2(x)$. Then $x + y = 1$, and the result follows directly. $\qquad \square$

Exercise 125: This exercise appears in many places, for example, [499, Prob. 31], where a brief solution outline is given.

For $n \geq 0$, let $A(n)$ be the assertion that

$$\cos(\alpha)\cos(2\alpha)\cos(4\alpha)\cdots\cos(2^n\alpha) = \frac{\sin(2^{n+1}\alpha)}{2^{n+1}\sin(\alpha)}.$$

BASE STEP: Since $\cos(\alpha) = \frac{\sin(2\alpha)}{2\sin(\alpha)}$ (just expand the numerator using identity (9.11)), $A(1)$ is seen to be true.

INDUCTIVE STEP: For some $k \geq 0$ suppose that

$$A(k): \quad \cos(\alpha)\cos(2\alpha)\cdots\cos(2^k\alpha) = \frac{\sin(2^{k+1}\alpha)}{2^{k+1}\sin(\alpha)}$$

is true. Then using the identity $\sin(A)\cos(A) = \sin(2A)/2$,

$$\cos(\alpha)\cos(2\alpha)\cdots\cos(2^k\alpha)\cos(2^{k+1}\alpha)$$

$$= \frac{\sin(2^{k+1}\alpha)}{2^{k+1}\sin(\alpha)}\cos(2^{k+1}\alpha) \quad \text{(by } A(k))$$

$$= \frac{\sin(2^{k+1}\alpha + 2^{k+1}\alpha)/2}{2^{k+1}\sin(\alpha)}$$

$$= \frac{\sin(2^{k+2}\alpha)}{2^{k+2}\sin(\alpha)}$$

shows that $A(k+1)$ is true. This completes the inductive step.

By MI, for every $n \geq 0$, $A(n)$ is true. $\qquad\qquad\qquad\qquad\qquad\qquad\square$

Exercise 126: (Dirichlet kernel) This exercise appears in many places, for example, [499, Prob. 34].

Let θ be an angle which is not an integer multiple of 2π; for each $n \geq 1$, denote the statement in the exercise by

$$P(n): \quad \frac{1}{2} + \sum_{j=1}^{n} \cos(jt) = \frac{\sin((2n+1)t/2)}{2\sin(t/2)}.$$

BASE STEP: To prove $P(1)$, use the identity

$$\sin(A+B) - \sin(A-B) = 2\cos A \sin B, \qquad\qquad (27.6)$$

which was proved in the solution of Exercise 118. Using $A = t$ and $B = t/2$, equation (27.6) yields

$$\sin(3t/2) - \sin(t/2) = 2\cos(t)\sin(t/2);$$

hence

$$\frac{1}{2} + \cos(t) = \frac{\sin(t/2) + 2\cos(t)\sin(t/2)}{2\sin(t/2)} = \frac{\sin(3t/2)}{2\sin(t/2)},$$

which proves $P(1)$.

INDUCTIVE STEP: For some fixed $k \geq 1$ assume that

$$P(k): \quad \frac{1}{2} + \sum_{j=1}^{k} \cos(jt) = \frac{\sin(\frac{(2k+1)t}{2})}{2\sin(\frac{t}{2})}$$

is true. To be proved is

$$P(k+1): \quad \frac{1}{2} + \sum_{j=1}^{k+1} \cos(jt) = \frac{\sin(\frac{(2k+3)t}{2})}{2\sin(\frac{t}{2})}.$$

Starting with the left side of $P(k+1)$,

$$\frac{1}{2} + \sum_{j=1}^{k+1} \cos(jt)$$

$$= \frac{1}{2} + \sum_{j=1}^{k} \cos(jt) + \cos((k+1)t)$$

$$= \frac{\sin(\frac{(2k+1)t}{2})}{2\sin(\frac{t}{2})} + \cos((k+1)t) \quad \text{(by } (P(k)))$$

$$= \frac{\sin(\frac{(2k+1)t}{2}) + 2\cos((k+1)t)\sin(\frac{t}{2})}{2\sin(\frac{t}{2})}$$

$$= \frac{\sin(\frac{(2k+1)t}{2}) + \sin((k+1)t + \frac{t}{2}) - \sin((k+1)t - \frac{t}{2})}{2\sin(\frac{t}{2})} \quad \text{(by eqn (27.6))}$$

$$= \frac{\sin(\frac{(2k+3)t}{2})}{2\sin(\frac{t}{2})},$$

this proves $P(k+1)$, and hence finishes the inductive step.

By mathematical induction, for all $n \geq 1$, $P(n)$ is true. $\qquad \square$

Exercise 127: (Fejér kernel) Using the notation from Exercise 126 put

$$K_N(t) = \frac{1}{N+1} \sum_{n=0}^{N} D_n(t).$$

For $N \geq 0$, let denote the statement to be proved by

$$S(N): \quad K_N(t) = \frac{\sin^2((N+1)t/2)}{2(N+1)\sin^2(t/2)}.$$

BASE STEP: When $N = 0$, there is only one summand in D_0, namely when $n = 0$; in this case, $K_0 = \frac{1}{2}$, and

$$\frac{\sin^2((0+1)t/2)}{2(0+1)\sin^2(t/2)} = \frac{1}{2}$$

as well, so $S(0)$ is true.

Instead of proving $S(n)$ by induction, a simpler but equivalent statement is proved. Since by Exercise 126,

$$K_N(t) = \frac{1}{N+1} \sum_{n=0}^{N} D_n(t) = \frac{1}{N+1} \sum_{n=0}^{N} \frac{\sin((2n+1)t/2)}{2\sin(t/2)},$$

$S(N)$ says

$$\frac{1}{N+1} \sum_{n=0}^{N} \frac{\sin\left(\frac{2n+1}{2}t\right)}{2\sin(t/2)} = \frac{\sin^2\left(\frac{N+1}{2}t\right)}{2(N+1)\sin^2(t/2)}.$$

Cancelling the term $\frac{1}{2(N+1)\sin(t/2)}$ shows that $S(N)$ is equivalent to

$$S^*(N): \sum_{n=0}^{N} \sin\left(\frac{2n+1}{2}t\right) = \frac{\sin^2\left(\frac{N+1}{2}t\right)}{\sin(t/2)}.$$

Proceed with the induction using S^* rather than S. (The base case for $S^*(N)$ is proved by $S(0)$.)

INDUCTIVE STEP: For some fixed $k \geq 0$, assume that

$$S^*(k): \sum_{n=0}^{k} \sin\left(\frac{2n+1}{2}t\right) = \frac{\sin^2\left(\frac{k+1}{2}t\right)}{\sin(\frac{t}{2})}$$

is true. To be proved is

$$S^*(k+1): \sum_{n=0}^{k+1} \sin\left(\frac{2n+1}{2}t\right) = \frac{\sin^2\left(\frac{k+2}{2}t\right)}{\sin(\frac{t}{2})}.$$

Starting with the left side of $S^*(k+1)$,

$$\sum_{n=0}^{k+1} \sin\left(\frac{2n+1}{2}t\right) = \left[\sum_{n=0}^{k} \sin\left(\frac{2n+1}{2}t\right)\right] + \sin\left(\frac{2k+3}{2}t\right)$$

$$= = \frac{\sin^2\left(\frac{k+1}{2}t\right)}{\sin(\frac{t}{2})} + \sin\left(\frac{2k+3}{2}t\right) \quad \text{(by } S*(k)\text{)}$$

$$= = \frac{\sin^2\left(\frac{k+1}{2}t\right) + + \sin\left(\frac{2k+3}{2}t\right)\sin(\frac{t}{2})}{\sin(\frac{t}{2})}$$

Using $\sin^2 B + \sin(A+B)\sin(A-B) = \sin^2 A$ (this is equation (27.4), proved in Exercise 120) with $A = \frac{k+2}{2}t$ and $B = \frac{k+1}{2}t$ in the last line above, obtain

$$\sum_{n=0}^{k+1} \sin\left(\frac{2n+1}{2}t\right) = \frac{\sin^2\left(\frac{k+2}{2}t\right)}{\sin(\frac{t}{2})},$$

which is precisely $S^*(k+1)$. Since this is equivalent to $S(k+1)$, the inductive step is completed.

Hence, by mathematical induction, for all $n \geq 0$, $S(N)$ is true.   □

Note: The statement $S^*(N)$ is equivalent to that in Exercise 120, using the replacement $t = 2\theta$.

Exercise 128: This exercise occurs in [499, Prob. 35], for example.

Let x be a real number which is not an integer multiple of 2π. For each $n \geq 1$ let $S(n)$ be the statement

$$\sum_{j=1}^{n} j \sin(jx) = \frac{(n+1)\sin(nx) - n\sin((n+1)x)}{4\sin^2(x/2)}.$$

BASE STEP: $S(1)$ says $\sin(x) = \frac{2\sin(x) - \sin(2x)}{4\sin^2(x/2)}$. Why is this true? To begin with, a standard identity is arrived at by applying identity (9.12) with $\alpha = \beta = \frac{x}{2}$,

$$\cos(x) = \cos^2(x/2) - \sin^2(x/2) - 1 - \sin^2(x/2) - \sin^2(x/2) = 1 - 2\sin^2(x/2).$$

It follows that

$$4\sin^2(x/2) = 2 - 2\cos(x). \tag{27.7}$$

Also, identity (9.11) with $\alpha = \beta = x$ yields $\sin(2x) = 2\sin(x)\cos(x)$. Using these identities,

$$\frac{2\sin(x)}{4\sin^2(x/2)} \cdot \frac{\sin(2x)}{} = \frac{2\sin(x) - 2\sin(x)\cos(x)}{2 - 2\cos(x)} = \sin(x)$$

shows that $S(1)$ is indeed true.

INDUCTIVE STEP: Fix $k \geq 1$, and suppose that

$$S(k) : \quad \sum_{j=1}^{k} j\sin(jx) = \frac{(k+1)\sin(kx) - k\sin((k+1)x)}{4\sin^2(x/2)}$$

is true. It remains to show that

$$S(k+1) : \quad \sum_{j=1}^{k+1} j\sin(jx) = \frac{(k+2)\sin((k+1)x) - (k+1)\sin((k+2)x)}{4\sin^2(x/2)}$$

follows. To prove $S(k+1)$, use the following identity

$$\sin(kx) - 2\sin((k+1)x)\cos(x) = -\sin((k+2)x), \tag{27.8}$$

which is perhaps most easily seen by using the trick $kx = (k+1)x - x$ and expanding $\sin(kx)$ using identity (9.11) as follows:

$$\begin{aligned}
&\sin(kx) - 2\sin((k+1)x)\cos(x) \\
&= \quad \sin((k+1)x - x) - 2\sin((k+1)x)\cos(x) \\
&= \quad \sin((k+1)x)\cos(-x) + \cos((k+1)x\sin(-x) - 2\sin((k+1)x)\cos(x) \\
&= \quad \sin((k+1)x)\cos(x) - \cos((k+1)x\sin(x) - 2\sin((k+1)x)\cos(x) \\
&= \quad -\sin((k+1)x)\cos(x) - \cos((k+1)x\sin(x)
\end{aligned}$$

$$= -\sin((k+2)x).$$

Starting with the left side of $S(k+1)$,

$$\sum_{j=1}^{k+1} j\sin(jx)$$

$$= \left[\sum_{j=1}^{k} j\sin(jx)\right] + (k+1)\sin((k+1)x)$$

$$= \frac{(k+1)\sin(kx) - k\sin((k+1)x)}{4\sin^2(x/2)} + (k+1)\sin((k+1)x) \quad \text{(by } S(k)\text{)}$$

$$= \frac{(k+1)\sin(kx) - k\sin((k+1)x) + 2(k+1)\sin((k+1)x)(1 - \cos(x))}{4\sin^2(x/2)}$$

$$\text{(by eq'n (27.7))}$$

$$= \frac{(k+1)\sin((k+1)x) + (k+1)\sin(kx) - 2(k+1)\sin((k+1)x)\cos(x)}{4\sin^2(x/2)}$$

$$= \frac{(k+1)\sin((k+1)x) + (k+1)[\sin(kx) - 2\sin((k+1)x)\cos(x)]}{4\sin^2(x/2)}$$

$$= \frac{(k+2)\sin((k+1)x) - (k+1)\sin((k+2)x)}{4\sin^2(x/2)} \quad \text{(by eq'n (27.8))},$$

which is the right side of $S(k+1)$. This completes the inductive step.

By MI, for all $n \geq 1$, $S(n)$ is true, completing the solution to Exercise 128. $\square$

Exercise 129: This exercise occurs in *e.g.*, [499, Prob. 36].

Let $x \in \mathbb{R}$ be fixed which is not an integer multiple of 2π. For each $n \geq 1$ let $S(n)$ be the statement

$$\sum_{j=1}^{n} \cos(jx) = \frac{(n+1)\cos(nx) - n\cos((n+1)x) - 1}{4\sin^2(x/2)}.$$

BASE STEP: $S(1)$ says $\cos(x) = \dfrac{2\cos(x) - \cos(2x) - 1}{4\sin^2(x/2)}$. Using

$$\cos(2x) = \cos^2(x) - \sin^2(x) = 2\cos^2(x) - 1,$$

and equation (27.7) [which says $4\sin^2(x/2) = 2 - 2\cos(x)$, proved in Exercise 128],

$$\frac{2\cos(x) - \cos(2x) - 1}{4\sin^2(x/2)} = \frac{2\cos(x) - (2\cos^2(x) - 1) - 1}{2 - 2\cos(x)} = \cos(x),$$

verifies that $S(1)$ indeed holds.

INDUCTIVE STEP: Let $k \geq 1$ be fixed and assume that

$$S(k): \quad \sum_{j=1}^{k} \cos(jx) = \frac{(k+1)\cos(kx) - k\cos((k+1)x) - 1}{4\sin^2(x/2)}$$

is true. To complete the inductive step, one must prove

$$S(k+1): \quad \sum_{j=1}^{k+1} \cos(jx) = \frac{(k+2)\cos((k+1)x) - (k+1)\cos((k+2)x) - 1}{4\sin^2(x/2)}$$

Similar to that used in Exercise 128, the following identity is used to accomplish the proof of $S(k+1)$:

$$(k+1)\cos(kx) - 2(k+1)\cos(x)\cos((k+1)x) = -(k+1)\cos((k+2)x). \quad (27.9)$$

The sequence of steps in proving equation (27.9) is very similar to that used to prove equation (27.8), so only an outline is given:

$$
\begin{aligned}
&\cos(kx) - 2\cos(x)\cos((k+1)x) \\
&= \ [\cos((k+1)x)\cos(x) + \sin((k+1)x)\sin(x)] - 2\cos(x)\cos((k+1)x) \\
&= \ -[\cos((k+1)x)\cos(x) - \sin((k+1)x)\sin(x)] \\
&= \ -\cos((k+2)x).
\end{aligned}
$$

Multiplication throughout by $(k+1)$ finishes the proof of (27.9). To prove $S(k+1)$,

$$\sum_{j=1}^{k+1} j\cos(jx)$$

$$= \ \left[\sum_{j=1}^{k} j\cos(jx)\right] + (k+1)\cos((k+1)x)$$

$$= \ \frac{(k+1)\cos(kx) - k\cos((k+1)x) - 1}{4\sin^2(x/2)} + (k+1)\cos((k+1)x) \quad \text{(by } S(k))$$

$$= \ \frac{(k+1)\cos(kx) - k\cos((k+1)x) - 1 + 2(1 - \cos(x))(k+1)\cos((k+1)x)}{4\sin^2(x/2)}$$

$$\text{(by eq'n 27.7)}$$

$$= \ \frac{(k+2)\cos((k+1)x) + (k+1)\cos(kx) - (k+1)\cos(x)\cos((k+1)x) - 1}{4\sin^2(x/2)}$$

$$= \ \frac{(k+2)\cos((k+1)x) - (k+1)\cos((k+2)x) - 1}{4\sin^2(x/2)} \quad \text{(by eq'n (27.9))},$$

finishing the proof of $S(k+1)$, and hence the inductive step.

By mathematical induction, for all $n \geq 1$, $S(n)$ is true, completing the solution to Exercise 129. $\qquad\square$

Exercise 130: This exercise occurred in, for example, [499, Prob. 37].

Suppose that $x \in \mathbb{R}$ is not an integer multiple of π. For each $n \geq 1$, let $T(n)$ be the statement

$$\sum_{j=1}^{n} \frac{1}{2^j} \tan\left(\frac{x}{2^j}\right) = \frac{1}{2^n} \cot\left(\frac{x}{2^n}\right) - \cot(x).$$

BASE STEP: Applying identity (9.12) with $\alpha = \beta = x/2$, one gets $\cos(x) = \cos^2(x/2) - \sin^2(x/2)$. Similarly, using equation (9.11), $\sin(x) = 2\cos(\frac{x}{2})\sin(\frac{x}{2})$. Using these, together with the definitions $\cot(A) = \frac{\cos(A)}{\sin(A)}$, and $\tan(B) = \frac{\sin(B)}{\cos(B)}$, one proof of $T(1)$ is:

$$
\begin{aligned}
\frac{1}{2}\tan(\frac{x}{2}) &= \frac{\sin(\frac{x}{2})}{2\cos(\frac{x}{2})} \\
&= \frac{\sin^2(\frac{x}{2})}{2\sin(\frac{x}{2})\cos(\frac{x}{2})} \\
&= \frac{\cos^2(\frac{x}{2}) - [\cos^2(\frac{x}{2}) - \sin^2(\frac{x}{2})]}{2\sin(\frac{x}{2})\cos(\frac{x}{2})} \\
&= \frac{\cos(\frac{x}{2})}{\sin(\frac{x}{2})} - \frac{\cos(x)}{\sin(x)} \\
&= \frac{1}{2}\cot\left(\frac{x}{2}\right) - \cot(x).
\end{aligned}
$$

Thus $T(1)$ is true.

INDUCTIVE STEP: For some fixed $k \geq 1$, suppose that

$$T(k) : \quad \sum_{j=1}^{k} \frac{1}{2^j} \tan\left(\frac{x}{2^j}\right) = \frac{1}{2^k} \cot\left(\frac{x}{2^k}\right) - \cot(x)$$

is true. It remains to prove

$$T(k+1) : \quad \sum_{j=1}^{k+1} \frac{1}{2^j} \tan\left(\frac{x}{2^j}\right) = \frac{1}{2^{k+1}} \cot\left(\frac{x}{2^{k+1}}\right) - \cot(x).$$

To save a little work, notice that in proving the base case, for any angle A which is not a multiple of π, the identity.

$$\frac{1}{2}\tan\left(\frac{A}{2}\right) = \frac{1}{2}\cot\left(\frac{A}{2}\right) - \cot(A),$$

was actually proved, or equivalently,

$$\cot(A) + \frac{1}{2}\tan\left(\frac{A}{2}\right) = \frac{1}{2}\cot\left(\frac{A}{2}\right). \tag{27.10}$$

In proving $T(k+1)$, use this with $A = \frac{x}{2^k}$, vastly simplifying calculations:

$$\sum_{j=1}^{k+1} \frac{1}{2^j}\tan\left(\frac{x}{2^j}\right)$$

$$= \left[\sum_{j=1}^{k} \frac{1}{2^j}\tan\left(\frac{x}{2^j}\right)\right] + \frac{1}{2^{k+1}}\tan\left(\frac{x}{2^{k+1}}\right)$$

$$= \frac{1}{2^k}\cot\left(\frac{x}{2^k}\right) - \cot(x) + \frac{1}{2^{k+1}}\tan\left(\frac{x}{2^{k+1}}\right) \quad \text{(by } T(k)\text{)}$$

$$= \frac{1}{2^k}\left[\cot\left(\frac{x}{2^k}\right) + \frac{1}{2}\tan\left(\frac{x}{2^{k+1}}\right)\right] - \cot(x)$$

$$= \frac{1}{2^k}\left[\frac{1}{2}\cot\left(\frac{x}{2^{k+1}}\right)\right] - \cot(x) \quad \text{(by eq'n 27.10)}$$

$$= \frac{1}{2^{k+1}}\cot\left(\frac{x}{2^{k+1}}\right) - \cot(x).$$

This proves $T(k+1)$, completing the inductive step.

By MI, for every $n \geq 1$, $T(n)$ is true, completing the solution to Exercise 130. $\square$

Exercise 131: This exercise occurs in, for example, [499, Prob. 38]. It serves as a good workout for understanding of inverse trigonometric functions (though all steps are found to be simple). Recall that if $\tan(\theta) = y$, then the tan inverse function $\tan^{-1}$ is defined by $\tan^{-1}(y) = \theta$. The reciprocal of $\tan(\theta)$ is $(\tan(\theta))^{-1} = \cot(\theta)$, though remember that $\tan^{-1}$ and $\cot$ are different functions (one is the inverse with respect to functions, and the other is the inverse with respect to multiplication).

For each $n \geq 1$ let $S(n)$ be the statement

$$\sum_{k=1}^{n} \cot^{-1}(2k+1) = \sum_{j=1}^{n} \tan^{-1}\left(\frac{j+1}{j}\right) - n\tan^{-1}(1).$$

BASE STEP: $S(1)$ says

$$\cot^{-1}(3) = \tan^{-1}(2) - \tan^{-1}(1).$$

To prove $S(1)$, take the tangent of each side. For the left side, let $\cot(\theta) = 3$; then $\tan(\theta) = \frac{1}{3}$, and $\theta = \cot^{-1}(3)$, so $\tan(\cot^{-1}(3)) = \tan(\theta) = \frac{1}{3}$. For the right side, an application of (9.15) gives

$$
\begin{aligned}
\tan\left(\tan^{-1}(2) - \tan^{-1}(1)\right) &= \frac{\tan(\tan^{-1}(2)) - \tan(\tan^{-1}(1))}{1 + \tan(\tan^{-1}(2))\tan(\tan^{-1}(1))} \\
&= \frac{(2) - (1)}{1 + (2)(1)} \\
&= \frac{1}{3},
\end{aligned}
$$

which agrees with the tangent of the left side. Since tangent is defined so that it is one-to-one, conclude that since the tangents of both sides of $S(1)$ agree, then $S(1)$ itself is true.

INDUCTIVE STEP: Fix some $m \geq 1$ and assume that

$$
S(m): \quad \sum_{k=1}^{m} \cot^{-1}(2k+1) = \left[\sum_{j=1}^{m} \tan^{-1}\left(\frac{j+1}{j}\right)\right] - m\tan^{-1}(1)
$$

is true. It remains to show that

$$
S(m+1): \quad \sum_{k=1}^{m+1} \cot^{-1}(2k+1) = \sum_{j=1}^{m+1} \tan^{-1}\left(\frac{j+1}{j}\right) - (m+1)\tan^{-1}(1)
$$

is true. Beginning with the left side of $S(k+1)$,

$$
\begin{aligned}
&\sum_{k=1}^{m+1} \cot^{-1}(2k+1) \\
&= \left\{\sum_{k=1}^{m} \cot^{-1}(2k+1)\right\} + \cot^{-1}(2m+3) \\
&= \left\{\left[\sum_{j=1}^{m} \tan^{-1}\left(\frac{j+1}{j}\right)\right] - m\tan^{-1}(1)\right\} + \cot^{-1}(2m+3) \quad \text{(by } S(m)\text{)} \\
&= \left[\sum_{j=1}^{m} \tan^{-1}\left(\frac{j+1}{j}\right)\right] - (m+1)\tan^{-1}(1) + \tan^{-1}(1) + \cot^{-1}(2m+3)
\end{aligned}
$$

So, to finish the proof of $S(m+1)$, it suffices to prove that

$$
\cot^{-1}(2m+3) = \tan^{-1}\left(\frac{m+2}{m+1}\right) - \tan^{-1}(1).
$$

This can be checked by taking tangents of each side:

$$\tan(\cot^{-1}(2m+3)) = \frac{1}{2m+3},$$

and

$$\tan\left(\tan^{-1}\left(\frac{m+2}{m+1}\right) - \tan^{-1}(1)\right) = \frac{\frac{m+2}{m+1} - 1}{1 + \frac{m+2}{m+1}(1)} = \frac{\frac{1}{m+1}}{\frac{2m+3}{m+1}} = \frac{1}{2m+3}$$

agree, and so $S(m+1)$ is true. This concludes the inductive step.

By MI, for each $n \geq 1$, $S(n)$ is true. $\qquad\square$

Exercise 132: This exercise occurs in Trim's calculus book [534, Ex. 40, p. A-6]— it received a three star rating (out of three). For $n \geq 0$, let $P(n)$ be the proposition that there exist constants $a_0, a_1, \ldots, a_n$, and $b_0, b_1, \ldots, b_n$ so that

$$\sin^n(x) = \sum_{r=0}^{n} [a_r \cos(rx) + b_r \sin(rx)].$$

BASE STEP: The exercise asks to prove $P(n)$ for $n \geq 2$, however, it appears to be true for even $n \geq 0$. If x is not a multiple of π, $\sin(x) \neq 0$, and then the result is true even for $n = 0$, since $\sin^0(x) = 1 = 1 \cdot \cos(0 \cdot x) + 0 \sin(0 \cdot x)$. When $n = 1$, pick $a_0 = 0$, $a_1 = 0$, b_0 can be anything, and $b_1 = 1$.

For $n = 2$, equations (9.12) and (9.13) yield $\cos(2x) = 1 - 2\sin^2(x)$, and hence

$$\sin^2 = \frac{1}{2} - \frac{1}{2}\cos(2x),$$

so $a_0 = \frac{1}{2}$, anything for b_0, $a_1 = b_1 = b_2 = 0$ and $a_2 = \frac{-1}{2}$ show that $S(2)$ is true.

INDUCTIVE STEP: Let $k \geq 2$ be fixed, and suppose that $P(k)$ is true, that is, assume for some fixed constants $a_0, a_1, \ldots, a_k$ and $b_0, b_1, \ldots, b_k$

$$\sin^k(x) = \sum_{r=0}^{k} [a_r \cos(rx) + b_r \sin(rx)].$$

To be shown is that $P(k+1)$ is true, that is, that there exist constants $\hat{a}_0, \hat{a}_1, \ldots, \hat{a}_{k+1}$ and $\hat{b}_0, \hat{b}_1, \ldots, \hat{b}_{k+1}$, so that

$$\sin^{k+1}(x) = \sum_{r=0}^{k+1} [\hat{a}_r \cos(rx) + \hat{b}_r \sin(rx)].$$

Starting with the left-hand side of this above equation,

$$\sin^{k+1}(x) = \sin(x)\sin^k(x).$$

Applying $P(k)$ to the factor $\sin^k(x)$ above, there exist a_i's and b_i's so that

$$\sin^{k+1}(x) = \sin(x)\left(\sum_{r=0}^{k}[a_r\cos(rx) + b_r\sin(rx)]\right)$$

$$= \sum_{r=0}^{k}[a_r\sin(x)\cos(rx) + b_r\sin(x)\sin(rx)].$$

It suffices to express, for each $r = 0, 1, \ldots, k$, each of $\sin(x)\cos(rx)$ and $\sin(x)\sin(rx)$ as linear combinations of $\cos(0)$, $\cos(x)$,..., $\cos((k+1)x)$, and $\sin(0)$, $\sin(x)$,..., $\sin((k+1)x)$. This is done via the two identities

$$\sin(x)\cos(rx) = \frac{1}{2}\sin((1+r)x) + \frac{1}{2}\sin((1-r)x),$$

$$\sin(x)\sin(rx) = \frac{1}{2}\cos((1-r)x) - \frac{1}{2}\cos((1+r)x).$$

(The proofs of these are simple: for the first, expand each of $\sin(A+B)$ and $\sin(A-B)$ using equation (9.11) and add the two equations; for the second, expand $\cos(A+B)$ and $\cos(A-B)$ using equation (9.12) and subtract the two equations.)

Notice also that when $r > 1$, one replaces $\sin((1-r)x)$ with $-\sin((r-1)x)$ and $\cos((1-r)x)$ with $\cos((r-1)x)$. So, using these replacements, $\sin^{k+1}(x)$ is indeed expressible as linear combinations of $\cos(0), \cos(x), \ldots, \cos((k+1)x)$, and $\sin(0), \sin(x), \ldots, \sin((k+1)x)$. This completes the proof of $P(k+1)$ and hence the inductive step.

By MI, for all $n \geq 2$, the statement is true (in fact, for all $n \geq 1$, and if $\sin(x) \neq 0$, for all $n \geq 0$). $\qquad\square$

Comments on Exercise 132: In the inductive step above, what are the $\hat{a}_i$'s and $\hat{b}_i$'s explicitly? I think that they work out as follows:

r	$\hat{a}_r$	$\hat{b}_r$
0	$\frac{1}{2}b_1$	$\frac{1}{2}a_1$
1	$\frac{1}{2}b_2$	$a_0 - \frac{1}{2}a_2$
2	$\frac{1}{2}(b_3 - b_1)$	$\frac{1}{2}(a_1 - a_3)$
3	$\frac{1}{2}(b_4 - b_2)$	$\frac{1}{2}(a_4 - a_2)$
$\vdots$		
$k-1$	$\frac{1}{2}(b_k - b_{k-2})$	$\frac{1}{2}(a_{k-2} - a_k)$
k	$-\frac{1}{2}b_{k-1}$	$\frac{1}{2}a_{k-1}$
$k+1$	$-\frac{1}{2}b_k$	$\frac{1}{2}a_k$

Actually, if one were to develop a recursion, perhaps $b_0 = 0$ should be declared in every case. Using the chart above recursively, one arrives at

$$\sin^3(x) = \frac{3}{4}\sin(x) - \frac{1}{4}\sin(3x)$$

or equivalently,
$$4\sin^3(x) = 3\sin(x) - \sin(3x),$$

which can be easily verified. One also notices that the $\hat{a}_i$'s depend only on previous b_j's and $\hat{b}_i$'s depend only on previous a_j's, and since for $n = 2$, all b_i's are zero, by induction, if n is even, $\sin^n(x)$ is expressible as a linear combination of only $\cos(rx)$'s and if n is odd, $\sin^n(x)$ is expressible only as a linear combination of $\sin(rx)$'s. It might be interesting to try and give a formula which would describe the coefficients explicitly for any n. Perhaps this has been done.

Exercise 133: This appeared in, *e.g.*, [161, 8.19, pp. 208, 215]. Fix x and α so that $x + \frac{1}{x} = 2\cos(\alpha)$. For every $n \geq 1$, denote the equality in the exercise by

$$E(n): \quad x^n + \frac{1}{x^n} = 2\cos(n\alpha).$$

BASE STEP: $E(1)$ is given. To see $E(2)$, $x^2 + \frac{1}{x^2} = (x + 1/x)^2 - 2 = 4\cos^2(\alpha) - 2 = 2\cos(2\alpha)$.

INDUCTIVE STEP: Suppose that for some $k \geq 2$, both $E(k)$ and $E(k-1)$ hold. To see that $E(k+1)$ follows, the identity

$$\cos(A+B) + \cos(A-B) = 2\cos(A)\cos(B) \qquad (27.11)$$

helps. [This follows from expanding $\cos(A+B)$ and $\cos(A-B)$ using equation (9.12) and adding the two equations.] Then

$$
\begin{aligned}
x^{k+1} &+ \frac{1}{x^{k+1}} \\
&= \left(x + \frac{1}{x}\right)\left(x^k + \frac{1}{x^k}\right) - x^{k-1} - \frac{1}{x^{k-1}} \\
&= 4\cos(\alpha)\cos(k\alpha) - 2\cos((k-1)\alpha) \quad \text{(by } E(1), E(k), E(k-1)) \\
&= 2\cos((k+1)\alpha) + 2\cos((k-1)\alpha) - 2\cos((k-1)\alpha) \quad \text{(by eqn (27.11))} \\
&= 2\cos((k+1)\alpha),
\end{aligned}
$$

which shows that $E(k+1)$ is true, completing the inductive step.

By induction, for each $n \geq 1$, the statement $E(n)$ is proved. $\qquad \square$

27.4   Solutions: Miscellaneous identities

Exercise 134: (Outline) Let $S(n)$ be the statement in the exercise. For $n = 1$, the result is clear. For $n = 2$, if $x_1 + x_2 = 0$ and say, $x_1 > 0$, then $x_2 < 0$, contrary to x_2 being non-negative, so $S(2)$ hold.

For the inductive step, use $S(2)$ and $S(k)$ to prove $S(k+1)$ as follows: $x_1 + \cdots + x_k + x_{k+1} = 0$ implies by $S(2)$ that both $x_1 + \cdots + x_k = 0$ and $x_{k+1} = 0$. Applying $S(k)$ to the first expression shows that $x_1 = \cdots = x_k = 0$ as well. $\quad\square$

Exercise 135: Fix x and b. Use the fact that for every $y \in \mathbb{R}^+$, $\log_b(xy) = \log_b(x) + \log_b(y)$. [The proof of this is quite simple, since $b^r b^s = b^{r+s}$.] Here is the proof of the result by induction on n:

BASE STEP: For $n = 1$, the statement is trivial.

INDUCTIVE STEP: Suppose that for some $k \geq 1$, $\log_b(x^k) = k \log_b(x)$. To be shown is that $\log_b(x^{k+1}) = (k+1) \log_b(x)$. Then

$$
\begin{aligned}
\log_b(x^{k+1}) &= \log_b(x \cdot x^k) \\
&= \log_b(x) + \log_b(x^k) \quad \text{(by fact above with } y = x^k) \\
&= \log_b(x) + k \log_b(x) \quad \text{(by induction hypothesis)} \\
&= (k+1) \log_b(x),
\end{aligned}
$$

concluding the inductive step.

Hence, by MI, the result is true for all $n \in \mathbb{Z}^+$. $\quad\square$

Exercise 136: Let $a_1, a_2, \ldots$ and $b_1, b_2, \ldots$ be real numbers, and let $P(n)$ denote the proposition

$$
P(n): \quad \sum_{i=1}^{n}(a_i + b_i) = \sum_{i=1}^{n} a_i + \sum_{j=1}^{n} b_j.
$$

BASE STEP ($n = 1$): Since $\sum_{i=1}^{1}(a_i + b_i) = a_1 + b_1$, and both $\sum_{i=1}^{1} a_i = a_1$ and $\sum_{j=1}^{1} b_i = b_1$, the statement

$$
P(1): \quad \sum_{i=1}^{1}(a_i + b_i) = \sum_{i=1}^{1} a_i + \sum_{j=1}^{1} b_j
$$

follows.

INDUCTIVE STEP ($P(k) \to P(k+1)$): For some fixed $k \geq 1$, assume that

$$
P(k): \quad \sum_{i=1}^{k}(a_i + b_i) = \sum_{i=1}^{k} a_i + \sum_{j=1}^{k} b_j
$$

holds. To be proved is that

$$
P(k+1): \quad \sum_{i=1}^{k+1}(a_i + b_i) = \sum_{i=1}^{k+1} a_i + \sum_{j=1}^{k+1} b_j
$$

follows. Beginning with the left side of $P(k+1)$,

$$\sum_{i=1}^{k+1}(a_i + b_i) \;=\; \left[\sum_{i=1}^{k}(a_i + b_i)\right] + (a_{k+1} + b_{k+1}) \quad \text{(by Def'n 2.5.6)}$$

$$=\; \left[\sum_{i=1}^{k}a_i + \sum_{j=1}^{k}b_j\right] + (a_{k+1} + b_{k+1}) \quad \text{(by ind. hyp. } P(k))$$

$$=\; \left(\sum_{i=1}^{k}a_i\right) + a_{k+1} + \left(\sum_{j=1}^{k}h_j\right) + b_{k+1})$$

$$=\; \left(\sum_{i=1}^{k+1}a_i\right) + \left(\sum_{j=1}^{k+1}b_j\right) \quad \text{(by Def'n 2.5.6)},$$

the right-hand side of $P(k+1)$, completing the inductive step.

Hence, by the principle of mathematical induction, for all $n \geq 1$, $P(n)$ is true. $\quad\square$

Exercise 137: (Telescoping sum) Let $a_1, a_2, a_3, \ldots$ be a sequence of real numbers. For each positive integer $n \geq 1$, let $T(n)$ be the claim that

$$\sum_{i=1}^{n}(a_i - a_{i+1}) = a_1 - a_{n+1}.$$

BASE STEP: When $i = 1$, there is only one summand and so $T(1)$ is trivially true.

INDUCTION STEP: For some fixed $k \geq 1$, assume that $T(k)$ is true, that is, assume that

$$\sum_{i=1}^{k}(a_i - a_{i+1}) = a_1 - a_{k+1}.$$

To prove $T(k+1)$, one needs to show

$$\sum_{i=1}^{k+1}(a_i - a_{i+1}) = a_1 - a_{k+2}.$$

Beginning with the left side of this equation,

$$\sum_{i=1}^{k+1}(a_i - a_{i+1}) \;=\; \left[\sum_{i=1}^{k}(a_i - a_{i+1})\right] + (a_{k+1} - a_{k+2})$$

$$=\; a_1 - a_{k+1} + (a_{k+1} - a_{k+2}) \quad \text{(by } T(k))$$

$$= a_1 - a_{k+2}$$

as desired. This completes the inductive step $T(k) \to T(k+1)$.

Hence, by MI, for all $n \in \mathbb{Z}^+$, the statement $T(n)$ is true. □

Exercise 138: This exercise appeared in, *e.g.*, [582, Prob. 22].
For $n \geq 1$, let $S(n)$ denote the statement

$$S(n): \quad \sum_{j=1}^{n}(3j^2 - j + 2) = n(n^2 + n + 2).$$

BASE STEP $(n = 1)$: Checking $S(1)$, it says $3(1^2) - 1 + 2 = 1(1^2 + 1 + 2)$, which boils down to $4 = 4$, so $S(1)$ is true.

INDUCTIVE STEP: For some fixed $k \geq 1$, assume the inductive hypothesis

$$S(k): \quad \sum_{j=1}^{k}(3j^2 - j + 2) = k(k^2 + k + 2).$$

To be shown is that

$$S(k+1): \quad \sum_{j=1}^{k+1}(3j^2 - j + 2) = (k+1)((k+1)^2 + (k+1) + 2).$$

follows. Beginning with the left side of $S(k+1)$,

$$\sum_{j=1}^{k+1}(3j^2 - j + 2)$$

$$= \left(\sum_{j=1}^{k}(3j^2 - j + 2) \right) + 3(k+1)^2 - (k+1) + 2$$

$$= k(k^2 + k + 2) + 3(k+1)^2 - (k+1) + 2 \quad \text{(by ind. hyp.)}$$

$$= k(k(k+1) + 2) + 3(k+1)^2 - (k+1) + 2$$

$$= k^2(k+1) + 2k + 3(k+1)^2 - (k+1) + 2$$

$$= k^2(k+1) + 3(k+1)^2 + k + 1$$

$$= (k+1)[k^2 + 3(k+1) + 1]$$

$$= (k+1)[k^2 + 3k + 4]$$

$$= (k+1)[k^2 + 2k + 1 + k + 1 + 2],$$

which is precisely the right side of $S(k+1)$. This concludes the inductive step $S(k) \to S(k+1)$.

Therefore, by the principle of mathematical induction, for all $n \geq 1$, $S(n)$ is true. $\square$

Exercise 140: Outline: Observe that $1 = \frac{2!}{2!}$, so for the purpose of this exercise, it suffices to consider 1 as a prime. Now suppose that $\frac{a}{b} \in \mathbf{Q}$ is some given rational number. Let p be the largest prime in the factorization of either a or b; without loss, suppose that for some $m \geq 1$, $\frac{a}{b} = \frac{p^m k}{b}$ (where p does not divide k). Use the induction assumption that for all fractions $\frac{x}{y}$ with x and y having all prime factors smaller than p, that $\frac{x}{y}$ is expressible as desired. Then, for example, when $m = 1$,

$$\frac{a}{b} = \frac{pk}{b} = \frac{p! k}{(p-1)! b},$$

and k, $p-1$, and b all have factorizations using primes smaller than p, so apply the induction hypothesis to $\frac{k}{(p-1)! b}$. When $m > 1$, a similar argument applies. $\square$

Exercise 141: This exercise appeared in [499, Problem 17]. Induct on n to show that for every $n \geq 1$, the proposition $P(n)$:

$$\frac{1}{x(x+1)} + \frac{1}{(x+1)(x+2)} + \cdots + \frac{1}{(x+n-1)(x+n)} = \frac{n}{x(x+n)}$$

is true.

BASE STEP: $P(1)$ says $\frac{1}{x(x+1)} = \frac{1}{x(x+1)}$, which is clearly true.

INDUCTION STEP: For some fixed $k \geq 1$, assume that $P(k)$ is true. To be shown is that $P(k+1)$:

$$\frac{1}{x(x+1)} + \frac{1}{(x+1)(x+2)} + \cdots + \frac{1}{(x+k)(x+k+1)} = \frac{k+1}{x(x+k+1)}$$

is true. Beginning with the left side, (writing in the second last term so that it is clear how to apply the inductive hypothesis)

$$\frac{1}{x(x+1)} + \cdots + \frac{1}{(x+k-1)(x+k)} + \frac{1}{(x+k)(x+k+1)}$$

$$= \frac{k}{x(x+k)} + \frac{1}{(x+k)(x+k+1)} \quad (\text{by } P(k))$$

$$= \frac{k(x+k+1)}{x(x+k)(x+k+1)} + \frac{x}{x(x+k)(x+k+1)}$$

$$= \frac{k(x+k) + k + x}{x(x+k)(x+k+1)}$$

$$= \frac{(k+1)(x+k)}{x(x+k)(x+k+1)}$$

$$= \frac{k+1}{x(x+k+1)},$$

arrive at the right side of $P(k+1)$, completing the inductive step.

Therefore, by MI, for all $n \geq 1$ the statement $P(n)$ is true. □

Exercise 142: This problem appeared in, *e.g.*, [499, Prob. 21]. To be shown is that for $n \geq 1$, the expression

$$E(n): \quad \frac{1}{1+x} + \frac{2}{1+x^2} + \cdots + \frac{2^n}{1+x^{2^n}} = \frac{1}{x-1} + \frac{2^{n+1}}{1-x^{2^{n+1}}}$$

is true.

BASE STEP: $E(1)$ says $\frac{1}{1+x} + \frac{2}{1+x^2} = \frac{1}{x-1} + \frac{4}{1-x^4}$; putting everything over a common denominator of $1 - x^4$ indeed shows that this is an equality. (Details are left to the reader.)

INDUCTION STEP: For some fixed $k \geq 1$, assume that $E(k)$ is true. To prove $E(k+1)$, use sigma notation for brevity:

$$\sum_{j=0}^{k+1} \frac{2^j}{1+x^{2^j}} = \sum_{j=0}^{k} \frac{2^j}{1+x^{2^j}} + \frac{2^{k+1}}{1+x^{2^{k+1}}}$$

$$= \frac{1}{x-1} + \frac{2^{k+1}}{1-x^{2^{k+1}}} + \frac{2^{k+1}}{1+x^{2^{k+1}}} \quad (\text{by } E(k))$$

$$= \frac{1}{x-1} + \frac{2^{k+1}(1+x^{2^{k+1}}) + 2^{k+1}(1-x^{2^{k+1}})}{1-x^{2^{k+2}}}$$

$$= \frac{1}{x-1} + \frac{2^{k+1} + 2^{k+1}}{1-x^{2^{k+2}}}$$

$$= \frac{1}{x-1} + \frac{2^{k+2}}{1-x^{2^{k+2}}},$$

giving the truth of $E(k+1)$, completing the inductive step.

By mathematical induction, for $n \geq 1$, the expression $E(n)$ holds. □

Exercise 143: This problem has appeared many places, *e.g.*, [499, Prob. 23].
For $n \geq 1$, let $S(n)$ denote the statement

$$1 - \frac{x}{1!} + \frac{x(x-1)}{2!} - \cdots + (-1)^n \frac{x(x-1)\cdots(x-n+1)}{n!}$$

$$= (-1)^n \frac{(x-1)(x-2)\cdots(x-n)}{n!}.$$

BASE STEP: $S(1)$ states $1 - x = (-1)^1 \frac{x-1}{1!}$ which is true.

INDUCTIVE STEP: Let $k \geq 1$ and assume that $S(k)$ is true; to be shown is $S(k+1)$:

$$1 - \frac{x}{1!} + \frac{x(x-1)}{2!} - \cdots + (-1)^k \frac{x(x-1)\cdots(x-k+1)}{k!}$$

$$+ (-1)^{k+1} \frac{x(x-1)\cdots(x-k)}{(k+1)!}$$

$$= (-1)^{k+1} \frac{(x-1)(x-2)\cdots(x-k)(x-k-1)}{(k+1)!}.$$

By the inductive hypothesis, the left side of $S(k+1)$ is equal to

$$(-1)^k \frac{(x-1)(x-2)\cdots(x-k)}{k!} + (-1)^{k+1} \frac{x(x-1)\cdots(x-k)}{(k+1)!}$$

$$= (-1)^{k+1} \frac{x(x-1)\cdots(x-k)}{(k+1)!} - (-1)^{k+1} \frac{(k+1)(x-1)(x-2)\cdots(x-k)}{(k+1)!}$$

$$= (-1)^{k+1} \frac{(x-1)(x-2)\cdots(x-k)(x-k-1)}{(k+1)!},$$

that which was desired. This completes the inductive step.

By mathematical induction, for all $n \geq 1$, the statement $S(n)$ holds.   □

Exercise 144: For each $n \geq 0$, denote the equality in the exercise by

$$\xi(n): \quad (1)(3)(5)\cdots(2n+1) = \frac{(2n+1)!}{2^n n!}.$$

[*Note:* "ξ" is the lower case Greek letter pronounced "ksee".]

BASE STEP: For $n = 0$, $\xi(n)$ reads $(1) = \frac{1!}{2^0 \cdot 0!}$; since $2^0 = 1$ and $0! = 0$, $\xi(0)$ is indeed an equality.

INDUCTIVE STEP: For some fixed $k \geq 0$, assume that

$$\xi(k): \quad (1)(3)(5)\cdots(2k+1) = \frac{(2k+1)!}{2^k k!}$$

is true. It remains to show that

$$\xi(k+1): \quad (1)(3)(5)\cdots(2(k+1)+1) = \frac{(2(k+1)+1)!}{2^{k+1}(k+1)!}$$

follows from the truth of $\xi(k)$. Rewriting the left side of $\xi(k+1)$ with the penultimate factor for clarity,

$$(1)(3)(5)\cdots(2k+1)(2(k+1)+1) = \frac{(2k+1)!}{2^k k!}(2k+3) \qquad \text{(by } \xi(k))$$

$$= \frac{(2k+3)(2k+2)(2k+1)!}{(2k+2)2^k k!}$$

$$= \frac{(2k+3)!}{(2)2^k(k+1)k!}$$

$$= \frac{(2k+3)!}{2^{k+1}(k+1)!},$$

thereby proving $\xi(k+1)$ and hence completing the inductive step.

Thus, by MI, for all $n \geq 0$, the statement $\xi(n)$ is true.   □

Exercise 145: This problem appeared in [583, 10.19]. Working out the first values for $n = 0, 1, 2, 3, 4$ gives 1, 5^2, 11^2, 19^2, and 29^2. The bases increase by 4, 6, 8, 10, respectively, so an expression for the base of the perfect square might be $2+4+6+\cdots+2(n+1)-1$. By Exercise 39, this last expression is $(n+1)(n+2)-1$. Thus the statement to try proving is

$$S(n): \quad 1 + n(n+1)(n+2)(n+3) = [(n+1)(n+2)-1]^2.$$

Simply multiplying out each side verifies $S(n)$; however, this is a book on induction, so an inductive proof is given:

The base case is done above, so assume that for some $m \geq 0$, $S(m)$ is true. Starting with the left side of $S(m+1)$,

$$1 + (m+1)(m+2)(m+3)(m+4)$$
$$= 1 + m(m+1)(m+2)(m+3) + 4(m+1)(m+2)(m+3)$$
$$= [(m+1)(m+2)-1]^2 + 4(m+1)(m+2)(m+3) \qquad \text{(by } S(m))$$
$$= [m^2+3m+1]^2 + 4(m^2+3m+2)(m+3)$$
$$= m^4 + 9m^2 + 1 + 6m^3 + 2m^2 + 6m + 4m^3 + 24m^2 + 44m + 24$$
$$= m^4 + 10m^3 + 35m^2 + 50m + 25$$
$$= (m^2 + 5m + 5)^2$$
$$= [(m+2)(m+3)-1]^2,$$

arrive at the right side of $S(m+1)$, as desired, completing the inductive step. By mathematical induction, for all $n \geq 0$, the statement $S(n)$ is true.   □

Exercise 146: This problem occurs in (at least) [462, p. 282]. For each $n \geq 1$, let $S(n)$ denote the statement

$$S(n): \quad \sum_{\emptyset \neq S \subseteq [n]} \frac{1}{\prod_{s \in S} s} = n.$$

BASE STEP: When $n = 1$, the set $[1]$ has only one non-empty subset, namely $\{1\}$, in which case the sum in the left side of $S(1)$ is simply $\frac{1}{1} = 1$; so $S(1)$ is true.

INDUCTIVE STEP: Fix some $k \geq 1$ and assume that

$$S(k): \quad \sum_{\emptyset \neq S \subseteq [k]} \frac{1}{\prod_{s \in S} s} = k$$

is true. To complete the inductive step, it suffices to show that

$$S(k+1): \quad \sum_{\emptyset \neq S \subseteq [k+1]} \frac{1}{\prod_{s \in S} s} = k+1$$

follows. Starting with the left side of $S(k+1)$,

$$\sum_{\emptyset \neq S \subseteq [k+1]} \frac{1}{\prod_{s \in S} s} = \sum_{\emptyset \neq S \subseteq [k]} \frac{1}{\prod_{s \in S} s} + \sum_{\emptyset \neq S \subseteq [k+1], k+1 \in S} \frac{1}{\prod_{s \in S} s}$$

$$= k + \sum_{\emptyset \neq S \subseteq [k+1], k+1 \in S} \frac{1}{\prod_{s \in S} s} \qquad \text{(by } S(k)\text{)}$$

$$= k + \frac{1}{k+1} \sum_{\emptyset \neq S \subseteq [k]} \frac{1}{\prod_{s \in S} s} + \frac{1}{k+1}$$

$$= k + \frac{1}{k+1} \cdot k + \frac{1}{k+1} \qquad \text{(by } S(k)\text{)}$$

$$= k+1,$$

shows that $S(k+1)$ is true, competing the inductive step.

By mathematical induction, for each $n \geq 1$, the statement $S(n)$ is true. □

Exercise 147: For every $n \geq 1$, let $S(n)$ be the statement that for every $x \in \mathbb{R} \setminus \mathbb{Z}$,

$$\frac{n}{x} + \frac{n(n-1)}{x(x-1)} + \frac{n(n-1)(n-2)}{x(x-1)(x-2)} + \cdots = \frac{n}{x-n+1}.$$

Note that in the above infinite sum, only the first n terms are non-zero.

BASE STEP: Let $x \in \mathbb{R} \setminus \mathbb{Z}$. Since $\frac{1}{x} = \frac{1}{x-1+1}$, $S(1)$ is true.

INDUCTIVE STEP: Let $k \geq 1$ and suppose that $S(k)$ holds. Fix $x \in \mathbb{R} \setminus \mathbb{Z}$. Then,

$$\frac{k+1}{x} + \frac{(k+1)k}{x(x-1)} + \cdots + \frac{(k+1)!}{x(x-1)\cdots(x-(k+1)-1)}$$

$$= \frac{k+1}{x}\left(1 + \frac{k}{x-1} + \cdots + \frac{k!}{(x-1)\cdots(x-1-k-1)}\right)$$

$$= \frac{k+1}{x}\left(1 + \frac{k}{x-1-k+1}\right) \quad \text{(by } S(k) \text{ with } x-1\text{)}$$

$$= \frac{k+1}{x}\left(\frac{x-k+k}{x-(k+1)+1}\right)$$

$$= \frac{k+1}{x-(k+1)+1}.$$

Thus $S(k+1)$ holds.

By MI, for all $n \geq 1$, $S(n)$ remains true.   □

Exercise 148: (Brief solution) This problem appeared in [280, Prob. 5], with a brief solution—though perhaps a bit too brief, since technically, two base cases are needed. This also appeared in [161, 8.22, p. 208]. If $n = 1$, then the sum of the products is $1 = 2! - 1$. If $n = 2$, the sum of the products is $1^2 + 2^2 = 5 = 3! - 1$. So assume the statement is true for $n = k - 1$ and $n = k$, and examine all subsets of $\{1, 2, \ldots, k+1\}$ that contain no two consecutive numbers. Of these, there are two kinds of subsets, those containing $k+1$, and those which don't. Of those which contain $k+1$, they can not contain k, and so by inductive hypothesis with $n = k-1$, the contribution to the sum of the squares of products for sets in $\{1, 2, \ldots, k-1\}$ is $k! - 1$. Since the product of numbers in the sets containing both $k+1$ and elements from $\{1, 2, \ldots, k-1\}$ have an additional factor of $(k+1)^2$, the total for such sets is $(k+1)^2[k! - 1]$. Together with the set $\{k+1\}$, the total for all suitable subsets containing $k+1$ is $(k+1)^2[k! - 1] + (k+1)^2$. For those sets not containing $k+1$, this reduces precisely to the case when $n = k$, and so by inductive hypothesis, the sum of the squares of the products of subsets not containing $k+1$ is $(k+1)! - 1$.

Therefore, in all, the sum of the squares is

$$(k+1)^2[k! - 1] + (k+1)^2 + (k+1)! - 1 = (k+1)^2 k! + (k+1)! - 1$$
$$= (k+2)! - 1.$$

This finishes the inductive step, and hence by MI, the solution.   □

Chapter 28

Solutions: Inequalities

Exercise 150: Let x and y be fixed positive real numbers with $x < y$. For each $n \geq 1$, let $P(n)$ denote the statement

$$P(n): \quad x^n < y^n.$$

BASE STEP: For $n = 1$, $P(1)$ says $x^1 < y^1$, which is assumed, so $P(1)$ is trivially true.

INDUCTIVE STEP: Suppose that for some $k \geq 1$,

$$P(k): \quad x^k < y^k,$$

or equivalently, suppose that

$$P'(k): \quad \frac{x^k}{y^k} < 1.$$

Needed to show is

$$P(k+1): \quad x^{k+1} < y^{k+1}.$$

Since

$$
\begin{aligned}
\frac{x^{k+1}}{y^{k+1}} &= \frac{x^k}{y^k}\frac{x}{y} \\
&< 1 \cdot \frac{x}{y} && \text{(by } P'(k)) \\
&< 1 && \text{(since } x < y),
\end{aligned}
$$

the statement $P(k+1)$ follows. This completes the inductive step.

By the principle of mathematical induction, for every $n \geq 1$, $P(n)$ is true. Since x and y were arbitrary (but fixed!) positive reals with $x < y$, the proof is complete. $\square$

Exercise 151: The inductive proof is presented first, then commented on, and finally a different proof not requiring induction is given.

For $n \geq 6$, let $P(n)$ be the statement

$$P(n): \quad 4n < n^2 - 7.$$

BASE STEP ($n = 6$): Since $24 < 36 - 7$, $S(6)$ holds.

INDUCTIVE STEP: Fix some $k \geq 6$ and suppose that the inductive hypothesis

$$P(k): \quad 4k < k^2 - 7$$

holds. To be shown is

$$P(k+1): \quad 4(k+1) < (k+1)^2 - 7.$$

Beginning with the left side of $P(k+1)$,

$$
\begin{aligned}
4(k+1) \ &= \ 4k + 4 \\
&< \ k^2 - 7 + 4 \quad \text{(by inductive hypothesis)} \\
&= \ (k^2 + 2k + 1) - 7 + 4 - (2k+1) \\
&< \ (k+1)^2 - 7 \quad \text{(since } k \geq 6, \ -2k + 3 \leq -9 < 0) \\
&= \ (k+1)^2 - 7,
\end{aligned}
$$

which is the right side of $P(k+1)$. This concludes the inductive step $P(k) \rightarrow P(k+1)$.

Therefore, by MI, for all $n \geq 6$, the proposition $P(n)$ is true. $\qquad\square$

Comments on the above proof: Note that the above proof was not very "tight"; the base step had room to improve as did the inductive step. Can this be improved to $4n < n^2 - 11$ for $n \geq 6$? In this case, the base step is still true (albeit with a lesser margin) and the inductive step is nearly identical. When a proof has so much "play", it is very natural to question whether or not the proof is a good one. It is often comforting to find another way to see the result of the exercise. One simple solution uses high school algebra:

Another solution of Exercise 151: If $x^2 - 7 > 4x$, then $x^2 - 4x - 7 > 0$, so examine the polynomial $p(x) = x^2 - 4x - 7$. If $p(x) > 0$ for every $x \geq 6$, then the statement in the exercise is true. By the quadratic formula, the roots of $p(x)$ are

$$x = \frac{4 \pm \sqrt{4^2 - 4(1)(-7)}}{2} = 2 \pm \sqrt{11}.$$

In any case, the roots are smaller than 6, and since the graph of $y = p(x)$ is a parabola opening upward (with vertex at $x = 2$), it follows that $p(x) > 0$ for all $x \geq 6$, and hence also $p(n) > 0$ for all natural numbers $n \geq 6$. $\qquad\square$

Exercise 152: For $n \geq 3$, let $Q(n)$ denote the statement in the exercise,

$$Q(n): \quad 2n + 1 < n^2.$$

BASE STEP ($n = 3$): Since $2 \cdot 3 < 3^2$, $Q(3)$ holds.

INDUCTIVE STEP: Fix some $k \geq 3$ and suppose that

$$Q(k): \quad 2k + 1 < k^2$$

is true. It suffices to prove

$$Q(k+1): \quad 2(k+1) + 1 < (k+1)^2.$$

This is done by starting with the left side of $Q(k+1)$, and deriving the right:

$$
\begin{aligned}
2(k+1) + 1 &= (2k+1) + 2 \\
&< k^2 + 2 && \text{(by ind. hyp. } Q(k)) \\
&< k^2 + 2k + 1 && \text{(since } k \geq 3,\ 2k > 1) \\
&= (k+1)^2,
\end{aligned}
$$

which is the right side of $Q(k+1)$. This concludes the inductive step $Q(k) \rightarrow Q(k+1)$.

Therefore, by MI, for all $n \geq 3$, the inequality $Q(n)$ is true. $\qquad\square$

Exercise 153: For $n \geq 2$, denote the inequality

$$S(n): \quad 4n^2 > n + 11.$$

BASE STEP ($n = 2$): The statement $S(2)$ says $16 > 2 + 11$, and so $S(2)$ holds.

INDUCTIVE STEP: For some fixed $k \geq 2$, suppose that the inductive hypothesis

$$S(k): \quad 4k^2 > k + 11$$

holds. To be shown is

$$S(k+1): \quad 4(k+1)^2 > k + 12.$$

Then

$$4(k+1)^2 = 4k^2 + 8k + 4 > k + 11 + 8k + 4 > k + 12,$$

where the first inequality follows from the inductive hypothesis and the second inequality is true because $k \geq 2$. So $S(k+1)$ is true, and thus this concludes the inductive step $S(k) \rightarrow S(k+1)$.

Therefore, by MI, for all $n \geq 2$, $S(n)$ is true. $\qquad\square$

Exercise 154: For $n \geq 3$, define the statement

$$P(n): \quad 2n < 2^n.$$

BASE STEP $(n = 3)$: $P(3)$ says $2 \cdot 3 < 2^3$, or $6 < 8$, which is true, proving the base case.

INDUCTIVE STEP: Suppose that for some $k \geq 3$,

$$P(k): \quad 2k < 2^k$$

is true. To be shown is

$$P(k+1): \quad 2(k+1) < 2^{k+1}.$$

Starting with the left side of $P(k+1)$,

$$
\begin{aligned}
2(k+1) &= 2k + 2 \\
&< 2^k + 2 && \text{(by } P(k)) \\
&< 2^k + 2^k && \text{(since } k \geq 3) \\
&= 2(2^k) \\
&= 2^{k+1},
\end{aligned}
$$

proving $P(k+1)$. This concludes the inductive step $P(k) \rightarrow P(k+1)$.

Therefore, by MI, for all $n \geq 3$, the proposition $P(n)$ is true. $\qquad \square$

Exercise 155: For each $n \geq 2$, define the statement

$$S(n): \quad 1 + 2^n < 3^n.$$

BASE STEP $(n = 2)$: The statement $S(2)$ says $1 + 2^2 < 3^2$, or $5 < 9$, which proves the base case.

INDUCTIVE STEP: Fix some $k \geq 2$, and assume that the inductive hypothesis

$$S(k): \quad 1 + 2^k < 3^k$$

is true. To be shown is the statement

$$S(k+1): \quad 1 + 2^{k+1} < 3^{k+1}.$$

Then

$$1 + 2^{k+1} = 1 + 2 \cdot 2^k$$

$$< 3 + 3 \cdot 2^k \qquad\qquad \text{(looking ahead)}$$
$$= 3(1 + 2^k)$$
$$< 3(3^k) \qquad\qquad \text{(by } S(k)\text{)}$$
$$= 3^{k+1},$$

which proves $S(k+1)$. This concludes the inductive step $S(k) \to S(k+1)$.

Therefore, by MI, for all $n \geq 2$, the statement $S(n)$ is true. □

Exercise 156: For each $n \geq 2$, define the proposition

$$P(n): \quad n + 1 < 2^n.$$

BASE STEP $(n = 2)$: $P(2)$ says $2 + 1 < 2^2$, which is true.

INDUCTIVE STEP: Suppose that for some fixed $k \geq 2$,

$$P(k): \quad k + 1 < 2^k$$

is true. Then applying the inductive hypothesis in the first step below, and the fact that $k \geq 2$ implies $1 < 2^k$,

$$(k+1) + 1 \overset{IH}{<} (2^k) + 1 < 2^k + 2^k = 2(2^k) = 2^{k+1},$$

and so

$$P(k+1): \quad (k+1) + 1 < 2^{k+1}$$

has been proved. This concludes the inductive step $P(k) \to P(k+1)$.

Therefore, by MI, for $n \geq 2$, the proposition $P(n)$ is true. □

Exercise 157: This is a special case of Bernoulli's inequality; see Exercise 198 for the general solution.

For each $n \geq 1$, define the statement

$$S(n): \quad \left(1 + \frac{1}{3}\right)^n \geq 1 + \frac{n}{3}.$$

BASE STEP $(n = 1)$: Since $S(1)$ merely says $1 + \frac{1}{3} = 1 + \frac{1}{3}$, the base case is true.

INDUCTIVE STEP: Fix some $k \geq 1$, and suppose that

$$S(k): \quad \left(1 + \frac{1}{3}\right)^k \geq 1 + \frac{k}{3},$$

is true. To be shown is that the statement

$$S(k+1): \quad \left(1 + \frac{1}{3}\right)^{k+1} \geq 1 + \frac{k+1}{3},$$

is also true. Beginning with the left side of $S(k+1)$,

$$
\begin{aligned}
\left(1 + \frac{1}{3}\right)^{k+1} &= \left(1 + \frac{1}{3}\right)\left(1 + \frac{1}{3}\right)^{k} \\
&\geq \left(1 + \frac{1}{3}\right)\left(1 + \frac{k}{3}\right) \qquad \text{(by } S(k)) \\
&= 1 + \frac{k}{3} + \frac{1}{3}\left(1 + \frac{k}{3}\right) \\
&> 1 + \frac{k}{3} + \frac{1}{3} \\
&= 1 + \frac{k+1}{3},
\end{aligned}
$$

and so $S(k+1)$ is true (actually, a slightly stronger statement where inequality is replace with strict inequality is true). This concludes the inductive step $S(k) \to S(k+1)$.

By MI, for all $n \geq 1$, the statement $S(n)$ holds true. $\qquad\square$

Exercise 158: For $n \geq 4$, denote the statement involving n by

$$S(n): \quad 2^n < n!.$$

BASE STEP $(n = 4)$: Since $2^4 = 16$ and $4! = 24$, the statement $S(4)$ is true.

INDUCTIVE STEP: Fix some $k \geq 4$ and assume that

$$S(k): \quad 2^k < k!$$

is true. To be shown is that

$$S(k+1): \quad 2^{k+1} < (k+1)!$$

follows. Beginning with the left side of $S(k+1)$,

$$
\begin{aligned}
2^{k+1} &= 2(2^k) \\
&< 2(k!) &&\text{(by } S(k)) \\
&< (k+1)(k!) &&\text{(since } k \geq 4) \\
&= (k+1)!,
\end{aligned}
$$

the right side of $S(k+1)$. This concludes the inductive step $S(k) \to S(k+1)$.

Therefore, by MI, for all $n \geq 4$, the inequality $S(n)$ is true. $\qquad\square$

Exercise 159: For $n \geq 5$, denote the inequality

$$S(n): \quad n^2 < 2^n.$$

BASE STEP ($n = 5$): Since $5^2 = 25 < 32 = 2^5$, $S(5)$ is true, completing the proof of the base step.

INDUCTIVE STEP: For some fixed $k \geq 5$, let the induction hypothesis

$$S(k): \quad k^2 < 2^k$$

be assumed to be true. To prove the inductive step, one needs to show that

$$S(k+1): \quad (k+1)^2 < 2^{k+1}$$

also holds. There are a number of sequences of inequalities which prove this—only one is given here. [There are often many ways to prove inequalities!]

Beginning with the left side of $S(k+1)$,

$$
\begin{aligned}
(k+1)^2 &= k^2 + 2k + 1 \\
&= 2^k + 2k + 1 && \text{(by } S(k)) \\
&< 2^k + k + 1 && \text{(since } k \geq 5 \geq 1) \\
&< 2^k + 2^k && \text{(by Ex. 156, since } k \geq 5 \geq 2) \\
&= 2(2^k) \\
&= 2^{k+1},
\end{aligned}
$$

the right side of $S(k+1)$, which proves $S(k+1)$. This concludes the inductive step $S(k) \to S(k+1)$.

Therefore, by MI, for all $n \geq 5$, the inequality $S(n)$ is true. $\qquad \square$

Exercise 160: For $\ell \geq 6$, let $S(\ell)$ denote the statement

$$S(\ell): \quad 6\ell + 6 < 2^\ell.$$

BASE STEP: When $\ell = 6$, the statement is true since $42 < 64$, proving the base case $S(6)$.

INDUCTIVE STEP: Assume that for some fixed $m \geq 6$,

$$S(m): \quad 6m + 6 < 2^m.$$

Then

$$
\begin{aligned}
6(m+1) + 6 &= 6m + 6 + 6 \\
&< 2^m + 6 && \text{(by } S(m)) \\
&< 2^m + 2^m && \text{(because } m \geq 6) \\
&= 2^{m+1}
\end{aligned}
$$

proves $S(m + 1)$, finishing the inductive step.

So, by MI, for all $\ell \geq 6$, the inequality $S(\ell)$ is true.   □

Exercise 161: For $k \geq 10$, denote by $S(k)$ the statement

$$S(k): \qquad 3k^2 + 3k + 1 < 2^k.$$

BASE STEP: $S(10)$, is true, because $331 < 1000$. [The base case $k = 8$ is actually true, and so the result in the exercise could be strengthened!]

INDUCTIVE STEP: Fix $\ell \geq 10$ and suppose that $S(\ell)$ is true. Then proving $S(\ell+1)$,

$$
\begin{aligned}
3(\ell + 1)^2 + 3(\ell + 1) + 1 &= 3\ell^2 + 6\ell + 3 + 3\ell + 3 + 1 \\
&= (3\ell^2 + 3\ell + 1) + 6\ell + 4 \\
&< 2^\ell + 6\ell + 4 &&\text{(by } S(\ell)) \\
&< 2^\ell + 2^\ell &&\text{(by Exercise 160)} \\
&= 2(2^\ell) \\
&= 2^{\ell+1},
\end{aligned}
$$

and so $S(\ell) \to S(\ell + 1)$, finishing the inductive step.

Hence, by the principle of mathematical induction, for all $k \geq 10$, the statement $S(k)$ is true.   □

Exercise 162: For $n \geq 10$, let $I(n)$ denote the inequality

$$I(n): \qquad n^3 < 2^n.$$

BASE STEP $(n = 10)$: The statement $I(10)$ says $10^3 < 2^{10}$, which is true because $10^3 = 1000$ whereas $2^{10} = 1024$.

INDUCTIVE STEP: Suppose that for some fixed $k \geq 10$,

$$I(k): \qquad k^3 < 2^k$$

holds. It remains to show that

$$I(k + 1): \qquad (k + 1)^3 < 2^{k+1}$$

follows. Beginning with the left side of $I(k + 1)$,

$$
\begin{aligned}
(k + 1)^3 &= k^3 + 3k^2 + 3k + 1 \\
&< 2^k + 3k^2 + 3k + 1 &&\text{(by } I(k)) \\
&< 2^k + 2^k &&\text{(by Exercise 161)}
\end{aligned}
$$

$$= 2(2^k)$$
$$= 2^{k+1},$$

the right side of $I(k+1)$. This proves that $I(k+1)$ is also true, and so concludes the inductive step $I(k) \rightarrow I(k+1)$.

Therefore, by MI, for all $n \geq 10$, the statement $I(n)$ is true. $\qquad\square$

Exercise 163: For $k \geq 4$, let $S(k)$ be the statement

$$S(k) : 3k^2 + 3k + 1 < 2(3^k).$$

One could prove this directly by checking the values $k = 4, 5, \ldots, 9$ directly and then apply Exercise 161 for the cases thereafter since 2^k is (by Exercise 150, say) less than $2(3^k)$, however, an inductive proof is given here.

BASE STEP: Since $3(4^2) + 3(4) + 1 = 61 < 162 = 2(3^4)$, the base case $S(4)$ is true. Similarly, $S(5)$ says $91 < 2 \cdot 2 \cdot 243$, which is true, and $S(6)$ says $127 < 2 \cdot 729$. (The extra base cases have been added only so as to use Exercise 160 below, avoiding having to prove yet another auxiliary result.)

INDUCTIVE STEP: For some $m \geq 6$, assume

$$S(m) : 3m^2 + 3m + 1 < 2(3^m).$$

To be proved is

$$S(m+1) : 3(m+1)^2 + 3(m+1) + 1 < 2(3^{m+1}).$$

Beginning with the left side and deriving the right side,

$$
\begin{aligned}
3(m+1)^2 + 3(m+1) + 1 &= (3m^2 + 3m + 1) + 6m + 6 \\
&< 2(3^m) + 6m + 6 && \text{(by ind. hyp. } S(m)) \\
&< 2(3^m) + 2^m && \text{(by Exercise 160)} \\
&< 2(3^m) + 3^m && \text{(by Exercise 150)} \\
&= 3(3^m) \\
&= 3^{m+1} \\
&< 2(3^{m+1}),
\end{aligned}
$$

finishing the inductive step.

By MI, for all $k \geq 4$, for all $k \geq 4$, the statement $S(k)$ is true. $\qquad\square$

Comment: This result above seems rather weak—perhaps the "2" on the right can be eliminated? The main feature of this exercise was the algebra behind $2 \cdot 3^m + 3^m = 3^{m+1}$.

Exercise 164: For $n \geq 4$, define the proposition

$$P(n): \quad n^3 < 3^n.$$

BASE STEP ($n = 4$): Since $4^3 = 64 < 81 = 3^4$, the statement $P(4)$ is true.

INDUCTIVE STEP: Suppose that for some fixed $k \geq 4$,

$$P(k): \quad k^3 < 3^k$$

holds. It remains to show that

$$P(k+1): \quad (k+1)^3 < 3^{k+1}$$

follows. Beginning with the left side of $P(k+1)$,

$$
\begin{aligned}
(k+1)^3 &= k^3 + 3k^2 + 3k + 1 \\
&< 3^k + 3k^2 + 3k + 1 && \text{(by } P(k)) \\
&< 3^k + 2(3^k) && \text{(by Exercise 163)} \\
&= 3(3^k) \\
&= 3^{k+1},
\end{aligned}
$$

the right side of $P(k+1)$. This proves that $P(k+1)$ is also true, and so concludes the inductive step $P(k) \to P(k+1)$.

Therefore, by MI, for all $n \geq 4$, $P(n)$ is true. □

Exercise 165: For $n > 6$, define the statement

$$Q(n): \quad 3^n < n!.$$

BASE STEP ($n = 7$): Since $3^7 = 2157$ and $7! = 5040$, the base case $Q(7)$ is true. (Note that $Q(6)$ is false, since $3^6 = 729$, yet $6! = 720$.)

INDUCTIVE STEP: Let $k \geq 7$ be fixed, and suppose that

$$Q(k): \quad 3^k < k!$$

holds. Then

$$
\begin{aligned}
3^{k+1} &= 3 \cdot 3^k \\
&< 3(3!) && \text{(by } Q(k)) \\
&< (k+1)(k!) && \text{(since } k \geq 7) \\
&= (k+1)!,
\end{aligned}
$$

shows that $Q(k+1)$ also holds. This concludes the inductive step $Q(k) \to Q(k+1)$.

Therefore, by mathematical induction, for each $n \geq 7$, the inequality $Q(n)$ is true. $\qquad\qquad\qquad\qquad\qquad\qquad\qquad\qquad\qquad\qquad\qquad\qquad\qquad\qquad$ $\square$

Exercise 167: The solution given here has a peculiarity in the inductive step; see comments after the proof.

For $n \geq 1$, define the statement

$$S(n): \quad n^2 \geq 2n - 1.$$

BASE STEP $(n = 1)$: $S(1)$ says $1^2 \geq 2(1) - 1$, which is true.

INDUCTIVE STEP: Suppose that for some fixed $k \geq 1$, the statement

$$S(k): \quad k^2 \geq 2k - 1$$

is true. Observe that $(k+1)^2 = k^2 + 2k + 1 > 2k + 1 = 2(k+1) - 1$, whence

$$S(k+1): \quad (k+1)^2 \geq 2(k+1) - 1$$

follows directly. This concludes the inductive step $S(k) \dashrightarrow S(k+1)$.

Therefore, by MI, for all $n \geq 1$, $S(n)$ is true. $\qquad\qquad\qquad\qquad\qquad$ $\square$

Comment on above solution: Notice that $S(k)$ was not used in the inductive step! Had $S(k)$ been employed, the sequence might have looked like $(k + 1)^2 = k^2 + 2k + 1 \geq (2k - 1) + 2k + 1) = 2k + 2k \geq 2k + 2 = 2(k+1) > 2(k+1) - 1$.

Exercise 168: For $n \geq 1$, define the proposition

$$P(n): \quad 2n + 1 \leq 3^n.$$

BASE STEP $(n = 1)$: $P(1)$ says $2(1) + 1 \leq 3^1$, which is true.

INDUCTIVE STEP: Fix some $k \geq 1$ and suppose that

$$P(k): \quad 2k + 1 \leq 3^k.$$

Needed to show is

$$P(k+1): \quad 2(k+1) + 1 \leq 3^{k+1}.$$

Starting with the left side of $P(k+1)$,

$$
\begin{aligned}
2(k+1) + 1 &= (2k+1) + 2 \\
&\leq 3^k + 2 && \text{(by } S(k)) \\
&< 3^k + 2 \cdot 2 \cdot 3^k && \text{(since } k \geq 1, \ 3^k \geq 3 > 1) \\
&= (1+2)3^k \\
&= 3 \cdot 3^k,
\end{aligned}
$$

which is equal to the right-hand side of $P(k+1)$; hence $P(k+1)$ is also true. This concludes the inductive step $P(k) \rightarrow P(k+1)$.

Therefore, by MI, for all $n \geq 1$, $P(n)$ is true . $\square$

Exercise 169: For $n \geq 3$, define the assertion

$$A(n): \quad n^n \geq (n+1)!.$$

BASE STEP $(n = 3)$: Since $3^3 = 27 \geq 24 = 4!$, $S(3)$ is true, proving the base case $A(3)$.

INDUCTIVE STEP: Fix $k \geq 3$ and suppose that $A(k): \quad k^k \geq (k+1)!$ is true. It remains to prove $A(k+1): \quad (k+1)^{k+1} \geq (k+2)!$. This is done by looking at the expansion of $(k+1)^{k+1}$ by the binomial theorem:

$$(k+1)^{k+1} = \sum_{i=0}^{k} +1 \binom{k+1}{i} k^{k+1-i}$$

$$= k^{k+1} + (k+1)k^k + \sum_{i=2}^{k} +1 \binom{k+1}{i} k^{k+1-i}$$

$$= k^k \left(k + (k+1) + \sum_{i=2}^{k} +1 \binom{k+1}{i} k^{1-i} \right)$$

$$\begin{aligned}
&> k^k(2k+1) &&\text{(since } k \geq 3) \\
&> k^k(k+2) &&\text{(since } k \geq 3) \\
&> (k+1)!(k+2) &&\text{(by } A(k)) \\
&= (k+2)!,
\end{aligned}$$

which completes the proof of $A(k+1)$, and hence the inductive step $A(k) \rightarrow A(k+1)$.

By mathematical induction, for all $n \geq 3$, $A(n)$ is true. $\square$

Note: In the sequence of inequalities above, the inequality $2k + 1 > k + 2$ was used, hardly an optimal inequality for large k, so one might guess that an even stronger inequality holds, namely

$$n^n > 2(n+1)!,$$

which indeed is true for $n \geq 4$. This can be proved in a nearly identical manner to that used above.

Exercise 170: This problem appears in, for example, [350].

For $n \geq 3$, define the statement

$$P(n): \quad n^{n+1} > (n+1)^n.$$

BASE STEP ($n = 3$): The statement $P(3)$ says $3^4 > 4^3$, which is equivalent to $81 > 64$, so $P(3)$ is true.

INDUCTIVE STEP: Fix some $k \geq 3$, and suppose that

$$P(k): \quad k^{k+1} > (k+1)^k$$

is true. Next to prove is

$$P(k+1): \quad (k+1)^{k+2} > (k+2)^{k+1}.$$

First, restate $P(k)$ in a form convenient to use later:

$$k^{k+1} > (k+1)^k \quad \Rightarrow \quad k \cdot k^k > (k+1)^k,$$

and so $P(k)$ can be rewritten as

$$P'(k): \quad \left(\frac{k+1}{k}\right)^k < k.$$

Observe that $(k+1)^2 = k^2 + 2k + 1 > k^2 + 2k = k(k+2)$, and so for positive k,

$$\frac{k+2}{k+1} < \frac{k+1}{k},$$

which implies (by Exercise 150)

$$\left(\frac{k+2}{k+1}\right)^{k+1} < \left(\frac{k+1}{k}\right)^{k+1}. \tag{28.1}$$

Starting with the right side of $P(k+1)$,

$$
\begin{aligned}
(k+2)^{k+1} &= \left(\frac{k+2}{k+1}\right)^{k+1} (k+1)^{k+1} \\[2mm]
&< \left(\frac{k+1}{k}\right)^{k+1} (k+1)^{k+1} && \text{(by eqn (28.1))} \\[2mm]
&= \left(\frac{k+1}{k}\right)^{k} \frac{k+1}{k} (k+1)^{k+1} \\[2mm]
&< k \frac{k+1}{k} (k+1)^{k+1} && \text{(by } P'(k))
\end{aligned}
$$

$$= (k+1)^{k+2}$$

arrive at the left side of $P(k+1)$, proving the required statement. This concludes the inductive step $P(k) \rightarrow P(k+1)$.

Therefore, by MI, for all $n \geq 3$, the statement $P(n)$ is true.   □

Exercise 171: For $n \geq 5$, define the statement

$$S(n): \quad (n+1)! > 2^{n+3}.$$

BASE STEP $(n = 5)$: $S(5)$ says $6! > 2^8$, which is true since $720 > 256$.

INDUCTIVE STEP: Fix some $k \geq 5$ and suppose that

$$S(k): \quad (k+1)! > 2^{k+3}$$

is true. It suffices to prove that

$$S(k+1): \quad (k+2)! > 2^{k+4}$$

holds. Beginning with the left side of $S(k+1)$,

$$
\begin{aligned}
(k+2)! &= (k+1)! \cdot (k+2) \\
&> 2^{k+3}(k+2) &&\text{(by ind. hyp. } S(k)) \\
&\geq 2^{k+3} \cdot 7 &&\text{(since } k \geq 5) \\
&> 2^{k+3} \cdot 2 \\
&= 2^{k+4},
\end{aligned}
$$

the right side of $S(k+1)$. Thus, $S(k) \rightarrow S(k+1)$, completing the inductive step.

Hence, by MI, for all $n \geq 5$, the statement $S(n)$ is true.   □

Exercise 172: This problem occurred in, for example, [350].

For $n \geq 3$, let $S(n)$ denote the statement

$$(n!)^2 > n^n.$$

Two proofs are provided here, an easy direct proof, and an inductive proof. An inductive proof following from Exercise 170 appears in [350], however, for interests sake, a slightly different proof is offered here.

Direct proof for Exercise 172: Interpret the expression $(n!)^2 = (n!)(n!)$ as the product of n terms, each term a pair of the form $(k+1)(n-k)$, where $k = 0, 1, \ldots, n-1$. For $1 < k \leq n-2$, each term is $(k+1)(n-k) = k(n-k) + n - k = n + k(n-k-1)$, which is greater than n (since $n \geq 3$). Thus, since there are n terms with the two outside terms equal to n and the middle terms all greater than n, the product is greater than n^n. $\qquad\square$

Inductive proof for Exercise 172:
BASE STEP: Since $(3!)^2 = 6^2 = 36 > 27 = 3^3$, $S(3)$ holds.

INDUCTIVE STEP: For some fixed $k \geq 4$, suppose that $S(k)$ is true. It remains to prove
$$S(k+1): \quad ((k+1)!)^2 > (k+1)^{k+1}.$$

First, an observation is made that will streamline the proof. For any $k \geq 3$ and each $2 \leq \ell \leq k$, $\binom{k}{\ell} < k^\ell$, and so, by the binomial theorem,

$$(k+1)^k = k^k + \binom{k}{1}k^{k-1} + \sum_{\ell=2}^{k} \binom{k}{\ell} k^{k-\ell} < (k+1)k^k.$$

To prove $S(k+1)$,

$$
\begin{aligned}
((k+1)!)^2 &= ((k+1)k!)^2 \\
&= (k+1)^2 (k!)^2 \\
&> (k+1)^2 \cdot k^k && \text{(by } S(k)) \\
&= (k+1)[(k+1)k^k] \\
&> (k+1)(k+1)^k && \text{(by above observation)} \\
&= (k+1)^{k+1}.
\end{aligned}
$$

This completes the inductive step $S(k) \to S(k+1)$.

By mathematical induction, for all $n \geq 3$, $S(n)$ is true. $\qquad\square$

Exercise 173: This exercise appeared in [350] (Exercise 13), together with two solutions. The first, apparently found in [282], is a rather clever application of the AM-GM inequality (Theorem 3.3.1) and the formula for the sum of the first n odd numbers (which itself has an inductive proof—see Exercise 38). The second proof given in [350] is that which one would arrive at naturally when trying to prove the result inductively. Both proofs are given here.

For $n \geq 2$, let $P(n)$ denote the statement

$$1 \cdot 3 \cdot 5 \cdots (2n-1) < n^n.$$

Proof using AM-GM: By Exercise 38, $1 + 3 + 5 + \cdots + (2n-1) = n^2$, and so by the AM-GM inequality,

$$(1 \cdot 3 \cdot 5 \cdots (2n-1))^{1/n} < \frac{1 + 3 + 5 + \cdots + (2n-1)}{n} = n.$$

Raising each side to the power of n (and applying Exercise 150) finishes the proof.

□

Proof by induction:

BASE STEP: Since $3 < 2^2$, $P(2)$ is true.

INDUCTIVE STEP: For some fixed $k \geq 2$, suppose that

$$P(k): \quad 1 \cdot 3 \cdot 5 \cdots (2k-1) < k^k$$

is true. It remains to prove the statement

$$P(k+1): \quad 1 \cdot 3 \cdot 5 \cdots (2k+1) < (k+1)^{k+1}.$$

Indeed,

$$
\begin{aligned}
1 \cdot 3 \cdot 5 \cdots (2k-1)(2k+1) &< k^k(2k+1) && \text{(by } P(k)\text{)} \\
&= k^{k+1} + (k+1)k^k \\
&= k^{k+1} + \binom{k+1}{1} k^k \\
&< (k+1)^{k+1},
\end{aligned}
$$

where the last inequality follows from the binomial theorem. Thus, $P(k+1)$ is true, finishing the inductive step.

Therefore, by MI, for each $n \geq 2$, the statement $P(n)$ is true.

□

Added note: In the inductive step above, it was essentially proved that $(k+1)^{k+1} > k^k(2k+1)$ (just as in Exercise 169, too). This can also be seen by applying the Bernoulli inequality (Exercise 198, which says $(1+x)^n > 1 + nx$ when $x > 0$) as follows:

$$(k+1)^{k+1} = k^{k+1}\left(1 + \frac{1}{k}\right)^{k+1} > k^{k+1}\left(1 + \frac{k+1}{k}\right) = k^k(2k+1).$$

Exercise 174: For $n \geq 5$, denote the inequality in the exercise by

$$P(n): \quad (2n)! < (n!)^2 4^{n-1}.$$

BASE STEP $(n = 5)$: One need only verify that $(10)! < (5!)^2 4^4$, which is true, since $10! = 3628800$ and $(5!)^2 \cdot 16 = 3686400$, so $P(5)$ holds.

INDUCTIVE STEP: Fix some $k \geq 5$ and suppose that $P(k)$ is true. Consider the statement

$$P(k+1): \quad (2k+2)! < ((k+1)!)^2 4^k.$$

Beginning with the left side of $P(k+1)$,

$$
\begin{aligned}
(2k+2)! &= (2k)!(2k+1)(2k+2) \\
&< (k!)^2 4^{k-1}(2k+1))(2k+2) \quad \text{(by } P(k)) \\
&< (k!)^2 4^{k-1}(2k+2)(2k+2) \\
&= (k!)^2 4^{k-1}(k+1)^2 \cdot 4 \\
&= ((k+1)!)^2 4^k,
\end{aligned}
$$

the right side of $P(k+1)$. So $P(k+1)$ also holds, completing the inductive step. By mathematical induction, for every $n > 5$, the statement $P(n)$ holds. $\square$

Exercise 175: This exercise appears in [350, Ex. 16]. For $n \geq 2$, denote the inequality in the exercise by

$$
S(n): \quad \frac{4^n}{n+1} < \frac{(2n)!}{(n!)^2}.
$$

BASE STEP: The left side of $S(2)$ is $16/3$ and the right side is 6, so $S(2)$ is true.

INDUCTIVE STEP: For some fixed $k \geq 2$, suppose that

$$
S(k): \quad \frac{4^k}{k+1} < \frac{(2k)!}{(k!)^2}
$$

is true. It remains to prove that

$$
S(k+1): \quad \frac{4^{k+1}}{k+2} < \frac{(2(k+1))!}{((k+1)!)^2}
$$

follows. To streamline the derivation of $S(k+1)$, the following fact is used (which holds for any $n \geq 1$):

$$
\frac{4n+4}{n+2} < \frac{4n+2}{n+1}. \tag{28.2}
$$

To see this, simply cross-multiply to obtain $4n^2 + 8n + 4 < 4n^2 + 10n + 4$.

Beginning with the left side of $S(k+1)$,

$$
\begin{aligned}
\frac{4^{k+1}}{k+2} &= \frac{4^k}{k+1} \cdot \frac{4(k+1)}{k+2} \\
&< \frac{(2k)!}{(k!)^2} \cdot \frac{4(k+1)}{k+2} \quad \text{(by } S(k)) \\
&< \frac{(2k)!}{(k!)^2} \cdot \frac{4k+2}{k+1} \quad \text{(by eq'n (28.2))} \\
&= \frac{(2k)!}{(k!)^2} \cdot \frac{(2k+2)(2k+1)}{(k+1)^2}
\end{aligned}
$$

$$= \frac{(2k+2)!}{((k+1)!)^2},$$

which is the right side of $S(k+1)$. This proves $S(k+1)$ also holds, and so completes the inductive step.

Consequently, by MI, for each $n \geq 2$, $S(n)$ is true. □

Exercise 176: For each positive integer n, denote the assertion in the exercise by

$$A(n): \quad n! > 4^n.$$

Observe that $A(8)$ is false since $8! = 40320$, yet $4^8 = 65536$. However, $A(9)$ is true, since $9! = 362880$ and $4^9 = 262144$. It should seem clear that for every $n > 9$, $A(n)$ is true since at each step, the previous term on the left is multiplied by $n + 1$, yet the right side increases only by a factor of 4. Nevertheless, here is the inductive proof that $A(n)$ is true for every $n \geq 9$:
BASE STEP: The case $n = 9$ was verified above.

INDUCTIVE STEP: For some $k \geq 9$, suppose that $A(k)$ is true. Then

$$
\begin{aligned}
(k+1)! &= (k+1) \cdot k! \\
&> (k+1) \cdot 4^k \quad \text{(by } A(k)) \\
&> 4 \cdot 4^k \quad \text{(since } k \geq 9) \\
&= 4^{k+1},
\end{aligned}
$$

proving that $A(k+1)$ follows. This completes the inductive step.

By mathematical induction, it follows that for all $n \geq 9$, $A(n)$ is true. □

Exercise 177: For each positive integer n, let $S(n)$ be the assertion that $\ln(n) < n$.

BASE STEP: Since $\ln(1) = 0 < 1$, $S(1)$ is true.

INDUCTIVE STEP: For some $k \geq 1$, assume that $S(k)$ is true, that is, $\ln(k) < k$, or equivalently, $k < e^k$. It remains to prove $S(k+1)$, that is, $\ln(k+1) < k = 1$, or equivalently, $k+1 < e^{k+1}$. To accomplish this, a little trick is used: since $2 < e$, it follows that $1 < e - 1$ and so $\frac{1}{e^k} < e - 1$. Hence $1 + \frac{1}{e^k} < e$ and multiplying each side e^k yields $e^k + 1 < e^{k+1}$. By induction hypothesis, $k < e^k$ and so the previous equation yields $k+1 < e^{k+1}$, the statement equivalent to $S(k+1)$. This completes the inductive step.

By MI, for all $n \geq 1$ the statement $S(n)$ is true . □

Exercise 179: For $n \geq 1$, let $S(n)$ denote the statement

$$1 + 2 + 3 + \cdots + n < \frac{1}{8}(2n+1)^2.$$

BASE STEP ($n = 1$): Since $1 < \frac{1}{8}3^2$, $S(1)$ holds.

INDUCTION STEP: Fix $k \geq 1$ and suppose that $S(k)$ holds. Consider the statement

$$S(k+1): \quad 1 + 2 + 3 + \cdots + k + (k+1) < \frac{1}{8}(2k+3)^2.$$

Beginning with the left side of $S(k+1)$,

$$
\begin{aligned}
1 + 2 + 3 + \cdots + k + (k+1) \ &< \ \frac{1}{8}(2k+1)^2 + (k+1) \quad \text{(by } S(k)) \\
&= \ \frac{1}{8}[(2k+1)^2 + 8k + 8] \\
&= \ \frac{1}{8}[4k^2 + 4k + 1 + 8k + 8] \\
&= \ \frac{1}{8}[4k^2 + 12k + 9] \\
&= \ \frac{1}{8}(2k+3)^2,
\end{aligned}
$$

completing the proof of $S(k+1)$, and hence the inductive step.

By the principle of mathematical induction, for all $n \geq 1$, the statement $S(n)$ is true. $\qquad \square$

Exercise 180: This exercise shows that the sum of the reciprocals of the squares converges to something at most 2; in fact, the series converges to $\frac{\pi^2}{6}$.

For $n \geq 1$, denote the statement in the exercise by

$$S(n): \quad 1 + \frac{1}{4} + \frac{1}{9} + \cdots + \frac{1}{n^2} \leq 2 - \frac{1}{n}.$$

BASE STEP ($n = 1$): Since $1 = 2 - \frac{1}{1}$, $S(1)$ holds.

INDUCTION STEP: Fix some $k \geq 1$ and suppose that $S(k)$ is true. It remains to show that

$$S(k+1): \quad 1 + \frac{1}{4} + \frac{1}{9} + \cdots + \frac{1}{k^2} + \frac{1}{(k+1)^2} \leq 2 - \frac{1}{k+1}$$

holds. Starting with the left side of $S(k+1)$,

$$
\begin{aligned}
1 &+ \frac{1}{4} + \cdots + \frac{1}{k^2} + \frac{1}{(k+1)^2} \\
&\leq 2 - \frac{1}{k} + \frac{1}{(k+1)^2} \qquad\qquad \text{(by } S(k)) \\
&= 2 - \frac{1}{k+1}\left(\frac{k+1}{k} - \frac{1}{k+1}\right)
\end{aligned}
$$

$$= 2 - \frac{1}{k+1}\left(\frac{k^2-k}{k(k+1)}\right)$$

$$\leq 2 - \frac{1}{k+1} \qquad\qquad \text{(since } k \geq 1,\ k^2 - k \geq 0\text{)},$$

the right side of $S(k+1)$. Thus $S(k+1)$ is true, thereby completing the inductive step.

By mathematical induction, for any $n \geq 1$, the statement $S(n)$ is true. $\qquad\square$

Exercise 181: For $n \geq 2$, denote the statement in the exercise by

$$S(n): \quad 2\left(1 + \frac{1}{8} + \frac{1}{27} + \cdots + \frac{1}{n^3}\right) < 3 - \frac{1}{n^2}.$$

This exercise shows that the series $\sum_{m=1}^{\infty} \frac{1}{m^3}$ converges to something at most $\frac{3}{2}$.

BASE STEP $(n = 2)$: Since $S(2)$ says $2(1+\frac{1}{8}) < 3 - \frac{1}{4}$, or equivalently, $\frac{9}{4} < \frac{11}{4}$, which is true.

INDUCTION STEP: Suppose that for some fixed $k \geq 2$, $S(k)$ is true. Consider the statement

$$S(k+1): \quad 2\left(1 + \frac{1}{8} + \frac{1}{27} + \cdots + \frac{1}{k^3} + \frac{1}{(k+1)^3}\right) < 3 - \frac{1}{(k+1)^2}.$$

Beginning with the left side of $S(k+1)$,

$$
\begin{aligned}
2\left(1 + \frac{1}{8} + \frac{1}{27} + \cdots + \frac{1}{k^3} + \frac{1}{(k+1)^3}\right) &= 3 - \frac{1}{k^2} + \frac{1}{(k+1)^3} \quad \text{(by } S(k)\text{)} \\
&= 3 - \frac{(k+1)^3 - k^2}{k^2(k+1)^3} \\
&= 3 - \frac{k^3 + 2k^2 + 3k + 1}{k^2(k+1)^3} \\
&< 3 - \frac{k^3 + k^2}{k^2(k+1)^3} \\
&= 3 - \frac{k^2(k+1)}{k^2(k+1)^3} \\
&= 3 - \frac{1}{(k+1)^2},
\end{aligned}
$$

arrive at the right side of $S(k+1)$. Thus, $S(k+1)$ is true, which completes the inductive step $S(k) \Rightarrow S(k+1)$.

By MI, one concludes that for every $n \geq 2$, $S(n)$ is true. $\qquad\square$

Exercise 182: This exercise is very popular, but I can not find an early source; it recently appeared in [534], for example. The result of this exercise shows that the power series expansion for e indeed converges, and to something at most 3.

For $n \geq 4$, denote the inequality in the exercise by

$$S(n): \quad 1 + \frac{1}{1!} + \frac{1}{2!} + \frac{1}{3!} + \cdots + \frac{1}{n!} < 3 - \frac{1}{n}.$$

BASE STEP: $S(4)$ reads $1 + \frac{1}{1} + \frac{1}{2} + \frac{1}{6} + \frac{1}{24} \leq 3 - \frac{1}{4}$, or $\frac{65}{24} < \frac{11}{4}$, and since $\frac{11}{4} = \frac{66}{4}$, $S(4)$ is true. (For $n = 1, 2, 3$, one has equality, so the statement with strict inequality must start at $n = 4$.)

INDUCTION STEP: Fix $m \geq 4$, and suppose that $S(m)$ is true. It remains to show that

$$S(m+1): \quad 1 + \frac{1}{1!} + \frac{1}{2!} + \frac{1}{3!} + \cdots + \frac{1}{m!} + \frac{1}{(m+1)!} \leq 3 - \frac{1}{m+1}$$

follows. In a sequence of inequalities below, the following inequality (which holds for $m \geq 1$) is used:

$$\frac{1}{m+1} \leq \frac{1}{m} - \frac{1}{(m+1)!}. \tag{28.3}$$

This holds because $m! \leq (m-1)!(m+1)$ (check for $m = 1$, then multiply out for larger m), and so

$$\frac{1}{m+1} + \frac{1}{(m+1)!} = \frac{m!+1}{(m+1)!} \geq \frac{(m-1)!(m+1)}{(m+1)!} = \frac{1}{m}.$$

With this in hand, beginning with the left side of $S(m+1)$,

$$1 + \frac{1}{1!} + \frac{1}{2!} + \cdots + \frac{1}{m!} + \frac{1}{(m+1)!} < 3 - \frac{1}{m} + \frac{1}{(m+1)!} \quad \text{(by } S(m)\text{)}$$

$$\leq 3 - \frac{1}{m+1} \quad \text{(by eq'n (28.3))},$$

which proves $S(m+1)$, thereby completing the inductive step.

By mathematical induction, for each $n \geq 4$, $S(n)$ holds. $\qquad \square$

Exercise 183: For $n \geq 1$, let $L(n)$ denote the inequality

$$\frac{1}{1!} + \frac{2}{3!} + \frac{3}{5!} + \cdots + \frac{n}{(2n-1)!} \leq 2 - \frac{1}{(2n)!}.$$

BASE STEP: For $n = 1$, $L(1)$ reads $\frac{1}{1} \leq 2 - \frac{1}{2}$, which is true.

INDUCTIVE STEP: For some fixed $k \geq 1$, let the inductive hypothesis be that $L(k)$ is true. The next step is to show that $L(k+1)$:

$$\frac{1}{1!} + \frac{2}{3!} + \frac{3}{5!} + \cdots + +\frac{k}{(2k-1)!} + \frac{k+1}{(2k+1)!} \leq 2 - \frac{1}{(2k+2)!}$$

is also true. Starting with the left side of $L(k+1)$,

$$
\begin{aligned}
\frac{1}{1!} + \frac{2}{3!} + \cdots + \frac{k}{(2k-1)!} + \frac{k+1}{(2k+1)!} \quad &\leq\quad 2 - \frac{1}{(2k)!} + \frac{k+1}{(2k+1)!} \quad \text{(by } L(k)) \\
&=\quad 2 - \frac{2k+1-(k+1)}{(2k+1)!} \\
&=\quad 2 - \frac{k}{(2k+1)!} \\
&<\quad 2 - \frac{1}{(2k+1)!} \\
&<\quad 2 - \frac{1}{(2k+2)!},
\end{aligned}
$$

which is the right side of $L(k+1)$. This completes the inductive step $L(k) \Rightarrow L(k+1)$.

By mathematical induction, for all $n \geq 1$, the statement $L(n)$ is true. □

Added notes: The inductive step above seems to have too much room, and hence indicates that a much tighter result is possible (or some simple mistake was made?). Furthermore, the above result shows that the infinite series $\sum_{i=1}^{\infty} \frac{i}{(2i-1)!}$ converges to a number which is at most 2. Can you find out what this series actually converges to?

Exercise 184: For $n \geq 1$, let $I(n)$ denote the inequality

$$\frac{1}{2!} + \frac{2}{3!} + \frac{3}{4!} + \cdots + \frac{n}{(n+1)!} \leq 1 - \frac{1}{(n+1)!}.$$

BASE STEP: $I(1)$ says $\frac{1}{2} \leq 1 - \frac{1}{2!}$ which is true because, in fact, equality is attained.

INDUCTION STEP: Let $k \geq 1$ and suppose that $I(k)$ holds. Then

$$
\begin{aligned}
\frac{1}{2!} + \frac{2}{3!} + \cdots + \frac{k}{(k+1)!} + \frac{k+1}{(k+2)!} \quad &\leq\quad 1 - \frac{1}{(k+1)!} + \frac{k+1}{(k+2)!} \quad \text{by } I(k) \\
&=\quad 1 - \frac{k+2}{(k+2)!} + \frac{k+1}{(k+2)!} \\
&=\quad 1 - \frac{1}{(k+2)!},
\end{aligned}
$$

proves that $I(k+1)$ also holds, completing the inductive step.

Therefore, by induction, for all $n \geq 1$, the statement $I(n)$ holds. $\qquad\square$

Exercise 185: For $n \geq 2$, let $S(n)$ be the statement

$$S(n): \quad \frac{1}{n+1} + \frac{1}{n+2} + \cdots + \frac{1}{2n} > \frac{13}{24}.$$

BASE STEP ($n = 2$): $S(2)$ says $\frac{1}{2+1} + \frac{1}{2+2} > \frac{13}{24}$, which is true since $\frac{1}{2+1} + \frac{1}{2+2} = \frac{14}{24}$.

INDUCTION STEP: Fix some $k \geq 2$, and suppose that

$$S(k): \quad \frac{1}{k+1} + \frac{1}{k+2} + \cdots + \frac{1}{2k} > \frac{13}{24}$$

is true (the induction hypothesis). The next step is to show that

$$S(k+1): \quad \frac{1}{k+2} + \frac{1}{k+3} + \cdots + \frac{1}{2k+2} > \frac{13}{24}$$

is true. Beginning with the left-hand side of $S(k+1)$, one must add and subtract an extra term so that the inductive hypothesis can be applied:

$$\frac{1}{k+2} + \frac{1}{k+3} + \cdots + \frac{1}{2k} + \frac{1}{2k+1} + \frac{1}{2k+2}$$

$$= \frac{1}{k+1} + \frac{1}{k+2} + \frac{1}{k+3} + \cdots + \frac{1}{2k} + \frac{1}{2k+1} + \frac{1}{2k+2} - \frac{1}{k+1}$$

$$> \frac{13}{24} + \frac{1}{2k+1} + \frac{1}{2k+2} - \frac{1}{k+1} \quad \text{(by ind. hyp.)}$$

$$= \frac{13}{24} + \frac{1}{2k+1} + \frac{1}{2k+2} - \frac{2}{2k+2}$$

$$= \frac{13}{24} + \frac{1}{2k+1} - \frac{1}{2k+2}$$

$$= \frac{13}{24} + \frac{1}{(2k+1)(2k+2)}$$

$$> \frac{13}{24},$$

proving $S(k+1)$, and completing the inductive step $S(k) \rightarrow S(k+1)$.

Hence, by mathematical induction, $S(n)$ holds for all $n \geq 2$. $\qquad\square$

Exercise 186: For each $n \geq 2$, let $R(n)$ denote the statement

$$2^n < \binom{2n}{n} < 4^n.$$

BASE STEP: $R(2)$ says $2^2 < \binom{4}{2} < 4^2$, a true statement since $4 < 6 < 16$.

INDUCTIVE STEP: Let $k \geq 2$ and suppose that $R(k)$ is true. It remains to prove

$$R(k+1): \quad 2^{k+1} < \binom{2k+2}{k+1} < 4^{k+1}$$

First note that

$$\binom{2k+2}{k+1} = \frac{(2k+2)!}{(k+1)!(k+1)!} = \frac{(2k+2)(2k+1)}{(k+1)(k+1)} \frac{(2k)!}{k!k!} = \frac{2(2k+1)}{k+1} \binom{2k}{k}.$$

To see the left inequality in $R(k+1)$,

$$\binom{2k+2}{k+1} = \frac{2(2k+1)}{k+1} \binom{2k}{k}$$
$$> \frac{2(2k+1)}{k+1} 2^k \qquad\qquad \text{(by } R(k)\text{)}$$
$$> 2 \cdot 2^k = 2^{k+1}.$$

To see the second inequality in $R(k+1)$, first note that $\frac{2(2k+1)}{k+1} = \frac{4k+2}{k+1} = 4 - \frac{2}{k+1} < 4$, and so

$$\binom{2k+2}{k+1} = \frac{2(2k+1)}{k+1} \binom{2k}{k}$$
$$< \frac{2(2k+1)}{k+1} 4^k \qquad\qquad \text{(by } R(k)\text{)}$$
$$< 4 \cdot 4^k = 4^{k+1}.$$

So both inequalities in $R(k+1)$ hold, completing the inductive step.

By mathematical induction, for each $n \geq 2$, $R(n)$ holds. $\qquad\qquad \square$

Exercise 187: For each $n \geq 1$, let the statement in the exercise be denoted by

$$C(n): \quad \binom{2n}{n} \geq \frac{2^{2n}}{2n}.$$

BASE STEP: $C(1)$ says $\binom{2}{1} \geq \frac{2^2}{2}$, or equivalently, $2 \leq 2$, and so $C(1)$ holds.
Inductive step: Fix some $m \geq 1$ and suppose that

$$C(m): \quad \binom{2m}{m} \geq \frac{2^{2m}}{2m}$$

is true. To be shown is that

$$C(m+1): \quad \binom{2m+2}{m+1} \geq \frac{2^{2m+2}}{2m+2}$$

follows. Using the equality mentioned in Exercise 186,

$$\binom{2m+2}{m+1} = \frac{2(2m+1)}{m+1}\binom{2m}{m}$$

$$\geq \frac{2(2m+1)}{m+1}\frac{2^{2m}}{2m} \qquad\qquad \text{(by } C(m)\text{)}$$

$$= \frac{2(2m+1)}{m} \cdot \frac{2^{2m}}{2(m+1)}$$

$$> 4 \cdot \frac{2^{2m}}{2m+2} = \frac{2^{2m+2}}{2m+2},$$

and so $C(m+1)$ indeed follows from $C(m)$, completing the inductive step.

By mathematical induction, for each $n \geq 1$, the inequality $C(n)$ is true. $\qquad\square$

Exercise 188: For $n \geq 1$, let $R(n)$ denote the inequality

$$\frac{1}{\sqrt{1}} + \frac{1}{\sqrt{2}} + \frac{1}{\sqrt{3}} + \cdots + \frac{1}{\sqrt{n}} \leq 2\sqrt{n} - 1.$$

BASE STEP: The statement $R(1)$ says $\frac{1}{1} \leq 2\sqrt{1} - 1$, and so is true.

INDUCTIVE STEP: Suppose that for some fixed $k \geq 1$,

$$R(k): \quad \frac{1}{\sqrt{1}} + \frac{1}{\sqrt{2}} + \frac{1}{\sqrt{3}} + \cdots + \frac{1}{\sqrt{k}} \leq 2\sqrt{k} - 1$$

holds. To complete this inductive step, it remains to show that

$$R(k+1): \quad \frac{1}{\sqrt{1}} + \frac{1}{\sqrt{2}} + \frac{1}{\sqrt{3}} + \cdots + \frac{1}{\sqrt{k}} + \frac{1}{\sqrt{k+1}} \leq 2\sqrt{k+1} - 1$$

follows. In doing so, the following inequality (which one discovers is needed only after one tries proving the inductive step) is helpful:

$$2\sqrt{k} + \frac{1}{\sqrt{k+1}} < 2\sqrt{k+1}, \quad \text{(for } k > 0\text{)} \qquad\qquad (28.4)$$

One proof of equation (28.4) is by the following reasoning: $4k(k+1) < (2k+1)^2$, and taking square roots yields $2\sqrt{k}\sqrt{k+1} < 2k+1$; then $2\sqrt{k}\sqrt{k+1}+1 < 2(k+1)$ and division by $\sqrt{k+1}$ now yields the result. Okay, with that in hand, beginning with the left side of $R(k+1)$,

$$\frac{1}{\sqrt{1}} + \frac{1}{\sqrt{2}} + \cdots + \frac{1}{\sqrt{k}} + \frac{1}{\sqrt{k+1}} \quad \leq \quad 2\sqrt{k} - 1 + \frac{1}{\sqrt{k+1}} \quad \text{(by } R(k)\text{)}$$

$$< \ 2\sqrt{k+1} - 1 \quad \text{(by equation (28.4))},$$

the right side of $R(k+1)$. This completes the inductive step $R(k) \to R(k+1)$.

By mathematical induction, for all $n \geq 1$, $R(n)$ holds. $\qquad\qquad$ □

Exercise 189: For $n \geq 1$, denote the given inequality by

$$S(n): \quad \frac{1}{\sqrt{1}} + \frac{1}{\sqrt{2}} + \frac{1}{\sqrt{3}} + \cdots + \frac{1}{\sqrt{n}} \geq \sqrt{n}.$$

BASE STEP: Since $1 \geq \sqrt{1}$, $S(1)$ holds.

Before proceeding with the inductive step, inequality is given (and proved) which streamlines calculations: For any $x \geq 1$,

$$\sqrt{x} + \frac{1}{\sqrt{x+1}} > \sqrt{x+1}. \tag{28.5}$$

To see this, observe that for $x \geq 1$, $\sqrt{x(x+1)} > x$, and so $\sqrt{x(x+1)} + 1 > x + 1$. Division by $\sqrt{x+1}$ proves equation (28.5).

INDUCTIVE STEP: For some fixed $k \geq 1$, suppose that $S(k)$ is true. Then

$$\frac{1}{\sqrt{1}} + \frac{1}{\sqrt{2}} + \cdots + \frac{1}{\sqrt{k}} + \frac{1}{\sqrt{k+1}} \ \geq \ \sqrt{k} + \frac{1}{\sqrt{k+1}} \quad \text{(by } S(k)\text{)}$$
$$> \ \sqrt{k+1} \quad \text{(by eqn 28.5)},$$

which shows that $S(k+1)$ is true as well. This completes the inductive step.

By the principle of mathematical induction, for all $n \geq 1$, $S(n)$ is true. $\qquad$ □

Exercise 190: For $n \geq 1$, let $S(n)$ denote the statement

$$2 + \frac{1}{\sqrt{1}} + \frac{1}{\sqrt{2}} + \frac{1}{\sqrt{3}} + \cdots + \frac{1}{\sqrt{n}} > 2\sqrt{n+1}.$$

BASE STEP: Since $2 + 1 > 2 \cdot 1$, $S(1)$ is clearly true.

INDUCTIVE STEP: For some fixed $k \geq 1$, suppose that $S(k)$ is true. To complete the inductive step, it remains to prove $S(k+1)$:

$$2 + \frac{1}{\sqrt{1}} + \frac{1}{\sqrt{2}} + \cdots + \frac{1}{\sqrt{n}} + \frac{1}{\sqrt{k+1}} > 2\sqrt{k+2}.$$

Beginning with the left side of $S(k+1)$,

$$2 + \frac{1}{\sqrt{1}} + \frac{1}{\sqrt{2}} + \cdots + \frac{1}{\sqrt{k}} + \frac{1}{\sqrt{k+1}} > 2\sqrt{k+1} + \frac{1}{\sqrt{k+1}} \quad \text{(by } S(k)\text{)}.$$

To finish the proof of $S(k+1)$, it suffices to prove

$$2\sqrt{k+1} + \frac{1}{\sqrt{k+1}} \geq 2\sqrt{k+2}.$$

Perhaps the easiest way to see this is by the following sequence of equations, each following from the previous:

$$
\begin{aligned}
(2k+3)^2 &> 4(k+2)(k+1); \\
2k+3 &> 2\sqrt{(k+2)(k+1)}; \\
\frac{2(k+1)+1}{\sqrt{k+1}} &> 2\sqrt{k+2}.
\end{aligned}
$$

This yields the desired inequality and so completes the proof of $S(k+1)$ and hence the inductive step.

By MI, for all $n \geq 1$, the statement $S(n)$ holds true. $\qquad\square$

Exercise 191: For each $n \geq 1$, let $S(n)$ be the statement

$$S(n): \quad \frac{1}{2n} \leq \frac{1 \cdot 3 \cdot 5 \cdots (2n-1)}{2 \cdot 4 \cdot 6 \cdots (2n)} \leq \frac{1}{\sqrt{n+1}}.$$

BASE STEP: Since $S(1)$ says $\frac{1}{2} \leq \frac{1}{2} \leq \frac{1}{\sqrt{2}}$, which is true.

INDUCTION STEP: Assume that for some $k \geq 1$,

$$S(k): \quad \frac{1}{2k} \leq \frac{1 \cdot 3 \cdot 5 \cdots (2k-1)}{2 \cdot 4 \cdot 6 \cdots (2k)} \leq \frac{1}{\sqrt{k+1}}$$

is true. The statement

$$S(k+1): \quad \frac{1}{2k+2} \leq \frac{1 \cdot 3 \cdot 5 \cdots (2k-1)(2k+1)}{2 \cdot 4 \cdot 6 \cdots (2k)(2k+2)} \leq \frac{1}{\sqrt{k+2}}$$

is to be shown. Each inequality is proved separately. For the first inequality, in $S(k+1)$, begin with the right side:

$$
\begin{aligned}
\frac{1 \cdot 3 \cdot 5 \cdots (2k-1)(2k+1)}{2 \cdot 4 \cdot 6 \cdots (2k)(2k+2)} &\geq \frac{1}{2k} \cdot \frac{2k+1}{2k+2} \quad \text{(by ind. hyp.)}, \\
&> \frac{1}{2k+2},
\end{aligned}
$$

the left-hand side of the first inequality in $S(k+1)$.

To prove the second equality in $S(k+1)$, begin with left side:

$$\frac{1 \cdot 3 \cdot 5 \cdots (2k-1)(2k+1)}{2 \cdot 4 \cdot 6 \cdots (2k)(2k+2)} \leq \frac{1}{\sqrt{k+1}} \cdot \frac{2k+1}{2k+2} \quad \text{(by ind. hyp.)}$$

$$= \frac{1}{\sqrt{k+2}} \frac{\sqrt{k+2}\,(2k+1)}{\sqrt{k+1}\,(2k+2)}$$

$$= \frac{1}{\sqrt{k+2}} \frac{\sqrt{(k+2)(2k+1)^2}}{\sqrt{(k+1)(2k+2)^2}}$$

$$= \frac{1}{\sqrt{k+2}} \frac{\sqrt{4k^3 + 12k^2 + 9k + 2}}{\sqrt{4k^3 + 12k^2 + 12k + 4}}$$

$$< \frac{1}{\sqrt{k+2}},$$

the right-hand side of the second inequality in $S(k+1)$. This completes the inductive step for both inequalities.

Hence, by mathematical induction, $S(n)$ holds for all $n \geq 1$. □

Exercise 192: This problem appears in many places (*e.g.* [161, p. 180, 7.16]). For $n \geq 1$, let $P(n)$ denote the proposition

$$\frac{1}{2} \cdot \frac{3}{4} \cdot \frac{5}{6} \cdots \frac{2n-1}{2n} \leq \frac{1}{\sqrt{3n+1}}.$$

BASE STEP: When $n = 1$, both sides of $S(1)$ are equal.

INDUCTIVE STEP: Only the details of the solution as found in [161, p. 188] are provided; the reader is left to fill in the steps. Suppose that for some $k \geq 1$, $S(k)$ is valid. To prove $S(k+1)$, it suffices to prove $\frac{2k+1}{2k+2} \leq \sqrt{\frac{3k+1}{3k+4}}$. The derivation is as follows (with the proof starting at the other end, of course):

$$\frac{2k+1}{2k+2} \leq \sqrt{\frac{3k+1}{3k+4}} \iff \left(\frac{2k+1}{2k+2}\right)^2 \leq \frac{3k+1}{3k+4}$$
$$\iff (4k^2 + 4k + 1)(3k + 4) \leq (4k^2 + 8k + 4)(3k + 2)$$
$$\iff 12k^3 + 28k^2 + 19k + 4 \leq 12k^3 + 28k^2 + 20k + 4$$
$$\iff 0 \leq k.$$

. □

Many authors (including [161] and [462]) point out that the inequality

$$\frac{1}{2} \cdot \frac{3}{4} \cdot \frac{5}{6} \cdots \frac{2n-1}{2n} \leq \frac{1}{\sqrt{3n}}$$

is more difficult to prove, even though it is a weaker statement.

Exercise 193 (Triangle inequality); For each $n \geq 1$, let $S(n)$ be the statement,

$$S(n): \quad \forall x_1, x_2, \ldots x_n \in \mathbb{R}, \ |x_1 + x_2 + \cdots + x_n| \leq |x_1| + |x_2| + \cdots + |x_n|.$$

To prove $S(n)$ for all $n \geq 1$, first prove $S(1)$ and $S(2)$ separately, then proceed by strong induction.

BASE STEP:

Base case $n = 1$: The statement $S(1)$ says that for any $x_1 \in \mathbb{R}$, $|x_1| \leq |x_1|$, which is trivially true.

Base case $n = 2$: Let x and y be any real numbers. Since

$$\begin{aligned}
|x+y|^2 &= (x+y)^2 \\
&= x^2 + 2xy + y^2 \\
&\leq x^2 + |2xy| + y^2 \\
&= |x|^2 + 2|x| \cdot |y| + |y|^2 \\
&= (|x| + |y|)^2,
\end{aligned}$$

and so by Lemma 10.0.2, with $b = |x| + |y|$ and $a = |x + y|$, one concludes that $|x + y| \leq |x| + |y|$, which is precisely the statement $S(2)$. The base step is done.

INDUCTIVE STEP: Fix some $k \geq 2$ and suppose that $S(2), S(3), \ldots, S(k)$ all hold. Yet to prove is

$$S(k+1): \quad |x_1 + x_2 + \cdots + x_k + x_{k+1}| \leq |x_1| + |x_2| + \cdots + |x_k| + |x_{k+1}|.$$

Beginning with the left-hand side of $S(k+1)$,

$$\begin{aligned}
|x_1 + x_2 + \cdots + x_k + x_{k+1}| &\leq |x_1 + x_2 + \cdots + x_k| + |x_{k+1}| \quad \text{(by } S(2)\text{)}, \\
&\leq |x_1| + |x_2| + \cdots + |x_k| + |x_{k+1}| \quad \text{(by } S(k)\text{)},
\end{aligned}$$

the right-hand side of $S(k+1)$, completing the inductive step.

By the principle of strong mathematical induction, $S(n)$ is true for all $n \geq 2$; together with the statement $S(1)$, this proves that for all $n \geq 1$, the statement $S(n)$ is true. □

Exercise 194: This is a famous inequality, appearing in many places (*e.g.*, see [499, Problem 50]). First a slightly stronger result for a special case is proved; the general result then follows quite easily.

Let $x + y > 0$, and $x \neq y$ and let $S(n)$ be the statement

$$S(n): \quad 2^{n-1}(x^n + y^n) > (x+y)^n.$$

Proving $S(n)$ for $n \geq 2$ is done by induction on n:

BASE STEP: $S(2)$ says $2(x^2 + y^2) > (x+y)^2$. Since $x \neq y$, $(x-y)^2 > 0$, and so adding $(x+y)^2$ to each side gives $S(2)$.

INDUCTIVE STEP: Suppose that for some $k \geq 2$, that

$$S(k): \quad 2^{k-1}(x^k + y^k) > (x+y)^k.$$

holds. It remains to prove

$$S(k+1): \quad 2^k(x^{k+1} + y^{k+1}) > (x+y)^{k+1}.$$

Proving $S(k+1)$ directly is a bit messy, so first observe that $x^k - y^k$ and $x - y$ are both the same sign (using Exercise 150) and so

$$(x^k - y^k)(x - y) > 0 \Rightarrow x^{k+1} + y^{k+1} > y^k x + x^k y.$$

Adding $x^{k+1} + y^{k+1}$ to each side of the last inequality gives

$$
\begin{aligned}
2(x^{k+1} + y^{k+1}) \ &> \ x^{k+1} + y^{k+1} + y^k x + x^k y \\
&= \ (x^k + y^k)(x + y) \\
&\geq \ \frac{1}{2^{k-1}}(x+y)^k (x+y) \quad \text{(by ind. hyp.)}, \\
&= \ \frac{1}{2^{k-1}}(x+y)^{k+1}
\end{aligned}
$$

and so division by 2 gives

$$x^{k+1} + y^{k+1} > \frac{1}{2^k}(x+y)^{k+1},$$

which is precisely $S(k+1)$. This completes the inductive step.

Hence, by MI, $S(n)$ holds. To finish the proof of the inequality in the exercise, one observes that equality holds when $x = y$, and the inequality is trivial when either $x = 0$ or $y = 0$.   □

Exercise 195: This problem appeared as [280, Challenge Problem 2] (without solution, but solutions most likely appeared in a later issue).

Exercise 196: The problem is to show that for any non-negative real numbers $x_1, x_2, \ldots, x_n$,

$$\frac{x_1^2 + x_2^2 + \cdots + x_n^2}{n} x_1 x_2 \cdots x_n \leq \left(\frac{x_1 + x_2 + \cdots + x_n}{n} \right)^{n+2}.$$

The renowned problem poser/solver Murray S. Klamkin gave this inequality as Problem 1324 in *Mathematics Magazine*, June 1989: In fact, the problem was actually proposed with the added condition $x_1 + x_2 + \ldots + x_n = 1$, and many solutions were received which used the method of Lagrange multipliers (a method from multivariate calculus often used to solve problems with such constraints); however, Klamkin gave a solution in [313] which was by induction on n; that solution is given here, but with just a few more details supplied. Some simple algebraic steps are still left to the reader.

Let $x_1, x_2, \ldots, x_n$ be non-negative reals, and let $S(n)$ be the statement

$$\frac{x_1^2 + x_2^2 + \cdots + x_n^2}{n} x_1 x_2 \cdots x_n \leq \left(\frac{x_1 + x_2 + \cdots + x_n}{n}\right)^{n+2}.$$

It is convenient to rewrite $S(n)$ as

$$(x_1^2 + x_2^2 + \cdots + x_n^2) x_1 x_2 \cdots x_n \leq (n) \left(\frac{x_1 + x_2 + \cdots + x_n}{n}\right)^{n+2}.$$

First observe that the inequality is trivial if any of the x_i's are 0, so will assume that each $x_i > 0$.

BASE STEPS: For $n = 1$, $S(1)$ says $x_1^3 \leq x_1^3$. For $n = 2$, $S(2)$ says

$$(x_1^2 + x_2^2) x_1 x_2 \leq \frac{1}{8}(x_1 + x_2)^4,$$

which reduces to $0 \leq (x_1 - x_2)^4$, which is certainly true.

INDUCTIVE STEP: Let $k \geq 2$ be fixed and suppose that $S(k)$ holds:

$$S(k): \quad (x_1^2 + x_2^2 + \cdots + x_k^2) x_1 x_2 \cdots x_k \leq (k) \left(\frac{x_1 + x_2 + \cdots + x_k}{k}\right)^{k+2}.$$

Yet to prove is (using $x = x_{k+1}$)

$$S(k+1): \quad (x_1^2 + \cdots + x_k^2 + x^2) x_1 x_2 \cdots x_k x \leq (k+1) \left(\frac{x_1 + \cdots + x_k + x}{k+1}\right)^{k+3}.$$

Put $A = \dfrac{x_1 + x_2 + \cdots + x_k}{k}$ and $P = x_1 x_2 \cdots x_k$. With this notation, the inductive hypothesis is

$$S(k): \quad (x_1^2 + x_2^2 + \cdots + x_k^2) P \leq k A^{k+2}$$

and would like to prove

$$S(k+1): \quad (x_1^2 + x_2^2 + \cdots + x_k^2 + x^2) P x \leq (k+1) \left(\frac{kA + x}{k+1}\right)^{k+3}.$$

The left-hand side of $S(k+1)$ is

$$(x_1^2 + x_2^2 + \cdots + x_k^2 + x^2) P x = (x_1^2 + x_2^2 + \cdots + x_k^2) P x + P x^3$$

$$\leq \quad kA^{k+2}x + Px^3 \quad \text{(by } S(k)\text{)}.$$

So to prove $S(k+1)$, it suffices to prove that

$$kA^{k+2}x + Px^3 \leq (k+1)\left(\frac{kA+x}{k+1}\right)^{k+3}.$$

By the AM-GM inequality (Theorem 3.3.1), $P \leq A^k$, so it suffices to prove

$$kA^{k+2}x + A^kx^3 \leq (k+1)\left(\frac{kA+x}{k+1}\right)^{k+3}.$$

The next idea is to restrict to the situation where the sum $x_1 + x_2 + \cdots + x_k + x$ is held constant, and prove the result with this added constraint. The general result then follows. Observe that for any constant c, the statement $S(n)$ holds for $x_1, \ldots, x_n$ if and only if it holds for $cx_1, \ldots, cx_n$ (the factor c^{n+2} appears on each side). So, consider only those $(x_1, \ldots, x_k, x) \in \mathbb{R}^{k+1}$ for which $x_1 + x_2 + \cdots + x_k + x = k+1$, that is,

$$kA + x = k + 1.$$

So, to prove $S(k+1)$, it suffices to show

$$kA^{k+2}x + A^kx^3 \leq k + 1.$$

The left-hand of the above inequality is a function of A (and $x = k+1-kA$, also a function of A), and so maximize the expression using calculus:

$$\frac{d}{dA}[kA^{k+2}x + A^kx^3]$$

$$= \quad k(k+2)A^{k+1}x + kA^{k+2}\frac{dx}{dA} + kA^{k-1}x^3 + A^k3x^2\frac{dx}{dA}$$

$$= \quad k(k+2)A^{k+1}x + kA^{k-1}x^3 - k(kA^{k+2} + A^k3x^2).$$

Putting $A = tx$, this expression becomes (after a bit of algebra)

$$(1-t)(kt^2 - 2t + 1)kt^{k-1}x^{k+2}.$$

Since $k \geq 2$, the above has roots at only $t = 0$ and $t = 1$, and so the derivative is positive for $0 < t < 1$ and negative for $t > 1$. Thus, $kA^{k+2}x + A^kx^3$ achieves a maximum when $t = 1$, that is, when $A = x = 1$. Hence,

$$kA^{k+2}x + A^kx^3 \leq k + 1,$$

and so $S(k+1)$ follows, completing the inductive step.

Thus, by mathematical induction, for all $n \geq 1$, the statement $S(n)$ is true. $\quad\square$

Exercise 197: This exercise occurred in, *e.g.*, [437], (Prob. 11-7) and [499] (Prob. 51).

Fix a positive real number x, and for each $n \geq 1$, let $S(n)$ denote the statement

$$x^n + x^{n-2} + x^{n-4} + \cdots + \frac{1}{x^{n-4}} + \frac{1}{x^{n-2}} + \frac{1}{x^n} \geq n+1.$$

Notice that when n is even, there are $n+1$ terms on the left side of $S(n)$, the middle of which is 1. For example, when $n = 4$, the left side is

$$x^4 + x^2 + x^0 + \frac{1}{x^2} + \frac{1}{x^4}.$$

On the other hand, when n is odd, there are also $n+1$ terms; for example, when $n = 3$, the left side of $S(n)$ is

$$x^3 + x^1 + \frac{1}{x^1} + \frac{1}{x^3}.$$

Before beginning the proof, another simple observation helps: For any real number y, $y^2 - 2y + 1 = (y-1)^2 \geq 0$; this implies $y^2 + 1 \geq 2y$, and when $y > 0$, yields the inequality

$$y + \frac{1}{y} \geq 2. \tag{28.6}$$

It should now be apparent as to the method of the solution: pair up terms in the sum, and proceed with two cases, n even, and n odd; this will mean that the inductive step will jump by two, and hence two base cases are needed.

BASE STEP: When $n = 1$, $S(1)$ says

$$x^1 + \frac{1}{x} \geq 2,$$

which was proved precisely in equation (28.6). For $n = 2$, $S(2)$ says

$$x^2 + 1 + \frac{1}{x^2} \geq 3,$$

which is again true by equation (28.6) with $y = x^2$.

INDUCTIVE STEP: $(S(k) \to S(k+2))$ Fix some $k \geq 1$ and assume that

$$S(k): x^k + x^{k-2} + x^{k-4} + \cdots + \frac{1}{x^{k-4}} + \frac{1}{x^{k-2}} + \frac{1}{x^k} \geq k+1$$

holds. One now wants to show that

$$S(k+2): x^{k+2} + x^k + x^{k-2} + \cdots + \frac{1}{x^{k-2}} + \frac{1}{x^k} + \frac{1}{x^{k+2}} \geq k+3$$

follows. Indeed,

$$x^{k+2} + x^k + x^{k-2} + \cdots + \frac{1}{x^{k-2}} + \frac{1}{x^k} + \frac{1}{x^{k+2}}$$

$$\geq \quad x^{k+2} + k + 1 + \frac{1}{x^{k+2}} \quad \text{(by } S(k)\text{)}$$

$$\geq \quad k + 1 + 2 \quad \text{(by eq'n (28.6) with } y = x^{k+2}\text{)}$$

proves $S(k+2)$ as desired.

By mathematical induction (actually, two inductive proofs are wrapped up in one, one for odd n and one for even n), for $n \geq 1$, $S(n)$ is true. $\qquad \square$

Exercise 198: (Bernoulli's inequality) Fix $x \in \mathbb{R}$ with $x > -1$ and $x \neq 0$. For each $n \geq 2$, let $S(n)$ be the statement that $(1+x)^n > 1 + nx$ holds.

BASE STEP: Since $x \neq 0$, $x^2 > 0$, and so $(1+x)^2 = 1 + 2x + x^2 > 1 + 2x$, $S(2)$ holds.

INDUCTIVE STEP: Fix $k \geq 2$, and suppose that $S(k)$ holds, that is, $(1+x)^k > 1+kx$. Before proving $S(k+1)$, a subtlety regarding inequalities is addressed: If $a > b$ and $c > 0$, then $ac > bc$; if $c < 0$, then $a > b$ implies $ac < bc$. Here is the proof of $S(k+1)$:

$$
\begin{aligned}
(1+x)^{k+1} &= (1+x)^k(1+x) \\
&> (1+kx)(1+x) \quad \text{(since } x + 1 > 0\text{)} \\
&= 1 + x + kx + kx^2 \\
&= 1 + (k+1)x + kx^2 \\
&> 1 + (k+1)x \quad \text{(because } x \neq 0\text{)}.
\end{aligned}
$$

So $S(k+1)$ is true, completing the inductive step.

By MI, for all $n \geq 2$, the statement $S(n)$ is true . $\qquad \square$

Exercise 199: This exercise occurs in, for example, [437]; that solution is very elegant; however, here the solution is presented in a slightly different manner.

For $n \geq 2$, let $S(n)$ be the statement that for any n positive real numbers, if their product is one, then their sum is at least n.

BASE STEP: Let $a_1 a_2 = 1$. Then $a_2 = \frac{1}{a_1}$, and then $a_1 + a_2 = a_1 + \frac{1}{a_1}$. Since the sum of a number and its reciprocal is at least 2 (see Exercise 197) then $a_1 + a_2 \geq 2$, proving the base step $S(2)$.

INDUCTIVE STEP: Suppose that for some $k \geq 2$, $S(k)$ holds. It remains to prove that $S(k+1)$ holds. Let $a_1 a_2 \cdots a_k a_{k+1} = 1$. Since $a_1 a_2 \ldots (a_k a_{k+1}) = 1$ is a product equal to one, by $S(k)$, $a_1 + a_2 + \cdots + a_k a_{k+1} \geq n$. One only needs to find i and j so that $a_i a_j \leq a_i + a_j - 1$. If $a_1 = a_2 = \cdots = a_{k+1} = 1$, then any i and j will do. If some $a_i \neq 1$, say $a_k > 1$, then some $a_j < 1$, say $a_{k+1} < 1$. In this case,

$(a_k - 1)(1 - a_{k+1} - 1) > 0$, which says $a_k - a_k a_{k+1} - 1 + a_{k+1} > 0$. Putting this all together (with $a_k > 1$ and $a_{k+1} < 1$),

$$
\begin{aligned}
a_1 + a_2 + \cdots + a_{k+1} &= (a_1 + a_2 + \cdots + a_k a_{k+1}) + a_k - a_k a_{k+1} + a_{k+1} \\
&\geq k + a_k - a_k a_{k+1} + a_{k+1} \quad (\text{by S(k)}) \\
&= (k+1) + a_k - a_k a_{k+1} - 1 + a_{k+1} \\
&= k + 1 + (a_k - 1)(1 - a_{k+1} - 1) \\
&\geq k + 1 \quad (\text{since } a_k > 1 \text{ and } a_{k+1} < 1).
\end{aligned}
$$

This proves $S(k+1)$, and hence finishes the inductive step.

Therefore, by induction, for $n \geq 1$, the statement $S(n)$ is true. $\qquad \square$

Remark on Solution to Exercise 199: From the statement $S(n)$, one can prove quite easily the AM-GM inequality (Theorem 3.3.1) without the complicated downward induction proof. Here is the proof (see, *e.g.*, [437]):
Let $g = (a_1 a_2 \cdots a_n)^{1/n}$ be the geometric mean. Then

$$
\left(\frac{a_1}{g} \cdot \frac{a_2}{g} \cdots \frac{a_n}{g} \right)^{1/n} = 1,
$$

and hence $\frac{a_1}{g} \cdot \frac{a_2}{g} \cdots \frac{a_n}{g} = 1$. Thus, by Exercise 199, $\frac{a_1}{g} + \frac{a_2}{g} + \cdots + \frac{a_n}{g} \geq n$, and so $\frac{a_1 + a_2 + \cdots + a_n}{n} \geq g$. $\qquad \square$

Exercise 200: (GM-HM inequality) To be proved is that for any positive real numbers $a_1, a_2, \ldots, a_n$,

$$
(a_1 a_2 \cdots a_n)^{1/n} \geq \frac{n}{\frac{1}{a_1} + \frac{1}{a_2} + \cdots + \frac{1}{a_n}}.
$$

Here is a direct (non-inductive) proof:
By the AM-GM inequality,

$$
\begin{aligned}
\frac{a_1^{-1} + a_2^{-1} + \cdots + a_n^{-1}}{n} &\geq (a_1^{-1} a_2^{-1} \cdots a_n^{-1})^{1/n} \\
&= (a_1 a_2 \cdots a_n)^{-1/n}.
\end{aligned}
$$

Taking reciprocals (which reverses the inequality) finishes the proof. $\qquad \square$

Exercise 201: The solution is outlined in [161, 8.24, pp. 208, 216], where the following inequality is used without proof:

Lemma 28.0.1. *Let* $n \geq 2$, *and let* $a_1, a_2, \ldots, a_n$ *be positive integers with*

$$
\frac{1}{a_1} + \frac{1}{a_2} + \cdots + \frac{1}{a_n} = 1
$$

(the a_i's don't need to be increasing). For $m < n$,

$$\frac{1}{a_1 a_2 \cdots a_m} \le 1 - \left(\frac{1}{a_1} + \cdots + \frac{1}{a_m}\right). \tag{28.7}$$

Proof: Let y be a positive rational number so that

$$\frac{1}{a_1} + \frac{1}{a_2} + \cdots + \frac{1}{a_m} + \frac{1}{y} = 1.$$

Simplifying,

$$
\begin{aligned}
y &= \frac{1}{1 - \left(\frac{1}{a_1} + \cdots + \frac{1}{a_m}\right)} \\
&= \frac{a_1 a_2 \cdots a_m}{a_1 a_2 \cdots a_m - \left(\sum_{i=1}^{m} \prod_{j \ne i} a_j\right)}.
\end{aligned}
$$

Since both sides above are positive, the denominator of the right-hand side is positive, and since this is an expression in positive integers, the the denominator is at least 1. Hence, $y \le a_1 a_2 \cdots a_m$, and inverting, equation (28.7) follows. $\qquad\square$

Here is the solution to the exercise: For $n \ge 2$, let $S(n)$ be the statement that for positive integers satisfying $1 \le a_1 \le a_2 \le \cdots \le a_n$, if $\frac{1}{a_1} + \frac{1}{a_2} + \cdots + \frac{1}{a_n} = 1$, then $a_n < 2^{n!}$.

Suppose, in hopes of a contradiction, for some fixed $n \ge 2$, that $S(n)$ fails, that is, the statement

$$P(n): \quad a_n \ge 2^{n!}$$

holds. Using $P(n)$ as the base step, by (strong) downward induction, prove that for each $j = 1, \ldots, n$, that $P(j): \quad a_j \ge 2^{j!}$ holds: Fix $1 \ge m < n$, and let the inductive hypothesis (IH) be that for each $k = m+1, m+2, \ldots, n$, $P(k)$ is true; it remains to prove $P(m)$. To this goal,

$$
\begin{aligned}
\frac{1}{a_m} &\le \left(\frac{1}{a_1 \cdots a_m}\right)^{1/m} \\
&\le \left(1 - \frac{1}{a_1} - \cdots - \frac{1}{a_m}\right)^{1/m} \\
&= \left(\frac{1}{a_{m+1}} + \cdots + \frac{1}{a_n}\right)^{1/m} \\
&\le \left(\sum_{k=m+1}^{n} \frac{1}{2^{k!}}\right)^{1/m} \qquad\qquad \text{(by IH)} \\
&\le \frac{1}{2^{m!}}.
\end{aligned}
$$

Hence $S(n)$ failing implies $P(1), P(2), \ldots, P(n)$ hold. It remains to be observed that

$$\frac{1}{2^{1!}} + \frac{1}{2^{2!}} + \cdots + \frac{1}{2^{n!}} < 1,$$

so $S(n)$ does not hold. $\qquad\square$

Exercise 202: This exercise, complete with solution can be found in [350, prob. 50].

Exercise 204: This appeared as Challenge Problem 1 in the article [280], but with the hint to see the solution to Exercise 205. In fact, this is only the special case $a = 0$ of Exercise 205, so only the more general solution is given here. $\qquad\square$

Exercise 205: This appeared complete with solution in [280, Problem 1]; it also appeared as a problem in the contest "Tournament of the Towns" (1987).

For $n \geq 1$, let $S(n)$ be the statement that for any non-negative real a,

$$\sqrt{a+1+\sqrt{a+2+\cdots+\sqrt{a+n}}} < a+3.$$

BASE STEP: $S(1)$ says $\sqrt{a+1} < a+3$, which is verifiable since $a+1 < (a+3)^2 \Leftrightarrow 0 < a^2 + 5a + 8$, which is true for $a \geq 0$.

INDUCTIVE STEP: Fix some $k \geq 0$, and suppose that for any non-negative a,

$$\sqrt{a+1+\sqrt{a+2+\cdots+\sqrt{a+k}}} < a+3$$

is true. To prove is that for every non-negative b,

$$\sqrt{b+1+\sqrt{b+2+\cdots+\sqrt{b+k+\sqrt{b+k+1}}}} < b+3.$$

Indeed, using $a = b+1$,

$$
\begin{aligned}
& \sqrt{b+1+\sqrt{b+2+\cdots+\sqrt{b+k+\sqrt{b+k+1}}}} \\
={} & \sqrt{b+1+\sqrt{a+1+\cdots+\sqrt{a+k-1+\sqrt{a+k}}}} \\
<{} & \sqrt{b+1+a+3} \quad \text{(by } S(k)\text{)} \\
={} & \sqrt{2b+5} \\
<{} & b+3,
\end{aligned}
$$

where the last inequality follows since $2b+5 < (b+3)^2 = b^2 + 6b + 9$ and $b^2 \geq 0$. This proves $S(k+1)$, concluding the inductive step.

Hence, by MI, for all $n \geq 1$, $S(n)$ is true. $\qquad\square$

Chapter 29

Solutions: Number theory

29.1 Solutions: Primes

Exercise 206: (FTOA, Outline) Use the result from Exercise 297 and strong induction on n, and at the inductive step, argue by contradiction, that is, suppose that n has two different factorizations. There are other proofs of this result that don't use induction; the standard one supposes that there are two different prime factorizations, then one argues that the primes must be the same (up to relabelling), and then one shows by contradiction that the exponents are the same.

Exercise 207: For $n \in \mathbb{Z}^+$, let $S(n)$ be the statement that there are at least n primes.
BASE STEP $(n = 1)$: Since 2 is a prime, $S(1)$ holds.
INDUCTIVE STEP: For some fixed $k \geq 1$, suppose $S(k)$, that is, suppose that there are k different primes, say $p_1, p_2, \ldots, p_k$. Examine the number

$$x = p_1 p_2 \cdots p_k + 1.$$

When dividing x by any one of the primes given, there is a remainder of 1, so none of the primes given are factors of x. That means that either x is itself a prime or a product of primes not yet listed. In either case, there is at least one more prime, that is, there are at least $k + 1$ primes, showing $S(k + 1)$, and thus completing the inductive step.

By induction, it is proved that for any (finite) $n \geq 1$, there are at least n primes. Hence, there are infinitely many primes. $\qquad \square$

Exercise 208: For each $k \geq 1$, let $S(k)$ be the statement that for any $n \geq 1$, if

$$n = p_1^{e_1} p_2^{e_2} \cdots p_k^{e_k}$$

is the prime power decomposition of n, then the sum of the divisors of n is

$$\sigma(n) = \sigma(p_1^{e_1}) \sigma(p_2^{e_2}) \cdots \sigma(p_k^{e_k}).$$

BASE STEP: When $k = 1$ there is nothing to prove, so $S(1)$ is trivially true.
INDUCTION STEP: Let $\ell \geq 1$ and suppose that $S(\ell)$ is true. Let $n = p_1^{e_1} p_2^{e_2} \cdots p_\ell^{e_\ell} p_{\ell+1}^{e_{\ell+1}}$ be the prime power decomposition of n, and for convenience, put

$$a = p_1^{e_1} p_2^{e_2} \cdots p_\ell^{e_\ell},$$

and put

$$b = p_{\ell+1}^{e_{\ell+1}};$$

so $n = ab$. Since a and b are relatively prime, any divisor of n is a product of a divisor of a and a divisor of b. For ease of notation, if z is a divisor of n, then write $z = xy$, where x is a divisor of a and y is a divisor of b. Also write $z \mid n$ to mean that z divides n. Summing all the divisors of n,

$$\sum_{z \mid n} z = \sum_{x \mid a} \sum_{y \mid b} xy = \sum_{x \mid a} x \sum_{y \mid b} y = \sigma(a)\sigma(b).$$

By induction assumption $S(\ell)$, $\sigma(a) = \sigma(p_1^{e_1})\sigma(p_2^{e_2}) \cdots \sigma(p_\ell^{e_\ell})$, so the equation above shows that $S(\ell+1)$ is true, completing the inductive step.

By mathematical induction, for each $k \geq 1$, the statement $S(k)$ is true.   □

Exercise 209 (Division lemma): Fix natural numbers m and n. The proof is divided into two parts, first showing existence of $q \geq 0$ and r with $0 \leq r < m$ so that $n = qm + r$, then showing the uniqueness of such a q and r.

(Existence) One needs to find integers q and r which satisfy

$$\left.\begin{array}{l} q \geq 0, \\ 0 \leq r < m, \\ n = qm + r. \end{array}\right\} \tag{29.1}$$

One might want to say "pick the largest q for which $qm < n$", however, to be formal, this is done in terms of well-ordering. There are at least two proofs based on well-ordering, probably the first of which is most common.

First proof of existence: (All variables are integers.) Examine the set $R = \{r \geq 0 :$ there exists $q \geq 0$ so that $n = qm + r\}$. Since $n \geq 0$, putting $q = 0$ shows that $n \in R$, and so R is non-empty. Since R is a non-empty well-ordered set, it contains a least element r^*, and so there is a $q \geq 0$ so that $n = qm + r^*$. Rewriting this, $n = (q+1)m + (r^* - m)$ shows that if $r^* - m \geq 0$, then $r^* - m \in R$, and in this case, $r*$ would not be the least element, so $r*-m < 0$, that is, $r^* < m$. By the definition of the set R, $r \geq 0$, so the existence of q and r satisfying (29.1) has been proved.

Second proof of existence: Examine the set

$$S = \{a \in \mathbb{Z} : n + am \geq 0\}.$$

Since $m > 0$, $n + (-n - 1)m \leq 0$, and hence any $a \in S$ must satisfy $a \geq -n$. Since $n > 0$, $n + 0 \cdot m > 0$ and so $0 \in S$. Thus, S is a well-ordered set which is non-empty, and so has a least element s. [*Note*: S is not precisely a subset of $\mathbb{N}$, so the well-ordering theorem can not be used directly, however S is a subset of integers each greater than or equal to $-n$, so it, too, is a well-ordered set.) Since $0 \in S$, $s \leq \min(S) = 0$. Putting $q = -s$, then $q \geq 0$ is the largest integer with $n - qm \geq 0$. Put $r = n - qm$.

Claim: $0 \leq r < m$. By the choice of elements in S, $n - qm = n + sm \geq 0$, so $r \geq 0$. If $r \geq m$, say, $r = m + k$ for some $k \geq 0$, then $r = n - qm = m + k$ implies $n - (q + 1)m = k \geq 0$, and so $s^* = (q + 1)$ is in S; but $s*$ is then smaller than s, contradicting the minimality of s—so $r \geq m$ leads to a contradiction. Hence, $r < m$ as desired, completing the proof of the claim, and the proof of (29.1). This completes the existence part of the proof.

(Uniqueness) Suppose that $n = q_1 m + r_1$ and $n = q_2 m + r_2$, where $q_1 \geq 0$, $0 \leq r_1 < m$, $q_2 \geq 0$, $0 \leq r_2 < m$. Subtracting these two equations, obtain $(q_1 - q_2)m = r_1 - r_2$. Without loss of generality, assume that $q_1 \geq q_2$. If $q_1 = q_2$, then $0 = r_1 - r_2$, which shows that $r_1 = r_2$ as well.

So examine the case when $q_1 > q_2$. Then $(q_1 - q_2)m = r_1 - r_2$ implies that $r_1 - r_2 \geq m$ and $r_1 = r_2 + m$. In this situation, it is impossible for both r_1 and r_2 to be in the interval $[0, m)$, so the case $q_1 > q_2$ is not possible.

Conclude that $q_1 = q_2$ and $r_1 = r_2$, that is, the q and r found in the existence proof are indeed unique. $\qquad\qquad\square$

Exercise 210: For each $n \geq 1$, let $C(n)$ be the claim that if $m_1, m_2, \ldots, m_n$ are pairwise relatively prime natural numbers, and y is a natural number so that each m_i divides y, then $m_1 m_2 \ldots m_n$ divides y.

BASE STEP: If m_1 divides y, then the product of only one m_1 trivially divides y, so $C(1)$ is true.

INDUCTIVE STEP: Let $k \geq 1$ and suppose that $m_1, m_2, \ldots, m_k, m_{k+1}$ are pairwise relatively prime. Assume that $C(k)$ is true, that is, if y is such that each of m_1, $m_2, \ldots, m_k$ divide y, then $m_1 m_2 \cdots m_k$ divides y as well. Suppose that each of $m_1, m_2, \ldots, m_k, m_{k+1}$ divides y. By $C(k)$, the product $M = m_1 \cdots m_k$ divides y. Since m_{k+1} is relatively prime to each of $m_1, \ldots, m_k$, conclude that M and m_{k+1} are relatively prime. Since M divides y and m_{k+1} divides y, there exist integers a and b so that $y = aM$ and $y = bm_{k+1}$; hence $aM = bm_{k+1}$. Since m_{k+1} is relatively prime to M, m_{k+1} divides a, say $a = \ell m_{k+1}$, and so $y = \ell m_{k+1} M$, giving that $m_{k+1} M = m_1 m_2 \cdots m_k m_{k+1}$ divides y, completing the inductive step.

By mathematical induction, for every $n \geq 1$, the statement $C(n)$ is true. $\qquad\square$

Exercise 211 (Euclidean Division Algorithm): Recall that Lemma 11.1.1 says for $m, n \in \mathbb{Z}+$, if q and r are integers satisfying $q \geq 0$, $0 \leq r < m$ and $n = mq + r$, then $gcd(m, r) = gcd(m, n)$.

Let m and n be positive integers, n not a multiple of m, and apply the division lemma repeatedly, producing quotients $q_1, q_2, \ldots, q_k$ ($k \geq 2$) and remainders $r_1, r_2, \ldots, r_k$, where

$$
\begin{aligned}
n &= q_1 m + r_1 & (0 < r_1 < m) \\
m &= r_1 q_2 + r_2 & (0 < r_2 < r_1) \\
r_1 &= r_2 q_3 2 + r_3 & (0 < r_3 < r_2) \\
r_2 &= r_3 q_4 + r_4 & (0 < r_4 < r_3) \\
&\ \ \vdots \\
r_{k-3} &= r_{k-2} q_{k-1} + r_{k-1} & (0 < r_{k-1} < r_{k-2}) \\
r_{k-2} &= r_{k-1} q_k & (0 = r_k).
\end{aligned}
$$

If $k = 1$, then put $r_1 = 0$, in which case m divides n, and $\gcd(m, n) = m$. Note that since the r_i's are decreasing positive integers, by well-ordering, there is a least k for which $r_k = 0$ (and so $r_{k-1} \neq 0$). One must prove that $r_{k-1} = \gcd(m, n)$. One can either induct upward or downward, making certain that certain gcd's are maintained; in either case, the proof is tantamount to proving the chain

$$\gcd(n, m) = \gcd(m, r_1) = \gcd(r_1, r_2) = \gcd(r_2, r_3) = \cdots = \gcd(r_{k-2}, r_{k-1}) = r_{k-1}.$$

The final equality is clear since r_{k-1} divides r_{k-2}. For any of the other equalities, say the j-th equality ($j < k$)

$$\gcd(r_{j-2}, r_{j-1}) = \gcd(r_{j-1}, r_j),$$

follows from the corresponding equation from the above algorithm,

$$r_{j-2} = r_{j-1} q_j + r_j,$$

and Lemma 11.1.1. By induction, equality holds throughout. □

Exercise 212: (Proof outline) For $n \geq 2$, let $S(n)$ be the statement that the product of all primes at most n is at most 2^{2n}.

To get ready for the induction step, make a couple of observations: The two middle terms $\binom{2m+1}{m}$ and $\binom{2m+1}{m+1}$ from the binomial expansion of $(1 + 1)^{2m+1}$ are equal, and so $2^{2m} \geq \binom{2m+1}{m}$. Also, for each prime p between m and $2m + 1$, $\binom{2m+1}{m}$ is divisible by p. Thus, the product of all primes between m and $2m + 1$ must divide $\binom{2m+1}{m}$ and hence be less than 2^{2m}. Apply induction on n, and examine two cases, even and odd. (The details are spelled out in [150, pp. 177-8]; this is one step in proving Bertrand's theorem, namely that for every $n \geq 2$, there is a prime between n and $2n$.)

Exercise 213: A solution can be found in, for example, [19, p. 82]; it is by induction on the number of prime factors of n. Note that part of this exercise (regarding $\sigma(n)$) is implicitly proved in Exercise 208, however, that solution is not relied upon here.

BASE STEP: If n has one prime factor, say $n = p^\alpha$, then the factors of n are $1, p, p^2, \ldots, p^\alpha$, of which there are $d(n) = \alpha + 1$ and their sum is (by Exercise 49) $\sigma(n) = \frac{p^{\alpha+1}-1}{p-1}$.

INDUCTIVE STEP: Let $m = np^\alpha$, where $n = p_1^{\alpha_1} \cdots p_s^{\alpha_s}$ and $p \notin \{p_1, \ldots, p_s\}$, and $\alpha \geq 1$. Assume that the two equalities (11.4) and (11.5) for $d(n)$ and $\sigma(n)$ hold true for n.

Any divisor d of m is of the form $d'p^\beta$, where d' is a divisor of n and $0 \leq \beta \leq \alpha$. By the induction hypothesis, $d(n) = (\alpha_1 + 1)(\alpha_2 + 1) \cdots (\alpha_s + 1)$ and the number of divisors of p^α is $\alpha + 1$, so

$$d(m) = (\alpha_1 + 1)(\alpha_2 + 1) \cdots (\alpha_s + 1)(\alpha + 1).$$

This completes the inductive step for the expression (11.4) for $d(n)$.

Partitioning the divisors of m into $\alpha + 1$ classes, depending on the power of p,

$$
\begin{aligned}
\sum_{d|m} d &= \sum_{d'|n} d' + \sum_{d'|n} d'p + \sum_{d'|n} d'p^2 + \cdots + \sum_{d'|n} d'p^\alpha \\
&= \sum_{d'|n} d' + p\sum_{d'|n} d' + p^2\sum_{d'|n} d'p^2 + \cdots + p^\alpha \sum_{d'|n} d' \\
&= \sum_{d'|n} d'(1 + p + p^2 + \cdots + p^\alpha) \\
&= \sigma(n)\frac{p^{\alpha+1}-1}{p-1} \\
&= \frac{p_1^{\alpha_1+1}-1}{p_1-1} \cdot \frac{p_2^{\alpha_2+1}-1}{p_2-1} \cdots \frac{p_s^{\alpha_s+1}-1}{p_s-1} \frac{p^{\alpha+1}-1}{p-1},
\end{aligned}
$$

where the last equality holds by induction hypothesis. Since this last expression is of the proper form (for $s + 1$ prime factors), this completes the inductive step for the expression (11.5) for $\sigma(n)$.

Therefore, by mathematical induction, both expressions are valid for all $s \geq 1$. This completes the solution to the exercise. $\qquad\square$

Exercise 214: This problem appears in, e.g., [150, Ex. 3, p. 130]. All variables here denote positive integers. To be shown is that if $r^2 = st$ and s and t are relatively prime, then both s and t are squares. One straightforward non-inductive proof is based on the following fact: if a has prime factorization $a = p_1^{\alpha_1} p_2^{\alpha_2} \cdots p_k^{\alpha_k}$, a is a perfect square if and only if each α_i is even. An inductive proof can be based around the same fact, yet it seems that such an inductive proof might be somewhat redundant. Two other inductive proofs are given—one using strong induction, and the other by infinite descent.

Proof by strong induction: For $r \geq 1$, let $S(r)$ be the statement of the exercise, and induct on r:

BASE STEP ($r = 1$): $S(1)$ says $1^2 = 1 \cdot 1$, in which case both factors on the right are perfect squares. (For $r = 2$, $2^2 = 1 \cdot 4$ or $2^2 = 4 \cdot 1$ are the only decompositions into two relatively prime numbers, and both 1 and 4 are perfect squares, so $S(2)$ is true as well.)

INDUCTION STEP: Fix $n \geq 1$, and assume that for all $r \leq n$, $S(n)$ is true. It remains to show that $S(n + 1)$ is true.

Let $(n+1)^2 = st$ with s and t relatively prime. For any prime divisor p of $n+1$, $p^2 \mid (n+1)^2$, and only one of $p \mid s$ or $p \mid t$ holds. Without loss of generality, suppose that p is a prime divisor of $n+1$ and $p \mid s$. Since $p^2 \mid (n+1)^2$, and p does not divide t, then $p^2 \mid s$, say $s = p^2 x$. Let $k = \frac{n+1}{p}$. Then

$$k^2 = \frac{(n+1)^2}{p^2} = \frac{st}{p^2} = xt,$$

where x and t are relatively prime. Applying the induction hypothesis to $k = (n+1)/p$ (since $k \leq n$), both x and t are perfect squares. Hence $s = p^2 x$ and t are both squares. This completes the inductive step.

Therefore, by the principle of strong mathematical induction, for all $r \geq 1$, $S(r)$ is true. $\qquad\qquad\square$

Proof by infinite descent: For positive integers r, s, and t, let $S(r, s, t)$ denote the statement that s and t are relatively prime with $r^2 = st$, where s and t are not both squares.

For the sake of argument, suppose that $S(r, s, t)$ holds. Without loss of generality, suppose that s is not a square. So $s \neq 0$ and $s \neq 1$. Hence s has a prime divisor p; write $s = kp$. Then p divides r^2 and hence r, say $r = \ell p$. Then $(\ell p)^2 = kpt$ implies $\ell^2 p = kt$. But $\gcd(s, t) = 1$, so p does not divide t and hence p divides k, say $k = mp$. Since $s = kp = mp^2$ is not a perfect square, neither is m. Furthermore, m divides s, and since $\gcd(s, t) = 1$, so also $\gcd(m, t) = 1$. Hence $S(\ell, m, t)$ holds, and $\ell < r$; repeated application of the argument above produces yet another, smaller, triple. By induction, this process continues forever, violating the well-ordering of positive integers, and so the original $S(r, s, t)$ must be false. By the method of infinite descent, both s and t must be perfect squares. $\qquad\square.$

Exercise 215: This problem appears in [161, 8.1, pp. 207, 211], (complete with typo), and is referred to as "this famous IMO 1988 problem". The solution given there is a proof credited to J. Campbell (Canberra), which is reproduced here in spirit.

For non-negative integers a and b, let $S(ab)$ be the statement that if $q = \frac{a^2+b^2}{ab+1}$ is an integer, then $q = (\gcd(a,b))^2$. The proof here is by induction on ab.

BASE STEP: When $ab = 0$, one of $a = 0$ or $b = 0$ holds, and so either $q = a^2$ or $q = b^2$. In either case, the result is true using $\gcd(a,0) = a$ or $\gcd(0,b) = b$.

INDUCTIVE STEP: Fix a, b, and for every c with $0 \leq c < b$, assume that $S(ac)$ holds. Put $q = \frac{a^2+b^2}{ab+1}$; it remains to show that $q = \gcd(a,b)^2$. Without loss of generality, assume that $a \leq b$. First seek a c giving the same q, that is, look for c so that

$$q = \frac{a^2+c^2}{ac+1}.$$

To accomplish this, use an old trick: if $\frac{A}{B} = \frac{C}{D} = q$, then $\frac{A-C}{B-D} = \frac{Bq-Dq}{B-D} = q$ as well. So putting

$$q = \frac{a^2+b^2}{ab+1} = \frac{a^2+c^2}{ac+1},$$

and subtracting both numerators and denominators as in the trick, get

$$q = \frac{b^2-c^2}{ab-ac} = \frac{b+c}{a},$$

and so $c = aq - b$ works.

Claim: $0 \leq c < b$. To see the upper bound,

$$q = \frac{a^2+b^2}{ab+1} < \frac{a^2+b^2}{ab} = \frac{a}{b} + \frac{b}{a},$$

which gives $aq < \frac{a^2}{b} + b \leq \frac{b^2}{b} + b = 2b$, and so $c = aq - b < b$. To see the lower bound, $q = \frac{a^2+c^2}{ac+1}$ implies that $ac+1 > 0$ and so $c \geq 0$. Thus, the claim is proved.

Since a suitable c less than b has been found, it remains to observe that

$$\gcd(a,c) = \gcd(a, aq-b) = \gcd(a,b),$$

and so by the induction hypothesis $S(ac)$, $q = \gcd(a,c)^2 = \gcd(a,b)^2$, completing the proof of $S(ab)$.

Thus, by induction, for all a, b, the statement $S(ab)$ holds. $\square$

Remark: The above proof could be translated into a proof by induction on b. It is not clear why the condition $a \leq b$ was necessary.

Exercise 216: This problem is adapted from [161, 8.33, p. 209]. For $n \geq 1$ let $S_n = \sum \frac{1}{xy} = 1$ where the sum is taken over all $x, y \leq n$ with $\gcd(x, y) = 1$, and $x + y > n$.

BASE STEP: $S_1 = 1$ since the only choices for x and y are $x = y = 1$.

INDUCTIVE STEP: For some $k \geq 1$, suppose that $S_k = 1$. Then all terms in S_k with $x + y > k + 1$ stay in the sum S_{k+1}. In S_{k+1}, the terms in S_k with $x + y \leq k + 1$ are gone; these are fractions of the form $\frac{1}{x(k+1-x)}$. For each such deleted fraction, two other fractions $\frac{1}{x(n+1)}$ and $\frac{1}{(k+1-x)(k+1)}$ are introduced. If x and $n + 1$ are relatively prime, then so too are $k + 1 - x$ and $k + 1$. Since

$$\frac{1}{x(n+1-x)} = \frac{1}{x(n+1)} + \frac{1}{(k+1-x)(k+1)},$$

it follows that $S_{k+1} = S_k$, which was by induction hypothesis, equal to 1. This completes the inductive step.

By mathematical induction, for all $n \geq 1$, $S_n = 1$. $\qquad \square$

Exercise 217: This problem (with solution) comes from [138], a paper entitled "Some beautiful arguments using mathematical induction."

Fix a prime p. The proof is by induction on n. Write both $m = m(n!)$ and $s = s(n)$ as functions. To streamline the proof, two observations are helpful.

Observation 1: Since p is prime, for all positive integers x, y, $m(xy) = m(x) + m(y)$; hence

$$m((n+1)!) = m(n!(n+1)) = m(n!) + m(n+1). \tag{29.2}$$

Observation 2: The value $m(n)$ is the number of zeros at the end of the p-ary representation of n. For example, when $p = 5$, $m(500) = m(4 \cdot 5^3) = 3$ and $500 = 4000_5$ has three zeros at the end. Also, $m(n + 1)$ is the number of times the digit $p - 1$ occurs at the end of the p-ary representation of n. Hence $s(n + 1) = s(n) - m(n+1)(p-1) + 1$, which implies

$$m(n+1) = \frac{s(n) + 1 - s(n+1)}{p-1}. \tag{29.3}$$

For any non-negative integer n, let the assertion of the exercise be denoted by

$$A(n): \quad m(n!) = \frac{n - s(n)}{p - 1}.$$

BASE STEP: When $n = 0$, $n! = 1$, which has base p representation 0 and so the sum of p-ary digits is $s(0) = 0$. On the other hand, $m(0!) = m(1) = 0$. Thus, $A(0)$ holds.

INDUCTION STEP: Fix some $k \geq 0$ and suppose that $A(k)$ holds. Consider $n = k+1$. Then

$$m((k+1)!) = m(k!) + m(k+1) \quad \text{(by (29.2))}$$

$$= \frac{k - s(k)}{p - 1} + m(k + 1) \quad (\text{by } A(k))$$

$$= \frac{k - s(k)}{p - 1} + \frac{s(n) + 1 - s(n + 1)}{p - 1}$$

$$= \frac{k + 1 - s(k + 1)}{p - 1},$$

shows that $A(k + 1)$ is also true, completing the inductive step.

By induction on n, for each $n \geq 0$, $A(n)$ is true.   □

Exercise 218: This problems appears with kind permission from José Espinosa's website [176, No. 9] on uncommon mathematical induction problems. Can you find an inductive proof? A non-inductive proof given (by Naoki Sato) is as follows:

The expression is congruent to $2(2^{n-1})^2$ modulo 13. Since 2 is not a square modulo 13, neither is the expression. In fact, as Espinosa points out, that it is not difficult to prove that the expression $4^{2n-1} + 9^{2n-1}$ is divisible by 13.   □

Exercise 219: This problems appears with kind permission from José Espinosa's website [176, No. 10] A hint Espinosa gives is that the expression is divisible by 13, but not by 13^2. [There may be a typo in his hint.]

Exercise 221: This problem appears in [161, p. 376]. Suppose that positive integers a, b, c, d satisfy equation (11.6). Since the right side (11.6) is even, so is the left side. Hence among a, b, c, d, the number of odd numbers is even. If all four are odd, the left side is divisible by 4 but the right is divisible by only two. If exactly two are odd, the left is divisible by only 2 (not 4) and the right is divisible by 8. Hence all of a, b, c, d are even, say $a = 2a_1$, $b = 2b_1$, $c = 2c_1$, $d = 2d_1$. Replacing these values into (11.6) gives (after a bit of simplification)

$$a_1^2 + b_1^2 + c_1^2 + d_1^2 = 8a_1 b_1 c_1 d_1. \tag{29.4}$$

Arguing as before, the left side of (29.4) is even; if all of a_1, b_1, c_1, d_1 are odd, then the left is divisible by only 4, whereas the right is divisible by 8; if just two are odd, the left is divisible by only 2. So all are even, say $a_1 = 2a_2$, $b_1 = 2b_2$, $c_1 = 2c_1$, and $d_1 = 2d_2$. Replacing these values in (29.4),

$$a_2^2 + b_2^2 + c_2^2 + d_2^2 = 32a_2 b_2 c_2 d_2. \tag{29.5}$$

Continuing inductively, for each $t \in \mathbb{Z}^+$, the numbers $a_t = a/2^t$, $,b_t = b/2^t$, $c_t = c/2^t$, and $d_t = d/2^t$ are positive integers so that

$$a_t^2 + b_t^2 + c_t^2 + d_t^2 = 2^{2t+1} a_t b_t c_t d_t. \tag{29.6}$$

However, this gives an infinite decreasing sequence of positive integers a_i's (for example), contrary to the well-ordering property of $\mathbb{Z}^+$. Hence no such solution a, b, c, d exists.   □

Exercise 223 occurred as Exercise 4 in [332, p.352] with no solution; however, I think the proof is standard; the result is attributed to Fermat.

Exercise 224 occurred as Exercise 8 in [332, p.352] with no solution; however, I think the proof is standard, only slightly more difficult than the last exercise; this result is attributed to Fermat as well.

Exercise 225: This is a standard exercise in many number theory books; see, e.g., [485, Thm 77, p. 174] for a complete proof. This can also be proved by infinite descent (see, e.g., [332, p.344]). The "smallest" solution (x_1, y_1) is called the *fundamental solution* to Pell's equation.

29.2   Solutions: Congruences

Exercise 226: This problem appeared as a question in the 1991 USA Mathematical Olympiad, and was reproduced in [357] complete with solution. The style of the following proof could be improved.

For $n \geq 1$, let $T(n)$ denote the statement that modulo n, the sequence of towers of 2's is eventually constant. One proof is by strong induction on n.

BASE STEP: When $n = 1$, the sequence is the constant sequence of 0s. For those not convinced that $n = 1$ is not a meaningful base step, when $n = 2$, the sequence is still the constant zero sequence.

INDUCTIVE STEP: Fix $k \geq 2$, and suppose that $T(1), \ldots, T(k-1)$ are all true. To complete the inductive step, it remains to show that $T(k)$ (that the tower sequence is eventually constant modulo k) is also true .

The proof is now separated into two cases, depending on whether k is even or odd. For convenience, denote the sequence

$$2, 2^2, 2^{2^2}, 2^{2^{2^2}}, \ldots, \quad (\text{mod } n)$$

by $a_1, a_2, a_3, a_4, \ldots$.

Case 1: k is even. Write $k = 2^s q$, where $s \geq 1$ and $q < k$ is odd. For large enough j, $a_{j-2} \geq s$, and for such j,

$$a_j = 2^{2^{a_j}}$$

is a multiple of 2^s; thus, for sufficiently large j, $a_j \equiv 0 \pmod{2^s}$. It follows that for large enough i, 2^s divides $(a_{i+1} - a_i)$.

By the induction hypothesis $T(q)$, the sequence $a_1, a_2, a_3, \ldots$ is eventually constant modulo q, and so for large enough i, q divides $(a_{i+1} - a_i)$.

Since q and 2^s are relatively prime, and both divide large $a_{i+1} - a_i$, their product k also divides $(a_{i+1} - a_i)$, for sufficiently large i. Hence, for large i, $a_{i+1} - a_i \equiv 0$

(mod k), that is, the sequence is eventually constant modulo k, completing the proof of $T(k)$ for even k.

$\quad$ *Case 2: k is odd.* When k is odd, $\gcd(2, k) = 1$, and so by Euler's theorem (Theorem 11.2.2)

$$2^{\phi(k)} \equiv 1 \pmod{k}.$$

Put $r = \phi(k)$. Since $r < k$, by the induction hypothesis $T(r)$, the sequence $a_1, a_2, a_3, \ldots$ is eventually constant modulo r, say congruent to c modulo r, that is, for large enough i, $a_i \cong c \pmod{r}$, or equivalently, for some m_i, $a_i = m_i r + c$. Then

$$a_{i+1} = 2^{a_i} = 2^{m_i r + c} = (2^r)^{m_i} 2^c \equiv (1)^{m_i} 2^c \equiv 2^c \pmod{k}$$

shows that the sequence $a_1, a_2, a_3, \ldots$ is eventually constant modulo k. This completes the proof of $T(k)$ for odd k.

$\quad$ Together, the two cases complete the inductive step.

$\quad$ By the principle of strong mathematical induction, for every $n \geq 1$, the result $T(n)$ holds. $\hspace{8cm}$ $\square$

Exercise 227: In the inductive step below, one might get stuck without the following simple fact:

Fact: When n is odd, then $n^2 \equiv n \pmod{2n}$.
To see this, write $n = 2m + 1$ and calculate

$$
\begin{aligned}
n^2 &= (2m + 1)^2 \\
&= 4m^2 + 4m + 1 \\
&= m(4m + 2) + 2m + 1 \\
&\equiv 2m + 1 \pmod{4m + 2},
\end{aligned}
$$

and since the last line is $n \pmod{2n}$, the fact is proved.

Let n be a fixed odd positive integer, and for $k \geq 1$, let $S(k)$ be the statement that $(n + 1)^k \equiv n + 1 \pmod{2n}$. Induction is on k.

BASE STEP: When $k = 1$, $S(1)$ reads $(n + 1)^1 \equiv n + 1 \pmod{2n}$, which is clearly true.

INDUCTIVE STEP: For some fixed $j \geq 1$, assume that $S(j)$ is true. Then

$$
\begin{aligned}
(n + 1)^{j+1} &= (n + 1)^j (n + 1) \\
&\equiv (n + 1)(n + 1) \pmod{2n} && \text{(by } S(j)) \\
&\equiv n^2 + 2n + 1 \pmod{2n} \\
&\equiv n^2 + 1 \pmod{2n} \\
&\equiv n + 1 \pmod{2n} && \text{(by above fact).}
\end{aligned}
$$

This proves $S(j + 1)$, completing the inductive step.

By the principle of mathematical induction, for all $k \geq 1$, $S(k)$ is true. $\qquad\square$

Exercise 228: Fix a prime p. The solution given here relies on the binomial theorem and the following simple fact: if p is prime, then for all $0 < \ell < p$, the binomial coefficient $\binom{p}{\ell}$ is divisible by p. [Note, this does not work if p is not prime: for example, if $p = 4$, then $\binom{4}{2} = 6$, which is not divisible by 4. The result in the exercise is not always true if p is not prime—for example, $3^4 - 3 = 78$ is not divisible by 4. However, it is sometimes true when p is not prime—for example, $5^4 - 5 = 620$ is divisible by 4.]

For $a \geq 0$, let $F(a)$ be the statement that $a^p - a$ is divisible by p.

BASE STEP: When $a = 0$, $a^p - a = 0$, which is divisible by p, so $F(0)$ is true.

INDUCTIVE STEP: Assume that for some $b \geq 0$, $F(b)$ holds, that is, $b^p - b$ is divisible by p. Then, by the binomial theorem,

$$(b+1)^p - (b+1) = b^p + \binom{p}{p-1}b^{p-1} + \cdots + \binom{p}{1}b^1 + 1 - (b+1)$$

$$= (b^p - b) + \sum_{0 < \ell < p} \binom{p}{\ell} b^\ell.$$

By induction hypothesis $F(b)$, the first term in the last line above is divisible by p, and by the comments preceding the proof, each term in the sum is also divisible by p; hence $(b+1)^p - (b+1)$ is also divisible by p. This proves $F(b+1)$, completing the inductive step.

Hence, by mathematical induction, for all $a \geq 0$, the statement $F(a)$ is true. To complete the proof, a way is needed to handle the negative a's. This can be done by induction downward (using $F(a) \to F(-a)$) in an identical manner to above, or one can take a more direct approach based on the truth of $F(a)$ for positive integers. If $p = 2$, $F(a)$ says $a^2 - a$ is divisible by 2, which is true regardless of the value of a, since $a^2 - a = a(a-1)$, and one of a or $a-1$ is divisible by 2, so also is their product. If $p > 2$, then p is odd, and if $a \geq 0$, then $(-a)^p - (-a) = -a^p + a = -(a^p - a)$ shows that $(-a)^p - (-a)$ is also divisible by p, proving $F(-a)$. $\qquad\square$

Exercise 229: For $p(x) = a_n x^n + a_{n-1} x^{n-1} + \cdots + a_1 x + a_0$, where each $a_i \in \mathbb{Z}$, if a prime p does not divide a_n then the congruence $p(x) \equiv 0 \pmod{p}$ has at most n distinct $\pmod{p}$ solutions. Here, strong induction on n is used, however, proving something which appears to be stronger: if $p(x)$ has more than n roots $\pmod{p}$, then it is the zero polynomial modulo p.

BASE CASES: For $n = 0$, if $a_0 \not\equiv 0$, then $p(x) = a_0$ has no solutions. For $n = 1$ and $p(x) = a_1 x + a_0$, if $a_1 \not\equiv 0$, then a_1^{-1} exists, and so $x = a_1^{-1}(-a_0)$ is the unique solution $\pmod{p}$. For $n = 1$, if p divides a_1, reduce to the case $n = 0$, in which case there are no roots, or all integers are solutions, and since $p \geq 2$, this gives either no solutions or two distinct solutions.

INDUCTIVE STEP: Fix $k \geq 1$ and suppose that the result holds for all $n < k$. Let $f(x) = a_k x^k + a_{k-1} x^{k-1} + \ldots + a_1 x + a_0$. Assume, that $f(x)$ has $k+1$ different roots modulo p, say $w_1, \ldots, w_k, w_{k+1}$. Examine

$$g(x) = f(x) - a_k(x - w_1)(x - w_2) \cdots (x - w_k).$$

Then $g(x)$ is a polynomial of degree less than k. Each of $w_1, \ldots, w_k$ is a root of $g(x)$, and so, by the induction hypothesis, $g(x)$ is the zero polynomial modulo p. Then also $g(w_{k+1}) \equiv 0$, giving

$$f(w_{k+1}) \equiv a_k(w_{k+1} - w_1)(w_{k+1} - w_2) \cdots (w_{k+1} - w_k) \pmod{p}.$$

However, $f(w_{k+1}) \equiv 0$, and for each $i = 1, \ldots, k$, $(w_{k+1} - w_i) \not\equiv 0 \pmod{p}$, which is impossible when $a_k \not\equiv 0$. Conclude that p divides a_k, that is, that $a_k \equiv 0 \pmod{p}$. Repeating this argument inductively shows that each a_i must be congruent to 0 modulo p, giving that if $f(x)$ has more than k roots, then $f(x)$ is the zero polynomial. This completes the inductive step.

Therefore, by strong mathematical induction, the statement is true for all $n \geq 0$. $\square$

Exercise 230: For each positive integer n, let $S(n)$ be the statement that 16^n ends in a 6.

BASE STEP: $(16)^1 = 16$, which ends in a 6.

INDUCTIVE STEP: For some $k \geq 1$, assume that $S(k)$ is true, that is, $(16)^k$ ends in a 6. Then for some a_k, $(16)^k = a_k \cdot 10 + 6$. Then $(16)^{k+1} = (16)^k \cdot 16 = (a_k \cdot 10 + 6)(10 + 6) = a_k \cdot 10^2 + (6a_k + 6)10 + 36$. Simplifying, $(16)^{k+1} = (16a_k + 9)10 + 6$, and so $(16)^{k+1}$ also ends in a 6, proving $S(k+1)$.

By MI, for all $n \geq 1$, the statement $S(n)$ is true. $\square$

Exercise 231: Let $S(n)$ be the statement that $10^n \equiv (-1)^n \pmod{11}$. Since $10 \equiv -1 \pmod{11}$, $S(n)$ is trivially true by raising each side of the congruence to the n-th power, and so induction is hardly necessary. Nevertheless, a purely inductive proof is given.

BASE STEP: Since $10^1 = 10 \equiv -1 \pmod{11}$, $S(1)$ is true.

INDUCTIVE STEP: For some fixed $k \geq 1$, assume that $S(k)$ is true. Then

$$
\begin{aligned}
10^{k+1} &\equiv 10^k \cdot 10 \pmod{11} \\
&\equiv 10^k \cdot (-1) \pmod{11} \\
&\equiv (-1)^k(-1) \pmod{11} \text{ (by } S(k)) \\
&\equiv (-1)^{k+1},
\end{aligned}
$$

proves $S(k + 1)$, finishing the inductive step.

Consequently, by mathematical induction, for all $n \geq 1$, $S(n)$ is true. $\qquad \square$

Exercise 232: The result in this exercise was given by Robert Haas as Lemma 3.1 in [249], where he labelled it "Finite Goldbach". Hint: induct on the number of prime factors of $2m$.

Exercise 233: (Chinese Remainder Theorem) The solution is by induction on n. BASE STEP: When $n = 1$, select $x = a_1$ and then the conditions are trivially satisfied.

INDUCTIVE STEP: For some $k \geq 1$, assume that the theorem is true for $n = k$, and let $m_1, m_2, \ldots, m_k, m_{k+1}$ and $a_1, a_2, \ldots, a_k, a_{k+1}$ be fixed, where each $m_i > 1$ and the m_i's are pairwise relatively prime. To be found is a solution to the $k + 1$ congruences

$$\left. \begin{array}{rcll} x & \equiv & a_1 & (\text{mod } m_1), \\ x & \equiv & a_2 & (\text{mod } m_2), \\ \vdots & & \vdots & \\ x & \equiv & a_n & (\text{mod } m_k), \\ x & \equiv & a_n & (\text{mod } m_{k+1}), \end{array} \right\} \tag{29.7}$$

and then prove that all solutions are those congruent to the given solution modulo $m_1 m_2 \cdots m_k m_{k+1}$.

Assuming for the moment that one can find a solution to (29.7), the last condition above is taken care of first (since induction is not required—in fact, this could have been proved before the inductive proof was begun). If x^* is a solution to (29.7) and $x' \equiv x^* \pmod{m_1 m_2 \cdots m_k m_{k+1}}$, then for some $b \in \mathbb{Z}$

$$x' = x^* + b m_1 m_2 \cdots m_k m_{k+1}$$

and so for each $i = 1, 2, \ldots, k, k + 1$,

$$x' = x^* + b m_1 m_2 \cdots m_k m_{k+1} \equiv x^* \equiv a_i \pmod{m_i},$$

and so x' is also a solution to (29.7). Suppose that x^* and x' are two solutions to (29.7); then for each $i = 1, 2, \ldots, k, k + 1$, $x^* - x'$ is divisible by m_i. Since the m_i's are relatively prime, by Exercise 210, $x^* - x'$ is divisible by the product $m_1 m_2 \cdots m_k m_{k+1}$, that is, $x' \equiv x^* \pmod{m_1 m_2 \cdots m_k m_{k+1}}$. This finishes the proof of the last condition stated above. So, it remains only to find one solution to (29.7).

By induction hypothesis, there is a x_0 which is a solution to the first k lines of (29.7), and w is a solution to these first k lines if and only if

$$w \equiv x_0 \pmod{m_1 m_2 \cdots m_k}.$$

The next idea is to find one of these solutions x^* which is also a solution to the last line, that is, find $y, z \in \mathbb{Z}$ so that

$$x^* = x_0 + y(m_1 m_2 \cdots m_k),$$

and

$$x^* = a_{k+1} + z m_{k+1}.$$

Combining these two equations, it suffices to find y and z so that

$$x_0 + y(m_1 m_2 \cdots m_k) = a_{k+1} + z m_{k+1}. \qquad (29.8)$$

Examine two situations, depending on the value of x_0.

If $x_0 = a_{k+1}$, then obvious choices for y and z are

$$y = m_{k+1} \quad \text{and} \quad z = m_1 m_2 \cdots m_k.$$

Then (29.8) is satisfied and the number x^* defined by

$$x^* = a_{k+1} + m_1 m_2 \cdots m_k m_{k+1}$$

satisfies each line of (29.7) because for each $i = 1, 2, \ldots, k, k+1$,

$$x^* = x_0 + m_1 m_2 \cdots m_k m_{k+1} \equiv x_0 \equiv a_i \pmod{m_i}.$$

Consider the case where $x_0 \neq a_{k+1}$. A little more "algebraic power" is needed to find a suitable y and z. This is where Bezout's Lemma comes in: Since m_{k+1} is relatively prime to each of $m_1, m_2, \ldots, m_k$, it is relatively prime to the product $m_1 m_2 \cdots m_k$, that is, $\gcd(m_1 m_2 \cdots m_k, m_{k+1}) = 1$, and so by Bezout's Lemma, there exist integers r and s so that

$$1 = r m_1 m_2 \cdots m_k + s m_{k+1}.$$

Multiplying this equation on each side by $(x_0 - a_{k+1})$ yields

$$x_0 - a_{k+1} = (x_0 - a_{k+1}) r m_1 m_2 \cdots m_k + (x_0 - a_{k+1}) s m_{k+1}. \qquad (29.9)$$

With the choices $y = (x_0 - a_{k+1})r$ and $z = (x_0 - a_{k+1})s$, (and a very little rearranging of terms) (29.9) becomes (29.8) as desired. So the number

$$x^* = x_0 + (x_0 - a_{k+1}) r m_1 \cdots m_k = a_{k+1} + (x_0 - a_{k+1}) s m_{k+1}$$

is a solution to (29.7). This completes the inductive step.

By mathematical induction, for all $n \geq 1$, the statement of the Chinese Remainder Theorem is true. $\qquad \square$

Exercise 234: For each positive integer n, let $C(n)$ be the statement that

$$2^n + 3^n \equiv 5^n \pmod 6.$$

First notice that a direct proof of $C(n)$ is fairly easy: by the binomial theorem, for any $k \geq 1$,

$$(2+3)^n = \sum_{i=0}^{n} \binom{n}{i} 2^{n-i} 3^i$$

and in the above sum, for each $i = 1, 2, \ldots, n-1$, each expression $2^i 3^{n-i}$ is divisible by 6, so

$$5^n = (2+3)^n \equiv 2^n + 3^n \pmod 6,$$

from which the result follows. Here is an attempt at an inductive proof:

BASE STEP: Since $2^1 + 3^1 = 5^1$, $C(1)$ is true.

INDUCTION STEP: Fix $m \geq 1$ and let the inductive hypothesis be that $C(m)$ holds, that is,

$$2^m + 3^m \equiv 5^m \pmod 6.$$

It remains to show that

$$C(m+1): \quad 2^{m+1} + 3^{m+1} \equiv 5^{m+1} \pmod 6$$

follows. Starting with the left side of $C(m+1)$, employ the trick of adding and subtracting the same terms in order to use $C(m)$:

$$
\begin{aligned}
2^{m+1} + 3^{m+1} &= 2 \cdot 2^m + 2 \cdot 3^m + 3 \cdot 2^m + 3 \cdot 3^m - (2 \cdot 3^m + 3 \cdot 2^m) \\
&= (2+3)(2^m + 3^m) - (2 \cdot 3^m + 3 \cdot 2^m) \\
&= 5(2^m + 3^m) - 6(3^{m-1} + 2^{m-1}) \\
&\equiv 5(2^m + 3^m) \pmod 6 \quad \text{(since } m \geq 1) \\
&\equiv 5 \cdot 5^m \pmod 6 \quad \text{(by ind. hyp. } C(m)) \\
&\equiv 5^{m+1} \pmod 6,
\end{aligned}
$$

arriving at the right side of $C(m+1)$, and so $C(m+1)$ is true. This completes the inductive step $C(m) \rightarrow C(m+1)$.

By mathematical induction, for each $n \geq 1$, $C(n)$ holds. $\qquad \square$

Exercise 235: For each positive integer n, let $M(n)$ be the statement that

$$16^n \equiv 1 - 10n \pmod{25}.$$

BASE STEP: $M(1)$ says that $16 \equiv -9 \pmod{25}$, which is true.

INDUCTIVE STEP: Fix $k \geq 1$ and assume that

$$M(k): \quad 16^k \equiv 1 - 10k \pmod{25}$$

holds. It remains to show that

$$M(k+1): \quad 16^{k+1} \equiv 1 - 10(k+1) \pmod{25}$$

follows. One can show this using just modular arithmetic, however, being a bit more pedantic, a few more steps are added for clarity. Since $M(k)$ holds, there is some integer y so that $16^k = 1 - 10k + 25y$. [Note that $y > 0$, but this fact is not needed.] Then

$$
\begin{aligned}
16^{k+1} &= 16^k \cdot 16 \\
&= (1 - 10k + 25y)16 \\
&= 16 - 160k + 400y \\
&= 1 + 15 - 160k + 400y \\
&= 1 + 15 - 10k - 150k + 400y \\
&= 1 + 15 - 10k + 25(16y - 6k) \\
&\equiv 1 + 15 - 10k \pmod{25} \\
&\equiv 1 + 15 + 10 - 10(k+1) \pmod{25} \\
&\equiv 1 - 10(k+1) \pmod{25},
\end{aligned}
$$

as desired, completing the proof of $M(k+1)$, and hence the inductive step.

By mathematical induction, for each $n \geq 1$, the statement $M(n)$ holds. □

Exercise 236: For each $n \geq 1$, let $S(n)$ be the statement that

$$3^n + 7^n \equiv 2 \pmod{8}.$$

BASE STEP: $S(1)$ says $3 + 7 \equiv 2 \pmod{8}$, which is true.

INDUCTIVE STEP: Fix some $k \geq 1$ and assume that $S(k)$ is true. Before showing $S(k+1)$, first note the following simple fact which helps: for any positive integer j, 3^j is odd and so $4 \cdot 3^j \equiv 4 \pmod{8}$. Starting with the left side of $S(k+1)$,

$$
\begin{aligned}
3^{k+1} + 7^{k+1} &= 3 \cdot 3^k + 7 \cdot 7^k \\
&\equiv 3 \cdot 3^k - 7^k \pmod{8} \\
&\equiv 4 \cdot 3^k - (3^k + 7^k) \pmod{8} \\
&\equiv 4 \cdot 3^k - 2 \pmod{8} && \text{(by } S(k)) \\
&\equiv 4 - 2 \pmod{8} && \text{(by above fact),}
\end{aligned}
$$

which is 2, concluding the demonstration of $S(k+1)$, and hence the inductive step $S(k) \to S(k+1)$.

By mathematical induction, for each $n \geq 1$, the statement $S(n)$ holds. □

Exercise 237: (Brief solution) For each non-negative integer n, let $T(n)$ be the assertion that

$$10^n \equiv (-1)^n \pmod{11}.$$

Since $10 \equiv -1 \pmod{11}$, $T(1)$ holds. For some $k \geq 1$, assuming that $T(k)$ holds,

$$10^{k+1} = 10 \cdot 10^k \equiv 10 \cdot (-1)^k \equiv -1 \cdot (-1)^k \equiv (-1)^{k+1} \pmod{11},$$

and so $T(k+1)$ follows. Hence by induction, for all $n \geq 1$, $T(n)$ holds. □

29.3   Solutions: Divisibility

Exercise 238: For any integer $n \geq 2$, let $P(n)$ be the statement that the product of n odd numbers is also odd.
Base step ($n = 2$): Let $a = 2k + 1$ and $b = 2\ell + 1$ be odd numbers (where k and ℓ are integers). Then

$$ab = 4k\ell + 2k + 2\ell + 1 = 2(2k\ell + k + \ell) + 1,$$

which is again odd.

INDUCTIVE STEP: Suppose that for some fixed $m \geq 2$, $P(m)$ holds. Let $a_1, a_2, \ldots,$ a_m, a_{m+1} be odd numbers, with, for each $i = 1, 2, \ldots, m + 1$, $a_i = 2k_i + 1$. By induction hypothesis, the product $a_1 a_2 \cdots a_m$ is odd, say $2s + 1$. Then

$$\begin{aligned} a_1 a_2 \cdots a_m a_{m+1} = (2s + 1)(2k_{m+1} + 1) &= 4sk_{m+1} + 2s + 2k_{m+1} + 1 \\ &= 2(2sk_{m+1} + s + k_{m+1}) + 1, \end{aligned}$$

and so the larger product is also odd. Thus, $P(m + 1)$ is true as well, completing the inductive step.

By MI, for each $n \geq 2$, the product of n odd numbers is again odd. □

Exercise 239: Let $S(n)$ be the statement that if n is odd, the sum of n odd numbers is odd.
BASE STEP: For $n = 1$, the sum of one odd number is trivially odd so $S(1)$ is true.

INDUCTIVE STEP ($S(k) \to S(k + 2)$): Let $k = 2m + 1$ be odd, where $k \geq 1$ (and so $m \geq 0$), and suppose that the sum of any k odd numbers is again odd. Since the next odd number is $k + 2 = 2m + 3$, it remains to show that $S(k + 2)$ follows, that is, that the sum of any $k + 2$ numbers is again odd. Let $a_1, a_2, \ldots, a_k, a_{k+1}, a_{k+2},$

where each $a_i = 2m_i + 1$ is odd. By induction hypothesis, $a_1 + a_2 + \ldots + a_k$ is odd, say $a_1 + a_2 + \ldots + a_k = 2\ell + 1$ for some integer ℓ. Then

$$
\begin{aligned}
a_1 + a_2 + \ldots + a_k + a_{k+1} + a_{k+2} &= 2\ell + 1 + a_{k+1} + a_{k+2} \\
&= 2\ell + 1 + 2m_{k+1} + 1 + 2m_{k+2} + 1 \\
&= 2(\ell + m_{k+1} + m_{k+2} + 1) + 1,
\end{aligned}
$$

and so the sum of the $k+2$ odd numbers is again odd. This completes the inductive step.

By mathematical induction, for all $n \geq 1$, $S(n)$ is true. $\qquad\square$

Exercise 240: This question appears in [462, 50, p.281], without solution. Here is one possible solution, perhaps not the intended one.

For each $n \geq 1$, let $S(n)$ be the statement that if $n + 1$ distinct numbers are selected from $[1, 2n] = \{1, 2, \ldots, 2n\}$, then one of these numbers must divide another.

BASE STEP: When $n = 1$, there are only $n + 1 = 2$ numbers in $[1, 2]$, one of which divides the other, so $S(1)$ is true.

INDUCTIVE STEP: Fix some $k \geq 1$, and assume that $S(k)$ holds. It remains to show $S(k + 1)$, that is, that among any $k + 2$ numbers chosen from $[1, 2k + 2]$, one divides the other. So let $T = \{t_1, \ldots, t_{k+2}\} \subset [1, 2k + 2]$ be given in order $(t_1 < t_2 < \cdots < t_{k+2}$. If $\{t_1, t_2, \ldots, t_{k+1}\} \subseteq [1, 2k]$, then by $S(k)$, T contains two elements one dividing the other. So suppose that $t_{k+1} > 2k$, so both $t_{k+1} = 2k + 1$ and $t_{k+2} = 2k + 2$.

Since $2k + 2 \in T$, one can eliminate, without loss of generality, the obvious divisors 1, 2, and $k + 1$ from T. Furthermore, assume that no two elements in $T' = \{t_1, \ldots, t_k\} \subset [1, 2k] \setminus \{k + 1\}$ divide one another. For the moment, add $k + 1$ to T' to give a set T'' of $k + 1$ numbers in $[1, 2k]$. By $S(k)$, there are two numbers in T'', one of which divides the other. Since T' has no such pair, and $k + 1$ is too large to divide any other number in $[1, 2k]$ it follows that for some $i \leq k$, t_i divides $k + 1$. Thus, t_i divides $2k + 2 = t_{k+2}$. So $S(k + 1)$ is true, completing the inductive step.

By mathematical induction, for all $n \geq 1$, $S(n)$ is true. $\qquad\square$

Comment on the above solution: When the extra $k + 1$ was added back in the set to produce a situation where the induction hypothesis could be used, one could have done the same with the number 2, proving (since 1 is forbidden) that 2 and some other even number was in $T' \subset [1, 2k]$.

Exercise 241: Repeat the solution of Exercise 240 (which is this problem when $r = 1$), introducing a dummy element when necessary.

Exercise 242: For every $n \geq 1$, let $P(n)$ be the statement that $n(n + 1)$ is even, that is, $2 \mid n(n + 1)$. (A direct proof is nearly trivial since for every integer n, one of n or $n+1$ is even, and so too is the product. Nevertheless, an inductive proof is presented.)

BASE STEP: For $n = 1$, since $1(1 + 1) = 2$ is even, $P(1)$ holds.

INDUCTION STEP: For some fixed $k \geq 1$, suppose that $P(k)$ is true, that is, $k(k+1)$ is even. Then
$$(k + 1)((k + 1) + 1) = (k + 1)k + (k + 1) \cdot 2,$$
and by the induction hypothesis, $(k + 1)k = k(k + 1)$ is even, and the second term $(k + 1) \cdot 2$ is also even. Since the sum of even numbers is again even, it follows that $(k + 1)((k + 1) + 1)$ is also even, proving $P(k + 1)$ is true as well. This completes the inductive step $P(k) \rightarrow P(k + 1)$.

By MI, for each $n \geq 1$, $P(n)$ is true. $\qquad\square$

Exercise 243: Showing that 3 divides $n^3 - n$ is done in a manner similar to that of Exercise 252.

Exercise 244: For every integer $n \geq 1$, let $D(n)$ be the statement that 3 divides $n^3 + 2n$.

BASE STEP: When $n = 1$, $n^3 + 2n = 3$, which is of course divisible by 3.

INDUCTION STEP: For some fixed $k \geq 1$, suppose that $D(k)$ is true, that is, for some integer m, $k^3 + 2k = 3m$. To see that $D(k + 1)$ is true,

$$
\begin{aligned}
(k + 1)^3 + 2(k + 1) &= k^3 + 3k^2 + 3k + 1 + 2k + 2 \\
&= k^3 + 2k + 3(k^2 + k + 1) \\
&= 3m + 3(k^2 + k + 1) \quad \text{by ind. hyp.} \\
&= 3[m + k^2 + k + 1],
\end{aligned}
$$

and so $(k + 1)^2 + 2(k + 1)$ is divisible by 3. This shows $D(k + 1)$, and so completes the inductive step $D(k) \rightarrow D(k + 1)$.

By induction, for each $n \geq 1$, $D(n)$ is true. $\qquad\square$

Exercise 245: For $n \geq 1$, let $S(n)$ be the statement that $3 \mid (2^{2n} - 1)$.

BASE STEP: Since $2^{2 \cdot 1} - 1 = 3$, the statement $S(1)$ is true.

INDUCTION STEP: For some fixed $k \geq 1$, suppose that $S(k)$ holds, that is, the induction hypothesis (IH) is that there is some integer m so that $2^{2k} - 1 = 3m$. Then

$$2^{2(k+1)} - 1 = 4 \cdot 2^{2k} - 1 = 4(2^{2k} - 1) + 3 \stackrel{\text{IH}}{=} 4(3m) + 3 = 3(4m + 1)$$

shows that $2^{2(k+1)} - 1$ is also divisible by 3, proving $S(k + 1)$. [*Note:* The notation $\stackrel{\text{IH}}{=}$ indicates where the Induction Hypothesis is used.]

Since $S(1)$ is true and $S(k) \rightarrow S(k+1)$, by mathematical induction, for any $n \geq 1$, the statement $S(n)$ holds.   □

Exercise 246: For $n \geq 1$, let $D(n)$ be the statement that $3 \mid (2^{2n+1} + 1)$.

BASE STEP: With $n = 1$, $2^{2n+1} + 1 = 9$ and 3 divides 9, so $D(1)$ holds.

INDUCTIVE STEP: Let $k \geq 1$ and suppose that $D(k)$ is true, say for some integer m, $2^{2k+1} + 1 = 3m$. It remains to show that $D(k+1)$ is also true, that is, that 3 divides $2^{2(k+1)+1} + 1$. Then

$$2^{2(k+1)+1} + 1 = 4 \cdot 2^{2k+1} + 1 = 4(2^{2k+1} + 1) - 3 = 4(3m) - 3 = 3(4m - 1),$$

(where the penultimate equality is true by induction hypothesis $D(k)$) which shows that $D(k+1)$ is also true. This completes the inductive step.

By mathematical induction, for each $n \geq 1$, the statement $D(n)$ holds.   □

Exercise 247: For every $n \geq 1$, let $S(n)$ denote the statement $3 \mid (5^{2n} - 1)$.

BASE STEP: Since $5^2 - 1 = 24$ is divisible by 3, $S(1)$ holds.

INDUCTION STEP: Let $k \geq 1$ and assume that $S(k)$ holds, that is, for some integer m, $5^{2k} - 1 = 3m$. Then

$$5^{2k+2} - 1 = 25 \cdot 5^{2k} - 25 + 24 = 25(5^{2k} - 1) + 24 \overset{S(k)}{=} 25 \cdot 3m + 24 = 3(25m + 8)$$

proves $S(k+1)$, completing the inductive step.

By mathematical induction, for every integer $n \geq 1$, $S(n)$ is true.   □

Exercise 248: For $n \geq 1$, let S_n be the statement

$$3 \mid \left(\frac{10^n + 5}{3} + 4^{n+2} \right).$$

BASE STEP: S_1 says that 3 divides 5+64 which is true.

INDUCTION STEP: Let $k \geq 1$ be fixed, and assume that

$$S_k : \quad 3 \mid \left(\frac{10^k + 5}{3} + 4^{k+2} \right)$$

is true, with say, $\frac{10^k + 5}{3} + 4^{k+2} = 3\ell$ for some positive integer ℓ. To be shown is that

$$S_{k+1} : \quad 3 \mid \left(\frac{10^{k+1} + 5}{3} + 4^{k+3} \right)$$

follows. With a little algebra,

$$
\begin{aligned}
\frac{10^{k+1}+5}{3} + 4^{k+3} &= \frac{10^k+5}{3} + 3 \cdot 10^k + 4^{k+2} + 3 \cdot 4^{k+2} \\
&= 3\ell + 3(10^k + 4^{k+2}),
\end{aligned}
$$

from which it follows that $3 \mid \left(\frac{10^{k+1}+5}{3} + 4^{k+3}\right)$, that is, S_{k+1} follows. This completes the inductive step.

By mathematical induction, for all $n \geq 1$, the statement S_n is true. $\qquad\square$

Exercise 249: This problem appears in, *e.g.*, [350]. For every $n \geq 0$, let S_n denote the statement $3 \mid (7^n + 2)$.

BASE STEP: S_0 says that 3 divides 3, which is true. Checking one more step, S_1 says that 3 divides 9, again true.

INDUCTION STEP: For some fixed $k \geq 0$, assume that S_k is true, that is, for some positive integer m, $7^k + 2 = 3m$. The statement $S_{k+1} : \ 3 \mid 7^{k+1} + 2$ remains to be proved. Noting that $7^{k+1} + 2 = 7 \cdot 7^k + 2 = 7(7^k + 2) - 12 = 7(3m) - 12 = 3(7m - 4)$, the statement S_{k+1} now evidently follows. This completes the inductive step.

By mathematical induction, for all $n \geq 0$, the statement S_n is true. $\qquad\square$

Exercise 250: The exercise is to show that for every $n \geq 1$,

$$
S_n : \quad 4 \mid n^2(n+1)^2
$$

is true. [Without using induction, the direct proof follows from the fact that among n and $n + 1$, one is even, and so one of n^2 and $(n + 1)^2$ has a factor of 4.]

BASE STEP: When $n = 1$, $n^2(n + 1)^2 = 4$, so S_1 holds.

INDUCTIVE STEP: Let $k \geq 1$ and let the induction hypothesis be that S_k holds; in particular, suppose that m is such that $k^2(k + 1)^2 = 4m$. It remains to show that

$$
S_{k+1} : \quad 4 \mid (k+1)^2(k+2)^2
$$

follows. Indeed,

$$
\begin{aligned}
(k+1)^2(k+2)^2 &= (k^2 + 4k + 4)(k+1)^2 \\
&= k^2(k+1)^2 + (4k+4)(k+1)^2 \\
&= 4m + 4(k+1)(k+1)^2 \quad \text{(by } S_k\text{)}
\end{aligned}
$$

shows that $(k+1)^2(k+2)^2$ is also divisible by 4, concluding the proof of S_{k+1} and hence the inductive step.

By the principle of mathematical induction, it is true that for any positive integer n, the number $n^2(n + 1)^2$ is divisible by 4. $\qquad\square$

Exercise 251: For each $n \geq 1$, let $S(n)$ be the statement "$4 \mid (6 \cdot 7^n - 2 \cdot 3^n)$".

BASE STEP ($n = 1$): $S(1)$ says $4 \mid (6 \cdot 7 - 2 \cdot 3)$, or $4 \mid 36$, which is clearly true.

INDUCTIVE STEP: For some fixed $k \geq 1$, assume that $S(k)$ is true, that is, that there is an ℓ so that $6 \cdot 7^k - 2 \cdot 3^k = 4\ell$. To be proved is

$$S(k+1): \quad 4 \mid (6 \cdot 7^{k+1} - 2 \cdot 3^{k+1}).$$

Calculating,

$$
\begin{aligned}
6 \cdot 7^{k+1} - 2 \cdot 3^{k+1} &= 3(2 \cdot 7^{k+1} - 2 \cdot 3^k) \\
&= 3(14 \cdot 7^k - 2 \cdot 3^k) \\
&= 3(8 \cdot 7^k + 6 \cdot 7^k - 2 \cdot 3^k) \\
&= 3(8 \cdot 7^k + 4\ell) \quad \text{by ind. hyp.)} \\
&= 4(6 \cdot 7^k + 3\ell),
\end{aligned}
$$

showing that $4 \mid (6 \cdot 7^{k+1} - 2 \cdot 3^{k+1})$, completing the inductive step.

By MI, for all $n \in \mathbb{Z}^+$, $S(n)$ holds. □

Exercise 252: Let $S(n)$ be the statement $5 \mid (n^5 - n)$.

BASE STEP ($n = 1$): The statement $S(1)$ says $5 \mid (1^5 - 1)$ or $5 \mid 0$, which is true.

INDUCTIVE STEP: Assume that for some fixed $k \geq 1$, $S(k)$ holds, that is, $k^5 - k = 5\ell$ for some $\ell \in \mathbb{Z}$. To be proved is

$$S(k+1): \quad 5 \mid (k+1)^5 - (k+1)$$

holds. Calculating

$$
\begin{aligned}
(k+1)^5 - (k+1) &= (k^5 + 5k^4 + 10k^3 + 10k^2 + 5k + 1) - (k+1) \\
&= (k^5 - k) + 5(k^4 + 2k^3 + 2k^2 + k) \\
&= 5\ell + 5(k^4 + 2k^3 + 2k^2 + k) \quad \text{(by ind. hyp)}, \\
&= 5[\ell + k^4 + 2k^3 + 2k^2 + k],
\end{aligned}
$$

a multiple of 5. Thus $5 \mid (k+1)^5 - (k+1)$, proving $S(k+1)$, completing the inductive step.

Hence by MI, for all $n \geq 1$, $S(n)$ holds. □

Exercise 253: For any $n \geq 1$, let $P(n)$ denote the proposition that $3^{2n} + 4^{n+1}$ is divisible by 5.

BASE STEP: $P(1)$ says that $3^2 + 4^2$ is divisible by 5, which is true.

INDUCTIVE STEP: Let $k \geq 1$ be fixed, and suppose that

$$P(k): \quad 3^{2k} + 4^{k+1} \text{ is divisible by 5}$$

holds, say, with $3^{2k} + 4^{k+1} = 5m$ for some integer m. To be shown is that

$$P(k+1): \quad 3^{2k+2} + 4^{k+2} \text{ is divisible by 5}$$

follows. Calculating,

$$
\begin{aligned}
3^{2k+2} + 4^{k+2} &= 9 \cdot 3^{2k} + 4 \cdot 4^{k+1} \\
&= 5 \cdot 3^{2k} + 4 \cdot 3^{2k} + 4 \cdot 4^{k+1} \\
&= 5 \cdot 3^{2k} + 4(3^{2k} + 4^{k+1}) \\
&= 5 \cdot 3^{2k} + 4(5m) \text{ by ind. hyp.} \\
&= 5(3^{2k} + 4m),
\end{aligned}
$$

showing that $3^{2k+2} + 4^{k+2}$ is also divisible by 5, proving $P(k+1)$, and hence concluding the inductive step.

By the principle of mathematical induction, for all $n \geq 1$, the statement $P(n)$ is true. $\qquad\square$

Exercise 255: For $n \geq 1$, let $S(n)$ denote the statement $6 \mid (n^3 - n)$.

BASE STEP: $S(1)$ says that 6 divides 0, which is true.

INDUCTIVE STEP: Let $k \geq 1$ and suppose that $S(k)$ is true, say $k^3 - k = 6m$. To show that $S(k+1)$ follows, show that 6 divides $(k+1)^3 - (k+1)$. To this purpose,

$$
\begin{aligned}
(k+1)^3 - (k+1) &= k^3 + 3k^2 + 3k + 1 - k - 1 \\
&= (k^3 - k) + (3k^2 + 3k) \\
&= 6m + 3k(k+1) \quad (\text{by } S(k)),
\end{aligned}
$$

and since one of k or $k+1$ is even, the last term is also divisible by 6, and so $S(k+1)$ is true, completing the inductive step.

By the principle of mathematical induction, for every $n \geq 1$, $S(n)$ is true. $\qquad\square$

Exercise 256: For every positive integer $n \geq 1$, let $M(n)$ denote the statement $6 \mid (7^n - 1)$.

BASE STEP: Since $6 \mid (7 - 1)$, $M(1)$ is true.

INDUCTIVE STEP: For some fixed $k \geq 1$, suppose that $M(k)$ is true; in particular, let m be an integer with $6m = 7^k - 1$. It remains to prove $M(k+1)$, namely, that 6 divides 7^{k+1}. Then

$$7^{k+1} - 1 = 7(7^k - 1) + 7 - 1 = 7(6m) + 6$$

shows that $7^{k+1} - 1$ is also divisible by 6, concluding the proof of $M(k+1)$, and hence the inductive step.

By MI, for every positive integer n, $M(n)$ is true. (In fact, $M(0)$ is also true.) $\qquad\square$

Exercise 257: For any positive integer n, let $D(n)$ denote the statement $6 \mid (n^3 + 5n)$.

BASE STEP: For $n = 1$, $n^3 + 5n = 6$, so $D(1)$ holds.

INDUCTIVE STEP: Let $k \geq 1$ be fixed, and suppose that $D(k)$ holds; in particular, let ℓ be an integer with $6\ell = k^3 + 5k$. Then

$$
\begin{aligned}
(k+1)^3 + 5(k+1) &= k^3 + 3k^2 + 3k + 1 + 5k + 5 \\
&= k^3 + 5k + 3k^2 + 3k + 6 \\
&= 6\ell + 3k(k+1) + 6 \quad \text{(by } D(k)\text{)}.
\end{aligned}
$$

Since one of k or $k+1$ is even, the term $3k(k+1)$ is divisible by 6, and so the last expression above is divisible by 6. This proves $D(k+1)$, and concludes the inductive step $D(k) \rightarrow D(k+1)$.

By MI, for each $n \geq 1$, the statement $D(n)$ is true. □

Exercise 258: This exercise has a direct solution because among n, $(n+1)$, and $(n+2)$, one is a multiple of 3 and at least one is even, yielding a factor of 6. An inductive proof is however also possible:

For each $n \geq 0$, let $S(n)$ denote the statement $6 \mid n(n+1)(n+2)$.

BASE STEP: Since 6 divides 0, $S(0)$ holds.

INDUCTIVE STEP: Fix $k \geq 0$, and suppose that $S(k)$ holds, that is, suppose that ℓ is an integer so that

$$
k(k+1)(k+2) = 6\ell.
$$

Then trying to prove $S(k+1)$,

$$
\begin{aligned}
(k+1)(k+2)(k+3) &= k(k+1)(k+2) + 3(k+1)(k+2) \\
&= 6\ell + 3(k+1)(k+2),
\end{aligned}
$$

and since one of $k+1$ or $k+2$ is even, 6 divides $3(k+1)(k+2)$, and so 6 divides $(k+1)((k+1)+1)((k+1)+2)$, proving $S(k+1)$, and concluding the inductive step $S(k) \rightarrow S(k+1)$.

By MI, for every $n \geq 0$, $S(n)$ holds. □

Exercise 259: For every $n \geq 0$, let $P(n)$ be the proposition that $n(n-1)(2n-1)$ is divisible by 6.

BASE STEP: $P(0)$ says that 0 is divisible by 6, a true statement.

INDUCTIVE STEP: Fix $k \geq 0$, and suppose that $P(k)$ is true, that is, fix an integer ℓ so that $k(k-1)(2k-1) = 6\ell$. To show that $P(k+1)$ is true, it remains to show that $(k+1)(k)(2(k+1)-1)$ is also divisible by 6. Calculating,

$$
(k+1)(k)(2k+1) = (k+1)k(2k-1) + 2(k+1)k
$$

$$
\begin{aligned}
&= \; k(k-1)(2k-1) + 2k(2k-1) + 2(k+1)k \\
&= \; 6\ell + 2k(3k) \quad (\text{by } P(k)) \\
&= \; 6(\ell + k^2),
\end{aligned}
$$

shows $P(k+1)$ is also true, completing the inductive step.

By MI, for every $n \geq 0$, the statement $P(n)$ is true.   □

Exercise 260: Show that 7 divides $n^7 - n$ in a manner similar to that done in the solution to Exercise 252.

Exercise 261: For each $n \geq 1$, let $A(n)$ be the assertion that $7 \mid (2^{n+2} + 3^{2n+1})$.

BASE STEP: $A(1)$ says that 7 divides $2^3 + 3^3 = 35$, which holds true.

INDUCTIVE STEP: Fix $k \geq 1$ and assume that $A(k)$ holds, say for some integer ℓ, $2^{k+2} + 3^{2k+1} = 7\ell$. To show that $A(k+1)$ follows, show that 7 also divides $2^{k+3} + 3^{2k+3}$. To this end,

$$
\begin{aligned}
2^{k+3} + 3^{2k+3} &= \; 2 \cdot 2^{k+1} + 9 \cdot 3^{2k+1} \\
&= \; 2(2^{k+1} + 3^{2k+1}) + 7 \cdot 3^{2k+1} \\
&= \; 2 \cdot 7\ell + 7 \cdot 3^{2k+1} \quad \text{by } A(k),
\end{aligned}
$$

which is divisible by 7, and thus $A(k+1)$ is also true, completing the inductive step.

By the principle of mathematical induction, for each $n \geq 1$, $A(n)$ holds.   □

Exercise 262: To be shown is that for every $n \geq 1$,

$$
S(n): \quad 7 \mid (11^n - 4^n).
$$

Before presenting the inductive proof, one might notice that a simple factorization shows that this result is true, and induction is not really necessary.

BASE STEP: $11^1 - 4^1 = 7$, and so $S(1)$ holds.

INDUCTIVE STEP: Fix $k \geq 1$ and suppose that $S(k)$ is true, where ℓ is some integer satisfying $11^k - 4^k = 7\ell$. To see that $S(k+1)$ is also true, compute:

$$
\begin{aligned}
11^{k+1} - 4^{k+1} &= \; 11 \cdot 11^k - 4 \cdot 4^k \\
&= \; 11(11^k - 4^k) + 7 \cdot 4^k \\
&= \; 11(7\ell) + 7 \cdot 4^k \quad (\text{by } S(k)).
\end{aligned}
$$

This last expression is divisible by 7, and so $S(k+1)$ is also true, completing the inductive step.

By MI, for each $n \geq 1$, $S(n)$ is true.   □

Exercise 263: For every $n \geq 1$, denote the assertion in the exercise by

$$A(n): \quad 7 \mid (23^{3n} - 1).$$

BASE STEP: $23^3 - 1 = (23 - 1)(23^2 + 23 + 1) = 22(553) = 22(79 \cdot 7)$, so $A(1)$ holds.

INDUCTIVE STEP: Fix $k \geq 1$ and suppose that $A(k)$ is true, where ℓ is some integer satisfying

$$23^{3k} - 1 = 7\ell.$$

Then

$$
\begin{aligned}
23^{3(k+1)} - 1 &= 23^3 \cdot 23^{3k} - 1 \\
&= 23^3(23^{3k} - 1) + 23^3 - 1 \\
&= 23^3(7\ell) + 22 \cdot 79 \cdot 7 \quad \text{(by } A(k))
\end{aligned}
$$

shows that 7 divides $23^{3(k+1)} - 1$, proving that $A(k+1)$ follows. This completes the inductive step.

By MI, for each $n \geq 1$, $A(n)$ is true. □

Exercise 264: For every $n \geq 1$, let $D(n)$ denote the statement "$3^{2n} - 1$ is divisible by 8".
BASE STEP: Since $3^2 - 1 = 8$, $D(1)$ holds.

INDUCTIVE STEP: Fix $k \geq 1$ and suppose that $D(k)$ is true, where ℓ is some integer satisfying

$$3^{2k} - 1 = 8\ell.$$

To see $D(k+1)$, show that $3^{2(k+1)} - 1$ is divisible by 8:

$$
\begin{aligned}
3^{2k+2} - 1 &= 9(3^2 k - 1) + 8 \\
&= 9 \cdot 8\ell + 8 \\
&= 8(9\ell + 1) \quad \text{(by } D(k)).
\end{aligned}
$$

This completes the proof of $D(k+1)$ and hence the inductive step.

By MI, for each $n \geq 1$, $D(n)$ is true. □

Exercise 265: For $n \geq 1$, denote the statement in the exercise by

$$A(n): \quad 8 \mid (3^n + 7^n - 2).$$

BASE STEP: Since $3^1 + 7^1 - 2 = 8$, $A(1)$ is true. (In fact, so is $A(0)$, so the result in this problem could be stated for $n \geq 0$.)

INDUCTIVE STEP $(A(k) \rightarrow A(k+1))$: Fix $k \geq 1$ and suppose that $A(k)$ is true, where ℓ is some integer satisfying

$$3^k + 7^k - 2 = 8\ell.$$

To show $A(k+1)$, if suffices to show that $3^{k+1} + 7^{k+1} - 2$ is also divisible by 8. Calculating,

$$
\begin{aligned}
3^{k+1} + 7^{k+1} - 2 &= 3 \cdot 3^k + 7 \cdot 7^k - 2 \\
&= 3(3^k + 7^k - 2) + 4 \cdot 7^k + 4 \\
&= 3 \cdot 8\ell + 4(7^k + 1) \quad (\text{by } A(k)).
\end{aligned}
$$

For $k \geq 1$, the number $7^k + 1$ is an even number, so the entire last expression above is divisible by 8, which proves $A(k+1)$, concluding the inductive step.

By MI, for each $n \geq 1$, $A(n)$ is true. $\square$

Exercise 266: For every $n \geq 1$, denote the statement by

$$S(n): \quad 8 \mid (5^{n+1} + 2 \cdot 3^n + 1).$$

BASE STEP: $S(1)$ says 8 divides $5^2 + 2 \cdot 3^1 + 1 = 32$, which is true.

INDUCTIVE STEP: Fix $k \geq 1$ and suppose that $S(k)$ is true, where ℓ is some integer satisfying

$$5^{k+1} + 2 \cdot 3^k + 1 = 8\ell.$$

Then to see $S(k+1)$,

$$
\begin{aligned}
5^{k+2} + 2 \cdot 3^{k+1} + 1 &= 5 \cdot 5^{k+1} + 3 \cdot 2 \cdot 3^k + 1 \\
&= 5(5^k + 2 \cdot 3^k + 1) + 4 \cdot 3^k - 4 \\
&= 5 \cdot 8\ell + 4(3^k - 1) \quad\quad\quad\quad\quad (\text{by } S(k)).
\end{aligned}
$$

Since for $k \geq 1$, 3^k is odd, $3^k - 1$ is divisible by 2, and so the last expression above is divisible by 8, proving $S(k+1)$, which concludes the inductive step.

By MI, for each $n \geq 1$, $S(n)$ is true. $\square$

Exercise 267: For $n \geq 0$, define the proposition

$$P(n): \quad 9 \mid (n^3 + (n+1)^3 + (n+2)^3).$$

BASE STEP $(n = 0)$: Since $0^3 + 1^3 + 8^3 = 9$, $P(0)$ is true. As an extra check, $1^3 + 2^3 + 3^3 = 1 + 8 + 27 = 36 = 4 \cdot 9$ shows $P(1)$ is true, too.

INDUCTIVE STEP: Let $k \geq 0$ be a fixed integer, and suppose that $P(k)$ is true, where for some integer ℓ,

$$k^3 + (k+1)^3 + (k+2)^3 = 9\ell.$$

To see that $P(k+1)$ is true,

$$
\begin{aligned}
(k+1)^3 + (k+2)^3 + (k+3)^3 &= k^3 + (k+1)^3 + (k+2)^3 + (k+3)^3 - k^3 \\
&= 9\ell + (k+3)^3 - k^3 \quad \text{(by } P(k)) \\
&= 9\ell + k^3 + 9k^2 + 27k + 27 - k^3 \\
&= 9\ell + 9(k^2 + 3k + 3),
\end{aligned}
$$

and so 9 divides $(k+1)^3 + ((k+1)+1)^3 + ((k+1)+2)^3$, proving $P(k+1)$. This completes the inductive step.

By MI, for each $n \geq 0$, the statement $P(n)$ holds. $\qquad \square$

Exercise 269: For every $n \geq 1$, denote the statement $10 \mid (n^5 - n)$ by $S(n)$. One could simply use Exercise 252 to guarantee a factor of 5, then observe that $n^5 - n$ is always even to yield the remaining factor of 2. However, an inductive proof is also possible:
BASE STEP: $S(1)$ says 10 divides $1^5 - 1 = 0$ which is true.

INDUCTION STEP: Let $k \geq 1$ and suppose that $S(k)$ is true, namely that $k^5 - k = 10m$ for some non-negative integer m. To show $S(k+1)$, one needs to show that 10 divides $(k+1)^5 - (k+1)$. Indeed,

$$
\begin{aligned}
(k+1)^5 - (k+1) &= k^5 + 5k^4 + 10k^3 + 10k^2 + 5k + 1 - k - 1 \\
&= k^5 - k + 10(k^3 + k^2) + 5(k^4 + k) \\
&= 10m + 10(k^3 + k^2) + 5(k^4 + k)
\end{aligned}
$$

and it only remains to observe that $k^4 + k$ is always even for one to conclude $S(k+1)$. This completes the inductive step.

By mathematical induction, conclude that for every $n \geq 1$, the number $n^5 - n$ is divisible by 10. $\qquad \square$

Exercise 270: For every $n \geq 1$, let

$$
P(n): \quad 15 \mid (4(47)^{4n} + 3(17)^{4n} - 7).
$$

denote the statement in the exercise.
BASE STEP: One can calculate

$$
4(47)^4 + 3(17)^4 - 7 = 19769280 = 15 \cdot 1317952,
$$

and observe that $P(1)$ is true. *Note:* One could have noticed that $P(0)$ is also true and more easily verified, and so proving the statement for all $n \geq 0$, is actually easier. The details of this suggestion are left to the reader.

INDUCTIVE STEP: Fix some $k \geq 1$ and assume that $P(k)$ holds. To prove $P(k+1)$, it suffices to show that 15 divides $4(47)^{4k+4} + 3(17)^{4k+4} - 7$. This is perhaps most easily done working modulo 15:

$$
\begin{aligned}
4(47)^{4k+4} + 3(17)^{4k+4} - 7 &= 4(47)^4(47)^{4k} + 3(17)^4(17)^{4k} - 7 \\
&\equiv 4(2^4)(47)^{4k} + 3(2^4)(17)^{4k} - 7 \pmod{15} \\
&\equiv 4(47)^{4k} + 3(17)^{4k} - 7 \pmod{15}, \\
&\equiv 0, \pmod{15} \quad \text{(by } P(k)\text{)}
\end{aligned}
$$

and so $P(k+1)$ follows.

By mathematical induction, for every $n \geq 1$ (or $n \geq 0$ if the suggestion in the base step above was followed), $P(n)$ is true. $\square$

Exercise 271: For every $n \geq 0$, let $S(n)$ denote the statement $15 \mid (2^{4n} - 1)$.
BASE STEP: The statement $S(0)$ says that 15 divides $2^4 - 1$, which is true.

INDUCTIVE STEP: Fix $k \geq 0$, and suppose $S(k)$ is true; in particular, let ℓ be an integer so that $2^{4k} - 1 = 15\ell$. To see that $S(k+1)$ follows,

$$
\begin{aligned}
2^{4(k+1)} - 1 &= 16 \cdot 2^{4k} - 1 \\
&= 16(2^{4k} - 1) + 15 \\
&= 16 \cdot 15\ell + 15 \\
&= 15(16\ell + 1).
\end{aligned}
$$

This completes the inductive step.

By MI, for each $n \geq 0$, the statement $S(n)$ holds. $\square$

Exercise 272: For every $n \geq 1$, let $T(n)$ be the statement that 16 divides $5^n - 4n - 1$.
BASE STEP: $T(1)$ holds because $5^1 - 4(1) - 1 = 0$, and 16 divides 0.

INDUCTIVE STEP: Fix some $k \geq 1$ and suppose that $T(k)$ holds, that is, suppose that there is some integer j so that $5^k - 4k - 1 = 16j$. It remains to show $T(k+1)$, that is, it remains to show that 16 also divides $5^{k+1} - 4(k+1) - 1$. With a little algebra,

$$
\begin{aligned}
5^{k+1} - 4(k+1) - 1 &= 5(5^k) - 4k - 5 \\
&= 5(5^k) - 20k - 1 + 16k \\
&= 5[5^k - 40k - 1] + 16k \\
&= 5[16j] + 16k \quad \text{(by } T(k)\text{)} \\
&= 16(5j + k),
\end{aligned}
$$

and so 16 divides $5^{k+1} - 4(k+1) - 1$, as desired. This concludes the inductive step $T(k) \to T(k+1)$.

By mathematical induction, for each $n \geq 1$, the statement $T(n)$ is true. $\qquad$ □

Exercise 273: For any $n \geq 0$, let $A(n)$ be the assertion that

$$17 \mid (3 \cdot 5^{2n+1} + 2^{3n+1}).$$

BASE STEP: When $n = 0$, the expression $3 \cdot 5^{2n+1} + 2^{3n+1}$ is equal to 17, so $A(0)$ is true.

INDUCTIVE STEP: Let $k \geq 0$ and let the induction hypothesis be that $A(k)$ holds. To show that $A(k+1)$ follows, one must show that $3 \cdot 5^{2n+3} + 2^{3n+4}$ is divisible by 17. Calculating,

$$
\begin{aligned}
3 \cdot 5^{2n+3} + 2^{3n+4} &= 25 \cdot 3 \cdot 5^{2k+1} + 8 \cdot 2^{3k+1} \\
&= 17 \cdot 3 \cdot 5^{2k+1} + 8(3 \cdot 5^{2k+1} + 8 \cdot 2^{3k+1})
\end{aligned}
$$

and since, by induction hypothesis $A(k)$, the last expression in parentheses is divisible by 17, it now follows that so too is $3 \cdot 5^{2n+3} + 2^{3n+4}$. This completes the inductive step.

By mathematical induction, for every $n \geq 0$, the statement $A(n)$ holds. $\qquad$ □

Exercise 275: This problem (with kind permission) is reproduced from José Espinosa's website [176, No. 15], with solution by Naoki Sato.
    For $n \geq 1$, put $f(n) = 2^{2^n} + 3^{2^n} + 3^{2^n}$, and let $C(n)$ be the claim that $f(n)$ is divisible by 19. The proof is by an alternative form of mathematical induction.
BASE STEP: As $f(1) = 38$ and $f(2) = 38 \cdot 19$, both $C(1)$ and $C(2)$ hold.

INDUCTIVE STEP: Fix some $k \geq 1$ and suppose that $C(k)$ is true. To see that $C(k+2)$ follows,

$$
\begin{aligned}
f(k+2) &= 2^{2^{k+2}} + 3^{2^{k+2}} + 5^{2^{k+2}} \\
&= 2^{4 \cdot 2^k} + 3^{4 \cdot 2^k} + 5^{4 \cdot 2^k} \\
&= 16^{2^k} + 81^{2^k} + 625^{2^k} \\
&\equiv 3^{2^k} + 5^{2^k} + 2^{2^k} \pmod{19} \\
&= f(k),
\end{aligned}
$$

and by induction hypothesis $C(k)$, $f(k)$ is divisible by 19, so also is $f(k+2)$. This completes the proof of $C(k+2)$ and hence the inductive step.

By mathematical induction, for all $n \geq 1$, $f(n)$ is divisible by 19. $\qquad$ □

Exercise 276: For $n \geq 1$, let $S(n)$ be the statement that

$$21 \mid (4^{n+1} + 5^{2n-1}).$$

BASE STEP: With $n = 1$, $4^{1+1} + 5^{2 \cdot 1 - 1} = 16 + 5 = 21$, so $S(1)$ is true.

INDUCTIVE STEP: Let $m \geq 1$ and assume that $S(m)$ is true, that is, $4^{m+1} + 5^{2m-1}$ is divisible by 21. To complete the inductive step, it remains to show that $S(m+1)$ is true, namely, that $4^{(m+1)+1} + 5^{2(m+1)-1}$ is also divisible by 21.

$$
\begin{aligned}
4^{(m+1)+1} + 5^{2(m+1)-1} &= 4 \cdot 4^{m+1} + 25 \cdot 5^{2m-1} \\
&= 4(4^{m+1} + 5^{2m-1}) + 21 \cdot 5^{2m-1},
\end{aligned}
$$

and since by induction hypothesis, $4^{m+1} + 5^{2m-1}$ is divisible by 21, so too is $4^{(m+1)+1} + 5^{2(m+1)-1}$, completing the inductive step $S(m) \to S(m+1)$.

Thus, by MI, for all $n \geq 1$, the statement $S(n)$ holds true. □

Exercise 277: For every positive integer n, let $T(n)$ be the statement that 24 divides $n(n^2 - 1)$. To be shown is that for every odd $n \geq 1$, $T(n)$ holds.
BASE STEP: Since $1(1^2 - 1) = 0$, and 24 trivially divides 0, $T(1)$ is true.

INDUCTIVE STEP: Fix an odd number $k \geq 1$, say $k = 2\ell + 1$ and suppose that $T(k)$ is true, that is, suppose that there is some integer m so that $k(k^2 - 1) = 24m$. The next odd number after k is $k + 2$, so it remains to prove $T(k+2)$, that is, that 24 divides $(k + 2)((k + 2)^2 - 1)$. To this end,

$$
\begin{aligned}
(k + 2)((k + 2)^2 - 1) &= k(k^2 + 4k + 3) + 2(k^2 + 4k + 3) \\
&= k(k^2 - 1) + k(4k + 4) + 2(k^2 + 4k + 3) \\
&= 24m + 4k(k + 1) + 2(k + 1)(k + 3) \quad \text{(by } T(k)\text{)} \\
&= 24m + (k + 1)(6k + 6) \\
&= 24m + (2\ell + 2)(12\ell + 12) \\
&= 24m + 24(\ell + 1)^2,
\end{aligned}
$$

which is evidently divisible by 24, concluding the inductive step $T(k) \to T(k+2)$ for k odd. [*Note*: The above sequence of equalities might not be the most efficient derivation of the result.]

By mathematical induction, for all odd positive integers n, the statement $T(n)$ is true. □

Exercise 278: This problem appears courtesy of José Espinosa and can be found on his website [176, No. 17]; the hint "Show that $f(n)$ has period 12 modulo 32." is provided by Naoki Sato.

The solution provided by Espinosa proceeds as follows: For $n = 1$, $f(3 \cdot 1) + f(3 \cdot 1 + 1) = f(3) + f(4) = 3(1 + 1) + 1 + 3(7 + 1) + 1 = 7 + 25 = 32$, which is trivially divisible by 32. For the inductive step,

$$f(3(k+1)) + f(3(k+1) + 1) = f(3k + 3) + f(3k + 4)$$

$$= f(3k+3) + 3[f(3k+3) + f(3k+2)] + 1$$
$$= 4f(3k+3) + 3f(3k+2) + 1$$
$$= 4[3(f(3k+2) + f(3k+1)) + 1] + 3f(3k+2) + 1$$
$$= 15f(3k+2) + 12f(3k+1) + 5$$
$$= 15(3(f(3k+1) + f(3k)) + 1) + 12f(3k+1) + 5$$
$$= 12f(3k+1) + 20 + 45(f(3k+1) + f(3k))$$
$$= 4(3f(3k+1) + 5) + 45(f(3k+1) + f(3k)).$$

So if one can prove that $3(f(3k+1)+5$ is divisible by 8, the proof can be completed. To this end, another proof by induction is helpful. For $k = 1$, $3f(4)+5 = 3\cdot25+5 = 80$, which is divisible by 8. Then

$$3f(3(k+1) + 1) + 5 = 3f(3k+4) + 5$$
$$= 3(3(f(3k+3) + f(3k+2)) + 1) + 5$$
$$= 3(3(3(f(3k+2) + f(3k+1)) + 1 + f(3k+2)) + 1) + 5$$
$$= 3(12f(3k+2) + 9f(3k+1) + 2) + 5$$
$$= 36f(3k+2) + 27f(3k+1) + 17$$
$$= 36f(3k+2) + 9(3f(3k+1) + 5) - 28$$
$$= 4(9f(3k+2) - 7) + 9(3f(3k+1) + 5).$$

To see that $9f(3k+2) - 7$ is divisible by 2, prove that $f(3k+2) - 1$ is divisible by 2. For $k = 1$, $f(3\cdot1+2) - 1 = f(5) - 2 = 97 - 1 = 96$, divisible by 2. Then

$$f(3(k+1) + 2) - 1 = f(3k+5)$$
$$= 3(f(3k+4) + f(3k+4)) + 1 - 1$$
$$= 3(3(f(3k+3) + f(3k+2)) + 1 + f(3k+3)$$
$$= 12f(3k+3) + 9f(3k+2) + 3$$
$$= 12f(3k+3) + 1) + 9(f(3k+2) - 1).$$

Therefore, if $f(3k+2) - 1$ is divisible by 2, then so is $f(3(k+1) + 2) - 1$. Hence $9f(3k+2) - 7$ is divisible by 2, and so $3(f(3k+1) + 5$ is divisible by 8, and also, for all positive integers n, $f(3k) + f(3k+1)$ is divisible by 32. □

Exercise 279: For each $n \geq 1$, let $D(n)$ be the statement that 43 divides $6^{n+1} + 7^{2n-1}$.

BASE STEP ($n = 1$): Since $6^{1+1} + 7^{2(1)-1} = 36 + 7$, $D(1)$ holds.

INDUCTIVE STEP: Suppose that for some $k \geq 1$, $D(k)$ holds, that is, $6^{k+1} + 7^{2k-1} = 43m$ for some positive integer m. To see that $D(k+1)$ is true, use the trick of adding and subtracting the expression $6 \cdot 7^{2k-1}$ as follows:

$$6^{(k+1)+1} + 7^{2(k+1)-1} = 6^{k+2} + 7^{2k+1}$$

$$
\begin{aligned}
&= 6^{k+2} + \underbrace{6 \cdot 7^{2k-1} - 6 \cdot 7^{2k-1}} + 7^{2k+1} \\
&= 6(6^{k+1} + 7^{2k-1}) + (-6 + 7^2)7^{2k-1} \\
&= 6 \cdot 43m + 43 \cdot 7^{2k-1} \quad \text{(by } D(k)\text{)} \\
&= 43(6m + 7^{2k-1}).
\end{aligned}
$$

Since $m \geq 1$ and $k \geq 1$, the number $6m + 7^{2k-1}$ is also a positive integer and so $6^{(k+1)+1} + 7^{2(k+1)-1}$ is divisible by 43, that is, $D(k+1)$ follows from $D(k)$. This concludes the inductive step.

By MI, for each $n \geq 1$, the statement $D(n)$ is true.   □

Exercise 280: This problem appeared in [161, 8.37, p. 209] with a nasty typo. The original problem read "1007, 10017, 10117, ..." from which it was impossible to determine what the sequence was. Use the fact that the successive differences are of the form 9010...0, a number divisible by 53.

Exercise 281: For every $n \geq 0$, denote the assertion by

$$
A(n): \quad 57 \mid (7^{n+2} + 8^{2n+1}).
$$

BASE STEP: Since $7^2 + 8 = 57$, $A(0)$ is true.

INDUCTIVE STEP: Let $k \geq 0$ be a fixed integer, and suppose that $A(k)$ is true with some integer ℓ satisfying

$$
7^{k+2} + 8^{2k+1} = 57\ell.
$$

Then,

$$
\begin{aligned}
7^{k+3} + 8^{2k+3} &= 7 \cdot 7^{k+2} + 64 \cdot 8^{2k+1} \\
&= 7(7^{k+2} + 8^{2k+1}) + 57 \cdot 8^{2k+1} \\
&= 7 \cdot 57\ell + 57 \cdot 8^{2k+1} \quad \text{(by } A(k)\text{)},
\end{aligned}
$$

whence it follows that 57 divides $7^{(k+1)+2} + 8^{2(k+1)+1}$, proving $A(k+1)$, and concluding the inductive step.

By MI, for each $n \geq 0$, $A(n)$ is true.   □

Exercise 282: This problem appears in, *e.g.*, [350, Problem 8]. The solution is a little tricky, as another inductive proof is used inside the inductive step of the main proof.

For each $n \geq 0$, denote the statement in the exercise by

$$
S(n): \quad 64 \mid (3^{4n+1} + 10 \cdot 3^{2n} - 13).
$$

BASE STEP $S(0)$: Using $n = 0$, $3^1 + 10 - 13 = 0$, and 64 divides 0, so $S(0)$ holds. As an extra check, with $n = 1$, $3^5 + 10 \cdot 3^2 - 13 = 243 + 90 - 13 = 320 = 64 \cdot 5$, so $S(1)$ holds, too.

INDUCTIVE STEP $(S(k) \rightarrow S(k+1))$: Fix some $k \geq 0$, and suppose that $S(k)$ is true, with some integer ℓ so that $3^{4k+1} + 10 \cdot 3^{2k} - 13 = 64\ell$. To show that $S(k+1)$ follows, show that $3^{4(k+1)+1} + 10 \cdot 3^{2(k+1)} - 13$ is also a multiple of 64:

$$
\begin{aligned}
3^{4k+5} + 10 \cdot 3^{2k+2} - 13 &= 81 \cdot 3^{4k+1} + 90 \cdot 3^{2k} - 13 \\
&= 9(3^{4k+1} + 3^{2k} - 13) + 72 \cdot 3^{4k+1} + 8 \cdot 13 \\
&= 9(64\ell) + 72 \cdot 3^{4k+1} + 8 \cdot 13.
\end{aligned}
$$

So to show $S(k+1)$, it suffices to show that $72 \cdot 3^{4k+1} + 8 \cdot 13$ is divisible by 64, or, equivalently, that $T(k) : \quad 9 \cdot 3^{4k+1} + 13$ is divisible by 8. The statement $T(k)$ is proved separately by induction:

BASE STEP $(T(0))$: $9 \cdot 3^1 + 13 = 40 = 8 \cdot 5$, so $T(0)$ holds.

INDUCTIVE STEP $(T(j) \rightarrow T(j+1))$: For some fixed $j \geq 0$, suppose that $T(j)$ holds with some integer z so that $9 \cdot 3^{4j+1} + 13 = 8z$. (Observe that $z \geq 5$.) Then to see that $T(j+1)$ holds,

$$
\begin{aligned}
9 \cdot 3^{4(j+1)+1} + 13 &= 9 \cdot 3^4 3^{4j+1} + 13 \\
&= 81(9 \cdot 3^{4j+1} + 13) - 80 \cdot 13 \\
&= 81 \cdot 8z - 80 \cdot 13 \\
&= 8(81z - 130).
\end{aligned}
$$

This completes the inductive step.

By MI, for each $k \geq 0$, $T(k)$ holds. Returning to the main body of the proof, since $T(k)$ holds, so does $S(k+1)$; this completes the inductive step $S(k) \rightarrow S(k+1)$.

By MI, for each $n \geq 0$, $S(n)$ holds. $\qquad \square$

Exercise 285: For each integer $n \geq 0$, denote the statement in the exercise by

$$
S(n) : \quad 73 \mid (8^{n+2} + 9^{2n+1}).
$$

BASE STEP: $8^2 + 9 = 73$, so $S(0)$ holds. As an extra check, $8^3 + 9^3 = 512 + 729 = 1251 = 73 \cdot 17$, so $S(1)$ is true, as well.

INDUCTIVE STEP: Fix some $k \geq 0$ and suppose that $S(k)$ holds, where for some integer ℓ, $8^{k+2} + 9^{2k+1} = 73\ell$. Then

$$
\begin{aligned}
8^{(k+1)+1} + 9^{2(k+1)+1} &= 8^{k+2} + 9^{2k+3} \\
&= 8 \cdot 8^{k+1} + 81 \cdot 9^{2k+1} \\
&= 8(8^{k+1} + 9^{2k+1}) + 73 \cdot 9^{2k+1} \\
&= 8 \cdot 73\ell + 73 \cdot 9^{2k+1} \quad \text{(by } S(k)) \\
&= 73(8 + 9^{2k+1})
\end{aligned}
$$

shows that 73 divides $8^{(k+1)+1} + 9^{2(k+1)+1}$, which is $S(k+1)$. This completes the inductive step.

By MI, for each $n \geq 0$, the statement $S(n)$ is true. □

Exercise 286: For each $n \geq 1$, let $P(n)$ denote the proposition that $80 \mid (3^{4n} - 1)$.

BASE STEP: Since $3^4 - 1 = 80$, $P(1)$ is true.

INDUCTIVE STEP: Fix some integer $k \geq 1$, and let the inductive hypothesis be that $P(k)$ is true; to be precise, assume that ℓ is an integer so that $3^{4k} - 1 = 80\ell$. The goal is to show that $P(k+1)$ is true, that is, that 80 divides $3^{4(k+1)} - 1$. Toward this goal,

$$
\begin{aligned}
3^{4k+4} - 1 &= 3^4 3^{4k} - 1 \\
&= 81(3^{4k} - 1) + 81 - 1 \\
&= 81 \cdot 80\ell + 80 \quad \text{(by } P(k)) \\
&= 80(81\ell + 1)
\end{aligned}
$$

shows that $P(k+1)$ indeed follows from $P(k)$.

By MI, for each $n \geq 1$, 80 divides $3^{4n} - 1$.

Exercise 287: This problem appears in [350] and many other places. For every $n \geq 0$, let $S(n)$ be the statement that 133 divides $11^{n+2} + 12^{2n+1}$.

BASE STEP: Since $11^2 + 12 = 133$, the statement $S(0)$ is true.

INDUCTION STEP: Fix $k \geq 0$ and suppose that $S(k)$ is true, say for some integer m, $11^{k+2} + 12^{2k+1} = 133m$. Then

$$
\begin{aligned}
11^{k+3} + 12^{2k+3} &= 11 \cdot 11^{k+2} + 144 \cdot 12^{2k+1} \\
&= 144(11^{k+2} + 12^{2k+1}) - 133 \cdot 11^{k+2} \\
&= 144(133m) - 133 \cdot 11^{k+2} \quad \text{(by } S(k)) \\
&= 133(144m - 11^{k+2}),
\end{aligned}
$$

and so $S(k+1)$ follows. This completes the inductive step.

By MI, for each $n \geq 0$, the statement $S(n)$ is true. □

Exercise 288: For every $n \geq 1$, let $S(n)$ be the statement that

$$
576 \mid (5^{2n+2} - 24n - 25).
$$

BASE STEP: When $n = 1$, $5^4 - 24 \cdot 1 - 25 = 625 - 49 = 576$, which is (of course) divisible by 576, so $S(1)$ is true.

INDUCTION STEP: Suppose that for some fixed $m \geq 1$, $S(m)$ is true, that is, there exists an integer k so that $5^{2m+2} - 24k - 25 = 576k$. Then

$$
\begin{aligned}
5^{2(m+1)+2} - 24(m+1) - 25 &= 25 \cdot 5^{2m+2} - 24m - 49 \\
&= 25 \cdot 5^{2m+2} - 25 \cdot 24m - 625 + 576m + 576 \\
&= 25[5^{2m+2} - 24m - 25] + 576(m+1) \\
&= 25 \cdot 576k + 576(m+1) \\
&= 576(25k + m + 1),
\end{aligned}
$$

which proves that $5^{2(m+1)+2} - 24(m+1) - 25$ is divisible by 576. Hence, $S(m+1)$ follows from $S(m)$, finishing the inductive step.

By mathematical induction, for each $n \geq 1$, $S(n)$ is true. $\qquad\square$

Exercise 290: This problem can be found in, *e.g.*, [350, Problem 88], however, the proof given there is essentially a direct proof—the inductive hypothesis is not required; however, one small lemma used in the main proof might be proved by induction. Note also that part of this proof also appears separately in Exercise 305.

To be shown is that for each $n \geq 1$, the proposition

$$
P(n): \quad 2^{n+1} \mid \lceil (1 + \sqrt{3})^{2n} \rceil
$$

holds. For example, $P(1)$ says that 4 divides

$$
\lceil (1 + \sqrt{3})^2 \rceil = \lceil 1 + 2\sqrt{3} + 3 \rceil = 4 + \lceil 2\sqrt{3} \rceil = 4 + 4,
$$

which is true. To deal with the ceiling function in general, observe that for any positive integer n,

$$
\begin{aligned}
(1 + \sqrt{3})^{2n} + (1 - \sqrt{3})^{2n} &= \sum_{i=0}^{2n} (1 + \sqrt{3})^i + \sum_{i=0}^{2b} (1 - \sqrt{3})^i \\
&= \sum_{i=0}^{2n} \binom{2n}{i} [(\sqrt{3})^i + (-\sqrt{3})^i] \\
&= \sum_{j=0}^{n} \binom{2n}{2j} 2(\sqrt{3})^{2j} \\
&= 2 \sum_{j=0}^{n} \binom{2n}{2j} 3^j
\end{aligned}
$$

is an (even) integer. Since $|(1 - \sqrt{3})^{2n}| < 1$ (a simple, but unnecessary, proof of which is by induction) it follows that for any positive integer n,

$$
\lceil (1 + \sqrt{3})^2 n \rceil = (1 + \sqrt{3})^{2n} + (1 - \sqrt{3})^{2n}.
$$

Thus,

$$
\begin{aligned}
\lceil (1+\sqrt{3})^{2n} \rceil &= (1+\sqrt{3})^{2n} + (1-\sqrt{3})^{2n} \\
&= (4+2\sqrt{3})^n + (4-2\sqrt{3})^n \\
&= 2^n[(2+\sqrt{3})^n + (2-\sqrt{3})^n] \\
&= 2^n\left[\sum_{i=0}^{n}\binom{n}{i}2^{n-i}(\sqrt{3})^i + \sum_{i=0}^{n}\binom{n}{i}2^{n-i}(-\sqrt{3})^i\right] \\
&= 2^n \cdot 2 \sum_{\substack{i=0 \\ i \text{ even}}}^{n}\binom{n}{i}2^{n-i}(\sqrt{3})^i \\
&= 2^{n+1}\sum_{j=0}^{\lfloor n/2\rfloor}\binom{n}{2j}2^{n-2j}3^j
\end{aligned}
$$

shows that $P(n)$ is true.   □

Remark: I have not yet found a purely inductive proof of $P(n)$ (*i.e.*, one that uses the inductive hypothesis in an inductive step).

Exercise 292: For $n \geq 1$, let $S(n)$ be the statement that the polynomial $x^{2n} - y^{2n}$ is divisible by $x^2 - y^2$. In fact, $S(n)$ is a special case of a more general statement: let $T(n)$ be the property that for any polynomials p and q, $p^n - q^n$ is divisible by $p - q$. (So $S(n)$ uses only $p = x^2$ and $Bq = y^2$.) One proof for $T(n)$ is direct by the factorization

$$
p^n - q^n = (p - q)(p^{n-1} + p^{n-2}q + \cdots + p^2 q^{n-3} + pq^{n-2} + q^{n-1}),
$$

however, for completeness, a proof of $T(n)$ by induction is included:

BASE STEP: Statement $T(1)$ says that for any polynomials p, q, $p^1 - q^1$ is divisible by $p - q$, which is true.

INDUCTIVE STEP: Fix $k \geq 1$ and suppose that $T(k)$ is true, with say, $p^k - q^k = (p - q)r$, where r is some polynomial. It remains to show that $T(k+1)$ follows, that is, that $p^{k+1} - q^{k+1}$ is also divisible by $p - q$. Toward this goal, calculate

$$
\begin{aligned}
p^{k+1} - q^{k+1} &= p \cdot p^k - q \cdot q^k \\
&= p(p^k - q^k) + (p - q)q^k \\
&= p(p - q)r + (p - q)q^k \quad \text{(by } T(k)\text{)} \\
&= (p - q)[pr + q^k],
\end{aligned}
$$

showing the desired result $T(k + 1)$. This completes the inductive step $T(k) \rightarrow T(k + 1)$.

By the principle of mathematical induction, for every positive integer n, the stronger result $T(n)$ holds, and hence $S(n)$ holds as well. $\square$

Exercise 293: For each $n \geq 0$, let $S(n)$ be the statement that the polynomial $x^{2n+1} + y^{2n+1}$ is divisible by $x + y$.

BASE STEP: $S(0)$ says that $x + y$ is divisible by $x + y$, and so $S(0)$ is valid.

INDUCTIVE STEP $(S(k) \to S(k+1))$: Fix some $k \geq 0$ and suppose that $S(k)$ holds, in particular, suppose that r is a polynomial so that

$$x^{2k+1} + y^{2k+1} = r(x + y).$$

Yet to be shown is that $S(k+1)$ holds, namely, that $x^{2k+3} + y^{2k+3}$ is divisible by $x + y$ as well:

$$
\begin{aligned}
x^{2k+3} + y^{2k+3} &= x^2 x^{2k+1} + y^2 y^{2k+1} \\
&= x^2(x^{2k+1} + y^{2k+1}) - x^2 y^{2k+1} + y^2 y^{2k+1} \\
&= x^2(x^{2k+1} + y^{2k+1}) + (y^2 - x^2)y^{2k+1} \\
&= x^2 \cdot r(x+y) + (y - x)(y + x)y^{2k+1}.
\end{aligned}
$$

from which it follows that $x + y$ is a factor, and so $S(k+1)$ follows, completing the inductive step.

By MI, for each $n \geq 0$, $S(n)$ holds. $\square$

Exercise 294: This problem appeared in [582, Prob. 28] without solution, but I seem to recall Bevan telling me that there was kind of a trick that made the problem simple. One such trick that almost works is to work modulo 5, a trick investigated below; another idea is to first prove

$$\sum_{i=1}^{n} i^5 = \frac{1}{12} n^2 (n+1)^2 (2n^2 + 2n - 1),$$

by induction, then use this formula to show that $\binom{n+1}{2}$ is indeed a factor—however, this idea does not seem to be in the spirit of the question.

For any $n \geq 1$, let $P(n)$ denote the proposition that $\binom{n+1}{2}$ divides $\sum_{j=1}^{n} j^5$.

BASE STEP: If $n = 1$, then $\binom{n+1}{2} = 1 = \sum_{j=1}^{n} j^5$, and so $P(1)$ is true.

INDUCTIVE STEP: For some $k \geq 1$, suppose that $P(k)$ holds. From Exercise 252, for any $n \geq 1$, 5 divides $n^5 - n$, and so, $n^5 \equiv n \pmod 5$. Then

$$
\begin{aligned}
\sum_{i=1}^{k+1} i^5 &\equiv \sum_{i=1}^{k+1} i \pmod 5 \\
&= \binom{k+2}{2} \pmod 5.
\end{aligned}
$$

So where does that get us? The inductive hypothesis has not been used yet.

Exercise 295: This problem appears with kind permission from José Espinosa [176, No. 5]. The solution outlines provided there by Naoki Sato were as follows:

For all n, let $s_n = a^{2n} + b^{2n} + c^{2n}$. Then $a^2 + b^2 + c^2 = 2(a^2 + ab + b^2)$ and since d is odd, d divides $a^2 + ab + b^2$. Also,

$$
\begin{aligned}
a^2 b^2 + a^2 c^2 + b^2 c^2 &= a^2 b^2 + (a^2 + b^2)(a + b)^2 \\
&= a^4 + 2a^3 b + 3a^2 b^2 + 2ab^3 + b^4 \\
&= (a^2 + ab + b^2)^2,
\end{aligned}
$$

so $a^2 b^2 + a^2 c^2 + b^2 c^2$ is divisible by d^2. Finally, by results on recursion, for all $n \geq 3$,

$$
s_n = (a^2 + b^2 + c^2)s_{n-1} - (a^2 b^2 + a^2 c^2 + b^2 c^2)s_{n-2} + a^2 b^2 c^2 s_{n-3}.
$$

(a) Note that $a^{6n-4} + b^{6n-4} + c^{6n-4} = s_{3n-2}$, and for all $n \geq 2$,

$$
s_{3n-2} = (a^2 + b^2 + c^2)s_{3n-3} - (a^2 b^2 + a^2 c^2 + b^2 c^2)s_{3n-4} + a^2 b^2 c^2 s_{3n-5}.
$$

For $n = 2$, $s_{3n-5} = s_1 = a^2 + b^2 + c^2$, which is divisible by d. Hence, by induction, for all $n \geq 1$, s_{3n-2} is divisible by d.

(b) Note that $a^{6n-2} + b^{6n-2} + c^{6n-2} = s_{3n-1}$, and for all $n \geq 2$,

$$
s_{3n-1} = (a^2 + b^2 + c^2)s_{3n-2} - (a^2 b^2 + a^2 c^2 + b^2 c^2)s_{3n-3} + a^2 b^2 c^2 s_{3n-4}.
$$

For $n = 2$,

$$
s_{3n-4} = s_2 = a^4 + b^4 + c^4 = 2a^4 + 4a^3 b + 6a^2 b^2 + 4ab^3 + 2b^4 = 2(a^2 + ab + b^2)^2,
$$

which is divisible by d^2. By part (a), s_{3n-2} is divisible by d. Also $a^2 + b^2 + c^2$ is divisible by d and $a^2 b^2 + a^2 c^2 + b^2 c^2$ is divisible by d^2. Hence, by induction, for all $n \geq 1$, s_{3n-1} is divisible by d^2.

(c) For all $n \geq 1$, 2^n is congruent to 2 or 4 modulo 6. The result then follows from parts (a) and (b).

(d) First observe that for $n \geq 1$, 4^n is congruent to 4 modulo 6. The result then follows from part (b). $\qquad\square$

Exercise 296: This problem essentially appears in [437, Prob. 12-25], where a slightly different question is posed, yet the proof presented there implicitly contains the proof of this result.

Suppose that A, B, and C are positive integers where $BC \mid (A - B - C)$. For each $n \geq 1$, let $S(n)$ be the statement that $BC \mid (A^n - B^n - C^n)$.

BASE STEP: $S(1)$ is just a restatement of the hypothesis, and so holds vacuously.

INDUCTIVE STEP: Fix some $k \geq 1$, and suppose that $S(k)$ is true, that is, suppose BC divides $A^k + B^k + C^k$, (and BC divides $A - B - C$). Let ℓ, m be integers with

$$A - B - C = BC\ell \quad \text{and} \quad A^k - B^k - C^k = BCm. \tag{29.10}$$

To show $S(k+1)$, it suffices to show that BC divides $A^{k+1} - B^{k+1} - C^{k+1}$. Observe that

$$
\begin{aligned}
A^{k+1} &= A \cdot A^k \\
&= (BC\ell + B + C)(BCm + B^k + C^k) \quad \text{(by eq'ns (29.10))} \\
&= (BC)^2 \ell m + B^{k+1} C\ell + BC^{k+1}\ell \\
&\qquad + B^2 Cm + B^{k+1} + BC^k + BC^2 m + B^k C + C^{k+1} \\
&= B^{k+1} + C^{k+1} + BC(BC\ell + B^\ell \ell + C^k \ell + Bm + C^{k-1} + Cm + B^{k-1}),
\end{aligned}
$$

and so BC divides $A^{k+1} - B^{k+1} - C^{k+1}$ as required. This completes the inductive step $S(k) \to S(k+1)$.

By MI, it follows that for each $n \geq 1$, $S(n)$ is true. $\qquad \square$

Exercise 297: Let p be a prime and $a_1, a_2, \ldots, a_n, \ldots$ be positive integers each larger than one. Let $A(n)$ be the assertion that if

$$p \mid (a_1 \cdot a_2 \cdot \ldots \cdot a_n),$$

then p divides some a_i.

BASE STEPS ($n = 1$, $n = 2$): The statement for $n = 1$ is trivially true.

Here is a standard proof for the case $n = 2$: If a prime p divides a product ab, and p does not divide a, then $\gcd(a, p) = 1$ and so (by the Euclidean division algorithm) there exist integers k and ℓ so that $1 = ka + \ell p$. Multiplying this equation by b gives $b = kab + \ell pb$; since p divides both terms on the right-hand side, p divides the left side, that is, $p \mid b$. Hence $A(2)$ has been proved.

INDUCTION STEP: Let $k \geq 2$ and suppose that $A(k)$ holds. To prove is that

$$A(k+1): \quad \text{``if } p \mid (a_1 a_2 \cdots a_n a_{k+1}) \text{ then } p \text{ divides one of the } a_i\text{''}.$$

So assume that $p \mid (a_1 a_2 \cdots a_n a_{k+1})$. Putting $a = a_1 a_2 \cdots a_k$ and $b = a_{k+1}$, by $S(2)$, either $p \mid a$ or $p \mid b$. If $p \mid b$, the antecedent of $A(k+1)$ is true, since then $p \mid a_{k+1}$. If p fails to divide b, then $p \mid a$, that is, $p \mid (a_1 a_2 \cdots a_k)$, and so by induction hypothesis $A(k)$, p divides some a_i, and again the antecedent of $A(k+1)$ is true. Thus $A(k+1)$ holds, completing the inductive step.

By MI, for every $n \geq 1$, the statement $A(n)$ is true. $\qquad \square$

Exercise 298: This result occurred in [280, Problem 2], complete with a solution. A few details are added to that solution.

BASE STEP: For $n = 1$, $N = 2$ suffices.

INDUCTIVE STEP: For some $k \geq 1$, suppose that N is divisible by 2^k, say, $N = a_{k-1}10^{k-1} + a_{k-2}10^{k-2} + \cdots a_1 10 + a_0$.

If N is also divisible by 2^{k+1}, then put $a_k = 2$ in front of N, getting a new number $N' = 2 \cdot 10^k + N$ which is divisible by 2^{k+1} since both $2 \cdot 10^k$ and N are.

If N is not divisible by 2^k, then put $a_k = 1$ in front of N, getting the new number $N' = 10^k + N$. Since 2^{k+1} does not divide N, $N/2^k$ is odd, and so $10^k + N = 2^k(5^k + N/2^k)$, and so $10^k + N$ is a product of 2^k and an even number, that is, 2^{k+1} divides $10^k + N$.

In either case, there is a $(k+1)$-digit number N' with all digits being 1 or 2, and which is divisible by 2^{k+1}. This concludes the inductive step.

By MI, for each $n \geq 1$, the result is true. $\square$

Exercise 299: This exercise has a very easy direct proof, accomplished simply by observing that in the expansion of $(2n)!$, there are n even numbers, each of which has a factor of (at least one) 2.

Here is the inductive proof: For any $n \geq 1$, let $S(n)$ denote the statement that $\dfrac{(2n)!}{2^n}$ is an integer.

BASE STEP: $S(1)$ says $2/2$ is an integer, so $S(1)$ is true.

INDUCTIVE STEP: Fix some $k \geq 1$ and suppose that $S(k)$ holds, say with some positive integer m satisfying $\dfrac{(2k)!}{2^k} = m$. Then

$$
\begin{aligned}
\frac{(2k+2)!}{2^{k+1}} &= \frac{(2k+2)(2k+1)((2k)!)}{2 \cdot 2^k} \\
&= \frac{(2k+2)(2k+1)}{2} \cdot \frac{(2k)!}{2^k} \\
&= \frac{(2k+2)(2k+1)}{2} \cdot m \\
&= (k+1)(2k+1)m,
\end{aligned}
$$

which is an integer, so $S(k+1)$ holds, completing the inductive step.

By MI, for each $n \geq 1$, $S(n)$ is true. $\square$

Exercise 300: For each integer $n \geq 1$, let $I(n)$ denote the statement that $\dfrac{(2n)!}{n!2^n}$ is an integer.

BASE STEP: Since $\frac{(2 \cdot 1)!}{1!2^1} = 1$, $I(1)$ holds.

INDUCTIVE STEP: For some fixed $k \geq 1$, suppose that $I(k)$ holds; in particular, let m be the integer $\frac{(2k)!}{k!2^k}$. To prove $I(k+1)$, it suffices to show that $\frac{(2k+2)!}{(k+1)!2^{k+1}}$ is also an integer. Rewriting this expression,

$$
\frac{(2k+2)!}{(k+1)!2^{k+1}} = \frac{(2k+2)(2k+1)[(2k)!]}{2(k+1)k!2^k}
$$

$$= m\frac{(2k+2)(2k+1)}{2(k+1)} \quad \text{(by } I(k)\text{)}$$
$$= m(2k+1),$$

which is an integer, so $I(k+1)$ also holds, completing the inductive step $I(k) \rightarrow I(k+1)$.

By MI, for each positive integer n, $I(n)$ holds. $\qquad \square$

Exercise 301: This exercise has appeared in many places, [582], for one.

For each positive integer n, let $P(n)$ be the proposition that

$$\frac{n^5}{5} + \frac{n^3}{3} + \frac{7n}{15}$$

is an integer.

BASE STEP: $\frac{1}{5} + \frac{1}{3} + \frac{7}{15} = 1$, so $P(1)$ is true.

INDUCTIVE STEP: Fix $k \geq 1$, and suppose that $S(k)$ is true; in particular, let $m = \frac{k^5}{5} + \frac{k^3}{3} + \frac{7k}{15}$ be an integer. To show that $S(k+1)$ holds, examine

$$\frac{(k+1)^5}{5} + \frac{(k+1)^3}{3} + \frac{7(k+1)}{15}$$
$$= \frac{k^5 + 5k^4 + 10k^3 + 10k^2 + 5k + 1}{5} + \frac{k^3 + 3k^2 + 3k + 1}{3} + \frac{7k + 7}{15}$$
$$= [\frac{k^5}{5} + \frac{k^3}{3} + \frac{7k}{15}] + \frac{5k^4 + 10k^3 + 10k^2 + 5k + 1}{5} + \frac{3k^2 + 3k + 1}{3} + \frac{7}{15}$$
$$= m + k^4 + 2k^3 + 2k^2 + k + k^2 + k + \frac{1}{5} + \frac{1}{3} + \frac{7}{15} \quad \text{(by } P(k)\text{)}$$
$$= m + k^4 + 2k^3 + 2k^2 + k + k^2 + k + 1,$$

which is an integer. Thus $S(k+1)$ is true, completing the inductive step.

So by MI, for each $n \geq 1$, the statement $S(n)$ is true. $\qquad \square$

Exercise 302: This problem, too, appeared in Youse's book [582] (however, that is the only place where I have seen it).

For each positive integer n, let $A(n)$ be the assertion that

$$\frac{n^7}{7} + \frac{n^3}{3} + \frac{11n}{21}$$

is an integer. Putting these three fractions over a common denominator gives

$$\frac{3n^7 + 7n^3 + 11n}{21},$$

so equivalently, to be shown is that 21 divides $3n^7 + 7n^3 + 11n$.

BASE STEP: With $n = 1$, $3n^7 + 7n^3 + 11n = 21$, so $A(1)$ is true.

INDUCTION STEP: Fix some $k \geq 1$ and suppose that $A(k)$ is true, that is, suppose that $3k^7 + 7k^3 + 11k$ is divisible by 21. To show that $A(k+1)$ is then true, show that $x_k = 3(k+1)^7 + 7(k+1)^3 + 11(k+1)$ is also divisible by 21. Expanding,

$$
\begin{aligned}
x_k &= 3(k^7 + 7k^6 + 21k^5 + 35k^4 + 35k^3 + 21k^2 + 7k + 1) \\
&\quad + 7(k^3 + 3k^2 + 3k + 1) + 11k + 11 \\
&= 3k^7 + 21k^6 + 63k^5 + 105k^4 + 105k^3 + 63k^2 + 21k + 3 \\
&\quad + 7k^3 + 21k^2 + 21k + 7 + 11k + 11 \\
&= (3k^7 + 7k^3 + 11k) + 21k^6 + 63k^5 + 105k^4 + 105k^3 + 84k^2 + 42k + 21 \\
&= (3k^7 + 7k^3 + 11k) + 21(k^6 + 3k^5 + 5k^4 + 5k^3 + 4k^2 + 2k + 1).
\end{aligned}
$$

In the last line above, by induction hypothesis, the first term is divisible by 21, and so the entire expression is also. This shows that x_k is divisible by 21, proving $A(k+1)$, and concluding the induction step.

By mathematical induction, for every $n \geq 1$, the statement $A(n)$ is true. $\square$

Comment: One might think that since $3n^7 + 7n^3 + 11n$ has a common factor of n, it might suffice to prove by induction that $3n^6 + 7n^2 + 11$ is always divisible by 21, however, already when $n = 3$, this expression is equal to 2261, not divisible by 21 (but divisible by 7).

Exercise 303: This exercise appeared in Trim's book [534].

Let $S(n)$ be the statement that if n is a positive integer, so is $(n^3 + 6n^2 + 2n)/3$.

BASE STEP: Since $1^3 + 6 \cdot 1^2 + 2 \cdot 1 = 9$, $S(1)$ is true.

INDUCTIVE STEP: Fix $k \geq 1$ and suppose that $S(k)$ is true, that is, $(k^3 + 6k^2 + 2k)/3$ is an integer. To show that $S(k+1)$ holds, show that $((k+1)^3 + 6(k+1)^2 + 2(k+1))/3$ is also an integer. Indeed,

$$
\begin{aligned}
& \frac{(k+1)^3 + 6(k+1)^2 + 2(k+1)}{3} \\
&= \frac{k^3 + 3k^2 + 3k + 1 + 6k^2 + 12k + 6 + 2k + 2}{3} \\
&= \frac{(k^3 + 6k^2 + 2k) + 3k^2 + 15k + 3}{3} \\
&= \frac{k^3 + 6k^2 + 2k}{3} + k^2 + 5k + 1,
\end{aligned}
$$

and by induction hypothesis, this is an integer as well. Thus $S(k+1)$ is verified, completing the inductive step $S(k) \to S(k+1)$.

By mathematical induction, for all positive integers n, the statement $S(n)$ is true. $\square$

Exercise 304: This exercise occurred in [280, Problem 6]. The solution is by double induction, but there are some subtleties, so the presentation here is rather pedantic.

Let x and y be non-zero reals so that $x + \frac{1}{x}$, $y + \frac{1}{y}$, and $xy + \frac{1}{xy}$ are integers. Let $S(m, n)$ be the statement that $x^m y^n + \frac{1}{x^m y^n}$ is an integer. The proof that $S(m, n)$ is true for all integers m and n is given in four stages:
   (i) For all $m \geq 0$, $S(m, 0)$ is true.
   (ii) For all $m \geq 0$, $S(m, 1)$ is true.
   (iii) For any fixed $m \geq 0$ and all $n \geq 0$, $S(m, n)$ is true.
   (iv) If either m or n (or both) are negative, then $S(m, n)$ is true.
Steps (i)–(iii) are proved below by induction, each with two base cases. Step (iv) uses a direct proof.

(i) For all $m \geq 0$, $S(m, 0)$ is true:

BASE STEP: Since $1 + \frac{1}{1}$ is an integer, $S(0, 0)$ is true, and by assumption, $S(1, 0)$ is true.
INDUCTIVE STEP: $(S(k - 1, 0) \wedge S(k, 0) \rightarrow S(k + 1))$
   For some fixed $k \geq 1$, assume that both $S(k - 1, 0)$ and $S(k, 0)$ are true. Then since

$$ x^{k+1} + \frac{1}{x^{k+1}} = \left(x + \frac{1}{x} \right) \left(x^k + \frac{1}{x^k} \right) - \left(x^{k-1} + \frac{1}{x^{k-1}} \right), $$

and the three bracketed terms on the right are integers by assumption, by $S(k)$, and $S(k - 1)$ respectively, so $S(k + 1, 0)$ holds as well.
   By an alternative form of mathematical induction, (or a restricted type of strong induction) for all $m \geq 0$, $S(m, 0)$ is true.

(ii) For all $m \geq 0$, $S(m, 1)$ is true:

BASE STEP: Both $S(0, 1)$ and $S(1, 1)$ are true by assumption.
INDUCTION STEP: $(S(k - 1, 1) \wedge S(k, 1) \rightarrow S(k + 1, 1))$
   For some fixed $k \geq 1$, suppose that $S(k - 1, 1)$ and $S(k, 1)$ be true. Then

$$ x^{k+1}y + \frac{1}{x^{k+1}y} = \left(x + \frac{1}{x} \right) \left(x^k y + \frac{1}{x^k y} \right) - \left(x^{k-1}y + \frac{1}{x^{k-1}y} \right), $$

and the three bracketed terms on the right are integers by assumption, $S(k, 1)$ and $S(k - 1, 1)$ respectively, showing that $S(k + 1, 1)$ is true.
   By an alternative form of induction, for all $m \geq 0$, the statement $S(m, 1)$ is true.

(iii) For any fixed $m \geq 0$ and all $n \geq 0$, $S(m, n)$ is true:

   Fix some $m \geq 0$.
BASE STEP: In parts (i) and (ii), $S(m, 0)$ and $S(m, 1)$ were shown to be true.
INDUCTIVE STEP: $(S(m, \ell - 1) \wedge S(m, \ell - 1) \rightarrow S(m, \ell + 1))$

For some fixed $\ell \geq 1$, suppose that both $S(m, \ell - 1)$ and $S(m, \ell)$ are true. Then

$$x^m y^{\ell+1} + \frac{1}{x^m y^{\ell+1}} = \left(x^m y^\ell + \frac{1}{x^m y^\ell}\right)\left(y + \frac{1}{y}\right) - \left(x^m y^{\ell-1} + \frac{1}{x^m y^{\ell-1}}\right),$$

and by $S(m, \ell)$, assumption, and $S(m, \ell - 1)$ respectively, the three bracketed expressions on the right are integers. This shows that $S(m, \ell + 1)$ is true, finishing the inductive step.

Therefore, by an alternative form of induction, $S(m, n)$ is true for the fixed $m \geq 0$ and all $n \geq 0$; since m was arbitrary, this shows that for all non-negative integers m and n, the statement $S(m, n)$ is true.

(iv) If either m or n is negative (or both), $S(m, n)$ is true:

If both m and n are positive, then since

$$x^m y^n + \frac{1}{x^m y^n} = \frac{1}{x^{-m} y^{-n}} + x^{-m} y^{-n}$$

is an integer, $S(-m, -n)$ is true as well.

Also, since

$$x^{-m} y^n + \frac{1}{x^{-m} y^n} = \left(x^{-m} + \frac{1}{x^{-m}}\right)\left(y^n + \frac{1}{y^n}\right) - \left(x^m y^n + \frac{1}{x^m y^n}\right)$$

and $x^{-m} + \frac{1}{x^{-m}} = x^m + \frac{1}{x^m}$, it follows that $x^{-m} y^n + \frac{1}{x^{-m} y^n}$ is an integer. Interchanging the roles of x and y, it follows that $x^m y^{-n} + \frac{1}{x^m y^{-n}}$ is also an integer. This concludes the proof of (iv) and hence the exercise.   □

Exercise 305: This exercise appears in [582, Problem 69] without solution. First, an (overly complicated) inductive solution is presented, then a much simpler direct, non-inductive proof (which was implicitly used in Exercise 290, as well) follows.

Let $I(n)$ denote the proposition that $(2 + \sqrt{3})^n + (2 - \sqrt{3})^n$ is an integers.

Inductive Proof of $I(n)$:
BASE STEP ($n = 1$): $(2 + \sqrt{3})^1 + (2 - \sqrt{3})^1 = 4$, which is an integer, so $I(1)$ is true.

INDUCTIVE STEP: For some $k \geq 1$, assume that $I(k)$ is true, that is, $(2 + \sqrt{3})^k + (2 - \sqrt{3})^k$ is an integer. Then

$$(2 + \sqrt{3})^{k+1} + (2 - \sqrt{3})^{k+1}$$

$$= (2 + \sqrt{3})(2 + \sqrt{3})^k + (2 - \sqrt{3})(2 - \sqrt{3})^k$$

$$= 2((2 + \sqrt{3})^k + (2 - \sqrt{3})^k) + \sqrt{3}((2 + \sqrt{3})^k - (2 - \sqrt{3})^k).$$

By $I(k)$, the first summand $2((2 + \sqrt{3})^k + (2 - \sqrt{3})^k)$ is an integer. It remains to show that $\sqrt{3}((2 + \sqrt{3})^k - (2 - \sqrt{3})^k)$ is an integer. This is shown by expanding both inner terms using the binomial theorem:

$$\sqrt{3}((2 + \sqrt{3})^k - (2 - \sqrt{3})^k$$

$$= \sqrt{3}\left(\sum_{i=0}^{k} \binom{k}{i} 2^{k-i}(\sqrt{3})^i - \sum_{i=0}^{k} \binom{k}{i} 2^{k-i}(-\sqrt{3})^i \right)$$

$$= \sqrt{3} \sum_{i=0}^{k} \binom{k}{i} 2^{k-i}((\sqrt{3})^i - (-\sqrt{3})^i).$$

When i is even, $(\sqrt{3})^i - (-\sqrt{3})^i = 0$, and when i is odd, $(\sqrt{3})^i - (-\sqrt{3})^i = 2(\sqrt{3}^i = 3^{(i-1)/2}2\sqrt{3}$, and together with the $\sqrt{3}$ term in front, produces an integer. Thus $(2 + \sqrt{3})^{k+1} + (2 - \sqrt{3})^{k+1}$ is an integer as well, that is, $I(k+1)$ is true, completing the inductive step.

Therefore, by mathematical induction, for all $n \geq 1$ the statement $I(n)$ is true.

$\square$

Direct Proof of $I(n)$:

$$[2 + \sqrt{3})^n + (2 - \sqrt{3})^n = \sum_{i=0}^{n} \binom{n}{i} 2^{n-i}(\sqrt{3})^i + \sum_{i=0}^{n} \binom{n}{i} 2^{n-i}(-\sqrt{3})^i]$$

$$= 2 \sum_{\substack{i = 0 \\ i \text{ even}}}^{n} \binom{n}{i} 2^{n-i}(\sqrt{3})^i$$

$$= 2 \sum_{j=0}^{\lfloor n/2 \rfloor} \binom{n}{2j} 2^{n-2j} 3^j,$$

which is an (even) integer. $\square$

Exercise 306: This problem occurred in the 1981 West Germany Mathematical Olympiad; the solution appears, *e.g.*, in [277, pp. 88–89], where it is also mentioned that a more general proof for all positive integers (not just powers of 2) was given by Ronald Graham in *Mathematical Intelligencer*, 1979, p. 250. Only the proof for the powers of 2 is given here. This also appeared in [161, 8.28, p.208].

BASE STEP: Since $n = 1$ is the first power of 2, note that the claim holds rather trivially (since there is only one number in the set). One might also observe that for the next case, $n = 2$, the claim also is true, because in any set of 3 numbers, two have the same parity, and so their sum is even (divisible by 2).

INDUCTION STEP: Suppose that the claim holds for $n = 2^{k-1}$ and let S be a set of $2(2^k) - 1$ integers. Select any $2 \cdot 2^{k-1} - 1$ elements. Apply the induction hypothesis, producing 2^{k-1} integers whose sum is divisible by 2^{k-1}, say, whose sum is $s_1 = 2^{k-1}a$. Delete these 2^{k-1} elements from S, producing S_1 with $2 \cdot 2^k - 1 - 2^{k-1} = 3 \cdot 2^{k-1} - 1$ integers left. Again pick any $2 \cdot 2^{k-1} - 1$ elements from S_1, apply the inductive hypothesis, and get 2^{k-1} more integers whose sum is divisible by 2^{k-1}, say with sum $s_2 = 2^{k-1}b$. Delete these from S_1, giving S_2 with $3 \cdot 2^{k-1} - 1 - 2^{k-1} = 2 \cdot 2^{k-1} - 1$ integers remaining. From S_2, again by induction hypothesis, pick 2^{k-1} with some equal to, say, $s_3 = 2^{k-1}c$.

Two of the integers a, b, c have the same parity, say a and b. In this case, the two subsets having sums s_1 and s_2 respectively contain 2^k integers in all, and have grand total $2^{k-1}(a + b)$. But $a + b$ is even, say $a + b = 2\ell$, so the grand total is $2^{k-1} \cdot 2\ell = 2^k\ell$, that is, the grand total is divisible by 2^k. This completes the inductive step, and hence the proof. □

Exercise 307: This exercise was found on the web [518]; I don't know the original source. The solution here might not be the intended one. For $n \geq 1$, let $A(n)$ be the assertion of the exercise.

BASE STEP: Since $2^{2^1} = 4 > 1$, at least one prime (namely, 2) is a divisor.

INDUCTION STEP: Suppose that for some $k \geq 1$, $A(k)$ is true. Then $2^{2^{k+1}} - 1 = (2^{2^k} - 1)(2^{2^k} + 1)$ and by $A(k)$, the first factor is divisible by at least k distinct primes $p_1, \ldots, p_k$. To prove $A(k+1)$, it remains to show that at least one prime factor of $2^{2^k} + 1$ is not in $\{p_1, \ldots, p_k\}$. In hopes of contradiction, suppose that p_1 divides $2^{2^k} + 1$. Putting $q = 2^{2^{k+1}} - 1$, p_1 would then divide both q and $q + 2$, and so would divide the difference 2, showing that p_1 must be 2 itself. However, $2^{2^{k+1}} - 1$ is odd, so $p_1 = 2$ is impossible. Thus, one must abandon the assumption that p_1 divides $2^{2^k} + 1$. Since this argument holds for any p_i, none of $p_1, \ldots, p_k$ divides $2^{2^k} + 1$, and so $2^{2^k} + 1$ contains yet one more prime factor. This completes the proof of $A(k+1)$ and hence the inductive step.

By mathematical induction, for each $n \geq 1$, the statement $A(n)$ holds. □

Exercise 308: This problem appears with kind permission from José Espinosa's website [176, No. 7] of "uncommon" [=hard?] mathematical induction problems. Three proofs are provided by Espinosa, occupying nearly three pages of dense mathematics, and so the reader is referred to the source for a complete proofs. One solution outlined (by Naoki Sato) does not appear to be inductive, though it might aid in finding an inductive proof:

First prove a lemma:

Lemma 29.3.1. *For any prime p and positive integer n not divisible by $p - 1$,*

$$\sum_{i=1}^{p-1} i^n \equiv 0 \pmod{p}.$$

Proof: Let s denote the given sum, and let g be a primitive root modulo p. Since n is not divisible by $p - 1$, $g^n \not\equiv 1 \pmod{p}$. Therefore,

$$g^n s = \sum_{i=1}^{p-1} (gi)^n \equiv \sum_{i=1}^{p-1} i^n \equiv s \pmod{p},$$

so $(g^n - 1)s \equiv 0 \pmod{p}$ implies $s \equiv 0 \pmod{p}$. $\qquad\square$

Let t denote the sum in the problem, and let u denote

$$
\begin{aligned}
\sum_{i=1}^{p-1} i^{2^n} &= \sum_{i=1}^{2k+1} i^{2^n} + \sum_{i=2k+2}^{4k+2} i^{2^n} \\
&= \sum_{i=1}^{2k+1} i^{2^n} + \sum_{i=1}^{2k+1} (p-i)^{2^n} \\
&\equiv 2t \pmod{p}.
\end{aligned}
$$

Since 2^n is not divisible by $p - 1 = 4k + 2 = 2(2k + 1)$, by Lemma 29.3.1, $u \equiv 0 \pmod{p}$, so $t \equiv 0 \pmod{p}$. $\qquad\square$

Exercise 309: This problem appears with kind permission from José Espinosa's website [176, No. 24]. An inductive proof is not given, but appears challenging. One solution provided (by Naoki Sato) does not appear to be by induction:

Let s denote the given sum, and let t denote

$$
\begin{aligned}
\sum_{i=1}^{p-1} i^{2 \cdot 3^n} &= \sum_{i=1}^{3k+2} i^{2 \cdot 3^n} + \sum_{i=3k+3}^{6k+4} i^{2 \cdot 3^n} \\
&= \sum_{i=1}^{2k+1} i^{2 \cdot 3^n} + \sum_{i=1}^{2k+1} (p-i)^{2 \cdot 3^n} \\
&\equiv 2s \pmod{p}.
\end{aligned}
$$

Hence $p - 1 = 6k + 4 = 2(3k + 2)$, which can not divide $2 \cdot 3^n$. Therefore, $t \equiv 0 \pmod{p}$, and so $s \equiv 0 \pmod{p}$. $\qquad\square$

The hint provided by Espinosa is to see the solution to Exercise 308, and to use the fact that if a and b are relatively prime, then both $a^2 - ab + b^2$ and $a^2 + ab + b^2$ are divisible by primes of the form $6k + 1$ or 3 (when b is relatively prime to 3).

Exercise 310: This problem appears with kind permission from José Espinosa's website [176, No. 29] without an inductive solution, but one (by Naoki Sato) that does not appear to be inductive:

Let s denote the given sum, and let t denote

$$
\begin{aligned}
\sum_{i=1}^{p-1} i^{4n+2} &= \sum_{i=1}^{2k} i^{4n+2} + \sum_{i=2k+1}^{4k} i^{4n+2} \\
&= \sum_{i=1}^{2k} i^{4n+2} + \sum_{i=1}^{2k} (p-i)^{4n+2} \\
&\equiv 2s \pmod{p}.
\end{aligned}
$$

Since $p - 1 = 4k$ does not divide $4n + 2$, $t \equiv 0 \pmod{p}$, so $s \equiv 0 \pmod{p}$. □

Note: A hint given for an inductive proof says: From Wilson's theorem, there exists an integer a so that $a^2 \equiv -1 \pmod{4k + 1}$; then prove that the integers $1, 2, \ldots, 2k$ can be arranged into k pairs so that the sum of the squares of each pair is divisible by $4k + 1$. By the well-known fact that $x^{2n+1} + y^{2n+1}$ is divisible by $x + y$, the result follows.

29.4   Solutions: Expressible as sums

Exercise 311: See proof of Theorem 3.4.1.

Exercise 312: For each integer $n \geq 8$ let $S(n)$ be the statement that n is expressible as a sum of 3's and 5's.

BASE STEP: Since $8 = 3 + 5$, $9 = 3 + 3 + 3$, $10 = 5 + 5$, $S(8)$, $S(9)$, and $S(10)$ are true.

INDUCTIVE STEP $(S(k) \to S(k+3))$: Let $k \geq 8$ and assume that $S(k)$ is true, where for some non-negative integers a, b, $k = a \cdot 3 + b \cdot 5$. Then $k + 3 = (a + 1) \cdot 3 + b \cdot 5$, a sum of 3's and 5's, so $S(k + 3)$ is true.

By MI (actually, by three cases of MI), for all $n \geq 8$, $S(n)$ is true.   □

Exercise 313: (Outline) To show that any positive integer $n \notin \{1, 3\}$ can be written as a sum of 2's and 5's, it suffices to check when $n = 2$, and when $n = 5$. The inductive step k to $k+2$ then covers all the cases: $2 \to 4 \to 6 \cdots$ and $5 \to 7 \to 9 \cdots$. [Follow the write up of Exercise 312.]   □

Exercise 314: (Outline) To show that any integer $n \geq 24$ can be written as a sum of 5's and 7's, first show that 24,25,26,27,28 all can be, and then induct from k to $k + 5$.   □

Exercise 315: (Outline) To show that any integer $n \geq 64$ can be expressed as a sum of 5's and 17's, first show that $64, 65, 66, 67, 68$ all can be, and then induct from k to $k + 5$.   □

29.5 Solutions: Egyptian fractions

Exercise 316: (Outline) Use the partial fractions identity

$$\frac{1}{A} = \frac{1}{A+1} + \frac{1}{A(A+1)}.$$

This expression can be used as many times as desired, each time applying it to the last term in the Egyptian fraction representation most recently obtained. Then a proof by mathematical induction shows that if any representation is given with k fractions, then for any $n > k$, there is an expression that uses at least n fractions.

Exercise 317: (Outline) The idea is a greedy algorithm, and although it may seem quite natural, this algorithm is credited to Leonardo Fibonacci in 1202 [191], and is sometimes called *Fibonacci's algorithm*. This algorithm was rediscovered by J. J. Sylvester [522, pp. 440–445]. The idea is to, recursively, find the largest unit fraction that is at most p/q, subtract off, and then repeat with the new remainder. Put $b = \lceil \frac{q}{p} \rceil$; then $\frac{1}{b} \le \frac{p}{q} \le \frac{1}{b-1}$, and so both $q \le \frac{1}{b}$ and $bp - p < q$. Thus $\frac{p}{q} - \frac{1}{b} = \frac{bp-q}{qb}$. The numerator of the remainder (without reducing the fraction) is $bp - q = bp - p + p - q < q + p - q = p$ is smaller than the original numerator. So $\frac{p}{q} = \frac{1}{b} + \frac{bp-q}{qb}$. Repeat this process with $\frac{p_1}{q_1} = \frac{bp-q}{qb}$. Eventually, the numerator of a remainder is 1 by the well-ordering of positive integers. □

Exercise 318: (Outline) One idea is by Exercise 395, use a sufficiently long series of harmonic numbers to get the original expression down to less than 1, and then apply Exercise 317. □

Note: Induction is not really necessary in the solution to Exercise 318, as the two main ingredients of the proof were already done by induction.

29.6 Solutions: Farey fractions

Exercise 319: If $\frac{p}{q} \le \frac{1}{2}$ is in $\mathcal{F}_n$, then so is $\frac{1}{2} + (\frac{1}{2} - \frac{p}{q}) = \frac{q-p}{q}$.

Exercise 320: Check the two statements $S_1(n)$ and $S_2(n)$ for $n = 1, 2, 3$, and then for the inductive hypothesis, assume that both $S_1(n)$ and $S_2(n)$ hold and prove both $S_1(n+1)$ and $S_2(n+2)$.

29.7 Solutions: Continued fractions

Exercise 323: First observe that $p_0 = a_0$ and $q_0 = 1$. Also, $C_1 = a_0 + \frac{1}{a_1} = \frac{a_0 a_1 + 1}{a_1}$, so $p_1 = a_0 a_1 + 1$ and $q_1 = a_1$. Let $S(k)$ be the two statements of equations (11.10) and (11.11).

BASE STEP: With simple algebra, $C_2 = \frac{a_2(a_0a_1+1)+a_0}{a_2a_1+1}$, so $p_2 = a_2(a_0a_1+1)+a_0 = a_2p_1 + p_0$ and $q_2 = a_2a_1 + 1 = a_2q_1 + q_0$ confirm $S(2)$.

INDUCTIVE STEP: Fix $m \geq 3$ and assume that for *any* rational continued fraction, the corresponding equations in $S(m-1)$ are true. It remains to show that $S(m)$ holds. Consider the rational continued fraction $D = [a_0, a_1, \ldots, a_{m-1} + \frac{1}{a_m}]$. By equation (11.8), $C_m = [a_0, a_1, \ldots, a_m] = D$, and for each $j = 0, 1, 2, \ldots, m-2$, each of C and D have the same j-convergents, that is, $C_j = D_j$. However, $C_m = D_{m-1}$. For each i, let $D_i = \frac{p'_i}{q'_i}$. Then

$$
\begin{aligned}
\frac{p_m}{q_m} &= C_m = D_{m-1} = \frac{p'_{m-1}}{q'_{m-1}} \\
&= \frac{(a_{m-1} + \frac{1}{a_m})p'_{m-2} + p'_{m-3}}{(a_{m-1} + \frac{1}{a_m})q'_{m-2} + q'_{m-3}} \quad \text{(by } S(m-1) \text{ applied to } D) \\
&= \frac{(a_{m-1} + \frac{1}{a_m})p_{m-2} + p_{m-3}}{(a_{m-1} + \frac{1}{a_m})q_{m-2} + q_{m-3}} \quad \text{(since } C_{m-2}) = D_{m-2}, C_{m-3} = D_{m-3} \\
&= \frac{a_m a_{m-1}p_{m-2} + p_{m-2} + a_m p_{m-3}}{a_m a_{m-1}q_{m-2} + q_{m-2} + a_m q_{m-3}} \\
&= \frac{a_m(a_{m-1}p_{m-2} + p_{m-3}) + p_{m-2}}{a_m(a_{m-1}q_{m-2} + q_{m-3}) + q_{m-2}} \\
&= \frac{a_m p_{m-1} + p_{m-2}}{a_m q_{m-1} + q_{m-2}} \quad \text{(by } S(m-1) \text{ applied to } C),
\end{aligned}
$$

and so comparing numerators and denominators, both equations in $S(m)$ hold true.

By mathematical induction, for every $k \geq 3$ and for any rational continued fraction, equations (11.10) and (11.11) are true. $\qquad \square$

Exercise 324: For each $k \geq 1$, let $S(k)$ denote the equality (11.12): $p_k q_{k-1} - p_{k-1}q_k = (-1)^{k-1}$.

BASE STEP: Using the values $p_0 = a_0$, $q_0 = 1$, $p_1 = a_0a_1 + 1$, and $q_1 = a_1$, $S(1)$ says

$$(a_0a_1 + 1) \cdot 1 - a_0a_1 = (-1)^{k-1},$$

which is true.

INDUCTIVE STEP: For some fixed $m \geq 2$, assume

$$S(m-1): \quad p_{m-1}q_{m-2} - p_{m-2}q_{m-1} = (-1)^{m-2}$$

holds. Then

$$p_m q_{m-1} - p_{m-1}q_m$$

$$
\begin{aligned}
&= (a_m p_{m-1} + p_{m-2}) q_{m-1} - p_{m-1}(a_m q_{m-1} + q_{m-2}) && \text{(by Thm 11.7.1)}\\
&= p_{m-2} q_{m-1} - p_{m-1} q_{m-2}\\
&= -(p_{m-1} q_{m-2} - p_{m-2} q_{m-1})\\
&= -(-1)^{m-2} && \text{(by } S(m-1))\\
&= (-1)^{m-1},
\end{aligned}
$$

shows that $S(m)$ is true, completing the inductive step.

By mathematical induction, for each $k \geq 1$, $S(k)$ is true. □

Exercise 325: The solution is very similar to that of Exercise 324, however, for the base case, one needs also the values $p_2 = a_0(a_1 a_2 + 1) + a_2$ and $q_2 = a_1 a_2 + 1$. The rest is left to the reader.

Exercise 326: (Outline) Let $C = [a_0, a_1, a_2, \ldots]$ be as in the hypothesis. Then $q_0 = 1$ and $q_1 = a_1 \geq 1$, and $q_2 = a_1 a_2 + 1 > q_1$ show that the first three terms are in increasing order. By equation (11.11), the result follows. If one were to use induction, one gets a slightly stronger result (after $i \geq 2$, the q_i's jump by at least two). □

Exercise 327: The solution is straightforward by induction using equation (11.11). Let $S(k)$ be the statement that $q_k \geq 2^{k/2}$.
BASE STEPS: When $k = 2$, $q_2 = a_1 a_2 + 1 \geq 2 = 2^{2/2}$, so $S(2)$ holds. A small calculation shows $q_3 = a_3(a_1 a_2 + 1) + a_1 \geq 3 > 2^{3/2}$, so $S(3)$ holds as well.

INDUCTIVE STEP $(S(m-2) \wedge S(m-1) \rightarrow S(m))$: Let $m \geq 4$ and suppose that $S(m-2)$ and $S(m-1)$ hold. Then

$$
\begin{aligned}
q_m &= a_m q_{m-1} + q_{m-2} && \text{(by eqn (11.11))}\\
&\geq q_{m-1} + q_{m-2} && \text{(since } a_m \geq 1)\\
&\geq 2^{\frac{m-1}{2}} + 2^{\frac{m-2}{2}} && \text{(by } S(m-1) \text{ and } S(m-2))\\
&= 2^{m/2}\left(2^{-1/2} + 2^{-1}\right)\\
&> 2^{m/2},
\end{aligned}
$$

shows $S(m)$ is true, completing the inductive step. [*Note:* It appears as if the result could be strengthened considerably, since $2^{-1/2} + 2^{-1} > 1.207$.]

By MI, for each $k \geq 2$, the statement $S(k)$ holds. □

Exercise 328: See [311, p. 7].

Chapter 30

Solutions: Sequences

30.1 Solutions: Difference sequences

Exercise 329: Fix some sequence $x = x_1, x_2, x_3, \ldots$, and let $(\Delta^k x)_n$ denote the n-th term of the k-th difference sequence. Let $S(k)$ denote the statement "For each $n \geq 1$,

$$(\Delta^k x)_n = \sum_{i=0}^{k} (-1)^i \binom{k}{i} x_{n+k-i}$$

holds." The proof is by induction on k.

BASE STEP: For $k = 0$, $(\Delta^k x)_n$ is simply x_n, and $\sum_{i=0}^{0}(-1)^i \binom{0}{i} x_{n-i} = x_n$ as desired. One might also check $k = 1$: $\sum_{i=0}^{1}(-1)^i \binom{1}{i} x_{n+1-i} = x_{n+1} - x_n$, which is precisely $(\Delta x)_n$.

INDUCTION STEP: For some $k \geq 0$, assume that $S(k)$ is true. To be shown is $S(k+1)$, namely, that for each $n \geq 1$,

$$(\Delta^{k+1} x)_n = \sum_{i=0}^{k+1} (-1)^i \binom{k+1}{i} x_{n+k+1-i}.$$

Beginning with the left side of $S(k+1)$, for each n,

$$
\begin{aligned}
&(\Delta^{k+1} x)_n \\
&= (\Delta^k x)_{n+1} - (\Delta^k x)_n \\
&= \sum_{i=0}^{k} (-1)^i \binom{k}{i} x_{n+1+k-i} - \sum_{i=0}^{k} (-1)^i \binom{k}{i} x_{n+k-i} \quad \text{(by } S(k)\text{)} \\
&= x_{n+1+k} - \binom{k}{1} x_{n+k} + \cdots + (-1)^k \binom{k}{k} x_{n+1}
\end{aligned}
$$

$$- \left[x_{n+k} - \binom{k}{1} x_{n+k-1} + \cdots + (-1)^{k-1} \binom{k}{k-1} x_{n+1} \right] - (-1)^k x_n$$

$$= x_{n+1+k} + \sum_{i=1}^{k} (-1)^i \left[\binom{k}{i} + \binom{k}{i-1} \right] x_{n+k-i} - (-1)^k x_n$$

$$= x_{n+1+k} + \sum_{i=1}^{k} (-1)^i \binom{k+1}{i} x_{n+k-i} - (-1)^{k+1} x_n \quad \text{(by Pascal's id.)}$$

$$= \sum_{i=0}^{k+1} (-1)^i \binom{k+1}{i} x_{n+k+1-i},$$

completing the inductive step.

By mathematical induction, for all $k \geq 0$, the statement $S(k)$ holds. $\qquad\square$

Exercise 330: According to Dickson [136, p. 60] it was shown by Schubert [480] that the k-th difference sequence of the k-th powers is the constant sequence $k!$. An inductive proof of this fact is in [77, p. 263]; it seems to parallel a proof of the fact that the k-derivative of x^k is $k!$.

To solve this exercise (by induction), one must first decide what to induct on. At first, it seems as if there are two choices, to induct on k, or to induct on the position in the sequence. Even after some initial experimentation, neither choice reveals itself to be natural. Indeed, one proof inducts on k, but the k from "k-th difference", and not the k from the exponents. First derive an expression for the k-th difference sequence of the sequence $1^n, 2^n, 3^n, \ldots$. If one wanted to be pedantic, this derivation can really be written up in the style of an inductive proof; however, since it is found via a recursive procedure, and the expressions are kind of hard to typeset, the formality is dispensed with here.

Fix some n and examine the sequence $s = 1^n, 2^n, 3^n, \ldots$. The m-th term of the first difference sequence Δs is $(m+1)^n - m^n$; denote this by simply Δ_m. Then, by the binomial theorem, (see Exercise 104)

$$\Delta_m = m^n + \binom{n}{n-1} m^{n-1} + \cdots + \binom{n}{1} m + 1 - m^n$$

$$= \sum_{i=1}^{n} \binom{n}{n-i} m^{n-i}.$$

Looking at the next term, $\Delta_{m+1} = \sum_{i=1}^{n} \binom{n}{n-i} (m+1)^{n-i}$. Their difference is the m-th term in the second difference sequence, denoted by simply Δ_m^2. Then, again applying the binomial theorem to each $(m+1)^{n-i}$, one obtains

$$\Delta_m^2 = \Delta_{m+1} - \Delta_m$$

$$= \sum_{i=1}^{n} \binom{n}{n-i} [(m+1)^{n-i} - m^{n-i}]$$

$$= \sum_{i=1}^{n} \binom{n}{n-i} \sum_{j=1}^{n-i} \binom{n-i}{n-i-j} m^{n-i-j}.$$

Continuing this procedure down to the k-th differences, (and replacing i with i_1, j with i_2, etc.) the m-th term of the k-th difference sequence Δ^k is $\Delta_m^k =$

$$\sum_{i_1=1}^{n} \binom{n}{n-i_1} \sum_{i_2=1}^{n-i_1} \binom{n-i_1}{n-i_1-i_2} \cdots \sum_{i_k=1}^{n-i_1-\ldots-i_{k-1}} \binom{n-i_1-\ldots-i_{k-1}}{n-i_1-\ldots-i_k} m^{n-i_1-\ldots-i_k}.$$

In particular, when $n = k$, it is kind of remarkable that the only term which survives is when $i_1 = i_2 = \cdots = i_k = 1$, in which case

$$\Delta_m^k = \binom{k}{k-1}\binom{k-1}{k-2} \cdots \binom{1}{0} = k(k-1)\cdots 1 = k!.$$

□

Note: The last expression for Δ_m^k above looks nothing like the expression from Exercise 329 when x_m's are replaced with m^n's. However, one should verify that both expressions indeed yield correct results. Can one find a more direct inductive proof that both expressions are indeed equal for the sequence here? This might be quite difficult to do directly; however, it would not be surprising to find that a more elegant proof exists.

Exercise 331: See [266, p.141]. The base case consists of four considerations, and the proof is not entirely trivial.

30.2 Solutions: Fibonacci numbers

Exercise 332: Let $S(k)$ be the statement that after k months, there are F_{k+2} pairs of rabbits. The solution involves proving a a stronger statement than $S(k)$, namely $T(k)$, the statement that after k months, there are F_k pairs of immature rabbits and F_{k+1} pairs of mature rabbits. Since $F_{k+2} = F_k + F_{k+1}$, $T(k)$ implies $S(k)$.

BASE STEP ($k = 0$): After 0 months there are $0 = F_0$ immature pairs, and $1 = F_1$ pair of mature rabbits, so the base case $T(0)$ holds.

INDUCTION STEP: Fix some $n \geq 0$, and suppose that $T(n)$ is true, that is, that after n months, there are F_n pairs of immature rabbits and F_{n+1} pairs of mature rabbits. During the $(n+1)$-st month, the mature pairs each have a pair, giving F_{n+1} new (immature) pairs, and the previously immature rabbits mature, giving

$F_n + F_{n+1} = F_{n+2}$ mature rabbit pairs. So at the end of the $(n+1)$-st month, there are F_{n+1} immature pairs and $F_{(n+1)+1}$ mature rabbit pairs, agreeing with $T(n+1)$. This completes the inductive step $T(n) \to T(n+1)$.

Hence, by mathematical induction, for all $k \geq 0$, the statement $T(k)$ holds. □

Exercise 333: For every $n \geq 6$, let $S(n)$ denote the statement

$$S(n): \quad F_n > \left(\frac{3}{2}\right)^{n-1}.$$

BASE STEPS ($n = 6, 7$): Since $F_6 = 8$ and $\left(\frac{3}{2}\right)^{6-1} = \frac{243}{32}$, which is less than 8, the case $S(6)$ is proved. (Note that the case $n = 5$ fails since $F_5 = 5$ and $\left(\frac{3}{2}\right)^{5-1} = \frac{81}{16} > 5$.) Also, $F_7 = 13$ and $\left(\frac{3}{2}\right)^{7-1} = \frac{729}{64} < 13$, showing $S(7)$.

INDUCTION STEP: Suppose that for some $k \geq 7$, both $S(k-1)$ and $S(k)$ hold. To be shown is

$$S(k+1): \quad F_{k+1} > \left(\frac{3}{2}\right)^k.$$

Beginning with the left side of $S(k+1)$,

$$
\begin{aligned}
F_{k+1} &= F_{k-1} + F_k \\
&> \left(\frac{3}{2}\right)^{k-2} + \left(\frac{3}{2}\right)^{k-1} \quad \text{(by } S(k-1) \text{ and } S(k)\text{)} \\
&= \left(\frac{3}{2}\right)^{k-2} \left[1 + \frac{3}{2}\right] \\
&> \left(\frac{3}{2}\right)^{k-2} \frac{9}{4} \\
&= \left(\frac{3}{2}\right)^{k-2} \left(\frac{3}{2}\right)^2 \\
&= \left(\frac{3}{2}\right)^k,
\end{aligned}
$$

which is the right side of $S(k+1)$, completing the proof of $S(k+1)$, and hence the inductive step $[S(k-1) \wedge S(k)] \to S(k+1)$.

Therefore, by mathematical induction, (a limited type of strong induction) for all $n \geq 6$, the statement $S(n)$ holds. □

Exercise 334: For every $n \geq 1$, let $S(n)$ denote the statement

$$S(n): \quad F_n \leq \left(\frac{7}{4}\right)^{n-1}.$$

BASE STEP: $S(1)$ says $F_1 \leq \left(\frac{7}{4}\right)^0$, and since $F_1 = 1$, this is true. $S(2)$ says $F_2 \leq \frac{7}{4}$, or $1 \leq \frac{7}{4}$; again this is true.

INDUCTION STEP: Fix some $k \geq 2$ and assume that both $S(k-1)$ and $S(k)$ are true. Then

$$
\begin{aligned}
F_{k+1} &= F_{k-1} + F_k \\
&\leq \left(\frac{7}{4}\right)^{k-2} + \left(\frac{7}{4}\right)^{k-1} \quad \text{(by } S(k-1) \text{ and } S(k)) \\
&= \left(\frac{7}{4}\right)^{k-2} \left(1 + \frac{7}{4}\right) \\
&= \left(\frac{7}{4}\right)^{k-2} \left(\frac{11}{4}\right) \\
&\leq \left(\frac{7}{4}\right)^{k-2} \left(\frac{49}{16}\right) \\
&= \left(\frac{7}{4}\right)^k,
\end{aligned}
$$

shows $S(k+1)$ is also true, completing the inductive step.

Therefore, by mathematical induction, for all $n \geq 1$, $S(n)$ holds. $\qquad \square$

Exercise 335: For each $n \geq 1$, let $S(n)$ denote the statement

$$
S(n): \quad F_n \leq \left(\frac{18}{11}\right)^{n-1}.
$$

BASE STEP: $S(1)$ says $F_1 \leq \left(\frac{18}{11}\right)^0$, and since $F_1 = 1$, this is true. $S(2)$ says $F_2 \leq \frac{18}{11}$, or $1 \leq \frac{7}{4}$; again this is true.

INDUCTION STEP: Fix some $k \geq 2$ and assume that both $S(k-1)$ and $S(k)$ are true. To be shown is that $S(k+1)$ is true.

$$
\begin{aligned}
F_{k+1} &= F_{k-1} + F_k \\
&\leq \left(\frac{18}{11}\right)^{k-2} + \left(\frac{18}{11}\right)^{k-1} \quad \text{(by } S(k-1) \text{ and } S(k)) \\
&= \left(\frac{18}{11}\right)^{k-2} \left(1 + \frac{18}{11}\right) \\
&= \left(\frac{18}{11}\right)^{k-2} \left(\frac{29}{11}\right) \\
&= \left(\frac{18}{11}\right)^{k-2} \left(\frac{319}{121}\right)
\end{aligned}
$$

$$< \left(\frac{18}{11}\right)^{k-2} \left(\frac{324}{121}\right)$$

$$= \left(\frac{18}{11}\right)^{k-2} \left(\frac{18^2}{11^2}\right)$$

$$= \left(\frac{18}{11}\right)^k,$$

which shows $S(k+1)$ is also true. This completes the inductive step.

Therefore, by mathematical induction, for all $n \geq 1$, $S(n)$ holds.   □

Exercise 336: For $n \geq 0$, denote the statement by

$$S(n): \quad F_n \leq \left(\frac{5}{3}\right)^{n-1}.$$

BASE CASES: $S(0)$ says $F_0 \leq \left(\frac{5}{3}\right)^{-1}$, and since $0 \leq \frac{3}{5}$, this is true. $S(1)$ says $F_1 \leq \left(\frac{5}{3}\right)^0$, or $1 \leq 1$; again a true statement.

INDUCTION STEP: Fix some $k \geq 1$ and assume that both $S(k-1)$ and $S(k)$ are true. To see that $S(k+1)$ is true, calculate

$$\begin{aligned}
F_{k+1} &= F_{k-1} + F_k \\
&\leq \left(\frac{5}{3}\right)^{k-2} + \left(\frac{5}{3}\right)^{k-1} \quad \text{(by } S(k-1) \text{ and } S(k)) \\
&= \left(\frac{5}{3}\right)^{k-2} \left(1 + \frac{5}{3}\right) \\
&= \left(\frac{5}{3}\right)^{k-2} \left(\frac{8}{3}\right) \\
&< \left(\frac{5}{3}\right)^{k-2} \left(\frac{25}{9}\right) \\
&= \left(\frac{5}{3}\right)^k,
\end{aligned}$$

which shows $S(k+1)$ is also true. This completes the inductive step.

By MI, for all $n \geq 0$, the statement $S(n)$ holds.   □

Exercise 337: For each $n \geq 0$, let $S(n)$ be the statement

$$F_n + 2F_{n+1} = F_{n+3}.$$

First observe that $S(n)$ has a trivial direct proof:

$$F_{n+3} = F_{n+1} + F_{n+2} = F_{n+1} + (F_n + F_{n+1}) = F_n + 2F_{n+1},$$

so it may seem rather silly to prove this by induction. Nevertheless, here is one such proof:

BASE STEP: $S(0)$ says $F_0 + 2F_1 = F_3$, which is true since $F_0 = 0$, $F_1 = 1$, and $F_3 = 2$.

INDUCTIVE STEP: For some fixed $k \geq 0$, assume that $S(k)$ is true. To be shown is that

$$S(k+1): \quad F_{k+1} + 2F_{k+2} = F_{k+4}$$

follows from $S(k)$. Note that $S(k+1)$ can be proved without the inductive hypothesis; however to formulate the proof as an inductive proof, following sequence of equalities uses the inductive hypothesis:

$$\begin{aligned}
F_{k+1} + 2F_{k+2} &= F_{k+1} + 2(F_k + F_{k+1}) \\
&= (F_{k+1} + F_k) + (F_k + 2F_{k+1}) \\
&= F_{k+2} + (F_k + 2F_{k+1}) \\
&= F_{k+2} + F_{k+3} \quad (\text{by } S(k)) \\
&= F_{k+4}.
\end{aligned}$$

This completes the inductive step $S(k) \rightarrow S(k+1)$.

Therefore, by the principle of mathematical induction, $S(n)$ is true for every $n \geq 0$. □

Exercise 338: For a non-negative integer n, let $S(n)$ be the statement that F_n is an even number if and only if n is divisible by 3. One can view $S(n)$ as saying if n is divisible by 3, then F_n is even, and if n is not divisible by 3, then F_n is odd.

BASE STEPS: Since 0 is divisible by 3, and $F_0 = 0$ is even, $S(0)$ is true. Since 1 is not divisible by 3 and $F_1 = 1$, $S(1)$ is true. Since 2 is not divisible by 3 and $F_2 = 1$, $S(2)$ is also true.

INDUCTIVE STEP: For some $k \geq 1$, suppose that both $S(k-1)$ and $S(k)$ are true. To prove $S(k+1)$ is true, there are three cases.

If $k+1$ is divisible by 3, then both $k-1$ and k are not divisible by 3, and by $S(k-1)$ and $S(k)$ respectively, both F_{k-1} and F_k are odd; thus $F_{k+1} = F_{k-1} + F_k$ is even.

When $k+1$ is not divisible by 3, there are two possibilities: $k+1 = 3\ell + 1$ for some $\ell \geq 1$, or $k+1 = 3\ell + 2$ for some $\ell \geq 0$.

Suppose that $k+1 = 3\ell + 1$; Then $k-1 = 3\ell - 1$ is not divisible by 3 and by $S(k-1)$, F_{k-1} is odd. Also, $k = 3\ell$ is divisible by 3, and by $S(k)$ respectively, F_k is even. Thus $F_{k+1} = F_{k-1} + F_k$ is odd plus even which is odd.

Finally, suppose $k+1 = 3\ell + 2$. Then $k-1$ is divisible by 3, so by $S(k-1)$, F_{k-1} is even. Also, k is not divisible by three, so by $S(k)$, F_k is even. Thus, $F_{k+1} = F_{k-1} + F_k$ is even plus odd, which is odd.

The inductive step $[S(k-1) \wedge S(k)] \to S(k+1)$ is now established in all three cases.

Therefore, by mathematical induction, for all $n \geq 0$, the statement $S(n)$ is true. $\square$

Exercise 339: For every $m \geq 0$, let $S(m)$ be the statement

$$F_{m+4} = 2F_m + 3F_{m+1}.$$

BASE STEP: When $m = 0$, $S(m)$ says $F_4 = 2F_0 + 3F_1$, which is true since $F_4 = 3$, $F_0 = 1$, and $F_1 = 1$.

INDUCTIVE STEP: Suppose that for some $n \geq 0$, $S(n)$ is true. To show that

$$S(n+1): \quad F_{n+5} = 2F_{n+1} + 3F_{n+2}$$

follows from $S(n)$, calculate

$$
\begin{aligned}
F_{n+5} &= F_{n+3} + F_{n+4} \\
&= F_{n+3} + (2F_n + 3F_{n+1}) \quad \text{(by } S(n)) \\
&= (F_{n+1} + F_{n+2}) + 2F_n + 3F_{n+1} \\
&= F_{n+1} + (F_n + F_{n+1}) + 2F_n + 3F_{n+1} \\
&= 2F_{n+1} + 3(F_n + F_{n+1}) \\
&= 2F_{n+1} + 3F_{n+2}.
\end{aligned}
$$

This completes the inductive step $S(n) \to S(n+1)$.

Hence, mathematical induction proves that for all $m \geq 0$, the statement $S(m)$ is true. $\square$

Exercise 340: For each $i \geq 0$, let $S(i)$ be the statement that F_{4i} is divisible by 3.

BASE STEP: $F_0 = 0$ and is divisible by 3, so $S(0)$ is true.

INDUCTION STEP $(S(k) \to S(k+1))$: Fix $k \geq 0$ and suppose that F_{4k} is divisible by 3. To be shown is that $F_{4(k+1)}$ is also divisible by 3. By Exercise 337, $F_{4k+4} = 2F_{4k} + 3F_{k+1}$. The term $3F_{k+1}$ is certainly divisible by 3, and since by induction hypothesis, F_{4k} is also divisible by 3, so too is $2F_{4k} + 3F_{k+1}$, that is, F_{4k+4} is also divisible by 3, concluding the inductive step.

Therefore, by mathematical induction, for every $i \geq 0$, $S(i)$ holds, that is, F_{4i} is divisible by 3. $\square$

Exercise 341: (Hint) To prove that every fifth Fibonacci number is divisible by 5, imitate the proof of Exercise 340. For the inductive step, one can either derive the identity

$$F_{n+5} = 3F_n + 5F_{n+1}$$

from Exercise 339 (in one easy step), prove it by induction, or prove it directly; here is a solution where this identity is derived directly.

For each $n \geq 0$, let $A(n)$ be the assertion that F_{5n} is divisible by 5.

BASE STEP: Since $F(0) = 0$, which is divisible by 5, $A(0)$ holds.

INDUCTIVE STEP: Fix some $k \geq 0$, and assume that

$$A(k): \quad F_{5k} \text{ is divisible by 5.}$$

To complete the inductive step, it remains to show

$$A(k+1): \quad F_{5k+5} \text{ is divisible by 5.}$$

Then

$$
\begin{aligned}
F_{5k+5} &= F_{5k+4} + F_{5k+3} \\
&= (F_{5k+3} + F_{5k+2}) + (F_{5k+2} + F_{5k+1}) \\
&= (F_{5k+2} + F_{5k+1}) + (F_{5k+1} + F_{5k}) + (F_{5k+1} + F_{5k}) + F_{5k+1} \\
&= ((F_{5k+1} + F_{5k}) + F_{5k+1}) + (F_{5k+1} + F_{5k}) + (F_{5k+1} + F_{5k}) + F_{5k+1} \\
&= 5F_{5k+1} + 3F_{5k}.
\end{aligned}
$$

By induction hypothesis, F_{5k} is divisible by 5, and since $5F_{5k+1}$ is also divisible by 5, so is their sum, and thus by calculations above, so is F_{5k+5}. This concludes the inductive step $A(k) \rightarrow A(k+1)$.

By mathematical induction, for each $n \geq 0$, $A(n)$ holds. $\qquad \square$

Exercise 342: The proof is by induction on n that for every $n \geq 0$, the statement

$$P(n): \quad F_{n+8} = 7F_{n+4} - F_n$$

is true.

BASE CASES: For $n = 0$, $F_8 = 21$, $F_4 = 3$, and $F_0 = 0$, so $F_8 = 7F_4 - F_0$ shows that $P(0)$ holds. For $n = 1$, $F_9 = 34$, $F_5 = 5$, and $F_1 = 1$, so $F_9 = 7F_5 - F_0$ holds true as well.

INDUCTIVE STEP: Fix some $k \geq 0$ and suppose that both $P(k)$ and $P(k+1)$ are true. To see that $P(k+2)$ follows,

$$
\begin{aligned}
F_{(k+2)+8} &= F_{k+8} + F_{(k+1)+8} \\
&= (7F_{k+4} - F_k) + (7F_{(k+1)+4} - F_{k+1}) \quad \text{(by } P(k) \text{ and } P(k+1)) \\
&= 7(F_{(k+1)+4} + F_{k+4}) - (F_{k+1} + F_k) \\
&= 7F_{(k+2)+4} - F_{k+2},
\end{aligned}
$$

and so $P(k+2)$ is true, concluding the inductive step.

By mathematical induction, for all $n \geq 0$, the statement $P(n)$ is true.   □

Exercise 343: (Hint) Imitate the proof of Exercise 340, using Exercise 342 in the inductive step to prove by induction that every 8th Fibonacci number is indeed divisible by 7.

Exercise 344: (Hint) This problem appears with kind permission from José Espinosa [176, No. 2]. The hint provided there (in solutions by Naoki Sato) was to show first that the expression has period 6 modulo 7.

Espinosa suggests to break the problem into three parts, for n of the form $n = 3m$, $n = 3m - 1$, and $n = 3m - 2$. Another suggested solution is of the form $S(k) \wedge S(k + 1) \to S(k + 2)$, where two base cases $S(1)$ and $S(2)$ are required.

Exercise 345: (Hint) To show by induction that for every $n \geq 0$, the proposition

$$P(n): \quad F_{n+10} = 11F_{n+5} + F_n$$

holds, repeat the proof technique as was used in Exercise 342.

Exercise 346: (Hint) Using Exercise 345, imitate the proof of Exercise 340.

Exercise 347: This problem comes from (with kind permission) José Espinosa's website [176, No. 26]. The hint given in accompanying solutions by Naiko Sato is to show that the expression has period 10 modulo 11. The hint given by Espinosa says to consider when n is even, and to use

$$F_{n+10} = 55F_{n+1} + 34F_n = 11(5F_{n+1} + 3F_n) + F_n.$$

Exercise 348: (Hint) For every non-negative integer n, F_{12n} is divisible by each of 6, 8, 9, and 12. Some of these follow from combinations of previous exercises. This exercise has a fairly substantial, however straightforward solution, and is left to the reader.

Exercise 350: Two proofs are given to show that every fifteenth Fibonacci number is divisible by 10; the first is not specifically by induction: By Exercise 338, every third Fibonacci is even, and by Exercise 341, every fifth Fibonacci number is divisible by 5. Since 15 is both a multiple of 3 and a multiple of 5, every fifteenth Fibonacci number is divisible by $2 \cdot 5 = 10$.   □

Inductive proof for Exercise 350: For each $i \geq 0$, let $S(i)$ be the statement that F_{15i} is divisible by 10.

BASE STEP: Checking $F_0 = 0$ shows $S(0)$ to be true.

INDUCTIVE STEP: For some fixed $k \geq 0$, suppose that $S(k)$ is true, that is, F_{15k} is divisible by 10. Then by Lemma 12.2.2 with $n = 15k$,

$$F_{15(k+1)} = F_{15k+15} = 10F_{15k+10} + 10F_{15k+5} + 10F_{15k+1} + 7F_{15k}.$$

Since the first three terms of the right-hand side are clearly divisible by 10, and by induction assumption, so is the last term, conclude that $F_{15(k+1)}$ is also divisible by 10. This completes the inductive step.

Therefore, by MI, every F_{15i} is divisible by 10. □

Checking the chart at the beginning of Section 12.2, $F_0 = 0$, $F_{15} = 610$, and $F_{30} = 832040$ are the only Fibonacci numbers shown that are multiples of 10.

Exercise 351: This problem appears with kind permission from José Espinosa's website [176, No. 3]. One hint provided (in solutions by Naoki Sato) was to prove that the expression has period 12 modulo 13.

Espinosa suggests to let

$$f(n) = 2(2^{2n} + 5^{2n} + 6^{2n}) + 3(-1)^{n+1}((-1)^{F_n} + 1).$$

He also suggests to prove that $f(n) - f(n+1) + f(n+2)$ is divisible by 13, or that $f(n)^2 + f(n+1)^2 + f(n+2)^2$ is divisible by 13, noting that if $f(n)^2$ is divisible by 13, then so is $f(n)$.

Exercise 352: For the moment, let m be fixed, and for every $n > m$, let $S_m(n)$ be the statement

$$F_{n-m+1}F_m + F_{n-m}F_{m-1} = F_n.$$

The proof of all such $S_m(n)$ is by induction on n.

BASE STEP: $(n = m+1, m+2)$ When $n = m+1$, $S_m(n)$ says

$$F_2F_m + F_1F_{m-1} = F_{m+1},$$

which is true because $F_2 = F_1 = 1$. Similarly, when $n = m+2$,

$$F_3F_m + F_2F_{m-1} = 2F_m + F_{m-1} = F_m + (F_m + F_{m-1}) = F_m + F_{m+1} = F_{m+2}$$

shows that $S_m(m+2)$ is also correct.

INDUCTIVE STEP: $([S_m(m+k-1) \wedge S_m(m+k)] \rightarrow S_m(m+k+1))$ Suppose that for some $k \geq 2$, both

$$
\begin{array}{rll}
S_m(m+k-1): & F_kF_m + F_{k-1}F_{m-1} &= F_{m+k-1}, \text{ and} \\
S_m(m+k): & F_{k+1}F_m + F_kF_{m-1} &= F_{m+k}
\end{array}
$$

are true. To show that

$$S_m(m + k + 1) : \quad F_{k+2}F_m + F_{k+1}F_{m-1} = F_{m+k+1}$$

follows, beginning with the left side of $S_m(m + k + 1)$,

$$
\begin{aligned}
F_{k+2}F_m &+ F_{k+1}F_{m-1} \\
&= (F_k + F_{k+1})F_m + (F_{k-1} + F_k)F_{m-1} \\
&= (F_k F_m + F_{k-1}F_{m-1}) + (F_{k+1}F_m + F_k F_{m-1}) \\
&= F_{m+k-1} + F_{m+k} \quad \text{(by } S_m(m + k - 1) \text{ and } S_m(m + k)) \\
&= F_{m+k+1},
\end{aligned}
$$

obtain the right side of $S_m(m + k + 1)$, completing the inductive step.

By an alternative form of mathematical induction, for all $n > m \geq 1$, $S_m(n)$ is true. Since $m \geq 1$ was arbitrary, this completes the solution.   □

Exercise 353: To be shown is that for any positive integers m and n, F_m divides F_{nm}. To accomplish this, fix $m \geq 1$ and induct on n. For each $n \geq 1$, let $S(n)$ denote the statement that F_m divides F_{mn}.

BASE STEP: For $n = 1$, F_m is identical to $F_{m \cdot 1}$, so the former divides the latter and $S(1)$ is then true.

INDUCTIVE STEP: For some fixed $k \geq 1$, suppose that $S(k)$ is true, that is, F_m divides F_{mk}, say, $qF_m = F_{mk}$. To be shown is $S(k + 1)$, namely, that F_m divides $F_{m(k+1)}$.

By Exercise 352, (with n replaced by $m(k + 1)$),

$$
\begin{aligned}
F_{m(k+1)} &= F_{m(k+1)-m+1}F_m + F_{m(k+1)-m}F_{m-1} \\
&= F_{mk+1}F_m + F_{mk}F_{m-1} \\
&= F_{mk+1}F_m + qF_m F_{m-1} \quad \text{(by } S(k)) \\
&= F_m(F_{mk+1} + qF_{m-1}),
\end{aligned}
$$

and so F_m divides $F_{m(k+1)}$ as well, proving $S(k + 1)$ and thereby completing the inductive step.

By mathematical induction, for all $n \geq 1$, the statement $S(n)$ is true. Since m was arbitrary, this completes the solution.   □

Exercise 354: (Cassini's identity) For every $n \geq 1$, let $S(n)$ denote the statement

$$F_{n-1}F_{n+1} = F_n^2 + (-1)^n.$$

BASE STEP: $S(1)$ says $F_0 F_2 = F_1 - 1$, which is true since both sides are 0.

INDUCTIVE STEP: For some fixed $k \geq 1$, suppose that

$$S(k): \quad F_{k-1}F_{k+1} = F_k^2 + (-1)^k$$

is true. To be proved is

$$S(k+1): \quad F_k F_{k+2} = F_{k+1}^2 + (-1)^{k+1}.$$

Starting with the left side of $S(k+1)$,

$$
\begin{aligned}
F_k F_{k+2} &= F_k(F_k + F_{k+1}) \\
&= F_k^2 + F_k F_{k+1} \\
&= (F_{k-1}F_{k+1} - (1)^k) + F_k F_{k+1} \quad \text{(by } S(k)) \\
&= (F_{k-1} + F_k)F_{k+1} - (-1)^k \\
&= F_{k+1}F_{k+1} - (-1)^k \\
&= F_{k+1}^2 + (-1)^{k+1}.
\end{aligned}
$$

This concludes the proof of $S(k+1)$ and hence the inductive step.

Hence, by MI, for all $n \geq 1$, the statement $S(n)$ is true. $\qquad \square$

Exercise 355: (Cassini converse) This exercise appeared in [230, 6.44, pp. 314, 553], with the following solution outline; details are added below. The credit given is originally due to Matiiasevich [369].

Outline: Replace the pair (ℓ, m) by $(-\ell, m)$, $(\ell, -m)$ or $(-\ell, -m)$ so that $0 \leq \ell \leq m$. The result is clear if $m = \ell$. If $m \geq \ell$, replace (ℓ, m) by $(m, m - \ell)$ and use induction to show that there is an integer n so that $\ell = F_n$ and $m = F_{n+1}$.

Details: Let $M(\ell, m)$ be the statement that if ℓ and m are integers such that $|m^2 - \ell m - \ell^2| = 1$, then there exists n so that $\ell = \pm F_n$ and $m = \pm F_{n+1}$. If $M(\ell, m)$ is true, then so are $M(-\ell, -m)$, $M(m, -m)$, and $M(-m, \ell)$ (notice the order is switched for the second two) so assume, without loss of generality, that $0 \leq \ell \leq m$.

$M(m, m)$ says that if $|m^2| = 1$, then for some n, $m = \pm F_n$ and $m = \pm F_{n+1}$. This is clear since if $|m^2| = 1$, then $m = \pm 1$; when $m = 1$, both $m = F_1$ and $m = F_2$ hold, and when $m = -1$, both $m = -F_1$ and $m = -F_2$ hold. So, for every m satisfying the hypothesis of $M(m, m)$, the conclusion of $M(m, m)$ holds.
INDUCTIVE STEP $M(k, m) \rightarrow M(m, m + k)$:. For some k and m with $0 \leq k \leq m$, suppose that $M(k, m)$ holds. Suppose that the hypothesis in $M(m, m + k)$ holds, that is, $|(m + k)^2 - m(m + k) - m^2| = 1$. Then $|m^2 - mk - k^2| = 1$ as well, and so by $M(k, m)$ (and since $0 \leq k \leq m$), there is an n so that $k = F_n$ and $m = F_{n+1}$. Then $k + m = F_{n+2}$ shows that $M(m, m + k)$ holds, finishing the inductive step.

So how does this prove the result? What is the base step? The inductive step above shows that to prove $M(\ell, m)$, with $0 \leq \ell < m$, it suffices to prove $M(m, m - \ell)$. Complete details are left to the reader.

Exercise 356: (Catalan's identity) For $1 \leq r \leq n$, let $C(n,r)$ denote the identity

$$F_n^2 - F_{n-r}F_{n+r} = (-1)^{n-r}F_r^2.$$

The solution for all r and n satisfying $1 \leq r \leq n$ is an induction of a special kind, essentially, inducting on $n+r$, but with the restriction that for each r, n must be at least r. The induction process can be viewed as follows: if one considers the pairs (n,r) as points in the Cartesian plane, the first step is prove the result when $n = r$, that is, on the line $y = x$. The next step is to prove the result for $r = 1$, that is, for points on the line $y = 1$. Then, prove the result recursively for points along $y = 2$, that is, for $(2,2),(3,2),(4,2),\ldots$, each time using the truth for points $(x,2),(x,1)$ and $(x+1,1)$ to prove it for $(x+1,2)$. Then continue along $y = 3$, starting with $(4,3)$ and each time, using the three points of the square to the lower left of each $(x,3)$. The induction then fills in each horizontal row consecutively.

The (equivalent) formulation of $C(n,r)$ preferred in the solution below is

$$C(n,r) : F_{n-r}F_{n+r} = F_n^2 - (-1)^{n-r}F_r^2.$$

BASE STEP: For any $n \geq 1$ and $r = 1$, $C(n,1)$ is precisely Cassini's (or Simson's, or Kepler's) identity, proved in Exercise 354. The case when $n = r$ says $0 = 0$, so for all $r \geq 1$, $C(r,r)$ is true.

INDUCTION STEP $C(k,s-1) \wedge C(k,s) \wedge C(k+1,s-1) \to C(k+1,s)$: Let k,s be fixed, with $2 \leq s \leq k$, and assume that each of

$$
\begin{aligned}
C(k,s-1) \quad &: \quad F_{k-(s-1)}F_{k+(s-1)} = F_k^2 - (-1)^{k-(s-1)}F_{s-1}^2, \\
C(k,s) \quad &: \quad F_{k-s}F_{k+s} = F_k^2 - (-1)^{k-s}F_s^2, \quad \text{and} \\
C(k+1,s-1) \quad &: \quad F_{k+1-(s-1)}F_{k+1+s-1} = F_{k+1}^2 - (-1)^{k-s+2}F_{s-1}^2
\end{aligned}
$$

hold. To complete the inductive step, one must show

$$C(k+1,s) : F_{k+1-s}F_{k+1+s} = F_{k+1}^2 - (-1)^{k+1-s}F_s^2.$$

Starting with the left side of $C(k+1,s)$,

$$
\begin{aligned}
F_{k+1-s}&F_{k+1+s} \\
&= \quad F_{k+1-s}(F_{k-1+s} + F_{k+s}) \\
&= \quad F_{k+1-s}F_{k-1+s} + F_{k+1-s}F_{k+s} \\
&= \quad F_k^2 - (-1)^{k-(s-1)}F_{s-1}^2 + F_{k+1-s}F_{k+s} \quad \text{(by } C(k,s-1)\text{)} \\
&= \quad F_{k-s}F_{k+s} + (-1)^{k-s}F_s^2 - (-1)^{k-(s-1)}F_{s-1}^2 + F_{k+1-s}F_{k+s} \quad \text{(by } C(k,s)\text{)} \\
&= \quad (F_{k-s} + F_{k+1-s})F_{k+s} + (-1)^{k-s}F_s^2 - (-1)^{k-(s-1)}F_{s-1}^2 \\
&= \quad F_{k+2-s}F_{k+s} + (-1)^{k-s}F_s^2 - (-1)^{k-(s-1)}F_{s-1}^2 \\
&= \quad F_{k+1-(s-1)}F_{k+1+s-1} + (-1)^{k-s}F_s^2 - (-1)^{k-(s-1)}F_{s-1}^2
\end{aligned}
$$

$$\begin{aligned} &= F_{k+1}^2 - (-1)^{k-s+2}F_{s-1}^2 + (-1)^{k-s}F_s^2 - (-1)^{k-(s-1)}F_{s-1}^2 \\ &\qquad\qquad\qquad\qquad\text{(by } C(k+1, s-1)) \\ &= F_{k+1}^2 - (-1)^{k-s+1}F_s^2, \end{aligned}$$

which is equal to the right side of $C(k+1, s)$. This completes the inductive step.

By mathematical induction on two variables, for all n and r satisfying $1 \le r \le n$, the statement $C(n, r)$ is true.   □

Exercise 358: Matijasevich's lemma and its proof can be found in [230, pp. 294–5]. One proof might begin with: For $n > 2$, let $M(n)$ denote the statement "F_m is a multiple of F_n^2 if and only if m is a multiple of nF_n." For each $k = 1, 2, 3, \ldots,$ examine $F_{kn} \pmod{F_n^2}$, and see when this is zero.

Exercise 359: For every $n \ge 0$, let $S(n)$ denote the statement

$$F_0 + F_1 + F_2 + \cdots + F_n = F_{n+2} - 1.$$

BASE STEP: $S(0)$ says $F_0 = F_2 - 1$, which is $0 = 1 - 1$, a true statement.

INDUCTIVE STEP: For some fixed $k \ge 0$, suppose that $S(k)$ is true. Then

$$\begin{aligned} F_0 + F_1 + F_2 + \cdots + F_k + F_{k+1} &= F_{k+2} - 1 + F_{k+1} \quad \text{(by } S(k)) \\ &= F_{k+1} + F_{k+2} - 1 \\ &= F_{k+3} \end{aligned}$$

verifies that $S(k+1)$ is also true, completing the inductive step.

Therefore, by MI, for all $n \ge 1$, the statement $S(n)$ is true.   □

Exercise 360: For every $n \ge 0$, let $P(n)$ denote the proposition

$$F_0^2 + F_1^2 + F_2^2 + \cdots + F_n^2 = F_n F_{n+1}.$$

BASE STEP: When $n = 0$, $P(0)$ reads $F_0^2 = F_0 F_1$, which is true since both sides are zero.

INDUCTIVE STEP: For some fixed $m \ge 0$, let the inductive hypothesis be that $P(m)$ is true. To be proved is

$$P(m+1): \quad F_0^2 + F_1^2 + F_2^2 + \cdots + F_{m+1}^2 = F_{m+1}F_{(m+1)+1}.$$

Starting with the left side of $P(m+1)$, (writing in one more term so that it is clear how to apply the inductive hypothesis)

$$\begin{aligned} F_0^2 + F_1^2 + F_2^2 + \cdots + F_m^2 + F_{m+1}^2 &= F_m F_{m+1} + F_{m+1}^2 \quad \text{(by } S(m)) \\ &= (F_m + F_{m+1})F_{m+1} \end{aligned}$$

$$\begin{aligned} &= F_{m+2}F_{m+1} \\ &= F_{m+1}F_{(m+1)+1}. \end{aligned}$$

This proves $S(m+1)$, concluding the inductive step.

From the principle of mathematical induction, one concludes that $P(n)$ holds for all $n \geq 0$. $\qquad\square$

Exercise 361: For each $n \geq 1$, let $S(n)$ denote the statement

$$F_1 + F_3 + \cdots + F_{2n-1} = F_{2n}.$$

BASE STEP: $S(1)$ says $F_1 = F_2$, which is correct, both being equal to one.

INDUCTIVE STEP: Fix some $k \geq 1$ and suppose that $S(k)$ is true. It remains to show that

$$S(k+1): \quad F_1 + F_3 + \cdots + F_{2k-1} + F_{2(k+1)-1} = F_{2(k+1)}$$

follows. Starting with the left-hand side of $S(k+1)$,

$$\begin{aligned} F_1 + F_3 + \cdots + F_{2k-1} + F_{2(k+1)-1} &= [F_1 + F_3 + \cdots + F_{2k-1}] + F_{2k+1} \\ &= F_{2k} + F_{2k+1} \quad \text{(by } S(k)\text{)} \\ &= F_{2k+2}, \end{aligned}$$

which is equal to the right side of $S(k+1)$, thereby finishing the inductive step $S(k) \to S(k+1)$.

By mathematical induction, for any positive integer n, the statement $S(n)$ is true. $\qquad\square$

Exercise 362: For each $n \geq 1$, let $A(n)$ be the assertion that

$$F_0 + F_2 + F_4 + \cdots + F_{2n} = F_{2n+1} - 1.$$

BASE STEP: $A(0)$ says $F_0 = F_1 - 1$ which is a true statement since $F_0 = 0$ and $F_1 = 1$.

INDUCTIVE STEP: Fix some $j \geq 0$ and suppose that

$$A(j): \quad F_0 + F_2 + F_4 + \cdots + F_{2j} = F_{2j+1} - 1$$

holds. To be shown is that

$$A(j+1): \quad F_0 + F_2 + F_4 + \cdots + F_{2(j+1)} = F_{2(j+1)+1} - 1$$

follows. Starting with the left side of $A(j+1)$ (and rewriting it slightly)

$$F_0 + F_2 + F_4 + \cdots + F_{2j} + F_{2j+2} = F_{2j+1} - 1 + F_{2j+2} \quad \text{(by } A(j)\text{)}$$

$$= F_{2j+3} - 1,$$

one arrives at the right side of $A(j+1)$. This concludes the inductive step.

Therefore, by mathematical induction, for all $n \geq 0$, $A(n)$ is true. □

Exercise 363: For every $n \geq 1$, denote the equality in the statement of the exercise by

$$S(n): \quad F_1 F_2 + F_2 F_3 + \cdots + F_{2n-1} F_{2n} = F_{2n}^2.$$

BASE STEP: When $n = 1$, $S(1)$ says $F_1 F_2 = F_2^2$, which is correct because $F_1 = F_2 = 1$.

INDUCTIVE STEP: Fix some $k \geq 1$ and assume the inductive hypothesis

$$S(k): \quad F_1 F_2 + F_2 F_3 + \cdots + F_{2k-1} F_{2k} = F_{2k}^2$$

to be true. The next statement to be proved is

$$S(k+1): \quad F_1 F_2 + F_2 F_3 + \cdots + F_{2k+1} F_{2k+2} = F_{2k+2}^2.$$

Starting from the left side of $S(k+1)$,

$$
\begin{aligned}
F_1 F_2 + F_2 F_3 &+ \cdots + F_{2k-1} F_{2k} + F_{2k} F_{2k+1} + F_{2k+1} F_{2k+2} \\
&= F_{2k}^2 + F_{2k} F_{2k+1} + F_{2k+1} F_{2k+2} \quad \text{(by } S(k)) \\
&= F_{2k}^2 + F_{2k} F_{2k+1} + F_{2k+1}(F_{2k} + F_{2k+1}) \\
&= F_{2k}^2 + F_{2k} F_{2k+1} + F_{2k} F_{2k+1} + F_{2k+1}^2 \\
&= (F_{2k} + F_{2k+1})^2 \\
&= F_{2k+2}^2,
\end{aligned}
$$

this completes the proof of $S(k+1)$, and hence the inductive step.

By the principle of mathematical induction, for every $n \geq 1$, $S(n)$ holds. □

Exercise 364: For $n \geq 1$, let $A(n)$ be the assertion in the exercise:

$$A(n): \quad F_1 F_2 + F_2 F_3 + \cdots + F_{2n} F_{2n+1} = F_{2n+1}^2 - 1.$$

BASE STEP: The statement $A(1)$ says

$$F_1 F_2 + F_2 F_3 = F_3^2 - 1,$$

which is true because $1 \cdot 1 + 1 \cdot 2 = 2^2 - 1$.

INDUCTIVE STEP: Fix $k \geq 1$ and suppose that $A(k)$ is true. Then

$$F_1 F_2 + F_2 F_3 + \cdots + F_{2k} F_{2k+1} + F_{2k+1} F_{2k+2} + F_{2k+2} F_{2k+3}$$

$$
\begin{aligned}
&= F_{2k+1}^2 - 1 + F_{2k+1}F_{2k+2} + F_{2k+2}F_{2k+3} \quad (\text{by } A(k)) \\
&= F_{2k+1}(F_{2k+1} + F_{2k+2}) + F_{2k+2}F_{2k+3} - 1 \\
&= F_{2k+1}F_{2k+3} + F_{2k+2}F_{2k+3} - 1 \\
&= (F_{2k+1} + F_{2k+2})F_{2k+3} - 1 \\
&= F_{2k+3}F_{2k+3} - 1
\end{aligned}
$$

proves $A(k+1)$, completing the inductive step.

By MI, for all $n \geq 1$, $A(n)$ is true. $\qquad\qquad\square$

Exercise 365: For $n \geq 0$, denote the equation in the exercise by

$$
P(n): \quad F_n^2 + F_{n+1}^2 = F_{2n+1}.
$$

BASE STEP: Since $F_0^2 + F_1^2 = 0^2 + 1^2 = 1^2 = F_1^2$, $P(0)$ is true.

INDUCTIVE STEP: For some fixed k, assume that $P(k)$ is true. To prove

$$
P(k+1): \quad F_{k+1}^2 + F_{k+2}^2 = F_{2k+3},
$$

begin with the left side:,

$$
\begin{aligned}
F_{k+1}^2 + F_{k+2}^2 &= F_{k+1}^2 + (F_k + F_{k+1})^2 \\
&= F_{k+1}^2 + F_k^2 + 2F_kF_{k+1} + F_{k+1}^2 \\
&= (F_k^2 + F_{k+1}^2) + 2F_kF_{k+1} + F_{k+1}^2 \\
&= F_{2k+1} + 2F_kF_{k+1} + F_{k+1}^2 \quad (\text{by } P(k)) \\
&= F_{2k+1} + F_{k+1}(F_{k+1} - 2F_k) \\
&= F_{2k+1} + F_{2k+2} \quad (\text{by Lemma 12.2.4 with } m = k+1) \\
&= F_{2k+3},
\end{aligned}
$$

which is the right side of $P(k+1)$. This completes the proof of $P(k+1)$ and hence the inductive step.

By mathematical induction, for all $n \geq 0$ the statement $P(n)$ is true. $\qquad\square$

Exercise 366: This problem is reproduced (with kind permission) from José Espinosa's website [176, No. 25]. The solutions by Naoki Sato give a hint to use the fact that $F_n^2 + F_{n+1}^2 = F_{2n+1}$ (which is Exercise 365 here). The inductive solution given by Espinosa relies on Exercise 360 here, and an identity similar to that in Exercise 365 here which itself has an inductive proof [in Espinosa's Problem 6—beware of a typo and different notation for Fibonacci numbers; both identities fail as written]. A direct proof is also available using Binet's formula.

Exercise 367: For every $n \geq 1$, let $S(n)$ denote the statement

$$
\frac{F_0}{2} + \frac{F_1}{4} + \frac{F_2}{8} + \cdots + \frac{F_{n-1}}{2^n} = 1 - \frac{F_{n+2}}{2^n}.
$$

BASE STEP: Since $F_0 = 0$, $F_1 = 1$, and $F_3 = 2$, the statement $S(1)$ says $0 + \frac{1}{4} = 1 - \frac{3}{4}$, a true statement.

INDUCTIVE STEP: Fix some $k \geq 1$ and suppose that

$$S(k): \quad \frac{F_0}{2} + \frac{F_1}{4} + \cdots + \frac{F_{k-1}}{2^k} = 1 - \frac{F_{k+2}}{2^k}$$

holds. To show

$$S(k+1): \quad \frac{F_0}{2} + \frac{F_1}{4} + \cdots + \frac{F_k}{2^{k+1}} = 1 - \frac{F_{k+3}}{2^{k+1}},$$

start with the left side of $S(k+1)$:

$$
\begin{aligned}
\frac{F_0}{2} + \frac{F_1}{4} + \cdots + \frac{F_{k-1}}{2^k} + \frac{F_k}{2^{k+1}} &= 1 - \frac{F_{k+2}}{2^k} + \frac{F_k}{2^{k+1}} \quad \text{(by } S(k)) \\
&= 1 + \frac{1}{2^{k+1}}(-2F_{k+2} + F_k) \\
&= 1 + \frac{1}{2^{k+1}}(-2F_{k+2} + F_{k+2} - F_{k+1}) \\
&= 1 + \frac{1}{2^{k+1}}(-F_{k+2} - F_{k+1}) \\
&= 1 - \frac{F_{k+3}}{2^{k+1}},
\end{aligned}
$$

which is the right side of $S(k+1)$. This completes the inductive step $S(k) \to S(k+1)$.

By mathematical induction, for all $n \geq 1$, the statement $S(n)$ holds. $\qquad\square$

Exercise 368: For $n \geq 1$, let $P(n)$ be the proposition

$$
\begin{bmatrix} 1 & 1 \\ 1 & 0 \end{bmatrix}^n = \begin{bmatrix} F_{n+1} & F_n \\ F_n & F_{n-1} \end{bmatrix}.
$$

BASE STEP: Since $F_2 = F_1 = 1$ and $F_0 = 0$, the base case $P(1)$ is true.

INDUCTIVE STEP: For some fixed $k \geq 1$, suppose that $P(k)$ is true. Then

$$
\begin{aligned}
\begin{bmatrix} 1 & 1 \\ 1 & 0 \end{bmatrix}^{k+1} &= \begin{bmatrix} 1 & 1 \\ 1 & 0 \end{bmatrix}^k \begin{bmatrix} 1 & 1 \\ 1 & 0 \end{bmatrix} \\
&= \begin{bmatrix} F_{k+1} & F_k \\ F_k & F_{k-1} \end{bmatrix} \begin{bmatrix} 1 & 1 \\ 1 & 0 \end{bmatrix} \quad \text{(by } P(k)) \\
&= \begin{bmatrix} F_{k+1} + F_k & F_{k+1} \\ F_{k+1} & F_k \end{bmatrix} = \begin{bmatrix} F_{k+2} & F_{k+1} \\ F_{k+1} & F_k \end{bmatrix},
\end{aligned}
$$

which proves $P(k+1)$, completing the inductive step.

Hence, by mathematical induction, for every $n \geq 1$, $P(n)$ is true. $\qquad\square$

Exercise 369: (Binet's formula for Fibonacci numbers) For every $n \geq 0$, let $A(n)$ be the assertion that

$$F_n = \frac{1}{\sqrt{5}} \left[\left(\frac{1 + \sqrt{5}}{2} \right)^n - \left(\frac{1 - \sqrt{5}}{2} \right)^n \right].$$

BASE STEPS: When $n = 0$, the expression on the right side $A(n)$ is 0, which is F_0. For $n = 1$, it is also not difficult to check that each side of $A(n)$ is 1.

INDUCTIVE STEP ($[A(k-1) \wedge A_k] \to A(k+1)$): Suppose that for some fixed k, that $A(k-1)$ and $A(k)$ are true. Calculating, (where the second equality below follows from $A(k-1)$ and $A(k)$),

$$
\begin{aligned}
F_{k+1} &= F_{k-1} + F_k \\
&= \frac{1}{\sqrt{5}} \left[\left(\frac{1+\sqrt{5}}{2} \right)^{k-1} - \left(\frac{1-\sqrt{5}}{2} \right)^{k-1} \right] + \frac{1}{\sqrt{5}} \left[\left(\frac{1+\sqrt{5}}{2} \right)^{k} - \left(\frac{1-\sqrt{5}}{2} \right)^{k} \right] \\
&= \frac{1}{\sqrt{5}} \left[\left(\frac{1+\sqrt{5}}{2} \right)^{k-1} \left(1 + \frac{1+\sqrt{5}}{2} \right) - \left(\frac{1-\sqrt{5}}{2} \right)^{k-1} \left(1 + \frac{1-\sqrt{5}}{2} \right) \right] \\
&= \frac{1}{\sqrt{5}} \left[\left(\frac{1+\sqrt{5}}{2} \right)^{k-1} \left(\frac{3+\sqrt{5}}{2} \right) - \left(\frac{1-\sqrt{5}}{2} \right)^{k-1} \left(\frac{3-\sqrt{5}}{2} \right) \right] \\
&= \frac{1}{\sqrt{5}} \left[\left(\frac{1+\sqrt{5}}{2} \right)^{k-1} \left(\frac{1+\sqrt{5}}{2} \right)^2 - \left(\frac{1-\sqrt{5}}{2} \right)^{k-1} \left(\frac{1-\sqrt{5}}{2} \right)^2 \right] \\
&= \frac{1}{\sqrt{5}} \left[\left(\frac{1+\sqrt{5}}{2} \right)^{k+1} - \left(\frac{1-\sqrt{5}}{2} \right)^{k+1} \right],
\end{aligned}
$$

so $A(k+1)$ is also true, completing the inductive step.

By mathematical induction, for each $n \geq 0$, $A(n)$ is true. $\square$

Exercise 370: For any non-negative integers m and n, let $P(m, n)$ be the proposition

$$\binom{n}{0} F_m + \binom{n}{1} F_{m+1} + \cdots + \binom{n}{n} F_{m+n} = F_{m+2n}.$$

For each n, $P(0, n)$ is true by Lemma 12.2.5. Here is the similar result when $m = 1$:

Lemma 30.2.1. *For $n \geq 0$,*

$$\binom{n}{0} F_1 + \binom{n}{1} F_2 + \cdots + \binom{n}{n} F_n = F_{2n+1}.$$

Proof:

$$\sum_{n=0}^{n} \binom{n}{i} F_{i+1} = \sum_{n=0}^{n} \binom{n}{i} \frac{1}{\sqrt{5}} [\phi_1^{i+1} - \phi_2^{i+1}]$$

$$= \frac{1}{\sqrt{5}} [\sum_{n=0}^{n} \binom{n}{i} \phi_1^{i+1} - \sum_{n=0}^{n} \binom{n}{i} \phi_2^{i+1}]$$

$$= \frac{1}{\sqrt{5}} [\phi_1 (1 + \phi_1)^n - \phi_2 (1 - \phi_2)^n]$$

$$= \frac{1}{\sqrt{5}} [\phi_1 (\phi_1^2)^n - \phi_2 (\phi_2^2)^n] \quad \text{(by eqn (12.1))}$$

$$= F_{2n+1}.$$

$\square$

All is now ready for the inductive proof; fix an arbitrary n, and induct on m.
BASE STEP: The cases $P(0, m)$ and $P(1, n)$ are handled above, by Lemmas 12.2.5 and 30.2.1.

INDUCTIVE STEP ($[P(m, n) \wedge P(m+1, n)] \rightarrow P(m+2, n)$): Fix some $m \geq 1$, and suppose that both

$$P(m, n): \quad \sum_{n=0}^{n} \binom{n}{i} F_{m+i} = F_{m+2n}$$

and

$$P(m+1, n): \quad \sum_{n=0}^{n} \binom{n}{i} F_{m+1+i} = F_{m+1+2n}$$

hold. Then

$$\sum_{n=0}^{n} \binom{n}{i} F_{m+2+i} = \sum_{n=0}^{n} \binom{n}{i} (F_{m+i} + F_{m+1+i})$$

$$= \sum_{n=0}^{n} \binom{n}{i} F_{m+i} + \sum_{n=0}^{n} \binom{n}{i} F_{m+1+i}$$

$$= F_{m+2n} + F_{m+1+2n} \quad \text{(by ind. hyp.)}$$

$$= F_{m+2+2n}$$

proves $P(m+2, n)$, completing the inductive step.

Since n was arbitrary, by mathematical induction, for all $m \geq 0$ and all $n \geq 0$, $P(m, n)$ is true. $\square$

Exercise 371: For $n \geq 1$, denote the statements by

$$S(n): \quad \sum_{i=0}^{n} \binom{2n - i}{i} = F_{2n+1},$$

and

$$T(n): \quad \sum_{i=0}^{n} \binom{n+1+i}{n-i} = F_{2n+2}.$$

Proving either one of these statements alone can be challenging, however, if one proves them simultaneously, the task is fairly simple.

BASE STEP: $S(1)$ says $\binom{2}{0} + \binom{1}{1} = F_3$, which is correct since both sides equal 2, and $T(1)$ says $\binom{3}{0} + \binom{2}{1} = F_4$, again correct since both sides are equal to 3.

INDUCTIVE STEPS: Fix some $k \geq 1$, and assume that both $S(k)$ and $T(k)$ hold. Showing that $S(k+1)$ and $T(k+1)$ hold is done in two steps, first $[S(k) \wedge T(k)] \rightarrow S(k+1)$, and then $[T(k) \wedge S(k+1)] \rightarrow T(k+1)$. Here are the equations showing the first step; beginning with the left side of $S(k+1)$, take off the first and last summand, and apply Pascal's identity:

$$
\begin{aligned}
\sum_{i=0}^{k+1} \binom{2k+2-i}{i} &= 1 + \left[\sum_{i=1}^{k} \binom{2k+2-i}{i} \right] + 1 \\
&= 1 + \left[\sum_{i=1}^{k} \left(\binom{2k+1-i}{i-1} + \binom{2k+1-i}{i} \right) \right] + 1 \\
&= \left[\sum_{j=0}^{k-1} \binom{2k-j}{j} \right] + 1 + \left(1 + \sum_{i=1}^{k} 2k+1-i \right) \\
&= \sum_{j=0}^{k} \binom{2k-j}{j} + \sum_{i=0}^{k} \binom{2k+1-i}{i} \\
&= F_{2k+1} + F_{2k+2} \quad \text{(by } S(k) \text{ and } T(k) \text{ resp.)} \\
&= F_{2k+3},
\end{aligned}
$$

giving the right side of $S(k+1)$, completing the step $[S(k) \wedge T(k)] \rightarrow S(k+1)$. The induction hypothesis can be extended to $S(k+1)$ and $T(k)$. Starting with the left side of $T(k+1)$, calculations are simpler than in the above step:

$$
\begin{aligned}
\sum_{i=0}^{k+1} \binom{2k+3-i}{i} &= \binom{2k+3}{0} + \sum_{i=1}^{k+1} \binom{2k+3-i}{i} \\
&= 1 + \sum_{i=1}^{k+1} \left(\binom{2k+2-i}{i-1} + \binom{2k+2-i}{i} \right) \\
&= \left[\sum_{j=0}^{k} \binom{2k+1-j}{j} \right] + 1 + \sum_{i=1}^{k+1} \binom{2k+2-i}{i} \\
&= \left[\sum_{j=0}^{k} \binom{2k+1-j}{j} \right] + \sum_{i=0}^{k+1} \binom{2k+2-i}{i}
\end{aligned}
$$

$$
\begin{aligned}
&= F_{2k+2} + F_{2k+3} \quad \text{(by } T(k) \text{ and } S(k+1), \text{ resp.)} \\
&= F_{2k+4}.
\end{aligned}
$$

This proves that $T(k+1)$ is true, completing the step $[T(k) \wedge S(k+1)] \to T(k+1)$.

Thus, by mathematical induction, for all $n \geq 1$, both statements $S(n)$ and $T(n)$ are true. $\qquad \square$

Comment: Is there an inductive proof of just one of these statements that does not rely on the other? For example, to prove $S(n)$, one might try expressing a Fibonacci number with an odd index as a sum of others with only odd indices, such as $F_9 = 2F_7 + F_5 + F_3 + F_1$, and proceed using strong induction.

Exercise 372: (Zeckendorf's theorem) Hint: Use a greedy algorithm.

Exercise 373: For $n \geq 1$, let $C(n)$ be the claim that the number of binary strings of length n not containing '11" as a substring is F_{n+2}. First note that the result in this exercise also follows from Exercise 380, (whose proof is considerably more complicated).

BASE CASES: Both binary strings 0 and 1 of length 1 do not contain "11", and since $F_{1+2} = F_3 = 2$, $C(1)$ is valid. The strings of length $n = 2$ not containing "11" are 00, 01, 10, of which there are $3 = F_4 = F_{2+2}$, and so too $C(2)$ is valid.

INDUCTIVE STEP: For some fixed $k \geq 1$, suppose that the claims $C(k)$ and $C(k+1)$ are true. To be proved is $C(k+2)$. A string of length $k+2$ can begin either with a 0 or a 1.

If a string of length $k+2$ begins with a 1, for it to not contain a "11" substring, then a 0 must occupy the second position; in this case, the remaining string of length k can not have 11 as a substring, and (by induction assumption $C(k)$) there are F_{k+2} such strings.

If a string of length $k+2$ begins with a 0, then the remaining string of length $k+1$ can be (by induction assumption $C(k+1)$) completed in F_{k+3} ways.

So, in all, there are $F_{k+2} + F_{k+3} = F_{k+4} = F_{(k+2)+2}$ strings of length $k+2$ which do not contain "11" as a substring; this is precisely the claim $C(k+2)$, finishing the inductive step.

Therefore, by mathematical induction, for every $n \geq 1$, $C(n)$ is valid. $\qquad \square$

Exercise 374: This exercise appeared in [161, 8.11, p. 207]. In the following solution, first a recursion is derived. A string beginning with 0 can be continued in a_{n-1} ways; a string beginning with 100 can be continued in a_{n-3} ways; a string beginning with 1100 can be continued in a_{n-4} ways. Thus

$$
a_n = a_{n-1} + a_{n-3} + a_{n-4}.
$$

Working out the first few values, $a_1 = 2$, $a_2 = 4 = F_2^2$, $a_3 = 6 = F_3 F_4$, $a_4 = 9 = F_4^2$, and $a_5 = F_4 F_5$. So one might conjecture that

$$a_{2m} = F_{m+2}^2, \quad a_{2m+1} = F_{m+2}F_{m+3}.$$

In fact, these two can be proved by induction on m. If this conjecture is true for all $k < 2m$, then

$$
\begin{aligned}
a_{2m} &= a_{2m-1} + a_{2m-3} + a_{2m-4} \\
&= F_{m+1}F_{m+2} + F_m F_{m+1} + F_m^2 \quad \text{(by IH)} \\
&= F_{m+1}F_{m+2} + F_m(F_{m+1} + F_m) \\
&= F_{m+1}F_{m+2} + F_m F_{m+2} \\
&= (F_{m+1} + F_m)F_{m+2} \\
&= F_{m+2}^2,
\end{aligned}
$$

and with similar calculations, one shows $a_{2m+1} = F_{m+2}F_{m+3}$. Hence the conjecture is true for $n = 2m$ and $n = 2m + 1$, completing the inductive step, and hence the proof of the conjecture in general. $\qquad\square$

Exercise 377: Use strong induction on k?

Exercise 378: Use Exercise 377 and Binet's formula. If $m = 1$, then $k = 0 = \log_\phi(1)$ and the result holds. If $m \geq 2$ then $k \geq 1$, and using the approximation $\frac{1-\sqrt{5}}{2} < 0.6 \leq (0.3)m$,

$$\phi^{k+1} < \sqrt{5}m + (0.3)m = (\sqrt{5} + 0.3)m.$$

Taking logarithms,

$$
\begin{aligned}
k + 1 &= \log_\phi(\phi^{k+1}) \\
&< \log_\phi((\sqrt{5} + 0.3)m) \\
&= \log_\phi(\sqrt{5} + 0.3) + \log_\phi(m) \\
&< 2 + \log_\phi(m),
\end{aligned}
$$

and so $k < \log_\phi(m) + 1$. $\qquad\square$

Exercise 379 (Lamé's Theorem): Proof outline: The proof is by induction on N. For $N = 1$, $m \leq 9$, and so by Exercise 378, the Euclidean algorithm computes $\gcd(m, n)$ in at most $\lfloor \log_\phi(9) \rfloor + 1 = 5$ steps.

To prove the inductive step, increasing the number of digits of $m \geq 9$ increases the number of steps by at most five since

$$\log_\phi(10m + 9) \leq \log_\phi(11m) = \log_\phi(11) + \log_\phi(m),$$

and $\log_\phi(11) < 5$. $\square$

Exercise 380: This problem appears in [6, Ex. 8, p. 98]. Note that the final conclusion follows directly from Exercise 373 because there is a one-to-one correspondence between binary words of length n and subsets of $\{1, 2, \ldots, n\}$. For example, 0111 corresponds to the subset $\{2, 3, 4\}$ of $\{1, 2, 3, 4\}$; the corresponding binary vector $(0, 1, 1, 1)$ is called the *characteristic vector* for the subset. So a binary word with consecutive 1's corresponds directly to a subset with consecutive integers. The solution provided here is slightly different than the one for Exercise 373.

Let $a_{n,k}$ be the number of k-subsets of $\{1, 2, \ldots, n\}$ which do not contain a pair of consecutive integers. Consider the statements

$$S(n, k) : \quad a_{n,k} = \binom{n - k + 1}{k},$$

and

$$S(n) : \quad \sum_{k \geq 0} a_{n,k} = F_{n+2}.$$

To see $S(n, k)$, proceed by induction on both n and k.

BASE CASES: When $n = 0$, there is only one 0-subset (namely the empty set), and for $k \geq 1$, there are no k-subsets, so $a_{0,0} = 1 = \binom{0-0+1}{0}$ and for $k \geq 1$, $a_{0,k} = 0 = \binom{-k+1}{k} = 0$. Also, for each $n \geq 1$, $a_{n,0} = 1 = \binom{n-0+1}{0}$ and $a_{n,1} = n = \binom{n+1-1}{1}$. So it is shown that for all $n \geq 0$, $k \geq 0$, $S(0, k)$, $S(1, k)$, $S(n, 0)$, and $S(n, 1)$ are all true.

INDUCTION STEP: For some $n \geq 2$ and $k \geq 1$, assume that both $S(n - 2, k - 1)$ and $S(n-1, k)$ hold. Examine a k-set $\{x_1, x_2, \ldots, x_k\}$ from $\{1, 2, \ldots, n\}$ that contains no consecutive integers. If $x_k = n$, then there are $a_{n-2,k-1}$ ways to select $x_1, \ldots, x_{k-1}$ since $n-1$ then can not be one of the remaining x_i's. If $n \notin \{x_1, \ldots, x_k\}$, then there are $a_{n-1,k}$ ways to select the x_i's. Hence,

$$\begin{aligned}
a_{n,k} &= a_{n-2,k-1} + a_{n-1,k} \\
&= \binom{n - 2 - (k - 1) + 1}{k - 1} + \binom{n - 1 - k + 1}{k} \qquad \text{(by ind. hyps.)} \\
&= \binom{n - k}{k - 1} + \binom{n - k}{k} \\
&= \binom{n - k + 1}{k}, \quad \text{(by Pascal's identity)}
\end{aligned}$$

proving $S(n, k)$, completing the inductive step $S(n-2, k-1) \wedge S(n-1, k) \to S(n, k)$.

By mathematical induction, $S(n, k)$ is true for all $n, k \geq 0$. $\square$

Note: The reader might check that the above base cases and inductive steps actually prove $S(n, k)$ for all n, k. The idea is to first induct on n for fixed k and

$k - 1$, then induct on k. For example, if one wanted to see that $S(6, 3)$ is true, proceed with implications as follows:

$$S(0, 1) \wedge S(1, 2) \quad \rightarrow \quad S(2, 2)$$
$$S(1, 1) \wedge S(2, 2) \quad \rightarrow \quad S(3, 2)$$
$$S(2, 1) \wedge S(3, 2) \quad \rightarrow \quad S(4, 2)$$
$$S(3, 1) \wedge S(4, 2) \quad \rightarrow \quad S(5, 2)$$

$$S(0, 2) \wedge S(1, 3) \quad \rightarrow \quad S(2, 3)$$
$$S(1, 2) \wedge S(2, 3) \quad \rightarrow \quad S(3, 3)$$
$$S(2, 2) \wedge S(3, 3) \quad \rightarrow \quad S(4, 3)$$
$$S(3, 2) \wedge S(4, 3) \quad \rightarrow \quad S(5, 3)$$
$$S(4, 2) \wedge S(5, 3) \quad \rightarrow \quad S(6, 3).$$

For $n \geq 0$, prove $S(n)$ by induction:

BASE STEP: For $n = 0$,

$$\sum_{k \geq 0} \binom{1 - k}{k} = \binom{1}{0} = 1 + 0 = 1 = F_2,$$

so $S(0)$ is true. For $n = 1$,

$$\sum_{k \geq 0} \binom{2 - k}{k} = \binom{2}{0} + \binom{1}{1} + \binom{0}{2} = 1 + 1 + 0 = 2 = F_3,$$

so $S(1)$ also holds.

INDUCTIVE STEP: Let $n \geq 2$ and assume that both $S(n - 2)$ and $S(n - 1)$ are true, that is, both

$$\sum_{j \geq 0} \binom{n - 2 - j + 1}{j} = F_n \quad \text{and} \quad \sum_{k \geq 0} \binom{n - 1 - k + 1}{k} = F_{n+1}.$$

Then

$$\sum_{k \geq 0} \binom{n - k + 1}{k} = \sum_{k \geq 0} \left[\binom{n - k}{k} + \binom{n - k}{k - 1} \right]$$

$$= \sum_{k \geq 0} \binom{n - k}{k} + \sum_{k \geq 0} \binom{n - k}{k - 1}$$

$$= \sum_{k \geq 0} \binom{n - k}{k} + \sum_{k \geq 1} \binom{n - k}{k - 1}$$

$$
\begin{aligned}
&= \sum_{k \geq 0} \binom{n-k}{k} + \sum_{j \geq 0} \binom{n-j-1}{j} \\
&= F_{n+1} + F_n \\
&= F_{n+2}
\end{aligned}
$$

shows that $S(n)$ also holds.

By induction, for each $n \geq 0$, the statement $S(n)$ is true. $\qquad$ $\square$

Exercise 381: This problem appears in [6, Ex. 10, p. 98]. To be shown is that $\sum_k \binom{n}{k} F_{m+k}$ is always a Fibonacci number. First make clear what the sum ranges over: If $0 \leq k \leq n$, then $\binom{n}{k}$ is non-zero; if $k < 0$, the binomial coefficient is undefined and if $k > n$, then $\binom{n}{k} = 0$, so one might try to show that $\sum_{k=0}^n \binom{n}{k} F_{m+k}$ is always a Fibonacci number.

One approach is to guess what Fibonacci number is obtained from this sum, and then prove this by induction. It seems natural to ask "Does one have to explicitly find which Fibonacci number is arrived at?" The answer is "perhaps not", however, I couldn't see any other way; this exercise might be possible to solve without answering this question.

With $n = 1$, the expression $\sum_{k=0}^n \binom{n}{k} F_{m+k}$ is $F_m + F_{m+1} = F_{m+2}$. With $n = 2$,

$$
\begin{aligned}
\binom{2}{0} F_m + \binom{2}{1} F_{m+1} + \binom{2}{2} F_{m+2} &= F_m + 2F_{m+1} + F_{m+2} \\
&= (F_m + F_{m+1}) + (F_{m+1} + F_{m+2}) \\
&= F_{m+2} + F_{m+3} \\
&= F_{m+4}.
\end{aligned}
$$

Similarly (with quite a few more lines of computation) with $n = 3$, one arrives at F_{m+6}. One might then conjecture that

$$
S(m, n): \quad \sum_{k=0}^n F_{m+k} = F_{m+2n}
$$

is the rule, as it holds for any $m \geq 0$ and $n = 1, 2, 3$. Inducting on n, a bit of a trick is involved. As is often the case with identities for Fibonacci numbers, two statements are needed in the inductive hypothesis, but rather than fixing an m and proving $S(m, n-1) \wedge S(m, n) \to S(m, n+1)$, use $S(m, n)$ and $S(m+1, n)$ to get $S(m, n+1)$:

BASE STEP: For all m, the statements $S(m, 1)$ and $S(m, 2)$ are true as verified above.

INDUCTION STEP: Let $p \geq 2$ and let $m \geq 1$ be arbitrary, but fixed. Assume that both $S(m, p)$ and $S(m+1, p)$ are true, that is, assume that $\sum_{k=0}^n \binom{n}{p} F_{m+p} = F_{m+2p}$

and $\sum_{k=0}^{p} \binom{p}{k} F_{m+1+p} = F_{m+1+2p}$. To show that $S(m, p+1)$ follows,

$$
\begin{aligned}
\sum_{i=0}^{p+1} \binom{p+1}{k} F_{m+i} &= F_m + \left(\sum_{i=1}^{p} \binom{p+1}{i} F_{m+i} \right) + F_{m+1+p} \\
&= F_m + \left(\sum_{i=1}^{p} \left[\binom{p}{i} + \binom{p}{i-1} \right] F_{m+i} \right) + F_{m+1+p} \\
&= F_m + \left(\sum_{i=1}^{p} \binom{p}{i} F_{m+i} \right) + \left(\sum_{i=1}^{p} \binom{p}{i-1} F_{m+i} \right) + F_{m+1+p} \\
&= \sum_{i=0}^{p} \binom{p}{i} F_{m+i} + \left(\sum_{j=0}^{p-1} \binom{p}{j} F_{m+j+1} \right) + F_{m+1+p} \\
&= \sum_{i=0}^{p} \binom{p+1}{i} F_{m+i} + \sum_{j=0}^{p} F_{m+1+j} \\
&= F_{m+2p} + F_{m+1+2p} \quad \text{(by } S(m,n) \text{ and } S(m+1,n)) \\
&= F_{m+2p+2},
\end{aligned}
$$

which proves $S(m, p+1)$, completing the inductive step $S(m,p) \wedge S(m+1,n) \rightarrow S(m,p+1)$.

Thus, by mathematical induction (inducting on n), for every $m \geq 0$ and $n \geq 1$, the sum $\sum_{k=0}^{n} \binom{n}{k} F_{m+k}$ is a Fibonacci number, in particular, F_{m+2n}. □

Exercise 382: For each $n = 1, 2, 3, \ldots$ let $S(n)$ be the statement that the net resistance R_n between points A and B in $FC(n)$ is

$$
R_n = \frac{F_{2n+1}}{F_{2n}}.
$$

BASE STEP: Since $FC(n)$ has only two resistors in series, $R_1 = 2 = \frac{2}{1} = \frac{F_3}{F_2}$, so $S(1)$ is true.

INDUCTIVE STEP: Fix some $k \geq 1$ and assume that $S(k)$ is true, namely that $R_k = \frac{F_{2k+1}}{F_{2k}}$. To prove $S(k+1)$, one must show that

$$
R_{k+1} = \frac{F_{2k+3}}{F_{2k+2}}.
$$

Let $FC(k)$ be drawn (or fixed) with points A and B as per the diagram in the question. Form the circuit $FC(k+1)$ by adding two resistors *on the left* of $FC(k+1)$, one resistor joining A and B, the other to the left of A, and extending the ground wire to the left, thereby determining new measuring points A' and B' with net resistance between A' and B' being $R(k+1)$.

The resistance now between A and B is (by Kirchoff's law)

$$\frac{1 \cdot R_k}{1 + R_k}.$$

Together with the new resistor on the left, the net resistance between A' and B' is

$$
\begin{aligned}
R_{k+1} &= 1 + \frac{1 \cdot R_k}{1 + R_k} \\
&= 1 + \frac{\frac{F_{2k+1}}{F_{2k}}}{1 + \frac{F_{2k+1}}{F_{2k}}} \\
&= 2 - \frac{1}{1 + \frac{F_{2k}}{F_{2k+1}}} \\
&= 2 - \frac{F_{2k}}{F_{2k} + F_{2k+1}} \\
&= 2 - \frac{F_{2k}}{F_{2k+2}} \\
&= \frac{2F_{2k+2} - F_{2k}}{F_{2k+2}} \\
&= \frac{F_{2k+3}}{F_{2k+2}}.
\end{aligned}
$$

which agrees with $S(k+1)$, completing the inductive step.

By mathematical induction, for all n, the statement $S(n)$ is true. $\qquad\square$

Exercise 384: See [328, Lemma 34.1, pp. 408–409] for a proof [which I have not seen] and [410] for related discussion.

30.3 Solutions: Lucas numbers

Exercise 385: For every $n \geq 1$, let $S(n)$ be the statement that $L_n = F_{n-1} + F_{n+1}$.

BASE STEP: Since $L_1 = 1 = 0 + 1 + F_0 + F_2$ and $L_3 = 3 = 1 + 2 = F_1 + F_3$, $S(1)$ and $S(2)$ hold.

INDUCTIVE STEP: Let $k \geq 2$ and suppose that both $S(k-1)$ and $S(k)$ hold. To show that $S(k+1)$ holds,

$$
\begin{aligned}
L_{k+1} &= L_{k-1} + L_k && \text{(by definition of } L_{k+1}) \\
&= F_{k-2} + F_k + F_{k-1} + F_{k+1} && \text{(by } S(k-1) \text{ and } S(k)) \\
&= (F_{k-2} + F_{k-1}) + (F_k + F_{k+1})
\end{aligned}
$$

$$= F_k + F_{k+2}.$$

Thus, $S(k+1)$ holds, completing the inductive step.

By mathematical induction, for each $n \geq 1$, the statement $S(n)$ is true. $\qquad\square$

Exercise 386: For every $n \geq 0$, let $S(n)$ be the statement that $F_n + L_n = 2F_{n+1}$.

BASE STEP: Since $F_0 + L_0 = 0 + 2 = 2 \cdot 1 = 2F_1$ and $F_1 + L_1 = 1 + 1 = 2 \cdot 1 = 2F_1$, $S(0)$ and $S(1)$ both hold.

INDUCTIVE STEP: For some fixed $k \geq 2$, suppose that both $S(k-1)$ and $S(k)$ holds. To show that $S(k+1)$ holds,

$$
\begin{aligned}
F_{k+1} + L_{k+1} &= F_k + F_{k-1} + L_k + L_{k-1} && \text{(by the recursive definitions)} \\
&= F_k + L_k + F_{k-1} + L_{k-1} \\
&= 2F_{k+1} + 2F_k && \text{(by } S(k-1) \text{ and } S(k)) \\
&= 2(F_{k+1} + F_k) \\
&= 2F_{k+2}.
\end{aligned}
$$

Thus, $S(k+1)$ holds, completing the inductive step.

By MI, for each $n \geq 0$, $S(n)$ holds. $\qquad\square$

Note: This result also has a non-inductive proof using Exercise 385: For $n \geq 1$, $F_n + L_n = F_n + F_{n-1} + F_{n+1} = F_{n+1} + F_{n+1} = 2F_{n+1}$. $\qquad\square$

Exercise 387: For every $n \geq 1$, let $S(n)$ be the statement that

$$L_{n-1}L_{n+1} - L_n^2 = 5(-1)^{n+1}.$$

BASE STEP: Since $L_0 L_2 - L_1^2 = 2 \cdot 3 - 1^2 = 5 = 5(-1)^{1+1}$, $S(1)$ holds.

INDUCTIVE STEP: Let $k \geq 1$ and assume that $S(k)$ holds. Then

$$
\begin{aligned}
L_k L_{k+2} - L_{k+1}^2 &= L_k L_{k+2} - L_k L_{k+1} + L_k L_{k+1} - L_{k+1}^2 \\
&= L_k(L_{k+2} - L_{k+1}) + L_{k+1}(L_k - L_{k+1}) \\
&= L_k(L_{k+1} + L_k - L_{k+1}) + L_{k+1}(L_k - L_k - L_{k-1}) \\
&= L_k L_k - L_{k+1}L_{k-1} \\
&= (-1)(L_{k+1}L_{k-1} - L_k^2) \\
&= (-1)5(-1)^{k+1} && \text{(by } S(k)) \\
&= 5(-1)^{k+2}.
\end{aligned}
$$

Thus, $S(k+1)$ holds, completing the inductive step.

By mathematical induction, for each $n \geq 1$, $S(n)$ is true. $\qquad\square$

Exercise 388: For $n \geq 1$, let $S(n)$ be the statement that $\sum_{i=1}^{n} L_i = L_{n+2} - 3$.

BASE STEP: Since $L_1 = 2 = 5 - 3 = L_3 - 3$, $S(1)$ holds.

INDUCTIVE STEP: Suppose that for some $k \geq 1$, $S(k)$ holds. Then,

$$\sum_{i=1}^{k+1} L_i = \sum_{i=1}^{k} L_i + L_{k+1}$$
$$= L_{k+2} - 3 + L_{k+1} \qquad \text{(by } S(k)\text{)}$$
$$= L_{k+3} - 3,$$

and hence $S(k+1)$ holds, completing the inductive step.

By mathematical induction, for each $n \geq 1$, the equality $S(n)$ holds. $\qquad\square$

Exercise 389: For $n \geq 1$, let $I(n)$ be the identity $L_n + 2L_{n+1} = L_{n+3}$.

BASE STEP: Since $L_1 + 2L_2 = 1 + 2 \cdot 3 = 7 = L_4$ and $L_2 + 2L_3 = 3 + 2 \cdot 4 = 11 = L_5$, both $I(1)$ and $I(2)$ hold.

INDUCTIVE STEP: Let $k \geq 2$ and suppose that both $I(k-1)$ and $I(k)$ hold. Then,

$$L_{k+1} + 2L_{k+2} = L_k + L_{k-1} + 2(L_{k+1} + L_k) \qquad \text{(by the recursive definition)}$$
$$= L_k + 2L_{k+1} + L_{k-1} + 2L_k$$
$$= L_{k+3} + L_{k+2} \qquad \text{(by } I(k) \text{ and } I(k-1)\text{)}$$
$$= L_{k+4},$$

and so $I(k+1)$ holds. The inductive step $[I(k-1) \wedge I(k)] \rightarrow I(k+1)$ is complete.

By mathematical induction, for each $n \geq 1$, $I(n)$ is correct. $\qquad\square$

Exercise 390: For $n \geq 1$, let $S(n)$ be the statement that $\sum_{i=1}^{n} L_i^2 = L_n L_{n+1} - 2$.

BASE STEP: Since $L_1^2 = 1^2 = 1 \cdot 3 - 2 = L_1 L_2 - 2$, the statement $S(1)$ holds.

INDUCTIVE STEP: Fix $k \geq 1$ and assume that $S(k)$ holds. Then,

$$\sum_{i=1}^{k+1} L_i^2 = \sum_{i=1}^{k} L_i^2 + L_{k+1}^2$$
$$= L_k L_{k+1} - 2 + L_{k+1}^2 \qquad \text{(by } S(k)\text{)}$$
$$= L_{k+1}(L_{k+1} + L_k) - 2$$
$$= L_{k+1} L_{k+2} - 2,$$

proving that $S(k+1)$ holds, completing the inductive step.

By the principle of mathematical induction, for each $n \geq 1$, $S(n)$ is true. $\qquad\square$

Exercise 391: For $n \geq 1$, let $A(n)$ be the assertion that

$$2 \sum_{i=0}^{n} L_{3i} = 1 + L_{3n+2}.$$

BASE STEP: Since $2(L_0 + L_3) = 2(2 + 4) = 12 = 1 + 11 = 1 + L_5$, $A(1)$ holds.

INDUCTIVE STEP: Let $k \geq 1$ and assume that $A(k)$ holds. Then

$$\begin{aligned}
2 \sum_{i=1}^{k+1} L_{3i} &= 2 \sum_{i=1}^{k} L_{3i} + 2L_{3k+3} \\
&= 1 + L_{3k+2} + L_{3k+3} + L_{3k+3} && \text{(by } A(k)) \\
&= 1 + L_{3k+4} + L_{3k+3} \\
&= 1 + L_{3k+5} = 1 + L_{3(k+1)+2},
\end{aligned}$$

and therefore $A(k+1)$ follows, completing the inductive step.

By MI, for each $n \geq 1$, the assertion $A(n)$ is true. $\qquad\square$

Exercise 392: For each $n \geq 1$, let $S(n)$ be the statement that $5F_{n+2} = L_{n+4} - L_n$.

BASE STEP: Since $5F_3 = 5 \cdot 2 = 10 = 11 - 1 = L_5 - L_1$ and $5F_4 = 5 \cdot 3 = 15 = 18 - 3 = L_6 - L_2$, both $S(1)$ and $S(2)$ hold.

INDUCTIVE STEP: Fix $k \geq 2$, and assume that both $S(k-1)$ and $S(k)$ hold. Then, starting with the right side of $S(k+1)$,

$$\begin{aligned}
L_{k+5} - L_{k+1} &= (L_{k+4} + L_{k+3}) - (L_k + L_{k-1}) \\
&= (L_{k+4} - L_k) + (L_{k+3} - L_{k-1}) \\
&= 5F_{k+2} + 5F_{k+1} && \text{(by } S(k-1) \text{ and } S(k)) \\
&= 5(F_{k+2} + F_{k+1}) \\
&= 5F_{k+3},
\end{aligned}$$

which is the left side of $S(k+1)$, completing the inductive step.

By mathematical induction, for each $n \geq 1$, the equality $S(n)$ holds. $\qquad\square$

30.4 Solutions: Harmonic numbers

Exercise 395: To see that the harmonic series diverges, it suffices to show that for any $n \in \mathbb{Z}^+$, there exists an m so that $H_m \geq n$. To see this, partition the series into groups, each group totalling at least $1/2$ as follows:

$$1 + \frac{1}{2} + \underbrace{\frac{1}{3} + \frac{1}{4}}_{\geq 1/2} + \underbrace{\frac{1}{5} + \cdots + \frac{1}{8}}_{\geq 1/2} + \underbrace{\frac{1}{9} + \cdots + \frac{1}{16}}_{\geq 1/2} + \cdots \,;$$

the number of groups can be made as large as necessary. To formalize this idea, for each $k \geq 1$,

$$\sum_{i=2^k+1}^{2^{k+1}} \frac{1}{i} \geq \sum_{i=2^k+1}^{2^{k+1}} \frac{1}{2^{k+1}} = (2^{k+1} - 2^k) \frac{1}{2^{k+1}} = \frac{1}{2}.$$

Claim: for each $p \geq 0$,

$$S(p): \quad H_{2^p} \geq 1 + \frac{p}{2}.$$

Proving this claim proves what is desired in this exercise, since for every n, choose p so large that $1 + \frac{p}{2} \geq n$, and then use $m = 2^p$. Here is an inductive proof of the claim:

BASE STEP: When $p = 0$, $H_1 \geq 1$ confirms that $S(0)$ is true.

INDUCTION STEP: Fix some $\ell \geq 0$, and assume that $S(\ell)$ is true. To show that $S(\ell + 1)$ follows,

$$
\begin{aligned}
H_{2^{\ell+1}} &= H_{2^\ell} + \sum_{i=2^\ell+1}^{2^{\ell+1}} \frac{1}{i} \\
&\geq H_{2^\ell} + \frac{1}{2} \quad \text{(by fact above)} \\
&\geq 1 + \frac{\ell}{2} + \frac{1}{2} \quad \text{(by } S(\ell)\text{)} \\
&= 1 + \frac{\ell + 1}{2},
\end{aligned}
$$

which completes the inductive step.

By mathematical induction, for every $p \geq 0$, the statement $S(p)$ holds true. $\quad\square$

So, for every n, there exists an m so that $H_m \geq n$. In other words, $\lim_{m \to \infty} H_m = \infty$.

Exercise 396: The following identity simplifies the inductive step below:

Lemma 30.4.1. *For any positive integer n,*

$$\sum_{i=1}^{n} \frac{(-1)^{i-1}}{i} \binom{n}{i-1} = \frac{1}{n+1}. \qquad (30.1)$$

Proof: Expanding $(1 - 1)^{n+1}$ using the binomial theorem,

$$
\begin{aligned}
0 &= (1 - 1)^{n+1} \\
&= \sum_{i=0}^{n+1} (-1)^i \binom{n+1}{i}
\end{aligned}
$$

$$= 1 + \sum_{i=1}^{n+1} (-1)^i \binom{n+1}{i}$$

$$= 1 + \sum_{i=1}^{n+1} (-1)^i \frac{n+1}{i} \binom{n}{i-1} \quad \text{(by eq'n (9.4))}$$

$$= 1 + (n+1) \sum_{i=1}^{n+1} \frac{(-1)^i}{i} \binom{n}{i-1},$$

so

$$1 = (n+1) \sum_{i=1}^{n+1} \frac{(-1)^{i-1}}{i} \binom{n}{i-1},$$

which gives (30.1).   □

For $n \geq 1$, consider the statement

$$S(n): \quad H_n = \sum_{i=1}^{n} \frac{(-1)^{i-1}}{i} \binom{n}{i}$$

BASE STEP: $S(1)$ says that $H_1 = \frac{(-1)^0}{1} \binom{1}{1}$, which is correct.

INDUCTIVE STEP: Fix some $k \geq 1$ and assume that

$$S(k): \quad H_k = \sum_{i=1}^{k} \frac{(-1)^{i-1}}{i} \binom{k}{i}$$

holds. To complete the induction step, it suffices to show

$$S(k+1): \quad H_{k+1} = \sum_{i=1}^{k+1} \frac{(-1)^{i-1}}{i} \binom{k+1}{i}.$$

Starting with the right-hand side of $S(k+1)$,

$$\sum_{i=1}^{k+1} \frac{(-1)^{i-1}}{i} \binom{k+1}{i}$$

$$= \left[\sum_{i=1}^{k} \frac{(-1)^{i-1}}{i} \binom{k+1}{i} \right] + \frac{(-1)^k}{k+1}$$

$$= \left[\sum_{i=1}^{k} \frac{(-1)^{i-1}}{i} \left(\binom{k}{i} + \binom{k}{i-1} \right) \right] + \frac{(-1)^k}{k+1} \qquad \text{(Pascal's id.)}$$

$$= \left[\sum_{i=1}^{k} \frac{(-1)^{i-1}}{i} \binom{k}{i} \right] + \left[\sum_{i=1}^{k} \frac{(-1)^{i-1}}{i} \binom{k}{i-1} \right] + \frac{(-1)^k}{k+1}$$

$$= H_k + \left[\sum_{i=1}^{k} \frac{(-1)^{i-1}}{i} \binom{k}{i-1} \right] + \frac{(-1)^k}{k+1} \qquad \text{(by } S(k)\text{)}$$

$$= H_k + \sum_{i=1}^{k+1} \frac{(-1)^{i-1}}{i} \binom{k}{i-1}$$

$$= H_k + \frac{1}{k+1} \qquad \text{(by (30.1))}$$

$$= H_{k+1},$$

which is the left side of $S(k+1)$, completing the inductive step $S(k) \to S(k+1)$.

By MI, for each $n \geq 1$, $S(n)$ is true. $\qquad \square$

Exercise 397: For each $n \geq 1$, denote the assertion in the exercise by

$$A(n): \quad H_1 + H_2 + \cdots + H_n = (n+1)H_n - n.$$

BASE STEP: Since $A(1)$ says $H_1 = 2(H_1 - 1)$; this is easily verified.

INDUCTIVE STEP: Suppose that for some $\ell \geq 1$, $A(\ell)$ is true. Then

$$
\begin{aligned}
\sum_{i=1}^{\ell+1} H_i &= (\ell+1)H_\ell - \ell + H_{\ell+1} \quad \text{(by } A(\ell)\text{)} \\
&= (\ell+1)\left[H_{\ell+1} - \frac{1}{\ell+1} \right] - \ell + H_{\ell+1} \\
&= (\ell+2)H_{\ell+1} - (\ell+1)
\end{aligned}
$$

shows that $A(\ell+1)$ is true also. This completes the inductive step.

So by MI, for every $n \geq 1$, the statement $A(n)$ holds. $\qquad \square$

Exercise 398: (Outline) This appeared as an exercise in [6, p. 99, 12(iv)]. For $n \geq 1$, denote the statement in the exercise by

$$S(n): \quad 1 + \frac{n}{2} \leq H_{2^n} \leq 1 + n.$$

Since the left-hand inequality in $S(n)$ was proved in Exercise 395 (by induction), it remains to prove the right-hand inequality.

Hint: In Exercise 395, one relied on the inequality

$$\sum_{i=2^k+1}^{2^{k+1}} \frac{1}{i} \geq \frac{1}{2}.$$

The remaining inequality in this exercise is based on a similar trick, something like

$$\sum_{i=2^k+1}^{2^{k+1}} \frac{1}{i} \leq 1.$$

Exercise 399: To be shown that for $n \geq 1$,

$$S(n): \quad \sum_{k=1}^{n} kH_k = \frac{n(n+1)}{2}H_{n+1} - \frac{n(n+1)}{4}.$$

BASE STEP: $S(1)$ says $H_1 = \frac{1 \cdot 2}{2}H_2 - \frac{1 \cdot 2}{4}$, that is, $1 = 1(3/2) - (1/2)$, which is correct.

INDUCTIVE STEP: Fix $m \geq 1$, and suppose that

$$S(m): \quad \sum_{k=1}^{m} kH_k = \frac{m(m+1)}{2}H_{m+1} - \frac{m(m+1)}{4}$$

is true. It remains to show that

$$S(m+1): \quad \sum_{k=1}^{m+1} kH_k = \frac{(m+1)(m+2)}{2}H_{m+2} - \frac{(m+1)(m+2)}{4}$$

follows. Beginning with the left side of $S(m+1)$,

$$
\begin{aligned}
\sum_{k=1}^{m+1} kH_k &= \sum_{k=1}^{m} kH_k + (m+1)H_{m+1} \\
&= \frac{m(m+1)}{2}H_{m+1} - \frac{m(m+1)}{4} + (m+1)H_{m+1} \quad \text{(by IH)} \\
&= \frac{(m+2)(m+1)}{2}H_{m+1} - \frac{m(m+1)}{4} \\
&= \frac{(m+1)(m+2)}{2}\left(H_{m+2} - \frac{1}{m+2}\right) - \frac{m(m+1)}{4} \\
&= \frac{(m+1)(m+2)}{2}H_{m+2} - \frac{m+1}{2} - \frac{m(m+1)}{4} \\
&= \frac{(m+1)(m+2)}{2}H_{m+2} - \frac{2(m+1) + m(m+1)}{4},
\end{aligned}
$$

which is indeed equal to the right side of $S(m+1)$. This concludes the inductive step $S(m) \to S(m+1)$.

By mathematical induction, for each $n \geq 1$, the statement $S(n)$ holds. $\square$

Exercise 400: This has appeared in many places, for example, [350, Ex. 66]. For $n \geq 2$, denote the statement in the exercise by

$$S(n): \quad \sum_{i=2}^{n} \frac{1}{i(i-1)} H_i = 2 - \frac{1}{n+1} - \frac{H_{n+1}}{n}.$$

BASE STEP: $S(2)$ says $\frac{1}{2}H_2 = 2 - \frac{H_3}{2} - \frac{1}{3}$, and since each side works out to $3/4$, $S(2)$ is true.

INDUCTIVE STEP: Let $k \geq 2$, and suppose that $S(k)$ is true. Then

$$
\begin{aligned}
\sum_{i=2}^{k+1} \frac{1}{(i-1)i} H_i &= \left(\sum_{i=2}^{k} \frac{1}{(i-1)i} H_i \right) + \frac{1}{k(k+1)} H_{k+1} \\
&= 2 - \frac{H_{k+1}}{k} - \frac{1}{k+1} + \frac{1}{k(k+1)} H_{k+1} \quad \text{(by IH)} \\
&= 2 - \left(\frac{1}{k} - \frac{1}{k(k+1)} \right) H_{k+1} - \frac{1}{k+1} \\
&= 2 - \frac{1}{k+1} H_{k+1} - \frac{1}{k+1} \\
&= 2 - \frac{1}{k+1} (H_{k+2} - \frac{1}{k+2}) - \frac{1}{k+1} \\
&= 2 - \frac{H_{k+2}}{k+1} + \frac{1}{(k+1)(k+2)} - \frac{1}{k+1} \\
&= 2 - \frac{H_{k+2}}{k+1} - \frac{1}{k+2}
\end{aligned}
$$

shows that $S(k+1)$ is also true, completing the inductive step.

By MI, for each $n \geq 2$, the statement $S(n)$ is true. □

Exercise 401: This has appeared as an exercise in [6, p. 99, 12(iii)] and [350, Ex. 65]. For $n \geq 1$ let $S(n)$ denote the statement that for any m satisfying $0 \leq m \leq n$,

$$\sum_{i=1}^{n} \binom{i}{m} H_i = \binom{n+1}{m+1} \left[H_{n+1} - \frac{1}{m+1} \right]. \tag{30.2}$$

Note that when $m > n$, both sides of (30.2) are zero, so from $S(n)$, the stronger statement that for all $m \geq 0$, equation (30.2) follows.

BASE STEP ($n = 1$): $S(1,0)$ says $\binom{1}{0} H_1 = \binom{2}{1}(H_2 - 1)$, which is correct and $S(1,1)$ says $\binom{1}{1} H_1 = \binom{2}{2}(H_2 - 1/2)$, which is again correct.

INDUCTIVE STEP: Fix $k \geq 1$, and suppose that $S(k)$ is correct, that is, for all $m \geq 0$, the statement

$$S(k,m): \quad \sum_{i=1}^{k} \binom{i}{m} H_i = \binom{k+1}{m+1} \left[H_{k+1} - \frac{1}{m+1} \right].$$

is true. To show $S(k+1)$, one must show that for any $p \geq 0$, the statement

$$S(k+1, p): \quad \sum_{i=1}^{k+1} \binom{i}{p} H_i = \binom{k+2}{p+1}\left[H_{k+2} - \frac{1}{p+1}\right]$$

follows. Fix p. In the sixth equality below, the identity $\binom{k+2}{p+1} = \frac{k+2}{p+1}\binom{k+1}{p}$ is used:

$$\sum_{i=1}^{k+1} \binom{i}{p} H_i$$

$$= \sum_{i=1}^{k} \binom{i}{p} H_i + \binom{k+1}{p} H_{k+1}$$

$$= \binom{k+1}{p+1}\left[H_{k+1} - \frac{1}{p+1}\right] + \binom{k+1}{p} H_{k+1} \quad \text{(by } S(k,p)\text{)}$$

$$= \binom{k+2}{p+1} H_{k+1} - \frac{1}{p+1}\binom{k+1}{p+1} \quad \text{(by Pascal's id.)}$$

$$= \binom{k+2}{p+1}\left[H_{k+2} - \frac{1}{k+2}\right] - \frac{1}{p+1}\binom{k+1}{p+1}$$

$$= \binom{k+2}{p+1} H_{k+2} - \frac{1}{k+2}\binom{k+2}{p+1} - \frac{1}{p+1}\binom{k+1}{p+1}$$

$$= \binom{k+2}{p+1} H_{k+2} - \frac{1}{p+1}\binom{k+1}{p} - \frac{1}{p+1}\binom{k+1}{p} H_{k+1}$$

$$= \binom{k+2}{p+1} H_{k+2} - \frac{1}{p+1}\left[\binom{k+1}{p} + \binom{k+1}{p+1}\right]$$

$$= \binom{k+2}{p+1} H_{k+2} - \frac{1}{p+1}\binom{k+2}{p+1} \quad \text{(by Pascal's id.)}$$

$$= \binom{k+2}{p+1}\left[H_{k+2} - \frac{1}{p+1}\right].$$

Thus $S(k+1, p)$ is also true, and so $S(k+1)$ holds, completing the inductive step.

By mathematical induction on n, for every $n \geq 1$, and all $m \geq 0$, the statement $S(m, n)$ is true.   □

30.5   Solutions: Catalan numbers

Only one solution to one of the classic exercises is given here. See [247] for these exercises, among a number of other examples regarding Catalan numbers and many references. Most combinatorics books seem to have solutions to one or all of the exercises in this section. The exercises given here either rely on being able to see the number of items being counted as an expression involving binomial coefficients, or on some kind of parenthesizing, or on Segner's formula and strong induction, as in the next solution:

Exercise 405: (Outline of inductive step) Let $S(n)$ be the claim that an $(n+2)$-gon can be triangulated in C_n ways. The proof is by strong induction on n.

Fix $m \geq 3$, and suppose that each of $S(3), \ldots, S(m)$ holds. Let P be a convex polygon with $m + 3$ vertices $x, y, p_1 \ldots, p_{m+1}$ in cyclic order, and for some $i = 1, \ldots, m + 1$, consider the triangle $T_i = \triangle xyp_i$.

If either $i = 1$ or $i = m + 1$, T_i lies on the outside of P, leaving a polygon on $m + 2$ vertices to be further triangulated; if $1 < i < m + 1$, T_i breaks P into two polygons, P_1 on vertices $y, p_1, p_2, \ldots, p_i$, and P_2 with vertices $y, p_i, p_{i+1}, \ldots, p_{m+1}$. So P_1 has $i + 1$ vertices and P_2 has $m + 3 - i$ vertices. By induction hypothesis, P_1 can be triangulated in C_{i-1} ways, and P_2 can be triangulated in C_{m+1-i} ways. In all, there are $C_{i-1}C_{m+1-i}$ ways to triangulate P that use the triangle T_i. So counting over all T_i, there are

$$\sum_{i=1}^{m+1} C_{i-1}C_{m+1-i} = \sum_{j=0}^{m} C_jC_{m-j}$$

possible ways to triangulate P. By Segner's recursion (12.2), this number is C_{m+1}, which proves $S(m + 1)$. $\qquad\square$

30.6   Solutions: Eulerian numbers

Exercise 407: The equation related to (12.5) is precisely (12.6), the one given in the exercise, which also follows from equation (12.4), by replacing k with $k - 1$. For any $1 \leq k \leq m$, let $S(m, k)$ denote the statement that the number $E_{m,k}$ derived from the recursion agrees with the definition. To prove that all $S(m, k)$ hold is done by simply inducting on m:

BASE STEP: When $m = 1$, there is only one possible value for k, namely $k = 1$, and $E_{1,1} = 1$ by the initial values given, which is correct by the definition (the number of length 1 permutations with 0 ascents is 1).

INDUCTIVE STEP ON m: Suppose that for that for some $m' \geq 2$, that for every $1 \leq \ell \leq m'$, $S(m' - 1, k')$ holds. Then $S(m', 1)$ holds by the initial values given and by the definition as in the base step. For any $1 < k' \leq m - 1$, by the inductive hypothesis, both $S(m' - 1, k' - 1)$ and $S(m' - 1, k')$ are true. But then equation (12.6) is precisely the recursion developed previous to the exercise, so $E_{m',k'}$ derived from the recursion is correct. Finally, $E_{m',m'}$ is correct by the initial values given and the fact that there is only permutation of length m' with $m' - 1$ ascents, namely the strictly increasing one. So for all $1 \leq k' \leq m$, $E_{m',k}$ derived from the recursion is correct, i.e., $S(m', k')$ is true, completing the inductive step.

By mathematical induction, for any $m \geq 1$ (and all $1 \leq k \leq m$), $S(m, k)$ is correct. □

Exercise 408: For each $m \geq 1$, let $P(n)$ denote the proposition that

$$\sum_{j=1}^{m} E_{m,j} = m!.$$

BASE STEP: $\sum_{j=1}^{1} E_{1,j} = E_{1,1} = 1 = 1!$, so $P(1)$ is true.

INDUCTION STEP: Fix $n \geq 1$ and assume $P(n)$ is true. Then

$$\sum_{k=1}^{n+1} E_{n+1,k} = 2 + \sum_{k=2}^{n} E_{n+1,k}$$

$$= 2 + \sum_{k=2}^{n} [(m+1-k+1)E_{n,k-1} + kE_{n,k}] \qquad \text{(by (12.6))}$$

$$= 2 + (n+1) \sum_{k=2}^{n} E_{n,k-1} + \sum_{k=2}^{n} kE_{n,k} - \sum_{k=2}^{n} (k-1)E_{n,k-1}$$

$$= 2 + (n+1) \left(-1 + \sum_{k=2}^{n+1} E_{n,k-1} \right) + \sum_{k=2}^{n} kE_{n,k} - \sum_{k=1}^{n-1} kE_{n,k}$$

$$= 2 + (n+1)(-1+n!) + \sum_{k=2}^{n} kE_{n,k} - \sum_{k=1}^{n-1} kE_{n,k} \qquad \text{(by $P(n)$)}$$

$$= 1 - n + (n+1)! + nE_{n,n} - E_{1,1}$$

$$= 1 - n + (n+1)! + n - 1$$

$$= (n+1)!$$

as desired, showing $P(n+1)$ is true.

By mathematical induction on m, for every $m \geq 1$, $P(m)$ holds. □

30.7 Solutions: Euler numbers

Exercise 411: By definition, $E_0 = 1$ (one might think of the "empty permutation" as vacuously satisfying the requirements for being alternating). There is only one permutation (1) on $\{1\}$, which is also trivially alternating, so $E_1 = 1$.

For $n \geq 1$, it remains to show

$$E_{n+1} = \frac{1}{2} \sum_{i=0}^{n} \binom{n}{i} E_i E_{n-i}.$$

Checking $n = 1$,

$$E_2 = \frac{1}{2}\left(\binom{1}{0}E_0E_1 + \binom{1}{1}E_1E_0\right) = \frac{1}{2}(1+1) = 1,$$

which is correct.

Some preliminary observations make the proof simple. By comments previous to the exercise, the number of up-alternating permutations and the number of down-alternating permutations is the same (by looking at the complement of a permutation). If an up-alternating permutation is reversed, an alternating permutation ending in a descent is created; similarly, a permutation ending in a descent, when reversed, begins with an ascent. So the number of alternating permutations on a k-set that end in a descent is also E_k. Note also that for any subset $J \subset \{1, \ldots, n\}$, with $|J| = j$ elements, there are as many up-alternating (or down) permutations on J as there are on $\{1, \ldots, j\}$. With these facts in hand, the proof can begin:

Consider a permutation σ on $\{1, 2, \ldots, n\}$ and for some $i = 0, \ldots, n$, let σ' be the permutation formed by inserting $n+1$ in between positions i and $i+1$ (between 0 and 1 is the position previous to $\sigma(1)$, and between n and $n+1$ is after the last entry). In order for σ' to be alternating, both $\tau_1 = (\sigma(1), \ldots, \sigma(i))$ and $\tau_2 = (\sigma(i+1), \ldots, \sigma(n))$ must also be alternating, where τ_1 ends in a descent and τ_2 begins with an ascent. There are E_i ways to have τ_1 as a permutation on $\{\sigma(1), \ldots, \sigma(i)\}$; similarly, there are E_{n-i} ways to have τ_2 on the remaining $n - i$ values. Since a subset $\{\sigma(1), \ldots, \sigma(i)\}$ can be chosen $\binom{n}{i}$ ways, there are $\sum_{i=0}^{n} \binom{n}{i} E_i E_{n-i}$ ways to create *all* alternating permutations of length $n+1$, and the number of up-alternating permutations is half of this. $\qquad\square$

30.8   Solutions: Stirling numbers

Exercise 412: The proof is by induction on n. For each $n \geq 1$, denote the proposition in the exercise by

$$P(n): \quad x^n = \sum_{k=1}^{n} S_{n,k} x(x-1)\cdots(x-k+1).$$

BASE STEP: When $n = 1$, the sum $\sum_{k=1}^{n} S_{n,k} x(x-1)\cdots(x-k+1)$ is simply $S_{1,1}x = x = x^1$, so $P(1)$ is true.

INDUCTIVE STEP: Fix some $m \geq 2$ and assume that $P(m-1)$ is true. The proof of $P(m)$ begins on the right side [for clarity, not all terms are shown]:

$$\sum_{k=1}^{m} S_{m,k} x(x-1)\cdots(x-k+1)$$

$$= \sum_{k=1}^{m}(S_{m-1,k-1}+kS_{m-1,k})x(x-1)\cdots(x-k+1) \quad \text{(by (12.7))}$$

$$= (S_{m-1,0}+1\cdot S_{m-1,1})x+(S_{m-1,1}+2S_{m-1,2})x(x-1)$$
$$\quad +(S_{m-1,2}+3S_{m-1,3})x(x-1)(x-2)+\cdots$$
$$\quad +(S_{m-1,m-1}+mS_{m-1,m})x(x-1)\cdots(x-n+1)$$
$$= S_{m-1,1}(x+x(x-1))+S_{m-1,2}(2x(x-1)+x(x-1)(x-2))$$
$$\quad +S_{m-1,3}(3x(x-1)(x-2)+x(x-1)(x-2)(x-3))+\cdots$$
$$= S_{m-1,1}x^2+S_{m-1,2}x^2(x-1)+S_{m-1,3}x^2(x-1)(x-2)+\cdots$$
$$= x[S_{m-1,1}x+S_{m-1,2}x(x-1)S_{m-1,3}x(x-1)(x-2)+\cdots]$$
$$= x\cdot x^{m-1} \quad \text{(by } P(m-1)),$$

which agrees with the left side of $P(m)$, completing the inductive step.

By mathematical induction, for each $n \geq 1$, $P(n)$ is true. $\qquad\square$

Exercise 413: For positive integers n and k, let $A(n,k)$ be the assertion

$$S_{n,k}=\frac{1}{k!}\sum_{i=1}^{k}(-1)^{k-i}\binom{k}{i}i^n$$

of the exercise. The proof here is by induction on n; the cases where $k=1$ have to be separated, but otherwise, k is arbitrary, and can even be fixed in advance—no induction on k is necessary.

BASE STEP: Since $S_{1,1}=1$, and

$$\frac{1}{1!}\sum_{i=1}^{1}(-1)^{1-i}\binom{1}{i}i^1=1,$$

the assertion $A(1,1)$ holds. For $k\geq 2$, $S_{1,k}=0$, and by Exercise 107 (or equation 9.7),

$$\frac{1}{k!}\sum_{i=1}^{k}(-1)^{k-i}\binom{k}{i}i^1=0,$$

so $A(1,k)$ is true.

INDUCTIVE STEP: Fix $m\geq 2$. Since $S_{m,1}=1$, and

$$\frac{1}{1!}\sum_{i=1}^{1}(-1)^{1-i}\binom{1}{i}i^m=1,$$

$A(m,1)$ holds. So let $k\geq 2$, and assume that both $A(m-1,k-1)$ and $A(m-1,k)$ are true.

Note regarding inductive hypothesis: The base step shows that for *all* k, $A(1, k)$ holds (one does not need to fix a k in advance to be carried into the inductive step). So the inductive hypothesis could have been that for all $\ell \geq 1$, $A(m - 1, \ell)$ holds. However, after fixing $k \geq 2$, the only values of ℓ needed are $\ell = k$ and $\ell = k - 1$.

Then for any $k \geq 2$,

$$
\begin{aligned}
S_{m,k} &= S_{m-1,k-1} + kS_{m-1,k} && \text{(by (12.7))}\\
&= \frac{1}{(k-1)!}\sum_{i=1}^{k-1}(-1)^{k-1-i}\binom{k-1}{i}i^{m-1} + k\frac{1}{k!}\sum_{i=1}^{k}(-1)^{k-i}\binom{k}{i}i^{m-1} && \text{(by IH)}\\
&= \frac{1}{(k-1)!}\left[\sum_{i=1}^{k-1}(-1)^{k-1-i}\binom{k-1}{i}i^{m-1} + \sum_{i=1}^{k}(-1)^{k-i}\binom{k}{i}i^{m-1}\right]\\
&= \frac{1}{(k-1)!}\left[\sum_{i=1}^{k-1}(-1)^{k-1-i}\binom{k-1}{i}i^{m-1} + \sum_{i=1}^{k-1}(-1)^{k-i}\binom{k}{i}i^{m-1} + k^{m-1}\right]\\
&= \frac{1}{(k-1)!}\left[\sum_{i=1}^{k-1}(-1)^{k-i}\left(\binom{k}{i}-\binom{k-1}{i}\right)i^{m-1} + k^{m-1}\right]\\
&= \frac{1}{(k-1)!}\left[\sum_{i=1}^{k-1}(-1)^{k-i}\binom{k-1}{i-1}i^{m-1} + k^{m-1}\right] && \text{(by Pascal's id.)}\\
&= \frac{1}{(k-1)!}\sum_{i=1}^{k}(-1)^{k-i}\frac{i}{k}\binom{k}{i}i^{m-1} && \text{(by Lemma 9.6.1)}\\
&= \frac{1}{k!}\sum_{i=1}^{k}(-1)^{k-i}\binom{k}{i}i^{m}
\end{aligned}
$$

shows that for any $k \geq 2$, $A(m, k)$ is also true, completing the inductive step.

By mathematical induction, for each n and k, $A(n, k)$ is true. $\qquad\square$

Exercise 414: The proof is in [6, p. 91]. For $n \geq 3$, it is not known whether or not $S_{n,k}$ always has a single maximum.

Exercise 415: Here is an outline of a proof as found in [549, p. 126]: For each $k \geq 1$, let $P(k)$ denote the equality

$$
\sum_{n=k}^{\infty} S(n, k)\frac{x^n}{n!} = \frac{1}{k!}(e^x - 1)^k. \tag{30.3}
$$

BASE STEP: When $k = 1$, for each $n \geq 1$, $S(n, 1) = 1$, and so $P(1)$ reads

$$
\sum_{n=1}^{\infty} \frac{x^n}{n!} = e^x - 1,
$$

which is true, so $P(1)$ holds.

INDUCTIVE STEP: For each $k \geq 1$, denote the left side of (30.3) as a function

$$f_k(x) = \sum_{n=k}^{\infty} S(n, k) \frac{x^n}{n!}.$$

Fix $m \geq 2$ and suppose that $P(m-1)$ holds, that is, equation (30.3) with $k = m-1$ holds. Taking derivatives,

$$
\begin{aligned}
f_m'(x) &= \sum_{n=m}^{\infty} S(n, m) \frac{x^{n-1}}{(n-1)!} \\
&= \sum_{n=m}^{\infty} [mS(n-1, m) + S(n-1, m-1)] \frac{x^{n-1}}{(n-1)!} \quad \text{(by (12.7)} \\
&= m \sum_{n=m}^{\infty} S(n-1, m) \frac{x^{n-1}}{(n-1)!} + \sum_{n=m}^{\infty} S(n-1, m-1) \frac{x^{n-1}}{(n-1)!} \\
&= m \sum_{n-1=m}^{\infty} S(n-1, m) \frac{x^{n-1}}{(n-1)!} + \sum_{n-1=m-1}^{\infty} S(n-1, m-1) \frac{x^{n-1}}{(n-1)!} \\
&\quad \text{(since } S(m-1, m) = 0) \\
&= mf_m(x) + f_{m-1}(x) \\
&= mf_m(x) + \sum_{n=m-1}^{\infty} \frac{1}{(m-1)!} (e^x - 1)^{m-1} \quad \text{(by } P(m-1))
\end{aligned}
$$

Solving this differential equation and using the condition $S(m, m) = 1$, arrive at the desired expression for $f_m(x)$. $\qquad \square$

Chapter 31

Solutions: Sets

31.1 Solutions: Properties of sets

Exercise 416: For this simple problem, see [181, p. 209, prob. 7.3.6(c)].

Exercise 419: The reader likely does not need to be reminded that an inductive proof is not the preferred method here; however, it may be an interesting exercise to show that induction can be used to prove the result nevertheless. Let "N-set" be an abbreviation for "set with N elements". A proof is by induction on n the statement $S(n)$: For every k satisfying $0 \leq k \leq n$, the number of distinct k-sets in an n-set is $\binom{n}{k}$.

BASE STEP: For $n = 0$, the only choice [no pun intended] for k is $k = 0$, and $\binom{0}{0} = \frac{0!}{(0-0)!0!} = 1$, and the number of 0-sets in a 0-set is 1 (the empty set is a subset of the empty set). When $n = 1$, there are two possibilities, $k = 0$ or $k = 1$. In each case, there is precisely one k-set in a set consisting of one element. Also, by definition, $\binom{1}{0} = \frac{1!}{0!1!} = 1$ and $\binom{1}{1} = \frac{1!}{1!0!} = 1$, so the statement is true for $n = 1$.

INDUCTION STEP $(S(m) \rightarrow S(m+1))$: For some $m \geq 1$, assume that $S(m)$ is true. Let A be an $(m+1)$-set. To show $S(m+1)$, one needs to prove that for every k satisfying $0 \leq k \leq m+1$, that the number of k-sets in A is $\binom{m+1}{k}$.

When $k = m+1$, there is precisely one k-set in A, namely A itself, and

$$\binom{m+1}{k} = \binom{m+1}{m+1} = \frac{(m+1)!}{0!(m+1)!} = 1.$$

Also, when $k = 0$,

$$\binom{m+1}{0} = \frac{(m+1)!}{(m+1)!0!} = 1,$$

the number of 0-sets in A.

So let k satisfy $1 \leq k \leq m$. The number of k-sets in A can be calculated as follows: Fix some $x \in A$. The number of k-sets in A which contain x is the number

651

of $(k-1)$-sets in $A\backslash\{x\}$, and since $|A\backslash\{x\}| = m$, by the induction hypothesis, there are $\binom{m}{k-1}$ such sets. The number of $(k-1)$-sets which do not contain x is (again by induction hypothesis) $\binom{m}{k}$. Thus, by Pascal's identity, the number of k-sets in A is

$$\binom{m}{k-1} + \binom{m}{k} = \binom{m+1}{k},$$

the desired number to complete the inductive step.

By mathematical induction, for all $n \geq 0$, the statement $S(n)$ is true. $\quad\square$

Exercise 426: (Functions defined with symmetric differences) Fix some set X. Let $A(n)$ be the assertion that for any n subsets of X, say $S_1, S_2, \ldots, S_n$, the set $D(S_1, S_2, \ldots, S_n)$ consists precisely of those elements in X belonging to an odd number of the S_i's. Let $C_X(S_1, \ldots, S_m)$ denote the set of all elements in X contained in exactly an odd number of the sets $S_1, \ldots, S_m$.

BASE STEP: With $n = 1$, there is only one set, S_1, and each element of S_1 occurs in S_1 precisely once, so $C_X(S_1) = S_1$, and by definition, $D(S_1) = S_1$, so these notions agree. [Just as an added check, for $n = 2$, the symmetric difference of two sets contain precisely the elements in one but not both sets, and one is odd.]

INDUCTION STEP: Fix some $k \geq 1$ and assume that $A(k)$ holds. Let $S_1, \ldots, S_{k+1}$ be subsets of X, put $D = D_{k+1}(S_1, \ldots, S_{k+1})$ and put $C = C_X(S_1, \ldots, S_{k+1})$. To complete the induction step, it suffices to show that $C = D$, which is accomplished by showing both $D \subseteq C$ and $C \subseteq D$.

First pick some $x \in D$. By definition, $D = D_k(S_1, \ldots, S_k) \triangle S_{k+1}$, and so either $x \in D_k(S_1, \ldots, S_k)$ or $x \in S_{k+1}$, and not both. If $x \in D_k(S_1, \ldots, S_k)$, then by the induction hypothesis $A(k)$, x is contained in an odd number of the sets $S_1, \ldots, S_k$, and since $x \notin S_{k+1}$, x is still contained in an odd number of the sets $S_1, \ldots, S_k, S_{k+1}$. If $x \in S_{k+1}$ and not in $D_k(S_1, \ldots, S_k)$, then by induction hypothesis $A(k)$, x is in an even number of the sets $S_1, \ldots, S_k$, so together with its membership in S_{k+1} again puts x in an odd number of the sets. In any case, $x \in C$, so $D \subseteq C$.

Pick $x \in C$. If $x \notin S_{k+1}$, then x is contained in an odd number of the sets $S_1, \ldots, S_k$, and so by $A(k)$, $x \in D_k(S_1, \ldots, S_k)$, and hence $x \in D$. If $x \in S_{k+1}$, then x is contained in an even number of the sets $S_1, \ldots, S_k$, and thus by $A(k)$, $x \notin D_k(S_1, \ldots, S_k)$; then conclude that $x \in D_k(S_1, \ldots, S_k) \triangle S_{k+1} = D$. In any case, $x \in D$, showing that $C \subseteq D$. This completes the inductive step.

By mathematical induction, for any $n \geq 1$, the assertion $A(n)$ is true. $\quad\square$

Exercise 429: (Tukey's lemma) Let $\mathcal{F}$ be a family of subsets of X with the property that $F \in \mathcal{F}$ if and only if every finite subset of F is in $\mathcal{F}$. To be shown is that $\mathcal{F}$ has a maximal member with respect to the partial order containment.

First of all, $\mathcal{F}$ is non-empty since $\emptyset \in \mathcal{F}$.

Let $\mathcal{C}$ be a chain in $\{\mathcal{F}, \subseteq\}$ and set $F_{\mathcal{C}} = \cup_{F \in \mathcal{C}} F$. In order to apply Zorn's lemma, it suffices to show that $F_{\mathcal{C}}$ is in $\mathcal{F}$, that is, to show that any finite subset of $F_{\mathcal{C}}$ is in $\mathcal{F}$.

Let $E \subset F_{\mathcal{F}}$ be finite, say $E = \{x_1, \ldots, x_n\}$. Since E is contained in a union, for each $i = 1, \ldots, n$, there exists $F_i \in \mathcal{C}$ so that $x_i \in F_i$. Since $\mathcal{C}$ is a chain in $\mathcal{F}$, $E^* = \cup_{i=1}^n F_i \in \mathcal{C}$ and so. The set E is a finite subset of E^* and $E^* \in \mathcal{F}$, so by the assumption on $\mathcal{F}$, it follows that $E \in \mathcal{F}$. Since E was an arbitrary finite subset of $F_{\mathcal{C}}$, conclude that $F_{\mathcal{C}} \in \mathcal{F}$.

By Zorn's lemma, $\mathcal{F}$ contains a maximal element. $\qquad\square$

Exercise 430: This proof is found in [95], however with natural numbers starting at 0, giving a slightly different base case (and perhaps in a more elegant style).

Let $S(n)$ be the statement "if $\{0, 1, \ldots, n-1\}$ and $\{0, 1, \ldots, m-1\}$ have the same cardinality, then $m = n$."

BASE STEP: When $n = 1$, $S(1)$ says that if $\{0\}$ and $\{0, 1, \ldots, m-1\}$ have the same cardinality, then $m = 1$. If there is a bijection between $\{0\}$ and $\{0, 1, \ldots, m-1\}$, then $\{0, 1, \ldots, m-1\}$ can have only one element, and this occurs precisely when $m = 1$, proving $S(1)$.

INDUCTIVE STEP: Assume that for some $k \geq 1$, $S(k)$ is true. To complete the inductive step, one proves $S(k+1)$, namely, that if there is a bijection between $\{0, 1, \ldots, k-1, (k+1)-1\}$ and $\{0, 1, \ldots, m-1\}$, then $m = k+1$. So suppose that $f : \{0, 1, \ldots, k-1, (k+1)-1\} \to \{0, 1, \ldots, m-1\}$ is a bijection.

If $f(k) = m-1$, then the function f restricted to $\{0, 1, \ldots, k-1\}$ is a bijection to $\{0, 1, \ldots, m-2\}$, and so by $S(k)$, $k = m-1$, giving $m = k+1$ as desired.

If $f(k) = \ell \neq m-1$, then create a new function g which differs from f in just two places. If w is so that $f(w) = m-1$, then put $g(w) = \ell$, $g(k) = m-1$, and for all $x \in \{0, 1, \ldots, k\} \backslash \{w, k\}$, put $g(x) = f(x)$. Then g is a bijection from $\{0, 1, \ldots, k\}$ to $\{0, 1, \ldots, m-1\}$ satisfying $g(k) = m-1$. As in the last paragraph, conclude that $k = m-1$ as desired, ending the inductive step.

By MI, for all $n \geq 1$ the statement $S(n)$ is true. $\qquad\square$

Note: The set $\{0, 1, \ldots, n-1\}$ is called the *ordinal n*. This exercise proves that two distinct finite ordinals have distinct cardinalities. See Section 4.2 for a brief discussion of ordinals; to give a complete theory of ordinals is well beyond the scope of this book, so the interested reader might look at nearly any book on set theory (*e.g.*, [289, 347]) for more on ordinals.

Exercise 431: Assume that both $|A| \leq |B|$ and $|B| \leq |A|$ with injections (one-to-one functions) $f : A \to B$ and $g : B \to A$ as witnesses. To prove the theorem, it suffices to find a bijection $h : A \to B$, injective and onto B. Two proofs are given, the first of which may be more transparent; an accompanying diagram is given for each method in Figure 31.1. The idea in both is the same: construct h inductively.

See also [316, p. 11] for a proof (with a version of the first diagram), or [160], where a similar proof occurs (and the same diagram is on its cover!).

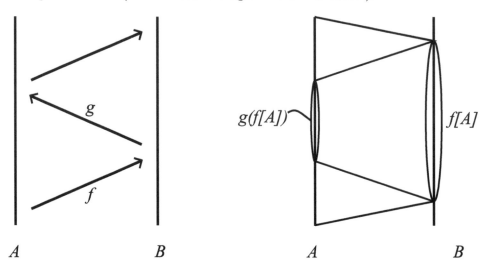

Figure 31.1: Two proofs for Cantor–Bernsetin–Schröder

Proof 1: (Due, essentially, to Ron Aharoni, personal communication.) Assume that $A \cap B = \emptyset$, since otherwise, construction of a bijection between the two non-intersecting parts is easily extended to a bijection by using the identity function on $A \cap B$. Consider $f = \{(x, f(x)) : x \in A\}$ and $g^{-1} = \{(g(y), y) : y \in B\}$ as sets of ordered pairs. Then $f \cup g$ can be seen as the edges of a bipartite graph on sets A and B, where any vertex is in at most two edges. Components of this graph are either finite cycles or infinite paths.

[*Comment:* There can be no infinite cycles, because in a cycle, each vertex has a unique predecessor and unique successor; for example, if $x_1, x_2, \ldots$, is an infinite sequence of vertices in a cycle (whose indices are $\mathbb{Z}^+$), the predecessor of x_1 is not identifiable; similarly, any "first infinite" vertex has no well-defined predecessor. *If* indeed "infinite cycles" could be defined, the idea in the proof below would be identical: take every (second) pair in this cycle whose first vertex is in A, that is, use only the f part of the cycle to give a bijection between one "half" of the cycle and the other half—or use only the g^{-1} part. To see that paths are infinite, argue by contradiction; if some path $x_1, \ldots, x_r$ is finite, if $x_r \in A$, $(x_r, f(x_r))$ extends the path, and if $x_r \in B$, $(g(x_r), x_r)$ extends the path.]

If C is a cycle on elements $x_1, x_2, \ldots, x_{2n}$, where $x_1 \in A$, for each $i = 1, \ldots n$, define $h(x_{2i-1}) = x_{2i}$.

Similarly, for every doubly infinite chain (with no circuits), take every second pair to be in h (there are two choices for h here, those pairs in f, or those pairs in g^{-1}; either works). For those infinite paths with a starting point, take the first and every subsequent pair to be in h, that is, for paths beginning in A, include in h all pairs in f from this path; for those paths beginning in B, include in h only those pairs from g^{-1}.

The function h created is a bijection. $\qquad\square$

Proof 2: (Based on Jech [289, p.23]) For any set $S \subseteq A$, use the standard shorthand $f[S] = \{f(s) : s \in S\}$; similarly define $g[\cdot]$. Instead of assuming A and B are disjoint, Jech argues as follows: Letting $B' = g[B] \subseteq A$, and $A_1 = g[f[A]]$, then since $f[A] \subseteq B$, $A_1 \subseteq B' \subseteq A$, and so $\alpha = g \circ f$ is an injection from A onto A_1, that is, $|A_1| = |A|$. Thus, it suffices to prove the theorem when $A_1 \subseteq B \subseteq A$ and α is an injection from A_1 onto A. Still yet to be proved is that $|A| = |B|$.

Inductively define sets A_n and B_n by setting $A_0 = A$, $B_0 = B$, and for each $n \geq 0$, define $A_{n+1} = \alpha[A_n]$ and $B_{n+1} = \alpha[B_n]$. Define the desired bijection h by

$$h(x) = \begin{cases} \alpha(x) & \text{if for some } n, \; x \in A_n \setminus B_n \\ x & \text{otherwise} \end{cases}$$

Then h is an injection from A onto B "as the reader will gladly verify". $\qquad\square$

Exercise 432: (Hint) Use the same trick of adjusting two values of a function as seen in the solution to Exercise 430. (This argument appears in Cameron's book [95, p. 22].)

Exercise 433: Let A be a set and suppose that it is Peano infinite. Select a sequence of elements $a_0, a_1, \ldots$ from A. Since A is Peano infinite, one can continue this sequence *ad infinitum*, because otherwise, if $A \backslash \{a_0, \ldots, a_{n-1}\}$ is empty at some point, then there would be a bijection from A to $\{0, 1, \ldots, n-1\}$ showing that A is Peano finite. (Selecting the infinite sequence all at once is where the Axiom of Choice is used.)

Let $f : A \to A$ be the function defined by $f(a_i) = a_{i+1}$ for each a_i in the sequence, and $f(a) = a$ for those not in the sequence. This function is a bijection from A to $A \backslash \{a_0\}$, a proper subset, and so A is Dedekind infinite. $\qquad\square$

Comment: The above proof is even easier if one well-orders A first (then f is just a shift operator, shifting everything down by one—as in the "Hilbert Hotel", a fanciful hotel containing infinitely many guests, where room is made for a new guest by simply shifting everyone up one room, leaving room 1 empty).

Exercise 434 (countable union): This standard exercise is solved in [160, pp. 159–160] or [289, p. 39].

Exercise 435: The proof is by induction on the size of $X = \cup_{i=1}^{k} A_i \cup B_i$. For each $n \geq 0$, let $S(n)$ be the statement that if $|X| = n$ and a family of subsets of X is cross-intersecting, then equation (13.3) holds (for the corresponding value of k). BASE STEP: When $n = 0$, necessarily $k = 1$ and $A_1 = B_1 = \emptyset$, in which case the left side of (13.3) is $\binom{0+0}{0} = 1$, so $S(0)$ is true.

INDUCTIVE STEP: For some fixed $m > 0$, suppose that $S(m-1)$ holds. Let X be a set with $|X| = m$ elements, and let $A_1, \ldots, A_k, B_1, \ldots, B_k$ be a cross-intersecting family of subsets of X.

It is not difficult to verify that for each $x \in X$, the family $\{A_i : x \notin A_i\} \cup \{B_i \setminus \{x\} : i \notin A_i\}$ is also cross-intersecting. Therefore, for each $x \in X$, (with $k' = |\{i : x \notin A_i\}|$ replacing k), the induction hypothesis $S(m-1)$ yields

$$\sum_{i:x\notin A_i} \left(\frac{|A_i| + |B_i \setminus \{x\}|}{|A_i|} \right)^{-1} \le 1. \tag{31.1}$$

Summing (31.1) over all $x \in X$, For each $i = 1, \ldots, k$,

$$\sum_{x\in X} \sum_{i:x\notin A_i} \left(\frac{|A_i| + |B_i \setminus \{x\}|}{|A_i|} \right)^{-1} \le m. \tag{31.2}$$

Using the following facts

$$|\{x \in X \setminus A_i : |B_i \setminus \{x\}| = |B_i| - 1\}| = |B_i|,$$
$$|\{x \in X \setminus A_i : |B_i \setminus \{x\}| = |B_i|\}| = m - |A_i| - |B_i|,$$

the left-hand side of (31.2) is

$$\sum_{i=1}^{k} \sum_{x\in X\setminus A_i} \left(\frac{|A_i| + |B_i \setminus \{x\}|}{|A_i|} \right)^{-1}$$

$$= \sum_{i=1}^{k} \left(\sum_{x\in X\setminus A_i, x\in B_i} \left(\frac{|A_i| + |B_i \setminus \{x\}|}{|A_i|} \right)^{-1} + \sum_{x\in X\setminus A_i, x\notin B_i} \left(\frac{|A_i| + |B_i \setminus \{x\}|}{|A_i|} \right)^{-1} \right)$$

$$= \sum_{i=1}^{k} \left(|B_i| \left(\frac{|A_i| + |B_i| - 1}{|A_i|} \right)^{-1} + (m - |A_i| - |B_i|) \left(\frac{|A_i| + |B_i \setminus \{x\}|}{|A_i|} \right)^{-1} \right).$$

To simplify the last expression, the equality

$$\frac{a+b)a!b!}{(a+b)!} = \frac{(b)a!(b-1)!}{(a+b-1)!},$$

implies that

$$(a+b) \binom{a+b}{a}^{-1} = b \binom{a+b-1}{a}^{-1},$$

which in turn yields

$$n \binom{a+b}{a}^{-1} = b \binom{a+b-1}{a}^{-1} + (n-a-b) \binom{a+b}{a}^{-1}.$$

Thus equation (31.2) becomes

$$\sum_{i=1}^{k} n \left(\frac{|A_i| + |B_i|}{|A_i|} \right)^{-1} \le n,$$

from which $S(m)$ follows, thereby concluding the inductive step.

By mathematical induction on $|X|$, for any set X and any cross-intersecting family of subsets of X, equation (13.3) holds.   □

Exercise 436: This solution can be also found in [250, Prob. 71]. For $n \ge 3$, let $S(n)$ be the statement that if the pairs of an n-set are partitioned into k classes so that every two pairs in the same class share a vertex, then $k \ge n - 2$.

BASE STEP: When $n = 3$, any partition of the three pairs of elements into either $k = 1, 2, 3$ classes has the desired property, so $S(3)$ holds trivially.

INDUCTION STEP: Fix $m \ge 3$ and suppose that $S(m)$ holds. Let $X = \{x_0, \ldots, x_m\}$ and let $[X]^2 = \{Y \subset X : |Y| = 2\}$ denote the pairs in X. Suppose that

$$[X]^2 = C_1 \cup \cdots \cup C_k$$

is a partition of the pairs in X so that for each i, any two pairs in the class C_i share a common point.

Case 1: There exists an element of X, say x_0 so that x_0 is contained in at least three pairs from the same class, say $\{x_0, x_1\}, \{x_0, x_2\}, \{x_0, x_3\} \in C_k$. Let $Z = X \backslash \{x_0\} = \{x_1, \ldots, x_m\}$. Since any pair in Z contains at most two of x_1, x_2, x_3, no pair in Z occurs in C_k. So $[Z]^2$ is partitioned into $k - 1$ classes, and by $S(m)$, $k - 1 \le m - 2$. Thus $k \le (m + 1) - 2$, and so $S(m + 1)$ holds in this case.

Case 2: There is no element of X as in Case 1. Then every class contains at most 3 pairs, and so $3k \ge \binom{m+1}{2}$, which gives $k \ge \frac{m(m+1)}{6}$, and for $m \ge 3$, it is not difficult to check that for $m \ge 3$, $k \ge m - 1$, again proving $S(m + 1)$. These two cases complete the inductive step.

Mathematical induction shows that for each $n \ge 3$, the statement $S(n)$ holds.   □

Comment: Note that the result in Exercise 436 can not be improved because a partition into $n - 2$ classes exists so that all pairs in the same class share a vertex; such a partition exists by a simple induction argument (put all three pairs from three vertices in one class; add a vertex, and put all newly formed pairs in the next class, and continue similarly adding each new vertex).

Exercise 437: This problem and solution can be found in the classic reference by Hadwiger, Debrunner, and Klee [250, Prob. 78]. The proof here is by induction on $p - q \ge 0$.

BASE STEP: For $2 \leq q = p$, let $\mathcal{F}$ be a family of segments with the (p, q) property. By Lemma 13.1.5, $\mathcal{F}$ has the $(2, 2)$ property, and so by Helly's theorem (Theorem 20.1.7 and comments after for infinite families), some point is common to all sets in $\mathcal{F}$. So when $p = q$, $p - q + 1 = 1$ and the unique class, all of $\mathcal{F}$, has non-empty intersection.

INDUCTIVE STEP: Fix $d \geq 1$ and suppose that the result is true whenever $p - q < d$, and fix some p_0 and q_0 so that $p_0 - q_0 = d$ and let $\mathcal{F}$ be a family with the (p_0, q_0) property.

First, suppose that $\mathcal{F}$ is finite. Let P denote the leftmost of the right endpoints of the segments. Let $\mathcal{F}' \subseteq \mathcal{F}$ be the family of those sets containing P and $\mathcal{F}''$ be those sets not containing P. [The idea is to now partition $\mathcal{F}''$ into $d - 1$ classes, and use $\mathcal{F}'$ as the d-th class.]

Claim: $\mathcal{F}''$ can be partitioned into at most $p_0 - q_0$ classes so that each class has non-empty intersection.

Proof of claim: If $\mathcal{F}''$ contains at most $p_0 - q_0$ sets, the result is trivial. So suppose that $\mathcal{F}''$ contains more than $p_0 - q_0$ sets. By Lemma 13.1.5 with $r = q_0 - 2$, $\mathcal{F}$ has the $(p_0 - q_0 + 2, 2)$ property. So any $p_0 - q_0 + 1$ sets from $\mathcal{F}''$ together with the segment ending in P has a pair intersecting, and since one of these cannot be the one ending in P, this pair resides in $\mathcal{F}''$. Thus $\mathcal{F}''$ has the $(p_0 - q_0 + 1, 2)$ property, and since $p_0 - q_0 + 1 - 2 < p_0 - q_0 = d$, by the inductive hypothesis, $\mathcal{F}''$ has a partition into $(p_0 - q_0 + 1) - (2) + 1 = p_0 - q_0$ classes so that each class has non-empty intersection, finishing the proof of the claim.

Together with $\mathcal{F}'$ as one more class (which has common intersection P), $\mathcal{F}$ is partitioned into $p_0 - q_0 + 1$ classes, each with non-empty intersection, and completing the inductive step.

When $\mathcal{F}$ is infinite, the solution is only outlined. Let $R \subset \mathbb{R}$ be the set of right-hand endpoints of segments in $\mathcal{F}$. Then R is bounded on the left (for if not, arbitrarily many, more than p_0, disjoint segments are present, contradicting the (p_0, q_0) property). Hence, there is a rightmost point P^* such that no point from R lies to the left of P^*. Now proceed as in the finite case to complete the inductive step.

By mathematical induction, for every $p \geq q \geq 2$, the assertion in the exercise is true. $\square$

Exercise 438: The proof is by induction on n and d. For $d \geq 0$ and $n \geq 1$, let $S(d, n)$ be the statement that if $\mathcal{H}$ is a set-system on $|X| = n$ elements with VC-dim $= d$, then

$$|\mathcal{H}| \leq \binom{n}{0} + \binom{n}{1} + \cdots + \binom{n}{d}.$$

BASE STEP: If either $d = 0$, or $n \leq d$, the inequality given in $S(d, n)$ is trivial. [So, if the d's are plotted on the x-axis, and the n's are plotted on the y-axis, the base step covers the y-axis and the (infinite) triangle at or below $x = y$.]

INDUCTION STEP ($[S(c-1, m-1) \land S(c, m-1)] \rightarrow S(c, m)$): Let $0 < c < m$ and suppose that both $S(c-1, m-1)$ and $S(c, m-1)$ hold. Let $\mathcal{H}$ be a set system on X with $|X| = m$ and VC-dim($\mathcal{H}$) = c. Fix any $x \in X$ and consider the two families $\mathcal{H}_1$ and $\mathcal{H}_2$ of sets defined on $X \backslash \{x\}$ by

$$\mathcal{H}_1 = \{H \backslash \{x\} : x \in H \in \mathcal{H}\},$$
$$\mathcal{H}_2 = \{H \in \mathcal{H} : x \notin H \text{ and } H \cup \{x\} \in \mathcal{H}\}.$$

If $H \in \mathcal{H}$ contains x and also $H \backslash \{x\} \in \mathcal{H}$, then $H \backslash \{x\}$ in $\mathcal{H}_1$ arises from two different $\mathcal{H}_1$ in $\mathcal{H}$. Hence $|\mathcal{H}| = |\mathcal{H}_1| + |\mathcal{H}_2|$.

Since any subset that is shattered in $\mathcal{H}$ remains shattered in $\mathcal{H}_1$, it follows that

$$\text{VC-dim}(\mathcal{H}_1) \leq \text{VC-dim}(\mathcal{H} = c,$$

and so the hypothesis $S(c, m-1)$ applies to $\mathcal{H}_i$. If some $Y \subset X \backslash \{x\}$ is shattered in $\mathcal{H}_2$, then $Y \cup \{x\}$ is shattered in $\mathcal{H}$, so VC-dim($\mathcal{H}_\in$) $\leq c-1$, and so $S(c-1, m-1)$ applies to $\mathcal{H}_2$.

Putting all of these facts together,

$$|\mathcal{H}| = |cal H_1| + |\mathcal{H}_2|$$

$$\leq \sum_{i=0}^{c} \binom{m-1}{i} + \sum_{i=0}^{c-1} \binom{m-1}{i} \qquad \text{(by } S(c, m-1), S(c-1, m-1))$$

$$= \binom{m-1}{0} + \sum_{i=1}^{c} \binom{m-1}{i} + \sum_{i=}^{c} \binom{m-1}{i-i}$$

$$= \binom{m}{0} + \sum_{i=1}^{c} \binom{m}{i} \qquad \text{(by Pascal's id.)}$$

$$= \sum_{i=0}^{c} \binom{m}{i}.$$

and so $S(c, m)$ holds, completing the inductive step.

By a form of double induction, for all $d \geq 0$ and $n \geq 1$, $S(d, n)$ holds.   □

Comment: One can think of the induction step as an outer induction on d, with the inner induction done on n. Another way to see that the inductive step above indeed accomplishes the task, is start with those indices corresponding to the line $y = x$ and $x = 0$ as mentioned in the base step, and then to consecutively prove the theorem for each of $y = x+1$, $y = x+2$, and so on.

31.2   Solutions: Posets and lattices

Exercise 439: (Dilworth's theorem) For any poset $(P, \leq)$, let $d(P)$ denote the minimum number of disjoint chains required to cover P, and let $\alpha(P)$ be the number of elements in a largest antichain. Since any antichain $A \subseteq P$ requires $|A|$ chains to cover A alone, $\alpha(P) \leq d(P)$. It remains to prove the reverse inequality, and this is achieved by strong induction on $|P|$. For each $n \geq 0$, let $S(n)$ denote the statement "For any poset $(P, \leq)$ with $|P| = n$, P can be covered by at most $\alpha(P)$ disjoint chains."

BASE STEP: If $|P| = 0$, then $P = \emptyset$, and both $\alpha(P) = d(P) = 0$, verifying that $P(0)$ holds.

INDUCTIVE STEP: Fix some $k \geq 1$ and assume that for all $k' \in \{0, 1, \ldots, k - 1\}$, $P(k')$ is true. Let $(P, \leq)$ be a poset with $|P| = k$ elements and let $\alpha = \alpha(P)$ be the length of the longest antichain in P. It remains to show that P can be covered by (at most) α chains.

Let C be any maximal chain in P.

Case 1: If every antichain in $P \setminus C$ contains less than α elements, then by $S(|P \setminus C|)$, $P \setminus C$ can be covered by at most $\alpha - 1$ disjoint chains, and so together with C, P can be covered by at most α disjoint chains, and $S(n)$ is satisfied.

Case 2: Suppose there exists an antichain $A = \{a_1, a_2, \ldots, a_\alpha\}$ in $P \setminus C$ with α elements. Longer antichains do not exist by the definition of α, and so every element in $P \setminus C$ is comparable to some element of A. Define

$$U = \{p \in P : \exists a_i \in A \text{ with } a_i \leq p\}$$

and

$$L = \{p \in P : \exists a_i \in A \text{ with } p \leq a_i\}.$$

Then $U \cup L = P$. The maximum element of C is not in L, for if it were, C could be extended by adding some a_i; but C is maximal. Similarly, the minimum element of C is not in U. Thus $U \neq P$ and $L \neq P$. By induction hypotheses $S(|U|)$ and $S(|L|)$, let each of U and L can be covered by at most α disjoint chains, say C_i^U's and C_j^L's. Since A is an antichain with α elements, each $a_i \in A$ is contained in precisely one chain in U and one in L; relabelling the C_i^U's and C_j^L's if necessary, suppose that for each i, $a_i \in C_i^U$ and $a_i \in C_i^L$.

Observe that for each i, both $a_i = \min C_i^U$ and $a_i = \max C_i^L$, for if, say, $a_i \neq \min C_i^U$, that is, for some $x \in C_i^U$, $x < a_i$, then for some a_j, $a_j \leq x < a_i$ contradicts A being an antichain. Thus for each i, form the chain $C_i = C_i^U \cup C_i^L$; the chains $C_1, \ldots, C_\alpha$ are disjoint and cover P, completing the inductive step $S(|P|) = S(n)$.

By strong mathematical induction on $|P|$, for each finite poset P, Dilworth's theorem holds. $\qquad\square$

Exercise 440: For those wanting a bit more of hint, see [59, p.98, prob 53]. For a complete proof, see, *e.g.*, [pp 61–62][386]; the Rado selection principle is used, which is given with a compactness argument using Tychonff's theorem (see [386, pp.52–55]).

Exercise 441: See [387].

Exercise 442: One solution to this exercise appears in, *e.g.*, [58, p. 18], however, "convex" was not mentioned (though it is a consequence of the proof given). The proof is by induction on $|X| = n$.

BASE STEP: When $n = 1$, $\mathcal{P}(X)$ is itself a (symmetric and convex) chain with two elements (X and $\emptyset$).

INDUCTION STEP: Let X be a set with $k \geq 2$ elements, and suppose that any set with $k - 1$ elements has a symmetric chain decomposition. Let $x \in X$, and $Y = X \backslash \{x\}$. By induction hypothesis, let

$$\mathcal{P}(Y) = \mathcal{C}_1 \cup \mathcal{C}_2 \cup \cdots \cup \mathcal{C}_s$$

be a partition of all subsets of Y into (disjoint) convex symmetric non-empty chains. [Note that since each chain contains precisely one set of $\lfloor (k-1)/2 \rfloor$ elements, $s = \binom{k-1}{\lfloor (k-1)/2 \rfloor}$.] In $\mathcal{P}(X)$, each such chain $\mathcal{C}_i$ is replaced by two chains (one of which may be empty). If some $\mathcal{C}_i$ consists of j sets $A_1 \subset A_2 \subset \cdots \subset A_j$, let

$$\mathcal{Q}_i = \{A_1, A_2, \ldots, A_{j-1}, A_j, A_j \cup \{x\}\}$$

and

$$\mathcal{R}_i = \{A_1 \cup \{x\}, A_2 \cup \{x\}, \ldots, A_{j-1} \cup \{x\}\}$$

(which may be empty if $j = 1$). Since $\mathcal{C}_i$ is, by the induction hypothesis, convex and symmetric, it is not difficult to verify that so are both $\mathcal{Q}_i$ and $\mathcal{R}_i$ (see the example above—examine the even and odd cases separately). Also, (excluding the empty $\mathcal{R}_i$'s)

$$X = \bigcup_{i=1}^{s} \mathcal{Q}_i \cup \bigcup_{i=1}^{s} \mathcal{R}_i$$

is a partition of $\mathcal{P}(X)$, concluding the inductive step.

By MI, for all $n \geq 1$, the power set of any n-element set can be partitioned into symmetric convex chains. □

Comment: One can check that the inductive step produces the correct number of sets, doubling from 2^{k-1} to 2^k, since j sets in $\mathcal{C}_i$ are replaced by $j + 1$ sets and $j - 1$ sets respectively. Also, as noted by Bollobás, the number of chains produced is correct as well. When $|X| = k = 2\ell$ is even, two chains are produced from each in $\mathcal{P}(Y)$, giving

$$2\binom{k-1}{\lfloor (k-1)/2 \rfloor} = 2\binom{2\ell - 1}{\ell - 1} = \binom{2\ell}{\ell},$$

the correct number. However, when $|X| = 2\ell + 1$ is odd, there are chains in $\mathcal{P}(Y)$ consisting of a single set—in the decomposition for Y, (where $|Y| = 2\ell$), there are $\binom{2\ell}{\ell}$ chains, of which only $\binom{2\ell-1}{\ell-1}$ have more than one set (this takes some thought), so the construction gives (by Pascal's identity)

$$\binom{2\ell}{\ell} + \binom{2\ell - 1}{\ell - 1} = \binom{2\ell + 1}{\ell}$$

chains, precisely the number desired.

31.3 Solutions: Countable Zorn's lemma for measurable sets

Exercise 443: The proof is by contradiction; suppose that $\mu(U \backslash A) > 0$. Since

$$\sup\{\mu(A_i) : i = 1, 2, \ldots\} = \mu(A),$$

pick n be so large that

$$\mu(A) - \mu(A_n) < \frac{\mu(U \backslash A)}{2}.$$

By construction,

$$
\begin{aligned}
\mu(A_{n+1}) &\geq \frac{\sup\{\mu(B) : B \in \Theta, A_n \precsim B\} + \mu(A_n)}{2} \\
&\geq \frac{\mu(U) + \mu(A_n)}{2} \qquad\qquad\qquad \text{(since } A_n \precsim U\text{)} \\
&= \frac{\mu(U \backslash A) + \mu(A) + \mu(A_n)}{2} \\
&\geq \frac{\mu(U \backslash A) + \mu(A_n) + \mu(A_n)}{2} \qquad \text{(because } A_n \precsim A\text{)} \\
&= \frac{\mu(U \backslash A)}{2} + \mu(A_n) \\
&> \mu(A) - \mu(A_n) + \mu(A_n) \\
&= \mu(A),
\end{aligned}
$$

and so $\mu(A_{n+1}) > \mu(A)$, contradicting $A_{n+1} \subseteq A$. Thus one concludes that $\mu(U \backslash A) = 0$, and so $U \approx A$.

Hence any upper bound $V \in \Theta$ for $\mathcal{A}$ satisfies $\mu(V) = \mu(U) = \mu(A)$, and so $\mu(U \triangle V) = 0$. Thus U is maximal in P. $\qquad\square$

31.4 Solutions: Topology

Exercise 444: This problem is an old classic (see *e.g.*, [220, p. 114], or more recently, [462, Ex. 69, p. 282]). For $n \geq 1$ let $P(n)$ be the proposition that if $(a_1, b_1), (a_2, b_2), \ldots, (a_n, b_n)$ are open intervals of real numbers that pairwise intersect, then $\cap_{i=1}^{n}(a_i, b_i) \neq \emptyset$. Note that the assumption that every pair of intervals intersect ensures the tacit assumption that each $a_i < b_i$.

Two different versions of the inductive step are given; they differ in notation and style, but rely on the same general principles. The second, (based on [220, p. 114]) is perhaps a little easier to read.

BASE STEP: When $n = 1$, the statement $P(1)$ is trivial. (The case $n = 2$ is also trivial.)

INDUCTIVE STEP, VERSION 1: Let $m \geq 1$ and suppose that $P(m)$ is true. Let $(a_1, b_1), (a_2, b_2), \ldots, (a_m, b_m), (a_{m+1}, b_{m+1})$ be open intervals that are pairwise intersecting. Without loss of generality, assume that $a_1 \leq a_2 \leq \cdots \leq a_m \leq a_{m+1}$. By $P(m)$, $\cap_{i=1}^{m}(a_i, b_i)$ is non-empty. Let $b_\alpha = \min\{b_1, \ldots, b_m\}$; then $\cap_{i=1}^{m}(a_i, b_i) = (a_m, b_\alpha) \neq \emptyset$, so $b_\alpha > a_m$.

Case 1: If $b_{m+1} \leq b_\alpha$, then

$$\bigcap_{i=1}^{m+1}(a_i, b_i) = (a_{m+1}, b_{m+1}),$$

which is non-empty by assumption.

Case 2: If $b_\alpha < b_{m+1}$, then

$$\bigcap_{i=1}^{m+1}(a_i, b_i) = (a_m, b_\alpha) \cap (a_{m+1}, b_{m+1}) = (a_{m+1}, b_{m+1}) \cap (a_\alpha, b_\alpha),$$

which is non-empty because intervals pairwise intersect.

In both cases, $\cap_{i=1}^{m+1} = (a_{m+1}, b_{m+1}) \neq \emptyset$, completing the proof of $P(m+1)$.

INDUCTIVE STEP, VERSION 2: Note that if two open intervals intersect, they do so in an open interval. Let $m \geq 1$ and assume that $P(m)$ is true. Let $I_1, I_2, \ldots, I_m, I_{m+1}$ be pairwise intersecting open intervals in $\mathbb{R}$. By $P(m)$, $I_1, \ldots, I_m$ mutually intersect, and the intersection $\cap_{i=1}^{m} I_i$ is an interval I. To prove that I intersects I_{m+1}, suppose the contrary, that is, suppose that $I \cap I_{m+1} = \emptyset$. Then there exists a point $X \in \mathbb{R}$ in between I and I_{m+1}. Each of the intervals $I_1, \ldots, I_m$ contains I, and by assumption, intersects with I_{m+1}, and so contains all points between, namely X. So X belongs to I. The contradiction shows that $I \cap I_{m+1} \neq \emptyset$, and all points in this intersection $I \cap I_{m+1}$ are contained all intervals $I_1, \ldots, I_{m+1}$.

By MI, for each $n \geq 1$, if n open intervals intersect pairwise, they mutually intersect. $\qquad \square$

Exercise 445: Suppose that for every $\epsilon > 0$, there exists $x \in A$ such that $|x-a| < \epsilon$. In particular, for each $n \in \mathbb{N}$, the set

$$X_n = \{x \in A : |x - a| < \frac{1}{n}\}$$

is non-empty. By the Axiom of Choice, $\prod_{n\in\mathbb{N}} X_n \neq \emptyset$, and so there exists a sequence $\{x_n\}_{n\in\mathbb{N}}$ so that each $x_n \in X_n$. Then each $x_n \in A$ and $x_n \to a$ as $n \to \infty$. $\square$

Exercise 446: This proof is well-known (see, *e.g.*, [463]).

Let X and Y be sets, and let $\mathcal{F} \subset X \times Y$ be a collection of functions from some subset of X into Y. Let $\mathcal{C}$ be a chain in $(\mathcal{F}, \preceq)$, and put $h = \cup_{f\in\mathcal{C}} f$. Put $X' = \operatorname{dom}(h) = \cup_{f\in\mathcal{C}}\operatorname{dom}(f)$. To be shown is that h is a function, that is, for any $x \in X'$, there is unique y so that $(x, y) \in h$. In hopes of a contradiction, suppose that both $(x, y_1) \in h$ and $(x, y_2) \in h$. Since h is defined as a union of functions, there are g_1 and g_2 in $\mathcal{C}$ so that $(x, y_1) \in g_1$ and $(x, y_2) \in g_2$. Without loss of generality, suppose that $g_1 \subset g_2$. Then $(x, y_1) \in g_2$ as well, and since g_2 is a function, $y_1 = y_2$.

It is rudimentary to verify that the domain and range of h is as in the statement of the exercise, and so this is left to the reader. $\square$

Note: By the above proof, chains in the above order have their union in the family, so Zorn's lemma applies to any subset of all functions from subsets of X to Y to give the existence of a maximal elements in the ordering given above. However, without Zorn's lemma, it is nearly trivial to notice that any function with domain X and range in Y is maximal in this sense. So Zorn's lemma doesn't yield anything interesting for the entire set of such functions. It might, however, imply something interesting for some special subset of functions whose domains are in X. (See solution to Exercise 449, Tychonoff's theorem.)

Exercise 447: (using Zorn's lemma) Let $\mathcal{A} \subseteq \mathcal{P}(X)$ be a family with FIP, and define

$$\mathfrak{A} = \{\mathcal{C} \subseteq \mathcal{P}(X) : \mathcal{A} \subseteq \mathcal{C} \text{ and } \mathcal{C} \text{ has FIP}\}.$$

For any $\mathcal{C} \in \mathfrak{A}$, if $\emptyset \neq \cap_{C\in\mathcal{C}}\overline{C}$, then $\cap_{C\in\mathcal{A}}\overline{C} \supseteq \cap_{C\in\mathcal{C}}\overline{C} \neq \emptyset$.

Let $\{\mathcal{C}_\alpha : \alpha \in A\} \subset \mathfrak{A}$ be a chain (ordered by inclusion), and set $\mathcal{C}' = \cup_{\alpha\in A}\mathcal{C}_\alpha$. Fix $C_1, \ldots, C_n \in \mathcal{C}'$. For each $i = 1, \ldots, n$, let $C_i \in \mathcal{C}_{\alpha_i}$ (the α_i's need not be distinct). Since the $\mathcal{C}_\alpha$'s form a chain, there exists $j \in \{1, \ldots, n\}$ so that $C_1, C_2, \ldots, C_n \in \mathcal{C}_{\alpha_j}$. Since $\mathcal{C}_{\alpha_j}$ has FIP, $C_1 \cap \cdots \cap C_n \neq \emptyset$. Thus, $\mathcal{C}'$ has FIP, and so is an upper bound for $\{\mathcal{C}_\alpha : \alpha \in A\}$. By Zorn's lemma, $\mathfrak{A}$ has a maximal element. $\square$

Exercise 448: [Thanks to KR for this proof.] For each $n \geq 1$, let $S(n)$ be the statement that if $X_1, X_2, \ldots, X_n$ are compact topological spaces, then $X_1 \times X_2 \times \cdots \times X_n$, equipped with the product topology, is compact.

BASE STEP $(S(1), S(2))$: $S(1)$ is immediate. To prove $S(2)$, let X_1, X_2 be compact topological spaces and let $\{U_\alpha \times V_\alpha : \alpha \in I\}$ be an open cover of $X_1 \times X_2$ by basic open sets.

For every $b \in X_2$, let $I_b = \{\alpha \in I : b \in V_\alpha\}$. Then $\{U_\alpha \times V_\alpha : \alpha \in I_b\}$ covers $X_1 \times \{b\}$ and hence $\{U_\alpha : \alpha \in I_b\}$ is an open cover of X_1. Since X_1 is compact, there is a finite subcover $\{U_{\alpha_{b,i}} : 1 \leq i \leq k_b\}$. Then $X_1 \times \{b\} \subseteq \cup_{i=1}^{k_b} U_{\alpha_{b,i}} \times V_{\alpha_{b,i}}$.

Set $V_b = \cap_{i=1}^{k_b} V_{\alpha_{b,i}}$. Then $b \in V_b$, V_b is open in X_2, and $\{V_b : b \in X_2\}$ is an open cover of X_2. Since X_2 is compact, there is a finite subcover $\{V_{b_1}, V_{b_2}, \dots, V_{b_t}\}$. For every $j = 1, \dots, t$, if $1 \leq i \leq k_{b_j}$, then $V_{b_j} \subseteq V_{\alpha_{b_j,i}}$ and hence $\cup_{j=1}^{t} \cup_{i=1}^{k_{b_j}} U_{\alpha_{b_j,i}} \times V_{\alpha_{b_j,i}}$ is a finite open cover of $X_1 \times X_2$.

INDUCTIVE STEP: Suppose that for some $\ell \geq 2$, $S(\ell)$ holds. Let $X_1, \dots, X_\ell, X_{\ell+1}$ be compact spaces. By the inductive hypothesis, $X_1 \times \cdots \times X_\ell$ is compact and by $S(2)$, $X_1 \times \cdots \times X_\ell \times X_{\ell+1} \simeq (X_1 \times \cdots \times X_\ell) \times X_\ell$ is compact, finishing the proof of $S(\ell+1)$ and so the inductive step.

By MI, for each $n \geq 2$, $S(n)$ holds. $\qquad\square$

Exercise 449: Two solutions are given, the first perhaps more elegant. Proofs for Tychonoff's theorem are also found in most topology texts (*e.g.*, Dugundji [151]). The first proof given here is really based on the idea of filters/ideals, without actually mentioning them, and is based on Lemma 13.3.2 (and can be found in [390].) [Thanks to KR for help writing this proof.] Both proofs assume the axiom of choice and hence Zorn's lemma.

Proof 1: Assume that for each $i \in I$, X_i is a compact topological space, and set $X = \prod_{i \in I} X_i$, with the product topology.

By Exercise 447, let $\mathcal{C}^*$ be a maximal family of subsets of X with FIP If $Y \subset X$ has the property that for every $C \in \mathcal{C}^*$, $C \cap Y \neq \emptyset$, then $Y \in \mathcal{C}^*$.

In each coordinate space X_i, the family $\{\pi(C) : C \in \overline{\mathcal{C}*}\}$ has FIP, and so by the compactness of each X_i,

$$\emptyset \neq \bigcap_{C \in \mathcal{C}*} \overline{\pi_i(C)}.$$

By AC, for each $i \in I$, pick $x_i \in \cap_{C \in \mathcal{C}*} \overline{\pi_i(C)}$ and define $\mathbf{x} = (x_i)_{i \in I} \in X$. It remains to show that $\mathbf{x} \in \cap_{C \in \mathcal{C}*} \overline{C}$.

Since each $\overline{C}$ is closed, to show that $\mathbf{x} \in \cap_{C \in \mathcal{C}*} \overline{C}$, it suffices to show that for every open set $U \subseteq X$ containing $\mathbf{x}$, and for every $C \in \mathcal{C}^*$, $U \cap C \neq \emptyset$.

For each $i \in I$, let $U_i \subseteq X_i$ be an open neighborhood of x_i. For every $C \in \mathcal{C}^*$, $x_i \in \overline{\pi_i(C)}$, and since U_i is an open neighborhood of x_i, it follows that $U_i \cap \pi_i(C) \neq \emptyset$. Thus $\emptyset \neq \pi_i^{-1}(U_i) \cap C$, and by the maximality of $\mathcal{C}^*$, $\pi_i^{-1}(U_i) \in \mathcal{C}^*$.

Let U be open in X with $\mathbf{x} \in U$. Approximating by basis elements of X, there are $i_1, \dots, i_n \in I$ and $U_1 \subseteq X_{i_1}, \dots, U_n \subseteq X_{i_n}$ so that

$$\mathbf{x} \in \bigcap_{j=1}^{n} \pi_j^{-1}(U_j) \subseteq U.$$

By the previous paragraph, since for $j = 1, \ldots, n$, U_j is an open neighborhood for x_{i_j}, $\pi_{i_j}^{-1}(U_j) \in \mathcal{C}^*$, and since $\mathcal{C}^*$ has FIP,

$$\bigcap_{j=1}^{n} \pi_j^{-1}(U_j) \cap C \neq \emptyset.$$

Thus, $U \cap C \neq \emptyset$.

Therefore, for every $C \in \mathcal{C}^*$, $\mathbf{x} \in \overline{C}$, and hence $\mathbf{x} \in \cap_{C \in \mathcal{C}_*} \overline{C} \neq \emptyset$. $\qquad\square$

The second proof given here is an adaptation of that found in [463], a proof for which Paul Chernoff (1992) receives credit. Recall that a *net* is a generalization of an infinite sequence (a sequence is a function f whose domain is a well-ordered set I, whereas a net is a function from a partially ordered set with least upper bounds—see [571, p.73]).

Proof 2: For each $i \in I$, let X_i be a compact topological space, and set $X = \prod_{i \in I} X_i$. To show that X is compact, it suffices to show that every net in X has a cluster point. Any element of X can be viewed as a function with domain I. Fix a net $\langle f_\alpha \rangle_{\alpha \in A}$.

For each non-empty subset $J \subseteq I$ and g in $\prod_{i \in J} X_i$, g is called a *partial cluster point* of $\langle f_\alpha \rangle_{\alpha \in A}$ iff g is a cluster point of the net $\langle f_\alpha \mid_J \rangle_{\alpha \in A}$.

When $J = \{i\}$ is a single index, $\langle f_\alpha \mid_J \rangle_{\alpha \in A}$ is a net in X_i, and since X_i is compact, X_i has a cluster point. Hence, partial cluster points exist. Let P be the non-empty set of all partial cluster points in $\langle f_\alpha \rangle_{\alpha in A}$. As in Exercise 446, let $\prec$ be the partial order on functions ordered by extension.

By Exercise 446, for any chain $\mathcal{C}$ in P $h_\mathcal{C} = \cup_{g \in \mathcal{C}} g$ is a function with domain $J_\mathcal{C} = \cup_{g \in \mathcal{C}} \mathrm{dom}(g)$.

For the moment, fix a chain $\mathcal{C}$ in $(P, \preceq)$; the claim is that $h_\mathcal{C}$ is a partial cluster point of the net $\langle f_\alpha \rangle_{\alpha \in A}$. To show this, it suffices to show that $h_\mathcal{C}$ is a cluster point of $\langle f_\alpha \mid_{J_\mathcal{C}} \rangle_{\alpha \in A}$. Let F be a finite subset of $J_\mathcal{C}$ and for each $i \in F$, let U_i be an open set in X_i. Put

$$W = \{h \in \prod_{i \in J_\mathcal{C}} X_i : \forall i \in F, h(i) \in U_i\},$$

a basic neighborhood of $h_\mathcal{C}$. Since $\mathcal{C}$ is a chain, there exists $\beta \in A$ so that for each $i \in F$, $f_\beta \mid_{J_\mathcal{C}} \in U_i$, and so $f_\beta \mid_{J_\mathcal{F}} \in W$. Since $\alpha \in A$ is arbitrary and the basic neighborhood W of $h_\mathcal{C}$ is arbitrary, $h_\mathcal{C}$ is a cluster point of $\langle f_\alpha \mid_{J_\mathcal{C}} \rangle_{\alpha in A}$. Thus, $h_\mathcal{C} \in P$.

So any chain in P has an upper bound in P, namely $h_\mathcal{C}$. By Zorn's lemma, P has a maximal element, say, h^*.

Let $\mathrm{dom}(h^*) = J^*$. If $J^* = I$, then $\langle f_\alpha \rangle_{\alpha \in A}$ has a cluster point in X and in this case, the theorem is proved. So suppose that $\mathrm{dom}(h^*) = J^* \neq I$, and let $k \in I \backslash J^*$. Since $h^* \in P$, some subnet $\langle f_\beta \mid_{J^*} \rangle_{\beta \in B}$ converges to h^*. Since X_k is compact, the net $\langle f_\beta \mid_{J^*} (k) \rangle_{\beta \in B}$ in X_k must have a cluster point p. Extend h^* to g on $J^* \cup \{k\}$ by putting $g = h^*$ on J^* and $g * (k) = p$. Then h is a partial cluster point of $\langle f_\alpha \rangle_{\alpha \in A}$, putting $h \in P$. However, $h^* \preceq g$, contradicting that h^* was maximal, so $J^* \neq I$ is not possible. $\qquad\square$

Exercise 450: See [308]; a solution is also given in several topology texts.

Exercise 451: ($\mathbb{Q}$ is dense in $\mathbb{R}$) First consider the case when $0 < x < y$. By the Archimedean property, there exists $q \in \mathbb{Z}^+$ so that

$$q > \frac{1}{y - x},$$

and so $1 < q(y - x)$. Again, by the Archimedean property, there exists an $n \in \mathbb{Z}^+$ so that $qx < n$; by the well-ordering of $\mathbb{Z}^+$, there is a least such n, say p. Then $p - 1 \leq qx < p$ (because if $p = 1$, $0 \leq qx$ which is true, and if $p \neq 1$, $p - 1 > qx$ would contradict the minimality of p). So $qx < p = (p-1)+1 < qx + q(y-x) < qy$; and division by q yields

$$x < \frac{p}{q} < y.$$

To prove the case where $x \leq 0$, use the Archimedean property to find $m \in \mathbb{Z}^+$ so that $-x < m$, and apply the above proof to find $p/q \in \mathbb{Q}$ satisfying

$$x + m < \frac{p}{q} < y + m,$$

giving $\frac{p}{q} - m$ as the desired rational number. $\qquad\square$

Exercise 452: Suppose that the conclusion is false, that is, suppose f is not continuous. Then there exists $\epsilon > 0$ so that for every $\delta > 0$, there is an x so that $|x - a| < \delta$, yet $|f(x) - f(a)| \geq \epsilon$. Fix such an epsilon. Using δ's of the form $1/n$, for each $n = 1, 2, 3, \ldots$ there is y_n so that $|y_n - a| < \frac{1}{n}$ and $|f(y_n) - f(a)| \geq \epsilon$. For $n = 1, 2, 3 \ldots$, let X_n be the set of reals x which satisfy $|x - a| < \frac{1}{n}$ and $|f(x) - f(a)| \geq \epsilon$; then each X_n is non-empty (for example, $y_n \in X_n$). By the Axiom of Choice, there exists a sequence $\{x_n\}$ so that $x_n \to a$ but $\{f(x_n)\}$ does not converge to $f(a)$. $\qquad\square$

Question: Was the Axiom of Choice really required here? Why couldn't one just use the sequence $\{y_n\}$ from the proof? Essentially, the Axiom of Choice was used to select a sequence $\{x_n\}$ all at once, and the sequence $\{y_n\}$ was just used to show one-by-one that all the X_n's were non-empty. It seems as if this proof could be restated using MI rather than AC.

31.5   Solutions: Ultrafilters

Exercise 453: ($\Rightarrow$) Let $\mathcal{F_E}$ be a proper filter, that is, $\emptyset \notin \mathcal{F_E}$. Since $\mathcal{F_E}$ is a filter, and the intersection of any two sets must also be in the filter, every pair of sets in $\mathcal{F}_E$ intersect non-trivially, and since $E \subseteq \mathcal{F_E}$, then any two sets in E intersect. It follows by induction that $\mathcal{E}$ has FIP. Hence the remaining direction of Lemma 13.4.2 holds. $\qquad\square$

Exercise 454: Let $\mathcal{H}$ be the set of proper filters on S containing $\mathcal{F}$, ordered by inclusion (where $\mathcal{F} \subseteq \mathcal{F}' \Leftrightarrow [A \in \mathcal{F} \Rightarrow A \in \mathcal{F}']$). To apply Zorn's lemma, one needs to show the union of a chain or proper filters is again a proper filter.

First check that the union of a chain of filters is again a filter: Let $\mathcal{F}_\alpha : \alpha < \lambda$ be a chain of filters where $\alpha < \alpha' \Rightarrow \mathcal{F}_\alpha \subset \mathcal{F}_{\alpha'}$. Set

$$\mathcal{F} = \cup_{\alpha < \lambda} \mathcal{F}_\alpha = \{A : A \in \mathcal{F}_\alpha \text{ for some } \alpha < \lambda\}.$$

For each α, $S \in \mathcal{F}_\alpha$, and so $S \in \mathcal{F}$. Now suppose $A \in \mathcal{F}$ and $A \subseteq B \subseteq S$. Then for some α, $A \in \mathcal{F}_\alpha$, and since $\mathcal{F}_\alpha$ is a filter, $B \in \mathcal{F}_\alpha$, and hence $B \in \mathcal{F}$. Suppose that $C, D \in \mathcal{F}$ with $C \in \mathcal{F}_\alpha$ and $D \in \mathcal{F}_\beta$. Then $C \cap D \in \mathcal{F}_{\max\{\alpha,\beta\}}$, and hence $C \cap D \in \mathcal{F}$. So $\mathcal{F}$ is a filter.

Furthermore, $\mathcal{F}$ is a proper filter since $\emptyset$ never occurred as an element in any $\mathcal{F}_\alpha$, so $\emptyset \notin \mathcal{F}$. So Zorn's lemma applies giving the desired result. $\qquad\square$

Exercise 455: For $k = 1$, $X_1 = X \in \mathcal{F}$, so assume that $k > 1$. If $X_k \notin \mathcal{F}$, then since $\mathcal{F}$ is an ultrafilter, $\bar{X}_k = X_1 \cup \cdots \cup X_{k-1} \in \mathcal{F}$. The result now follows by induction on k. $\qquad\square$

Chapter 32

Solutions: Logic and language

32.1 Solutions: Sentential logic

Exercise 456: For each $m \in \mathbb{Z}^+$, let $S(m)$ be the statement that for $m + 1$ statements p_i,

$$[p_1 \rightarrow p_2] \wedge [p_2 \rightarrow p_3] \wedge \ldots \wedge [p_m \rightarrow p_{m+1}] \Rightarrow [(p_1 \wedge p_2 \wedge \ldots \wedge p_m) \rightarrow p_{m+1}].$$

BASE STEP: The statement $S(1)$ says

$$[p_1 \rightarrow p_2] \Rightarrow [(p_1 \wedge p_2) \rightarrow p_2], \tag{32.1}$$

which is true (since the right side is a tautology).

INDUCTIVE STEP: Fix $k \geq 1$, and assume that for any statements $q_1, \ldots, q_{k+1}$, both

$$S(1): \quad [q_1 \rightarrow p_2] \Rightarrow [(q_1 \wedge q_2) \rightarrow q_2]$$

and

$$S(k): \quad [q_1 \rightarrow q_2] \wedge [q_2 \rightarrow q_3] \wedge \ldots \wedge [q_k \rightarrow q_{k+1}] \Rightarrow [(q_1 \wedge q_2 \wedge \ldots \wedge q_k) \rightarrow q_{k+1}]$$

hold. It remains to show that for any statements $p_1, p_2, \ldots, p_k, p_{k+1}, p_{k+2}$,

$$S(k+1): [p_1 \rightarrow p_2] \wedge [p_2 \rightarrow p_3] \wedge \ldots \wedge [p_{k+1} \rightarrow p_{k+2}] \Rightarrow [(p_1 \wedge \ldots \wedge p_{k+1}) \rightarrow p_{k+2}]$$

follows. Begin with the left side of $S(k+1)$:

$$[p_1 \rightarrow p_2] \wedge \ldots \wedge [p_{k+1} \rightarrow p_{k+2}] \wedge [p_{k+1} \rightarrow p_{k+2}]$$
$$\Downarrow \quad \text{(definition of conjuction)}$$
$$[[p_1 \rightarrow p_2] \wedge [p_2 \rightarrow p_3] \wedge \ldots \wedge [p_{k+1} \rightarrow p_{k+2}]] \wedge [p_{k+1} \rightarrow p_{k+2}]$$
$$\Downarrow \quad \text{(by } S(k) \text{ with each } q_i = p_i)$$
$$[(p_1 \wedge p_2 \wedge \ldots \wedge p_k) \rightarrow p_{k+1}] \wedge [p_{k+1} \rightarrow p_{k+2}]$$

$\Downarrow$ (by $S(1)$ with $q_1 = p_1 \wedge \ldots \wedge p_k$) and $q_2 = p_{k+1}$)

$[[(p_1 \wedge p_2 \wedge \ldots \wedge p_k) \wedge p_{k+1}] \to p_{k+1}] \wedge [p_{k+1} \to p_{k+2}]$

$\Downarrow$ (by definition of conjuction)

$[(p_1 \wedge p_2 \wedge \ldots \wedge p_k \wedge p_{k+1}) \to p_{k+1}] \wedge [p_{k+1} \to p_{k+2}]$

$\Downarrow$ (since $a \wedge b \to b$ with $b = [p_{k+1} \to p_{k+2}]$)

$[(p_1 \wedge p_2 \wedge \ldots \wedge p_k \wedge p_{k+1}) \to p_{k+2}] \wedge [p_{k+1} \to p_{k+2}]$

$\Downarrow$ (since $a \wedge b \to a$)

$[p_1 \wedge p_2 \wedge \ldots \wedge p_k \wedge p_{k+1}] \to p_{k+2},$

precisely the right side of $S(k+1)$, which completes the inductive step.

By mathematical induction, for each $n \geq 1$, $S(n)$ holds. $\square$

Exercise 458: For each $n \geq 3$, let $C(n)$ be the claim that for each r satisfying $1 \leq r < n$,

$$(p_1 \wedge \cdots \wedge p_r) \wedge (p_{r+1} \wedge \cdots \wedge p_n) \Leftrightarrow p_1 \wedge \cdots \wedge p_r \wedge p_{r+1} \wedge \cdots \wedge p_n.$$

BASE STEP: Let $n = 3$; the cases $r = 1$ and $r = 2$ are, respectively equations (14.2) and (14.1), declared just before the statement of the exercise, so $C(3)$ holds.

INDUCTIVE STEP: Fix $k \geq 3$ and suppose that $C(k)$ is true. To show $C(k+1)$, one has to show that for each $s \in \{1, \ldots, k\}$,

$$(p_1 \wedge \cdots \wedge p_s) \wedge (p_{s+1} \wedge \cdots \wedge p_{k+1}) \Leftrightarrow p_1 \wedge \cdots \wedge p_s \wedge p_{s+1} \wedge \cdots \wedge p_{k+1}.$$

Applying each of the results of Exercise 457

$$
\begin{aligned}
&(p_1 \wedge \cdots \wedge p_s) \wedge (p_{s+1} \wedge \cdots \wedge p_{k+1}) \\
&\Leftrightarrow (p_1 \wedge \cdots \wedge p_s) \wedge [(p_{s+1} \wedge \cdots \wedge p_k) \wedge p_{k+1})] \\
&\Leftrightarrow (p_1 \wedge \cdots \wedge p_s) \wedge (p_{s+1} \wedge \cdots \wedge p_k) \wedge p_{k+1} && \text{(by } C(3)) \\
&\Leftrightarrow [(p_1 \wedge \cdots \wedge p_s) \wedge (p_{s+1} \wedge \cdots \wedge p_k)] \wedge p_{k+1} && \text{(by } C(3)) \\
&\Leftrightarrow [p_1 \wedge \cdots \wedge p_s \wedge p_{s+1} \wedge \cdots \wedge p_k] \wedge p_{k+1} && \text{(by } C(k)) \\
&\Leftrightarrow p_1 \wedge \cdots \wedge p_s \wedge p_{s+1} \wedge \cdots \wedge p_k \wedge p_{k+1},
\end{aligned}
$$

which is the right side of $C(k+1)$, completing the inductive step.

By mathematical induction, for each $n \geq 3$, the claim $C(n)$ holds. $\square$

Exercise 459: For $n \geq 2$, define recursively the *disjunction* of $n + 1$ statements $p_1, p_2, \ldots, p_n, p_{n+1}$ by

$$p_1 \vee \cdots \vee p_n \vee p_{n+1} \Leftrightarrow (p_1 \vee \cdots \vee p_n) \vee p_{n+1}.$$

$\square$

Exercise 460: To show that

$$(p_1 \vee \cdots \vee p_r) \vee (p_{r+1} \vee \cdots \vee p_n) \Leftrightarrow p_1 \vee \cdots \vee p_r \vee p_{r+1} \vee \cdots \vee p_n,$$

repeat the solution to Exercise 458, however replacing $\wedge$ with $\vee$.

Exercise 464: Having q defined as in the question, for $n \geq 2$, let $I(n)$ denote the implication

$$q(x_1, x_2, \ldots, x_n) \rightarrow (x_1 \rightarrow x_n).$$

BASE STEP: The implication $I(2)$ says $q(x_1, x_2) \rightarrow (x_1 \rightarrow x_2)$, which is the first part of the definition of q.

INDUCTIVE STEP: Fix $k \geq 2$, and suppose that

$$I(k): \quad q(x_1, x_2, \ldots, x_k) \rightarrow (x_1 \rightarrow x_k).$$

is true. It remains to prove

$$I(k+1): \quad q(x_1, x_2, \ldots, x_{k+1}) \rightarrow (x_1 \rightarrow x_{k+1}).$$

Beginning with the left side of $I(k+1)$ and applying the recursive definition of q,

$$
\begin{aligned}
q(x_1, x_2, \ldots, x_{k+1}) &= q(x_1, x_2, \ldots, x_k) \wedge (x_k \rightarrow x_{k+1}) \\
&\rightarrow (x_1 \rightarrow x_k) \wedge (x_k \rightarrow x_{k+1}) && \text{(by } I(k)) \\
&\Rightarrow (x_1 \rightarrow x_{k+1}),
\end{aligned}
$$

which completes the proof of $I(k+1)$ and so the inductive step.

By mathematical induction, for each $n \geq 2$, the implication $I(n)$ holds. $\qquad \square$

32.2 Solutions: Well-formed formulae

Exercise 465: The proof given here is in two parts. Set $W_0 = A$ and for each $n \geq 0$, set

$$W_{n+1} = W_n \cup \{(\neg p : p \in W_n\} \cup \{(p \wedge q) : p, q \in W_n\},$$

and $W' = \cup_{n \geq 0} W_n$. Since W' contains A and is closed under $\neg$ and $\wedge$, $W \subseteq W'$. Claim: $W = W'$. For every $n \geq 0$, let $S(n)$ be the statement that $W_n \subseteq W$.

BASE STEP: $S(0)$ is true since $W_0 = A \subseteq W$.

INDUCTIVE STEP: Let $k \geq 0$, and suppose that $S(k)$ is true. Let $x \in W_{k+1}$. If $x \in W_k$, then $x \in W$ by the induction hypothesis. If there is a $p \in W_k$ so that $x = (\neq p)$ or if there are p and $q \in W_k$ such that $x = (p \wedge q)$, then $x \in W$ by the induction hypothesis and since W is closed under $\neq$ and $\wedge$. Thus $W_{k+1} \subseteq W$.

By MI, for every $n \geq 0$, $W_n \subseteq W$ and hence $W' = \cup_{n \geq 0} W_n \subseteq W$.

For each $n \geq 0$, let $E(n)$ be the statement that for every $x \in W_n$, x has an even number of parentheses.

BASE STEP: If $x \in W_0$, then for some $i \geq 1$, $x = a_i \in A$ has no parentheses, an even number.

INDUCTIVE STEP: Fix $k \geq 0$, suppose that $E(k)$ holds, and fix $x \in W_{k+1}$. If $x \in W_k$, then x has an even number of parentheses by $E(k)$. If there is a $p \in W_k$ so that $x = (\neq p)$, then by $E(k)$, p has an even number of parenthesis, say 2ℓ. Then x has $1 + 2\ell + 1 = 2(\ell + 1)$ parenthesis. Otherwise, there are $q, w \in W_k$ such that $x = (q \wedge w)$. By the induction hypothesis, there are $s, t \geq 0$ so that q has $2s$ parentheses and w has $2t$ parentheses. Then x has $1 + 2s + 2t + 1 = 2(s + t + 1)$ parentheses.

By MI, for every $n \geq 0$, every word in W_n has an even number of parentheses. $\square$

Chapter 33

Solutions: Graphs

33.1 Solutions: Graph theory basics

Exercise 470: This exercise has a simple direct proof—since E is a subset of the unordered pairs of vertices, $|E| \leq \binom{|V|}{2}$. Here is an inductive proof: For each $n \geq 0$, let $A(n)$ be the assertion that any graph G on n vertices has at most $\binom{n}{2}$ edges.

BASE STEP: When $n = 0$, there are no vertices and hence no edges, and $\binom{0}{2} = 0$, so $S(0)$ holds. Similarly, when $n = 1$ there are no edges and $\binom{1}{2} = 0$, so $S(1)$ holds. The first non-trivial case (actually, it is trivial, too) is for $n = 2$. The maximum number of edges on two vertices is one, and since $\binom{2}{2} = 1$, $S(2)$ holds as well. Three base cases were not actually necessary, however, often it doesn't hurt to be sure to prove the first base case that actually says something.

INDUCTION STEP: Fix some $k \geq 2$, and suppose that $S(k)$ is true. Examine a graph G on $k + 1$ vertices and fix a vertex $x \in V(G)$. There are at most k edges incident with x, and in the remaining graph, by $S(k)$, there at most $\binom{k}{2}$ edges. Thus, in all, G contains at most

$$\binom{k}{2} + k = \frac{k(k-1)}{2} + k = \frac{k^2 - k + 2k}{2} = \frac{(k+1)k}{2} = \binom{k+1}{2}$$

edges, which shows that $S(k+1)$ is true as well. This completes the inductive step.

By mathematical induction, for each $n \geq 0$, the statement $S(n)$ is true. $\qquad\square$

Exercise 471: As pointed out, there is no need for a proof by induction, since each edge contributes 2 to the degree sum, making the result immediate. Two inductive proofs are available, one by induction on $|V(G)|$, and the other by induction on $|E(G)|$. The first natural choice might be to induct on the number of vertices, so this proof is presented here. As the reader can verify, however, inducting on the number of edges is far easier.

For $n \geq 1$, let $S(n)$ be the statement that for any graph G on n vertices, $\sum_{x \in V(G)} \deg(x) = 2|E(G)|$.

BASE STEP: For $n = 1$, the only graph is a single vertex x; both $\deg(x) = 0$ and $2|E(G)| = 0$, so $S(1)$ holds.

INDUCTIVE STEP: Let $k \geq 1$ and suppose that $S(k)$ is true. Let G be a graph on $k + 1$ vertices, and let $v \in V(G)$. Suppose that $\deg(v) = d$, and let $y_1, \ldots, y_d$ be the neighbors of x. Form G' by deleting v (and all incident edges, of course). Then G' has k vertices, and so by induction hypothesis $S(k)$, $\sum_{x \in V(G')} \deg_{G'}(x) = 2|E(G')|$. Since the degree of each y_i is precisely one less in G' than in G,

$$
\begin{aligned}
\sum_{x \in V(G)} \deg_G(x) &= d + \left[\sum_{x \in V(G')} \deg_{G'}(x) \right] + \deg_G(v) \\
&= d + 2|E(G')| + d \quad (\text{by } S(k)) \\
&= 2(d + |E(G')|) \\
&= 2|E(G)| \quad (d \text{ edges in } G \text{ contain } v).
\end{aligned}
$$

This shows that $S(k + 1)$ is true, completing the inductive step.

By MI, for any $n \geq 1$, $S(n)$ holds. □

Exercise 472: [Handshake problem] This problem is a classic; for example, it appears with solution (and non-solution) in, e.g., [566, pp. 481–2].

Translating the problem to one of graph theory, for the moment, fix a party with the set of people $V = \{x_1, y_1, x_2, y_2, \ldots, x_n, y_n\}$, where for each $i = 1, \ldots, n$, $\{x_i, y_i\}$ is the i-th couple. Let $G = (V, E)$ be the graph on V be defined by $\{v, w\} \in E(G)$ iff v shakes hands with w.

For $n \geq 1$, let $H(n)$ be the statement that for any party of n couples, with host x_1 and hostess y_1, if in the graph for the party, the degrees of all vertices but the host are different, then $\deg(y_1) = n - 1$.

If no partners shake hands, then for any $v \in V$, $0 \leq \deg(v) \leq 2n - 2$.

BASE STEP: When $n = 1$, there is only one couple, no handshakes, and the hostess shakes $n - 1 = 1 - 1 = 0$ hands as required.

INDUCTIVE STEP: Fix some $m \geq 1$ and suppose that $H(m)$ holds. Consider a party with $m + 1$ couples $\{x_i, y_i\}$, G being its graph, no couple shaking hands, x_1 the host, y_1 the hostess, where the degrees in G of all but the host are different. Since any vertex has degree at most $2(m + 1) - 2 = 2m$, the degrees of all vertices but the host are $0, 1, 2, \ldots, 2m$.

The person other than the host with degree $2m$ shook hands with everyone else except the person of degree 0, so these two form a couple. Since the host is not part of this couple, neither is the hostess, so let x_{m+1}, y_{m+1} be the couple with degrees $0, 2m$ respectively.

Consider the party where the couple x_{m+1}, y_{m+1} is deleted, with G' its graph on $2m$ vertices $V' = \{x_1, y_1, x_2, y_2, \ldots, x_m, y_m\}$, where x_1, y_1 are host, hostess, respectively. Because $\deg_G(x_{m+1}) = 2m$ and $\{x_{m+1}, y_{m+1}\} \notin E(G)$, x_{m+1} was connected in G to all vertices in V', so the degrees of vertices in $V' \backslash x_1$ are reduced by one in G'; thus, the degrees of vertices in $V' \backslash \{x_1\}$ are $0, 1, 2, \ldots, 2m - 2$. So G' corresponds to a party with m couples satisfying $H(m)$, and so $\deg_{G'}(y_1) = m - 1$. Adjoining the two deleted vertices shows that $\deg_G(y_1) = m$, the required degree of the hostess. This concludes the inductive step $H(m) \to H(m + 1)$.

By mathematical induction, for every $n \geq 1$, $H(n)$ is true. $\qquad \square$

Exercise 473: (Eulerian graphs) Induct on $m = |E(G)|$.
BASE STEP: For $m = 0$, the graph is a single vertex, in which case the theorem is trivially true.
INDUCTIVE STEP: Fix $\ell \geq 0$, and suppose that the theorem is true for all graphs with at most ℓ edges. Without loss of generality, let G be connected, with all degrees even, and with $\ell + 1$ edges.

Since $d(x) = 0$ is impossible, $\delta(G) \geq 2$, and so by Lemma 15.1.2, G contains a cycle C. Delete edges of C from G, giving a graph H that still has even degrees. Let $H_1, \ldots, H_k$ be the components of H. Since each H_i has fewer than ℓ edges, apply the IH to each to get an Eulerian circuit C_i in each H_i. Splice these circuits together with C to get an Eulerian circuit for G (traverse C, and at the first vertex of each H_i in C, stop, traverse C_i all the way (returning to the point in C where the diversion began), then continue on with C). Hence the theorem holds for graphs on $\ell + 1$ edges, completing the inductive step.

By MI, the theorem holds for graphs with any number of edges. $\qquad \square$

Exercise 474: For $n \geq 3$, let $S(n)$ be the statement that for any graph G with n vertices, if G has at least n edges, then G contains a cycle. An equivalent formulation of $S(n)$ is $S'(n)$: if G is a graph on n vertices, if G contains no cycle, then G has at most $n - 1$ edges.

BASE STEP: For $n = 3$, the only graph with 3 edges is K_3, which is a 3-cycle, so $S(3)$ holds.

INDUCTIVE STEP: Let $\ell \geq 3$, and suppose that $S(\ell)$ is true. Let G be a graph on $\ell + 1$ vertices. It suffices to show that if G contains no cycle, then G has at most ℓ edges. So suppose that G has no cycle.

Divide the proof into two cases. First suppose that G has a vertex x with degree either 0 or 1. Deleting x produces an acyclic graph G' with ℓ vertices, and so by $S'(\ell)$, G' contains less than ℓ edges. Putting back x shows that G has less than $\ell + 1$ edges, proving $S'(\ell + 1)$, completing the inductive step. Second, consider the case where G has no vertex of degree 0 or 1, that is, $\delta(G) \geq 2$; in this case, it is shown that G has a cycle, regardless of the edge count. Let P be a maximal path in G, between, say, vertices u and v. Since P is maximal, any neighbor of u appears in P. Let x be the vertex in P that is farthest from u. Then the path from u to x contains at least $\delta(G) + 1 \geq 3$ vertices, and the edge $\{u, x\}$ completes a cycle in G. This completes the proof of $S'(\ell + 1)$.

By MI, for all $n \geq 3$, $S(n)$ holds. $\qquad\qquad\qquad\qquad\qquad\qquad\qquad\qquad\square$

Exercise 475: [Erdős-Gallai] This result appeared in [169, Thm 2.7]; another proof by Woodall appears in [575] (and reproduced in [566, p.416]). Woodall's proof (given below) relies on Dirac's theorem [143] (Theorem 15.4.1 here). Bollobás [59] also discusses this problem, but leaves it as an exercise.

Let $S(n)$ be the statement that for any constant $c \geq 3$, for any $n \geq c$, if G is a graph with more than $(c-1)(n-1)/2$ edges, then the length of the longest cycle in G is at least c.

Proof: Fix $c \geq 3$ and use strong induction on n.

BASE STEP ($S(c)$): Since $\frac{(n-1)^2}{2} = \binom{n}{2} - (n-1)/2$, any graph with more than $\frac{(n-1)^2}{2}$ edges has $\delta(G) \geq n/2$, and so G is hamiltonian by Dirac's theorem. Thus $S(c)$ is true.

INDUCTIVE STEP: Fix $m \geq c$ and suppose that each of $S(c), S(c+1), \ldots, S(m)$ is true. The inductive step is to show that $S(m+1)$ follows. To this end, let G be a graph on $m+1$ vertices with $|E(G)| > (c-1)(m+1-1)/2 = (c-1)m/2$ edges.

Case 1: $\delta(G) \leq (c-1)/2$. Fix $x \in V(G)$ with $d(x) \leq (c-1)/2$. Deleting x produces a graph with $|E(G \backslash x)| \geq (c-1)m/2 - (c-1)/2 = (c-1)(m-1)/2$; by induction hypothesis, $G \backslash x$ contains a cycle of length c, hence so does G.

Case 2: $\delta(G) > (c-1)/2$. Without loss of generality, assume that G is connected, for if G is disconnected, then some component will have more than the average number of edges, in which case the induction hypothesis applies to that component.

A natural tactic used in inductive proofs for graphs is to find a set $W \subseteq V(G)$ that has few edges incident with it, delete W, obtain a smaller graph with a higher concentration of edges, and apply the inductive hypothesis to the smaller graph.

Among all longest paths in G pick a path $P = v_1 \ldots v_\ell$ with $d = \deg(v_1)$ maximum. Observe that if $\ell < m + 1$, $\{v_1, v_\ell\} \notin E(G)$ because otherwise, any edge leaving P would be in a longer path, and since G is connected (and $\ell < m+1$), such an edge exists.

Let $W = \{v_i : \{v_1, v_{i+1}\} \in E(G)\}$. Then $|W| = d$. Since P is maximal, every neighbor of v_1 lies in P and so $\deg(v_1) \leq \ell - 1$. If for some $v_k \in W$ and $j \geq k+1 \geq c$, the edge $\{v_k, v_j\}$ is present, then the cycle $v_1 v_2 \ldots v_k v_j v_{j-1} \ldots v_{k+1} v_1$ has length c, satisfying the theorem. So assume that for any $w_k \in W$ and $j > k+1 \geq c$, then $\{v_k, v_j\} \notin E(G)$.

For any $v_k \in W$, the path $v_k v_{k-1} \ldots v_1 v_{k+1} v_{k+2} \ldots v_\ell$ also has ℓ vertices, so $\deg(v_k) \leq d$ and as this new path is maximum length ℓ, hence maximal, it follows that $N(v_k) \subseteq V(P)$.

Let $r = \min\{\ell, c - 1\}$ and put $X = \{v_1, \ldots, v_r\}$. Since for each $v_k \in W$, $N(v_k) \subseteq X$, and $\deg_G(v_k) \leq d$. For the moment, suppose that $r = c - 1 \leq \ell$.

The number of edges in $G[W]$ (the graph induced by W) is at most

$$\frac{1}{2} \sum_{v \in W} \deg_W(v) \leq \frac{1}{2}d^2.$$

Let H be the bipartite subgraph consisting of edges between W and $Z \backslash W$. Then $|E(H)| \leq |W| \cdot |X \backslash W| = d(r - d)$. Let d_W, d_X denote degrees in the graphs induced by W, X respectively, and put $d_H = \deg_H$. Then

$$
\begin{aligned}
|E(G[W])| &= \frac{1}{2} \sum_{w \in W} d_W(w) \\
&= \frac{1}{2} \sum_{w \in W} d_W(w) + \frac{1}{2} \sum_{w \in W} d_H(w) - \frac{1}{2} \sum_{w \in W} d_H(w) \\
&= \frac{1}{2} \sum_{w \in W} (d_W(w) + d_H(w)) - \frac{1}{2} \sum_{w \in W} d_H(w) \\
&= \frac{1}{2} \sum_{w \in W} d_X(w) - \frac{1}{2} \sum_{w \in W} d_H(w) \\
&= \frac{1}{2} \sum_{w \in W} d_X(w) - \frac{1}{2}|E(H)| \\
&\leq \frac{1}{2}d^2 - \frac{1}{2}|E(H)|.
\end{aligned}
$$

Adding $|E(H)|$ to each side of the above equation shows that the number of edges incident with W is

$$
\begin{aligned}
|E(G[W])| + |E(H)| &\leq \frac{1}{2}d^2 + \frac{1}{2}|E(H)| \\
&\leq \frac{1}{2}d^2 + \frac{1}{2}d(r - d) \\
&= \frac{1}{2}dr.
\end{aligned}
$$

Upon deleting W and all edges incident with W, obtain a graph on $m - d$ vertices with more than

$$\frac{1}{2}(m - 1)(c - 1) - \frac{1}{2}dr \geq \frac{1}{2}(m - d - 1)(c - 1)$$

edges. By the induction hypothesis $S(m - d)$, this remaining graph contains a cycle of length at least c, hence so does G, proving $S(m + 1)$.

When $r = \ell < c - 1$, even fewer edges are incident with W, and the induction hypothesis again applies.

By (strong) mathematical induction, for all $n \geq c$, $S(n)$ is true. $\qquad \square$

33.2 Solutions: Trees and forests

Exercise 476: For $n \geq 1$, let $C(n)$ be the claim that for any connected graph G on n vertices, the sum of the distances from any fixed v to all other vertices of B, $\sum_{w \in V(G)} d(v, w)$, is at most $\binom{n}{2}$.

BASE STEP: When $n = 1$, the sum of distances from the single vertex is 0, and $\binom{n}{2} = \binom{1}{2} = 0$ as well, so $C(1)$ holds.

INDUCTIVE STEP: For some $m \geq 1$, suppose that $C(m)$ holds, and let G be a graph on $m + 1$ vertices with some vertex v fixed. Pick $x \in V(G) \backslash \{v\}$, and consider the graph G' formed by deleting x (and all edges incident with x). Since G is connected, $d(v, x) \leq m$. Then

$$
\begin{aligned}
\sum_{y \in V(G)} d(v, y) &= \left[\sum_{y \neq x} d(v, y) \right] + d(v, x) \\
&\leq \left[\sum_{y \neq x} d(v, y) \right] + m \\
&\leq \binom{m}{2} + m \quad \text{(by } C(m)\text{)} \\
&= \binom{m + 1}{2},
\end{aligned}
$$

showing that $C(m + 1)$ follows, completing the inductive step.

By mathematical induction, for every $n \geq 1$, $C(n)$ holds. $\qquad \square$

Remark: Here is one idea for a non-inductive proof. Let G be connected on n vertices and fix some vertex v. The sum of the distances from v is maximized when there is one vertex at maximal distance $n - 1$, in which case the vertices are a path, with distances from v being $1, 2, \ldots, n - 1$ whose sum is $\binom{n}{2}$.

Exercise 477: For $n \geq 1$ consider only graphs G on n vertices, and denote the three statements in the exercise by
$A(n)$: G is connected and acyclic (*i.e.*, G is a tree);
$B(n)$: G is connected and has $n - 1$ edges;
$C(n)$: G is acyclic and has $n - 1$ edges.

Of the three implications, only the proof of $A(n) \rightarrow B(n)$ given here is by induction; three such proofs are given.
$A(n) \rightarrow B(n)$: By induction on n: For $n \geq 1$, let $S(n)$ be the statement that any tree on n vertices has $n - 1$ edges.
Proof 1:

BASE STEP: For $n = 1$, the only tree with a single vertex has $1 - 1 = 0$ edges, so $S(1)$ holds.

INDUCTIVE STEP: Proof 1 (simple induction): For some fixed $k \geq 1$, suppose that $S(k)$ is true, and let T be a tree on $k + 1$ vertices. By Lemma 15.2.1, let $x \in V(T)$ be a leaf, and form the tree T' by deleting x (and the edge incident with x). Then $|V(T')| = k$, and so by $S(k)$, T' has $k - 1$ edges. Together with the edge removed, this shows that $|E(T)| = k = (k + 1) - 1 = |V(T)| - 1$, showing that $S(k + 1)$ follows, completing the inductive step.

Proof 2 (strong induction): Fix $k \geq 1$ and suppose that $S(1), \ldots, S(k)$ are all true. Let T be a tree on $k + 1$ vertices. By Lemma 15.2.3, let $\alpha = \{x, y\}$ be a bridge in T; removal of α produces two trees, say T_1 and T_2, each with at most k vertices. By the inductive hypotheses $S(|V(T_1)|)$ and $S(|V(T_2)|)$, T_1 has $|V(T_1)| - 1$ edges and T_2 has $|V(T_2)| - 1$ edges. Since T_1 and T_2 together have $k + 1$ vertices,

$$|E(T)| = |E(T_1)| + |E(T_2)| + 1 = |V(T_1)| - 1 + |V(T_2)| - 1 + 1 = k + 1 - 1 - 1 + 1 = k,$$

completing the proof of $S(k + 1)$.

By mathematical induction, for every $n \geq 1$, $S(n)$ holds. $\qquad \square$

A third proof of $A(n) \rightarrow B(n)$ is by well-ordering (and is the preferred method by some, e.g., see [71]). As pointed out above, $S(1)$ is true. Let $X = \{x \in \mathbb{Z}^+ : S(x)$ is false$\}$. Suppose, for the moment, that $X \neq \emptyset$; then X is well-ordered, and so has a least element, call it ℓ, where $\ell > 1$. Let T be a tree on ℓ vertices for which $S(\ell)$ fails, that is, $|E(T)| \neq \ell - 1$. Removing a leaf from T produces a tree T' with $\ell - 1$ vertices, and one fewer edge than in T; so $|E(T')| = |E(T)| - 1 \neq \ell - 2 = |V(T')| - 1$, contradicting the minimality of ℓ. Thus, $X = \emptyset$. $\qquad \square$

Note: One can not use Lemma 15.2.4 in any of the above proofs as Lemma 15.2.4 uses the result from this exercise!

$B(n) \rightarrow C(n)$: (For a connected graph G on n vertices, if G has $n - 1$ edges, then G is acyclic.)

Let G be connected on n vertices with $n - 1$ edges. Let $G' \subseteq G$ be any acyclic connected subgraph of G (delete edges from cycles in G, which doesn't disconnect the graph) on $V(G) = V(G')$. By $A(n) \rightarrow B(n)$, G' has $n - 1$ edges. Since $|E(G)| = n - 1$ was assumed, $G' = G$, and so G is acyclic.

$C(n) \rightarrow A(n)$: Let G be an acyclic graph on n vertices, $n - 1$ edges. One needs only to show that G is connected. Let $G_1, \ldots, G_k$ be the connected components of G. By $A(n) \rightarrow B(n)$, each component satisfies $|E(G_i)| = |V(G_i)| - 1$. Summing all components, $|E(G)| = n - k$. But $|E(G)| = n - 1$ was given, so $k = 1$, and thus G is connected. (So both $A(n)$ and $B(n)$ follow.) $\qquad \square$

Exercise 478: This problem is an old standard, and can be found in many places. See, *e.g.*, [566, p.70, Prop. 2.1.8] for a proof and further references.

For $k \geq 0$, let $A(k)$ be the assertion that if G is a graph with $\delta(G) \geq k$, then G contains every tree with k edges as a (weak) subgraph.

BASE STEP: The only tree with 0 edges is a single vertex, which is contained as a weak subgraph of any graph (with at least one vertex). Thus $A(0)$ is true.

INDUCTIVE STEP: Suppose that $t \geq 0$ and that $A(t)$ holds. Let G be a graph with $\delta(G) \geq t+1$. It remains to show that G contains every tree on $t+1$ vertices. Let T be a tree with $t+1$ vertices. Because $t+1 \geq 1$, T contains a leaf (vertex with degree 1) x, say, attached to $y \in V(T)$; form T' on t vertices by deleting x. By $A(t)$, since $\delta(G) \geq t+1 > t$ G contains a copy of T'. Let $z \in V(G)$ be the vertex corresponding to y in T'. Since $\delta(G) \geq t+1$, y is adjacent to some vertex w not in the copy of T'; adjoining w to the copy of T' produces a copy of T. Thus, G contains a copy of T, ending the inductive step.

By the principle of mathematical induction, for any $k \geq 0$, $A(k)$ holds. $\square$

Exercise 479: This was observed in 1869, proved by C. Jordan [294] and is found in, *e.g.*, [566, p. 72].

One has to show that the center of a tree is either a vertex or a single edge. Recall that a vertex u is in the center of a graph G iff the eccentricity of u, $\epsilon_G(u) = \max_{v \in V(G)} \deg_G(u, v)$, is minimal.

For $n \geq 1$, let $J(n)$ be the statement that the center of a tree on n vertices is either a vertex or a single edge. The theorem is proved by strong induction on n.

BASE STEP: When $n = 1$ or $n = 2$, there is nothing to prove as the center of a tree on 1 or 2 vertices is the tree itself.

INDUCTIVE STEP: Let $n \geq 2$ and suppose that for each $i \leq n$, $J(i)$ holds. Let T be a tree on $n + 1 \geq 3$ vertices. Form T' by deleting all leaves of T. Then T' is a tree with at least one vertex, and so by inductive hypothesis, the center of T' is either a vertex or a single edge. It suffices to show that the center of T is also the center of T'.

For any vertex $u \in V(T')$, another vertex at maximum distance from u is a leaf in T (otherwise one could extend the path from u farther). Since all leaves have been deleted and edges between other vertices remain, the length of any maximal path from a vertex $v \in V(T')$ is shortened by 1. Hence, for any $u \in V(T')$, $\epsilon_{T'}(u) = \epsilon_T(u) - 1$. The eccentricity of any leaf in T is greater than its neighbor, so the vertices minimizing eccentricity in T' also minimize eccentricity in T. Hence, the center of T' is the center of T, as required. This completes the inductive step.

By MI, for every $n \geq 1$, the statement $J(n)$ is true. $\square$

Exercise 483: For any non-negative integer h, let $S(h)$ be the statement that if T is any full binary tree with height h, then $|V(T)| = 2^{h+1} - 1$.

BASE STEP: When $h = 0$, there is only one full binary tree of height 0, namely a single vertex, and $2^0 - 1 = 1$, so $S(0)$ is true.

INDUCTIVE STEP: Fix $\ell \geq 0$ and suppose $S(\ell)$ is true. Let T be the full binary tree with height $\ell + 1$. Deleting the root of T produces two binary trees T_L and T_R, each of height ℓ, and so by $S(\ell)$, $|V(T_1)| = |V(T_2)| = 2^{\ell+1} - 1$. Thus the number of vertices in T is

$$1 + |V(T_1)| + |V(T_2)| = 1 + 2(2^{\ell+1} - 1) = 2^{\ell+2} - 1,$$

and so $S(\ell + 1)$ is true.

By mathematical induction, for all $h \geq 0$, $S(h)$ is true. □

Exercise 484: Repeat the proof for Exercise 483, and bound each $|V(T_i)|$ by the maximum of the two.

Exercise 485: (Outline) When $n = 0$, there is only one increasing tree, and $0! = 1$, so the base case is true. To see the inductive step, first an observation is made. Let T be an increasing tree on $\{0, 1, \ldots, k\}$. Since T is increasing, k is a leaf and removal of k creates another increasing tree T'. Furthermore, if T' is increasing, the attaching of k by an edge to any $j \in V(T')$ creates an increasing tree on $\{0, 1, \ldots, k\}$. Since there are k vertices $0, 1, \ldots, k - 1$ in T' where the new vertex k can be attached, the number of increasing trees on $\{0, 1, \ldots, k\}$ is k times the number of increasing trees on $\{0, 1, \ldots, k - 1\}$. If one assumes the inductive hypothesis that for any $k \geq 1$, there are $(k - 1)!$ increasing trees on $\{0, 1, \ldots, k - 1\}$, then there are $k(k - 1)! = k!$ increasing trees on $\{0, 1, \ldots, k\}$. □

Exercise 487: By an easy induction on i, each G_i is a tree, and so G_n is a spanning tree. It remains to prove that G_n is a minimal.

For $k \geq 1$, let $S(k)$ be the statement that G_k is a subtree of a minimum weight spanning tree.

BASE STEP: G_1 is a single vertex, which is a subtree of every spanning tree, so $S(1)$ is true.

INDUCTION STEP: Fix $k \geq 1$, and suppose $S(k)$, that G_k is a subtree of a minimum weight spanning tree T. Consider G_{k+1}; if G_{k+1} is a subtree of T, there is nothing to show. So suppose that G_{k+1} is not a subtree of T.

Let $e = \{x, y\}$ be the edge chosen by the algorithm with $x \in V_k = V(G_k)$, $y \in V \setminus V_k$, $G_{k+1} = G_k \cup \{e\}$ and $e \notin E(T)$.

Consider the graph $T \cup \{e\}$; since T is a tree containing x and y, $T \cup \{e\}$ has a unique cycle containing e, and so there is another edge e' of T from some vertex in $V(G_k) = V_k$ to a vertex outside of V_k. Because e' was not chosen at step k,

$w(e) \leq w(e')$. Examine the graph U formed by deleting e' from T and inserting e. Since e and e' are on the same cycle in $T \cup \{e\}$, U is a tree spanning $V_k \cup \{y\}$ with

$$w(U) = w(T) - w(e') + w(e) \leq w(T).$$

But T has minimum weight, so $w(U) = w(T)$, and $w(e) = w(e')$. Then U is a spanning tree with minimum weight. Hence, G_{k+1} is a subtree of the minimum spanning tree U, proving $S(k+1)$.

By mathematical induction, G_n produced by the algorithm is a tree contained in a minimum spanning tree—so G_n is a minimum spanning tree. □

33.3 Solutions: Connectivity, walks

Exercise 489: This theorem might be called Whitney's theorem (due to Whitney [568]) however Whitney is famous for many theorems. This result can also be found in, *e.g.*, [566].

To prove: any graph with at least 3 vertices is 2-connected if and only every pair of vertices are connected by two disjoint paths (vertex disjoint, except for endpoints, of course; in other words, every pair of vertices lie on a common cycle).

First, suppose that any two vertices x and y in a graph G are connected by disjoint paths. Then x and y can not be separated by removing a single vertex, and so G can not be disconnected by the removal of a vertex, and so is 2-connected.

Second, suppose that G is a graph with $|V(G)| \geq 3$, and that G is 2-connected. Since G is 2-connected, G is connected, and so for any $u, v \in V(G)$, the distance $d(u, v)$ is finite. Induction on $d(u, v)$ shows that there are two disjoint $u - v$ paths: BASE STEP: When $d(u, v) = 1$, the pair $\{u, v\}$ is an edge of G, and since removal of either u or v does not disconnect the graph, removing the edge uv does not disconnect G. Hence there is a $u - v$ path in G that does not use the edge uv; thus are there two disjoint $u - v$ paths (one of which is $\{u, v\}$).
INDUCTIVE STEP: Let $k \geq 1$ and suppose that for any 2-connected graph H (with at least 3 vertices), if two vertices $x, y \in V(H)$ have $d(x, y) \leq k$, then there exists two disjoint $x - y$ paths. Fix a 2-connected graph G and two vertices u, v with $d_G(u, v) = k + 1$, and let P be a (shortest) $u - v$ path with length $k + 1$. Let v_1 be the vertex in P next to v. Then $d_G(u, v_1) \leq k$, and so by induction hypothesis, there are two disjoint $u - v_1$ paths R_1 and R_2. Since $G \backslash v_1$ is connected (G is 2-connected), there exists a $u - v$ path S not containing v_1. If S is disjoint from either R_1 or R_2, then there are two disjoint $u - v$ paths (S and the R_i extended by $\{v_1, v\}$). So suppose that S intersects both R_1 and R_2 in an internal vertex. If y is the last vertex on S (closest to v) that lies on $R_1 \cup R_2$, say, on R_1, then the $u - v$ path that starts with R_1 and continues with $\{y, v\}$ is disjoint from the path $R_2 \cup \{v_1, v\}$. This completes the inductive step, showing two disjoint $u - v$ paths.

By mathematical induction, for any 2-connected graph G, and any two vertices $u, v \in V(G)$, (regardless of their distance) there are two disjoint $u - v$ paths. □

Exercise 490: Let $A = [a_{ij}]$ be the $n \times n$ adjacency matrix of a graph G on vertices $v_1, \ldots, v_n$. For each $k \geq 1$, let $S(k)$ be the statement that the (i, j)-entry of A^k is the number of walks of length k from v_i to v_j. The proof given here is by induction on k.

BASE STEP: For $k = 1$, $A^1 = A$ and $a_{ij} = 1$ iff $\{v_i, v_j\} \in E(G)$, and $a_{ij} = 0$ otherwise, and since a walk of length 1 is simply an edge, $S(1)$ holds.

INDUCTIVE STEP: Fix some $k \geq 1$ and suppose that $S(k)$ holds. Fix a graph on vertices $\{v_1, v_2, \ldots, v_n\}$ with adjacency matrix A. It remains to show $S(k+1)$, that is, the (i, j) entry of A^{k+1} is the number of walks of length $k + 1$ from v_i to v_j.

Put $A^k = B = [b_{ij}]$. Then $A^{k+1} = A^k A = BA$ and so the (i, j) entry of A^{k+1} is $\sum_{\ell=1}^{n} b_{i,\ell} a_{\ell,j}$. A walk of length $k + 1$ consists of a walk of length k first, to some v_ℓ, then one final edge from v_ℓ to v_j. For each $\ell = 1, \ldots, n$, by induction hypothesis, the number of walks of length k from v_i to v_ℓ is $b_{i,\ell}$. Such a walk can be completed to a walk of length $k + 1$ to v_j iff $\{v_\ell, v_j\} \in E(G)$, that is, iff $a_{\ell,j} = 1$. Counting all walks of length $k + 1$ from v_i to v_j then shows the total to be precisely the sum above, the (i, j) entry of A^{k+1}. This completes the inductive step.

By MI, for any $k \geq 1$, $S(k)$ holds. □

Exercise 491: As mentioned, this problem has an instant solution given the existence of a cyclic Gray code. The inductive solution given below essentially employs the recursive procedure used (see Exercise 564) to create a reflected Gray code, (just using a different language) and so shows the existence of Gray codes as well.

Let Q^n denote the graph of the n-dimensional unit cube; to be specific,

$$V(Q^n) = \{0, 1\} = \{(a_1, a_2, \ldots, a_n) : \forall i, a_i \in \{0, 1\}\},$$

and $\{(a_1, \ldots, a_n), (b_1, \ldots, b_n)\} \in E(Q^n)$ iff there exists $i \in \{1, \ldots, n\}$ so that $a_i = b_i + 1 \pmod{2}$ and for each $j \neq i$, $a_j = b_j$, that is, if the two vertices differ in precisely one coordinate.

For $n \geq 2$, let $A(n)$ be the assertion that Q^n is hamiltonian.

BASE STEP: For $n = 2$, the graph of Q^2 is simply a 4-cycle, which is itself hamiltonian.

INDUCTIVE STEP: Fix some $k \geq 2$ and suppose that $A(k)$ holds, that is, Q^k is hamiltonian. Examine the vertices of Q^{k+1} partitioned into two classes, $V_1 \cup V_2$, where V_i is the set of vertices with i as the first coordinate. Each V_i induces a graph isomorphic to Q^k (as only k coordinates vary), call it H_i. By the inductive hypothesis, each H_i has a hamiltonian cycle, say C_i; take these to be the same (up to labelling of first coordinate). Without loss of generality, suppose that in each

H_i, the hamiltonian cycle C_i contains consecutive vertices $u_i = (i, 0, a_2, \ldots, a_n)$ and $v_i = (i, 1, a_2, \ldots, a_n)$. The following is a hamiltonian cycle in Q^{k+1}: begin at $v_0 \in C_0$, traverse C_0 the long way around to u_0, then take the edge $\{u_0, u_1)\}$ to C_1, traverse C_1 the long way around to v_1 (in the opposite direction as in C_1), then finally close the cycle along the edge $\{v_1, v_0\}$ to v_0.

By MI, for every $n \geq 2$, $A(n)$ is true. $\qquad\square$

Exercise 492: Proceed by induction on n, using the fact that the n-cube is formed by taking two disjoint copies of the $(n-1)$-cube and adding an edges between corresponding vertices of each smaller cube. See, *e.g.*, [566, p.150].

33.4 Solutions: Matchings

Exercise 494: For $n \geq 1$, let $P(n)$ be the statement that the number of perfect matchings in K_{2n} is $\frac{(2n)!}{2^n(n!)}$.

BASE STEP: K_2 has only one perfect matching, and $\frac{(2)!}{2^1(2!)} = 1$, so $P(1)$ is true.

INDUCTIVE STEP: For some $k \geq 1$, suppose that $P(k)$ is true. Examine a copy of K_{2k+2}, and fix one vertex, say x. If an edge containing x is fixed, there are $2k$ remaining vertices to be matched, and by $P(k)$, this can be done in $\frac{(2k)!}{2^k(k!)}$ ways. As there are $2k + 1$ edges containing x, in all there are

$$(2k+1)\frac{(2k)!}{2^k(k!)} = \frac{(2k+1)(2k+2)}{2(k+1)}\frac{(2k)!}{2^k(k!)} = \frac{(2k+2)!}{2^{k+1}(k+1)!}$$

different perfect matchings in K_{2k+2}, proving $P(k+1)$, and completing the inductive step.

By MI, for every $n \geq 1$, the statement $P(n)$ is true. $\qquad\square$

Exercise 495: [Hall's theorem] The following proof is essentially due to Easterfield [156], rediscovered by Halmos and Vaughn [257].

The condition $|N(S)| \geq |S|$ is necessary, so it remains to show sufficiency. The proof of sufficiency is by induction on $m = |X|$.

BASE STEP: When $m = 1$, the condition is equivalent to saying that there is at least one edge leaving x.

INDUCTIVE STEP: Fix $k \geq 1$ and suppose that the condition is sufficient for all $m = 1, \ldots, k$. Let $G = (X \cup Y, E)$ be bipartite with $|X| = k + 1$.

Case 1: Suppose that any $j \in \{1, 2, \ldots, k\}$, if any j elements from X are adjacent to at least $j + 1$ elements of Y. Then pick one $x \in X$ and $y \in N(x)$ and delete the edge $\{x, y\}$ together with the vertices; the remaining graph satisfies the induction hypothesis for $m = k$, so all $k + 1$ elements of X can be matched.

Case 2: Suppose that for some j, $1 \leq j \leq k$, there are j vertices in X with precisely j neighbors in Y. By induction hypothesis, these j vertices can be matched. Remove these j vertices from X and their matches in Y. If some $\ell \leq k + 1 - j$ of the remaining vertices in X are adjacent to less than ℓ vertices in Y, then these ℓ together with the original k vertices in X would have less than $k + \ell$ neighbors in Y, violating the condition, thus the remaining graph satisfies the condition, and so remaining vertices in X can be matched by induction hypothesis (with $m = k+1-j$). These two matchings match all of X, concluding the inductive step.

By mathematical induction, the condition is sufficient for all bipartite graphs.

□

Remark: Given a set X and a family $\mathcal{S} = \{S_1, S_2, \ldots, S_n\}$ of subsets of X. *a system of distinct representatives* (denoted "SDR") for $\mathcal{S}$ is a collection of distinct elements $x_1, x_2, \ldots, x_n \in X$ so that for each $i = 1, 2, \ldots, n$, $x_i \in S_i$. The above proof can be adapted to prove the following theorem (see, *e.g.*, [7, p. 146]):

Theorem 33.4.1. *A family $\mathcal{S}$ has an SDR iff for each $m = 1, 2, \ldots, |\mathcal{S}|$, the union of any m sets in $\mathcal{S}$ contains at least m elements.*

33.5 Solutions: Stable matchings

Exercise 496: If in the first round every woman receives a proposal, the algorithm terminates. In this case, since each man proposed to exactly one woman, and all women received proposals, each man is married to his favorite woman. Such a marriage is stable just because no man would want to switch partners (though the women might not get their top picks).

Suppose that after round j some woman has still not received one proposal. Each man either proposed in round j, or was some woman's maybe at the end of round $j-1$. In either case, at the end of round j, each man is either freshly rejected, remains a "maybe", or freshly receives maybe status. Since not all women have a maybe on hold, there is at least one man that is freshly rejected in round j.

If some man receives a rejection and still has women he has not yet proposed to, the algorithm continues to the next round. Is it possible that the algorithm continues to the point where some man has been rejected by everyone and has no further proposals to make? No, because if some man has received n rejections, then he has proposed to all women (since he can only propose to each woman at most once), in which case no woman has been passed over, contrary to there existing a woman with no proposals. So the algorithm continues if and only if there are women with no proposals (or there are newly rejected men).

Why must the algorithm terminate? One easy way to see this is that since each man proposes to n women, so (by the pigeonhole principle) after at most $n^2 - n + 1$ rounds, some man will have exhausted all his choices, in which case each woman will have had at least one proposal. [In fact, with a closer analysis, at most $n^2 - 2n + 2$ rounds are necessary, and one can design a scenario requiring this many steps.]

To see that when the algorithm terminates a stable marriage ensues, suppose that the algorithm terminates (each woman has her "maybe"), and consider a man m and a woman w whom m prefers over the wife given him by the algorithm. It suffices to show that w does not prefer m over the husband given to her by the algorithm. Since m prefers w over his mate, m must have proposed to w at some earlier iteration and was either outright rejected, or was w's maybe, but later rejected when w accepted the proposal from a another man she prefers to m. In either case, w prefers her mate to m. $\qquad\square$

Exercise 497: For a solution, see [241, p.45–48], or for those who can read pseudocode, see [326, p. 50].

33.6 Solutions: Graph coloring

Exercise 498: Let $S(n)$ be the statement that if G is a graph on n vertices with maximum degree $\Delta(G) = k$, then G is $(k+1)$-colorable, that is, $\chi(G) \leq \Delta(G) + 1$. Proceed by induction on the number of vertices of G (not k).

BASE STEP: If $n = 1$, G is a single vertex with $\Delta(G) = 0$, and is 1-colorable, so $S(1)$ holds.

INDUCTION STEP: Suppose that for some $m \geq 1$, $S(m)$ holds, and let H be a graph on $m+1$ vertices $v_1, v_2, \ldots, v_m, v_{m+1}$. Examine the graph G formed by deleting v_{m+1} (and all edges incident with v_{m+1}). Put $\Delta(H) = k$ and note that $\Delta(G) \leq k$. By induction hypothesis $S(m)$, G is $(\Delta(G)+1)$-colorable, and hence is $(k+1)$-colorable, so fix a good $(k+1)$-coloring of G.

It remains to color v_{m+1}. Since $\Delta(H) = k$, v_{m+1} has at most k neighbors (among $v_1, v_2, \ldots, v_m$), already colored, so choose some remaining color for v_{m+1}, thereby producing a good $(k+1)$-coloring of H. This completes the inductive step.

By mathematical induction on the number of vertices n of graph, any graph G on n vertices is $(\Delta(G) + 1)$-colorable. $\qquad\square$

Comment 1: The *greedy coloring algorithm* is based on the above idea, however it colors vertices consecutively, each vertex with the *least available* color (colors are ordered, say, $1, 2, \ldots, k+1$). In this setting, it is worth noting that recursion and mathematical induction are essentially the same notion.

Comment 2: Can you find a proof using induction on $k = \Delta(G)$?

Comment 3: Don't confuse the result of this exercise with *Vizing's theorem,* which states that $\chi'(G) \leq \Delta(G) + 1$, where χ' is the edge-chromatic number, or *chromatic index,* the least number of colors required to color the *edges* so that all edges incident at a vertex receive different colors. See, *e.g.,* [566, pp. 275–278] for details.

Exercise 499: This result was proved in 1956 by Gaddum and Nordhaus [203], and appears as an exercise in [566, 5.1.41, p. 202].

Let $S(n)$ be the statement that if G is a graph with $|V(G)| = n$ vertices, then $\chi(G) + \chi(\overline{G}) \leq n + 1$.

BASE STEP: When $n = 1$, both G and $\overline{G}$ consist of a single vertex, each with chromatic number 1, so $S(1)$ is true. Checking $n = 2$, either G is a pair of isolated vertices or G is a single edge (and so $\overline{G}$ is either an edge or a pair of isolated vertices, resp.) and so the sum of chromatic numbers is 1+2=3, showing that $S(2)$ is also true. (This second base case is not needed for the proof, however.)

INDUCTION STEP: Fix $k \geq 1$ and suppose that $S(k)$ is true. Fix a graph H on $|V(H)| = k + 1$ vertices, and identify one vertex $x \in V(H)$. To show that $S(k + 1)$ holds, it suffices to show that

$$\chi(H) + \chi(\overline{H}) \leq (k+1) + 1.$$

Let G be the graph formed by deleting x (and all edges incident with x) from H. Then $|V(G)| = k$, and so by $S(k)$,

$$\chi(G) + \chi(\overline{G}) \leq k + 1.$$

Since H is only one vertex larger than G, it is clear that $\chi(H) \leq \chi(G) + 1$ and $\chi(\overline{H}) \leq \chi(\overline{G}) + 1$. If either $\chi(H) = \chi(G)$ or $\chi(\overline{H}) = \chi(\overline{G})$, (or both) then

$$\begin{aligned} \chi(H) + \chi(\overline{H}) &\leq \chi(G) + \chi(\overline{G}) + 1 \\ &\leq (k+1) + 1 \quad \text{(by } S(k)\text{)} \end{aligned}$$

confirms $S(k + 1)$.

There is only one case left: assume that both $\chi(H) = \chi(G) + 1$ and $\chi(\overline{H}) = \chi(\overline{G}) + 1$. Let there be g edges in H between x and G, and let there be g' edges in $\overline{H}$ between x and $\overline{G}$. Since $|V(G)| = k$, $g + g' = k$. The key observation is that since $\chi(H) = \chi(G) + 1$, necessarily $g \geq \chi(G)$ (since otherwise, one could extend a coloring of G to a coloring of H by coloring x with one of the colors not used by neighbors of x). Similarly, $g' \geq \chi(\overline{G})$. Hence,

$$k = g + g' \geq \chi(G) + \chi(\overline{G}),$$

and so $\chi(G) + \chi(\overline{G}) \leq k$. Then

$$\chi(H) + \chi(\overline{H}) = \chi(G) + \chi(\overline{G}) + 2 \leq k + 2 = (k+1) + 1$$

confirms $S(k + 1)$ in this case as well. This completes the inductive step $S(k) \to S(k + 1)$.

By mathematical induction, for all $n \geq 1$, $S(n)$ is true. $\qquad\qquad\qquad\square$

Exercise 500: For $n \geq 3$, let $S(n)$ be the statement that for any onto n-coloring of $E(K_n)$, there exists a rainbow colored triangle.
BASE STEP: All six onto 3-colorings of $E(K_3)$ are indeed injective.

INDUCTION STEP: Fix some $k > 3$, and assume that $S(k)$ holds. Let $\Delta : E(K_{k+1}) \to [k + 1]$ be onto. Pick any $x \in V(K_{k+1})$, and let $G = K_{k+1} \setminus \{x\}$, which is a copy of K_k. If $E(G)$ is k-colored, then by $S(k)$, G contains a rainbow K_3. So consider the cases when G is not k-colored. If $E(G)$ is $(k+1)$-colored, again G contains a rainbow K_3 (identify two colors, apply $S(k)$ with this new coloring). So assume that $E(G)$ is $(k - 1)$-colored. Then two of the colors are missing in G, so two edges incident with x receive these two colors; the third edge induced by these two receives yet another color, producing the rainbow triangle. Hence, $S(k + 1)$ is true.

By MI, for each $n \geq 3$, $S(n)$ is true. $\qquad\qquad\qquad\square$

Exercise 501: Proofs appears in many places besides the original, *e.g.*, a proof by L. Pósa appears in [354, 9.14, pp.67, 398–9] the first part of which is given here. First, a lemma proves an interesting property of infinite k-colorable graphs that have a maximal set of edges. Such a graph is then shown to exist by Zorn's lemma.

Lemma 33.6.1. *Suppose that G is an infinite (simple) graph whose every finite subgraph is k-colorable, but if any two non-adjacent vertices in G are joined by an edge, then there exists a finite subgraph that fails to be k-colorable. Then G is k-colorable.*

Note: The existence of such a graph is not yet established, but this question is answered below.

Proof of Lemma 33.6.1: First observe that since any finite subgraph of G is k-colorable, G does not contain any K_{k+1} (for such a graph is not k-colorable). Hence, if there exists such a desired partition, there are at most k (non-empty) classes.

To show that G is k-colorable, it suffices to show that $V(G)$ has a partition into at most k classes such that two points are in the same class iff they are non-adjacent (equivalently, they are adjacent iff they belong to different classes). To prove the lemma, it then suffices to prove that "non-adjacency" is an equivalence relation.

Since G is simple, both reflexivity and symmetry of "non-adjacency" are trivial. Suppose that $(a, b) \notin E(G)$ and $(b, c) \notin E(G)$. By maximality, there exists a finite subgraph H so that $(V(H), E(H) \cup \{\{a, b\}\}) = H + \{a, b\}$ is not k-colorable. Similarly, there exists a finite subgraph K so that $K + \{b, c\}$ is not k-colorable. Consider the graph $B = H \cup K + \{a, c\}$. It is now shown that B is not k-colorable. In hopes of a contradiction, suppose that B has a good k-coloring, say $\beta : V(B) \to \{1, 2, \ldots, k\}$.

Since under any good k-coloring of H (in particular, the one induced by β) the edge (a, b) can not be added, it follows that $\beta(a) = \beta(b)$. Similarly, $\beta(b) = \beta(c)$, and by transitivity of equality, $\beta(a) = \beta(c)$, which contradicts β being a good k-coloring of $V(B)$. So B is not k-colorable, and hence is not a subgraph of G (yet $H \cup K$ clearly is). Thus, $\{a, c\} \notin E(G)$, completing the proof of transitivity, and so, too, that non-adjacency is an equivalence relation, thereby proving the lemma. □

Note that Lemma 33.6.1 says that any such graph G satisfying its hypothesis is k-colorable. The goal is now to show that such a graph exists; this is done using Zorn's lemma.

Proof of Theorem 15.7.1: Let G be an infinite graph whose every finite subgraph is k-colorable. Consider the class of supergraphs of G,

$$\mathcal{G} = \{G \cup G' : V(G') = V(G), \text{all finite subgraphs of } G \cup G' \text{ are } k\text{-colorable}\},$$

and order $\mathcal{G}$ by inclusion. [*Note:* $G \in \mathcal{G}$, using G' to be the empty graph on $V(G)$.] Let $\mathcal{C}$ be a chain in $\mathcal{G}$, and let U be the union of the chain. Every finite subgraph H of U is k-colorable, because if not, H "appeared" in some $G_i \in \mathcal{G}$, $G_i \subseteq G$, whose every finite subgraph is assumed to be k-colorable. Hence, for every chain in $\mathcal{G}$, the union of the chain is also in $\mathcal{G}$. By Zorn's lemma, $\mathcal{G}$ contains a maximal element. Hence, without loss of generality, one can assume that G is maximal, that is, G satisfies the hypothesis of Lemma 33.6.1, and so G is k-colorable. □

Note: The proof in [354] does not state Zorn's lemma explicitly; a second ending is given that uses Tychonoff's theorem after establishing a notion of "closed sets" of colorings. This above form of compactness can also be translated to hypergraphs, and to theorems for other relational structures, and even to model theory (where, so I am told, some prefer to create a metric and a topology and then apply Tychonoff's compactness result for product spaces).

33.7   Solutions: Planar graphs

Exercise 503: The equation $v + f = e + 2$ for planar connected graphs is due to Euler [179]. There are many inductive proofs of this formula, some inducting on v, some on e. Perhaps the easiest is to induct on e.

Before beginning the proof, note that this result holds for general multigraphs, that is, those containing loops (edges of the form $\{x, x\}$) or multiple edges, (where a pair $\{x, y\}$ occurs as an edge more than once) yet the restriction here is to simple graphs. Perhaps the result for more general graphs is more easily proved by induction on v (see [566, p. 241], for example).

For $e \geq 0$, let $S(e)$ be the statement that if G is a connected planar graph with e edges, (with a planar embedding) then $v + f = e + 2$ (where $v = |V(G)|$ and f is number of faces or regions).

BASE STEP: When $e = 0$, any connected graph with no edges must be a single vertex, and there is only one face (the outer one). Then $v + f = 1 + 1 = 0 + 2 = e + 2$, so $S(0)$ holds. As an added check, if $e = 1$, since the graph must be connected, there are only $v = 2$ vertices and exactly $f = 1$ face, so $v + f = 2 + 1 = 1 + 2 = e + 2$ holds, showing $S(1)$, too.

INDUCTIVE STEP: Fix some $e \geq 1$, and assume that $S(e - 1)$ is true, that is, for a graph with $e' = e - 1$ edges, v' vertices, and f' faces, $v' + f' = e' + 2 = e + 1$. Let G be a planar connected graph (with a planar embedding) with e edges, v vertices and f faces. To be shown is that $S(e)$ holds, namely that $v + f = e + 2$.

If G contains no cycle, then since G is connected, G is a tree. In this case, there is only one face, and (by Exercise 477) $e = v - 1$, so $v + f = e + 2$ holds (without any induction).

So now suppose that G contains a cycle; fix one edge $\{x, y\}$ in that cycle and delete it, forming G' with $e' = e - 1$ edges, $v' = v$ vertices, and f' faces. Deletion of one edge in a cycle does not disconnect a graph, so G' is connected (and still planar). Since removal of an edge on a cycle joins one face inside the cycle to one outside, G' has $f' = f - 1$ faces. By $S(e - 1)$, $v' + f' = e' + 2$, which implies $v + (f - 1) = (e - 1) + 2$ and so $v + f = e + 2$ as desired. This completes the inductive step.

By MI, for every $e \geq 0$, Euler's formula for connected planar graphs with e edges is true. $\qquad\square$

Comment: Try an inductive proof which inducts on the number of vertices; however, be careful that when a vertex is deleted, the remaining graph is connected.

Exercise 504 (Euler's formula for planar graphs with n components): To be shown is that if a planar graph has n components, then

$$v - e + f = n + 1.$$

The proof is by induction on n.

BASE STEP: The base case $n = 1$ is true by Euler's formula for connected planar graphs (Exercise 503).

INDUCTIVE STEP: Suppose that the formula holds for a planar graph with $k \geq 1$ components, and consider a planar graph G with $k + 1$ connected components, say $C_0, C_1, \ldots, C_k$. Let G' be the graph obtained from G by removing C_0. If G' has v' vertices, e' edges and f' faces, then by the induction hypothesis,

$$v' - e' + f' = k + 1.$$

If the single component C_0 has v_0 vertices, e_0 edges and f_0 faces, then by Euler's formula (for a connected planar graph),

$$v_0 - e_0 + f_0 = 2.$$

Then G has $v = v' + v_0$ vertices, $e = e' + e_0$ edges, and $f = f' + f_1 - 1$ faces (since the infinite face of G' is a face of C_0), and hence

$$
\begin{aligned}
v - e + f &= (v_0 - e_0 + f_0) + (v' - e' + f') - 1 \\
&= 2 + (k + 1) - 1 \quad \text{(by Euler's formula and ind. hyp)} \\
&= (k + 1) + 1,
\end{aligned}
$$

proving the formula for graphs with $k + 1$ components, completing the inductive step.

Therefore, by induction, for every $n \geq 1$, the result holds for all graphs with n components. $\qquad \square$

Exercise 505: Let $S(n)$ be the statement that every planar graph with n vertices is 6-colorable.

BASE STEP: Every planar graph on six or fewer vertices is trivially 6-colorable (color each vertex with a different color) and so $S(n)$ is true for $n = 0, 1, 2, 3, 4, 5, 6$.

INDUCTION STEP: Let $k \geq 6$ and suppose that $S(k)$ is true. Let G be a planar graph with $k + 1$ vertices. By Lemma 15.8.2, let $x \in V(G)$ be a vertex of degree at most 5. Delete x (and all edges incident with x), forming a graph H. Since G was planar, so is H, and thus by $S(k)$, fix a good 6-coloring of H. Since x was of degree at most 5, there are at most 5 vertices in $V(H)$ connected to x in G, so color x with a color unused for these 5 (and color remaining vertices of G as in H). This represents a good 6-coloring of G, and so shows that $S(k + 1)$ is true.

By mathematical induction, for every $n \geq 0$, $S(n)$ holds, and so the statement of the exercise is true. $\qquad \square$

Exercise 506: The proof that every planar graph is 5-colorable relies on what are called "Kempe chains". See [327] for a discussion of Kempe chains, and how the 5-color proof arose from a failed 4-color proof by Kempe. It was Heawood who discovered that Kempe's ideas actually showed that every planar graph is 5-colorable. For a proof of the 5-color theorem by induction, see [64, pp. 291-2], where two more proofs are outlined.

33.8 Solutions: Extremal graph theory

Exercise 508: For $n \geq 2$, let $S(n)$ be the statement that the maximum number of edges in a disconnected simple n-vertex graph is $\binom{n-1}{2}$, with equality only for $K_1 + K_{n-1}$.

BASE STEP: The only disconnected graph on $n = 2$ vertices is a pair of isolated vertices, *i.e.*, $K_1 + K_1$, so $S(2)$ is true.

INDUCTION STEP: Let $k \geq 2$ and suppose that $S(k)$ is true. Let H be a disconnected graph with $k+1$ vertices, and suppose that among all disconnected graphs on $k+1$ vertices, H has the maximum number of edges. Fix a vertex $x \in V(H)$ and delete it, producing a graph G on k vertices.

If x is an isolated vertex, then $|E(G)| \leq \binom{k}{2}$ edges, and when $G = K_k$, equality is attained, in which case $H = K_1 + K_k$.

So assume that x is not isolated. Then G is disconnected, and $\deg_H(x) \leq k-1$. Thus an upper bound for the number of edges in H is $k-1$ plus the maximum number of edges in a disconnected graph on k vertices. By $S(k)$, this total is

$$\binom{k-1}{2} + k - 1 = \frac{(k-1)(k-2)}{2} + k - 1 = \binom{k}{2}.$$

In order to achieve this bound, one requires both $\binom{k-1}{2}$ edges in G and $\deg_H(x) = k-1$. The first condition says (by $S(k)$) that $G = K_1 + K_{k-1}$, and in this case, the only way to have to $\deg_H(x) = k-1$ (and still have H disconnected) is to have x connected to every vertex in the component K_{k-1} of G. Thus $K_1 + K_k$ is the unique disconnected graph on $k+1$ with maximal number of edges. This proves $S(k+1)$ and hence the inductive step.

Therefore, by mathematical induction, for every $n \geq 2$, the statement $S(n)$ is true. $\qquad\square$

Exercise 509 (Mantel's theorem): This result, apparently proved in 1905, was published in 1907 [362] and is the simplest case of Turán's theorem (published in 1941 (see Exercise 512)). This problem occurs in many graph theory texts; it even appears in disguise in [161, 8.1, p. 207].

Let $S(n)$ be the statement that if a simple graph G on n vertices has more than $n^2/4$ edges, then G contains a triangle.

BASE STEP: Since any graph on 1 or 2 vertices is triangle-free, consider the case $n = 3$. In this case $n^2/4 = 2.25$, so $S(3)$ says that any graph on three vertices with at least 3 edges has a triangle, a true statement.

There are two proofs of the inductive step. The first, rather short and sweet, is the proof that $S(k) \to S(k+2)$, and since there are two base cases done, it will suffice. The second proof is $S(k) \to S(k+1)$ and needs only a bit more care; however, it is often the first proof taught to students.

INDUCTIVE STEP $S(k) \to S(k+2)$: suppose that for some $k \geq 3$, $S(k)$ holds, and let H be a graph on $k+2$ edges with no triangle. Consider some edge $e = \{x,y\}$ in H, and the graph $G = H\backslash\{x,y\}$. There can be at most k edges from e to G, for otherwise a triangle is formed. By $S(k)$, G has at most $k^2/4$ edges, so H has at most $k^2/4 + k + 1 = (k+2)^2/4$ edges, thereby confirming $S(k+2)$.

INDUCTIVE STEP $S(k) \to S(k+1)$: Suppose that for some $k \geq 3$, $S(k)$ is true. Let H be a graph on $k+1$ vertices. The idea is to delete a vertex in H with smallest possible degree to create G on k vertices, and then show there are still lots of edges left, enough to apply $S(k)$.

Let H have more than $(k+1)^2/4$ edges. Note that if H has a triangle, then any graph with additional edges will also, so suppose, without loss of generality, that H has as few edges as possible, but still more than $k+1^2/4$. It is convenient to break the proof that H has a triangle into two cases, k even, and k odd.

First suppose that $k = 2m$. Then $\frac{(k+1)^2}{4} = \frac{4m^2+4m+1}{4}$, and so assume that H has $m^2 + m + 1$ edges. The average degree of vertices in H is $\frac{2(m^2+m+1)}{2m+1} = m + \frac{m+1}{2m+1}$, and so there is a vertex $x \in V(H)$ with degree at most m. Delete x (and all edges incident with x) to give a graph G on $k = 2m$ vertices and with at least $m^2 + m + 1 - m = m^2 + 1 = k^2/4 + 1$ edges. Thus by $S(k)$, G contains a triangle, and hence so did H.

Suppose that $k = 2m + 1$. Then $\frac{(k+1)^2}{4} = \frac{(2m+2)^2}{4} = (m+1)^2$, so assume that H has $(m+1)^2 + 1 = m^2 + 2m + 2$ edges. Then vertices in H have average degree $\frac{2(m^2+2m+2)}{2m+2} = m+1+\frac{2}{2m+2}$, so there is a vertex x with degree at most $m+1$. Delete x to give G with k vertices and at least

$$m^2 + 2m + 2 - (m+1) = m^2 + m + 1 > m^2 + m + \frac{1}{4} = \frac{(2m+1)^2}{4} = \frac{k^2}{4}$$

edges remaining. Thus by $S(k)$, G contains a triangle, and hence H also.

So in either case, k even or odd, $S(k+1)$ is true. This completes the inductive step $S(k) \rightarrow S(k+1)$.

Thus by mathematical induction, for all $n \geq 3$, $S(n)$ is true. □

Comment: Among all triangle-free graphs on n vertices, there is only one with the most number of edges, namely the complete bipartite graph $K_{\lfloor n/2 \rfloor, \lceil n/2 \rceil}$, where the partite sets are chosen as equal as possible in size.

Exercise 510: This exercise (without solution) occurred in [226, p. 320, Q. 15], and was generalized in [168]. Here, the term *bow tie* is used to describe the graph on 5 vertices consisting of two triangles with precisely one vertex in common (see Figure 33.1).

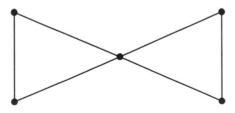

Figure 33.1: The bow tie graph

Let $S(n)$ be the statement that if a simple graph G on $n \geq 5$ vertices has more than $\frac{n^2}{4} + 1$ edges, then G contains a bow tie (as a weak subgraph).

BASE STEP: For $n = 5$, $\frac{n^2}{4} + 1 = 7.25$, so it suffices to show that a graph on 5 vertices and 8 edges contains a bow tie. (If such a graph contains a bow tie, then any graph with 5 vertices and 9 or 10 edges does, too.) It is not difficult to verify that there are only two such graphs; take the complete graph K_5 and remove two edges—these edges can be touching, or not. (One has vertices with degrees 4,4,3,3,3, and the other has degrees 4,3,3,3,3). Each has a bow tie, where the center of the bow tie is a vertex of degree 4.

INDUCTIVE STEP: Let $p \geq 5$ and suppose that $S(p)$ holds, that is, any graph with p vertices and more than $\frac{p^2}{4} + 1$ edges contains a bow tie. Let G be a graph with $p + 1$ vertices and q edges, where q is the smallest number larger than $\frac{(p+1)^2}{4} + 1$. Split the proof that G contains a bow tie into two cases. [Recall that $G \backslash x$ denotes a graph G with vertex x (and all edges incident with x) deleted.]

First consider the case when p is odd, say $p = 2m + 1$ (and so $m \geq 2$). Then

$$\frac{(p+1)^2}{4} + 1 = \frac{(2m+2)^2}{4} + 1 = m^2 + 2m + 2,$$

so $q = m^2 + 2m + 3$. The average degree of a vertex in G is

$$\frac{2(m^2 + 2m + 3)}{2m + 2} = m + 1 + \frac{4}{2m + 2} < m + 2,$$

and so there is a vertex $x \in V(G)$ with $\deg_G(x) \leq m + 1$. Deleting x leaves

$$
\begin{aligned}
|E(G \backslash x)| \quad &\geq \quad m^2 + 2m + 3 - (m + 1) \\
&= \quad m^2 + m + 2 \\
&= \quad \frac{4m^2 + 4m + 8}{4} \\
&= \quad \frac{(2m + 1)^2 + 7}{4} \\
&> \quad \frac{p^2}{4} + 1
\end{aligned}
$$

edges, so by $S(p)$, $G \backslash x$ contains a bow tie, and hence so does G.

Next consider when p is even, say $p = 2m$. Since $p \geq 5$, it follows that $m \geq 3$. Then

$$\frac{(p+1)^2}{4} + 1 = \frac{(2m+1)^2}{4} + 1 = m^2 + m + \frac{5}{4},$$

so $q = m^2 + m + 2$. The average degree of a vertex in G is

$$\frac{2(m^2 + m + 2)}{2m + 1} = m + \frac{m + 4}{2m + 1},$$

and so there is a vertex $x \in V(G)$ with $\deg_G(x) \leq m + \frac{m+4}{2m+1}$. Now there is a bit of a problem. In general, $\frac{m+4}{2m+1}$ might be as large as 1 when $m = 3$, but deleting a vertex of degree $m+1$ leaves $m^2 + 1 = \frac{p^2}{4} + 1$ edges, not quite enough to use the inductive hypothesis $S(p)$. It turns out, however, that when $m > 3$, this trick works, so let's take care of this first, then return to $m = 3$. Suppose for now that $m > 3$. Then $\frac{m+4}{2m+1} < 1$, making the average degree less than $m + 1$, so there is a vertex x with degree at most m. Delete x, leaving $m^2 + 2 = \frac{p^2}{4} + 2$ edges in $G \backslash x$, and so by $S(p)$, the graph $G \backslash x$ contains a bow tie, and hence so does G.

What remains is the annoying case $m = 3$, that is, $p = 7$. In this case $\frac{p^2}{4} + 1 = 13.25$, so assume that G has 7 vertices and 14 edges. If some vertex x in G has $\deg_G(x) \leq 3$, then delete x, leaving a graph on 6 vertices and at least $11 = \frac{6^2}{4} + 2$ edges, so by $S(6)$, [which follows from the above inductive step where $p = 5$ is odd] G contains a bow tie.

So suppose that every vertex has degree at least 4. Note that since the sum of the degrees ($\geq 7 \cdot 4$) is exactly twice the number of edges (14), every vertex has precisely degree 4 (that is, G is 4-regular). Thus, it remains to show that any 4-regular graph on 7 vertices contains a bow tie. This can be done by an exhaustive analysis, however it is made slightly easier by first deleting a vertex, getting a graph on 6 vertices with 10 edges, and vertex degrees 4,4,3,3,3. Up to isomorphism, there are only three such graphs (whose complements are $C_4 + K_2$, P_5, and $K_3 + P_2$, where P_i denotes a path of length i), two of which are easily seen to contain a bow tie. For the remaining one, re-affix the deleted vertex and the bow tie is easily spotted. This concludes the inductive step where $p = 2m$, and hence the inductive step in general.

By mathematical induction, for each $n \geq 5$, the statement $S(n)$ is true. $\square$

Exercise 511: This appears in [161, 8.23, p. 208]. For each $n \geq 0$ let $S(n)$ be the statement that if G is a graph on n vertices with no tetrahedron, then G contains at most $n^3/3$ edges.

BASE STEP: Since a tetrahedron contains 4 vertices, any graph on $n = 0, 1, 2, 3$ vertices contains no tetrahedron. The number of edges in each case is bounded above by 0, 0, 1, or 3, respectively, each of which is less than $n^3/3$,

INDUCTIVE STEP $S(k) \rightarrow S(k+3)$: Suppose for some $k \geq 3$, $S(k)$ is true. Consider a graph G with $k + 3$ vertices. If G does not contain a triangle, then it contains no tetrahedron, and by Mantel's theorem (see Exercise 509), G contains at most $(k+3)^2/4 < (k+3)^2/3$ edges. So suppose that three vertices in G form a triangle. These three cannot be connected to another common neighbor, so there are at most at most $2k$ additional edges to the remaining vertices. Since the remainder of the graph contains no tetrahedron, by $S(k)$ that part of the graph contains at most $k^2/3$ edges. In all, the maximum number of edges is $k^2/3 + 2k + 3 = (k+3)^2/3$, completing the proof of $S(k+3)$.

By mathematical induction (actually three inductive proofs rolled in one), for all $n \geq 0$, the statement $S(n)$ is true. $\square$

Exercise 512: (Turán's theorem) There are many solutions (see, for example, [7, pp. 183–187] for five different proofs); here is one by induction on n. Fix $k \geq 1$ and for each $n \geq 1$, let $S(n)$ be the statement that if G is a K_{k+1}-free graph on n vertices and $\mathrm{ex}(n; K_{k+1})$ edges, then $G = T(n, k)$.

BASE CASES: For each $i = 0, 1, \ldots, k$, the graph with the most edges on i vertices is $K_i = T(i, k)$, so $S(i)$ holds.

INDUCTIVE STEP: Fix some $m \geq k$ and suppose that $S(m - k)$ holds. Let G be a K_{k+1}-free graph on m vertices with $\mathrm{ex}(m, K_{k+1})$ edges. As G is extremal for K_{k+1}, G contains a copy of K_k, call it H, on vertices $A = \{a_1, \ldots, a_k\}$. Put $B = V(G)\backslash A$, and let G^* be the graph induced on B.

Since G is K_{k+1}-free, each vertex in B is adjacent to at most $k - 1$ vertices of A. Hence

$$
\begin{aligned}
|E(G)| &\leq \binom{k}{2} + |B|(k - 1) + |E(G^*)| \\
&\leq \binom{k}{2} + (m - k)(k - 1) + \mathrm{ex}(m - k; K_{k+1}) \quad \text{(since } G^* \text{ is } K_{k+1}\text{-free)} \\
&\leq \binom{k}{2} + (m - k)(k - 1) + t(m - k, k) \quad \text{(by IH, } S(m - k)) \\
&= t(m, k) \quad \text{(to see this, look at structure of } T(m, k)).
\end{aligned}
$$

Hence $|E(G)| \leq t(m, k)$. Also, since $T(m, k)$ is K_{k+1}-free and G has an extremal number of edges, $t(m, k) \leq |E(G)|$. Thus $|E(G)| = t(m, k)$, forcing equality in the equations above. Then each vertex in B is joined to exactly $k - 1$ vertices of A.

For each $i = 1, \ldots, k$, put $W_i = \{x \in V(G) : \{x, a_i\} \notin E(G)\}$. Note that $a_i \in W_i$ and the W_i's partition $V(G)$ since every vertex in B is not adjacent to one of the a_i's. Each W_i is an independent set, since if some $x, y \in W_i$ were adjacent, x, y and $A\backslash\{a_i\}$ form a K_{k+1}. Hence, G is k-partite.

Since $T(m, k)$ is the unique k-partite graph with as many edges as possible, $G = T(m, k)$. This completes the inductive step $S(m - k, k) \rightarrow S(m, k)$.

By mathematical induction, for all $n \geq 0$, $S(n)$ is true. $\qquad\qquad\square$

Exercise 513: For each $r \geq 1$, let $A(r)$ be the assertion (both statements) in the theorem. The proof is by induction on r.

BASE STEP: When $r = 1$, G has no edges, so is "1-partite".

INDUCTIVE STEP: Suppose that $s \geq 2$ and that $A(s - 1)$ is true. Let G contain no K_{s+1}. Pick a vertex $x \in V(G)$ of maximal degree in G, and put $Y = N_G(x)$, the neighborhood of x, and put $X = V \setminus Y$ (so $x \in X$). Then the graph $G^* = G[Y]$ induced by vertices of Y is K_s-free (otherwise $G^* + x = K_{s+1}$). Applying $A(s-1)$, get an $(s-1)$-partite graph H^* on vertex set Y, where for every $y \in Y$, $d_{G^*}(y) \leq d_{H^*}(y)$, and if G^* is not complete $(s - 1)$-partite, there exists a vertex in Y with strict inequality. Form the graph H by adding the vertices of $V \setminus Y$ to H^*, connecting all vertices in W to all in $V \setminus Y$.

For $v \in X$, $d_G(v) \leq d_G(x) = |Y| = d_H(x) = d_H(v)$. For $z \in Y$, $d_G(v) \leq d_{G^*}(v) + |X| \leq d_{H^*} + |X| = d_H(v)$. In any case, $d_G(v) \leq d_H(v)$, as required.

To show the second statement in $A(s)$, it suffices to show that if $d_G(z) < d_H(z)$ never holds, then G is a complete s-partite graph. So assume that for every $v \in V$, $d_G(v) = d_H(v)$. Counting degrees in Y,

$$\sum_{v \in Y} d_{H^*}(v) + |X| \cdot |Y| = \sum_{v \in Y} d_H(v) = \sum_{v \in Y} d_G(v) \leq \sum_{v \in Y} d_{G^*}(v) + |X| \cdot |Y|, \quad (33.1)$$

and so $\sum_{v \in Y} d_{H^*}(v) \leq \sum_{v \in Y} d_{G^*}(v)$. However, H^* was chosen so that for each $y \in Y$, $d_{G^*}(y) \leq d_{H^*}(y)$, so $d_{G^*}(y) = d_{H^*}(y)$ and equality holds in (33.1). Hence

$$\sum_{v \in Y} d_G(v) = \sum_{v \in Y} d_{G^*}(v) + |X| \cdot |Y|,$$

which says that the number of edges leaving Y in G is maximized, that is, each vertex in Y is adjacent to all of X. Since $|E(G)| = |E(H)|$, $|E(G^*)| = |E(H^*)|$, and all edges in H contain at least one vertex in Y, it follows from equality in (33.1) that $|E(G[X])| = 0$, and so G is a complete s-partite graph, proving the second statement in $A(s)$. The inductive step is complete.

By mathematical induction, for each $r \geq 1$, $A(r)$ is true. $\qquad \square$

33.9   Solutions: Digraphs and tournaments

Exercise 514: Repeat the proof of Euler's theorem for graphs—the version for digraphs is nearly identical. Begin by showing that if all vertices have minimum outdegree 1, there exists a directed cycle (look at a maximal directed path). $\qquad \square$

Exercise 515: This problem of showing a king exists was first proved by H. G. Landau in 1953 [338], and continues to appear in graph theory texts and puzzle books ever since. For example, in [161, 8.3, p. 207], it appears as a problem on one-way streets in Sikinia (where ever that is!). First a "standard" analysis is given that leads to an observation (the claim below) which has a simple proof of a stronger claim that implies every tournament contains a king; two purely inductive proofs of the original statement are then given.

Analysis: Let $T = (V, E)$ be a tournament with $|V|$ finite, and consider some vertex x. If x is not a king, then there is some other vertex y so that there is no path of length at most two from x to y. This means that $(y, x) \in E$ and for every other $z \in V$, if $(x, z) \in E$, then $(y, z) \in E$. But this implies that $d^+(y) > d^+(x)$. So repeat this argument replacing x with y, a vertex with higher outdegree. This process can not continue *ad infinitum*, and so when it stops, a king is located. Such a king will have maximum outdegree, and so perhaps proving a stronger result is easier:

Claim: In a tournament, any vertex with maximum outdegree is a king.

Proof of claim: Let v be a vertex of maximum outdegree in a tournament T. Let $X = \{x \in V(T) : (v, x) \in E(T)$, and $Y = \{y \in V(T) : (y, v) \in E(T)\}$. (Then $X \cap Y = \emptyset$, $V(T) = \{v\} \cup X \cup Y$, and $d^+(v) = |X|$.) In hopes of contradiction, suppose that v is not a king, that is, there exists some $z \in V(T)$ so that there is no path of length of at most two from v to z. Then $z \in Y$, and for every $x \in X$, $(z, x) \in E(T)$. Since $(z, v) \in E(T)$ as well, $d^+(z) = |X| + 1$, contradicting that v has maximum outdegree. □

Here is an inductive proof of the original exercise:

Inductive proof: Induct on the number of vertices. Let $C(n)$ be the claim that any tournament on n vertices contains a king. BASE STEP: If $n = 1$ or $n = 2$, the claim holds trivially.

INDUCTIVE STEP: Suppose that for some $k \geq 2$, $C(k)$ is true, and let $T = (V, E)$ be a tournament on $k + 1$ vertices, and fix some vertex $x \in V$. The remaining k vertices form a tournament $T' = T\backslash\{x\}$, and so by $C(k)$, T' contains a king w.

There are two cases. If $(w, x) \in E(T)$, then w is a king of T as well, and nothing is left to show. If there is a path of length two from w to x, again, w is a king in T.

So suppose that $(x, w) \in E(T)$ and that there is no path of length two from w to x. Thus for every y with $(w, y) \in E$, one has $(x, y) \in E$ (for if not, $(w, y), (y, x)$ is a path of length two witnessing w as a king). Since all remaining vertices are reachable from w through such y's, so are they similarly reachable from x. Hence x is a king. This concludes the inductive step.

Therefore, by mathematical induction, for any $n \geq 1$, the claim $C(n)$ is true. □

Remark: The inductive proof did not immediately reveal the much stronger statement derived in the first proof. It also might be of interest to know that there can be more than one king, and in fact, it is easy to construct a tournament with two kings. Maurer [374] proved that if $1 \leq k \leq n$, there is a tournament on n vertices with precisely k kings—except when $k = 2$ or $n = k = 4$. So in large tournaments, everyone can be king!

A simpler proof uses strong induction, and this inductive step is only outlined. [Thanks to Liji Huang for reminding me of this proof.] Let T be a tournament, and let x be any vertex in T. Let A be the set of all those vertices dominated by x, and let B be the set of all those vertices that dominate x. By inductive hypothesis, the tournament induced by vertices of B contains a king, say y. If $B = \emptyset$, then x is a king of T, so assume $B \neq \emptyset$. For any $b \in B$, since y is a king of B, there is a directed path from y to b of length at most two, and for any vertex in $a \in A$, there is a path (through x) from y to a of length 2. □

Exercise 516: This appeared in [566, 1.4.35, p. 65] marked as a harder problem. The reference supplied was [338]. Good luck!

Exercise 517: Let $S(n)$ be the statement that if a tournament is held among n players, where every pair of players meets precisely once, then there is a listing of all the players $a, b, c, \ldots$, so that a beat b, b beat c, and so on, continuing until the last player. Call such a listing a *ranking*.

Note: The term "ranking" is often used in a different sense; one could define a *ranking* of any n-element set V to be simply a bijection $\sigma : V \to [n]$. For example, see [503, p. 6]. The ordering of vertices in this present context might more aptly be called a "chain ranking", or some such.

Base step: If $n = 0$, the empty list is vacuously a ranking. If $n = 1$, the list consisting of a single player is also vacuously a ranking. When $n = 2$, there is one match, and so simply list the winner first and the loser second.

Inductive step: Use strong induction on k. Fix some $k \geq 2$, and suppose for each $i = 0, 1, 2, \ldots, k$ that $S(i)$ is true, and that a tournament T has been held with $k+1$ players. Fix any player p, and consider the two groups, A made of players who beat p, and B, those whom p beat. [Think: A=above, B=below] Since $|A| \leq k$, by induction hypothesis $S(|A|)$, a ranking of players in A exists; similarly a ranking for B exists. Since p was beaten by any member of A, in particular, by the last member of A, and p beat every member of B, in particular, the first member of B, the new listing formed by concatenating the ranking of A, followed by p, then followed by the ranking of B, is a ranking of the $k + 1$ players as desired. So $S(k + 1)$ is true, completing the inductive step.

By mathematical induction, for each $n \geq 0$, any tournament with n players has a ranking.   □

Exercise 518: For $n \geq 3$, let $P(n)$ denote the proposition "if a tournament T has a directed cycle on n vertices, then it contains a directed cycle on three vertices."

BASE STEP: When $n = 3$, there is nothing to prove.

INDUCTIVE STEP: Fix $k \geq 4$, and assume $P(k-1)$ holds, that is, if T is a tournament with a cycle on $k - 1$ vertices, then T contains a directed cycle.

Let $T = (V, D)$ be a tournament containing a cycle on k vertices, say $x_0, x_1, \ldots, x_{k-1} \in V$, where for each $i = 0, 1, \ldots, k-1$, $(x_i, x_{i+1}) \in D$ (addition in indices is done modulo k). If for any i, $(x_{i+2}, x_i) \in D$, then x_i, x_{i+1}, x_{i+2} is a directed triangle. So suppose that all such pairs (two apart on the cycle) are directed in the same direction as the cycle, *i.e.*, $(x_i, x_{i+2}) \in D$. Then, *e.g.*, the vertices $x_0, x_2, x_3, \ldots, x_{k-1}$ form a directed cycle with $k - 1$ vertices. Thus, by $P(k-1)$, T contains a directed triangle, and so $P(k)$ is true.

Thus, by MI, for each $n \geq 3$, $P(n)$ is true.   □

Exercise 519: (Outline) See [90] for more details. The right side of the equation in Exercise 519 is the sum of the first $n-1$ squares (see Exercise 54). Thus the critical observation is that there is no directed triangle if and only if all outdegrees are different (needs proof), in which case they must be $0, 1, 2, \ldots, n-1$. Then prove that the sum of the squares is maximized when all degrees are different, and this is done by induction.

33.10   Solutions: Geometric graphs

Exercise 520: For each $n \geq 2$, let $A(n)$ be the assertion that if the edges of a complete geometric graph on n vertices are 2-colored, then there exists a monochromatic plane spanning tree. The proof is by strong induction on n.
BASE STEP: $A(2)$ holds trivially as there is only one edge in a spanning tree.

INDUCTION STEP: Let $k \geq 3$ and suppose that $A(2)$, ..., $A(k-1)$ are all true. Consider points $P = \{p_1, \ldots, p_k\}$ in general position, let G denote the geometric graph on P, and let a red-blue coloring of $E(G)$ be given.

Case 1: At least two edges on the border of the convex hull of P receive different colors. Then there exist consecutive points, say m, p, q, on the border of conv(P) so that the segments mp and pq are colored differently. By the induction hypothesis $A(k-1)$ on $P' = P\backslash\{p\}$, the geometric graph induced by P' has a monochromatic plane spanning tree, and together with one of mp or pq a monochromatic plane spanning tree for G is formed.

Case 2: All edges on the border of conv(P) are colored identically, say red. Without loss of generality, assume that the x-coordinates of all p_i's are strictly increasing, and so the points are ordered $p_1, \ldots, p_k$ left to right. For each $1 < i < k$, let G_i^ℓ and G_i^r be the graphs induced by $\{p_1, p_2, \ldots, p_i\}$ and $\{p_i, \ldots, p_k\}$ respectively. For each such i, by the inductive hypotheses $A(i)$ and $A(k-i)$, each of G_i^ℓ and G_i^r has a monochromatic plane spanning tree, say T_i^ℓ and T_i^r, respectively. If for some i, both T_i^ℓ and T_i^r are colored the same, then their union forms a monochromatic plane spanning tree for G, so assume that each such pair of trees has different colors. Furthermore, if either T_2^r or T_{k-1}^ℓ is red, a red edge on the border of conv(P) joining either p_1 to T_2^r or T_{k-1}^ℓ to p_k produces a red plane spanning tree for G, so assume that both T_2^r and T_{k-1}^ℓ are blue. Then the sequence of left-right color pairs begins red-blue, and ends in blue-red; hence there exists an $i \in \{2, \ldots, k-2\}$ so that $T_i\ell$ is red, T_i^r is blue, T_{i+1}^ℓ is blue, and T_{i+1}^r is red.

Adjoining any red edge from the border of conv(P) that joins T_i^ℓ to T_{i+1}^r (which crosses a vertical line between p_i and p_{i+1}) yields a red plane spanning tree for G. In any case, $A(k)$ is true, completing the inductive step.

By strong mathematical induction, for each $n \geq 2$, the assertion $A(n)$ is true. $\square$

Exercise 521: See [304, Thm. 1.2] for the brief solution.

Chapter 34

Solutions: Recursion and algorithms

34.1 Solutions: Recursively defined sets

Exercise 524: Let AP$_3$ be short for "3-term arithmetic progression". This exercise appeared as problem 5 of the 1983 IMO competition, where it asked if it is possible to choose 1983 distinct positive integers in $[1, 10^5]$ that is AP$_3$-free. The solution given in [342] uses S_n and T_n defined as follows: Let $S_1 = \{1\} = T_1$. For $n > 1$, recursively define

$$S_n = \{3^{n-2} + x : x \in T_{n-1}\},$$
$$T_n = S_n \cup T_{n-1}.$$

Let $A(n)$ be the combined assertions that T_n does not contain any AP$_3$. $|T_n| = 2^{n-1}$, and the largest element in T_n being $(3^{n-1} + 1)/2$. The proof is by induction on n. To get an idea of why this result is true, examine the first few T_i's and S_i:

$$
\begin{aligned}
S_1 = T_1 &= \{1\} \\
S_2 &= \{2\} \\
T_2 &= \{1, 2\} \\
S_3 = \{3 + 1, 3 + 2\} &= \{4, 5\} \\
T_3 = S_3 \cup T_2 &= \{1, 2, 4, 5\} \\
S_4 = \{9 + 1, 9 + 2, 9 + 4, 9 + 5\} &= \{10, 11, 13, 14\} \\
T_4 = S_4 \cup T_3 &= \{1, 2, 4, 5, 10, 11, 13, 14\}.
\end{aligned}
$$

BASE STEP $A(1)$: The singleton $T_1 = \{1\}$ certainly contains no AP$_3$, $|T_1| = 1 = 2^{1-1}$, and the largest element in T_1 is $1 = (3^{1-1} + 1)/2$, so $A(1)$ holds.

701

INDUCTIVE STEP $A(k) \to A(k+1)$: Let $k \geq 1$ and suppose that $A(k)$ holds. Then $T_{k+1} = S_{k+1} \cup T_k$. By induction hypothesis $A(k)$, T_k contains no AP3, and T_k contains 2^{k-1} elements, the largest of which is $(3^{k-1}+1)/2$. Since T_k contains no AP3, neither does S_{k+1}, as S_{k+1} is just a copy of T_k shifted by 3^{k-1}.

Suppose, for the moment, that T_{k+1} contains an AP3, say, $a, a+d, a+2d$. Then not all three terms are in the first half of T_{k+1} (that is, from S_{k+1}), nor could they be all from the second half (that is, from T_k). Thus d must be greater than, the size of the gap between S_{k+1} and T_k, which is

$$\min(S_{k+1}) - \max(T_k) = 3^{k-1} + 1 - \frac{3^{k-1}+1}{2} = \frac{3^{k-1}+1}{2}.$$

With such a value for d, it is impossible to have both a and $a+d$ in T_k, or to have both $a+d$ and $a+2d$ in S_{k+1}. Hence, T_{k+1} contains no AP3.

To finish the inductive step, it remains to observe that T_{k+1} contains 2^k elements and that the largest element in T_{k+1} is

$$\max(T_{k+1}) = \max(S_{k+1}) = 3^{k-1} + \max(T_k) = 3^{k-1} + \frac{3^{k-1}+1}{2} = \frac{3^k+1}{2}.$$

Therefore, by mathematical induction, for each $n \geq 1$, the assertion $A(n)$ is true. □

Comment: Starting with a large set $[1, n]$, one way to form an AP3-free set is to delete the middle third, which forces any AP3 into one end, then in each of the remaining thirds, delete the middle thirds again, continuing (as in the construction of the Cantor set) until this is no longer possible. This can be done "optimally" when n is a power of 3.

34.2 Solutions: Recursively defined sequences

Exercise 526: Fix the interest rate r and principal P. For any integer $n > 0$, let $S(n)$ be the statement that the amount in the account is $P(1+r)^n$.

BASE STEP: The value $n = 0$ corresponds to the time of the deposit, so there is P in the account, and since $P = P(1+r)^0$, $S(0)$ is true.

INDUCTIVE STEP: For some $k \geq 0$, suppose that $S(k)$ is true, that is, after k years, there is $P(1+r)^k$ in the account. After one more year, interest accrued is $rP(1+r)^k$. Together with the amount in the account, after $k+1$ years, there is $P(1+r)^k + rP(1+r)^k = (1+r)P(1+r)^k = P(1+r)^{k+1}$ in the account, which proves $S(k+1)$. This concludes the inductive step.

By mathematical induction, for all $n \geq 0$, the statement $S(n)$ is true. □

34.2.1 Solutions: Linear homogeneous recurrences of order 2

Exercise 527: See [150, Prob. 8, p. 209]; the solution is straightforward.

34.2.2 Solutions: Applying the method of characteristic roots

Exercise 528:. (Outline) The inductive proof is nearly since even plus odd is odd. The characteristic equation is $x^2 - x - 2 = 0$, which has roots $\alpha = 2$, $\beta = -1$. By Theorem 16.3.1, the general solution is

$$a_n = 2^{n+1} + (-1)^n.$$

The proof of this solution by induction is straightforward, using two base cases and two cases for the induction hypothesis (the reader is invited to write this proof formally):

When $n = 0$, $2^1 + 1 = 3 = a_0$, and when $n = 1$, $2^2 - 1 = 3 = a_1$, concluding the base step.

For the inductive step, let $k/geq1$ and for the induction hypothesis (IH), assume that both $a_{k-1} = 2^k + (-1)^{k-1}$ and $a_k = 2^{k+1} + (-1)^k$ are correct. Then by the recurrence relation $a_n = 2a_{n-2} + a_{n-1}$,

$$
\begin{aligned}
a_{k+1} &= 2a_{k-1} + a_k \\
&= 2(2^k + (-1)^{k-1}) + 2^{k+1} + (-1)^k \qquad \text{(by IH)} \\
&= 2^{k+1} + 2^{k+1} + (-1)^{k-1}(2 - 1) \\
&= 2^{k+2} + (-1)^{k-1} \\
&= 2^{k+2} + (-1)^{k+1},
\end{aligned}
$$

which is a form for a_{k+1} which agrees with the solution (when $n = k+1$), completing the inductive step.

Then by MI, for all $n \geq 0$, the solution above is verified. □

Exercise 529: The characteristic function is $x^2 - x - 1 = 0$, $\alpha = \frac{1+\sqrt{5}}{2}$ (the golden ratio) and $\beta = \frac{1-\sqrt{5}}{2}$, so Binet's formula follows directly from Theorem 16.3.1. □

Exercise 530: This example appears in [292, Ex. 21, p. 244], where the answer is $a_n = 2(-4) + 3n(-4)^n$. □

Exercise 531: The induction proof asked for is nearly trivial. To solve the recursion, the characteristic equation is $x^2 - 2x + 2 = 0$, with roots $\alpha = 1 + i$ and $\beta = 1 - i$. Using $\theta = \pi/4$, Theorem 16.3.1 gives

$$a_n = 2^{n/2}[\cos(\frac{n\pi}{4}) + \sin(\frac{n\pi}{4})].$$

This recurrence is also solved directly in [239, Ex. 10.21, pp. 465–466].

Exercise 532: Define $a_1 = 1$, $a_2 = 3$ and for each $k \geq 2$, define $a_{k+1} = 3a_k - 2a_{k-1}$.
For $n \geq 1$, let $P(n)$ be the proposition that $a_n = 2^n - 1$.

BASE STEPS: Since $a_1 = 1 = 2^1 - 1$ and $a_2 = 3 = 2^2 - 1$, both $P(1)$ and $P(2)$ hold.

INDUCTION STEP: Let $\ell \geq 3$ and suppose that $P(\ell - 2)$ and $P(\ell - 1)$ hold, that is,
$a_{\ell-2} = 2^{\ell-1} - 1$ and $a_{\ell-1} = 2^{\ell-1} - 1$. Then

$$
\begin{aligned}
a_\ell &= 3a_{\ell-1} - 2a_{\ell-2} \\
&= 3(2^{\ell-1} - 1) - 2(2^{\ell-2} - 1) \\
&= 6 \cdot 2^{\ell-2} - 3 - 2 \cdot 2^{\ell-2} + 2 \\
&= 4 \cdot 2^{\ell-2} - 1 = 2^\ell - 1,
\end{aligned}
$$

and so $P(\ell)$ is true, completing the inductive step.

Therefore, by (an alternative form of) MI, for all $n \geq 1$, $P(1)$ holds.   □

Exercise 533: The characteristic equation is $x^2 - bx + b^2$, with roots $\alpha = b(\frac{1+\sqrt{3}}{2})$,
$\beta = b(\frac{1-\sqrt{3}}{2})$. and so (with further simplification as in the case of complex roots,
using $\theta = \pi/3$) by Theorem 16.3.1,

$$
a_n = b^n[\cos(n\pi/3) + (\frac{1}{\sqrt{3}})\sin(n\pi/3)].
$$

This same recursive definition is also solved in detail in [239, Ex. 10.22, pp. 466–
7], where the sequence is a sequence of determinants. The inductive proof of this
solution is left to the reader (and may be made easier by translating back to numbers
without angles).   □

Exercise 534: For $n \geq 1$, let $P(n)$ be the statement that equation 16.5 holds.

BASE STEPS: $P(0)$ says $a_0 = \frac{2+1}{3} = 1$, and since by definition, $a_0 = 1$, $P(0)$ is true.
$P(1)$ says $a_1 = \frac{2^2-1}{3} = 1$, which is true by definition, so $P(1)$ holds.

INDUCTIVE STEP: Fix $k \geq 1$, and suppose that both

$$
P(k-1): \quad a_{k-1} = \frac{2^k + (-1)^{k-1}}{3}
$$

and

$$
P(k): \quad a_k = \frac{2^{k+1} + (-1)^k}{3}.
$$

It remains to show that

$$
P(k+1): \quad a_{k+1} = \frac{2^{k+2} + (-1)^{k+1}}{3}
$$

follows. Starting with the left-hand side of $P(k+1)$,

$$
\begin{aligned}
a_{k+1} &= a_k + 2a_{k-1} \\
&= \frac{2^{k+1} + (-1)^k}{3} + 2 \cdot \frac{2^k + (-1)^{k-1}}{3} \qquad \text{(by } S(k) \text{ and } S(k-1)) \\
&= \frac{2^{k+1} + (-1)^k + 2^{k+1} + 2 \cdot (-1)^{k-1}}{3} \\
&= \frac{2^{k+2} + (-1)^{k-1}(-1+2)}{3} \\
&= \frac{2^{k+2} + (-1)^{k-1}}{3} \\
&= \frac{2^{k+2} + (-1)^{k+1}}{3},
\end{aligned}
$$

which is the right side of $P(k+1)$, concluding the inductive step.

By mathematical induction (an alternate form), for each $n \geq 0$, $P(n)$ holds. $\qquad \square$

Remark: This result is for $n \geq 0$, slightly more than what the question asked for.

Exercise 535: A version of this problem appears in [499, Problem 18]. Define $a_1, a_2, a_3, \ldots$ by $a_1 = 2$, $a_2 = 3$ and for each $k \geq 2$,

$$
a_{k+1} = 3a_k - 2a_{k-1}.
$$

For each $n \geq 1$, let $C(n)$ be the claim that $a_n = 2^{n-1} + 1$.

BASE STEPS: Since $a_1 = 2 = 2^0 + 1$, and $a_2 = 2^1 + 1$, both $C(1)$ and $C(2)$ hold.

INDUCTIVE STEP: Fix $k \geq 1$ and suppose that both $C(k) : a_k = 2^{k-1} + 1$ and $C(k+1) : a_{k+1} = 2^k + 1$ hold. It remains to show $C(k+2) : a_{k+2} = 2^{k+1} + 1$. Then

$$
\begin{aligned}
a_{k+2} &= 3a_{k+1} - 2a_k \\
&= 3(2^k) - 2(2^{k-1} + 1) \quad \text{by } C(k+1) \text{ and } C(k)) \\
&= 3 \cdot 2^k + 3 - 2^k - 2 \\
&= 2 \cdot 2^k + 1 \\
&= 2^{k+1} + 1,
\end{aligned}
$$

as desired. This completes the inductive step $[C(k) \wedge C(k+1)] \to C(k+2)]$.

By MI (an alternative form) for each $n \geq 1$, $C(n)$ is true. $\qquad \square$

34.2.3   Solutions: Linear homogeneous recurrences of higher order

Exercise 536: This problem appeared in [18, Prob. 4, pp. 85, 209]. One notices that the first few values for x_n are 1, 1, 4, 9, 25, 64, ..., squares of the Fibonacci numbers. So, let $S(n)$ be the statement that $x_n = F_n^2$, and try to prove this by induction.

BASE STEPS: Since $x_1 = 1 = 1^2 = F_1^2$, $x_2 = 1 = 1^2 = F_2^2$, and $x_3 = 4 = 2^2 = F_3^2$, statements $S(1)$, $S(2)$, and $S(3)$ hold.

INDUCTIVE STEP $(S(k) \wedge S(k+1) \wedge S(k+2) \rightarrow S(k+3))$: Fix some $k \geq 1$, and assume that $S(k)$, $S(k+1)$, and $S(k+2)$ hold. One shows that $S(k+3) : \quad x_{k+3} = F_{k+3}^2$ is true by

$$
\begin{aligned}
x_{k+3} &= 2x_{k+2} + 2x_{k+1} - x_k \\
&= 2F_{k+2}^2 + 2F_{k+1}^2 - F_k^2 \quad \text{(by } S(k+2), S(k+1), S(k), \text{ resp.)} \\
&= F_{k+2}^2 + F_{k+1}^2 + F_{k+2}^2 + F_{k+1}^2 - F_k^2 \\
&= F_{k+2}^2 + F_{k+1}^2 + F_{k+2}^2 + (F_{k+1} + F_k)(F_{k+1} - F_k) \\
&= F_{k+2}^2 + F_{k+1}^2 + F_{k+2}^2 + F_{k+2}(F_{k+1} - F_k) \\
&= F_{k+2}^2 + F_{k+1}^2 + F_{k+2}(F_{k+2} + F_{k+1} - F_k) \\
&= F_{k+2}^2 + F_{k+1}^2 + F_{k+2}((F_{k+1} + F_k) + F_{k+1} - F_k) \\
&= F_{k+2}^2 + 2F_{k+2}F_{k+1} + F_{k+1}^2 \\
&= (F_{k+2} + F_{k+1})^2 \\
&= F_{k+3}^2.
\end{aligned}
$$

Thus $S(k+3)$ holds, finishing the inductive step.

By MI, for all $n \geq 1$, $S(n)$ holds.   □

Exercise 538: Define $s_0 = 1$, $s_1 = 2$, $s_2 = 3$, and for $n \geq 3$, define $s_n = s_{n-3} + s_{n-2} + s_{n-1}$.

For $n \geq 0$, let $C(n)$ be the claim that $s_n \leq 3^n$.

BASE STEPS: $s_0 = 1 = 3^0$, $s_1 = 2 \leq 3^1$, and $s_2 = 3 \leq 3^2$, so $C(0)$, $C(1)$, and $C(2)$ hold.

INDUCTION STEP $([C(k-3) \wedge C(k-2) \wedge C(k-1)] \rightarrow C(k))$: Fix $k \geq 3$, and suppose that $C(k-3)$, $C(k-2)$, and $C(k-1)$ all hold. Then

$$
\begin{aligned}
s_k &= s_{k-3} + s_{k-2} + s_{k-1} \\
&\leq 3^{k-3} + 3^{k-2} + 3^{k-1} \\
&= 3^{k-3}(1 + 3 + 9) \\
&< 3^{k-3}3^3 = 3^k.
\end{aligned}
$$

shows that $C(k) : s_k \leq 3^k$ holds (in fact, strict inequality holds), finishing the inductive step.

By (an alternative form of) MI, for all $n \geq 0$, $C(n)$ is true. □

Exercise 539: (Outline) Define $s_1 = s_2 = s_3 = 1$, and for $n \geq 1$, $s_{n+3} = s_n + s_{n+1} + s_{n+2}$. To show is that for each $n \geq 1$, $s_n < 2^n$. Repeat, (essentially) the solution from Exercise 538, where instead of using $1 + 3 + 9 < 3^3$, use $1 + 2 + 4 < 2^3$. □

Exercise 540: Define $s_1 = 2$, $s_2 = 4$, $s_3 = 7$, and for $n \geq 1$, define $s_{n+3} = s_n + s_{n+1} + s_{n+2}$. The s_i's are called the "tribonacci numbers". Let t_n be the number of binary strings that do not contain the substring "000". s_n is the number of binary strings of length n that do not contain the substring "000". For each $n \geq 1$, let $C(n)$ be the claim that $t_n = s_n$. To show that for all $n \geq 1$, $C(n)$ is true, is by induction:

BASE STEPS: Since $t_1 = 2$, $t_2 = 4$, and $t_3 = 7$ are easily verified, $C(1)$, $C(2)$, and $C(3)$ are true.

INDUCTION STEP: For some $j \geq 1$, suppose that $C(j)$, $C(j+1)$, and $C(j+2)$ are true, and let $f = (f(1), f(2), \ldots, f(j+3)) \in \{0, 1\}^{j+3}$ denote a binary word of length $j + 3$. Suppose that f does not contain three consecutive 0s. If $f(1) = f(2) = 0$, then $f(3) = 1$, and so there are t_j ways to complete f. If $f(1) = 0$ and $f(1) = 1$, then there are t_{j+1} ways to complete f. If $f(1) = 1$, then there are t_{j+2} ways to complete f. All three patterns in the first three letters of f fall into one of the above categories, so

$$
\begin{aligned}
t_{j+3} &= t_j + t_{j+1} + t_{j+2} \\
&= s_j + s_{j+1} + s_{j+2} \quad \text{(by } C(j), C(j+1), \text{ and } C(j+2)) \\
&= s_{j+3} \quad \text{(by def'n)},
\end{aligned}
$$

showing $C(j+3)$, completing the inductive step.

By (an alternative form of) MI, for all $n \geq 1$, $C(n)$ is true. □

Comment: For the general formula for tribonacci number s_n, let a, b, c be the (real) roots of $x^3 - x^2 - x - 1$; then

$$
s_n = \frac{a^{n+1}}{(a-b)(a-c)} + \frac{b^{n+1}}{(b-a)(b-c)} + \frac{c^{n+1}}{(c-a)(c-b)}.
$$

Proving this expression by induction is cumbersome.

34.2.4 Solutions: Non-homogeneous recurrences

Exercise 541: This problem comes from (with kind permission) José Espinosa's website [176, No. 20] without solution, however the solution is straightforward. Espinosa points out that the form of induction required is to show $S(1)$ and $S(2)$, then show that $S(k) \wedge S(k+1) \rightarrow S(k+2)$.

Exercise 542: See the solution to Exercise 543.

Exercise 543: Thanks to Stephanie Portet for providing this problem and solution. For each $t \in \mathbb{Z}^+$, define the statement

$$P_t : \quad x_t = \left[\prod_{i=0}^{t-1} a_i \right] x_0 + b_{t-1} + \sum_{i=0}^{t-2} \left[\prod_{r=i+1}^{t-1} a_r \right] b_i$$

BASE STEPS: At rank $t = 1$: using the difference equation, $x_1 = a_0 x_0 + b_0$, so P_1 is true. At rank $t = 2$: $x_2 = a_1 x_1 + b_1 = a_1 a_0 x_0 + b_1 + a_1 b_0$, so P_2 is true.

INDUCTIVE STEP: For some $k \geq 1$, assume that P_k holds true, *i.e.*, assume that

$$x_k = \left[\prod_{i=0}^{k-1} a_i \right] x_0 + b_{k-1} + \sum_{i=0}^{k-2} \left[\prod_{r=i+1}^{k-1} a_r \right] b_i,$$

and express x_{k+1}:

$$x_{k+1} = a_k x_k + b_k$$

$$= a_k \left\{ \left[\prod_{i=0}^{k-1} a_i \right] x_0 + b_{k-1} + \sum_{i=0}^{k-2} \left[\prod_{r=i+1}^{k-1} a_r \right] b_i \right\} + b_k$$

$$= \left[a_k \prod_{i=0}^{k-1} a_i \right] x_0 + a_k b_{k-1} + \sum_{i=0}^{k-2} \left[a_k \prod_{r=i+1}^{k-1} a_r \right] b_i + b_k$$

$$= \left[\prod_{i=0}^{k} a_i \right] x_0 + b_k + a_k b_{k-1} + \sum_{i=0}^{k-2} \left[\prod_{r=i+1}^{k} a_r \right] b_i$$

$$= \left[\prod_{i=0}^{k} a_i \right] x_0 + b_k + \sum_{i=0}^{k-1} \left[\prod_{r=i+1}^{k} a_r \right] b_i.$$

Thus P_{k+1} holds, completing the inductive step.

By the principle of mathematical induction conclude that, for all $t \in \mathbb{Z}^+$, P_t is true. $\square$

Exercise 544: Proof outline: $a_2 = a_1 + 1$, $a_4 = a_2 + 2 = a_1 + 3$, and so $a_3 = a_1 + 2$ (the sequence is increasing). Then prove that for each $n \geq 1$, $a_n = a_1 + n - 1$ as follows: For $n = 2^k$, use induction on k. Then for $2^k < n < 2^{k+1}$,

$$a_1 + 2^k - 1 = a_{2^k} < a_{2^k+1} < \cdots < a_n < \cdots a_{2^{k+1}} = a_1 + 2^{k+1} - 1,$$

which is only possible when $a_n = a_1 + n + 1$.

Next prove that $a_1 = 1$ by contradiction: if $p < q$ are consecutive primes, $a_1 < p$, then $a_q - a_1 + 1 = a_1 + q - a + 1 = q$. and so $q - a_1 + 1$ is also prime. Then

$q-a_1+1 \leq p$, and so $q-p \leq a_1-1$. But the numbers $(a_1+1)!+2,\ldots,(a_1+1)!+a_1+1$ are all composite, so if p and q are consecutive primes with $p < (a_1+1)!+2 < (a_1+1)!+a_1+1 < q$, then $q-b > a_1+1$, a contradiction. Thus $a_1 = 1$ and so for each n, $a_n = n$. □

34.2.5 Solutions: Non-linear recurrences

Exercise 546: This problem appeared in [582, Prob. 54]. Fix $c \in (0,1]$. For each $n \geq 1$, let $A(n)$ be the assertion that $s_n < s_{n+1}$. Observe that each s_n is positive, since $c > 0$ and $x^2 \geq 0$.

BASE STEP: Since $s_2 = \frac{(c/2)^2+c}{2} > \frac{c}{2} = s_1$, $A(1)$ is true.

INDUCTION STEP: Fix some $k \geq 1$ and suppose that $A(k)$ is true, that is, $s_k < s_{k+1}$. To show that $A(k+1) : s_{k+1} < s_{k+2}$,

$$s_{k+2} - s_{k+1} = \frac{s_{k+1}^2 + c}{2} - \frac{s_k^2 + c}{2} = \frac{1}{2}(s_{k+1}^2 - s_k^2),$$

and since (by $A(k)$) $s_k < s_{k+1}$ and $s_{k+1} > 0$, the expression above is also positive. Thus $A(k+1)$ is true, completing the inductive step.

By mathematical induction, each $A(n)$ is true, that is, the sequence is increasing.

To see that each s_n is at most 1, again induction can be used: For each $n \geq 1$, let $B(n)$ be the statement that $s_n \leq 1$.

BASE STEP: Since $c \leq 1$, $s_1 = \frac{c}{2} < 1$, so $B(1)$ holds.

INDUCTIVE STEP: Fix $k \geq 1$ and assume that $B(k) : s_k \leq 1$ holds. Then

$$B(k) : s_k \leq 1 \Rightarrow s_k^2 \leq 1 \Rightarrow s_k^2 + c \leq 2 \Rightarrow s_{k+1} = \frac{s_k^2 + c}{2} \leq 1,$$

and so $B(k+1)$ follows, completing the inductive step.

By MI, for all $n \geq 1$, $B(n)$ holds. □

Note: One might observe that since $s_1 < 1$, the statement $B(n)$ could have been proved with strict inequality. So, a strictly increasing sequence bounded above by 1 has a limit, say L, which depends on c. To find L, where $0 < L \leq 1$, solve $L = \frac{L^2+c}{2}$ to get $L = 1 - \sqrt{1-c}$.

Exercise 547: Define $a_1 = 1$ and for each $n \geq 1$, define $a_{n+1} = \sqrt{a_n + 5}$. For $n \geq 1$, let $A(n)$ be the claim that $a_n < 3$ and let $B(n)$ be the claim that $a_{n+1} > a_n$. BASE STEP $A(1)$: Since $a_1 = 1 < 3$, $A(1)$ holds.

INDUCTIVE STEP $A(k) \to A(k+1)$: Fix $k \geq 1$, and assume that $A(k) : a_k < 3$ holds. Then $a_{k+1} = \sqrt{a_k + 5} \leq \sqrt{3+5} < 3$ shows that $A(k+1)$ holds.

By MI, for all $n \geq 1$, $A(n)$ holds.

BASE STEP $B(1)$: Since $a_2 = \sqrt{6} > 1 = a_1$, $B(1)$ holds.

INDUCTIVE STEP: $B(k) \rightarrow B(k+1)$: Fix $k \geq 1$, and assume that $B(k) : a_{k+1} > a_k$ holds. Then

$$a_{k+1} + 5 > a_k + 5 \Rightarrow \sqrt{a_{k+1} + 5} > \sqrt{a_k + 5} \Rightarrow a_{k+2} > a_{k+1}$$

shows that $B(k+1)$ follows, completing the inductive step.

By MI, for all $n \geq 1$, $B(n)$ holds, and so the sequence is increasing.

The sequence is bounded above and increasing, and so has a limit; let

$$\lim_{n \to \infty} a_n = L.$$

To find L, suppose that n is so large that a_n is "close" to L, but calculate as if $a_n = a_{n+1} = L$: then $L = \sqrt{L + 5}$ implies that $L^2 - L - 5 = 0$, and so by the quadratic formula, $L = \frac{1 \pm \sqrt{21}}{2}$. Since each $a_i > 0$, the limit must be non-negative, and so $L = \frac{1 + \sqrt{21}}{2}$, (and as a check, L is near 2.7913, less than 3). $\qquad \square$

Exercise 548: Define $a_1 = 1$ and for $n \geq 1$, define $a_{n+1} = 1 + \sqrt{a_n + 5}$. For $n \geq 1$, let $P(n)$ be the proposition that $a_n > 4 - \frac{4}{n}$.

BASE STEP: $a_1 = 1 > 0 = 4 - \frac{4}{1}$, so $P(1)$ holds.

INDUCTIVE STEP: Fix $j \geq 1$, and suppose that $P(j) : a_j > 4 - \frac{4}{j}$ holds. It remains to show that $P(j+1) : a_{j+1} > 4 - \frac{4}{j+1}$ also holds.

By $P(j)$, $a_{j+1} = 1 + \sqrt{a_j + 5} > 1 + \sqrt{4 - \frac{4}{j} + 5} = \sqrt{9 - \frac{4}{j}}$, so it remains to show that

$$1 + \sqrt{9 - \frac{4}{j}} > 4 - \frac{4}{j+1},$$

or equivalently, (by subtracting 1 from each side and squaring, since terms are positive)

$$9 - \frac{4}{j} > (3 - \frac{4}{j+1})^2 = 9 - \frac{24}{j+1} + \frac{16}{(j+1)^2}.$$

Subtracting 9 from each side, and multiplying by $-j(j+1)^2$, the last inequality is equivalent to

$$4(j+1)^2 < 24j(j+1) - 16j,$$

which simplifies to $4 < 20j^2$, which is true for every $j \geq 1$.

By MI, for each $n \geq 1$, $P(n)$ holds. $\qquad \square$

Note: The sequence appears to have a limit by solving $L = 1 + \sqrt{L+5}$ (for $L > 0$) giving $L = 4$; perhaps one can prove by induction that the sequence is strictly increasing, and since each term (by the above solution) is bounded above by $4 - \frac{4}{n} < 4$, this would be an example of where a limit actually attains the upper bound.

Exercise 549: Define $a_1 = 1$ and for each $n \geq 1$, define $a_{n+1} = \sqrt{2a_n + 1}$. For each $n \geq 1$, let $U(n)$ be the claim that $a_n < 4$ and let $V(n)$ be the claim that $a_{n+1} > a_n$.
BASE STEP $U(1)$: Since $a_1 = 1 < 4$, so $U(1)$ holds.

INDUCTIVE STEP $U(k) \rightarrow U(k+1)$: Fix $k \geq 1$, and suppose that $U(k) : a_k < 4$ holds. Then $a_{k+1} = \sqrt{2a_k + 1} < \sqrt{2 \cdot 4 + 1} = 3 < 4$, so $U(k+1)$ holds. [Note: this bound is rather sloppy, and so the 4 could have easily been replaced by, say, 3.]

By MI, for each $n \geq 1$, $U(n)$ holds.

BASE STEP $V(1)$: Since $a_2 = \sqrt{3} > 1 = a_1$, $V(1)$ holds.

INDUCTIVE STEP $V(\ell) \rightarrow V(\ell+1)$: Fix $\ell \geq 1$, and assume that $V(\ell) : a_{\ell+1} > a_\ell$ holds. Then $V(\ell)$ implies

$$2a_{\ell+1} > 2a_\ell \Rightarrow 2a_{\ell+1} + 1 > 2a_\ell + 1 \Rightarrow \sqrt{2a_{\ell+1} + 1} > \sqrt{2a_\ell + 1} \Rightarrow a_{\ell+2} = a_{\ell+1},$$

which shows $V(\ell+1)$ is true.

By MI, for each $n \geq 1$, $V(n)$ is true.

So the sequence $a_1, a_2, a_3, \ldots$ is bounded above and is strictly increasing, therefore, converges to some limit L. Replacing both a_n and a_{n+1} with L gives $L = \sqrt{2L - 1}$, and since $L > 0$, the quadratic formula gives $L = 1 + \sqrt{2}$. $\square$

Exercise 550: Solution outline (for a more thorough write-up, repeat the format given in the solution of Exercise 549):
Define $a_1 = 4$ and for each $n \geq 1$, define $a_{n+1} = \sqrt{a_n + 2}$.
To see that each $a_n > 2$, the base step is $a_1 = 4 > 2$, and for $k \geq 1$, assuming that $a_k > 2$, $a_{k+1} = \sqrt{a_k + 2} > \sqrt{2 + 2} = 2$. To see that the sequence is decreasing, $a_2 = \sqrt{6} < 4 = a_1$, and for $k \geq 1$,

$$a_{k+1} < a_k \Rightarrow a_{k+1} + 2 < a_k + 2 \Rightarrow \sqrt{a_{k+1} + 2} < \sqrt{a_k + 2} \Rightarrow a_{k+2} < a_{k+1}.$$

By MI, each of the desired claims is proved. Since the sequence is bounded below and decreasing, it has a limit $L \geq 2$. Setting $L = \sqrt{L + 2} > 0$ implies $L^2 - L - 2 = 0$, and the quadratic formula gives $L = 2$ as the only positive solution. $\square$

Exercise 551: Solution outline (for a more thorough write-up, repeat the format given in the solution of Exercise 549):
Define $a_1 = 2$ and for each $n \geq 1$, define $a_{n+1} = 1 + \sqrt{a_n + 5}$.

To see that each $a_n < 4$, $a_1 = 2 < 4$, and for $k \geq 1$, assuming that $a_k < 4$, $a_{k+1} = 1 + \sqrt{a_k + 5} < 1 + \sqrt{9} = 4$. To see that the sequence is increasing, $a_2 = 1 + \sqrt{7} > 2$, and for $k \geq 1$, $a_{k+1} > a_k$ implies

$$a_{k+1} + 5 > a_k + 5 \Rightarrow \sqrt{a_{k+1} + 5} > \sqrt{a_k + 5} \Rightarrow 1 + \sqrt{a_{k+1} + 5} > 1 + \sqrt{a_k + 5},$$

and so $a_{k+2} > a_{k+1}$. Thus, by MI, each of the desired properties follows. Since the sequence is bounded above and is increasing, it has a limit L, where $0 < a_1 < L \leq 4$. Putting $L = 1 + \sqrt{L + 5}$ and solving, $L = 4$ is the only positive solution. □

Exercise 552: This exercise appears in [161, 7.24, pp. 181, 189]. For each $n \geq 1$, denote the expression in the exercise by

$$E(n): \quad \sqrt{n} \leq x_n \leq \sqrt{n} + 1.$$

BASE STEP: For $n = 1$, $\sqrt{1} \leq 1 \leq \sqrt{1} + 1$ is a true statement.

INDUCTION STEP: For some $m \geq 1$, suppose that

$$E(m): \quad \sqrt{m} \leq x_m \leq \sqrt{m} + 1$$

is true. To be proved is

$$E(m+1): \quad \sqrt{m+1} \leq x_{m+1} \leq \sqrt{m+1} + 1.$$

To see the first inequality,

$$
\begin{aligned}
x_{m+1} &= 1 + m/x_m \\
&\geq 1 + \frac{m}{\sqrt{m} + 1} \quad \text{(by 2nd ineq. of } E(m)) \\
&> 1 + \frac{m+1-1}{\sqrt{m+1} + 1} \\
&= 1 + \sqrt{m+1} - 1 \\
&= \sqrt{m+1},
\end{aligned}
$$

and the second inequality is proved by

$$
\begin{aligned}
x_{m+1} &= 1 + m/x_m \\
&\leq 1 + \frac{m}{\sqrt{m}} \quad \text{(by 1st ineq. of } E(m)) \\
&= 1 + \sqrt{m} \\
&< \sqrt{m+1} + 1.
\end{aligned}
$$

This completes the inductive step.

By MI, for each $n \geq 1$, the expression $E(n)$ is true. In fact, for $n \geq 2$, strict inequality holds in $E(n)$. □

Exercise 553: Define $s_1 = 0$ and for $n \geq 2$, define $s_n = 1 + s_{\lfloor n/2 \rfloor}$. For every $n \in \mathbb{Z}^+$, let $A(n)$ be the assertion that $s_n = \lfloor \log_2(n) \rfloor$. The proof of $A(n)$ is divided into two parts: when n is a power of 2, then another step for values of n between consecutive powers of two. The first few values of s_n are $s_1 = 0$, $s_2 = s_3 = 1$, $s_4 = s_5 = s_6 = s_7 = 2$, $s_8 = s_9 = \cdots = s_{15} = 3$, so the pattern might be evident.

FIRST BASE STEP: $s_1 = 1 = 1 + 0 = 1 + \log_2(1)$, so $A(1) = A(2^0)$ holds.

FIRST INDUCTION STEP $(A(2^k) \to A(2^{k+1}))$: Fix some $k \geq 0$, and assume that $A(2^k)$ is true. To see that $A(2^{k+1})$ follows,

$$
\begin{aligned}
s_{2^{k+1}} &= 1 + s_{\lfloor 2^{k+1}/2 \rfloor} \\
&= 1 + s_{\lfloor 2^k \rfloor} \\
&= 1 + s_{2^k} \\
&= 1 + \lfloor \log_2(2^k) \rfloor \quad \text{by } A(2^k)) \\
&= 1 + k \\
&= \lfloor \log_2(2^{k+1}) \rfloor.
\end{aligned}
$$

So by MI, for each $k \geq 0$, $A(2^k)$ is true.

For each $k \geq 0$, let $C(k)$ be the claim that for all values of $m \in \{0, 1, \ldots, 2^k - 1\}$, that $s_{2^k} = s_{2^k + m}$. (So, e.g., $C(3)$ says that $s_8 = s_9 = \cdots = s_{15}$.) The sequence of claims $C(0), C(1), C(2), C(3), \ldots$ is proved by induction:

BASE STEP: When $k = 0$, the only value of m is $m = 0$, so $C(0)$ says only $s_1 = s_1$. [When $k = 1$, the values of m are 0,1, so $C(1)$ says $s_2 = s_3$, and since $s_2 = 1 + s_1 = s_3$, $C(1)$ also holds.]

INDUCTION STEP $(C(r) \to C(r+1))$: Fix some $r \geq 0$, and suppose that $C(r)$ holds, namely, that $s_{2^r} = s_{2^r+1} = \cdots = s_{2^r + 2^r - 1}$. To show that $C(r+1)$ holds, one must show for any $m \in \{0, 1, \ldots, 2^{r+1} - 1\}$, that $s_{2^{r+1}} = s_{2^{r+1}+m}$. Indeed, let $0 \leq m \leq 2^{r+1} - 1$; then

$$
\begin{aligned}
s_{2^{r+1}+m} &= 1 + s_{\lfloor (2^{r+1}+m)/2 \rfloor} \\
&= 1 + s_{2^r + \lfloor m/2 \rfloor} \\
&= 1 + s_{2^r} \quad \text{(by } C(r) \text{, since } \lfloor m/2 \rfloor \leq 2^r - 1) \\
&= s_{2^{r+1}},
\end{aligned}
$$

as desired. This concludes the inductive step $(C(r) \to C(r+1))$.

Hence, by MI, for all $k \geq 0$, $C(k)$ is true.

To complete the proof of all $A(n)$ between powers of 2, let $n = 2^k + i$ for some $0 \leq i < 2^k$, and note that $s_n = \lfloor \log_2(2^k + i) \rfloor = \lfloor \log_2 2^k \rfloor = k = s_{2^k}$. □

Exercise 554: This problem appeared as question A-5 of the 1983 Putnam examination, and a solution prepared by Loren Larson and Bruce Hanson, with the assistance of the St. Olaf College Problem Solving Group, appeared in *Math. Magazine* **57** (May 1984), 188–189 (published by MAA).

Exercise 555: This problem appears in [499, Problem 19].

Exercise 556: This problem appeared in [161, 8.13, p. 208] from TT (Tournament of the Towns) where it was asked to show that a_{10} had more than 1000 nines in decimal notation.

Exercise 557: This problem appears in [437, Problem 3-1].

Exercise 558: Hint: First prove that $t_{n+1} = t_n^2 - t_n + 1$.

Exercise 559: This solution outline is from [411, Problem 60]. To prove the result, set $y_n = x_n^2 - 2n$ and obtain

$$y_{n+1} = y_n + \frac{1}{y_n + 2n}.$$

This shows that $y_n < y_{n+1} < y_n + \frac{1}{2n}$. Then by induction show that $0 < y_n < \frac{1}{2}\log n$. The result then follows.   □

34.2.6   Solutions: Towers of Hanoi

Exercise 562: Let $f(n)$ be the minimum number of moves necessary to move n disks from one peg to some other peg. One can verify that $f(1) = 1$ and $f(2) = 3$. Let $n \geq 3$ and suppose that $f(n-1)$ is known. Suppose that there are n disks. Before the bottom disk can be moved, the top $n-1$ disks must be moved all to one other peg, taking $f(n-1)$ moves. After the bottom disk is moved, the remaining $n-1$ disks must be returned, and since the algorithm can be executed to move $n-1$ disks to either of the remaining pegs, $f(n-1)$ more moves are required. Since this sequence of moves just described succeeds in moving all n disks, $f(n) = f(n-1) + 1 + f(n-1) = 2f(n-1) + 1$.

Now a simple proof by induction shows that $f(n) = 2^n - 1$ because $2(2^{n-1} - 1) + 1 = 2^n - 1$.   □

Exercise 563: This solution outline is also given in [16]. The base cases for $n = 1, 2$ have been confirmed just before the statement of the exercise. Suppose the number $3^k - 1$ is correct for k disks, and consider the game with $k+1$ disks, labelled with say, $1, 2, \ldots, k+1$, with $k+1$ being the largest. By induction hypothesis, it takes $3^k - 1$ moves to move $1, \ldots, k$ from A to C, one move to take $k+1$ from A to B, an additional $3^k - 1$ moves to bring $1, \ldots, k$ back from C to A (using the induction hypothesis for the reverse game, unaffected by the large disk on B), one move to

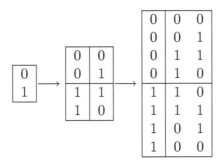

Table 34.1: The Gray codes C_1, C_2, and C_3

take $k + 1$ to C, and finally another $3^k - 1$ moves to return $1, \ldots, k$ to C. In all, there are

$$(3^k - 1) + 1 + (3^k - 1) + 1 + (3^k - 1) = 3(3^k - 1) + 2 = 3^{k+1} - 1,$$

confirming the number of moves for $k + 1$ disks, and so the inductive step. □

34.3 Solutions: Data structures

Exercise 564: The Gray code given by the following outline is called a "binary reflected Gray code", which also turns out to be *cyclic*, that is, also the first and last word differ in precisely one bit as well.

Since $B_1 = \{0, 1\}$, $C_1 = (0, 1)$ is a Gray code for B_1. Let $k \geq 1$, and suppose that a Gray code has been constructed for B_k, say $C_k = (\mathbf{w}_1, \ldots, \mathbf{w}_{2^k})$. Create a new code C_{k+1} by affixing a 0 in front of every word in C_k, and then affixing a 1 in front of a reversed copy of C_k, and juxtapose these two new lists (see Table 34.3 for the first few examples). By induction hypothesis, no element in B_{k+1} is repeated, and there are $2^k + 2^k = 2^{k+1}$ words in C_{k+1}.

Exercise 565 (hypercube): (Outline) First note that a very simple counting argument proves the result: For each collection of $n - i$ coordinates and each of the 2^{n-i} binary words on such coordinates, there is a unique i-facet and there are $\binom{n}{n-i} = \binom{n}{i}$ ways to choose the collection of $n - i$ such coordinates.

To prove this result by induction, some work must be done. The cases $n = 1$ and $i = 0$ (and any n) are trivial: $v_1^1 = 1$ and $v_0^n = 2^n$.

Before proceeding with the inductive argument, first prove the recursion

$$v_i^n = 2v_i^{n-1} + v_{i-1}^{n-1}.$$

Then the inductive argument (inducting on n) might go something like

$$v_i^n = 2v_i^{n-1} + v_{i-1}^{n-1}$$

$$= 2 \cdot 2^{n-1-i} \binom{n-1}{i} + 2^{n-1-(i-1)} \binom{n-1}{i-1} \qquad \text{(by ind. hyp.)}$$

$$= 2^{n-i} \binom{n-1}{i} + 2^{n-i} \binom{n-1}{i-1}$$

$$= 2^{n-i} \binom{n}{i} \qquad \text{(by Pascal's id.).}$$

It remains for the reader to prove the recursion, then to write up the proof formally.

$\square$

34.4   Solutions: Complexity

Exercise 568: Hint: Use Exercise 567 and induction.

Chapter 35

Solutions: Games and recreation

35.1 Solutions: Tree games

Exercise 569: Hint: Use induction on the height of the tree. Look at a tree game with a tree of height n. Look at the subtrees not including any starting position; apply the inductive hypotheses to these. For more details, see [42, p. 323].

Exercise 570: [December 31 game] Reference, hint and partial solution: See [124, Ex. 3.62], where a hint is to first list winning dates. For example, any player naming 30 November can win (because the next player can pick from only 30 December, or 31 November, from which 31 December is a "legal" move). The hint also suggests using strong induction to describe the "winning dates". If (day, month) is a winning date, then perhaps (day-1, month-1) is also? Perhaps the second player need only pick from among 20 Jan., 21 Feb., 22 Mar., 23 Apr., 24 May, 25 June, 26 July, 27 Aug., 28 Sept., 29 Oct., 30 Nov., or 31 Dec., depending on which is a legal move after the first player moves.

35.1.1 Solutions: The game of NIM

Exercise 571: This solution is based on notes given by Eric Milner in PMAT 340, 1987–88.

For $n \geq 0$, let $A(n)$ be the assertion that the position $(1, n, n + 1)$ is a losing position (P-position) in NIM if and only if n is even. Use the notation $(a, b, c) \in P$ to indicate that the position (a, b, c) is a P-position; similarly use $(a, b, c) \in N$ for N-positions.

BASE STEP: When $n = 0$, the position $(1, 0, 1)$ is a P position, and n is even.

INDUCTION STEP: Fix $r > 0$ and suppose that for every $m < r$, that $A(m)$ is true. To accomplish the inductive step, one needs to prove that $A(r)$ is true, that is, (1) if r is even, then $(1, r, r + 1) \in P$, and (2) if r is odd, then $(1, r, r + 1) \in N$.

(1) Let r be even. It suffices to show that whatever move the first player makes, there is a response that results in a P-position. The possible moves by the first player from $(1, r, r + 1)$ are to $(0, r, r + 1)$, $(1, k, r + 1)$ where $k < r$, and $(1, r, k)$ where $k \leq r$. The responses:

$$
\begin{aligned}
(0, r, r + 1) \to &(0, r, r) \in P \\
(1, k, r + 1) \to &(1, k, k + 1) \in P \quad \text{if } k \text{ is even (by } A(k)\text{), or} \\
&(1, k, k - 1) \in P \quad \text{if } k \text{ is odd (by switching piles and } A(k)\text{)} \\
(1, r, k) \to &(r, k - 1, k) \in P \quad \text{if } k \text{ is odd (by } A(k - 1)\text{), or} \\
&(1, k + 1, k) \in P \quad \text{if } k \text{ is even, } r \neq k \text{ (by switching and } A(k)\text{), or} \\
&(0, r, r) \in P \qquad\;\; \text{if } k \text{ is even, } r = k.
\end{aligned}
$$

[Note: If $k < r$ and r is even, then $k + 1 < r$.]

(2) When r is odd, the move $(1, r, r + 1) \to (1, r, r - 1)$ leads to a P-position (by $A(r - 1)$ with a different order of piles), and so when r is odd, $(1, r, r + 1) \in N$.

These two cases complete the inductive step $A(0) \wedge \cdots \wedge A(r - 1) \to A(r)$.

By mathematical induction, for each $n \geq 0$, the assertion $A(n)$ holds. $\qquad\square$

Exercise 573: For each $n \geq 0$, let $S(n)$ be the statement that whenever $n = n_1 + \cdots + n_t$, a position $(n_1, \ldots, n_t)$ in NIM(2) is a losing position (a P-position) if and only if it is satisfactory.

BASE STEP: When $n = 0$, since each $n_i \geq 0$, the only solution to $n_1 + \cdots + n_t = 0$ is when each $n_i = 0$. The position $(0, 0, \ldots, 0)$ is a losing (actually, lost) position, and each $\sigma_i = 0 \equiv 0 \pmod 3$.

INDUCTIVE STEP: Let $m > 0$ and suppose that for all $m' < m$, $S(m')$ is true. Consider the position $(n_1, \ldots, n_t)$, where $m = n_1 + \cdots n_t$.

Case 1: Suppose that $(n_1, \ldots, n_t)$ is satisfactory, *i.e.*, for each i, $\sigma_i \equiv 0 \pmod 3$. It remains to show that any move $(n_1, \ldots, n_t) \to (n'_1, \ldots, n'_t)$ that decreases at most two piles results in an N-position, which, by the inductive hypothesis, is an unsatisfactory position.

Without loss of generality, suppose that $n'_1 < n_1$ and $n'_2 < n_2$, and for all $i = 3, \ldots, t$, $n'_i = n_i$. Letting $b_i(n)$ denote the i-th binary digit of n, there is some i so that $b_i(n'_1) + b_i(n'_2) < b_i(n_1) + b_i(n_2)$ (*e.g.*, take the largest i so that there is inequality). Thus, for some $\rho \in \{1, 2\}$,

$$
\sigma_i(n_1, \ldots, n'_t) = \sigma_i(n_1, \ldots, n_t) - \rho.
$$

Thus $\sigma_i(n'_1, \ldots, n'_t) \not\equiv 0 \pmod 3$, and so by induction hypothesis, $(n'_1, \ldots, n'_t) \in N$, which proves that $(n_1, \ldots, n_t) \in P$.

Case 2: Suppose that $(n_1, \ldots, n_t)$ is not satisfactory, *i.e.*, for some i, $\sigma_i \not\equiv 0$ (mod 3). To be shown is that $(n_1, \ldots, n_t) \in N$, that is, there is some move to a P-position. Put

$$I = \{i : \sigma_i \not\equiv 0 \pmod{3}\} \neq \emptyset.$$

Let $\mu = \max I$, and for each $\rho \in \{1, 2\}$, set

$$I^\rho = \{i \in I : \sigma_i \equiv \rho \pmod{3}\}.$$

Thus $I = I^1 \cup I^2$. Also, for $\varepsilon \in \{0, 1\}$, put

$$I^\rho_\varepsilon = \{i \in I^\rho : b_i(n_p) \equiv 2 \pmod{3}\},$$

where p is chosen so that $b_\mu(n_p) = 1$. Thus I^ρ is the disjoint union of I^ρ_0 and I^ρ_1.

Case 2a: $I^1_0 = \emptyset$ and $I^2_1 = \emptyset$, *i.e.*, for all i, $\sigma_i + b_i(n_p) \equiv 2 \pmod 3$. In this case, make the move

$$n_p \to n'_p = n_p - \left[\sum_{i \in I^1} 2^i - \sum_{i \in I^2} 2^i\right] = n_p - x,$$

and for $j \neq p$, $n'_j = n_j$ (remove stones from the p-th pile, and leave all others untouched). Note that $\mu \in I^1$ (since $b_\mu(n_p) = 1$ and $I^2_1 = \emptyset$), so that $x \geq 2^\mu - (2^{\mu-1} + 2^{\mu-1} + \cdots + 1) = 1$ (*i.e.*, the removal of x stones from the p-th pile is a proper move). Moreover, for each $i \in I$, if $\sigma_i = 1$, then $b_i(n'_p) = b_i(n_p) - 1$, and if $\sigma_i = 2$, then $b_i(n'_p) = b_i(n_p) + 1$. Thus for $i \in I$, it follows that $\sigma_i(n'_1, \ldots, n'_t) \equiv 0$ (mod 3). For $i \notin I$, $\sigma_i(n'_1, \ldots, n'_t) \equiv 0 \pmod 3$ since $\sigma_i(n'_1, \ldots, n'_t) = \sigma_i(n_1, \ldots, n_t)$. Thus the position $(n'_1, \ldots, n'_t)$ is satisfactory, and by the induction hypothesis, is a P-position.

Case 2b: $I^1 \cup I^2_1 \neq \emptyset$. Put $\lambda = \max(I^1 \cup I^2)$. It is possible that $\lambda = \mu$, but in this case, $\mu \in I^2_1$ ($\mu \notin I^1_0$ since $b_m(p) = 1$), *i.e.*, $\sigma_\mu \equiv 2 \pmod 3$, and so there is $q \neq p$ so that $b_\lambda(n_q) = 1$. If $\lambda \neq \mu$ then, since $\sigma_\lambda(n_p) \pmod 3$, there is again $q \neq p$ such that $b_\lambda(n_q) = 1$. For $\rho \in \{1, 2\}$ and $\varepsilon, \delta \in \{0, 1\}$, let

$$I^\rho_{\varepsilon\delta} = \{i \in I^\rho_\varepsilon : b_i(n_q) = \delta\}.$$

Then, for example, I^1_0 is the disjoint union of I^1_{00} and I^1_{01}. Put

$$x = \sum_{i \in I^1_1 \cup I^2_{11}} 2^i - \sum_{i \in I^1_{00} \cup I^2_{01} \cup I^2_{00}} 2^i,$$

$$y = \sum_{i \in I^1_{01} \cup I^2_{11}} 2^i - \sum_{i \in I^1_{00} \cup I^2_{10}} 2^i,$$

and consider the move

$$n_p \to n'_p = n_p - x,$$

$$n_q \to n_q' = n_q - y,$$
$$n_j \to n_j' = n_j \qquad\qquad\qquad \text{for } j \notin \{p, q\}.$$

Since $\mu = \max(I) \in I_1^\rho$, note that $x \geq 2^\mu - (2^{\mu-1} + 2^{\mu-2} + \cdots + 1) = 1$, and since $\lambda \in I_{01}^1 \cup I_{11}^2$, $y \geq 2^\lambda - (2^{\lambda-1} + 2^{\lambda-2} + \cdots + 1) = 1$, the removal of x stones from pile p and y stones from pile q is a proper move. It remains to verify that for each i,

$$\sigma_i' = \sigma_i(n_1', \ldots, n_t') \equiv 0 \pmod 3,$$

so that $(n_1', \ldots, n_t')$ is satisfactory and is hence a P-position by the induction hypothesis.

When $i \notin I$, $\sigma_i' = \sigma_i \equiv 0 \pmod 3$. So suppose that

$$i \in I = I_1^1 \cup (I_{00}^1 \cup I_{01}^1) \cup (I_{00}^2 \cup I_{01}^2 \cup I_{10}^2).$$

Only a couple of cases are given here, the remainder being left to the reader as an exercise. For example, if $i \in I_1^1$, then $b_i(n_p') = b_i(n_p) - 1$, $b_i(n_q') = b_i(n_q)$, and so $\sigma_i' = \sigma_i - 1 \equiv 0 \pmod 3$. If $i \in I_{00}^1$, then $b_i(n_p') = b_i(n_p) + 1$ and $b_i(n_q') = b_i(n_q) + 1$, so $\sigma_i' = \sigma_i + 2 \equiv 0 \pmod 3$. In each case, when $i \in I_{\epsilon\delta}^\rho$, one must check that $\sigma_i' \equiv 0 \pmod 3$, which shows $S(m)$, concluding the inductive step.

By mathematical induction, for each $n \geq 0$, the assertion $S(n)$ is true. $\square$

35.1.2   Solutions: Chess

Exercise 575: See [161, 8.30, pp. 209, 217] for details; only an outline is given here. Let $f(n)$ denote the number of squares that the knight can land on after precisely n moves. Verifying $f(0) = 1$ and $f(1) = 8$ is direct (see Figure 17.2.3), and from Figure 35.1.2, one sees that $f(2) = 33$ and $f(3) = 76$ (68 new squares labelled 3 and the 8 labelled 1). After an odd number of moves, the knight is on a white square, and after an even number of moves, the knight lands on a black square.

Observe that after 3 moves, all reachable squares are within an octagon with 4 white cells per side. By induction, one can prove that after n moves, the reachable cells fill an octagon with $n+1$ squares per side (all of the same color). To finish the proof, one need only count the squares of one color in such an octagon. One idea is to complete the octagon to a square of size $4n + 1$ and remove the four corners, leaving

$$\frac{(4n+1)^2 + 1}{2} - n^2 = 7n^2 + 4n + 1.$$

(One needs to consider the two cases when n is even or when n is odd.) $\square$

Exercise 576: This is a popular (and easy) exercise (see, e.g., [462, 51, p. 281]). For $m \geq 1$ and $n \geq 1$, let $S(m + n)$ denote the statement that starting at $(1, 1)$,

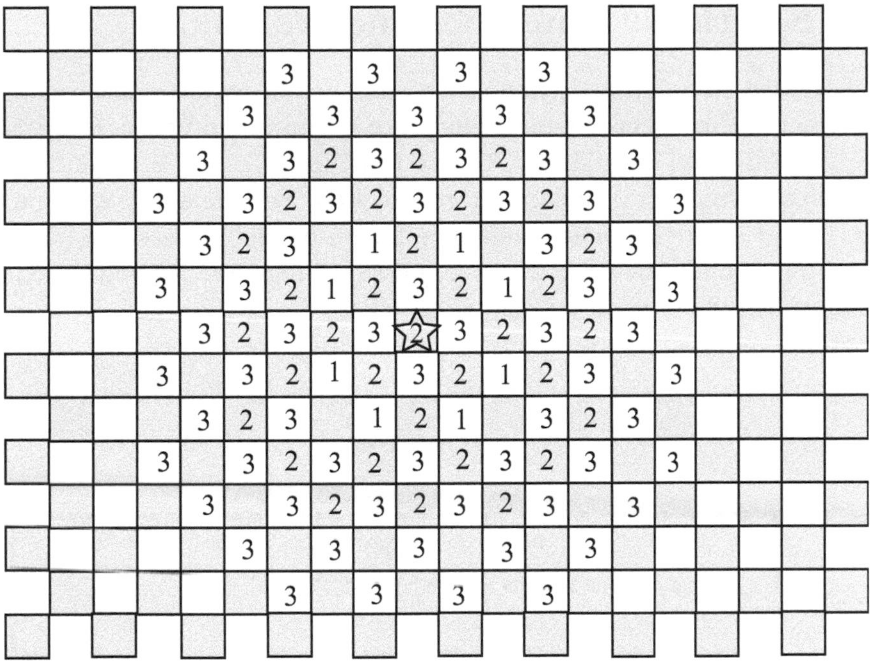

Figure 35.1: Squares reachable by knight after 2 or 3 moves

the square (m, n) can be reached in a finite number of moves. The induction is on $m + n$.

BASE STEP: Starting at $(1, 1)$, after no moves, one arrives at $(1, 1)$, so $S(2)$ is true.

INDUCTIVE STEP: Fix positive integers s, t, and suppose that (s, t) can be reached in a finite number of moves, that is, assume that $S(s + t)$ holds. It remains to show that $S(s + t + 1)$ holds, that is, that both $(s + 1, t)$ and $(s, t + 1)$ can be reached in a finite number of moves.

To see that $(s, t + 1)$ can be reached, consider the moves

$$(s, t) \to (s + 2, t + 1) \to (s + 1, t + 3) \to (s, t + 1).$$

By $S(s + t)$, (s, t) can be reached in a finite number of moves, and together with the three moves just given, $(s, t + 1)$ can be reached in a finite number of steps. To see that $(s + 1, t)$ can be similarly reached, use moves

$$(s, t) \to (s + 1, t + 2) \to (s + 3, t + 1) \to (s + 1, t).$$

Hence, $S(s + t + 1)$ is true, completing the inductive step.

By mathematical induction, for all $k \geq 2$, $S(k)$ is true. $\qquad\qquad\square$

35.2   Solutions: Dominoes and trominoes

577: For each $n \geq 1$, let d_n denote the number of ways to place dominoes in a $2 \times n$ array is a Fibonacci number. The proof is tantamount to showing that the d_i's satisfy the Fibonacci recursion. Observe that both $d_1 = 1 = F_2$ and $d_2 = 2 = F_3$ are Fibonacci numbers, completing the base step. For some $k \geq 2$, assume that not only d_{k-1} and d_k are Fibonacci numbers, but in particular, assume both $d_{k-1} = F_k$ and $d_k = F_{k+1}$, and consider a $2 \times (k+1)$ array. There are two ways the last column of the array can be covered by dominoes—either by one vertical domino, or by two horizontal dominoes that occupy column k as well (see Figure 35.2).

Figure 35.2: Two ways to finish the $k+1$ column

In the first case, there are d_k ways to complete the first k columns, and in the second way, there are d_{k-1} ways to complete the first $k-1$ columns; in all, there are

$$d_{k+1} = d_k + d_{k-1} = F_{k+1} + F_k = F_{k+2}$$

ways to cover the $2 \times (k+1)$ array. So d_{k+1} is the desired Fibonacci number, thereby completing the inductive step. The general result holds by mathematical induction on the number of columns. □

Exercise 578: Let m and n be multiples of 2 and 3 respectively. It is then easy to partition an $m \times n$ checkerboard (with no squares missing) into 2×3 cells, each of which can be covered with two L-shaped trominoes (see Figure 35.3). So, induction is not really needed. □

Exercise 579: (Hint:) For any $n \geq 2$, a $6 \times n$ checkerboard can be covered by L-shaped trominoes. If n is a multiple of either 2 or 3, the result in Exercise 578 guarantees the conclusion. It remains to give a construction whenever $n > 2$ is

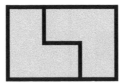

Figure 35.3: The 2×3 configuration

congruent 1 or 5 modulo 6. If one can find a tiling for both the 6×5 rectangle and the 6×7 rectangle, then a simple induction (adding 6×6 squares) gives the result. In fact, these two patterns have tilings formed by combinations in Figure 35.4.

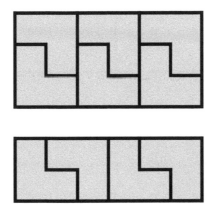

Figure 35.4: Ways to tile an $n \times 6$ board

Exercise 580: This result is due to Golomb [218] and was also stated as a problem in the West German Mathematical Olympiad (1981, number 3, first round); this proof also appears in [277, p. 85].

To be shown is that for each $n \in \mathbb{Z}^+$, a $2^n \times 2^n$ checkerboard with any one square removed can be covered with L-shaped trominoes.

For $n = 1$, the remaining squares on the checkerboard are precisely in the shape of one L-shaped tromino, so the base case holds.

For the remainder of the solution, consider Figure 35.5 and the quote from Honsberger [277]:

"Quartering a $2^n \times 2^n$ board gives 4 squares $2^{n-1} \times 2^{n-1}$. The deleted square must occur in one of these quarters. Placing an L-tromino appropriately at the center yields a board in which each quarter has a single square that needs no further attention. Now it is obvious that the conclusion can be obtained easily by induction."
$\square$

Exercise 581: (Outline) This result appeared in [147, p. 86]. In fact, the following theorem is proved in [107, p. 38] by induction:

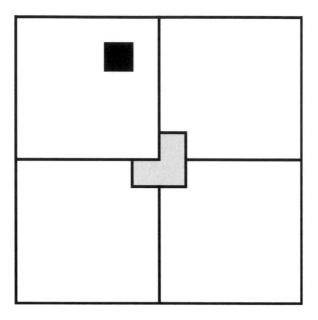

Figure 35.5: The idea behind Golomb's inductive step

Theorem 35.2.1. *If $n > 5$ is odd and 3 does not divide n, then any $n \times n$ board with one square removed can be tiled with L-shaped trominoes.*

To establish the base case $n = 7$, one first observes that a 5×5 board with one corner removed can be tiled with L-shaped trominoes. Using this fact, one finds a tiling for each of the cases (by symmetry, if (i, j) is the missing square, one need only consider $1 \le i \le j \le 4$). Another base case $n = 11$ requires looking at one 7×7 corner, and partitioning the remainder of the 11×11 board into a 4×6 rectangle, a 6×4 rectangle, and a 5×5 square with a corner missing.

The induction proceeds similarly to that of obtaining the tiling of the 11×11 board from the 7×7 board: break off an $(n - 6) \times (n - 6)$ corner or break off an $(n - 3) \times (n - 3)$ corner, and get two more rectangles and a square with a corner missing. See the references for the full details.

Exercise 582: The condition that 3 divides $n^2 - 1$ is equivalent to 3 not dividing n since 3 divides $(n_1)(n + 1)$ iff 3 divides one of the factors. See [107] for a proof that essentially relies on some cases with separate proofs and the result from Exercise 581.

35.2.1   Solutions: Muddy children

Exercise 583: Fix $n \in \mathbb{Z}^+$, and for each $k \in \{1, 2, \ldots, n\}$, let $S(k)$ be the statement of the proposition for k muddy children.

Base step: When $k = 1$, the child with the muddy forehead sees no other muddy foreheads, and knowing that there is at least one muddy forehead, raises a hand upon the first announcement. Since all other children can see the one child with a muddy forehead raising a hand upon the first announcement, these clean children can deduce that the only way that the one muddy child can know, is when there are no others, and so on the second announcement, all can safely raise their hand. Hence $S(1)$ is true.

Inductive step: Fix $\ell \in \{1, 2, \ldots, n - 1\}$, and suppose that $S(1), S(2), \ldots, S(\ell)$ are true—and that all children know that each of these statements is true. Suppose that there are precisely $\ell + 1$ muddy children.

Because there are $\ell + 1$ muddy faces, by the inductive hypotheses, nobody raises their hands after each of the first ℓ announcements. A muddy child sees ℓ other muddy faces, and since nobody raised a hand after the ℓth announcement, by $S(\ell)$ a muddy child concludes that there is at least one more—him/herself, and so can raise a hand upon the $(\ell + 1)$st announcement.

A non-muddy child (if there are any) sees $\ell + 1$ other muddy foreheads, and so by each of $S(1), \ldots, S(\ell)$, knows that his or her own state cannot be determined until at least after the $(\ell + 1)$st announcement. Since any non-muddy child sees $\ell + 1$ hands raise upon the $(\ell + 1)$st announcement, and since each non-muddy child knows that those with hands raised used reasoning based upon seeing only ℓ other muddy faces, each mudddy child correctly concludes that theirs was not one of these ℓ other muddy faces.

So in either case (muddy or not), each child makes the correct deduction on step $\ell + 1$ or $\ell + 2$. Thus $S(\ell + 1)$ is true, completing the inductive step.

By strong mathematical induction, for each $k = 1, \ldots, n$, $S(k)$ is true. $\qquad \square$

35.2.2   Solutions: Colored hats

Exercise 584: See references provided before the exercise. It is also amazing that this puzzle was on the on-line version of *New York Times*, in a blog called "Tierney Lab"; see [532] for the solution and a link to the original puzzle (also see the 3 hat puzzle, first posted 16 March 2009).

35.3   Solutions: Detecting counterfeit coin

See also Exercise 18 for related discussion of coin weighing.

Exercise 585: For each positive integer n, let $A(n)$ be the assertion that 3^n is the largest number of coins so that if exactly one coin is lighter, n weighings can find the light coin. The solution is a very simple induction on n.

BASE STEP: Consider three coins. Put one coin in each pan of the balance, and set the third coin aside. If the scales tip, then the counterfeit coin is on the side that raises. If the scales balance, the third coin must then be the counterfeit. Hence the counterfeit coin can be determined in precisely one weighing. Now consider four coins: any weighing uses either a 1-1 or 2-2 coin pattern. In the first case, if the scales balance, one of the two remaining coins is counterfeit, but nothing more can be concluded; in any 2-2 weighing, the scales will tip, but the only conclusion is that the raised pan contains the counterfeit; again, there is not enough information to determine the bad coin. Hence, $A(1)$ is true.

INDUCTIVE STEP: Fix $k \geq 1$ and suppose that $A(k)$ holds. There are two sides of $A(k+1)$ to be shown: $k+1$ weighings are sufficient for 3^{k+1} coins, but $k+1$ weighings are insufficient for more coins.

Consider 3^{k+1} coins. Divide these coins into three groups, each group with 3^k coins. Put the first group in the left pan and the second group in the right pan. If the scales balance, the counterfeit coin is in the third group, and by $A(k)$, it can be located with an additional k weighings ($k+1$ in all). If the scales tip, the group in the rising pan contains the counterfeit coin, and again, the induction hypothesis $A(k)$ applies. Thus, if one light counterfeit coin is among 3^{k+1} coins, $k+1$ weighings can find it.

On the other hand, suppose that $3^{k+1} + 1$ coins are given. Any weighing with at most 3^k coins in each pan leaves more than 3^k coins untested, and by $A(k)$, the unweighed coins then requires (in general) at least $k+1$ additional weighings to find a counterfeit coin. If the counterfeit coin is in a group of more than 3^k coins in one pan, then one weighing identifies the group; by $A(k)$, k additional weighings are in general insufficient to find the coin among that group. In any case, for more than 3^{k+1} coins, $k+1$ weighings are insufficient to accurately identify the lighter coin.

So $A(k+1)$ is true, completing the inductive step.

By mathematical induction, for each $n \geq 1$, $A(n)$ is true. $\square$

Exercise 586: INDUCTIVE STEP: Fix $k \geq 1$ and assume that $C(k)$ is true. Consider a collection of 3^{k+1} coins, one of which is counterfeit. Assume that there is one more black than white (the same argument works if there is one more white than black), so suppose that $\frac{3^{k+1}}{2}$ are black and $\frac{3^{k}-1}{2}$ are white. Partition the coins into three groups, G_1, G_2, and G_3, each with 3^k coins, and each with a near balance of black and white: In G_1, put $\frac{3^k-1}{2}$ black and $\frac{3^k+1}{2}$ white coins. In both G_2 and G_3, put $\frac{3^k+1}{2}$ black and $\frac{3^k-1}{2}$ white coins.

Putting G_1 aside, weigh G_2 on the left against G_3 on the right. If the scales balance, the counterfeit coin is in G_1, and $C(k)$ applies to G_1, finding the fake coin in an additional k weighings, $k+1$ in all, and so $C(k+1)$ is true in this case. If the scales go down on the left, either one of the $\frac{3^k+1}{2}$ black coins from G_2 is heavy, or one of the $\frac{3^k-1}{2}$ white coins from G_3 is light; these

$$\frac{3^k+1}{2} + \frac{3^k-1}{2} = 3^k$$

coins satisfy the hypothesis for $C(k)$, and so the coin is found in $k+1$ weighings. The analogous argument works when the right pan lowers.

By mathematical induction, for each positive integer n, $C(n)$ is true. □

Exercise 587: [Pennies in boxes] If any two boxes have the same number of pennies, one can be emptied, so assume that all boxes have a different number of pennies, no box empty. The first step is to show that among any three boxes, it is possible to empty one.

Claim: Suppose that boxes A, B, C contain a, b, c pennies respectively, where $0 < a < b < c$. By the division algorithm, let q and r be so that $b = aq + r$, with $0 \leq r < a$. Then by moving pennies from boxes B and/or C into A, it is possible to leave box B with r pennies.

Proof of Claim: Any move from B or C to A doubles the number of pennies in A, so after any i such iterations, the number of pennies in A is of the form $2^i a$. Write q in binary representation,

$$q = q_0 + q_1 \cdot 2 + q_2 \cdot 2^2 + \cdots + q_k \cdot 2^k,$$

where $q_k \neq 0$, and so

$$b = aq_0 + aq_1 \cdot 2 + aq_2 \cdot 2^2 + \cdots + aq_k \cdot 2^k + r.$$

For each $i = 0, 1, \ldots, k$, if $q_i \neq 0$, then move $a \cdot 2^i$ pennies from box B to A; if $q_i = 0$, then move $a \cdot 2^i$ pennies from C to A. This proves the claim.

So applying the claim to three boxes, one arrives at a distribution between three boxes where one box (in this case, B) has fewer pennies than the previous "smallest" box (in this case, A). Since the non-negative integers are well-ordered, eventually one of the three boxes is emptied. [By induction, get a sequence of decreasing r's, which must stop.]

Repeatedly applying the claim to any three boxes, (say, $n-2$ times) leaves at most two boxes non-empty, finishing the proof of the first part of the exercise.

To answer the second part, assume that there are 2^m pennies, and the result is true for 2^{m-1} pennies. Then the number of boxes with an odd number of pennies is even, so pair up "odd boxes"; then between odd boxes in a pair, one transfer of pennies produces an even number in each. So one may assume that all boxes have an even number of pennies. In each box, pair up pennies, giving 2^{m-1} "penny-pairs" in all, and applying the inductive hypothesis for half as many penny-pairs shows that all can be transferred into one box. □

Exercise 588: (Outline) For each $n \geq 2$, let $S(n)$ be the statement that if $n = 2^a + t$ (with $0 \leq t < 2^a$), then $J(n, 2) = 2t + 1$. When $t = 0$, each turn around the circle knocks out half of the soldiers, leaving yet another power of two (it is only important that an even number of soldiers remains after each pass), and the next pass again starts with the second position. In each case, the position 1 survives, and when only $n = 2$ remain, position 1 is still the last standing. So by downward induction, $J(2^a, 2) = 1 = 2 \cdot 0 + 1$, and so $S(2^a)$ is true.

When $t > 0$, apply equation (17.1) t times (by upward induction), showing that $J(2^a + t, 2) = J(2^a, 2) + 2t$. Hence, $J(2^a + t, 2) = 2t + 1$ and so $S(2^a + t)$ holds. □

Remark 1: See [230] for a slightly different inductive proof, where equation (17.1) for $q = 2$ is a consequence, and also discussion of Josephus permutations and binary representations. This problem is also related to shuffling and fixed points (see [481]). See also [34, 32–36] for more history and related problems.

Remark 2: In general, if n is a power of q, then one might think that $J(n, q) = 1$ since each pass around the circle, a multiple of q remain and the last person in the circle to be killed is the one just before soldier 1. However, it is easy to check by hand that $J(3, 3) = 2$, $J(4, 4) = 2$, $J(5, 5) = 2$, and $J(6, 6) = 4$. The problem is that in the case $q = 2$, when $n = 2$, there is only one person left; however, when q is larger, the "last round" after $n = q$ remain leaves q standing, and although number 1 survives up until this point, it is not clear who will be left after the next $q - 1$ suicides.

Exercise 589 [Gossip problem]: (Outline) If n people share all secrets after $2n - 4$ phone calls using four central organizers, a new person added to the community could first call one of the organizers (before the $2n - 4$ calls are made), and then have that same organizer phone back at the end, giving 2 more phone calls, that is, $2(n + 1) - 4$ calls in total. □

Remark: To prove that $2n - 5$ phone calls are never sufficient is difficult. Essentially, it boils down to proving that some 4-cycle is present. Standard references to the gossip problem, in chronological order, are [531], [32], [252], [317], and [84]. For a recent review of the literature (and discussion of the origin of the problem), see [492].

Exercise 590: This problem is famous, occurring in, *e.g.*, [161, 8.2, pp. 207,210]. Fix a circular track of circumference C. For $n \geq 1$, let $S(n)$ be the statement that if n cars are placed anywhere on the track with exactly enough gas among them to complete one lap, there is one car that can make its way around the track traveling clockwise by collecting gas from the other cars along the way.

BASE STEP: $S(1)$ is true since the one car will have enough gas to complete one lap.

INDUCTIVE STEP: Let $k \geq 1$, suppose that $S(k)$ holds and let $k + 1$ cars be placed on the track with exactly enough gas among them to complete one lap. Label the cars

$1, 2, \ldots, k, k+1$ consecutively clockwise around the track and for $1 \le i \le k+1$, let g_i be the distance car i can travel with the gas in its tank and let d_i be the distance from car i to car $i+1$ (addition modulo $k+1$). Since

$$g_1 + g_2 + \cdots + g_{k+1} = C = d_1 + d_2 + \cdots d_{k+1},$$

there is an i so that $g_i \ge d_i$. Without loss of generality, suppose that $g_k \ge d_k$.

Consider the distribution of k cars on the track in exactly the same positions as cars $1, 2, \ldots, k$ and all with the same amount of gas as before, except for car k which has enough gas to travel a distance $g_k + g_{k+1}$ (it has the gas from car k and car $k+1$ in the previous arrangement). By the induction hypothesis, there is a car, say i_0, that can travel clockwise and complete one lap of the track, collecting gas from the other cars as it goes.

In the original distribution of $k+1$ cars on the track, if car i_0 begins traveling clockwise around the track, collecting gas as it goes, when it reaches car k, car i_0 collects enough gas to travel distance $g_k \ge d_k$ and so can make it to car $k+1$ and collect its gas and complete the lap. Thus $S(k+1)$ is true, completing the inductive step.

By mathematical induction, the proof is complete. $\qquad\square$

Chapter 36

Solutions: Relations and functions

36.1 Solutions: Binary relations

Exercise 591: For each $n \geq 1$, let $C(n)$ be the claim that R^n is symmetric on A.

BASE STEP: Since R^1 is equal to R, which is assumed to be symmetric on A, $C(1)$ is true.

INDUCTIVE STEP: Fix some $k \geq 1$ and suppose that $C(k)$ is true. To prove $C(k+1)$, one has to show that R^{k+1} is symmetric. Suppose that $(x, y) \in R^{k+1} = R^k \circ R$. Then there exists $z \in A$ so that $(x, z) \in R^k$ and $(z, y) \in R$. By $C(k)$, $(z, x) \in R_k$ and by $C(1)$, $(y, z) \in R$. Then $(y, x) \in R^k \circ R = R^{k+1}$, completing the proof of $C(k+1)$, and the inductive step.

By mathematical induction, for each $n \geq 1$, R^n is symmetric. $\square$

36.2 Solutions: Functions

Exercise 593: This problem appears in that wonderful article by Hrimiuc [280, Prob. 4], complete with solution. For $n = 1$, by definition, $2 = f(2) = 1 + f(1)$, and so $f(1) = 1$. For $n = 2$,

$$f(3) = 1 + f(1) + 2f(2) = 1 + 1 + 2 \cdot 2 = 6.$$

One can work out that $f(4) = 24$ and $f(5) = 120$, and propose that for each $n \geq 1$,

$$S(n): \quad f(n) = n!.$$

The proof is by induction on n.

BASE STEP: The cases $n = 1, 2, 3, 4, 5$ can serve as base cases.

INDUCTIVE STEP: Fix $k \geq 1$ (say) and suppose that each of $S(1)$, $S(2)$, ..., $S(k)$ are all true, that is, that $f(1) = 1!$, $f(2) = 2!$, ..., $f(k) = k!$ are all true. To complete the inductive step, it remains to show that $S(k+1)$: $f(k+1) = (k+1)!$ also holds. By the recursion above, then it is sufficient to show that

$$(k+1)! = 1 + 1 \cdot 1! + 2 \cdot 2! + \cdots k \cdot k!. \tag{36.1}$$

However, equation (36.1) has an easy inductive proof, done here as Exercise 70. The proof boils down to assuming the inductive hypothesis

$$(k)! = 1 + 1 \cdot 1! + 2 \cdot 2! + \cdots + (k-1) \cdot (k-1)!,$$

and then starting with the right side of equation (36.1),

$$1 + 1 \cdot 1! + 2 \cdot 2! + \cdots + k \cdot k! = k! + k \cdot k! \qquad \text{(by ind. hyp.)}$$
$$= (1+k)k! = (k+1)!.$$

So accepting that (36.1) is true completes the inductive step.

By mathematical induction, the only f that satisfies the requirements of the question is defined by $f(n) = n!$. □

Exercise 594: Suppose that $f : A \to B$ is surjective (onto B). For any $b \in B$, let $f^{-1}(b) \subset A$ denote the set $\{a \in A : f(a) = b\}$. Then for each $b \in B$, $f^{-1}(b) \neq \emptyset$; furthermore, $\cup_{b \in B} f^{-1}(b) = A$. By the Axiom of Choice, let $\gamma : \{f^{-1}(b) : b \in B\} \to A$ be a choice function. Define $g : B \to A$ by $g(b) = \gamma(f^{-1}(b))$. For the moment, fix $b \in B$ and let $\gamma(f^{-1}(b) = x$; then $f(x) = b$ and so $f \circ g(b) = f(g(b)) = f(\gamma(f^{-1}(b)) = f(x) = b$. Since b was arbitrary, this shows that $f \circ g = I_B$. □

Exercise 595: Let $X = \{x_1, x_2, \ldots, x_m,\}$ and $Y = \{y_1, y_2, \ldots, y_n\}$. For some x not in X, put $X^* = X \cup \{x\}$. Any surjection $f : X \to Y$ can be extended to $f^* : X^* \to Y$ by assigning $f(x)$ to any y_i. There are $O_{m,n}$ onto functions $f : X \to Y$ and n ways to extend each to $f^* : X^* \to Y$, so there are $nO_{m,n}$ ways to extend an existing surjection. How else can one make a surjection from X^* onto Y? For each $i = 1, \ldots, n$, one could extend a surjection of the form $g : X \to Y \backslash \{y_i\}$ by setting $g^*(x) = y_i$. For each fixed i, such a g is a surjection from an m-element onto an $(n-1)$-element set, and there are n such i's, so there are $nO_{m,m-1}$ more surjections from X^* onto Y. This proves the recursion

$$O_{m+1,n} = nO_{m,n} + nO_{m,n-1}.$$

Before proving inductively the formula for $O_{m,n}$, here is a simple identity used in the proof.

$$\binom{n}{i} - \binom{n-1}{i} = \binom{n}{i} - \frac{n-i}{n}\binom{n}{i} = \frac{i}{n}\binom{n}{i}. \tag{36.2}$$

For each $m \geq 0$, let $P(m)$ be the proposition that for every $n \geq 0$, the statement

$$S(m, n): \quad O_{m,n} = \sum_{i=0}^{n} \binom{n}{i} (-1)^{n-i} i^m$$

holds.

BASE STEP: If $m = 0$, there are no onto functions from the empty set, so $O_{0,n} = 0$ for each n. On the other hand, the right-hand side of $S(0, n)$ is

$$\sum_{i=0}^{n} \binom{n}{i} (-1)^{n-i},$$

which is equal to 0 by Exercise 100, finishing the proof of the base case $P(0)$.

INDUCTIVE STEP: Suppose that for some $k \geq 0$, $P(k)$ holds, that is, for every n,

$$S(k, n): \quad O_{k,n} = \sum_{i=0}^{n} \binom{n}{i} (-1)^{n-i} i^k$$

holds. To be proved is $P(k+1)$, namely that for every n,

$$S(k+1, n): \quad O_{k+1,n} = \sum_{i=0}^{n} \binom{n}{i} (-1)^{n-i} i^{k+1}.$$

Beginning with the left-hand side, (and using the inductive hypothesis twice on the second line),

$$
\begin{aligned}
O_{k+1,n} &= n(O_{k,n} + O_{k,n-1}) \quad \text{(by recursion established above)} \\[2mm]
&= n\left(\sum_{i=0}^{n} \binom{n}{i} (-1)^{n-i} i^k + \sum_{i=0}^{n-1} \binom{n-1}{i} (-1)^{n-1-i} i^k \right) \\[2mm]
&= n\left(n^k + \sum_{i=0}^{n-1} \binom{n}{i} (-1)^{n-i} i^k - \sum_{i=0}^{n-1} \binom{n-1}{i} (-1)^{n-i} i^k \right) \\[2mm]
&= n\left(n^k + \sum_{i=0}^{n-1} \left[\binom{n}{i} - \binom{n-1}{i} \right] (-1)^{n-i} i^k \right) \\[2mm]
&= n\left(n^k + \sum_{i=0}^{n-1} \left[\binom{n}{i} \frac{i}{n} \right] (-1)^{n-i} i^k \right) \quad \text{(by eqn (36.2) above)} \\[2mm]
&= n^{k+1} + \sum_{i=0}^{n-1} \binom{n}{i} (-1)^{n-i} i^{k+1}
\end{aligned}
$$

$$= \sum_{i=0}^{n} \binom{n}{i}(-1)^{n-i} i^{k+1},$$

which is the right side of $S(k+1, n)$. This proves $P(k+1)$, finishing the inductive step.

Hence, by MI, for all $m \geq 1$, $P(m)$ is true, that is, the formula for $O_{m,n}$ is true for all $m, n \geq 0$. $\square$

Remark on Exercise 595: By Exercise 107, whenever $m < n$, the formula for $O_{m,m}$ gives zero, which is reasonable, since there are then no surjections from an m-element set onto an n-element set.

Exercise 597: For each $n \geq 1$, let $S(n)$ be the statement that $f^n(x) = a^n(x+b)-b$.
BASE STEP: Since $f^1(x) = f(x) = a(x+b) - b = a^1(x+b) - b$, $S(1)$ is true.

INDUCTIVE STEP: Fix $k \geq 1$, and suppose that $S(k)$ is true. To be shown is

$$S(k+1): \quad f^{k+1}(x) = a^{k+1}(x+b) - b.$$

Beginning with the left side of $S(k+1)$,

$$
\begin{aligned}
f^{k+1}(x) &= f(f^k(x)) \\
&= f(a^k(x+b) - b) && \text{(by } S(k)\text{)} \\
&= a(a^k(x+b) - b + b) - b \\
&= a(a^k(x+b) - b \\
&\quad a^{k+1}(x+b) - b,
\end{aligned}
$$

which is the right side of $S(k+1)$. So the inductive step $S(k) \to S(k+1)$ is complete.

By mathematical induction, for each $n \geq 1$, $S(n)$ is true. $\square$

Exercise 598: Fix a function $f : X \to X$. For each $m, n \geq 1$, let $A(m, n)$ be the assertion that $f^m \circ f^n = f^{m+n}$. An inductive proof is not really necessary, but an inductive proof is easy by fixing one variable and inducting on the other. Depending on which variable is fixed, one could define a statement of one variable to be proved, but leaving the notation $A(m, n)$ helps to remember what is being done.
     Fix $n \in Z^+$.

BASE STEP $A(1, n)$: Since $f^1 \circ f^n = f \circ f^n = f^{n+1}$, $A(1, n)$ is true.

INDUCTIVE STEP $A(k, n) \to A(k+1, n)$: Fix $k \geq 1$, and assume that

$$A(k, n): \quad f^k \circ f^n = f^{k+n}$$

is true. To be shown is that

$$A(k+1, n): \quad f^{k+1} \circ f^n = f^{k+1+n}.$$

Starting with the left side of $A(k+1, n)$,

$$f^{k+1} \circ f^n = (f \circ f^k) \circ f^n$$
$$= f \circ (f^k \circ f^n) \qquad \text{(associativity of composition)}$$
$$= f \circ f^{k+n} \qquad \text{(by } A(k, n))$$
$$= f^{k+n+1},$$

which agrees with the right side of $A(k+1, n)$, completing the inductive step.

By mathematical induction on m, for each $n \geq 1$, and for all $m \geq 1$, $A(m, n)$ is true. □

Exercise 600: This problem was a question in the 1983 International Mathematical Olympiad, and its solution appeared in [342, pp. 121–2].

Let $x \in \mathbb{R}^+$. By property (i), $f(xf(x)) = xf(x)$, and so $f(f(xf(x))) = f(xf(x)) = xf(x)$, and so with $x = 1$,

$$f(f(f(1))) = f(1). \tag{36.3}$$

Putting $w = f(1)$, $f(f(w)) = f(1 \cdot f(w)) = wf(1) = f(1)f(1)$, and so

$$f(f(f(1))) = f(1)f(1). \tag{36.4}$$

From equations (36.3) and (36.4), $f(1) = 1$.

Suppose that $z \in \mathbb{R}^+$ satisfies $f(z) = z$. Then

$$zf(\frac{1}{z}) = f(\frac{1}{z}f(z)) = f(1) = 1,$$

and so $f(\frac{1}{z}) = \frac{1}{z})$.

Claim: $f(z^n) = z^n$.

Proof of this claim is by induction on n. When $n = 1$, there is nothing to prove, so suppose that for some $m \geq 1$, the claim is true when $n = m$. Then

$$f(z^{m+1}) = f(z^m \cdot z)$$
$$= f(z^m f(z))$$
$$= zf(z^m) \quad \text{(by (i) with } x = z^m, y = z)$$
$$= z \cdot z^m \quad \text{(by ind. hyp.)}$$
$$= z^{m+1},$$

showing that the claim is true for $n = m + 1$, completing the inductive step. Thus, by MI, for each $n \geq 1$, the claim is true.

Since $\frac{1}{z}$ is also a fixed point, by the claim above, for each positive integer n, $f(\frac{1}{z^n}) = \frac{1}{z^n}$.

If $z > 1$, then $f(z^n) = z^n \to \infty$, or if $z < 1$, then $f(\frac{1}{z^n}) \to \infty$ as well, violating condition (ii). Hence, the only number fixed by f is 1. Hence, for any x, the equation $f(xf(x)) = xf(x)$ implies $xf(x) = 1$. Thus the only function satisfying (i) and (ii) is $f(x) = \frac{1}{x}$. □

Exercise 602: Let f be convex on $[a, b]$. Let $S(n)$ be the statement that for any reals $x_1, x_2, \ldots$ in $[a, b]$,

$$f\left(\frac{\sum_{i=1}^n x_i}{n}\right) \leq \frac{\sum_{i=1}^n f(x_i)}{n}.$$

The left-hand side makes sense, since (by Exercise 757, an easy inductive proof), if $x_1, x_2, \ldots, x_n$ are in $[a, b]$, then $\frac{\sum_{i=1}^n x_i}{n}$ is in $[a, b]$, and so f is defined for this average value of x_i's.

The proof of $S(n)$ for all $n \geq 1$ is by "downward induction"; first proving $S(n)$ when n is any power of 2, then proving that if $S(k)$ is true, so too is $S(k-1)$.

BASE CASES: $S(2^0) = S(1)$ says only that for any $x \in [a, b]$, $f\left(\frac{x}{1}\right) \leq \frac{f(x)}{1}$, which is true. $S(2^1)$ says that for any $x, y \in [a, b]$, $f\left(\frac{x+y}{2}\right) \leq \frac{f(x)+f(y)}{2}$ which is true because f is convex on $[a, b]$.

INDUCTIVE STEP $(S(2^k) \to S(2^{k+1}))$: Assume that for some $k \geq 1$, $S(2^k)$ is true, that is, assume that for any $x_1, x_2, \ldots, x_{2^k}$ in $[a, b]$,

$$f\left(\frac{\sum_{i=1}^{2^k} x_i}{2^k}\right) \leq \frac{\sum_{i=1}^{2^k} f(x_i)}{2^k}.$$

Consider $S(2^{k+1})$: for any $y_1, \ldots, y_{2^{k+1}}$ in $[a, b]$,

$$f\left(\frac{\sum_{i=1}^{2^{k+1}} y_i}{2^{k+1}}\right) \leq \frac{\sum_{i=1}^{2^{k+1}} f(y_i)}{2^{k+1}}.$$

Beginning with the left side of $S(2^{k+1})$,

$$f\left(\frac{\sum_{i=1}^{2^{k+1}} y_i}{2^{k+1}}\right)$$

$$= f\left(\frac{\frac{\sum_{i=1}^{2^k} y_i}{2^k} + \frac{\sum_{i=2^k+1}^{2^{k+1}} y_i}{2^k}}{2}\right)$$

$$\leq \frac{f\left(\frac{\sum_{i=1}^{2^k} y_i}{2^k}\right) + f\left(\frac{\sum_{i=2^k+1}^{2^{k+1}} y_i}{2^k}\right)}{2} \quad \text{(by } S(2))$$

$$\leq \frac{\frac{\sum_{i=1}^{2^k} f(y_i)}{2^k} + \frac{\sum_{i=2^k+1}^{2^{k+1}} f(y_i)}{2^k}}{2} \quad \text{(by } S(2^k), \text{ twice)}$$

$$= \frac{\sum_{i=1}^{2^{k+1}} f(y_i)}{2^{k+1}},$$

which finishes the proof of $S(2^{k+1})$ and the upward inductive step.

Therefore by mathematical induction, whenever n is a power of 2, the statement $S(n)$ is true.

DOWNWARD INDUCTIVE STEP: Suppose that for some $m \geq 2$,

$$f\left(\frac{\sum_{i=1}^{m} y_i}{m}\right) \leq \frac{\sum_{i=1}^{m} f(y_i)}{m}$$

is true for any numbers $y_1, \ldots, y_m$ in $[a, b]$. To be proved is

$$f\left(\frac{\sum_{i=1}^{m-1} y_i}{m-1}\right) \leq \frac{\sum_{i=1}^{m-1} f(y_i)}{m-1}.$$

To achieve this, two facts are relied upon. First, if $y_1, \ldots, y_{m-1}$ are in $[a, b]$, then so too is $\frac{y_1 + y_2 + \cdots + y_{m-1}}{m-1}$. The second fact used is the equality

$$\frac{\sum_{i=1}^{m-1} y_i}{m-1} = \frac{(m-1)+1}{m}\left(\frac{\sum_{i=1}^{m-1} y_i}{m-1}\right) = \frac{\sum_{i=1}^{m-1} y_i + \frac{\sum_{i=1}^{m-1} y_i}{m-1}}{m}.$$

Then

$$f\left(\frac{\sum_{i=1}^{m-1} y_i}{m-1}\right) = f\left(\frac{\sum_{i=1}^{m-1} y_i + \frac{\sum_{i=1}^{m-1} y_i}{m-1}}{m}\right)$$

$$\leq \frac{\sum_{i=1}^{m-1} f(y_i) + f\left(\frac{\sum_{i=1}^{m-1} y_i}{m-1}\right)}{m} \quad \text{(by } S(m)).$$

The term $f\left(\frac{\sum_{i=1}^{m-1} y_i}{m-1}\right)$ appears on both sides of the above inequality, and so solving for this term gives the inequality,

$$\frac{m-1}{m} f\left(\frac{\sum_{i=1}^{m-1} y_i}{m-1}\right) \leq \frac{\sum_{i=1}^{m-1} f(y_i)}{m},$$

which implies

$$f\left(\frac{\sum_{i=1}^{m-1} y_i}{m-1}\right) \leq \frac{\sum_{i=1}^{m-1} f(y_i)}{m-1},$$

the desired inequality. This completes the downward step.

Therefore, by downward induction, for all $n \geq 1$, $S(n)$ is true. □

Exercise 603: (Jensen's inequality) A downward induction proof is available, however, a simple inductive proof is shorter.

Let f be a convex function on $[a, b]$. For each $n \geq 1$, let $S(n)$ be the statement that for all $p_1, p_2, \ldots, p_n \in \mathbb{R}^+$ and $x_1, x_2, \ldots, x_n \in [a, b]$,

$$f\left(\frac{\sum_{i=1}^n p_i x_i}{\sum_{i=1}^n p_i}\right) \leq \frac{\sum_{i=1}^n p_i f(x_i)}{\sum_{i=1}^n p_i}.$$

BASE STEP: $S(1)$ says that for every $p_1 \in \mathbb{R}^+$ and $x_1 \in [a, b]$,

$$f\left(\frac{p_1 x_1}{p_1}\right) \leq \frac{p_1 f(x_1)}{p_1},$$

which is true.

INDUCTIVE STEP: Fix $k \geq 1$ and suppose that $S(k)$ holds. Let $p_1, \ldots, p_k, p_{k+1} \in \mathbb{R}^+$ and $x_1, x_2, \ldots, x_k, x_{k+1} \in [a, b]$. Note that since $x_k, x_{k+1} \in [a, b]$,

$$\frac{p_k x_k + p_{k+1} x_{k+1}}{p_k + p_{k+1}} \in [a, b].$$

Thus,

$$f\left(\frac{\sum_{i=1}^{k+1} p_i x_i}{\sum_{i=1}^{k+1} p_i}\right)$$

$$= f\left(\frac{\left(\sum_{i=1}^{k-1} p_i x_i\right) + (p_k + p_{k+1})\frac{p_k x_k + p_{k+1} x_{k+1}}{p_k + p_{k+1}}}{\left(\sum_{i=1}^{k-1} p_i\right) + (p_k + p_{k+1})}\right)$$

$$\leq \frac{\left(\sum_{i=1}^{k-1} p_i f(x_i)\right) + (p_k + p_{k+1}) f\left(\frac{p_k x_k + p_{k+1} x_{k+1}}{p_k + p_{k+1}}\right)}{\left(\sum_{i=1}^{k-1} p_i\right) + (p_k + p_{k+1})} \qquad \text{(by } S(k)\text{)}$$

$$\leq \frac{\sum_{i=1}^{k-1} p_i f(x_i) + (p_k + p_{k+1})\left(\frac{p_k}{p_k + p_{k+1}} f(x_k) + \frac{p_{k+1}}{p_k + p_{k+1}} f(x_{k+1})\right)}{\left(\sum_{i=1}^{k-1} p_i\right) + (p_k + p_{k+1})} \qquad \text{(by convexity)}$$

$$= \frac{\sum_{i=1}^{k+1} p_i f(x_i)}{\sum_{i=1}^{k+1} p_i}$$

and hence $S(k + 1)$ holds.

Therefore, by MI, for all $n \geq 1$, $S(n)$ is true for all $n \geq 1$. □

Exercise 604: The proof is rather interesting; see [373, pp. 124-126].

Exercise 606: This problem has appeared in many places, [499, Problem 41], for one.

Exercise 607: (Hahn–Banach) The idea is to extend f one dimension at a time, use Zorn's lemma, and then "look back". (The proof presented here is roughly based on one given in lectures by P. Waltman, Emory University, 1992; the nearly identical proof is also found on pages 261–263 of Bridges' analysis text [74].)

Let X be a real normed linear space and without loss of generality, let $\|f\| = 1$. Pick $x_0 \in X\backslash M$ and set

$$M_1 = \mathrm{span}(M, x_0) = \{x + \lambda x_0 : x \in M, \lambda \in \mathbb{R}\}.$$

If for any fixed $\alpha \in \mathbb{R}$, define $f_1 : M_1 \to \mathbb{R}$ by

$$f_1(x + \lambda x_0) = f(x) + \lambda \alpha.$$

It is now shown that f_1 is again a bounded linear functional. The boundedness is trivial, so it remains to check linearity:

$$
\begin{aligned}
f_1(x + \lambda x_0 + k(y + \mu x_0)) &= f_1(x + ky + (\lambda + k\mu)x_0) \\
&= f(x + ky) + (\lambda + k\mu)\alpha \\
&= f(x) + kf(y) + \lambda\alpha + k\mu\alpha \\
&= f(x) + \lambda\alpha + k(f(y) + \mu\alpha) \\
&= f_1(x + \lambda x_0) + kf_1(y + \mu x_0),
\end{aligned}
$$

and so f_1 is linear.

The next goal is to choose α so that $\|f_1\| = \|f\| = 1$, that is, so that

$$\|f_1\| = \sup_{x+\lambda x_0} \frac{|f(x + \lambda x_0)|}{\|x + \lambda x_0\|} = \sup_{x+\lambda x_0} \frac{|f(x) + \lambda\alpha|}{\|x + \lambda x_0\|} = 1.$$

This is satisfied if for every $x \in M$ and $\lambda \in \mathbb{R}$,

$$|f(x) + \lambda\alpha| \le \|x + \lambda x_0\|,$$

or if any of the equivalent statements hold:

$$
\begin{aligned}
\forall x \in M, \forall \lambda \in \mathbb{R}, \quad & |f(-\lambda x) + \lambda\alpha| \le \| - \lambda x + \lambda x_0\|, \\
\forall x \in M, \forall \lambda \in \mathbb{R}, \quad & |-\lambda f(x) + \lambda\alpha| \le \| - \lambda x + \lambda x_0\|, \\
\forall x \in M, \quad & |f(x) - \alpha| \le \|x - x_0\|, \\
\forall x \in M, \quad & f(x) - \|x - x_0\| \le \alpha \le f(x) + \|x - x_0\|.
\end{aligned}
$$

So if for every $x, y \in M$, $f(x) - \|x - x_0\| \le f(y) + \|y - x_0\|$, then one can find such an α; the following shows this to be true:

$$f(x) - f(y) = f(x - y)$$

$$\leq \quad \|x - y\|$$
$$\leq \quad \|x - x_0\| + \|x_0 - y\|$$
$$= \quad \|x - x_0\| + \|y - x_0\|.$$

So there is such an α, and hence there is norm preserving extension $f_1 : M_1 \rightarrow \mathbb{R}$ of f. Such an extension need not be unique, so one introduces some notation to depict all such extensions. As long as there are still points not yet in the domain of a bounded linear functional extending f, one can repeat the above construction getting further extensions.

Define the set $\mathcal{A} \subseteq X \times \mathbb{R}$ with elements

$$[G, g] = \left\{ \begin{array}{c} (x, g(x)) : \quad x \in G, M \subseteq G \subseteq X, g \text{ a bounded linear functional} \\ \text{on } G \text{ extending } f \text{ with } \|g\| = \|f\| = 1 \end{array} \right\}.$$

For example, $[M, f]$ and $[\text{span}\{M, x_0\}, f_1]$ are elements in $\mathcal{A}$. Define a partial order $\preceq$ on $\mathcal{A}$ by $[G, g] \preceq [H, h]$ if and only if $G \subset H$ and h extends g (that is, for every $x \in G$, $h(x) = g(x)$).

At this point, there are two ways to go: Zorn's lemma or Hausdorff's Maximality Principle. First, here is the Zorn's lemma finale.

First check that the union of a chain exists in $\mathcal{A}$: If $\mathcal{C}$ is a chain in $\mathcal{A}, \preceq)$, say $\{(G_\alpha, g_\alpha) : \alpha \in I\}$, (where I is some well-ordered set) then the union of the chain is some pair (G^*, g^*), where $G^* = \cup_{\alpha \in I} G_\alpha$. To define the function $g^*(x)$, if $x \in G^*$, then there is some μ so that $x \in G_\mu$; in this case, define $g^*(x) = g_\mu(x)$. It is easy to check that g^* is a functional, (since this value $g_\mu(x)$ is fixed throughout all extensions of g_μ.) One has only to check that g^* is linear: this is easy because $g(x + ky)$ is calculated in sets which contain both x and y. So the union of a chain in $(\mathcal{A}, \preceq)$ is again an element of $\mathcal{A}$.

By Zorn's lemma, there exists a $[G, g]$ which is maximal with respect to $\preceq$. First observe that $G = X$, for if not, there is a point $x_0 \in X \backslash G$ and the above construction can be used to extend g to g_1 on $G \cup \{x_0\}$; in this case, g would not be maximal, so $G = X$. Then $g = F$ is the linear functional desired in the statement of the theorem.

One could also use Hausdorff's Maximality Principle: *Every poset has a maximal (totally ordered) chain.* In particular, $(\mathcal{A}, \preceq)$, has a maximal chain, say $\{(B_\beta, g_\beta) : \beta \in J\}$. Take the union of this chain and check that $\cup_{\beta \in J} B_\beta$ is indeed X and that the functional F thereby naturally defined is indeed linear. $\qquad \square$

Exercise 611: For each $n \geq 2$, let $S(n)$ be the equation (18.2) in the statement of the exercise. Assume the product rule (18.1) holds.

BASE STEP: The statement $S(2)$ is equation (18.1), which is assumed to be true.

INDUCTION STEP: For some fixed $k \geq 2$, let $f_1, \ldots, f_k, f_{k+1}$ be differentiable functions and assume $S(k)$ is true. By the product rule $S(2)$, applied with $f = f_1 f_2 \cdots f_k$

and $g = f_{k+1}$,

$$(f_1 f_2 \cdots f_k f_{k+1})' = (f_1 \cdots f_k)' f_{k+1} + (f_1 \cdots f_k) f'_{k+1}.$$

Expanding the term $(f_1 \cdots f_k)'$ by $S(k)$ and then distributing f_{k+1} over this sum gives the desired form of the expression required to prove $S(k+1)$, thereby concluding the inductive step.

By mathematical induction, for every $n \geq 2$, the general product rule (18.2) for derivatives holds. $\square$

Exercise 614: (Outline, from Pólya [435, Prob.3.85, pp. 191–2, vol I]) Let the function f be defined $f(x) = \frac{\ln x}{x}$, and for each $n \geq 1$, let $S(n)$ be the statement that the nth derivative of the function is of the form

$$f^{(n)}(x) = (-1)^n \frac{n! \ln x}{x^{n+1}} + (-1)^{n-1} \frac{c_n}{x^{n+1}},$$

where c_n is an integer depending only on n (and not on x). To prove this, it suffices to find explicitly c_n in terms of n.

The base case $S(1)$ is easy to verify (as is $S(0)$, using $c_0 = 0$). Assuming the desired form for $f^{(n)}(x)$, that is, assuming $S(n)$, differentiating once more shows

$$f^{(n+1)}(x) = (-1)^{n+1}! \frac{\ln x}{x^{n+2}} + (-1)^n \frac{n! + (n+1)c_n}{x^{n+2}}.$$

So let $c_{n+1} = n! + (n+1)c_n$, giving the recurrence relation

$$\frac{c_{n+1}}{(n+1)!} = \frac{c_n}{n} + \frac{1}{n+1}.$$

Using $c_1 = 1$, find [by induction, using the above recurrence]

$$c_n = n! \left(1 + \frac{1}{2} + \frac{1}{3} + \cdots + \frac{1}{n}\right).$$

$\square$

Exercise 616: See Pólya [435, Prob.3.82, pp. 190, vol I].

Exercise 618: This solution appears courtesy of Julien Arino.

Suppose that the initial data is of class C^p on $[-1, 0]$ and that f is C^1. Fix $n \geq 1$, and assume the induction hypothesis

P_n : For $t \in [n-1, n]$, the solution to (18.5) exists, is unique and of class C^{p+n}.

To show P_1, one constructs the solution to (18.5) on the interval $[0, 1]$ by using the integral form of the solution to the ordinary differential equation initial value problem

$$x' = f(t, x(t))$$
$$x(0) = x_0,$$

given for $t \geq 0$ by

$$x(t) = x_0 + \int_0^t f(s, x(s))ds.$$

Write (18.5) as

$$x' = f(t, x(t), \phi_0(t - 1))$$
$$x(0) = \phi_0(0), \quad 0 \leq t \leq 1 \tag{36.5}$$

Then consider (18.5) as a nondelayed initial value problem on the interval $[0, 1]$. Indeed, on this interval, one can consider (36.5). That the latter is a nondelayed problem is obvious if the differential equation is rewritten as

$$x'(t) = g(t, x(t)) \tag{36.6}$$

with $g(t, x(t)) = f(t, x(t), \phi_0(t - 1))$, which is well defined on the interval $[0, 1]$ since for $t \in [0, 1]$, $t - 1 \in [-1, 0]$, on which the function ϕ_0 is defined.

Then use the integral form to construct the solution on the interval $[0, 1]$,

$$x(t) = x(0) + \int_0^t g(s, x(s))ds$$
$$= \phi_0(0) + \int_0^t f(s, x(s), \phi_0(s - 1))ds.$$

Now consider the nature of the function f. As problem (36.5) is an ordinary differential equations initial value problem, existence and uniqueness of solutions on the interval $[0, 1]$ follow the usual scheme. To discuss the required properties on f and ϕ_0, the best strategy is to use (36.6). Recall that a vector field has to be continuous both in t and in x for solutions to exist. Thus to have existence of solutions to the equation (36.6), g must be continuous in t and x. This implies that $f(t, x, \phi_0(t - 1))$ must be continuous in t, x. Thus ϕ_0 has to be continuous on $[-1, 0]$.

For uniqueness of solutions to (36.6), it is required that g is Lipschitz in x, *i.e.*, the same property is required from f. Note that this does not affect either ϕ_0 or the way f depends on ϕ_0.

If $\phi_0 \in C^p$ on $[-1, 0]$ and f is integrable on $[-1, 0]$, then on $[0, 1]$, $x(t)$ is C^{p+1}, since it is given by

$$x(t) = \phi_0(0) + \int_0^t f(s, x(s), \phi_0(s - 1))ds.$$

Therefore, P_1 holds true.

Now assume that P_k holds true. Let $\phi_k(t)$ be the solution of (18.5) on the interval $[k-1, k]$. Using the same method as for P_1, write the problem as, for $t \in [k, k+1]$,

$$
\begin{aligned}
x' &= f(t, x(t), \phi_k(t-1)) \\
x(k) &= \phi_k(k), \quad k-1 \le t \le k.
\end{aligned}
\tag{36.7}
$$

Since f is integrable, and as $\phi_k \in C^{k+p}$, the solution to (36.7) exists on $[k, k+1]$. $\quad\Box$

Exercise 619: (Brief) The initial condition $I_0 = 1 - \frac{1}{e}$ is equal to

$$
0! - \frac{(0+1)!}{e} \sum_{j=0}^{0} \frac{1}{0!},
$$

and so satisfies (18.6). Let $k \ge 0$ and assuming that I_k is of the proper form,

$$
\begin{aligned}
I_{k+1} &= (k+1)I_k - \frac{1}{e} \\
&= (k+1)\left[k! - \frac{k!}{e} \sum_{j=0}^{k} \frac{1}{j!} \right] - \frac{1}{e} \qquad \text{(ind. hyp.)} \\
&= (k+1)! - \frac{(k+1)!}{e} \left[\left(\sum_{j=0}^{k} \frac{1}{j!} \right) - \frac{1}{(k+1)!} \right] \\
&= (k+1)! - \frac{(k+1)!}{e} \sum_{j=0}^{k+1} \frac{1}{j!}
\end{aligned}
$$

shows that I_{k+1} is also of the proper form.

By mathematical induction, I_n is of the form in 18.8. $\quad\Box$

Exercise 625: This exercise occurs in [161, 8.39, p. 209]. For each $n \ge 0$, let $S(n)$ denote the statement y_n and y_{n+1} are relatively prime integers. First observe that α and β are non-zero. Viete's relations are used in this proof: if $x^2 + px - 1 = (x - \alpha)(x - \beta)$, then $p = -(\alpha + \beta)$ and $-1 = \alpha\beta$.

BASE STEPS: When $n = 0$, $y_0 = 2$, which is an integer. When $n = 1$, (by Viete's relations) $\alpha + \beta = -p$, an integer relatively prime to 2 since p is odd. So $S(0)$ is true.

INDUCTIVE STEP: Let $k \ge 0$ and assume that $S(k)$ is true, that is, assume that y_k and y_{k+1} are relatively prime integers. Then

$$
y_{k+2} = \alpha^{k+2} + \beta^{k+2}
$$

$$= (\alpha^{k+1} + \beta^{k+1})(\alpha + \beta) - \alpha\beta(\alpha^k + \beta^k)$$
$$= -y_{k+1}p + y_k \qquad\qquad \text{(by Viete's relations)},$$

which is an integer. Furthermore, if d is a common divisor of y_{k+2} and y_{k+1}, then by the above equation, d also divides y_k. By $S(k)$, $\gcd(y_{k+1}, y_k) = 1$, and so $d = \pm 1$, showing that y_{k+2} and y_{k+1} are relatively prime. The statement $S(k+1)$ is true, completing the inductive step.

By mathematical induction, for each $n \geq 0$, y_n and y_{n+1} are relatively prime integers. $\qquad\square$

Exercise 631: This problem appeared in [411, Prob. 77, pp. 14, 36, 91], complete with solution. One proof is by induction on n:

For a fixed m, let $S_m(n)$ be the statement that for any polynomial $P(x, y)$ with x-degree m and y-degree n, then $P(x, e^x)$ can have at most $mn + m + n$ real zeros. Throughout, fix $m \geq 0$. For this solution, recall Rolle's theorem: If $f : \mathbb{R} \to \mathbb{R}$ is a differentiable function on $[a, b]$ with $f(a) = f(b) = 0$, then there exists $c \in (a, b)$ so that $f'(c) = 0$.

BASE STEP: When $n = 0$, $P(x, 0)$ is a polynomial in x of degree m, which has at most m zeros; $S_m(0)$ follows.

INDUCTIVE STEP: Fix $k \geq 0$ and suppose that $S_m(k)$ holds. Let $P(x, y)$ be a polynomial with x-degree m and y-degree $k + 1$. For the moment, suppose that $P(x, y)$ has N zeros. Then by Rolle's theorem (repeated $m + 1$ times, by induction, if necessary) the $(m+1)$-st derivative of $P(x, e^x)$ has at least $N - (m+1)$ zeros. Furthermore, this derivative is of the form $e^x Q(x, e^x)$ where $Q(x, y)$ has x degree m and y degree k.

By the induction hypothesis $S(k)$, $Q(x, e^x)$ has at most $mk + m + k$ zeros, and so $e^x Q(x, e^x)$ also has at most $mk + m + k$ zeros. Thus $N - (m+1) \leq mk + m + k$ implies $N \leq m(k+1) + m + (k+1)$, proving $S(k+1)$.

By mathematical induction, for any non-negative integers m and n, the statement $S_m(n)$ holds. $\qquad\square$

Chapter 37

Solutions: Linear and abstract algebra

37.1 Solutions: Linear algebra

Exercise 637: A reduced row echelon form of a matrix is unique. This proof is by induction on the number of columns, as found in an article by Yuster [584]. Another proof is in [273, p. 58], but it is not by induction.

Let $U(n)$ be the statement that for any $m \geq 1$, the RREF of an $m \times n$ matrix A is unique.

BASE STEP: For $n = 1$, the proof is direct.

INDUCTIVE STEP: Fix $k \geq 1$ and suppose that $U(k)$ is true. Let A be an $m \times (k+1)$ matrix with two RREF's $B = [b_{i,j}]$ and $C = [c_{i,j}]$. Let A' be the $m \times k$ matrix obtained from A by deleting the last column of A; similarly, let B' and C' be formed from B and C respectively by deleting last columns.

Any sequence of elementary row operations that puts A in RREF also puts A' in RREF. Thus, by induction hypothesis $U(k)$, $B' = C'$; it remains to show that $B = C$, and for this, it suffices to show that the last columns of B and C agree.

For the moment, suppose that $B \neq C$, and let $i \in \{1, 2, \ldots, m\}$ be so that $b_{i,k+1} \neq c_{i,k+1}$. (This assumption is contradicted by showing that $B = C$ follows, from which one concludes $B = C$ in any case. This proof strategy seems strange, so the reader is invited to find a more direct proof.)

Let $\mathbf{x} \in \mathbb{R}^{k+1}$ satisfy $B\mathbf{x} = \mathbf{0}$. Since elementary row operations do not affect the solution space, $C\mathbf{x} = \mathbf{0}$ as well, and hence $(B - C)\mathbf{u} = \mathbf{0}$. The first k columns of $B - C$ are columns of zeros. Calculating the i-th coordinate of $B\mathbf{x}$,

$$(b_{i,1} - c_{i,1})x_1 + \cdots + (b_{i,k+1} - c_{i,k+1})x_{k+1} = 0;$$

as the first k columns of $B - C$ are zeros, and $b_{i,k+1} - c_{i,k+1} \neq 0$, it follows that $x_{k+1} = 0$. Thus any solution to $B\mathbf{x} = \mathbf{0}$ or $C\mathbf{x} = \mathbf{0}$ must have $x_{k+1} = 0$.

Claim: Both the $(k+1)$-th columns of B and C contain leading 1's.

Proof of Claim: In the system $A\mathbf{x} = \mathbf{0}$, for $\mathbf{x} = (x_1 \cdots x_{k+1})^T$, if the $(k+1)$-th column of B or C does not contain a leading 1, then x_{k+1} is a free variable (not necessarily 0), contradicting $x_{k+1} = 0$, proving the claim.

As the first k columns of B and C agree, the leading 1's in the $(k+1)$-th columns of B and C occurs in the same position, namely the first zero row of $B' = C'$.

Because the remaining entries in the $(k+1)$-th columns of B and C must all be zero, these columns agree, giving $B = C$. This contradicts the assumption $B \neq C$, so $B = C$. This proves $U(k+1)$, finishing the inductive step.

By MI, for any $n \geq 1$, $U(n)$ holds, and since m was arbitrary, the RREF of any $m \times n$ matrix is unique. $\square$

Exercise 638: (Marked matrix) This appeared in [161, 8.26, pp. 208, 216] but is an old problem. For positive integers m and n, let $P(m,n)$ be the statement of the problem: In any $m \times n$ matrix of real numbers, if at least p of the largest numbers in each column ($p \leq m$), and at least q of the largest numbers in each row are marked, then at least pq of the numbers are marked at least twice. The result is trivial if either $p = 0$ or $q = 0$, so assume $p, q > 0$. Proof is by induction on $m + n$.

BASE STEP: If either $m = 1$ (then $p = 1$) or $n = 1$ (then $q = 1$) the statement is easily verified.

INDUCTIVE STEP: Fix m, n, both at least 2, and for some $1 \leq p \leq m$, $1 \leq q \leq n$, and assume that both $P(m, n-1)$ and $P(m-1, n)$ hold. If every entry that is marked is marked twice, then pq entries are marked twice. So assume that some entries that are marked are marked only once; among those marked only once, let M be the largest and suppose M is entry (i, j).

Suppose for the moment that M is among the p largest in column j. M is not among the q largest in row i, but since M is largest, these q largest in row i are marked twice. Delete the row i. Then the $(m-1) \times n$ matrix that remains has at least $p - 1$ largest entries marked in each column, and q largest entries in each row are marked. By induction hypothesis $P(m-1, n)$, at least $(p-1)q$ entries in the smaller matrix are marked twice. Together with the q entries from the original row i that are marked twice, there are $p(q-1) + q = pq$ entries in the original matrix marked twice. Repeat the argument when M is among the q largest in row i, and conclude that $P(m, n)$ holds.

By MI, for all $m, n \geq 1$, $P(m, n)$ holds. $\square$

Exercise 639: For each positive integer n, let $P(n)$ be the proposition that if A and B are matrices with $AB = BA$ then $(AB)^n = A^n B^n$.

BASE STEP: When $n = 1$, there is nothing to prove, so $P(1)$ holds.

INDUCTIVE STEP: Fix some $k \geq 1$ and suppose that $P(k)$ holds. Then,

$$
\begin{aligned}
(AB)^{k+1} &= (AB)(AB)\cdots(AB) \\
&= A(BA)^k B \quad \text{(by associativity)} \\
&= A(AB)^k B \\
&= A(A^k B^k)B \quad \text{(by } P(k)\text{)} \\
&= A^{k+1}B^{k+1},
\end{aligned}
$$

and so $P(k+1)$ also holds, completing the inductive step.

By MI, for all $n \geq 1$, the statement $P(n)$ is true. $\qquad\square$

Exercise 640: For each positive integer n, let $S(n)$ be the statement that for any matrices $A_1, A_2, \ldots, A_n$ of the same size, then

$$
(A_1 + A_2 + \cdots + A_n)^T = A_1^T + A_2^T + \cdots + A_n^T.
$$

BASE STEPS: $S(1)$ holds trivially. To see $S(2)$, let $A = [a_{ij}]$ and $B = [b_{ij}]$ be $\ell \times m$ matrices. Then the (i,j)-th entry of $(A+B)^T$ is the (j,i) entry of $A+B$, namely, $a_{ji} + b_{ji}$, which is precisely the (i,j) entry of $A^T + B^T$, so $S(2)$ holds as well.

INDUCTIVE STEP: For some $k \geq 2$, suppose that $S(k)$ holds, and let $A_{1,}, \ldots, A_k, A_{k+1}$ be matrices of the same size. Then

$$
\begin{aligned}
(A_1 + A_2 + \ldots + A_k + A_{k+1})^T &= ([A_1 + A_2 + \ldots + A_k] + A_{k+1})^T \\
&= [A_1 + A_2 + \ldots + A_k]^T + A_{k+1}^T \quad \text{(by } S(2)\text{)} \\
&= A_1^T + A_2^T + \cdots + A_k^T + A_{k+1} \quad \text{(by } S(k)\text{)}
\end{aligned}
$$

proving that $S(k+1)$ holds, completing the inductive step.

By the principle of mathematical induction, for all $n \geq 1$, $S(n)$ holds. $\qquad\square$

Exercise 641: For a positive integer n, let $P(n)$ denote the statement that if $A_1, A_2, \ldots, A_n$ are square matrices of the same size, then

$$
(A_1 A_2 \cdots A_n)^T = A_n^T \cdots A_2^T A_1^T.
$$

As in Exercise 640, most of the work is done in proving $P(2)$.

BASE STEPS: When $n = 1$, there is nothing to prove, so examine the case $n = 2$. For notational convenience, instead of using A_1 and A_2, suppose that $A = (a_{ij})$ and $B = (b_{ij})$ are two $m \times m$ matrices. To show $P(2)$, one needs to prove that

$$
(AB)^T = B^T A^T. \tag{37.1}
$$

For $k, \ell \in \{1, \ldots, m\}$; the (k, ℓ) entry of $(AB)^T$ is the (ℓ, k) entry of AB, namely

$$a_{\ell,1}b_{1,k} + a_{\ell,2}b_{2,k} + \cdots + a_{\ell,m}b_{m,k},$$

which is precisely the (k, ℓ) entry of $B^T A^T$, which proves equation (37.1), and so the base case $P(2)$.

INDUCTIVE STEP: Fix some $q \geq 2$ and suppose that $P(q)$ is true. Let $A_1, \ldots, A_q$, A_{q+1} be square matrices of the same size. Then

$$\begin{aligned}
(A_1 \cdots A_q A_{q+1})^T &= ((A_1 A_2 \cdots A_q) A_{q+1})^T \\
&= A_{q+1}^T (A_1 A_2 \cdots A_q)^T &\text{(by } P(2)) \\
&= A_{q+1}^T A_q^T \cdots A_2^T A_1^T &\text{(by } P(p))
\end{aligned}$$

proves that $P(p+1)$ also holds, completing the inductive step.

By MI, for each $n \geq 1$, $P(n)$ holds. $\qquad\square$

Exercise 642: The only difference between this exercise and Exercise 641 is that $A_1, A_2, \ldots, A_n$ are matrices that might not be square. One need only repeat the proof of Exercise 641 being careful to note that if the sizes of the matrices are such that the product $A_1 A_2 \cdots A_n$ is defined, then so is the product $A_n^T \cdots A_2^T A_1^T$. $\quad\square$

Exercise 643: If A and M are $m \times m$ matrices and one wants to show that $M = A^{-1}$, by definition, one needs to show that $AM = I_m = MA$. In fact, it suffices to show only one of $AM = I$ or $MA = I$; one proof of this fact uses the fact that M is invertible iff for any column matrix X, $MX = \mathbf{0}$ implies $X = 0$. (Suppose that $AM = I$ and put $MX = \mathbf{0}$. Multiplying on the left by A, $AMX = \mathbf{0}$, and so $X = 0$, which shows that M^{-1} exists. Multiplying $AM = I$ on the right by M^{-1} then gives $A = M^{-1}$, so A is invertible with $A^{-1} = M$.)

For $n \geq 1$, let $C(n)$ be the claim that if $A_1, A_2, \ldots, A_n$ are invertible matrices of the same size, then

$$(A_1 A_2 \cdots A_n)^{-1} = A_n^{-1} \cdots A_2^{-1} A_1^{-1}.$$

BASE STEPS: The claim $C(1)$ says $A_1^{-1} = A_1^{-1}$, a trivially true statement. To prove $C(2)$, one must show that $(AB)^{-1} = B^{-1}A^{-1}$. To this end, observe that, by associativity of multiplication,

$$AB(B^{-1}A^{-1}) = A(BB^{-1})A^{-1} = AIA^{-1} = AA^{-1} = I.$$

INDUCTIVE STEP: Fix $k \geq 2$ and suppose that $C(k)$ is true. Let $B_1, \ldots, B_k, B_{k+1}$ be invertible matrices of the same size. By $C(k)$, the matrix $B_1 B_2 \cdots B_k$ is invertible, with

$$(B_1 B_2 \cdots B_k)^{-1} = B_k^{-1} \cdots B_2^{-1} B_1^{-1}.$$

By $C(2)$, with $A_1 = B_1 \cdots B_k$ and $A_2 = B_{k+1}$, the product $B_1 B_2 \cdots B_k B_{k+1}$ is also invertible with

$$
\begin{aligned}
(B_1 B_2 \cdots B_k B_{k+1})^{-1} &= B_{k+1}^{-1}(B_1 \cdots B_k)^{-1} \\
&= B_{k+1}^{-1} B_k^{-1} \cdots B_1^{-1},
\end{aligned}
$$

which proves $C(k+1)$, completing the inductive step.

By MI, for each $n \geq 1$, the claim $C(n)$ is true. $\qquad \square$

Exercise 644: This proof is outlined by Foulds and Johnston in [196], where totally unimodular matrices are related to integer programming problems.

Let A be an $m \times n$ matrix with the following two properties:

(i) Every column of A contains at most two non-zero entries.

(ii) Every entry in A is either -1, 0, or 1.

Furthermore, assume that there is a partition of the rows of A into disjoint sets R_1 and R_2 so that the next two properties hold:

(iii) For $i \neq j$, if a_{ik} and a_{jk} are non-zero and have the same sign, then row i is in R_1 and row j is in R_2, or row i is in R_2 and row j is in R_1.

(iv) For $i \neq j$, if a_{ik} and a_{jk} are non-zero and have different signs, then row i and row j are both in R_1, or row i and row j are both in R_2.

To show that A is totally unimodular, one applies induction on the size of the square submatrices (and show each has determinant -1, 0, or 1). Let $U(p)$ denote the statement that any non-singular $p \times p$ submatrix of A is unimodular.

BASE STEP: Since A satisfies (ii), each 1×1 submatrix of A is either singular or has determinant -1 or 1, and so $U(1)$ holds.

INDUCTIVE STEP: Fix some $\ell \geq 1$, and suppose that every non-singular $\ell \times \ell$ submatrix of A is unimodular, that is, $U(\ell)$ holds, and let B be a $(\ell + 1) \times (\ell + 1)$ submatrix of A. If B has a row of zeros, B is singular.

If B has a column, say the k-th column, with precisely one non-zero entry, say, $a_{ik} \in \{-1, 1\}$ then the determinant of B can be expanded along column j, and $U(\ell)$ applies to the $\ell \times \ell$ cofactor of a_{ij}, giving B a determinant equal to -1 or 1.

Suppose that every column of B has two non-zero entries. Then for every column k, (iii) or (iv) applies, and thus

$$
\sum_{i \in R_1} a_{ik} = \sum_{i \in R_2} a_{ik}.
$$

With a little work, one can see that this equation implies that some linear combination of rows is zero, and so B is singular.

In all cases, B is either singular or unimodular, concluding the inductive step $U(\ell) \to U(\ell + 1)$.

By mathematical induction, for each $\ell \geq 1$, $U(\ell)$ holds; in other words, every non-singular square submatrix of A is unimodular, and so A is totally unimodular. □

Exercise 645: This exercise has a direct proof without induction—simply look at the cofactor expansion along a row of zeros. If one expands the determinant along any column, induction can be used:

For each $n \geq 1$, let $P(n)$ be the proposition that if A is an $n \times n$ matrix containing a row of zeros, then $\det(A) = 0$.

BASE STEP: The 1×1 matrix $[0]$ trivially consists of one row of zeros, and has determinant 0, so $P(1)$ holds.

INDUCTIVE STEP: Fix some $r \geq 1$ and suppose that $P(r)$ holds. Let $M = [m_{ij}]$ be an $(r+1) \times (r+1)$ matrix with row z all zeros. Examine the cofactor expansion along the first column, and for each row i, let M_i be the $r \times r$ principal submatrix determined by deleting the first column and row i. Since A had a row of zeros, for each $i \neq z$, M_i contains a row of zeros and by induction hypothesis $P(r)$, $\det(M_i)=0$. Expanding along the first column,

$$\det(M) = \left(\sum_{i \neq z} m_{i1} \det(M_i) \right) + m_{z1} \det(M_z)$$

$$= \left(\sum_{i \neq z} m_{i1} \cdot 0 \right) + 0 \cdot \det(M_z),$$

and so $\det(M) = 0$, proving $P(r+1)$ and concluding the inductive step.

By mathematical induction, for each $n \geq 1$, $P(n)$ holds. □

Exercise 646: For each integer $n \geq 2$, let $P(n)$ denote the statement that if A is an $n \times n$ matrix with two columns equal, then $\det(A) = 0$. (In the proof below, straight bars outside of a matrix indicate the determinant of the matrix.)

BASE STEP: $\begin{vmatrix} a & a \\ b & b \end{vmatrix} = ab - ab = 0$ shows $P(2)$ is true.

INDUCTIVE STEP: Let $k \geq 2$ and suppose $P(k)$ holds. Let A be a $(k+1) \times (k+1)$ matrix with two columns identical. Fix two of the columns that are identical. Use the cofactor expansion along any of the other columns; all the $k \times k$ submatrices used in the expansion will have identical columns, to which the inductive hypothesis $P(k)$ applies, giving zero cofactors, and hence zero determinant.

By MI, for each $n \geq 2$, the proposition $P(n)$ is true. □

Remark: If, in the inductive step above, one tries to expand along a row, one gets all but at most two cofactors being zero, but then one needs to show that the last two are negative of one another, and for this, one needs to know about the affect on the determinant of switching two columns—however, if one had that fact, the result of the exercise is trivial because switching two identical columns gives the same matrix and $-\det(A) = \det(A)$ implies $\det(A) = 0$.

Exercise 647: The proof given here is for $\beta = \gamma = 1$; the general proof is similar but notationally a bit more cumbersome. Let $S(n)$ be the statement of the result for $n \times n$ matrices. The following notation is convenient: if M is a matrix, the matrix produced by deleting the i-th row and j-th column is denoted M_{ij} (in contrast with m_{ij}, the (i, j) entry of M). The proof proceeds by induction on n.

BASE STEP: For $n = 1$, the result is easily verified (as it is also for $n = 2$).

INDUCTIVE STEP: Fix $m \geq 2$ and suppose that $S(m - 1)$ is true. Let A, B, C be $m \times m$ matrices so that the k-th column of A is the sum of the k-th columns of B and C. Examine the Laplace expansion of A along the first row:

$$\det(A) = a_{11} \det A_{11} - \cdots + (-1)^{1+j} a_{1j} \det A_{1j} + \cdots + (-1)^{1+n} a_{1m} \det A_{1m}. \quad (37.2)$$

Examine the terms above, considering two cases:

1. When $j \neq k$, by $S(m - 1)$, since A_{1j}, B_{1j}, and C_{1j} are identical except that one column of $A_{1,j}$ is the sum of the corresponding columns in B_{1j} and C_{1j},

$$\det A_{1j} = \det B_{1j} + \det C_{1j}.$$

Also, $a_{1j} = b_{1j} = c_{1j}$.

2. When $j = k$, $a_{1k} = b_{1k} + c_{1k}$ and $A_{1k} = B_{1k} = C_{1k}$.

Thus (37.2) becomes

$$
\begin{aligned}
\det(A) &= \sum_{j \neq k} (-1)^{1+j} a_{1j} \det A_{1j} + (-1)^{1+k} a_{1k} \det A_{1k} \\
&= \sum_{j \neq k} (-1)^{1+j} a_{1j} [\det B_{1j} + \det C_{1j}] + (-1)^{1+k} (b_{1k} + c_{1k}) \det A_{1k} \\
&= \sum_{j \neq k} (-1)^{1+j} a_{1j} \det B_{1j} + \sum_{j \neq k} (-1)^{1+j} a_{1j} \det C_{1j} \\
&\qquad + (-1)^{1+k} b_{1k} \det A_{1k} + (-1)^{1+k} c_{1k} \det A_{1k} \\
&= \sum_{j \neq k} (-1)^{1+j} b_{1j} \det B_{1j} + \sum_{j \neq k} (-1)^{1+j} c_{1j} \det C_{1j} \\
&\qquad + (-1)^{1+k} b_{1k} \det B_{1k} + (-1)^{1+k} c_{1k} \det C_{1k} \\
&= \det(B) + \det(C),
\end{aligned}
$$

thereby proving $S(m)$. This concludes the inductive step.

By MI, for all $m \geq 1$, the result $S(m)$ holds (for $\beta = \gamma = 1$). $\qquad\qquad\square$

Exercise 648: (Brief) For $n \geq 1$, let $C(n)$ be the statement that for any $n \times n$ matrix A, $\det(A^T) = \det(A)$.

BASE STEP: For $n = 1$, $A^T = A$ and so $C(1)$ is direct.

INDUCTIVE STEP: For some $k \geq 1$, assume that $C(k)$ holds, and let A be $(k+1) \times (k+1)$. Expand the determinant of A along any row, and compute the determinant of A^T by expanding along the corresponding column. All the corresponding minors are "transposes" of one another, and so by $C(k)$, have the same determinant. Thus the expansions are identical, showing $C(k+1)$.

Thus, by induction, for each $n \geq 1$, $C(n)$ is true. $\qquad\qquad\square$

Exercise 649: For each $n \in \mathbb{Z}^+$, let $S(n)$ be the statement that for any $n \times n$ triangular matrix, its determinant is the product of the entries on the main diagonal.

BASE STEP: Since all 1×1 matrices are triangular, and the determinant of such a matrix $T = [t]$ is t, a trivial product of the diagonal elements, so $S(1)$ is true.

INDUCTIVE STEP: Fix $k \geq 1$, and suppose that $S(k)$ is true. Let T be an $(k+1) \times (k+1)$ triangular matrix with diagonal entries $d_1, \ldots, d_{k+1}$. Without loss of generality, assume that T is upper triangular (for a similar proof holds when T is lower triangular). Let the $(1,1)$ minor of T (obtained by deleting the first row and first column of T) be denoted by $T_{1,1}$. Then expanding the determinant of T along the first column, only one term survives, namely $\det(T) = d_1 \det(T_{1,1})$. Since the $k \times k$ matrix $T_{1,1}$ is still triangular, by $S(k)$, $\det(T_{1,1})$ is the product of the remaining d_i's, and so $\det(T)$ is the product of all its diagonal elements, proving $S(k+1)$.

By mathematical induction, for every $n \geq 1$, $S(n)$ is true. $\qquad\qquad\square$

Exercise 650: S. H. Lui kindly gave this solution in the form of source code from his class notes [360]; I have slightly modified some details.

When $n = 1$, the result is trivial, so let $n \geq 2$.

To prove one direction, suppose that an $n \times n$ matrix A has the LU-decomposition $A = LU$, and let A_k be the $k \times k$ leading principal submatrix of A. If one writes

$$L = \begin{bmatrix} L_{11} & \mathbf{0} \\ L_{21} & L_{22} \end{bmatrix}, \quad \text{and} \quad U = \begin{bmatrix} U_{11} & U_{12} \\ \mathbf{0} & U_{22} \end{bmatrix}.$$

where L_{11} and U_{11} are the $k \times k$ leading principal submatrices of L and U respectively (and the $\mathbf{0}$-matrices are of appropriate sizes), then by block multiplication, $A_k = L_{11}U_{11}$.

Since A is non-singular, every diagonal entry of U and hence of U_{11} is non-zero; thus U_{11} is invertible, and so A_k is invertible. This proves one direction of the theorem.

Suppose that for each $k \in \{1, 2, \ldots, n\}$, the $k \times k$ leading principal submatrix A_k of A is non-singular. The proof that A has an LU-decomposition is achieved by showing inductively that each $A_1, A_2, A_3, \ldots$ has an LU-decomposition, and ultimately, that $A = A_n$ has an LU-decomposition.

The case $k = 1$ is trivial. Fix some $k \geq 2$ and let the inductive hypothesis be that A_{k-1} has an LU-decomposition $L_{k-1}U_{k-1} = A_{k-1}$, where L_{k-1} is an invertible lower triangular $(k-1) \times (k-1)$ matrix, and U_{k-1} is upper triangular. (By assumption both A_{k-1} and A_k are also invertible.) Write

$$A_k = \begin{bmatrix} A_{k-1} & B \\ C & a_{kk} \end{bmatrix}$$

for some $(k-1) \times 1$ matrix B and $1 \times (k-1)$ matrix C. It is not too difficult to verify that A_k has an LU factorization $A_k = L_k U_k$ where

$$L_k = \begin{bmatrix} L_{k-1} & \mathbf{0} \\ CU_{k-1}^{-1} & 1 \end{bmatrix} \quad \text{and} \quad U_k = \begin{bmatrix} U_{k-1} & L_{k-1}^{-1}B \\ \mathbf{0} & a_{kk} - c^T A_{k-1}^{-1}B \end{bmatrix}.$$

Of course L_k is unit lower triangular while U_k is upper triangular. Note that $a_{kk} - CA_{k-1}^{-1}B \neq 0$ since A_k is non-singular. $\square$

Exercise 651: The proof is by induction on $n \geq 1$. For each $n \in \mathbb{Z}^+$, let $P(n)$ be the proposition for $n \times n$ matrices.

BASE STEP: The statement $P(1)$ is trivial because any 1×1 matrix is already upper triangular, so use $U = [1] = I_1$.

INDUCTION STEP: Fix a positive integer k and assume that $P(k)$ is true. Let $A \in M_{(k+1) \times (k+1)}(\mathbb{C})$. Let λ be an eigenvalue of A and let $\mathbf{v} \neq \mathbf{0}$ be an eigenvector associated with λ. Without loss of generality, assume that $\|\mathbf{v}\| = 1$. Extend $\mathbf{v}$ to an orthonormal basis $\{\mathbf{v}, \mathbf{v}_1, \ldots, \mathbf{v}_k\}$ of $\mathbb{C}^{k+1}$. Viewing these vectors as column vectors, form the matrix

$$U_1 = [\mathbf{v} \mid \mathbf{v}_1 \mid \cdots \mid \mathbf{v}_k] \in M_{(k+1) \times (k+1)}(\mathbb{C}).$$

It is not difficult to verify that U_1 is unitary. (In fact, any unitary matrix with $\mathbf{v}$ as its first column is all that is needed.) Let $W \in M_{(k+1) \times (k)}(\mathbb{C})$ be the principal submatrix of U_1 formed by deleting the first column, i.e., $W = [\mathbf{v}_1 \mid \cdots \mid \mathbf{v}_k]$ or $U_1 = [\mathbf{v} \mid W]$. Define

$$A_1 = U_1^* A U_1 = \begin{bmatrix} \mathbf{v}^* \\ W^* \end{bmatrix} A[\mathbf{v} W] = \left[\begin{array}{c|c} \mathbf{v}^* A \mathbf{v} & \mathbf{v}^* A W \\ \hline W^* A \mathbf{v} & W^* A W \end{array} \right].$$

Since $A\mathbf{v} = \lambda \mathbf{v}$ and $\mathbf{v}^* \mathbf{v} = 1$, then $\mathbf{v}^* A \mathbf{v} = \lambda$. Also, since the columns of W are all orthogonal to $\mathbf{v}$, $W^* \mathbf{v} = \mathbf{0}_{k \times 1)}$, and so $W^* A \mathbf{v} = \lambda W^* \mathbf{0} = \lambda \mathbf{0} = \mathbf{0}$. Thus

$$A_1 = \left[\begin{array}{c|c} \lambda & \mathbf{v}^* A W \\ \hline \mathbf{0}_{k \times 1} & W^* A W \end{array} \right].$$

Then $W^*AW \in M_{k \times k}(\mathbb{C})$, and so by the induction hypothesis $P(k)$, there exists a unitary $U_2 \in M_{k \times k}(\mathbb{C})$ so that $T_1 = U_2^*(W^*AW)U_2 \in M_{k \times k}(\mathbb{C})$ is upper triangular.

Form the matrix $\hat{U}_2 = \left[\begin{array}{c|c} 1 & \mathbf{0}_{k \times 1} \\ \hline \mathbf{0}_{1 \times k} & U_2 \end{array}\right] \in M_{(k+1) \times (k+1)}(\mathbb{C})$. Then

$$(\hat{U}_2)^* \hat{U}_2 = \left[\begin{array}{c|c} 1 & \mathbf{0}_{k \times 1} \\ \hline \mathbf{0}_{1 \times k} & (U_2)^* U_2 \end{array}\right] = \left[\begin{array}{c|c} 1 & \mathbf{0}_{k \times 1} \\ \hline \mathbf{0}_{1 \times k} & I_k \end{array}\right] = I_{k+1},$$

and so $\hat{U}_2$ is unitary. Also, by simple block multiplication,

$$(\hat{U}_2)^* A_1 \hat{U}_2 = \left[\begin{array}{c|c} \lambda & \mathbf{v}AWU_2 \\ \hline \mathbf{0}_{1 \times k} & (U_2)^*(W^*AW)U_2 \end{array}\right] = \left[\begin{array}{c|c} \lambda & \mathbf{v}AWU_2 \\ \hline \mathbf{0}_{1 \times k} & T_1 \end{array}\right],$$

which is upper triangular. Then $P(k+1)$ is satisfied with upper triangular $T = (\hat{U}_2)^* A \hat{U}_2 = \hat{U}_2^* U_1^* A U_1 \hat{U}_2 \in M_{(k+1) \times (k+1)}(\mathbb{C})$ and unitary $U = U_1 \hat{U}_2$. This completes the inductive step $P(k) \to P(k+1)$.

By mathematical induction, for each $n \geq 1$, $P(n)$ is true. $\square$

Exercise 652: S. H. Lui kindly gave this solution in the form of source code from his class notes [360]; aside from minor typesetting changes, the proof is verbatim. The statement given by Lui was for real matrices, however I think his proof is for complex matrices.

($\Rightarrow$) Let $A = PDP^*$ with $P^*P = I$ and D diagonal. Now

$$\begin{aligned} AA^* &= PDP^*(PDP^*)^* = PDP^*PD^*P^* \\ &= PDD^*P^* = PD^*DP^* = PD^*P^*PDP^* \\ &= AA^* \end{aligned}$$

since D is diagonal and so $DD^* = D^*D$.

($\Leftarrow$) True for $n = 1$. Suppose true for $n-1$. Let A be a normal $n \times n$ matrix. From the Schur decomposition theorem [see Exercise 651], there exists a unitary U such that $U^*AU = T$, where T is upper triangular. Let t be an $(n-1)$ row vector, $\tau \in \mathbb{C}$, and $S \in \mathbb{R}^{(n-1) \times (n-1)}$ be upper triangular such that

$$T = \left[\begin{array}{c|c} \tau & t \\ \hline 0 & S \end{array}\right].$$

Since A is normal, T is also normal.

$$T^*T = TT^* \implies \left[\begin{array}{c|c} \bar{\tau} & 0 \\ \hline t^* & S^* \end{array}\right]\left[\begin{array}{c|c} \tau & t \\ \hline 0 & S \end{array}\right] = \left[\begin{array}{c|c} \tau & t \\ \hline 0 & S \end{array}\right]\left[\begin{array}{c|c} \bar{\tau} & 0 \\ \hline t^* & S^* \end{array}\right]$$

$$LHS = \left[\begin{array}{c|c} |\tau|^2 & \bar{\tau}t \\ \hline t^*\tau & |t|_2^2 + S^*S \end{array}\right], \quad RHS = \left[\begin{array}{c|c} |\tau|^2 + |t|_2^2 & tS^* \\ \hline St^* & SS^* \end{array}\right].$$

Therefore, $|\tau|^2 = |\tau|^2 + |t|_2^2$ which implies that $t = 0$. Now $S^*S = SS^* \implies S$ is normal.

Induction hypothesis implies that there exists unitary V such that $V^*SV = E \in \mathbb{R}^{(n-1)\times(n-1)}$, a diagonal matrix. Let $Q = U \left[\begin{array}{c|c} 1 & \\ \hline & V \end{array} \right]$. Then

$$
\begin{aligned}
Q^*AQ &= \left[\begin{array}{c|c} 1 & \\ \hline & V^* \end{array} \right] U^*AU \left[\begin{array}{c|c} 1 & \\ \hline & V \end{array} \right] \\
&= \left[\begin{array}{c|c} 1 & \\ \hline & V^* \end{array} \right] \left[\begin{array}{c|c} \tau & 0 \\ \hline 0 & S \end{array} \right] \left[\begin{array}{c|c} 1 & \\ \hline & V \end{array} \right] \\
&= \left[\begin{array}{c|c} \tau & \\ \hline & E \end{array} \right].
\end{aligned}
$$

Note that Q is unitary since both U and V are. $\qquad\square$

Exercise 653: S. H. Lui kindly gave this solution in the form of source code from an early draft of his class notes [360]; I have only slightly modified some typesetting, otherwise not altering his notes.

Induction on n. $n = 1$ is fine.

Suppose true for $n - 1$. Let $A \in \mathbb{R}^{n\times n}$ be symmetric positive definite.

$$
A = \left[\begin{array}{c|c} B & \mathbf{a} \\ \hline \mathbf{a}^T & a_{nn} \end{array} \right], \qquad B \in \mathbb{R}^{n-1\times n-1} \text{ symmetric}, \ \mathbf{a} \in \mathbb{R}^{n-1}.
$$

Note that $a_{nn} = e_n^T A e_n > 0$ since A is positive definite. Let $\mathbf{y} \in \mathbb{R}^{n-1}$, $\mathbf{y} \neq \mathbf{0}$, show $y^T By > 0$. Since A is positive definite,

$$
\left[\begin{array}{c} \mathbf{y} \\ 0 \end{array} \right]^T A \left[\begin{array}{c} \mathbf{y} \\ 0 \end{array} \right] = \left[\begin{array}{c} \mathbf{y} \\ 0 \end{array} \right]^T \left[\begin{array}{c|c} B & \mathbf{a} \\ \hline \mathbf{a}^T & a_{nn} \end{array} \right] \left[\begin{array}{c} \mathbf{y} \\ 0 \end{array} \right] = y^T By > 0.
$$

Induction hypothesis implies that there exists a unique upper triangular S with positive diagonal entries such that $B = S^T S$. Let

$$
A = \left[\begin{array}{c|c} S^T S & \mathbf{a} \\ \hline \mathbf{a}^T & a_{nn} \end{array} \right] = \left[\begin{array}{c|c} S^T & \\ \hline \mathbf{b}^T & c \end{array} \right] \left[\begin{array}{c|c} S & \mathbf{b} \\ \hline & c \end{array} \right] = R^T R
$$

where $c \in \mathbb{R}$, $\mathbf{b} \in \mathbb{R}^{n-1}$. From the above system, $\mathbf{a} = S^T \mathbf{b}$ and $a_{nn} = \mathbf{b}^T \mathbf{b} + c^2$. Since S is nonsingular, $\mathbf{b} = S^{-T}\mathbf{a}$. If it can be shown that $a_{nn} - \mathbf{b}^T \mathbf{b} > 0$, then one is able to define $c = \sqrt{a_{nn} - \mathbf{b}^T \mathbf{b}}$.

By a direct calculation, $a_{nn} - \mathbf{b}^T\mathbf{b} = a_{nn} - \mathbf{a}^T B^{-1}\mathbf{a}$. Define $\gamma = B^{-1}\mathbf{a}$. Since A is positive definite,

$$
0 < [\gamma^T, -1] \left[\begin{array}{c|c} B & \mathbf{a} \\ \hline \mathbf{a}^T & a_{nn} \end{array} \right] \left[\begin{array}{c} \gamma \\ -1 \end{array} \right]
$$

$$\begin{aligned} &= a_{nn} - \mathbf{a}^T B^{-1} \mathbf{a} \\ &= a_{nn} - \mathbf{b}^T \mathbf{b}. \end{aligned}$$

This is what was wanted to show.

Finally, to show uniqueness: Let $A = R^T R = \tilde{R}^T \tilde{R}$ where $\tilde{R}$ is upper triangular with positive diagonal entries. Then $\tilde{R}^{-T} R^T = \tilde{R} R^{-1}$. Note that the inverse of an upper triangular matrix is upper triangular and the product of two upper triangular matrices is upper triangular. The same remark applies to lower triangular matrices. Hence $\tilde{R}^{-T} R^T = \tilde{R} R^{-1} = D$, a diagonal matrix. Look at the (i, i) entry of $\tilde{R} = DR$ and of $R^T = \tilde{R}^T D$ to obtain $\tilde{r}_{ii} = d_{ii} r_{ii}$ and $r_{ii} = \tilde{r}_{ii} d_{ii}$ or $d_{ii} = \pm 1$. Since r_{ii} and $\tilde{r}_{ii}$ are positive, $d_{ii} = 1$ for every i and so D is the identity matrix which implies that $R = \tilde{R}$. $\qquad\square$

Exercise 654: Fix a constant c, and for each $n \geq 1$, let $C(n)$ be the claim that

$$\begin{bmatrix} 1 & c \\ 0 & 1 \end{bmatrix}^n = \begin{bmatrix} 1 & cn \\ 0 & 1 \end{bmatrix}.$$

BASE STEP: $C(1)$ is trivially true.

INDUCTIVE STEP: Fix $k \geq 1$ and let the inductive hypothesis be that $C(k)$ is true. It remains to prove that

$$C(k+1) : \begin{bmatrix} 1 & c \\ 0 & 1 \end{bmatrix}^{k+1} = \begin{bmatrix} 1 & c(k+1) \\ 0 & 1 \end{bmatrix}.$$

Starting with the LHS of $C(k+1)$,

$$\begin{aligned} \begin{bmatrix} 1 & c \\ 0 & 1 \end{bmatrix}^{k+1} &= \begin{bmatrix} 1 & c \\ 0 & 1 \end{bmatrix}^k \begin{bmatrix} 1 & c \\ 0 & 1 \end{bmatrix} \\ &= \begin{bmatrix} 1 & ck \\ 0 & 1 \end{bmatrix} \begin{bmatrix} 1 & c \\ 0 & 1 \end{bmatrix} \qquad \text{(by } C(k)\text{)} \\ &= \begin{bmatrix} 1 & c + ck \\ 0 & 1 \end{bmatrix}, \end{aligned}$$

which is equal to the RHS of $C(k+1)$. This completes the inductive step $C(k) \rightarrow C(k+1)$.

By mathematical induction, for each $n \geq 1$, $C(n)$ is true. $\qquad\square$

Exercise 655: Fix constants a, b. For each $n \geq 1$, let $C(n)$ be the claim that
$$\begin{bmatrix} a & 0 \\ 0 & b \end{bmatrix}^n = \begin{bmatrix} a^n & 0 \\ 0 & b^n \end{bmatrix}.$$

BASE STEP: $C(1)$ says $\begin{bmatrix} a & 0 \\ 0 & b \end{bmatrix}^1 = \begin{bmatrix} a^1 & 0 \\ 0 & b^1 \end{bmatrix}$, which is true.

INDUCTIVE STEP: Fix some $k \geq 1$ and suppose that $C(k)$ is true. Then

$$
\begin{aligned}
\begin{bmatrix} a & 0 \\ 0 & b \end{bmatrix}^{k+1} &= \begin{bmatrix} a & 0 \\ 0 & b \end{bmatrix}^k \begin{bmatrix} a & 0 \\ 0 & b \end{bmatrix} \\
&= \begin{bmatrix} a^k & 0 \\ 0 & b^k \end{bmatrix}\begin{bmatrix} a & 0 \\ 0 & b \end{bmatrix} \quad \text{(by } C(k)) \\
&= \begin{bmatrix} a^{k+1} & 0 \\ 0 & b^{k+1} \end{bmatrix}
\end{aligned}
$$

shows that $C(k+1)$ is also true, completing the inductive step.

By mathematical induction, for each $n \geq 1$, $C(n)$ is true. □

Exercise 656: For each $n \geq 2$, let $A(n)$ be the assertion that

$$
\begin{bmatrix} 4 & 2 \\ 2 & 1 \end{bmatrix}^n = 5^{n-1} \begin{bmatrix} 4 & 2 \\ 2 & 1 \end{bmatrix}.
$$

In fact, even when $n = 1$, $A(n)$ is true, however, in keeping with the exercise, this proof starts at $n = 2$.

BASE STEP $(n = 2)$: Since

$$
\begin{bmatrix} 4 & 2 \\ 2 & 1 \end{bmatrix}\begin{bmatrix} 4 & 2 \\ 2 & 1 \end{bmatrix} = \begin{bmatrix} 20 & 10 \\ 10 & 5 \end{bmatrix} = 5 \begin{bmatrix} 4 & 2 \\ 2 & 1 \end{bmatrix},
$$

$A(2)$ holds.

INDUCTIVE STEP: Fix some $k \geq 2$ and assume that $A(k)$ is true. Then

$$
\begin{aligned}
\begin{bmatrix} 4 & 2 \\ 2 & 1 \end{bmatrix}^{k+1} &= \begin{bmatrix} 4 & 2 \\ 2 & 1 \end{bmatrix}^k \begin{bmatrix} 4 & 2 \\ 2 & 1 \end{bmatrix} \\
&= 5^{k-1} \begin{bmatrix} 4 & 2 \\ 2 & 1 \end{bmatrix}\begin{bmatrix} 4 & 2 \\ 2 & 1 \end{bmatrix} \quad \text{(by } A(k)) \\
&= 5^{k-1} \begin{bmatrix} 20 & 10 \\ 10 & 5 \end{bmatrix} \\
&= 5^{k-1} \cdot 5 \begin{bmatrix} 4 & 2 \\ 2 & 1 \end{bmatrix} = 5^k \begin{bmatrix} 4 & 2 \\ 2 & 1 \end{bmatrix}
\end{aligned}
$$

shows that $A(k+1)$ is also true, completing the inductive step.

By mathematical induction, for each $n \geq 2$, $A(n)$ holds. □

Exercise 657: (rotation matrix) For any $\theta \in \mathbb{R}$ and $n \geq 1$, let $S(n)$ be the statement

$$\begin{bmatrix} \cos(\theta) & -\sin(\theta) \\ \sin(\theta) & \cos(\theta) \end{bmatrix}^n = \begin{bmatrix} \cos(n\theta) & -\sin(n\theta) \\ \sin(n\theta) & \cos(n\theta) \end{bmatrix}.$$

BASE STEP: The statement $S(1)$ holds trivially.

INDUCTIVE STEP: Fix some $k \geq 1$ and suppose that $S(k)$ holds. Then

$$\begin{bmatrix} \cos(\theta) & -\sin(\theta) \\ \sin(\theta) & \cos(\theta) \end{bmatrix}^{k+1} = \begin{bmatrix} \cos(\theta) & -\sin(\theta) \\ \sin(\theta) & \cos(\theta) \end{bmatrix}^k \cdot \begin{bmatrix} \cos(\theta) & -\sin(\theta) \\ \sin(\theta) & \cos(\theta) \end{bmatrix}$$

$$= \begin{bmatrix} \cos(k\theta) & -\sin(k\theta) \\ \sin(k\theta) & \cos(k\theta) \end{bmatrix} \cdot \begin{bmatrix} \cos(\theta) & -\sin(\theta) \\ \sin(\theta) & \cos(\theta) \end{bmatrix} \quad \text{(by } S(k)\text{)}$$

$$= \begin{bmatrix} \cos(k\theta)\cos(\theta) - \sin(k\theta)\sin(\theta) & -\cos(k\theta)\sin(k\theta) - \sin(k\theta)\cos(\theta) \\ \sin(k\theta)\cos(\theta) + \cos(k\theta)\sin(\theta) & -sin(k\theta)sin(\theta) + \cos(k\theta)\cos(\theta) \end{bmatrix}$$

$$\text{(by eq'ns (9.11) and (9.12))}$$

$$= \begin{bmatrix} \cos((k+1)\theta) & -\sin((k+1)\theta) \\ \sin((k+1)\theta) & \cos((k+1)\theta) \end{bmatrix}$$

which shows that $S(k+1)$ holds as well, completing the inductive step.

By the principle of mathematical induction, for each $n \geq 1$, $S(n)$ holds. $\square$

Comment: There is a connection between this exercise and DeMoivre's formula (see Exercise 115).

Exercise 658: Let $n \geq 1$ be fixed and let $M_n = J_n - I_n$. For $k \geq 1$, let $S(k)$ denote the statement that

$$M_n^k = \left(\frac{(n-1)^k - (-1)^k}{n}\right)J_n + (-1)^k I_n.$$

BASE STEP: When $k = 1$, $S(1)$ reads $M_n = \left(\frac{n-1-(-1)}{n}\right)J_n + (-1)I_n$ and the right-hand side is $J_n - I_n$, so $S(1)$ is true.

INDUCTIVE STEP $(S(j) \rightarrow S(j+1))$: Fix some $j \geq 1$ and suppose that $S(j)$ is true. One needs to show that $S(j+1)$

$$M_n^{j+1} = \left(\frac{(n-1)^{j+1} - (-1)^{j+1}}{n}\right)J_n + (-1)^{j+1}I_n$$

follows. Beginning with the left side of $S(j+1)$,

$$\begin{aligned} M_n^{j+1} &= (J_n - I_n)^{j+1} \\ &= (J_n - I_n)^j (J_n - I_n) \end{aligned}$$

$$
\begin{aligned}
&= \left(\left(\left(\frac{(n-1)^j - (-1)^j}{n}\right)J_n + (-1)^j I_n\right)(J_n - I_n) \quad \text{(by } S(j)) \\
&= \left(\frac{(n-1)^j - (-1)^j}{n}\right)J_n(J_n - I_n) + (-1)^j I_n(J_n - I_n) \\
&= \left(\frac{(n-1)^j - (-1)^j}{n}\right)(J_n^2 - J_n) + (-1)^j(J_n - I_n) \\
&= \left(\frac{(n-1)^j - (-1)^j}{n}\right)(nJ_n - J_n) + (-1)^j(J_n - I_n) \\
&= \left(\frac{(n-1)^j - (-1)^j}{n}\right)(n-1)J_n + (-1)^j J_n + (-1)^{j+1} I_n \\
&= \left(\frac{(n-1)^{j+1} - (-1)^j(n-1)}{n} + (-1)^j\right)J_n + (-1)^{j+1} I_n \\
&= \left(\frac{(n-1)^{j+1} - (-1)^j(n-1)}{n} + (-1)^j\right)J_n + (-1)^{j+1} I_n \\
&= \left(\frac{(n-1)^{j+1} - (-1)^{j+1}}{n}\right)J_n + (-1)^{j+1} I_n,
\end{aligned}
$$

one arrives at the right side of $S(j+1)$, concluding the inductive step.

By MI, for each $k \in \mathbb{Z}^+$, $S(k)$ is true. $\square$

Exercise 659: One natural construction of Hadamard matrices is an easy recursion, making larger ones from four smaller ones each time—this construction can be thought of as some special kind of product of two matrices, a 2×2 Hadamard matrix and an $n \times n$ Hadamard matrix, giving a $2n \times 2n$ matrix. Starting with $H_2 = \begin{bmatrix} 1 & 1 \\ 1 & -1 \end{bmatrix}$ one can recursively generate

$$
H_{k+1} = \begin{bmatrix} H_k & H_k \\ H_k & -H_k \end{bmatrix},
$$

often denoted by a "tensor product" $H_{k+1} = H_k \otimes H_2$; this construction is due to Sylvester and such matrices are said to be of Sylvester type.

Note: There are other ways to create Hadamard matrices. The Sylvester construction only works to create Hadamard matrices whose order is a power of 2; in general, it is not known if a Hadamard matrix exists for every n divisible by 4 (a necessary condition for a Hadamard matrix to exist). I think that Hadamard matrices have been found for all possible orders up to 264, but no Hadamard matrix of order 268 has yet been found.

Exercise 661: (Vandermonde determinant) Let $S(n)$ be the statement that for any matrix of the form

$$M = \begin{bmatrix} 1 & c_0 & c_0^2 & \cdots & c_0^n \\ 1 & c_1 & c_1^2 & \cdots & c_1^n \\ \vdots & \vdots & \vdots & & \vdots \\ 1 & c_n & c_n^2 & \cdots & c_n^n \end{bmatrix}$$

then $\det(M) = \prod_{0 \le i < j \le n}(c_j - c_i)$. The proof is by induction on n.

BASE STEP: When $n = 1$, $M = \begin{bmatrix} 1 & c_0 \\ 1 & c_1 \end{bmatrix}$, and $\det(M) = c_1 - c_0$, so $S(1)$ is true.

INDUCTION STEP: Fix some $k \ge 1$, and assume that $S(k)$ holds. Let $d_0, \ldots, d_{k+1}$ be scalars and let

$$M = \begin{bmatrix} 1 & d_0 & d_0^2 & \cdots & d_0^{k+1} \\ 1 & d_1 & d_1^2 & \cdots & d_1^{k+1} \\ \vdots & \vdots & \vdots & & \vdots \\ 1 & d_{k+1} & d_{k+1}^2 & \cdots & d_{k+1}^{k+1} \end{bmatrix}.$$

By subtracting from each column d_0 times the previous column,

$$\begin{vmatrix} 1 & d_0 & d_0^2 & \cdots & d_0^{k+1} \\ 1 & d_1 & d_1^2 & \cdots & d_1^{k+1} \\ \vdots & \vdots & \vdots & & \vdots \\ 1 & d_{k+1} & d_{k+1}^2 & \cdots & d_{k+1}^{k+1} \end{vmatrix} = \begin{vmatrix} 1 & 0 & 0 & \cdots & 0 \\ 1 & d_1 - d_0 & d_1^2 - d_1 d_0 & \cdots & d_1^{k+1} - d_1^k d_0 \\ 1 & d_2 - d_0 & d_2^2 - d_1 d_0 & \cdots & d_2^{k+1} - d_2^k d_0 \\ \vdots & \vdots & \vdots & & \vdots \\ 1 & d_{k+1} - d_0 & d_{k+1}^2 - d_{k+1}d_0 & \cdots & d_{k+1}^{k+1} - d_{k+1}^k d_0 \end{vmatrix}.$$

Expanding along the first row and factoring, the above determinant becomes

$$\begin{vmatrix} d_1 - d_0 & d_1(d_1 - d_0) & \cdots & d_1^k(d_1 - d_0) \\ d_2 - d_0 & d_2(d_2 - d_0) & \cdots & d_2^k(d_2 - d_0) \\ \vdots & \vdots & & \vdots \\ d_{k+1} - d_0 & d_{k+1}(d_{k+1} - d_0) & \cdots & d_{k+1}^k(d_{k+1} - d_0) \end{vmatrix}$$

$$= (d_1 - d_0)(d_2 - d_0)\cdots(d_{k+1} - d_0)\begin{vmatrix} 1 & d_1 & \cdots & d_1^k \\ 1 & d_2 & \cdots & d_2^k \\ \vdots & \vdots & & \vdots \\ 1 & d_{k+1} & \cdots & d_{k+1}^k \end{vmatrix},$$

and by the induction hypothesis (using $c_0 = d_1$, $c_1 = d_2$, ..., $c_k = d_{k+1}$), the above determinant is

$$(d_1 - d_0)(d_2 - d_0)\cdots(d_{k+1} - d_0) \prod_{1 \le i < j \le k+1}(d_j - d_i) = \prod_{0 \le i < j \le k+1}(d_j - d_i),$$

showing that $S(k + 1)$ holds, completing the inductive step.

By MI, for each positive integer n, $S(n)$ holds. □

Exercise 662: Let $x \neq 1$, and for each $n \geq 1$, let $S(n)$ denote the statement

$$\begin{bmatrix} +1 & -1 \\ 0 & x \end{bmatrix}^n = \begin{bmatrix} +1 & \frac{x^n-1}{1-x} \\ 0 & x^n \end{bmatrix}.$$

BASE STEP: $S(1)$ is trivially true as $x^1 = x$.

INDUCTIVE STEP: Fix some $k \geq 1$ and suppose that

$$S(k): \quad \begin{bmatrix} +1 & -1 \\ 0 & x \end{bmatrix}^k = \begin{bmatrix} +1 & \frac{x^k-1}{1-x} \\ 0 & x^k \end{bmatrix}.$$

Then

$$\begin{bmatrix} +1 & -1 \\ 0 & x \end{bmatrix}^{k+1} = \begin{bmatrix} +1 & -1 \\ 0 & x \end{bmatrix}^k \begin{bmatrix} +1 & -1 \\ 0 & x \end{bmatrix}$$

$$= \begin{bmatrix} +1 & \frac{x^k-1}{1-x} \\ 0 & x^k \end{bmatrix} \begin{bmatrix} +1 & -1 \\ 0 & x \end{bmatrix} \qquad \text{(by } S(k)\text{)}$$

$$= \begin{bmatrix} +1 & -1 + \left(\frac{x^k-1}{1-x}\right)x \\ 0 & x^{k+1} \end{bmatrix}$$

$$= \begin{bmatrix} +1 & \frac{x^{k+1}-1}{1-x} \\ 0 & x^{k+1} \end{bmatrix},$$

proves $S(k+1)$, finishing the inductive step.

By mathematical induction, for each $n \geq 1$, the statement $S(n)$ is true. □

Exercise 663: For each $n \geq 1$ and the $n \times n$ matrix

$$A_n = \begin{bmatrix} 2 & -1 & 0 & \cdots & 0 & 0 \\ -1 & 2 & -1 & \cdots & 0 & 0 \\ 0 & -1 & 2 & \cdots & 0 & 0 \\ 0 & 0 & -1 & \cdots & 0 & 0 \\ \vdots & \vdots & \vdots & \ddots & \vdots \\ 0 & 0 & 0 & \cdots & -1 & 2 \end{bmatrix},$$

let $P(n)$ be the proposition that $\det(A_n) = n+1$. Since the inductive step given here uses two previous cases, two base cases are required.

BASE STEPS $(n = 1,2)$: The 1×1 matrix A_1 is simply the number 2, which has $\det[2] = 2 = 1+1$, so $P(1)$ is true. For $n = 2$, $A_2 = \begin{bmatrix} 2 & -1 \\ -1 & 2 \end{bmatrix}$, and $\det(A_2) = 3 = 2+1$ shows that $P(2)$ is true.

INDUCTIVE STEP $(P(k) \to P(k+1))$: Fix some $k \geq 2$ and assume that the inductive hypotheses $P(k-1)$ and $P(k)$ is true. To show $P(k+1)$, one must show that $\det(A_{k+1}) = k+2$. Expanding along the first row of A_{k+1} (where vertical bars around the $k \times k$ matrix below indicate determinant)

$$\det(A_{k+1}) = 2\det(A_k) + \begin{vmatrix} -1 & 0 & 0 & \cdots & 0 & 0 \\ -1 & 2 & -1 & \cdots & 0 & 0 \\ 0 & -1 & 2 & \cdots & 0 & 0 \\ 0 & 0 & -1 & \cdots & 0 & 0 \\ \vdots & \vdots & \vdots & \ddots & \vdots \\ 0 & 0 & 0 & \cdots & -1 & 2 \end{vmatrix}$$

$$= 2\det(A_k) + (-1)\det(A_k - 1) \quad \text{(expanding along 1st row)}$$
$$= 2(k+1) - (k-1+1)$$
$$= k+2,$$

shows that $P(k+1)$ also holds, completing the inductive step.

By MI, for each $n \geq 1$, the statement $P(n)$ holds. □

Exercise 664: (Companion matrix) One solution is by induction on $k \geq 2$.

BASE STEP: When $k = 2$, $A = \begin{bmatrix} 0 & -a_0 \\ 1 & -a_1 \end{bmatrix}$, in which case,

$$\det(A - tI) = \begin{vmatrix} -t & -a_0 \\ 1 & -a_1 - t \end{vmatrix} = t^2 + a_1 t + a_0,$$

which proves the only base case.

INDUCTIVE STEP: Suppose that for some fixed $n \geq 2$, the result holds for any constants. Let $a_0, \ldots, a_n$ be scalars, and put

$$g(t) = (-1)^{n+1}(a_0 + a_1 t + \cdots + a_{n-1}t^{n-1} + a_n t^n + t^{n+1}).$$

Let

$$B = \begin{bmatrix} 0 & 0 & \cdots & 0 & -a_0 \\ 1 & 0 & \cdots & 0 & -a_1 \\ 0 & 1 & \cdots & 0 & -a_2 \\ \vdots & \vdots & & \vdots & \vdots \\ 0 & 0 & \cdots & 0 & -a_{n-1} \\ 0 & 0 & \cdots & 1 & -a_n \end{bmatrix}.$$

Expanding along the first row, and letting $M_{i,j}$ denote the principal submatrix formed by deleting the i-th row and j-th column of M,

$$\det(A - tI_{n+1}) = (-t)\det(M_{1,1}) + (-1)^{n+2}a_0\det(M_{1,n+1}).$$

The matrix $M_{1,1}$ is a matrix of the same form as $A - tI_n$, so by induction hypothesis, $\det(M_{1,1}) = (-1)^n(a_1 + a_2 t + \cdots + a_n t^{n-1} + t^n)$. Also, since $M_{1,n+1}$ is upper triangular with 1's on the main diagonal, $\det(M_{1,n+1}) = 1$. Putting these together,

$$
\begin{aligned}
\det(A - tI_{n+1}) &= (-t)\det(M_{1,1}) + (-1)^{n+2}(-a_0)\det(M_{1,n+1}) \\
&= (-t)(-1)^n(a_1 + a_2 t + \cdots + a_n t^{n-1} + t^n) + (-1)^{n+2}(-a_0) \\
&= (-1)^{n+1}t(a_1 + a_2 t + \cdots + a_n t^{n-1} + t^n) + (-1)^{n+1}a_0 \\
&= (-1)^{n+1}[a_0 + a_1 t + a_2 t^2 + \cdots + a_n t^n + t^{n+1}],
\end{aligned}
$$

showing that the result holds for $k = n + 1$, completing the inductive step.

By MI, for any integer $k \geq 2$, the result holds. $\qquad\square$

Exercise 665: (Leontief) The following is essentially the solution in [580], which leaves a few details to be verified. Let $S(n)$ be the statement that for any $n \times n$ Leontief matrix H, conditions (1), (2), and (3) are equivalent.

BASE STEP: Let $H = [h]$ be a 1×1 matrix (which is Leontief for any h). Each of the three above conditions merely says that $h > 0$.

INDUCTIVE STEP: Fix some k and suppose that $S(k)$ is true. Let H be a $(k+1) \times (k+1)$ Leontief matrix.

$(1) \Rightarrow (2)$: Let $\mathbf{x}' = \begin{bmatrix} \mathbf{x} \\ \delta \end{bmatrix}$ be a column vector with $H\mathbf{x}' > \mathbf{0}$. Because H is Leontief, there exists a permutation matrix P and a Leontief $k \times k$ matrix G, non-negative A and B ($k \times 1$ and $1 \times k$, respectively) and $\gamma \in \mathbb{R}^+$ with

$$
P^{-1}HP = \begin{bmatrix} G & -A \\ -B & \gamma \end{bmatrix},
$$

for if not, one must have $H\mathbf{x}' \leq \mathbf{0}$ (why?). By Lemma 19.1.4, one may assume without loss of generality that H itself can be expressed as

$$
H = \begin{bmatrix} G & -A \\ -B & \gamma \end{bmatrix},
$$

and

$$
H\begin{bmatrix} \mathbf{x} \\ \delta \end{bmatrix} = \begin{bmatrix} G\mathbf{x} - \delta A \\ -B\mathbf{x} + \gamma\delta \end{bmatrix} > \mathbf{0}.
$$

To show that H is a positive matrix, by Lemma 19.1.3, one only needs to show that $G - \frac{1}{\gamma}AB$ is a positive matrix. As $G - \frac{1}{\gamma}AB$ is seen to be Leontief,

$$
(G - \frac{1}{\gamma}AB)\mathbf{x} = \underbrace{G\mathbf{x} - \delta A}_{>0} + A\underbrace{(-B\mathbf{x} + \gamma\delta)}_{>0}/\gamma > \mathbf{0}.
$$

By $S(k)$, $G - \frac{1}{\gamma}AB$ is a positive $k \times k$ matrix.

(2)$\Rightarrow$ (3): Suppose that

$$H = \begin{bmatrix} G & -A \\ -B & \gamma \end{bmatrix}$$

is a positive matrix. By Lemma 19.1.3, $A \geq 0$, $B \geq 0$, $\gamma > 0$, and $G - \frac{1}{\gamma}AB$ is a positive $k \times k$ matrix. By $S(k)$, all entries of $[G - \frac{1}{\gamma}AB]^{-1}$ are non-negative. Using $K = G - \frac{1}{\gamma}AB$, verify that

$$B^{-1} = \begin{bmatrix} K^{-1} & \frac{1}{\gamma}K^{-1}A \\ \frac{1}{\gamma}BK^{-1} & \frac{1}{\gamma} + \frac{1}{\gamma^2}BK^{-1}A \end{bmatrix},$$

and one can check that all entries of B^{-1} are non-negative.

(3)$\Rightarrow$ (1) is trivial, so the circle of implications is complete for $(k+1) \times (k+1)$ matrices, completing the inductive step.

By mathematical induction, for each $n \geq 1$, $S(n)$ is true. $\qquad\square$

Exercise 666: Fix $n \geq 2$ and let M be an $m \times n$ latin rectangle M with $m < n$. For each $m \leq m' \leq n$, let $P(m')$ be the proposition that for any there is an $m' \times n$ latin rectangle that extends M.

BASE STEP: When $m' = m$, M itself satisfies $P(m')$.

INDUCTIVE STEP: Let M be an $m \times n$ latin rectangle with $m < n$. Fix $p \in \{m, m+1, \ldots, n-1\}$, suppose that $P(p)$ is true, and let M' be a witness to $P(p)$; that is, M' is an $p \times n$ latin rectangle that extends M. To show that $P(p+1)$ holds, it suffices to show that one can add another row to M' producing an $(p+1) \times n$ latin rectangle. Put $X = \{1, 2, \ldots, n\}$, and for each $j \in \{1, 2, \ldots, n\}$, let $T_j \subset S$ be those elements in X not in column j of M', and put $\mathcal{T} = \{T_1, \ldots, T_n\}$. To find one more row to add to M', it suffices to show that there is a matching (in the obvious bipartite graph) from X into $\mathcal{T}$, that is, to show the existence of a transversal in $\mathcal{T}$.

By Hall's theorem (Theorem 15.5.1) or its equivalent form for SDR's (Theorem 33.4.1), it suffices to show that the union of *any* k of the T_i's contains at least distinct k elements. Such a union contains, with repetition, $k(n-p)$ elements, and if such a union were to contain fewer than k distinct elements, then one element $x \in X$ is repeated more than $n-p$ times—meaning that x appears in less than p rows, contrary to M being a latin rectangle. Hence, M can be extended by adding one more row to an $(m'+1) \times n$ latin rectangle, showing $P(p+1)$ is true.

By finite mathematical induction, for any $m' \in \{m, m+1, \ldots, n\}$, $P(m')$ is true; in particular, $P(n)$ is true—if $m < n$, any $m \times n$ latin rectangle can be completed to a latin square. $\qquad\square$

Exercise 667: See [161, 8.32, p. 209]; this problem also occurred in the 1988 Tournament of the Towns. Hint: build a sequence of square matrices, each time doubling in size, and take the limit.

Exercise 668: This solution was kindly written for inclusion here by Michael Doob. [I have edited only a very little.] For more on Hankel matrices and Hankel transforms, see, *e.g.*, [344]. It appears as if something similar to the following solution first appeared in 2006 [293], although I can not confirm this [thanks to Liji Huang for supplying this reference].

For $n = 1$, get $H_1 = [a_0]$ for the first sequence and $H'_1 = [a_1]$ for the second one, and so the determinant condition implies $a_0 = a_1 = 1$. For $n = 2$ the two Hankel matrices are $H_2 = \begin{bmatrix} 1 & 1 \\ 1 & a_2 \end{bmatrix}$ and $H'_2 = \begin{bmatrix} 1 & a_2 \\ a_2 & a_3 \end{bmatrix}$. The assumption that the two determinants are one implies that $a_2 = 2$ and then $a_3 = 5$.

It may now be clear how to obtain further values. The two $n \times n$ Hankel matrices H_n and H'_n can be used to determine a_{2n-2} and a_{2n-1}. In each case, all but one element of the matrix is already determined, and the determinant condition yields the last element. The case $n = 3$ gives $a_4 = 14$ and $a_5 = 42$.

From here on, it is convenient to use lower case c_n to denote the Catalan number C_n, as a matrix with certain catalan number entries is to be called C.

Consider the $n \times n$ matrix C where the i, j entry of C is c_{i+j}, and C' where the i, j entry is c_{i+j+1}. The next goal is to show for all $n \geq 1$ that $\det(C) = \det(C') = 1$. To do this, define a lower triangular matrix B in the following way:

1. For the first row, $B_{0,0} = 1$ and $B_{0,j} = 0$ for $j > 1$.

2. Having defined rows $0, \ldots, i - 1$, let

   - $B_{i,0} = B_{i-1,0} + B_{i-1,1}$
   - $B_{i,j} = B_{i-1,j-1} + 2B_{i-1,j} + B_{i-1,j+1}$ otherwise.

(when $j = m$, interpret $B_{i-1,j+1}$ to be 0).

For example, when $n = 7$, get the following matrix:

$$B = \begin{bmatrix} 1 & 0 & 0 & 0 & 0 & 0 & 0 \\ 1 & 1 & 0 & 0 & 0 & 0 & 0 \\ 2 & 3 & 1 & 0 & 0 & 0 & 0 \\ 5 & 9 & 5 & 1 & 0 & 0 & 0 \\ 14 & 28 & 20 & 7 & 1 & 0 & 0 \\ 42 & 90 & 75 & 35 & 9 & 1 & 0 \\ 132 & 297 & 275 & 154 & 54 & 11 & 1 \end{bmatrix}$$

Then B is lower triangular and for each $k = 1, \ldots, n$, $B_{k,k} = 1$.

Define a matrix B' in a manner similar to B:

1. For the first row, $B'_{0,0} = 1$ and $B'_{0,j} = 0$ for $j > 1$.

2. Having defined rows $0, \ldots, i - 1$, let

   - $B'_{i,0} = 2B'_{i-1,0} + B'_{i-1,1}$

- $B'_{i,j} = B'_{i-1,j-1} + 2B'_{i-1,j} + B'_{i-1,j+1}$ otherwise.

(when $j = m$, interpret $B'_{i-1,j+1}$ to be 0).

For example, when $n = 7$, get the following matrix:

$$B' = \begin{bmatrix} 1 & 0 & 0 & 0 & 0 & 0 & 0 \\ 2 & 1 & 0 & 0 & 0 & 0 & 0 \\ 5 & 4 & 1 & 0 & 0 & 0 & 0 \\ 14 & 14 & 6 & 1 & 0 & 0 & 0 \\ 42 & 48 & 27 & 8 & 1 & 0 & 0 \\ 132 & 165 & 110 & 44 & 10 & 1 & 0 \\ 429 & 572 & 429 & 208 & 65 & 12 & 1 \end{bmatrix}$$

Claim 37.1.1. $BB^T = C$ and $B'B'^T = C'$

The proof relies on Claim 37.1.1, however, before proving Claim 37.1.1, a few observations are in order.

Consequence of Claim 37.1.1: Since B is 1 on the diagonal, then $\det(B) = \det(B^T) = 1$ and hence $\det(C) = 1$. Similarly $\det(B') = \det(B'^T) = 1$ and hence $\det(C') = 1$.

Note: The matrices B and B^T are the LU-decomposition of C, and B' and B'^T are the LU-decomposition of C'.

It is a standard application that the number of solutions to the ballot problem with $2n$ votes is just the Catalan number C_n, which is denoted here by c_n for convenience. (See Exercise 765.)

There is a straightforward but useful observation to be made. If at some point in the drawing the number of ballots for A is not less than the number of ballots for B, then in the remaining ballots the number of ballots for B is not less than the number of ballots for A. Hence if one starts with a solution to the ballot problem, reverses the roles of A and B, and then draws the ballots in the opposite order, one gets a (possibly identical) solution to the ballot problem. Call this the *reversing principle*.

Extend the ballot problem to the case where the numbers of ballots for each candidate are not necessarily equal. If there are m votes for candidate A and n for candidate B, let $c_{m,n}$ be the number of ways the the ballots can be drawn so that candidate A is never behind; then $c_{m,0} = 1$, $c_{n,n} = c_n$ and for $m < n$, $c_{m,n} = 0$. When counting all the ballots to compute $c_{m,n}$, there are two cases: the last vote is A (so the previous ballots are $c_{m-1,n}$ in number), or the last vote is B (so the previous ballot draws where $c_{m,n-1}$ in number). This gives:

Theorem 37.1.2. *If m and n are positive, $m \geq n$, then*

$$c_{m,n} = c_{m-1,n} + c_{m,n-1}$$

Define the matrix C' whose (m, n) entry is $c_{m,n}$ (limiting its size appropriately); then

$$C' = \begin{bmatrix} 1 & 0 & 0 & 0 & 0 & 0 & 0 & 0 \\ 1 & 1 & 0 & 0 & 0 & 0 & 0 & 0 \\ 1 & 2 & 2 & 0 & 0 & 0 & 0 & 0 \\ 1 & 3 & 5 & 5 & 0 & 0 & 0 & 0 \\ 1 & 4 & 9 & 14 & 14 & 0 & 0 & 0 \\ 1 & 5 & 14 & 28 & 42 & 42 & 0 & 0 \\ 1 & 6 & 20 & 48 & 90 & 132 & 132 & 0 \\ 1 & 7 & 27 & 75 & 165 & 297 & 429 & 429 \end{bmatrix}$$

Next relate the matrix C' given above with the matrix B defined earlier. In fact the entries of B run down alternate skew diagonals of C'.

Theorem 37.1.3. *The entries in B and B' satisfy*

$$b_{i,j} = c_{i+j,i-j} \text{ and}$$
$$b'_{i,j} = c_{i+j+1,i-j}$$

To prove Theorem 37.1.3, use induction on the rows of B and B'. Theorem 37.1.2 and the recursive definition of B validate the result. There is now enough information to give:

Proof of Claim 37.1.1: Compute the (r, s) entry of BB^T: On the one hand,

$$\sum_{k=0}^{n-1} b_{r,k} b_{s,k} = \sum_{k=0}^{n-1} c_{r+k,r-k} c_{s+k,s-k}.$$

On the other hand, one can consider c_{r+s} and the associated ballot problem: in this case there are $r + s$ votes for candidate A and the same number for candidate B. Break the counting into two stages: the drawing of the first $2r$ ballots and the drawing of the remaining $2s$ ballots. Let $2k \geq 0$ be the lead of candidate A after the first $2r$ ballots are drawn. There are $c_{r+k,r-k}$ possible ways this can happen. So one can then ask, for each of these possible ways, how many ways can the sequence be completed to one that is counted by c_{r+s}? Using the reversing principle, get $c_{s+k,s-k}$ such completions. This gives (by Segner's recursion (12.2)),

$$c_{r+s} = \sum_{k=0}^{n-1} c_{r+k,r-k} c_{s+k,s-k}.$$

Essentially the same argument gives $B'B'^T = C'$. Hence the claim is verified. ☐

Exercise 669: See [266, pp. 145–147] for two proofs, one by induction.

37.2 Solutions: Groups and permutations

Exercise 670: (Brief solution) Assume that G is abelian, that is, for every $x, y \in G$, $xy = yx$ holds. For each positive integer n, let $S(n)$ denote the statement "for every $a, b \in G$, $(ab)^n = a^n b^n$ holds". The base case $S(1)$ is trivial, so assume that for some fixed $m \geq 1$, $S(m)$ holds. Then for any $a, b \in G$,

$$
\begin{aligned}
(ab)^{m+1} &= (ab)(ab)^m \\
&= (ab)(a^m b^m) && \text{(by } S(m)) \\
&= a(ba^m)b^m && \text{(by associativity in a group)} \\
&= a(a^m b)b^m && \text{(since } G \text{ is abelian)} \\
&= (a \cdot a^m)(b \cdot b^m) && \text{(by associativity)} \\
&= a^{m+1} b^{m+1},
\end{aligned}
$$

which completes the proof of $S(m+1)$, and hence the inductive step. By MI, $S(n)$ holds for all $n \geq 1$. □

Exercise 671: Induct on $|G|$, and use Cauchy's theorem. See, *e.g.* [198, p. 239] or [152, pp.142–4] for proofs.

Exercise 673: This exercise appears in [6, p. 97].

Exercise 674: (Solution outline) For each $r \geq 2$, let $S(r)$ denote the statement "every cycle of length $r \geq 2$ can be written as a product of transpositions, as in

$$(x_1, \ldots, x_r) = (x_1, x_2) \circ (x_1, x_3) \circ \cdots \circ (x_1, x_r)."$$

For $r = 2$, (x_1, x_r) is already in desired form, so assume that for some fixed $m \geq 2$, $S(m)$ holds and let $(x_1, x_2, \ldots, x_m, x_{m+1})$ be a cycle. Direct computations show that

$$(x_1, x_2, \ldots, x_m, x_{m+1}) = (x_1, x_2, \ldots, x_m) \circ (x_1, x_{m+1}).$$

Use the inductive hypothesis and substitute the cycle $(x_1, x_2, \ldots, x_m)$ with $(x_1, x_2) \circ (x_1, x_3) \circ \cdots \circ (x_1, x_m)$. □

Exercise 675: (Hint for induction step) If a given permutation τ on n elements is not itself a cycle, then inside find a cycle σ (by taking an arbitrary element and following its path like in the decomposition of Example 19.2.4) and express τ as the composition of two permutations, say $\tau = \sigma \circ \rho$, where ρ is a permutation on fewer than n elements. Then apply the induction hypothesis to ρ.

Exercise 676: (Hint for inductive step) One could prove a more general lemma first, namely, that if τ and σ are bijections from A to A then so is $\tau \circ \sigma$. Then apply the lemma with $\tau = \sigma^m$.

Exercise 677: (Hint) First prove the recursion

$$D_n = nD_{n-1} + (-1)^n.$$

Also, by simple inclusion-exclusion, one arrives at a more direct solution by establishing

$$D_n = \sum_{i=0}^{n} (-1)^i \binom{n}{i} (n-i)!.$$

Exercise 679: Let p_n be the number of permutations with the desired property. For $n = 1$, there is only one permutation, namely the trivial one, so $p_1 = 1$; there are two permutations of two elements, each of which has the desired property, so $p_2 = 2$.

Let $n \geq 3$ be fixed. Of those permutations π of $\{1, 2, \ldots, n\}$ which have the desired property, either $\pi(n) = n$ or $\pi(n) = n - 1$. There are p_{n-1} of these that fix n. For those with $\pi(n) = n - 1$, $\pi(n-1) = n$, so the number of these is p_{n-2}. Hence, for $n \geq 3$, $p_n = p_{n-1} + p_{n-2}$. Thus p_n satisfies the recursion for the Fibonacci numbers. $\square$

Exercise 680: (Comments) This problem appears in [6, p. 25, Ex. 14, p. 98 Ex. 11] without solution. Here are some comments toward to what might constitute a proof: For any $f, g \in S_n$, define $\delta(f, g) = \max_{1 \leq i \leq n} |f(i) - g(i)|$. For $n, r \in \mathbb{Z}^+$, and $f \in S_n$, define

$$a_f(n, r) = |\{g \in S_n : \delta(f, g) \leq r\}|.$$

To be shown is that the numbers $a_f(n, r)$ are independent of f, so that $a(n, r) = a_f(n, r)$ is well defined.

Let both $f, h \in S_n$ be fixed; one must show $a_f(n, r) = a_h(n, r)$.

Let $\pi \in S_n$ be defined by $f \circ \pi = h$, that is, for each i, $h(i) = f(\pi(i))$. Put $g' = g \circ \pi$. Then for each i, $f(\pi(i)) - g(\pi(i)) = h(i) - g'(i)$. Hence, the set of differences $\{f(j) - g(j) : j = 1 \ldots, n\}$ is the same as $\{h(i) - g'(i) : i = 1, \ldots, n\}$. Thus, for every g with $\delta(f, g) \leq r$, there is a (unique) $g' = g \circ \pi$ with $\delta(h, g') \leq r$. Hence, $a_f(n, r)$ is independent of f, so $a(n, r) = a_f(n, r)$ is well defined.

It remains to prove that for $n \geq 6$,

$$a(n, 2) = 2a(n-1, 2) + 2a(n-3, 2) - a(n-5, 2),$$

and that $a(n, 1) = F_{n+1}$, the $(n+1)$-st Fibonacci number. Does this lead to an inductive solution?

37.3   Solutions: Rings

Exercise 682: Assume ZFC. As in, *e.g.*, [95, p. 120], commutativity is not assumed for this proof. See also [416, p. 525] (where commutativity is assumed).

Let $\mathcal{I}$ be the set of proper ideals of R, and consider $\mathcal{I}$ as partially ordered set ordered by inclusion.

Let $\mathcal{C} \subseteq \mathcal{I}$ be a chain in $\mathcal{I}$. If $\mathcal{C} = \emptyset$, then the ideal $\{0\}$ is an upper bound, so suppose $\mathcal{C} \neq \emptyset$. Put $I = \cup \mathcal{C}$.

One verifies that $I \in \mathcal{I}$ (which then shows I is an upper bound for $\mathcal{C}$):

Let $x_1, x_2 \in I$, with say, $x_1 \in I_1 \in \mathcal{C}$ and $x_2 \in I_2 \in \mathcal{C}$, and without loss of generality, assume that $I_1 \subseteq I_2$. Then $x_1, x_2 \in I_2$ and so $x_1 + x_2 \in I_2$ (because I_2 is an ideal, and hence a subring) and thus $x_1 + x_2 \in I$. In a similar manner, one sees that $x \in I$ and $r \in R$ implies that both $rx \in I$ and $xr \in I$. Since an ideal containing 1 is all of R, and all ideals in $\mathcal{C}$ are proper, none contain 1, and so neither does their union I. Thus I is a proper ideal, that is, $I \in \mathcal{I}$.

By Zorn's lemma, $\mathcal{I}$ contains a maximal element, namely a maximal ideal of R. □

Exercise 683: This appears in many abstract algebra texts; *e.g.*, see [469, p.269]. Let $f(x) = x^n + b_{n-1}x^{n-1} + \cdots + b_0 \in \mathbb{Z}[x]$ have α as a root. Put

$$G = \langle \alpha^{n-1}, \ldots, \alpha^2, \alpha, 1 \rangle.$$

To answer the exercise, it suffices to show that for each $k \geq n$, $\alpha^k \in G$. The proof is by induction on k.

BASE STEP: When $k = n$, using the fact that f is monic,

$$\alpha^n = -(b_{n-1}x^{n-1} + \cdots + b_0) \in G.$$

INDUCTIVE STEP: Fix $k \geq n$, and assume that $\alpha^k \in G$, with constants $c_i \in \mathbb{Z}$ so that

$$\alpha^k = c_{n-1}\alpha^{n-1} + \cdots + c_1\alpha + c_0.$$

Then

$$\alpha^{k+1} = c_{n-1}\alpha^n + \cdots + c_1\alpha^2 + c_0\alpha$$
$$= c_{n-1}[-(b_{n-1}x^{n-1} + \cdots + b_0)] + \cdots + c_1\alpha^2 + c_0\alpha,$$

which lies in G.

By mathematical induction, for each $k \geq n$, $\alpha^k \in G$, and so G is finitely generated. □

Exercise 629: See [117].

37.4   Solutions: Fields

Exercise 684: Let p be a prime and let $\mathbb{F} = \mathrm{GF}(p)$. For $0 \le L \le p$, let $S(L)$ be the statement that if $x_0 \in \mathbb{F}$ and $f(x) \in \mathbb{F}[x]$ are so that for $0 \le \ell < L$, $f^{(\ell)}(x_0) = 0$, then x_0 is a zero of f with multiplicity at least L. The proof given here is a pedantic version of that found in [60, p. 349], proceeding by induction on L.

BASE STEP: For $L = 0$, there is nothing to prove.

INDUCTION STEP: Let $N > 0$ and suppose that $S(0), S(1), \dots, S(N-1)$ hold. To be shown is that $S(N)$ holds. Let $f(x) \in \mathbb{F}[x]$ and suppose that for $0 \le \ell < N$, $f^{(\ell)}(x_0) = 0$ holds. By induction hypothesis $S(N-1)$, x_0 is root of $f(x)$ with multiplicity at least $N-1$; thus one can write $f(x) = (x - x_0)^{N-1} r(x)$ for some polynomial $r(x) \in \mathbb{F}[x]$. Write $r(x) = (x - x_0)s(x) + c$ for some $s(x) \in \mathbb{F}[x]$ and some $c \in \mathbb{F}$. Then $f(x) = (x - x_0)^N s(x) + c \cdot (x - x_0)^{N-1}$. It is not too difficult to verify that $f^{(N-1)}(x_0) = c(L-1)!$, and since $f^{(N-1)}(x_0) = 0$, one concludes that $c = 0$. Hence $f(x) = (x - x_0)^N s(x)$, and so x_0 is a root of f with multiplicity at least N.

By the principle of mathematical induction, for all $L \ge 0$, $S(L)$ is true. $\square$

Note: By Fermat's little theorem, in $\mathrm{GF}(p)$, $x^p = x$, and so the pth derivative is zero in any case—the reason for assuming $\ell < L \le p$.

Exercise 685: (Brief) Induct on m. For $m = 1$, simply pick $a_1 = f$. Suppose $k > 1$ and suppose that the result holds when $m = k - 1$. Since g_1 is relatively prime to the product $g_2 \cdots g_m$, by the comments just before the exercise, there exist $s, t \in K[x]$ so that $1 = sg_1 + t(q_2 \cdots q_m)$. Then

$$
\begin{aligned}
\frac{f}{g} &= (sg_1 + tg_2 \cdots g_m)\frac{f}{g} \\
&= \frac{sg_1 f}{g} + \frac{tg_2 \cdots g_m f}{g} \\
&= \frac{sf}{g_2 \cdots g_m} + \frac{tf}{g_1}.
\end{aligned}
$$

Applying the induction hypothesis to the first fraction in the last line above shows that the statement is true when $m = k$, completing the inductive step.

By mathematical induction, for each $m \ge 1$, the statement in the exercise is true. $\square$

Exercise 686: This problem appears in [469, pp. 263–4]. Let $f(x) \in K[x]$. Since $\deg(b) \ge 1$, there exists $m \ge 0$ with

$$
\deg(f) < (m+1)\deg(b) = \deg(b^{m+1}).
$$

For each $m \geq 0$, let $S(m)$ be the statement that if

$$\deg(b^m) \leq \deg(f) < \deg(b^{m+1}),$$

then f has an expression $d_m b^m + d_{m-1} b^{m-1} + \cdots d_0$ of the desired form.

BASE STEP: If $m = 0$, $0 = \deg(b^0) \leq \deg(f) < \deg(b)$, so define $d_0 = f$, so $S(0)$ holds.

INDUCTIVE STEP: Fix $n \geq 1$ and suppose that for all $k \in 1, \ldots, n-1$, $S(k)$ is true. If $m > 0$,

$$\deg(b^n) \leq \deg(f) < \deg(b^{n+1},$$

and the division algorithm gives d_n and r so that $f = d_n b^n + r$, where either $r = 0$ or $\deg(r) < \deg(b^n)$. Notice that $d_n \neq 0$ [since otherwise, $\deg(f) \geq b^{n+1}$]. If $r = 0$, define $d_{n-1} = \cdots = d_0 = 0$ and $f = d_n b^n$ is an expression of the desired form. If $r \neq 0$, then $S(\deg(r))$ shows that r, and hence f has an expression of the desired form.

By mathematical induction, for each $m \geq 0$, $S(m)$ is true. $\qquad\square$

Exercise 687: See, *e.g.*, [416, p. 527].

Exercise 688: For a finite positive integer k, let $S(k)$ denote the statement that every extension of K formed by adding k algebraic (over K) elements is finite and therefore algebraic over K. The proof is by induction on k.
BASE STEP $(k = 1)$: The statement $S(1)$ is Theorem 19.4.2.

INDUCTIVE STEP: Fix $m \geq 1$ and suppose that for all $j \leq m$, $S(m)$ holds. Let $a_1, \ldots, a_{m+1}$ be elements not in K, but algebraic over K. Let $K(a_1, \ldots, a_{m+1})$ be the corresponding field extension. By induction hypothesis $S(m)$, $K' = K(a_1, \ldots, a_m)$ is algebraic over K, so let $\{\nu_1, \ldots, \nu_r\}$ be a basis (over K) for K'.

If $K' = K'(a_{m+1})$, then there is nothing to prove, so assume $K' \neq K'(a_{m+1})$. Since a_{m+1} is algebraic over K, there is a polynomial $q(x) \in K[x] \subseteq K'[x]$ so that $q(a_{m+1}) = 0$. Thus $K'(a_{m+1})$ is algebraic over K'. Again, by Theorem 19.4.2, $K'' = K'(a)$ has a basis over K', say $\{\theta_1, \ldots, \theta_\ell\}$. For every $b \in K''$, there exist constants $c_1, \ldots, c_\ell$ so that $b = c_1 \theta_1 + \cdots + c_\ell \theta_\ell$. Also, for each $i = 1, \ldots, \ell$, let $d_{ij} \in K$ be so that

$$c_i = d_{i1}\nu_1 + \cdots + d_{ir}\nu_r.$$

Hence, $b = \sum_{i,j} d_{ij}\nu_j\theta_i$. The $r \cdot \ell$ elements $\nu_j\theta_i$ are linearly independent since $b = 0$ implies that the c_i's are all zero and so each d_{ij} is zero. Since $b \in K''$ was arbitrary, the vectors $\nu_j\theta_i$ span K', and so these vectors are a basis for K'' over K. Thus K'' is a finite dimensional vector space over K, concluding the inductive step.
Thus, by MI, for all $k \geq 1$, $S(k)$ holds. $\qquad\square$

Exercise 689: The proof is by induction on $\deg(f)$. If $\deg(f) = 1$, f is linear, so choose $E = K$. So assume that the result is true for all polynomials with degree smaller than $\deg(f)$. If $\deg(f) > 1$, write $f(x) = p(x)q(x)$, where p is irreducible. If $p(x)$ is linear, then $f(x)$ factors as products of linears whenever $q(x)$ does; moreover, such a factorization of $q(x)$ (over a larger field) does exist, by induction hypothesis, because $\deg(q) < \deg(f)$.

By the fact just before the statement of the exercise, there exists a field F containing K and a root z of $p(x)$, which implies that p factors as $p(x) = (x - z)h(x) \in F[x]$. By induction hypothesis, there exists a field E containing F (and hence K) so that $h(x)q(x)$ and hence $f(x) = (x - z)h(x)q(x)$ is a product of linear factors in $E[x]$. This completes the induction and so, by mathematical induction, the solution. $\square$

37.5 Solutions: Vector spaces

Exercise 690: (Distinct eigenvalues) The proof is by induction on the number of distinct eigenvalues (not the size of the matrix).

For each positive integer r, let $C(r)$ be the claim that if A is a matrix with r distinct eigenvalues $\lambda_1, \ldots, \lambda_r$ and for each $i = 1, \ldots, r$ if $\mathbf{v}_i$ is an eigenvector associated with λ_i, then the set $\{\mathbf{v}_1, \ldots, \mathbf{v}_r\}$ is linearly independent.

BASE STEP ($r = 1$): Let λ be an eigenvalue for A, with associated eigenvector $\mathbf{v}$. By definition of an eigenvector, $\mathbf{v} \neq \mathbf{0}$, and so $\{\mathbf{v}\}$ is linearly independent.

INDUCTION STEP: For some fixed $k \geq 1$, suppose that $C(k)$ is true, and let A be a matrix with distinct eigenvalues $\lambda_1, \ldots, \lambda_{k+1}$ with associated eigenvectors $\mathbf{v}_1, \ldots, \mathbf{v}_{k+1}$ respectively. To show that $C(k + 1)$ holds, one must show that the set $\{\mathbf{v}_1, \ldots, \mathbf{v}_{k+1}\}$ is linearly independent. To this end, suppose that $a_1, \ldots, a_{k+1}$ are so that

$$a_1\mathbf{v}_1 + \cdots + a_k\mathbf{v}_k + a_{k+1}\mathbf{v}_{k+1} = \mathbf{0}. \tag{37.3}$$

Multiplying equation (37.3) on the left by A,

$$a_1\lambda_1\mathbf{v}_1 + \cdots + a_l\lambda_k\mathbf{v}_k + a_{k+1}\lambda_{k+1}\mathbf{v}_{k+1} = \mathbf{0}. \tag{37.4}$$

Multiplication of equation (37.3) by λ_{k+1} and subtracting from equation (37.4) yields

$$a_1(\lambda_1 - \lambda_{k+1})\mathbf{v}_1 + \cdots + a_k(\lambda_k - \lambda_{k+1})\mathbf{v}_k = \mathbf{0}.$$

By the induction hypothesis $C(k)$, the set $\{\mathbf{v}_1, \ldots, \mathbf{v}_k\}$ is linearly independent and so

$$a_1(\lambda_1 - \lambda_{k+1}) = \cdots = a_k(\lambda_k - \lambda_{k+1}) = 0.$$

Since the λ_i's are distinct, this implies that $a_1 = \cdots = a_k = 0$. Thus equation (37.3) now reads $a_{k+1}\mathbf{v}_{k+1} = \mathbf{0}$, and using the fact that $\mathbf{v}_{k+1} \neq \mathbf{0}$ one concludes that $a_{k+1} = 0$ as well. Hence all of the a_i's are zero, showing that the set $\{\mathbf{v}_1, \ldots, \mathbf{v}_{k+1}\}$ is linearly independent, completing the proof of $C(k + 1)$.

By mathematical induction, for any $r \geq 1$, the claim $C(r)$ is true. $\qquad\square$

Remark: Notice that the above proof can be formulated for any linear operator T on any vector space V (see, *e.g.*, [202, p. 261]), not just the matrix operator which is implicitly defined on a finite dimensional vector space.

Exercise 691: This exercise appears as problem 23 in [202, p. 324] (without solution) with the hint to induct on k.

Exercise 693: Fix n and $A \in M_{n \times n}(\mathbb{F})$. It suffices to show that for any $k \in \mathbb{Z}^+$, $A_k \in \mathrm{span}(I_n, A, A^2, \ldots, A^{n-1})$. Call this statement $S(k)$. One proof is by induction on k.

BASE STEP: For $k = 1$, $A^1 = A$ and so $S(1)$ holds.

INDUCTIVE STEP: Assume that for a fixed $k \geq 1$, $S(k)$ holds, that is, assume that

$$A^k \in \mathrm{span}\{I_n, A, A^2, \ldots, A^{n-1}\},$$

that is, there exist constants $a_0, a_1, \ldots, a_{n-1}$ so that

$$A^k = a_0 I_n + a_1 A + a_2 A^2 + \cdots + a_{n-1} A^{n-1}.$$

To complete the inductive step, it remains to show that

$$A^{k+1} \in \mathrm{span}(I_n, A, A^2, \ldots, A^{n-1}).$$

Applying $A^{k+1} = A \cdot A^k$,

$$A^{k+1} = a_0 A + a_1 A^2 + a_2 A^3 + \cdots + a_{n-1} A^n,$$

and by the Cayley-Hamilton theorem, A^n is a linear combination of I_n, A, A^2, ..., A^{n-1}, and hence so is A^{k+1}.

By mathematical induction, for any $k \in \mathbb{Z}^+$, $S(k)$ holds, completing the proof. $\qquad\square$

Exercise 694: (Minkowski's inequality) Fix some non-negative integer n. Let $M(m)$ be the statement that for every collection of vectors $\mathbf{v}_1, \mathbf{v}_2, \ldots, \mathbf{v}_m$ in $\mathbb{R}^n$,

$$\|\mathbf{v}_1 + \mathbf{v}_2 + \cdots + \mathbf{v}_m\| \leq \|\mathbf{v}_1\| + \|\mathbf{v}_2\| + \cdots + \|\mathbf{v}_m\|.$$

BASE STEP ($m = 1, 2$): When $m = 1$, the result is trivial, since $M(1)$ says only that $\|\mathbf{v}\| \leq \|\mathbf{v}\|$.

Let $\mathbf{v} = (v_1, \ldots, v_n)$ and $\mathbf{w} = (w_1, \ldots, w_n)$. Then

$$\|\mathbf{v} + \mathbf{w}\|^2 = \sum_{i=1}^{n} (v_i + w_i)^2$$

$$= \sum_{i=1}^{n} v_i^2 + 2v_i w_i + w_i^2$$

$$= \sum_{i=1}^{n} v_i^2 + \sum_{i=1}^{n} w_i^2 + 2 \sum_{i=1}^{n} v_i w_i$$

$$= \|\mathbf{v}\|^2 + \|\mathbf{w}\|^2 + 2 \sum_{i=1}^{n} v_i w_i$$

$$\leq \|\mathbf{v}\|^2 + \|\mathbf{w}\|^2 + 2\|\mathbf{v}\|\|\mathbf{w}\| \quad \text{(by Cauchy–Schwarz)}$$

$$= (\|\mathbf{v}\| + \|\mathbf{w}\|)^2,$$

and so taking square roots shows

$$\|\mathbf{v} + \mathbf{w}\| \leq \|\mathbf{v}\| + \|\mathbf{w}\|,$$

proving $M(2)$.

INDUCTIVE STEP: For some fixed $k \geq 2$, assume that $M(k)$ holds. Let $\mathbf{v}_1, \mathbf{v}_2, \ldots, \mathbf{v}_k$ and $\mathbf{v}_{k+1}$ be vectors in $\mathbb{R}^n$. Then

$$\|\mathbf{v}_1 + \mathbf{v}_2 + \cdots + \mathbf{v}_k + \mathbf{v}_{k+1}\|$$

$$= \|(\mathbf{v}_1 + \mathbf{v}_2 + \cdots + \mathbf{v}_k) + \mathbf{v}_{k+1}\|$$

$$\leq |\mathbf{v}_1 + \mathbf{v}_2 + \cdots + \mathbf{v}_k)\| + \|\mathbf{v}_{k+1}\| \quad \text{(by } M(2))$$

$$\leq \|\mathbf{v}_1\| + \|\mathbf{v}_2\| + \cdots + \|\mathbf{v}_k\| + \|\mathbf{v}_{k+1}\| \quad \text{(by } M(k))$$

shows that $M(k+1)$ is true as well. This completes the inductive step.

By mathematical induction, for every $m \geq 1$, $M(m)$ holds. $\qquad\square$

Exercise 695: Two things are to be proved, the p-norm triangle inequality (which is often called Minkowski's inequality)

$$\|\mathbf{x} + \mathbf{y}\|_p \leq \|\mathbf{x}\|_p + \|\mathbf{y}\|_p,$$

and its generalization to larger sums,

$$\|\mathbf{x}_1 + \mathbf{x}_2 + \cdots + \mathbf{x}_n\|_p \leq \|\mathbf{x}_1\|_p + \|\mathbf{x}_2\|_p + \cdots + \|\mathbf{x}_n\|_p.$$

Note that when $p = 2$, this is Minkowski's inequality for two vectors.

Before proving the first inequality, here is a simple observation: if $p > 1$ and $\frac{1}{p} + \frac{1}{q} = 1$, then $q = (1 - \frac{1}{p})^{-1} = \frac{p}{p-1}$ and so $(p-1)q = p$. Putting $\mathbf{x} = (x_1, \ldots, x_n)$ and $\mathbf{y} = (y_1, \ldots, y_n)$, apply Hölder's inequality with y_i's replaced by $|x_i + y_i|^{p-1}$ as follows:

$$\sum_{i=1}^{n} |x_i + y_i|^p$$

$$= \sum_{i=1}^{n} |x_i + y_i| \cdot |x_i + y_i|^{p-1}$$

$$\leq \sum_{i=1}^{n} (|x_i| + |y_i|) \cdot |x_i + y_i|^{p-1} \quad \text{(by the triangle inequality)}$$

$$= \sum_{i=1}^{n} |x_i| \cdot |x_i + y_i|^{p-1} + \sum_{i=0}^{n} |y_i| \cdot |x_i + y_i|^{p-1}$$

$$\leq \|\mathbf{x}\|_p \left(\sum_{i=1}^{n} (|x_i + y_i|^{p-1})^q \right)^{1/q} + \|\mathbf{y}\|_p \left(\sum_{i=1}^{n} (|x_i + y_i|^{p-1})^q \right)^{1/q}$$

$$\text{(by Holder's inequality, twice)}$$

$$= \|\mathbf{x}\|_p \left(\sum_{i=1}^{n} |x_i + y_i|^{p} \right)^{1/q} + \|\mathbf{y}\|_p \left(\sum_{i=1}^{n} |x_i + y_i|^{p} \right)^{1/q}$$

$$= (\|\mathbf{x}\|_p + \|\mathbf{y}\|_p) \left(\sum_{i=1}^{n} |x_i + y_i|^{p} \right)^{1/q}.$$

Dividing each side by $(\sum_{i=1}^{n} |x_i + y_i|^p)^{1/q}$ leaves

$$\left(\sum_{i=1}^{n} |x_i + y_i|^{p} \right)^{1/p} \leq \|\mathbf{x}\|_p + \|\mathbf{y}\|_p,$$

which proves the first inequality in the exercise.

The second inequality now is a straightforward inductive argument completely analogous to that used to prove the triangle inequality, (see solution to Exercise 193) and so is left to the reader. $\qquad \square$

Exercise 696: (Gram–Schmidt) Throughout, all vectors are be assumed to be from some inner product space V. For $n \geq 1$, let $C(n)$ be the claim that for any linearly independent set $S = \{\mathbf{w}_1, \ldots, \mathbf{w}_n\}$ of vectors in V, the vectors $T = \{\mathbf{v}_1, \ldots, \mathbf{v}_n\}$ defined by $\mathbf{v}_1 = \mathbf{w}_1$, and for each $k = 2, \ldots, n$,

$$\mathbf{v}_k = \mathbf{w}_k - \sum_{j=1}^{k-1} \frac{\langle \mathbf{w}_k, \mathbf{v}_j \rangle}{\|\mathbf{v}_j\|^2} \mathbf{v}_j. \tag{37.5}$$

form an orthogonal set of non-zero vectors with $\mathrm{span}(T) = \mathrm{span}(S)$.

BASE STEP: When $n = 1$, $S = \{\mathbf{w}_1\} = \{\mathbf{v}_1\} = T$ is linearly independent.

INDUCTIVE STEP: Suppose that for some $m \geq 1$, the claim $C(m)$ holds, and let $S = \{\mathbf{w}_1, \ldots, \mathbf{w}_m\}$ be linearly independent. By $C(m)$, suppose that $T = \{\mathbf{v}_1, \ldots, \mathbf{v}_m\}$

is orthogonal (produced according to the recursion (37.5). Suppose $\mathbf{w}_{m+1}$ is so that $S' = S \cup \{\mathbf{w}_{m+1}\}$ is linearly independent, and put

$$\mathbf{v}_{m+1} = \mathbf{w}_{m+1} - \sum_{j=1}^{m} \frac{\langle \mathbf{w}_m, \mathbf{v}_j \rangle}{\|\mathbf{v}_j\|^2} \mathbf{v}_j.$$

Then for each $i = 1, \ldots, m$,

$$
\begin{aligned}
\langle \mathbf{v}_{m+1}, \mathbf{v}_i \rangle &= \langle \mathbf{w}_{m+1}, v_i \rangle - \sum_{j=1}^{m} \frac{\langle \mathbf{w}_{m+1}, \mathbf{v}_j \rangle}{\|\mathbf{w}_j\|^2} \langle \mathbf{v}_j, \mathbf{v}_i \rangle \\
&= \langle \mathbf{w}_{m+1}, v_i \rangle - \frac{\langle \mathbf{w}_{m+1}, \mathbf{v}_i \rangle}{\|\mathbf{w}_i\|^2} \langle \mathbf{v}_i, \mathbf{v}_i \rangle \text{ (see below for reason)} \\
&= 0
\end{aligned}
$$

since by $C(m)$, T is orthogonal, and so for $i \neq j$, $\langle \mathbf{v}_j, \mathbf{v}_i \rangle = 0$. Hence, $\mathbf{v}_{m+1}$ is orthogonal to all vectors in T, and so the set $T' = T \cup \{\mathbf{v}_{m+1}\}$ is orthogonal. By Lemma 19.5.5, T' is linearly independent.

By $C(m)$, it is assumed that $\mathrm{span}(S) = \mathrm{span}(T)$, and since $\mathbf{v}_{m+1} \in \mathrm{span}(S')$, then $\mathrm{span}(T') \subseteq \mathrm{span}(S')$. However, since both S' and T' are linearly independent, $m + 1 = \dim(\mathrm{span}(S')) = \dim(\mathrm{span}(T'))$; thus (do you know why?) $\mathrm{span}(T') = \mathrm{span}(S')$, finishing the proof of $C(m + 1)$ and hence the inductive step.

By mathematical induction, for each $n \geq 1$, the claim $C(n)$ is true.

If V is a finite dimensional vector space with an inner product, apply the Gram–Schmidt process to any basis for V, and obtain an orthogonal basis. Normalizing vectors in an orthogonal basis then gives an orthonormal basis. □

Chapter 38

Solutions: Geometry

Exercise 698: This problem occurred in [161, 7.26, pp. 181, 189].

BASE STEP: Since $c > a$ and $c > b$, $c^3 = c(a^2 + b^2) > a^3 + b^3$.

INDUCTIVE STEP: For some $k \geq 3$, suppose that $c^k > a^k + b^k$. Then $c^{k+1} = c \cdot c^k > c(a^k + b^k) > a^{k+1} + b^{k+1}$ proves the result for $k + 1$.

By mathematical induction, the statement is true for every $n > 2$. $\square$

Exercise 699: (Brief) Begin by constructing a right triangle with legs each of length 1; its hypotenuse has length $\sqrt{2}$. Supposing that for some $k \geq 2$, that a segment of length $\sqrt{k}$ has been constructed. From one end of this segment, extend a perpendicular 1 unit long, determining another right triangle with legs of length 1 and $\sqrt{k}$. By Pythagoras's theorem, the new hypotenuse has length $\sqrt{k+1}$. See Figure 38.1. $\square$

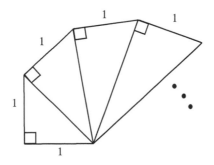

Figure 38.1: Constructing $\sqrt{n}$

Exercise 700: [This solution also occurs in [250, Prob. 52].] For each $n \geq 1$, let $A(n)$ be the assertion that among n points in the plane, the greatest distance between points is realized by at most n different pairs of points.

779

BASE STEP: Each of $A(1)$, $A(2)$, and $A(3)$ hold trivially since for each $n = 1, 2, 3$, the total number of pairs is at most n, so the pairs realizing a maximum distance is also.

INDUCTIVE STEP: Fix some $m \geq 4$ and assume that $A(m-1)$ is true. Consider a set of m points $p_1, \ldots, p_m$ and without loss of generality let the diameter of this set be 1 (that is, the maximum distance between any two points is 1). The proof now splits into two cases, the first of which does not require the induction hypothesis.

Case 1: Assume that each p_i is at distance 1 from at most two other points. Then by the handshaking lemma (Lemma 15.1.1), or simple double counting, there are at most m pairs with distance 1, so $A(m)$ is true.

Case 2: Assume that some point, say p_1 is at distance 1 from three other points, say p_2, p_3, and p_4.

Claim: At least one of p_2, p_3, p_4 has only p_1 at distance 1.

Since any two of p_2, p_3, p_4 are at distance at most one, all of the angles of the form $\angle p_i p_1 p_j$ are acute; suppose that these three vertices occur in order around p_1 (that is, p_3 is in the sector subtended by the acute angle $\angle p_2 p_1 p_4$). It is now shown that indeed p_3 satisfies the claim.

Suppose that for some k, p_k is at distance 1 from p_3. Then the segment $p_3 p_k$ is not disjoint from either $p_2 p_1$ or $p_4 p_1$ because of the following fact, whose proof is left to the reader:

Fact: If two line segments of length 1 are disjoint, two endpoints are at distance greater than 1.

Thus $p_3 p_k$ intersects both $p_2 p_1$ and $p_4 p_1$, and so $p_k = p_1$; thus p_3 satisfies the claim.

Deleting the one point guaranteed by the claim (in the above proof, p_3) gives a collection of $m-1$ points and by $A(m-1)$, the number of times distance 1 occurs is at most $m-1$; together with the deleted point gives at most m pairs of points at distance 1, proving $A(m)$ and completing the inductive step.

By mathematical induction, for each $n \geq 1$, the assertion $A(n)$ holds.   □

Comment: If n is odd, the maximum distance in a regular n-gon occurs twice at each vertex, so the bound in Exercise 700 is attained.

Exercise 701: The solution outline of the inductive step provided in [421, Ex. 6.1] goes as follows: Let points $c_1, \ldots, c_n$ be contained in a ball $B \subset \mathbb{R}^d$ of radius $\sqrt{2}$. First show that all points can be moved to the boundary of B without decreasing any pairwise distances. By rotating B appropriately, assume that $c_1 = (\sqrt{2}, 0, \ldots, 0)$. Let B' be the intersection of B with the (hyper)plane consisting of all points whose first coordinate is 0. Then show that for each $i = 2, \ldots, n$, the point c_i can be mapped to a point c_i' on the boundary of B' so that each $|c_i' - c_j'| > 2$. Apply the induction hypothesis to B' and $c_2', \ldots, c_n'$. [Added note: With the given position of c_1, by Pythagoras, all remaining points lie on the other side of B', and can be shifted "out" to meet the boundary of B' just as they were shifted in B.]   □

Exercise 703: (Main idea for proof) Given two squares with side lengths a and b, where $a \leq b$, cut the large square with two cuts at right angles to each other, where the cuts partition each edge in proportions $\frac{b-a}{2}$ and $\frac{b+a}{2}$. Do not cut the small square. Reassemble the five pieces into a larger square as in Figure 38.2. [This actually gives a standard proof of Pythagoras' theorem!]

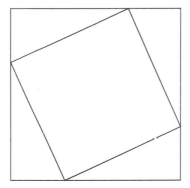

Figure 38.2: The square reassembled

Argue by induction using the above step.   □

38.1   Solutions: Convexity

Exercise 704: Fix $n \in \mathbb{Z}^+$ and for each $m \in \mathbb{Z}^+$, let $S(m)$ be the statement that if m points lie in a convex set, then any convex linear combination of these points also lies in the convex set.

BASE STEP: For $m = 1$, the result is trivial; the statement for $m = 2$ is simply the definition of convexity.

INDUCTION STEP: Fix $k \geq 2$ and suppose that $S(k-1)$ is true. Let $x_1, \ldots, x_k$ be points in a convex set C, and let $\alpha_1, \ldots, \alpha_k \in [0, 1]$ be real numbers with $\sum_{i=1}^{k} \alpha_i = 1$. To complete the proof of $S(k)$, it remains only to show that

$$\sum_{i=1}^{k} \alpha_i \mathbf{x}_i \in C.$$

To see this, begin by rewriting this sum:

$$\sum_{i=1}^{k} \alpha_i \mathbf{x}_i = \left(\sum_{i=1}^{k-1} \alpha_i \mathbf{x}_i \right) + \alpha_k \mathbf{x}_k$$

$$= (1 - \alpha_k)\left(\sum_{i=1}^{k-1} \frac{\alpha_i}{1-\alpha_k}\mathbf{x}_i\right) + \alpha_k\mathbf{x}_k. \tag{$*$}$$

Then each $\frac{\alpha_i}{1-\alpha_k} \in [0,1]$ and $\sum_{i=1}^{k-1}\frac{\alpha_i}{1-\alpha_k} = 1$, so by $S(k-1)$, $\sum_{i=1}^{k-1}\frac{\alpha_i}{1-\alpha_k}\mathbf{x}_i \in C$. Then by $S(2)$, the expression $(*)$ is also in C, completing the inductive step.

By the principle of mathematical induction, for each $m \geq 1$, the statement $S(m)$ is true. $\qquad\square$

Exercise 705: This exercise appears in [58, p. 86], where the hint says to apply induction on k and to show that for some $1 \leq i < j \leq k+1$ there is a point $\mathbf{x}_{ij} = \lambda\mathbf{x}_i + \mu\mathbf{x}_j$ with $\lambda, \mu \geq 0$, $\lambda + \mu = 1$, in the affine plane determined by $X' = X\backslash\{\mathbf{x}_i, \mathbf{x}_j\}$.

Exercise 706: (Helly's theorem) This solution appears in many places, *e.g.*, [61, p. 90] or [343, p. 48]. For each $r \geq n+1$, let $S(r)$ be the statement that if any of $n+1$ of convex sets $C_1, \ldots, C_r$ have non-empty intersection, then $\cap_{i=1}^r C_i \neq \emptyset$. Base step: The statement $S(n+1)$ holds trivially.

Inductive step: Let $k \geq n+1$ and assume that $S(k)$ is true. Let $C_1, \ldots, C_{k+1}$ be convex sets so that $n+1$ of them have non-empty intersection. By $S(k)$, for each $j = 1, \ldots, k+1$, there exists

$$\mathbf{x}_j \in C_1 \cap \cdots \cap C_{j-1} \cap C_{j+1} \cap \cdots \cap C_{k+1}.$$

Since $k+1 \geq n+2$, by Radon's theorem applied with the set $S = \{\mathbf{x}_1, \ldots, \mathbf{x}_{k+1}\}$, there exists a partition $S = S_1 \cup S_2$ so that $\text{conv}(S_1) \cap \text{conv}(S_2) \neq \emptyset$. Without loss of generality, let $S_1 = \{\mathbf{x}_1, \ldots, \mathbf{x}_p\}$ and $S_2 = \{\mathbf{x}_{p+1}, \ldots, \mathbf{x}_{k+1}\}$, and let $\mathbf{y} \in \text{conv}(S_1) \cap \text{conv}(S_2)$. Since $\{\mathbf{x}_1, \ldots, \mathbf{x}_p\} \subseteq C_{p+1} \cap \cdots \cap C_{k+1}$, which is convex, any convex combination of $\mathbf{x}_1, \ldots, \mathbf{x}_p$ is also in $C_{p+1} \cap \cdots \cap C_{k+1}$, namely

$$\mathbf{y} \in C_{p+1} \cap \cdots \cap C_{k+1}.$$

Arguing similarly, $\{\mathbf{x}_{p+1}, \ldots, \mathbf{x}_{k+1}\} \subseteq C_1 \cap \cdots \cap C_p$, and so

$$\mathbf{y} \in C_1 \cap \cdots \cap C_p.$$

Together, $\mathbf{y} \in \cap_{i=1}^{k+1} C_i$, concluding the proof of $S(k+1)$.

By mathematical induction on r, Helly's theorem is proved. $\qquad\square$

38.2 Solutions: Polygons

Exercise 714: See [266, Ex. 2.6(vi)].

Exercise 715: This problem is discussed in [433, p. 115].

Exercise 716: (Art gallery problem, outline) The original question was posed by Victor Klee in 1973, and solved by Václav Chvátal, [108] published in 1975. (See [419] for further generalizations and details.) The original proof was a delicate inductive argument, breaking off a piece of the polygon and reattaching it. A simpler proof (outlined below) was given by Steve Fisk [192] in 1978; Fisk's proof also relies on induction. For another exposition, see O'Rourke's book [420] on computational geometry.

Let $g(n)$ be the minimum number of guards required to guard the interior of any n-gon. It is not difficult to check that $g(3) = g(4) = g(5) = 1$. To see that $g(6) \leq 2$, use a diagonal (which exists by Lemma 20.2.1) to split the hexagon into two areas, one of which is a quadrilateral, and apply $g(3) = g(4) = 1$. Solving the case $n = 6$ is a prelude to the general inductive step.

To prove the general result, first (by Exercise 711, say, a result proved by induction) triangulate an n-gon into $n - 2$ triangles. At least one (in fact, at least two) of these triangles have two sides common to the polygon (another result proved by induction, see Exercise 712). Then prove by induction that the vertices of the polygon can then be 3-colored so that each triangle receives all three colors (or use Exercise 713). Put guards at all vertices using the least used color. This shows that $g(n) \leq \lfloor n/3 \rfloor$.

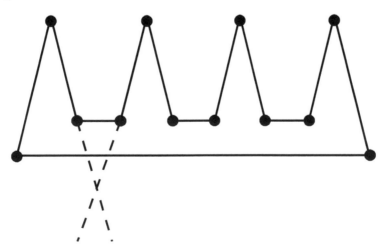

Figure 38.3: An art gallery with 12 vertices requiring 4 guards

To see that $\lfloor n/3 \rfloor$ guards are necessary for $n \geq 6$, when $n = 3k$, consider polygons of the form in Figure 38.3, (where $k = 4$). The diagram shows that as long as the extensions of adjacent acute angled walls do not meet inside the gallery, one guard is required for each of the saw-tooth peaks, and it suffices to have only one of these guards close to the bottom wall. ☐

Exercise 717: Pick's theorem is discussed in many popular sources, *e.g.*, [31, p. 17], [122], [212, p. 215], [240], [415], [508, pp. 96–98], and [565, pp. 183–4]. Pick's theorem also occurs as an exercise in many textbooks; *e.g.*, [462, 19, p. 292] contains a solution.

See Figure 38.4 for the idea behind step (iii); to measure the area of triangle ABC, calculate the area of the rectangle $BDEF$ and subtract off the area of the three outer triangles, each being a right triangle.

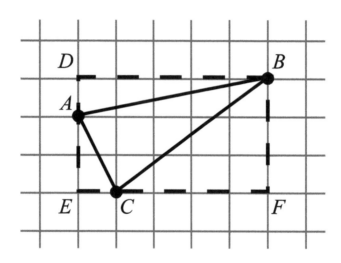

Figure 38.4: Finding area of arbitrary triangle

Exercise 718: The proof is by infinite descent. Let $n \geq 5$ and suppose that lattice points $P_1, \ldots, P_n$ form the vertices of a regular n-gon (in that order). All vectors $\overrightarrow{P_1P_2}$, $\overrightarrow{P_2P_3}$, ..., $\overrightarrow{P_{n-1}P_n}$, and $\overrightarrow{P_nP_1}$ also have integer coordinates.

As in Figure 38.5, attach the vector $\overrightarrow{P_2P_3}$ to P_1, $\overrightarrow{P_3P_4}$ to P_2, ..., and vector $\overrightarrow{P_1P_2}$ to P_n. These new segments do not overlap unless $n = 5$, in which case they just touch (try it!). All new endpoints thereby formed also have integer coordinates, yet form another n-gon with smaller side length (whose square is an integer, as in the proof of Lemma 20.2.3). Repeating this process of getting a smaller n-gon on gives an infinite sequence of n-gons on lattice points whose squares of the respective side lengths is a decreasing sequence of positive integers. This sequence violates the well-ordering of the positive integers, and so the original polygon does not exist. ☐

Another proof of Exercise 718: Oleksiy Klurman pointed out that the following standard result (whose proof is included for completeness) can also be used.

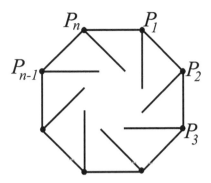

Figure 38.5: Making a smaller n-gon

Theorem 38.2.1. *For each integer $n \geq 5$, $\tan(\frac{\pi}{n})$ is irrational.*

Proof of Theorem 38.2.1: The proof is by contradiction and has two steps. In the first step, a simple pigeonhole argument shows that the the sequence $\{\tan(2^m \frac{\pi}{n})\}_{m=0}^{\infty}$ repeats. In the second step, assuming that $\tan(\frac{\pi}{n})$ is rational, then the remaining terms in this sequence are also rational with increasing denominators (and so never repeating), providing the necessary contradiction.

Step 1 begins with the following well-known lemma:

Lemma 38.2.2. *For any positive integer n, there exist positive integers $k < \ell$ so that n divides $2^\ell - 2^k$.*

Proof: Fix n, and consider the sequence $2, 2^2, 2^3, \ldots$ (mod n). By the pigeonhole principle, this sequence must eventually repeat. Then $2^k \equiv 2^\ell$ (mod n) is equivalent to n dividing $2^\ell - 2^k$. □

Continuing Step 1 in the proof of Theorem 38.2.1, by Lemma 38.2.2, let $k < \ell$ be so that for some $d \in \mathbb{Z}^+$, $nd = 2^\ell - 2^k$. Then

$$a_\ell - a_k = \tan(2^\ell \frac{\pi}{n}) - \tan(2^k \frac{\pi}{n})$$
$$= \frac{\sin(2^\ell \frac{\pi}{n} - 2^k \frac{\pi}{n})}{\cos(2^\ell \frac{\pi}{n}) \cos(2^k \frac{\pi}{n})}$$
$$= \frac{\sin(nd\frac{\pi}{n})}{\cos(2^\ell \frac{\pi}{n}) \cos(2^k \frac{\pi}{n})}$$
$$= 0,$$

and so $a_k = a_\ell$ with $k < \ell$, that is, the sequence $\{a_m\}$ repeats, completing the first step of the proof of Theorem 38.2.1.

Step 2: For some $n \geq 5$, suppose that $\tan(\frac{\pi}{n})$ is rational, that is, for some $p, q \in \mathbb{Z}^+$ with $\gcd(p, q) = 1$, $\tan(\frac{\pi}{n}) = \frac{p}{q}$. Since $n \geq 5$, $\frac{\pi}{n} < \frac{\pi}{4}$, and so $\frac{p}{q} < 1$; that is, $0 < p < q$.

For $m = 0, 1, 2, 3, \ldots$, put

$$a_m = \tan(2^m \frac{\pi}{n}).$$

Then $a_0 = \frac{p}{q}$, and

$$a_{m+1} = \frac{2\tan(2^m \frac{\pi}{n})}{1 - \tan^2(2^m \frac{\pi}{n})} = \frac{2a_m}{1 - a_m^2}.$$

Since a_0 is assumed to be rational, this last equation above and a simple inductive argument shows that each a_m is rational; for each $m \geq 0$, write $a_m = \frac{p_m}{q_m}$ where $(p_m, q_m) = 1$ (and $p = p_0$ and $q = q_0$). As above, for each m, $1 \leq p_m < q_m$. Then

$$\frac{p_{m+1}}{q_{m+1}} = \frac{2\frac{p_m}{q_m}}{1 - (\frac{p_m}{q_m})^2} = \frac{2p_m q_m}{q_m^2 - p_m^2}. \tag{38.1}$$

CLAIM: For each $m \geq 0$, $q_m < q_{m+1}$.

PROOF OF CLAIM: [The proof submitted by Klurman was by induction, however his calculations show that MI is not needed.] Fix $m \geq 0$. Since $(p_m, q_m) = 1$, it follows that $(p_m, q_m^2 - p_m^2) = (q_m, q_m^2 - p_m^2) = 1$ and so $(p_m q_m, q_m^2 - p_0^2) = 1$.

Case 1: Suppose that $(2, q_m^2 - p_m^2) = 1$. Then $(2p_m q_m, q_m^2 - p_m^2) = 1$, and thus by equation (38.1), both $p_{m+1} = 2p_m q_m$ and

$$\begin{aligned}
q_{m+1} &= q_m^2 - p_m^2 \\
&\geq q_m^2 - (q_m - 1)^2 &&\text{(since } p_m < q_m\text{)} \\
&= 2q_m - 1 \\
&> q_m &&\text{(since } q_m > 1\text{)},
\end{aligned}$$

finishing Case 1.

Case 2: Suppose that $(2, q_m^2 - p_m^2) \neq 1$. Then $(2, q_m^2 - p_m^2) = 2$ and by equation (38.1), $\frac{p_{m+1}}{q_{m+1}} = \frac{p_m q_m}{(q_m^2 - p_m^2)/2}$. Since $(p_m q_m, q_m^2 - p_m^2) = 1$, so also $(p_m q_m, (q_m^2 - p_m^2)/2) = 1$ and and so both $p_{m+1} = p_m q_m$ and $q_{m+1} = (q_m^2 - p_m^2)/2)$ (where $(p_{m+1}, q_{m+1}) = 1$). Since $p_m \equiv q_m \pmod{2}$, $p_m \leq q_m - 2$. If $q_m = 2$, then $p_m = 0$, which is impossible (because this gives $\tan(\frac{\pi}{n}) = 0$), so $q_m > 2$ and

$$\begin{aligned}
q_{m+1} &= (q_m^2 - p_m^2)/2 \\
&\geq (q_m^2 - (q_m - 2)^2)/2 &&\text{(since } p_m \leq q_m - 2\text{)} \\
&= 2q_m - 2 \\
&> q_m &&\text{(since } q_m > 2\text{)},
\end{aligned}$$

finishing Case 2; this completes the proof of the claim.

By the claim, $a_0, a_1, a_2, \ldots$ are all distinct, contradicting the result in Step 1, and completing the proof of Theorem 38.2.1. $\square$

With Theorem 38.2.1 in hand, another popular solution to Exercise 718 (provided here by Klurman) is as follows:

Let $n \geq 5$ and, to reach a contradiction, assume that P is a regular n-gon whose vertices are lattice points (in $\mathbb{Z}^+$). Let r be the circumradius of P, and let s be the side-length of P. Let O be the center of P, let X and Y be adjacent vertices of P, and consider the triangle $\triangle OXY$ where $m\angle O = \frac{2\pi}{n}$, $r = \|OX\| = \|OY\|$ and $s = \|XY\|$. Let the area of $\triangle OXY$ be A. By Pick's theorem (see Exercise 717), A is rational, and, since X and Y are rational points, $s^2 = \|XY\|^2$ is also rational.

By the cosine law (and simple calculation) $s^2 = r^2 + r^2 - 2r^2 \cos(\frac{2\pi}{n}) = 4r^2 \sin^2(\frac{\pi}{n})$. Also, $A = \frac{1}{2} r^2 \sin(\frac{2\pi}{n}) = r^2 \cos(\frac{\pi}{n}) \sin(\frac{\pi}{n})$. Combining these two facts,

$$\tan(\frac{\pi}{n}) = \frac{\sin(\frac{\pi}{n})}{\cos(\frac{\pi}{n})} = \frac{r^2 \sin^2(\frac{\pi}{n})}{r^2 \sin(\frac{\pi}{n}) \cos(\frac{\pi}{n})} = \frac{\frac{1}{4} s^2}{A},$$

which is rational, contradicting Theorem 38.2.1. $\square$

Exercise 719: (Outline) In the article [341], several ways to look at this problem are given, as well as many references for this exercise, which dates back to 1962. The idea using recursion is simple, however a bit messy because of different cases for n even and odd; these calculations are only outlined here.

Let $f(n)$ denote the number of triangles in a triangular grid T_n with side length n. To count $f(n)$, count all triangles that use a vertex from the bottom row, and add this number to $f(n-1)$, which counts the number of triangles not touching the bottom of the T_n, that is, $f(n-1)$ is the number of triangles contained in a T_{n-1} that forms the upper part of T_n. Of those triangles touching the bottom, there are two types, those whose orientation is the same as the big triangle, and those that are pointed upside-down with only a vertex touching the base line. Let $G(n)$ be the number of upward triangles touching the bottom of T_n (see left side of Figure 38.6) and $H(n)$ the number of upside-down triangles touching the bottom of T_n (as in the right side of Figure 38.6).

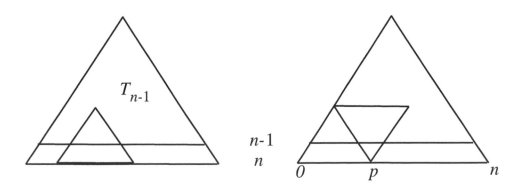

Figure 38.6: Counting triangles in T_n

Computing $G(n)$ is easy, as every point in T_{n-1} determines an upward triangle whose base is on T_n, and there are

$$G(n) = 1 + 2 + \cdots + n = \frac{n(n+1)}{2}$$

such points. Hence the number of upward triangles in all is

$$G(1) + G(2) + \cdots + G(n) = \sum_{v=1}^{n} \binom{v+1}{2} = \binom{n+2}{3},$$

where the last equality is proved in Exercise 88.

Counting $H(n)$ is only a little clumsier. For each vertex $p \in \{1, \ldots, n-1\}$ on the bottom row of T_n, the largest possible upside-down triangle than can be formed with p as its vertex has side length the smaller of p or $n - p$. So for each $p \in \{1, 2, \ldots, \lfloor n/2 \rfloor\}$, there are p such triangles, and by symmetry,

$$H(n) = \begin{cases} 2 \displaystyle\sum_{p=1}^{\frac{n-1}{2}} p & \text{when } n \text{ is odd} \\[2em] \left(2 \displaystyle\sum_{p=1}^{\frac{n}{2}} p \right) - \dfrac{n}{2} & \text{when } n \text{ is even.} \end{cases}$$

In [341], the notation $\delta(n)$ is used, which is 0 when n is even, and 1 when n is odd. With this notation, and a bit of fiddling, find that

$$H(n) = \frac{n^2}{4} - \frac{\delta(n)}{4},$$

whence it follows by induction that there are

$$\frac{n(n+2)(2n-1)}{24} - \frac{\delta(n)}{8}$$

upside down triangles in T_n. Together, these two counts (for upward and downward triangles) give the desired answer. $\qquad\square$

Exercise 720: See [266, p. 229] for two solutions, the first of which is by mathematical induction.

38.3 Solutions: Lines, planes, regions, and polyhedra

Exercise 721: This "now standard" solution seems to have initially appeared in 1948 [79]). For each $n \geq 3$, let $A(n)$ be the assertion that among n points not all on a line, there exist n different lines joining them.

BASE STEP: If three points are not all on one line, they determine a triangle of three lines, so $A(3)$ is true.

INDUCTION STEP: Fix $k \geq 3$ and suppose that $A(k)$ is true. Consider a placement of $k+1$ points $p_1, \ldots, p_k, p_{k+1}$, not all on a line.

By Lemma 20.3.1, there exists a line ℓ containing only two points, say p_k and p_{k+1}. Delete the point p_{k+1}. If the remaining points $p_1, \ldots, p_k$ are all on a line m, then the lines joining p_{k+1} to each of $p_1, \ldots, p_k$ are distinct, and together with m, there are $k+1$ lines, proving $A(k+1)$. If $p_1, \ldots, p_k$ are not all on a line, by $A(k)$, there are k distinct lines through these, and in addition to ℓ, there are $k+1$ lines, again proving $A(k+1)$.

By mathematical induction, for each $n \geq 3$, the statement $A(n)$ is true. $\square$

Exercise 722: (The result in this exercise is contained in [161, 8.31(a), p. 209], among many popular sources.)

The proof is by induction on ℓ for arbitrary p. Fix an arbitrary convex region R in the plane. For each $\ell \geq 0$, let $A(\ell)$ be the assertion that for each $p \in \{0, 1, \ldots, \binom{\ell}{2}\}$, if ℓ lines that cross R, with p intersection points inside of R, then the number of regions created inside R is $r = \ell + p + 1$.

BASE STEP: When no lines intersect R, $p = 0$, $r = 1$, and so $r = \ell + p + 1$, proving $A(0)$.

INDUCTION STEP: Fix some $k \geq 0$ and suppose that $A(k)$ holds for k lines and some $p \geq 0$ with $k + p + 1$ regions. Consider a collection $\mathcal{C}$ of $k+1$ lines each crossing R (not just touching), select some line $L \in \mathcal{C}$, and apply $A(k)$ to $\mathcal{C} \backslash \{L\}$ with some p intersection points inside R and $r = k + p + 1$ regions.

Let s be the number of lines intersecting L inside of R. As one draws a $(k+1)$-st line L, starting outside of R, a new region is created when L first crosses the border of R, and whenever L crosses a line inside of L. Hence, the number of new regions is $s + 1$. So the number of regions determined by the $k+1$ lines is

$$r + s + 1 = (k + p + 1) + s + 1$$
$$\text{(by } A(k))$$
$$= (k+1) + (p+s) + 1,$$

where $p + s$ is the total number of intersection points inside R; hence $A(k+1)$ is true.

By mathematical induction, for each $\ell \geq 0$, the assertion $A(\ell)$ is true. $\square$

Exercise 723: The solution follows directly from the result in Exercise 722. Since a circle is convex, and each intersection point is determined by a unique 4-tuple of points, there are $\binom{n}{4}$ intersection points and $\binom{n}{2}$ chords. $\square$

Exercise 725: This problem is discussed in [433], one of many sources. The first published solution [507] in 1826 is due to the Swiss mathematician, Jacob Steiner.

One proof is based on the following recursion:

Lemma 38.3.1. *Let r_n denote the maximum number of regions determined by n lines. Then $r_{n+1} = r_n + n + 1$.*

Proof of Lemma 38.3.1: Consider a system S of n lines, and let ℓ be a line not in S and let $S' = S \cup \{\ell\}$. Without loss of generality, assume that ℓ is not vertical (for if it is, rotate the system slightly). In the most left part of the plane, there are two regions, one above and one below ℓ both comprising a single region in S. As one follows ℓ to the right, if ℓ is not parallel to any line in S, then ℓ crosses all lines of S. As ℓ intersects each line in S, it cuts a region of S into two new regions. So when ℓ is not parallel to any line in S, then S' has $n + 1$ more regions than does S (if ℓ is parallel to any lines in S, then there are fewer). Hence, $r_{n+1} \leq r_n + n + 1$, with equality when ℓ is not parallel to any previous line.   □

To give the inductive proof of the exercise, for $n \geq 0$, let $R(n)$ be the statement that number of regions determined by n lines in general position is $1 + \binom{n+1}{2}$.

BASE STEP: When $n = 0$, there is only $1 = 1 + \binom{0+1}{2}$ region. Also, when $n = 1$, there are $2 = 1 + \binom{1+1}{2}$ regions.

INDUCTIVE STEP: Let $m \geq 0$ and suppose that $R(m)$ is true. Let S be a system of $m + 1$ lines in general position. Denoting the number of regions in S by r_{m+1}, by Lemma 38.3.1,

$$
\begin{aligned}
r_{m+1} &= r_m + (m + 1) + 1 \\
&= 1 + \binom{m+1}{2} + m + 1 \quad (\text{by } R(m)) \\
&= 1 + \frac{m(m+1)}{2} + (m + 1) \\
&= 1 + \frac{m(m+1)}{2} + \frac{2(m+1)}{2} \\
&= 1 + \frac{(m+2)(m+1)}{2} \\
&= 1 + \binom{m+2}{2},
\end{aligned}
$$

showing that the conclusion of $R(m + 1)$ holds, completing the inductive step $R(m) \rightarrow R(m+1)$.

By MI, for any number n of lines in general position in the plane, the number of regions determined is $1 + \binom{n+1}{2}$.   □

Exercise 726: (Outline) The base case $n = 1$ is clear since 1 circle creates 2 regions. For the inductive step, if k circles are present, any new circle added must intersect other circles in $2k$ distinct points; this means that there are $2k$ new arcs, each of which splits an old region into two parts, so the number of regions for $k + 1$ circles is $2k$ more than for k circles.   □

Exercise 727: See [230, pp 7–8, 19] for solution. The recursion needed is that Z_n is $2n$ fewer than the number of regions created by $2n$ lines.

Exercise 728: This problem appeared as [230, Ex.13, p. 19], with outline in the solution section.

Exercise 729: (Outline) The result is true for $n = 0$ since S_0 is the single point $(0, 0)$, which is covered by a single line. Supposing the result is true for $n = k$, since the line $x + y = k + 1$ is not covered by any lines covering S_k, at least one more line is required to cover the points on $x + y = k + 1$.   □

Exercise 730: (Brief) For one line, the result is clear. For the inductive step, assuming the result is true for k lines, show that the result is true for $k + 1$ lines by taking the 2-coloring of the regions determined by k lines, and reversing the colors on one side of the new line; see Figure 38.7.   □

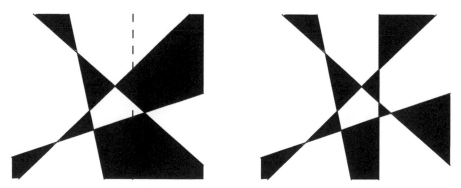

Figure 38.7: The new coloring after a line is added

Exercise 732: This problem is in [437, Prob. 6-8, Challenge 1].

Exercise 734: This problem (with solution) appears in [220, pp. 119–120]. Let $S(n)$ be the statement that if n half-planes cover $\mathbb{R}^2$, then some two or three of these cover $\mathbb{R}^2$.
BASE STEP: If either 2 or 3 planes cover the plane, then the result is trivial; so $S(2)$ and $S(3)$ hold.

INDUCTIVE STEP: Fix $k \geq 3$ and suppose that $S(k)$ holds. Let $H_1, \ldots, H_{k+1}$ be half-planes, combined which cover the plane. For each half-plane H_i, let ℓ_i be its boundary line.

If ℓ_{k+1} is entirely contained in one of the half-planes, say H_k, then ℓ_k and ℓ_{k+1} are parallel. If these two half-planes face opposite directions, then they cover the plane. If these two half-planes face the same direction, then one is contained in the other, making one superfluous, and so by $S(k)$, some two or three cover the entire plane.

For the remaining case, suppose that ℓ_{k+1} is not contained entirely in any of the half-planes $H_1, \ldots, H_k$. Since the entire plane is covered by the half-planes, so is all of ℓ_{k+1}. The intersection of each other half-plane with ℓ_{k+1} is a ray, and so all these rays cover ℓ_{k+1}. By Lemma 20.3.3, some two of these rays cover ℓ_{k+1}, say those rays determined by H_1 and H_2. There are two possible relative locations for the half-planes H_1, H_2 and H_{k+1}. (a) H_{k+1} contains the intersection of lines ℓ_1 and ℓ_2; in this case, these three half-planes cover the plane. (b) H_{k+1} does not contain the intersection between ℓ_1 and ℓ_2. In this case, the plane is covered by $H_1, H_2, \ldots, H_k$, and so again $S(k)$ applies, giving at most two half-planes covering the plane.

In any case, $S(k+1)$ is shown to follow from $S(k)$, completing the inductive step.

By mathematical induction, for each $n \geq 2$, $S(n)$ is true. $\qquad \square$

Exercise 735: This problem appeared, for example, in [499, Prob. 30].

Exercise 736: This problem is discussed in [433], one of many sources. Hint: If s_n is the maximum number of regions determined by n planes, then first prove that

$$s_{n+1} = s_n + r_n,$$

where r_n is defined in the solution to Exercise 725.

Exercise 737: Hint: duplicate the result for intersecting circles (Exercise 726) to show that the maximum number of regions n circles can partition a *sphere* into is $n^2 - n + 2$. If $f(n)$ is the answer to this question, show the recurrence

$$f(n+1) = f(n) + n^2 - n + 2$$

by looking at the pattern made on the $(n+1)$-th sphere by the remaining spheres.

Exercise 739 [Euler's formula for polyhedra]: (Outline) Form the "graph" of a polyhedron by imagining that one face is removed, and then the remaining structure is stretched out to lay flat. Since the graph of any simple (faces are non-intersecting and there are no holes) polyhedron is connected and planar, the result in this exercise follows directly from Exercise 503, the proof of which is a simple (strong) induction. This graph theory approach is the standard way to solve this exercise and it answers a slightly more general problem, since not every planar graph gives rise to a polyhedron (*e.g.*, a planar graph can have vertices of degree 0, 1, or 2, and polyhedra have no such corresponding vertices).

Inductive proofs using actual polyhedra can be tricky, since all simple polyhedra need to be inductively constructed. There are two approaches that seem to work. The base case for each is the tetrahedron.

First let P be polyhedron with only triangular faces. The number of edges per face is 3, and each edge is on 2 faces, and so counting all edge-triangle pairs, $2e = 3f$. Multiplying Euler's formula by 3 and making this replacement shows that it suffices to prove for triangle polyhedra (also called *deltahedra*), that $e = 3v - 6$, and this can be done by induction. For example, adding a new vertex to the center of any face, and joining it the three corners of the triangle, increases the number of vertices by 1, the number of edges by 3, and the number of faces by 2, and so each side of $v + f = e + 2$ is increased by 1+2=3. But there are other ways to create a triangulated polygon; an analysis is required for each case. To get the formula for a general polyhedron, subdivide each face into triangles, and observe that again Euler's formula is preserved upon each successive subdivision; induction from the triangulated polyhedron back to the general polyhedron does the rest.

For a second approach, use "truncation", a process by which one cuts off a corner of a polyhedron to get, if possible, a polyhedron with one fewer vertex (and so a proof is by induction on the number of vertices). For truncation to work one needs a small observation. Suppose that v is a vertex of a polyhedron P, and let $v_1, \ldots, v_d$ be those $d \geq 3$ vertices for which $\{v, v_i\}$ is an edge of P. If P is convex, then the v_i's can be shifted so that they all lie on one plane (while preserving all combinatorial properties of P). Then removal of v together with all edges touching v produces another polyhedron P' with $d - 1$ fewer faces, (d faces gone, and one new face with vertices $v_1, \ldots, v_d$), d fewer edges, and 1 fewer vertex, and so if Euler's formula is true for P', then it is true for P.

Exercise 740: This exercise occurs in [161, 8.21, pp. 208, 215].

38.4   Solutions: Finite geometries

Exercise 741: Let $S(v)$ be the statement that for any near linear space $\mathcal{S}$ on v points, (using the notation in Theorem 20.4.1) that if

$$\sum_{i=1}^{b} \binom{k_i}{2} \geq \binom{v}{2},$$

then $\mathcal{S}$ is linear.
Base step: If $v = 1$, then $b \leq 1$, and in any case $\mathcal{S}$ is linear.

Induction step: Fix $v \geq 2$ and suppose that $S(v-1)$ is true. Let $\mathcal{S} = (P, \mathcal{L})$ be a near linear space with $|P| = v$ points, and lines labelled $L_1, L_2, \ldots, L_b$, where each L_i contains $|L_i| = k_i$ points, and suppose that

$$\sum_{i=1}^{b} \binom{k_i}{2} \geq \binom{v}{2},$$

holds. Pick any point $p \in P$ and delete it—producing a new geometry $\mathcal{S}' = (P', \mathcal{L}')$, where $P = P \backslash \{p\}$, and lines in $\mathcal{L}'$ are those in $\mathcal{L}$, except those lines $L \in \mathcal{L}$ containing p are now $L' = L \backslash \{p\}$. One can verify that the restricted geometry $\mathcal{S}$ is still a near linear space. Let $v' = v - 1$ and for each $i = 1, \ldots, b$, put $k_i' = |L_i'|$. The first step is to show that the corresponding inequality also holds for $\mathcal{S}'$. In this derivation, there is no need to separate cases for small k_i if one interprets $\binom{0}{2} = \binom{1}{2} = 0$. The identity $\binom{n}{2} - (n-1) = \binom{n-1}{2}$ is used twice:

$$\sum_{i=1}^{b} \binom{k_i'}{2} = \sum_{i:p \notin L_i} \binom{k_i'}{2} + \sum_{i:p \in L_i} \binom{k_i'}{2}$$

$$= \sum_{i:p \notin L_i} \binom{k_i}{2} + \sum_{i:p \in L_i} \binom{k_i - 1}{2}$$

$$= \sum_{i:p \notin L_i} \binom{k_i}{2} + \sum_{i:p \in L_i} \left[\binom{k_i}{2} - (k_i - 1) \right]$$

$$= \sum_{i=1}^{b} \binom{k_i}{2} - \sum_{i:p \in L_i} (k_i - 1)$$

$$= \binom{v}{2} - \sum_{i:p \in L_i} (k_i - 1)$$

$$\leq \binom{v}{2} - (v - 1)$$

$$= \binom{v-1}{2}.$$

By $S(v-1)$, $\mathcal{S}'$ is linear; to show that $\mathcal{S}$ is linear, it remains to show that for every $q \in P \backslash \{p\}$, there is a unique line containing p and q. To see this, let r be any other point in P, and delete it; by the same argument used above for $\mathcal{S}'$, this new geometry is also linear, and so p and q are contained on a unique line. Hence, $\mathcal{S}$ is linear.

By mathematical induction, for each $v \geq 1$, $S(v)$ holds, finishing the proof of the other direction in Theorem 20.4.1. $\square$

Chapter 39

Solutions: Ramsey theory

Exercise 742: The proof is by induction on n. For $n \geq 1$, let $S(n)$ be the statement that if $\mathcal{P}([n])$ is n-colored, there are sets $A, B \in \mathcal{P}([n])$ monochromatic with $A \subsetneq B$. BASE STEP: When $n = 1$, if $\mathcal{P}([1]) = \{\emptyset, \{1\}\}$ is 1-colored, then $\emptyset$ and $\{1\}$ are the same color and $\emptyset\{1\}$.

INDUCTIVE STEP: Fix $k \geq 1$, and suppose $S(k)$ is true. Let

$$\Delta : \mathcal{P}([k+1]) \to [k+1]$$

be a given $(k+1)$-coloring. If for any $A \subsetneq [k+1]$, if $\Delta(A) = \Delta([k+1])$, then $B = [k+1]$ are as needed. So suppose that all proper subsets of $[k+1]$ are colored differently from $[k+1]$; thus all proper subsets of $[k+1]$ are k-colored. In particular, all subsets of $[k]$ are k-colored by Δ, and so by induction hypothesis $S(k)$, there exist two sets $A, B \subset \mathcal{P}([k])$, $A \subsetneq B$, that are colored the same, and since $\mathcal{P}([k])$ is contained in $\mathcal{P}([k+1])$, the statement $S(k+1)$ is again confirmed.

By mathematical induction, for each $n \geq 1$, $S(n)$ is true. $\qquad\square$

Exercise 743: (PHP: Brief solution) The proof is by induction on k. Fix some $n \geq 0$. For $k = 1$, the statement is trivial, so suppose that the result is true for $k = p$, that is, for any set with $np + 1$ elements partitioned into p parts, one part contains at least $n + 1$ elements. To show that the result is true for $k = p + 1$, let X be a set with $n(p+1) + 1$ elements partitioned into parts $X_1, X_2, \ldots, X_p, X_{p+1}$. If X_{p+1} does not contain $n + 1$ elements, then $X_1 \cup X_2 \cup \cdots \cup X_p$ is a set with at least $n(p+1) + 1 - n = np + 1$ elements, and so, by induction hypothesis, for some $i = 1, \ldots, p$, X_i contains $n + 1$ elements. This completes the inductive step, and hence the proof by induction. $\qquad\square$

Exercise 745 (Ramsey's theorem for 2-colorings): For $k \geq 1$, let $A(k)$ be the assertion that for every $a, b \geq 1$, $R_k(a, b)$ exists. The proof is by induction on k, and for each k, by induction on $a + b$.

BASE CASES ($k = 1, 2$): The statement $A(1)$ is the pigeonhole principle and $A(2)$ is Corollary 21.2.3. [The case $k = 2$ is not really required, because as is seen below, the proof for $k = 2$ below is nearly the same as for the Erdős–Szekeres recursion (Theorem 21.2.2).]

INDUCTIVE STEP ($A(j-1) \to A(j)$): Fix some $j \geq 1$ and suppose that $A(j-1)$ holds, that is, for every $a, b \geq 1$, that $R_{j-1}(a, b)$ exists (and is finite). It remains to show for each $a, b \geq 1$ that $R_j(a, b)$ exists. For each $m \geq 2$, let $B_j(m)$ be the statement that if $a + b = m$, then $R_j(a, b)$ exists. The proof of $B_j(m)$ is by induction on m.

BASE STEP ($\min\{a, b\} \leq j$): If $\min\{a, b\} < j$, then $R_j(a, b) = \min\{a, b\}$. If $\min\{a, b\} = j$, then $R_j(a, b) = R_j(b, a) = \max\{a, b\}$.

INDUCTIVE STEP ($B_j(m) \to B_j(m + 1)$): Let s and t satisfy $1 < j \leq \min\{s, t\}$, and suppose that both $R_j(s - 1, t)$ and $R_j(s, t - 1)$ exist, that is, for $m = s + t - 1$, that $B_j(m)$ holds. To prove $B_j(m + 1)$, it suffices to show that $R_j(s, t)$ exists, and this follows if one can show

$$R_j(s, t) \leq R_{j-1}(R_j(s - 1, t), R_j(s, t - 1)) + 1. \tag{39.1}$$

since, by $A(j - 1)$, the righthand side of equation (39.1 is a finite number; call this number N. Let X be a set with N elements, and consider any coloring of the form

$$\Delta : [X]^j \to \{\text{red}, \text{blue}\}.$$

Fix a vertex $x \in X$, put $Y = X \setminus \{x\}$ and consider the induced 2-coloring of the $(j - 1)$-subsets of Y,

$$\Delta^* : [Y]^{j-1} \to \{\text{red}, \text{blue}\}$$

defined, for each $W \in [Y]^{j-1}$, by $\Delta^*(W) = \Delta(W \cup \{x\})$. By the choice of N, there exists at least one of two kinds of subsets of Y: a set $Y_0 \subset Y$ with $R_j(s - 1, t)$ elements so that Δ^* restricted to $[Y_0]^{j-1}$ is monochromatic red, or a set $Y_1 \subset Y$ with $R_j(s - 1, t)$ elements so that Δ^* restricted to $[Y_1]^{j-1}$ is monochromatic blue.

Suppose that such a Y_0 exists. Then either Y_0 contains $s - 1$ vertices S_0 so that $[S_0]^j$ is monochromatic red in which case $[S_0 \cup \{x\}]^j$ is also, or Y_0 contains t vertices T_0 so that $[T_0]^t$ is blue. The similar argument shows that if Y_1 exists, there is either an s-set whose j-subsets are all red, or a t-subset, all of whose j-subsets are blue. This shows equation (39.1), and hence concludes the proof of the inductive step $B_j(m) \to B_j(m + 1)$.

Thus the inductive step $A(j - 1) \to A(j)$ is complete as well.

By mathematical induction, for $k \geq 1$, $A(k)$ is true. $\qquad \square$

Exercise 746: (Outline) Duplicate the solution in Exercise 745, replacing equation (39.1) with Theorem 21.2.4, and induct on the sum of the p_i's.

Exercise 748: For each $r \geq 2$, let $P(r)$ be the proposition that the theorem is true for this value of r. Induct on r.

BASE STEP: $P(2)$ is simply Theorem 21.2.5.

INDUCTIVE STEP: Fix $m \geq 2$ and suppose that $P(m)$ is true. Let T be an infinite set and let $\Delta : [T]^2 \to [1, m+1]$ be a coloring. Consider the two colors m and $m+1$ as one color c; then Δ induces an m-coloring $\Delta^* : [T]^2 \to [1, m-1] \cup \{c\}$. Then by $P(m)$, there exists an infinite set S so that $[S]^2$ is monochromatic under Δ^*. If $[S]^2$ is color $i \in [1, m-1]$, then $[S]^2$ is monochromatic under Δ. If the color of $[S]^2$ is c, then $[S]^2$ is 2-colored under Δ. Applying $P(2)$ (Theorem 21.2.5) with S playing the role of T, there exists another infinite set $S' \subseteq S \subseteq T$ so that $[S']^2$ is monochromatic. In either case, the infinite monochromatic set exists, proving $P(m+1)$. This completes the inductive step.

By mathematical induction, for every $r \geq 2$, $P(r)$ holds. □

Note: The above technique of grouping colors is frequently used to prove that a 2-coloring Ramsey-type theorem implies the corresponding r-coloring theorem.

Exercise 749: For $k, \ell \geq 2$, let $S(k, \ell)$ be the statement that if $\binom{k+\ell-4}{k-1} + 1$ points with no same x coordinates are chosen in $\mathbb{R}^2$, then some k of these points contains a k-cup or some ℓ of these points form an ℓ-cap. The proof is by induction on $k+\ell$.

BASE STEP: When $k = 2$ and $\ell = 2$, any two points form both a 2-cup and a 2-cap, and since $\binom{2+2-4}{2-2} + 1 = 2$, the case $S(2, 2)$ is true.

INDUCTIVE STEP: Fix $s, t \geq 3$, and suppose that both $S(s-1, t)$ and $S(s, t-1)$ hold. It remains to show that $S(s, t)$ holds. Put

$$N = \binom{s+t-4}{s-2} + 1 = \binom{s+t-4}{t-2} + 1,$$

and let $P = \{p_1, \ldots, p_N\}$ be points listed in order with strictly increasing x-coordinates. In hope of contradiction, suppose that no subset of P forms either an s-cup or a t-cap. Since

$$N \geq \binom{s-t-5}{s-3} + 1$$

by $S(s-1, t)$, the set P contains $(s-1)$-cups. In fact, P contains *many* $(s-1)$-cups: Let L be the set of rightmost points of all $(s-1)$-cups. $P \setminus L$ is a set with no $(s-1)$-cups, so by $S(s-1, \ell)$,

$$|P \setminus L| \leq \binom{s+t-5}{s-3}.$$

Hence

$$\|L\| \geq N - \binom{s+t-5}{s-3}$$

$$= \binom{s+t-4}{t-2} + 1 - \binom{s+t-5}{t-2}$$

$$= \binom{s+t-5}{t-3} + 1 \qquad\qquad \text{(Pascal's id.)}$$

and since L contains no s-cups, by $S(s, t-1)$, L contains at least one $(t-1)$-cap, say $\{q_1, \ldots, q_{t-1}\}$, given in left-right order, where each q_i is the rightmost point of some $(s-1)$-cup. Suppose that q_1 is the right endpoint of the $(s-1)$-cup on points $p_1, \ldots, p_{s-2}, p_{s-1} = q_1$, where p_{s-2} is the second last point. If the slope of $p_{s-2}p_{s-1}$ is greater than or equal to that of q_1q_2, then the points p_{s-2}, $p_{s-1} = q_1$, q_2, $\ldots$, q_{t-1} form a t-cap [I could have interchanged cups and caps and got a "t-cup" here?] Now argue depending on the relative position of q_2. If the slope of $p_{s-2}p_{s-1}$ is less than that of q_1q_2, then $p_1, p_2, \ldots, p_{s-2} = q_1$, q_2 forms an s-cup. Either position of q_2 contradicts the initial assumption, so $S(s, t)$ holds, thereby completing the inductive step.

By mathematical induction on $k + \ell$, for every $k, \ell \geq 2$, $S(k, \ell)$ holds. □

Exercise 751 (Hilbert's affine cube lemma):
A simple fact used below is that for some d and r, if $h(d, r)$ exists, then for *any* set S of $h(d, r)$ consecutive positive integers (not just $S = [1, h(d, r)]$) if $\chi : S \to [1, r]$ is given, S contains a monochromatic affine d-cube (just shift the x_0 in the cube).

Fix $r \geq 1$. For each positive integer d, let $C(d)$ be the claim that $h(d, r)$ exists. BASE STEP: Since a 1-dimensional affine cube consists of only two integers, by the pigeonhole principle, $h(1, r) = r + 1$, proving that $C(1)$ is true.

INDUCTIVE STEP $(C(p) \to C(p+1))$: Fix some $p \geq 1$ and suppose that $n = h(p, r)$ exists. Put $N = r^n + n$ and let $\chi : [1, N] \to [1, r]$ be an arbitrary r-coloring of $[1, N]$. For each $i = 0, 1, \ldots, r^n$, consider sets of n consecutive integers in $[1, N]$ of the form

$$S_i = \{i+1, i+2, \ldots, i+n\}.$$

Each of the $r^n + 1$ sets S_i can be colored in at most r^n ways, so by the pigeonhole principle there are $j < k$ so that the two sequences

$$S_j = \{j+1, j+2, \ldots, j+n\} \text{ and } S_k = \{k+1, k+2, \ldots, k+n\}$$

receive the same color pattern. As S_j contains n consecutive integers and is r-colored, the inductive hypothesis $C(p)$ applies, yielding a monochromatic affine p-cube $H_1 = (x_0, x_1, \ldots, x_p)$ in S_j. Put $k - j = x_{p+1}$. Then the corresponding cube $H_2 = H(x_0 + x_{p+1}, x_1 + x_{p+1}, \ldots, x_p + x_{p+1})$ in S_k is monochromatic and is the same color as H_1. Thus, the $(p+1)$-cube $H_1 \cup H_2 = H(x_0, x_1, \ldots, x_p, x_{p+1})$ is monochromatic, showing that $h(p+1, r)$ exists and so $C(p+1)$ is true. This completes the inductive step.

By mathematical induction on d, for each $d \geq 1$, $h(d, r)$ exists. Since r was fixed but arbitrary, for any $r \geq 1$ and $d \geq 1$, $h(d, r)$ exists. □

Comment: The above proof typifies many coloring arguments in Ramsey theory. In writing such a proof, one might elect to use more formal notation, perhaps like: "χ induces a coloring $\chi^* : \{S_0, S_1, \ldots, S_{r^n}\} \to r^n$ by $\chi * (S_i) = (\chi(i+1), \chi(i+2), \ldots, \chi(i+n))$" to record each coloring pattern as a vector. The efficacy of this notation becomes clear in more complicated arguments.

Exercise 752: The proof here is by induction on $|B|$, as in, *e.g.*, [296, p. 330].

BASE STEP: When $|B| = 1$, for every subgroup H,

$$|A + B| = |A| = |A| + |B| - 1 \geq |A| + |B| - |H|.$$

INDUCTIVE STEP: Let $|B| > 1$ and suppose that the theorem holds for all pairs of finite non-empty subsets A, B' with $|B'| < |B|$.

Case 1: For all $a \in A$ and all $b, c \in B$, $a + b - c \in A$. In this case, for all $b, c \in B$, $A + b - c \in A$. Let H be the subgroup of G generated by all elements of the form $b - c$ where $b, c \in B$. Then $|B| \leq |H|$ and $A + H = A \neq G$. Thus, H is a proper subgroup of G and

$$|A + B| \geq |A| \geq |A| + (|B| - |H|).$$

Case 2: For some $a \in A$ and $b, c \in B$, suppose that $a + b - c \notin A$. Setting $x = a - c$, define

$$A' = A \cup (B + x) \quad \text{and}$$
$$B' = B \cap (A - x).$$

If $b \in B'$, then for some $\hat{a} \in A$, $b = \hat{a} - x = \hat{a} - a + c$, implying

$$a + b - c = a + \hat{a} - a + c - c = \hat{a} \in A,$$

contrary to assumption, so $b \notin B'$. Since $0 \in A - a$, $c \in (A - a) + c = A - (a - c)$, and since also $c \in B$, conclude that $c \in B'$. Together, $b \notin B'$ and $c \in B'$ show that B' is a non-empty proper subset of B. By induction hypothesis (applied with A' and B'), there exists a proper subgroup H of G so that

$$|A' + B'| \geq |A'| + |B'| - |H|. \tag{39.2}$$

Then

$$\begin{aligned}
A' + B' &= [A \cup (B + x)] + [B \cap (A - x)] \\
&\subseteq (A \cup B) \cup [(B + x) + (A - x)] = A + B.
\end{aligned}$$

and

$$\begin{aligned}
|A'| + |B'| &= |A \cup (B + x)| + |B \cap (A - x)| \\
&= |A \cup (B + x)| + |(B + x) \cap A|
\end{aligned}$$

$$= \ |A| + |B + x|$$
$$= \ |A| + |B|,$$

together with equation (39.2) yield $|A+B| \geq |A|+|B|-|H|$, completing the inductive step.

By mathematical induction, the theorem is true for any B.   □

Exercise 753: (Using van der Waerden's theorem) The proof is by induction on r. For $r \in \mathbb{Z}^+$, let $P(r)$ be the statement that for each $k \geq 1$, $SB(k;r)$ exists.
BASE STEP: When $r = 1$, the result is trivial (for any k), so $P(1)$ holds.
INDUCTION STEP: Fix $t > 1$ and suppose that $P(t-1)$ holds, that is, that for each $k \geq 1$, $SB(k;t-1)$ exists. Consider the van der Waerden number

$$n = W(k \cdot SB(k;t-1) + 1, t),$$

and let $\Delta : [1,n] \to [1,t]$ be a t-coloring. By the choice of n, there exists an arithmetic progression

$$\{a + id : i \in [0, k \cdot SB(k;t-1) - 1]\} \subset [1,n].$$

that is monochromatic, say, with color i. If for any $j < SB(k,t-1)$, dj is also colored i, then $\{dj\} \cup \{a + i(jd) : i \in [0, k-1]\}$ is monochromatic, showing that $SB(k;t)$ exists. Otherwise, $\{d, 2d, \dots, SB(k, t-1)d\}$ is $(t-1)$-colored, and then by the inductive hypothesis $P(t-1)$, (with the same k), there exists a' and d' so that $\{d'd\} \cup \{(a' + id')d : i \in [0, k-1]\}$ is monochromatic. This is a k-term arithmetic progression together with difference $d'd$ in the same color, again confirming that $SB(k;t)$ exists, thereby completing the proof of $S(t)$, and hence the inductive step.
Thus, by MI, for all $r \in \mathbb{Z}^+$, $P(r)$ is true.   □

Exercise 754: This proof occurs in [231], and relies on van der Waerden's theorem. Only the beginning of the proof is given; the remainder follows as in Exercise 753.
For $r \geq 1$, let $P(r)$ be the proposition that for any positive integers k, s, such an $n = n(k, s, r)$ exists. The proof is by induction on r.
BASE STEP: For any k, s, let $n = \max\{k, s\}$. If $k \geq s$, then the progression $1, 2, \dots, k = n$ satisfies the theorem with $a = d = 1$. If $s > k$, then $1, 2, \dots, k, \dots, s$ also contains both the AP_k with $a = d = 1$ and the element s. No smaller n works, so $n(k, s, 1) = n = \max\{k, s\}$ exists, proving the base case $P(1)$.

INDUCTIVE STEP: Fix $R \geq 1$ and suppose that $P(r)$ holds, that is, for all k' and s', $n(k', s', R)$ exists. Fix k and s.
Claim: $n(k, s, R+1) = s \cdot W(k \cdot n(k, s, R), R+1)$. Proof of claim: Let $n = s \cdot W(k \cdot n(k, s, R), R+1)$ and let $\Delta : [1, n] \to [1, R+1]$ be given. Now continue as in Exercise 753.

Exercise 755: (Tarry-Escott problem) This solution appeared in many places (*e.g.*, see [85] for solution and many references). The desired partition is arrived at recursively. To describe this recursion, for a number x and a set S, the notation $x + S = \{x + s : s \in S\}$ is convenient.

Put $A_0 = \{0\}$ and $B_0 = \{1\}$. For $m \geq 0$, having defined A_m and B_m, put $A_{m+1} = A_m \cup (2^{m+1} + B_m)$ and $B_{m+1} = B_m \cup (2^{m+1} + A_m)$. To confirm, the few first partitions are: $0 \mid 1$; $0, 3 \mid 1, 2$; $0, 3, 5, 6 \mid 1, 2, 4, 7$; and

$$0, 3, 5, 6, 9, 10, 12, 15 \mid 1, 2, 4, 7, 8, 11, 13, 14.$$

For each $n \geq 1$, let $S(n)$ be the statement that A_n and B_n satisfy, for every $j = 0, 1, \ldots, n$

$$\sum_{a \in A_n} a^j = \sum_{b \in B_n} b^j.$$

BASE STEP: $A_0 = \{0\}$ and $B_0 = \{1\}$ satisfies $0^0 = 1^0$, so $S(0)$ is true.

INDUCTIVE STEP: For some $m \geq 0$, assume that $S(m)$ holds. Then for each (fixed) $j \in \{0, 1, \ldots, m + 1\}$,

$$\sum_{a \in A_{m+1}} a^j - \sum_{b \in B_{m+1}} b^j$$

$$= \left[\sum_{a \in A_m} a^j + \sum_{b \in B_m} (2^{m+1} + b)^j \right] - \left[\sum_{b \in B_m} b^j + \sum_{a \in A_m} (2^{m+1} + a)^j \right]$$

$$= \left[\sum_{a \in A_m} a^j - \sum_{b \in B_m} b^j \right] - \left[\sum_{a \in A_m} (2^{m+1} + a)^j - \sum_{b \in B_m} (2^{m+1} + b)^j \right]$$

$$= \left[\sum_{a \in A_m} a^j - \sum_{b \in B_m} b^j \right] - \sum_{k=0}^{j} \binom{j}{k} 2^{(m+1)(j-k)} \left[\sum_{a \in A_m} a^k - \sum_{b \in B_m} b^k \right].$$

When $0 \leq j \leq m$, by induction hypothesis $S(m)$, each term above is 0. When $j = m + 1$, the above becomes

$$\left[\sum_{a \in A_m} a^{m+1} - \sum_{b \in B_m} b^{m+1} \right] - \sum_{k=0}^{m+1} \binom{m+1}{k} 2^{(m+1)(m+1-k)} \left[\sum_{a \in A_m} a^k - \sum_{b \in B_m} b^k \right],$$

and again by $S(m)$, for each $k = 0, 1, \ldots, m$, the expression in the last brackets is again 0, so the entire expression becomes 0. This shows that $S(m + 1)$ is true as well, completing the inductive step.

By MI, for all non-negative integers n, the statement $S(n)$ is true. $\square$

Exercise 756: For each $n \geq 2$, let $A(n)$ be the assertion that G_n is triangle-free and $\chi(G_n) = n$.

BASE STEP: It is trivial to check that G_2 is triangle-free and $\chi(G_2) = 2$, so $A(2)$ is true. [It is also easy to check manually that both $S(3)$ and $S(4)$ hold as well.]

INDUCTIVE STEP: Fix $k \geq 2$ and suppose that $A(k)$ holds, that is, G_k is triangle-free and $\chi(G_k) = k$. Put $V = V(G_k) = \{v_1, \ldots, v_m\}$, (where m can be calculated recursively: $m = 2|V(G_{k-1})| + 1$). Define $U = \{u_1, \ldots, u_m\}$ and x as per the construction of G_{k+1} on vertex set $V \cup U \cup \{x\}$. It is not difficult to verify that G_{k+1} is again triangle-free; to show $A(k+1)$, it remains to verify that $\chi(G_{k+1}) = k+1$. To see that $\chi(G_{k+1}) \leq k+1$, consider any good k-coloring $c : V(G_k) \to \{1, \ldots, k\}$ of G_k, and extend this coloring to c' of G_{k+1} by defining, for each $i = 1, \ldots, m$, $c'(v_i) = c'(u_i) = c(v_i)$, and assigning x its own color. Since each u_i and v_i share the same neighbors, c' is a good $(k+1)$-coloring.

To see that G_{k+1} is not k-chromatic, consider any k-coloring c of $V(G_{k+1})$. Suppose that c restricted to the G_k portion of G_{k+1} is a good k-coloring. By $A(k)$, G_k is not $(k-1)$-colorable and so in each color class $c^{-1}(i)$ of $V(G_k)$, there exists a vertex v_j connected to vertices of all remaining $k-1$ colors. Since u_j is connected to the same vertices in G_k as v_j is, if c is to be a good coloring, u_j must receive the same color as v_j, and hence all of U receives all k colors—but x is adjacent to every vertex in U, and so this leaves no color available for x, so c is not a good k-coloring of G_{k+1}. So $\chi(G_{k+1}) = k+1$, proving $A(k+1)$ and completing the inductive step.

By mathematical induction, for each $n \geq 2$, $A(n)$ is true.   □

Chapter 40

Solutions: Probability and statistics

Exercise 757: Suppose that $x_1, x_2, \ldots$ are all real numbers in the interval $[a, b]$. Let $S(n)$ be the statement that

$$a \le \frac{x_1 + x_2 + \cdots + x_n}{n} \le b.$$

BASE STEP: For $n = 1$, $S(1)$ merely says that $a \le \frac{x_1}{1} \le b$, which is true by the assumption made on all the x_i's.

INDUCTIVE STEP: For some fixed $k \ge 1$, assume that $S(k)$ is true, namely

$$S(k): \quad a \le \frac{x_1 + x_2 + \cdots + x_k}{k} \le b.$$

To be shown is

$$S(k+1): \quad a \le \frac{x_1 + x_2 + \cdots + x_k + x_{k+1}}{k+1} \le b.$$

The two inequalities in $S(k+1)$ are proved separately.

Since $a \le \frac{x_1 + x_2 + \cdots + x_k}{k}$, it follows that $ak \le x_1 + x_2 + \cdots + x_k$ and since $a \le x_{k+1}$,

$$a(k+1) = ak + a \le ak + x_{k+1} \le x_1 + x_2 + \cdots + x_k + x_{k+1};$$

division by $k+1$ yields the lower bound of $S(k+1)$.

Similarly, since $\frac{x_1 + x_2 + \cdots + x_k}{k} \le b$, it follows that $x_1 + x_2 + \cdots + x_k \le bk$, and since $x_{k+1} < b$, then

$$x_1 + x_2 + \cdots + x_k + x_{k+1} \le bk + b = b(k+1).$$

Upon dividing by $k+1$, one has the upper bound in $S(k+1)$. This concludes the inductive step.

By mathematical induction, for all $n \geq 1$, $S(n)$ holds. $\qquad\qquad\square$

Exercise 759: See [411, Problem 52].

Exercise 760: Thanks to Brad Johnson (University of Manitoba, Dept. of Statistics), who provided this question and proof for inclusion here. **(a)** Let $X_0 = 1$ and note that, just prior to the nth draw there is a total of $n+1$ balls in the urn. Then

$$E(X_1) = E\left[E(X_1 \mid X_0)\right]$$
$$= E\left[(X_0 + 1)\frac{1}{2} + X_0\frac{1}{2}\right]$$
$$= E\left[X_0 + \frac{1}{2}\right] = \frac{3}{2}.$$

(b)

$$E(X_2) = E\left[E(X_2 \mid X_1)\right]$$
$$= E\left[(X_1 + 1)\frac{X_1}{3} + X_1\frac{3 - X_1}{3}\right]$$
$$= E\left[\frac{4X_1}{3}\right] = \left(\frac{4}{3}\right)\left(\frac{3}{2}\right) = 2.$$

(c)

$$E(X_3) = E\left[E(X_3 \mid X_2)\right]$$
$$= E\left[(X_2 + 1)\frac{X_2}{4} + X_2\frac{4 - X_2}{3}\right]$$
$$= E\left[\frac{5X_2}{4}\right] = \frac{5}{4} \cdot 2 = \frac{5}{2}.$$

(d) One might guess that $E(X_n) = (n+2)/2$.

(e) For each integer $n \geq 0$, let $S(n)$ denote the equality $E(X_n) = \frac{n+2}{2}$.

BASE STEP: $S(0)$ says $E[X_0] = 1$, which is true because $X_0 = 1$.

INDUCTIVE STEP: Fix $k \geq 1$, and suppose that

$$S(k-1): \quad E_{k-1} = \frac{k+1}{2}$$

is true. Then

$$
\begin{aligned}
E(X_k) &= E\left[E(X_k \mid X_{k-1})\right] \\
&= E\left[(X_{k-1}+1)\frac{X_{k-1}}{n+1} + X_{k-1}\frac{k+1-X_{k-1}}{k+1}\right] \\
&= E\left[X_{k-1}\left(1+\frac{1}{k+1}\right)\right] \\
&= \left(\frac{k+1}{2}\right)\left(\frac{k+2}{k+1}\right) \qquad\qquad\qquad \text{(by } S(k-1)) \\
&= \frac{k+2}{2},
\end{aligned}
$$

confirming $S(k)$, thereby completing the inductive step.

Therefore, by mathematical induction, for every $n \geq 0$, $E(X_n) = (n+2)/2$.

(**f**) Using the same arguments as above, $E(X_1) = 3/2$, $E(X_2) = 2$ and $E(X_3) = 5/2$ and it appears that $E(X_n) = (n+2)/2$ again. As in the inductive step of part (e), assume $S(k-1)$. Then

$$
\begin{aligned}
E(X_k) &= E\left[E(X_k \mid X_{k-1})\right] \\
&= E\left[X_{k-1}\frac{X_{k-1}}{n+1} + (X_{k-1}+1)\frac{k+1-X_{k-1}}{k+1}\right] \\
&= E\left[X_{k-1}\frac{n}{n+1} + 1\right] \\
&= \frac{k+2}{2} \qquad\qquad\qquad\qquad \text{(by } S(k-1)\text{, simplify)}.
\end{aligned}
$$

Therefore, by mathematical induction, for each $n \geq 0$, $E(X_n) = (n+2)/2$. Note, however, that the distribution of X_n is different in these two cases. $\qquad\square$

Exercise 761: The proof given here follows that in [467] or [90]. For each $n \geq 1$, let $S(n)$ be the statement that for any $a \in (0,1]$,

$$
\text{Prob}(M_a > n) = \frac{a^n}{n!}.
$$

BASE STEP: When $n = 1$,

$$
\text{Prob}(M_a > 1) = \text{Prob}(X_1 \leq a) = a = \frac{a^1}{1!},
$$

so $S(1)$ is true.

INDUCTIVE STEP: Fix an integer $k \geq 1$, and suppose that $S(k)$ is true. Conditioning on X_1,

$$
\begin{aligned}
\mathrm{Prob}(M_a > k + 1) &= \int_0^1 \mathrm{Prob}(M_a > k + 1 \mid X_1 = x)\, dx \\
&= \int_0^a \mathrm{Prob}(M_a > k + 1 \mid X_1 = x)\, dx \\
&= \int_0^a \mathrm{Prob}(M_{a-x} > k)\, dx \\
&= \int_0^a \frac{(a - x)^k}{k!}\, dx \qquad\qquad \text{(by } S(k) \text{ with } a - x) \\
&= \int_0^a \frac{u^k}{k!}\, du \\
&= \frac{a^{k+1}}{(k+1)!},
\end{aligned}
$$

shows $S(k + 1)$ is true, completing the inductive step.

By mathematical induction, for each $n \geq 1$, the statement $S(n)$ is true, so the Exercise is solved for any fixed a. □

Exercise 762: This exercise is adapted from [411, Problem 4].

Let x be the probability desired. The colony can last forever only if the microbe first splits, and only if at least one of the daughters leads to an everlasting colony. Thus one needs to solve $x = p(1 - x)^2$. The solutions to this quadratic are $x = 0$ and $x = 2 - \frac{1}{p}$. If $p \leq 1/2$, then $p - \frac{1}{2} \leq 0$. Assume $p > \frac{1}{2}$; there are still two possible solutions.

So, let p_n be the probability that the population lasts n generations. Then

$$
p_{n+1} = p(1 - (1 - p_n)^2).
$$

Since probabilities p_n are decreasing (they are dependent), it suffices to show that for each n,

$$
p_n > 2 - \frac{1}{p}.
$$

This is done by induction. Since $p_1 = 1$, the base case is trivial. For the inductive hypothesis, for some $k \geq 1$, assume that

$$
p_k \geq 2 - \frac{1}{p}.
$$

Then

$$
p_{k+1} \;=\; p(1 - (1 - p_k)^2)
$$

$$> \; p \left(1 - \left[1 - \left(2 - \frac{1}{p} \right) \right]^2 \right) \quad \text{(by ind. hyp.)}$$

$$= \; p \left(1 - \left[1 - \frac{2p-1}{p} \right]^2 \right)$$

$$= \; p \left(1 - \left[\frac{1-p}{p} \right]^2 \right)$$

$$= \; p \left(\frac{p^2 - (1-p)^2}{p^2} \right)$$

$$= \; \frac{p^2 - (1 - 2p + p^2)}{p}$$

$$= \; \frac{2p - 1}{p}$$

$$= \; 2 - \frac{1}{p}.$$

This completes the inductive step.

Hence, by induction, for each $n \geq 1$, the result $p_n > 2 - \frac{1}{p}$ holds as required. $\quad\square$

Exercise 763: Let $\tau \geq 0$ also satisfy

$$\tau = \sum_{j=0}^{\infty} \tau^j P_j.$$

For each $n \geq 0$, let $S(n)$ be the statement that $\text{Prob}(X_n = 0) \leq \tau$.
BASE STEP: For $n = 0$, $\text{Prob}(X_0 = 0) = 0 \leq tau$, so $S(0)$ is true. Also, for $n = 1$,

$$\text{Prob}(X_1 = 0) = P_0 = \tau^0 P_0 \leq \sum_{j=0}^{\infty} \pi^j P_j = \tau,$$

so $S(1)$ is also true.

INDUCTIVE STEP: Fix $k \geq 0$ and assume that $S(k)$ is true. Then

$$\text{Prob}(X_{k+1} = 0) = \sum_{j=0}^{\infty} \text{Prob}(X_{k+1} = 0 \mid X_i = j) P_j$$

$$= \sum_{j=0}^{\infty} [\text{Prob}(X_k = 0)]^j P_j$$

$$\leq \sum_{j=0}^{\infty} \tau^j P_j \qquad\qquad (\text{by } S(k))$$

$$= \tau,$$

proving $S(k+1)$.

By mathematical induction, for each $n \geq 0$, $\text{Prob}(X_n = 0) \leq \tau$.

Taking limits,

$$\tau \geq \lim_{n\to\infty} \text{Prob}(X_n = 0) = \text{Prob}(\text{population dies out}) = \pi.$$

So indeed, π is the smallest number satisfying the equality in the theorem.   □

Exercise 764: (Outline) Induct on $a + b \geq 1$. For each $a > 0$, $N(a, 0) = 1$, and for each b, $N(b, b) = 0$. Consider the two possibilities for the last vote. If the last vote is for A, there are $N(a - 1, b)$ such distributions; if the last vote is for B, then there are $N(a, b - 1)$ such distributions. Hence the recursion

$$N(a, b) = N(a - 1, b) + N(a, b - 1).$$

Applying the inductive hypothesis to each of the terms on the right-hand side, and using the identity $\binom{n-1}{k-1} = \frac{k}{n}\binom{n}{k}$,

$$
\begin{aligned}
N(a, b) &= \frac{a - 1 - b}{a - 1 + b}\binom{a - 1 + b}{a - 1} + \frac{a - (b - 1)}{a + b - 1}\binom{a + b - 1}{a} \\
&= \frac{a - 1 - b}{a - 1 + b}\frac{a}{a + b}\binom{a + b}{a} + \frac{a - b + 1}{a + b - 1}\frac{b}{a + b}\binom{a + b}{b} \\
&= \frac{(a - 1 - b)a + (a - b + 1)b}{(a + b - 1)(a + b)}\binom{a + b}{a} \\
&= \frac{a - b}{a + b}\binom{a + b}{a},
\end{aligned}
$$

as desired.   □

Exercise 768: For $\alpha + \beta = n \geq 2$, let $S(n)$ be the statement that stakes should be divided between B and A in the ratio

$$\sum_{i=0}^{\alpha-1}\binom{n-1}{i} \;:\; \sum_{j=\alpha}^{n-1}\binom{n-1}{j}.$$

or that A's share is

$$\frac{P}{2^{n-1}}\sum_{j=\alpha}^{n-1}\binom{n-1}{j}$$

and B's share is

$$\frac{P}{2^{n-1}} \sum_{i=0}^{\alpha-1} \binom{n-1}{i}.$$

BASE CASES ($n = 2, 3$): For the situation $n = 2 = \alpha + \beta$, each player lacks one point, (*i.e.*, $\alpha = \beta = 1$) and so the so each is as likely to win as the other, so the share should be divided equally. $S(2)$ says that the shares should be divided in the ratio $\binom{1}{0} : \binom{1}{1}$, namely the ratio $1 : 1$, so $S(2)$ is true.

Although the base case $n = 2$ has been shown, it is instructive to examine the case $n = 3$. For the case $n = 3$, either $\alpha = 1$ and $\beta = 2$ or $\alpha = 2$ and $\beta = 1$. Without loss of generality, suppose that $\alpha = 1$. If the play were to continue one more point and A wins, he is entitled to all of P, however if A loses, by virtue of $S(1)$, he is entitled to $P/2$. Thus, since A is as likely to win or lose the next point, A is entitled to $\frac{P + P/2}{2} = \frac{3}{4}P$ and so the ratio should be $1 : 3$. Indeed, $S(2)$ says the ratio should be

$$\binom{2}{0} : \left(\binom{2}{1} + \binom{2}{2} \right),$$

the ratio $1 : 3$, so $S(3)$ is true.

INDUCTIVE STEP ($S(m) \to S(m+1)$): Fix $m \geq 3$, and assume that if A is lacking α and B is lacking β, and $\alpha + \beta = m$, the share should be divided between B and A in the ratio

$$\sum_{i=0}^{\alpha-1} \binom{n-1}{i} : \sum_{j=\alpha}^{m-1} \binom{n-1}{j}.$$

Assume that A is lacking k and B is lacking ℓ, where $k + \ell = m + 1$. If the play were to continue and A wins the next point, then A would lack $k - 1$ points and B would lack ℓ points and $(k - 1) + \ell = m$, so the inductive hypothesis would apply with $\alpha = k - 1$ and $\beta = \ell$, giving a ratio

$$\sum_{i=0}^{k-2} \binom{m-1}{i} : \sum_{j=k-1}^{m-1} \binom{m-1}{j},$$

that is, B's share would be

$$\frac{P}{2^{m-1}} \sum_{i=0}^{k-2} \binom{m-1}{i}.$$

But if A were to lose, then B would win and the inductive hypothesis would hold with $\alpha = k$, giving B's share to be

$$\frac{P}{2^{m-1}} \sum_{i=0}^{k-1} \binom{m-1}{i}.$$

The arithmetic mean of these two figures for B's share is

$$\frac{P}{2^m} \left[\sum_{i=0}^{k-2} \binom{m-1}{i} + \sum_{i=0}^{k-1} \binom{m-1}{i} \right]$$

$$= \frac{P}{2^m} \left[2 \left(\sum_{i=0}^{k-2} \binom{m-1}{i} \right) + \binom{m-1}{k-1} \right]$$

$$= \frac{P}{2^m} \left[\binom{m-1}{0} + \sum_{i=0}^{k-2} \left(\binom{m-1}{i} + \binom{m-1}{i+1} \right) \right]$$

$$= \frac{P}{2^m} \left[\binom{m-1}{0} + \sum_{i=0}^{k-2} \binom{m}{i+1} \right] \quad \text{(by Pascal's identity)},$$

$$= \frac{P}{2^m} \left[\binom{m}{0} + \sum_{i=1}^{k-1} \binom{m}{i} \right],$$

precisely what $S(m+1)$ says B's share should be. This completes the inductive step.

Hence, by MI, the statement $S(n)$ is true for all $n \geq 2$. □

Exercise 769: The claim in the proof of Local Lemma was: For any $S \subsetneq [n]$ and any $i \notin S$,

$$P \left(A_i \mid \bigwedge_{j \in S} \overline{A_j} \right) \leq x_i.$$

As suggested in the exercise, the proof is by induction on $|S|$.

BASE STEP: For $|S| = 0$, considering the vacuous product to be 1,

$$P \left(A_i \mid \bigwedge_{j \in S} \overline{A_j} \right) = P(A_i) \leq x_i \prod_{j \in S} (1 - x_j) = x_i,$$

and so the claim is true when $|S| = 0$.

INDUCTION STEP: Fix some S and assume that the claim holds for all smaller S'. Fix some $i \notin S$. Put $S_1 = N(i) \cap S$ (the neighborhood of i in G). and $S_2 = S \backslash S_1$. Then (with some reasons given below due to room constraints),

$$P \left(A_i \mid \bigwedge_{j \in S} \overline{A_j} \right) = P \left(A_i \mid \left(\bigwedge_{j \in S_1} \overline{A_j} \right) \wedge \left(\bigwedge_{k \in S_2} \overline{A_k} \right) \right)$$

$$= \frac{P\left(A_i \wedge (\bigwedge_{j\in S_1}\overline{A_j})\,\big|\,\bigwedge_{k\in S_2}\overline{A_k}\right)}{P\left(\bigwedge_{j\in S_1}\overline{A_j}\,\big|\,\bigwedge_{k\in S_2}\overline{A_k}\right)} \quad \text{(by Lemma 22.6.1)}$$

$$\leq \frac{P\left(A_i\,\big|\,\bigwedge_{k\in S_2}\overline{A_k}\right)}{P\left(\bigwedge_{j\in S_1}\overline{A_j}\,\big|\,\bigwedge_{k\in S_2}\overline{A_k}\right)}$$

$$= \frac{P(A_i)}{P\left(\bigwedge_{j\in S_1}\overline{A_j})\,\big|\,\bigwedge_{k\in S_2}\overline{A_k}\right)}$$

$$\leq \frac{x_i\,\prod_{\{i,j\}\in E}(1-x_j)}{P\left(\bigwedge_{j\in S_1}\overline{A_j})\,\big|\,\bigwedge_{k\in S_2}\overline{A_k}\right)}.$$

(The third and fourth line above follow by the reasons: no control over dependent events, so ignore them; because A_i is independent of A_j when $j \in S_2$.)

If $S_1 = \emptyset$, the denominator is equal to 1, and the claim holds. Without loss of generality, let $S_i = \{1, \ldots, r\}$, $r \geq 1$, and $S_2 = \{r+1, r+2, \ldots, s\}$, where $i \notin S_1 \cup S_2$. Rewriting above,

$$P\left(A_i\,\Big|\,\bigwedge_{j\in S}\overline{A_j}\right) \leq \frac{x_i\displaystyle\prod_{\{i,j\}\in E}(1-x_j)}{P\left(\overline{A_1}\wedge\overline{A_2}\wedge\cdots\wedge\overline{A_r}\,\Big|\,\bigwedge_{k=r+1}^{s}\overline{A_k}\right)}. \tag{40.1}$$

Applying $P(A \wedge B \mid C) = P(A \mid B \wedge C) \cdot P(B \mid C)$ recursively, the denominator above becomes

$$P\left(\overline{A_1}\,\Big|\,\bigwedge_{k=2}^{r}\overline{A_k}\wedge\bigwedge_{k=r+1}^{s}\overline{A_k}\right)P\left(\bigwedge_{j=2}^{r}\overline{A_j}\,\Big|\,\bigwedge_{k=r+1}^{s}\overline{A_k}\right)$$

$$= P\left(\overline{A_1}\,\Big|\,\bigwedge_{k=2}^{s}\overline{A_k}\right)P\left(\overline{A_2}\,\Big|\,\bigwedge_{k=3}^{s}\overline{A_k}\right)P\left(\bigwedge_{j=3}^{r}\overline{A_j}\,\Big|\,\bigwedge_{k=r+1}^{s}\overline{A_k}\right)$$

$$\vdots$$

$$= P\left(\overline{A_1}\,\Big|\,\bigwedge_{k=2}^{s}\overline{A_k}\right)P\left(\overline{A_2}\,\Big|\,\bigwedge_{k=3}^{s}\overline{A_k}\right)\cdots P\left(\overline{A_r}\,\Big|\,\bigwedge_{k=r+1}^{s}\overline{A_k}\right)$$

$$= \left(1-P\left(A_1\,\Big|\,\bigwedge_{k=2}^{s}\overline{A_k}\right)\right)\left(1-P\left(A_2\,\Big|\,\bigwedge_{k=3}^{s}\overline{A_k}\right)\right)\cdots\left(1-P\left(A_r\,\Big|\,\bigwedge_{k=r+1}^{s}\overline{A_k}\right)\right)$$

Applying the induction hypothesis for each $k = 1, \ldots, r$ with $S' = S\backslash[k]$, the denominator in (40.1) is bounded below by

$$(1 - x_1)(1 - x_2) \cdots (1 - x_r) = \prod_{j \in S_1} (1 - x_j) \geq \prod_{\{i,j\} \in E} (1 - x_j)$$

because $N(i) \backslash S_1$ might be non-empty, and each $x_j \geq 0$ so $(1 - x_j) \leq 1$. Thus,

$$P \left(A_i \, \middle| \, \bigwedge_{j \in S} \overline{A_j} \right) \leq x_i.$$

This proves the claim for S, finishing the inductive step.

By mathematical induction, for all $S \subsetneq [n]$ and $i \notin S$, the claim is true. $\qquad \square$

Part IV

Appendices

Appendix A: ZFC axiom system

In Section 2.5, the theory of natural numbers from Peano's axioms is developed. The arithmetic that arises is called Peano arithmetic, denoted PA. Gödel's incompleteness theorem says that if PA is consistent, then the consistency of PA is not provable (using a finite sequence of mechanical steps) from Peano's axioms. A larger, somehow richer axiomatic system is necessary. This leads to the popular axiomatic system of set theory called ZFC (from which it is provable that Peano arithmetic is consistent!). See [95] or any of many recent standard logic books for more discussion.

Axioms for set theory rely on two abstract concepts, "element" and "set", and a binary relation $\in$ denoting "is a member of". Usually, one uses capital letters to denote sets and small letters to denote elements, and writes $x \in A$ to denote "the element x is a member of the set A". The usage of upper and lower case in these situations is often a bit too restrictive, since, for example, one might have a set whose elements are themselves sets. In set theory, a collection of axioms was assembled by Ernst Zermelo (and apparently improved by suggestions from Fraenkel). Depending on the source, this collection varies considerably (compare, *e.g.*, Cameron's list of ten in [95], with Jech's list of nine in [289]). Presented here is the list of ten based on [95]. The first nine form what is known as the Zermelo-Fraenkel system, denoted ZF, and together with the tenth, the Axiom of Choice, the system is denoted by ZFC. See, for example, [95] or for discussion as to what each entails.

ZF1 (Extensionality) If two sets have the same members, they are equal.

ZF2 (Empty set) There exists a set $\emptyset$ with no members.

ZF3 (Pairing) If A and B are sets, then there is a set $\{A, B\}$ whose members are precisely A and B.

ZF4 (Union) If A is a set, there is a set $\cup A$ whose members are the members of members of A.

ZF5 (Infinity) There is a set A such $\emptyset$ is a member of A, and if $x \in A$, then so too $\{x\} \in A$. (Such a set has to be infinite.)

ZF6 (Power set) If A is a set, there is a set $\mathcal{P}(A)$ whose members are all subsets of A.

ZF7 (Selection, Separation, or Specification) If A is a set and $S(x)$ is a sentence (with only one free variable x) that is either true or false for any $x \in A$, then there exists a set B consisting of all elements in A for which S is true.

ZF8 (Replacement) If f is a function, then for any X, there exists a set $Y = \{F(x) : x \in X\}$. (See Exercise 592 for a different formulation.)

ZF9 (Foundation) Every set A contains a member B that is disjoint from A. (This prevents Russell's paradox of "the set of all sets".)

In 1939, Kurt Gödel [217](1885-1978) showed that AC is consistent with ZF, (that is, adding AC to ZF did not lead to contradictions) and in 1963, Paul Cohen [112] proved that the negation of AC is also consistent with ZF. In this sense, AC is independent of ZF. In this book, as might be expected, ZFC is assumed.

Appendix B: Inducing you to laugh?

In many mathematics journals and popular journals, there often appear quips that use the principle of induction in humor.

Here is one that appeared in [429, p. 11] (attributed to Quantum Seep):

> *My mother is a mathematician, so she knows how to induce good behavior. "If I've told you n times, I've told you $n + 1$ times..."*

Here is another (found in, *e.g.*, [429, p. 7]) old joke employing well-ordering:

Statement: *All positive integers are interesting.*

Proof: *If not all positive integers are interesting, then the set of all non-interesting integers (by well-ordering) has a least integer, making it interesting—a contradiction.*

A sister joke is the following: [This is not original, but I cannot recall a reference.]
Statement: *All integers are boring.*

Proof: If not, there would be a least not-boring integer. Who cares?

This next one is so old that it seems as if most mathematicians have heard it [but I can not track down the original source].

Question: *What's yellow and equivalent to the Axiom of Choice?*

Answer: *Zorn's Lemon.*

Here is another whose origin is unclear:

> *Induction doesn't amount to a hill of beans (because 1 bean is not a hill, and if n beans did not form a hill, then neither does $n + 1$).*

In [110] Richard Cleveland gave some limericks named with Zermelo's axioms. Here are a few relevant ones:

1. *Extensionality*
We assume that our sets are extensional,
As opposed to their being intensional,
So the name of the game
Is that sets are the same
If they have the same members—that's sensible!

9. *Choice*
Zermelo had one more in store,
And this one we mustn't ignore:
By which anyone may
Take a messy array
And make it well-ordered once more.

OR

9. *Choice*
There once was maiden named Emma,
Who had a peculiar dilemma:
She had so many beaus,
That to choose, heaven knows,
She had to appeal to Zorn's lemma.

The following quote is from Edgar Allen Poe, as in [183, p.289], about two writers, Cornelius Mathews (referred to here as Mr. M.) and Willeam Ellery Channing [Mr. C.]:

> To speak algebraically: Mr. M. is execrable, but Mr. C. is $(x + 1)$=ecrable.

This next quip is from Martin Gardner as it appeared in *Penrose tiles to trapdoor ciphers*[1] [214, p. 143].

> Because mathematical induction often takes the from of "reducing to the preceding case," I close with an old joke. For a college freshman who cannot decide between physics and mathematics as his major subject the following two-part test has been devised. In the first part the student is taken to a room that contains a sink, a small stove with one unlighted burner and an empty kettle on the floor. The problem is to boil water. The student passes this part of the test if he fills the kettle at the sink, lights the burner and puts the kettle on the flame.
>
> For part two the same student is taken to the same room, but now the kettle is filled and on the unlighted burner. Again the problem is to boil water. The potential physicist simply lights the burner. The

---

[1]Used with permission; Copyright 1997 by the Mathematical Association of America, Washington, DC.

potential mathematician first empties the kettle and puts it on the floor. This reduces the problem to the preceding case, which he has already solved.

Appendix C: The Greek alphabet

lower case	upper case	name
α	A	alpha
β	B	beta
γ	Γ	gamma
δ	Δ	delta
ϵ	E	epsilon
ζ	Z	zeta
η	H	eta
θ	Θ	theta
ι	I	iota
κ	K	kappa
λ	Λ	lambda
μ	M	mu
ν	N	nu
ξ	Ξ	xi
o	O	omicron
π	Π	pi
ρ	P	rho
σ	Σ	sigma
τ	T	tau
υ	Υ	upsilon
ϕ	Φ	phi
χ	X	chi
ψ	Ψ	psi
ω	Ω	omega

References

[*]   **NOTE:** Bold numbers following each entry are page numbers in this book where the entry is cited. **xxiii**

[1]   N. H. Abel, Beweis eines Ausdrucks von welchem die Binomial-Formel ein einzelner Fall ist, *J. Reine Angew. Math.* (also called "Crelle's journal") **1** (1826), 159-160. **141**

[2]   H. G. Abeledo and U. G. Rothblum, Courtship and linear programming, *Linear algebra and its applications* **216** (1995), 111–124. **252**

[3]   F. Acerbi, Plato: Parmenides 149a7-c3, A proof by Complete Induction?, *Archive for History of Exact Sciences*, **55**(1) (2000), 57–76. **12**

[4]   A. D. Aczel, *Probability 1, Why there must be intelligent life in the universe*, Harcourt Brace & Company, New York, 1998. **146, 396**

[5]   J. Aczel, Ungleichungen und ihre Verwendung zur elementaren Lösung von Maximum und Minimumaufgaben, *L'Enseignment Math.*, **7** (1961), 214–249. **39**

[6]   M. Aigner, *Combinatorial theory*, (the 1979 edition, Vol 234 of "Grundlehren der mathematischen Wissenschaften") reprinted in "Classics in mathematics", Springer-Verlag, Berlin, Heidelberg, New York, 1997. **17, 210, 631, 633, 641, 643, 649, 768, 769**

[7]   M. Aigner and G. M. Ziegler, *Proofs from The Book (2nd ed.)*, Springer, 2001. [A 3rd edition is now available.] **81, 82, 403, 422, 685, 696**

[8]   M. O. Albertson and J. P. Hutchinson, *Discrete mathematics with algorithms*, John Wiley & Sons, 1988. **16**

[9]   B. Aldershoff and O. M. Carducci, Stable matchings with couples, *Discrete Applied Mathematics* **68** (1996), 203–207. **252**

[10] G. L. Alexanderson, R. M. Grassl, and A. P. Hillman, *Discrete and combinatorial mathematics*, Collier MacMillan Publishers, London, 1987. **16**

[11] R. E. Allardice and A. Y. Fraser, La Tour d'Hanoi, *Proc. Edinb. Math. Soc.* **2** (1884), 50–53. **276**

[12] J. P. Allouche and J. Shallit, The ring of k-regular sequences, *Theoret. Comput. Sci.* **98** (1992), 163–197. **277**

[13] N. Alon and J. Spencer, *The probabilistic method*, 2nd ed., Wiley-Interscience Series in Discrete Mathematics and Optimization, John Wiley & Sons, 2000. **385, 397**

[14] R. Alter, Some remarks and results on Catalan numbers, *Proc. 2nd Louisianna Conf. Comb. Graph Th. and Comput.* (1971), 109–132. **203**

[15] R. Alter and K. K. Kubota, Prime and prime power divisibility of Catalan numbers, *J. Combin. Theory, Ser. A* **15** (1973), 243–256. **204**

[16] S. Althoen, Towers of Hanoi revisited, *Coll. Math. J.* **40** (no. 3, May 2009), 225. **277, 714**

[17] E. Altman, On a problem of P. Erdős, *Amer. Math. Monthly* **70** (1963), 148–157. **349**

[18] T. Andreescu and B. Enescu, *Mathematical Olympiad Treasures*, Birkhäuser, Boston, 2004. **706**

[19] G. E. Andrews, *Number Theory*, Dover Publications, New York, 1994. [Originally published by W. B. Saunders Company, Philadelphia, 1971.] **161, 557**

[20] T. M. Apostol, "The principle of mathematical induction" in *Calculus, (2nd ed.) Vol 1, One-variable calculus, with an introduction to linear algebra*, Blaisdell, Waltham Ma, 1967, p. 34. **17**

[21] K. Appel and W. Haken, Every planar map is four colorable, Part I: Discharging, *Illinois J. of Math.* **21** (1977), 429–90. **255**

[22] K. Appel, W. Haken, J. Koch, Every planar map is four colorable, Part II: Reducibility, *Illinois J. of Math.* **21** (1977), 491–567. **255**

[23] J. W. Archibald, *Algebra*, 4th ed., Pitman Paperbacks, 1970. **321**

[24] R. G. Archibald, Goldbach's theorem, *Scripta Mathematica* **3** (1955), 44–50 and 153–161. **83**

[25] I. K. Argyros, On Newton's method and nondiscrete mathematical induction, *Bull. Austral. Math. Soc.* **38** (1988), no. 1, 131–140. **18**

[26] I. K. Argyros, On the secant method and nondiscrete mathematical induction, *Anal. Numér. Théor. Approx.* **18** (1989), no. 1, 27–28. **18**

[27] I. K. Argyros, Newton-like methods and nondiscrete mathematical induction, *Studia Sci. Math. Hungar.* **28** (1993), no. 3–4, 417–426. **18**

[28] R. Aubin, Mechanizing structural induction, *Theoret. Comput. Sci.* **9** (1979), 329–362. **18**

[29] R. Audi, (general editor), *The Cambridge dictionary of philosophy*, Cambridge University Press, 1995. **23**

[30] D. F. Bailey, Counting arrangements of 1's and -1's, *Math. Mag.* **69** (1996), 128–131. **203**

[31] D. Bailey and J. Borwein, *Mathematics by experiment: Plausible reasoning in the 21st century*, 2nd ed., A. K. Peters, Ltd., Wellesley, Massachusetts. **784**

[32] B. Baker and R. Shostak, Gossips and telephones, *Discrete Math.* **2** (1972), 191–193. **728**

[33] V. K. Balakrishnan, *Introductory discrete mathematics*, Prentice Hall, 1991. **16**

[34] W. W. R. Ball and H. S. M. Coxeter, *Mathematical Recreations and Essays*, 13th ed., Dover Publications, 1987. **728**

[35] I. Bárány, A short proof of Kneser's conjecture, *J. Combin. Theory Ser. A* **25** (1978), 325–326. (1978). **221**

[36] E. Barbeau (editor), Fallacies, flaws, and flimflam, *The College Math. J.* **22** (1991), 133. **73**

[37] É. Barbier, Generalisation du probleme resolu par M. J. Bertrand [Generalization of a problem solved by Mr. J. Bertrand], *Comptes Rendus de l'Academie des Sciences*, Paris **105** (1887), p. 407. **395**

[38] W. Barnier and J. B. Chan, *Discrete mathematics*, West Publishing Company, 1989. **16**

[39] J. Barwise, Scenes and other situations, *J. Philosophy* **78** (1981), 369–397. **298**

[40] L. M. Batten, *Combinatorics of finite sets*, 2nd ed., Cambridge University Press, 1997. **363**

[41] J. D. Beasley *The mathematics of games*, Oxford University Press, Oxford and New York, 1989. [Paperback edition released in 1990.] **289**

[42] A. Beck, M. N. Bleicher, and D. W. Crowe, *Excursions into mathematics*, Worth Publishers, Inc., 1969. **36, 176, 276, 291, 292, 717**

[43] E. F. Beckenbach and R. Bellman, *Inequalities*, Springer-Verlag, 1965. **39**

[44] E. T. Bell, Gauss, Prince of Mathematicians, in, *Men of Mathematics*, Simon and Schuster, Inc., 1937. **8**

[45] E. T. Bell, Gauss, Prince of Mathematicians, in *The world of mathematics*, vol. 1 (J. R. Newman, ed.), p. 298, 1956. **8**

[46] R. E. Bellman and K. L. Cooke, *Differential-difference equations*, Academic Press, 1963. **316**

[47] A. Benjamin and J. Quinn, *Proofs that really count*, MAA, 2003. **17**

[48] M. de Berg, O. Cheong, M. van Kreveld, and M. Overmars, *Computational geometry: algorithms and applications*, Springer, 2008. **287**

[49] E. R. Berlekamp, J. H. Conway, and R. K. Guy, *Winning ways for your mathematical plays*, 2 Vols, Academic Press, London, 1982. **291**

[50] J. Bertrand, Solution d'un problem [Solution of a problem], *Comptes Rendus de l'Academie des Sciences*, Paris **105** (1887), p. 369. **394**

[51] G. S. Bhat and C. D. Savage, Balanced Gray codes, *Electronic Journal of Combinatorics*, **3** (1996), # R25, 11 pages. **281**

[52] N. L. Biggs, *Discrete mathematics*, revised ed., Oxford Science Publications, Clarendon Press, Oxford, 1994. **16**

[53] J. Binet, Mémoire sur l'intégration des équations linéaires aux différences finies d'un ordre quelconque à coefficients variables, *Comptes Rendus hebdomadaires des séances de l'Académie des Sciences* (Paris) **17** (1843), 559–567. **196**

[54] T. Bisztriczky, K. Böröczky, Jr., and D. S. Gunderson, Cyclic polytopes, hyperplanes, and Gray codes, *J. Geom.* **78** (2003), 25–49. **281**

[55] K. P. Bogart, *Discrete Mathematics*, D. C. Heath and Company, 1988. **16**

[56] J. Boleman, *Physics: an introduction*, Prentice–Hall, 1985. **1**

[57] B. Bollobás, On generalized graphs, *Acta Math. Acad. Sci. Hungar.* **16** (1965), 447–452. **221**

[58] B. Bollobás, *Combinatorics: set systems, hypergraphs, families of vectors and combinatorial probability*, Cambridge University Press, 1986. **17, 94, 220, 353, 661, 782**

[59] B. Bollobás, *Modern graph theory*, Graduate texts in mathematics **184**, Springer, New York, 1998. **239, 369, 660, 676**

[60] B. Bollobás, *Random Graphs*, 2nd ed., Cambridge studies in advanced mathematics **73**, Cambridge University Press, 2001. **771**

[61] B. Bollobás, *The art of mathematics: coffee time in Memphis*, Cambridge University Press, 2006. **17, 353, 782**

[62] M. Bóna, *Combinatorics of permutations*, Chapman & Hall/CRC, Boca Raton, 2004. **210, 212**

[63] J. A. Bondy, Basic graph theory: Paths and circuits, in *Handbook of Combinatorics* Vol. 1, pp. 3–110, R. Graham, M. Grötschel and L. Lovász eds., Elsevier Science, 1995. **248**

[64] J. A. Bondy and U. S. R. Murty, *Graph theory*, Graduate texts in mathematics, Springer, 2008. **239, 691**

[65] R. D. Borgersen, *Topics in finite graph Ramsey theory*, Master's thesis, U. of Manitoba, 2007, *xiii*+247 pages. Available at http://mspace.lib.umanitoba.ca/dspace/handle/1993/2998. **386**

[66] A. Born, C. A. J. Hurkens, and G. J. Woeginger, The Freudenthal problem and its ramifications (Part III), *Bulletin of the EATCS* **95** (June 2008), 201–219. **295, 300**

[67] J. Borwein and D. Bailey, *Mathematics by experiment: plausible reasoning in the 21st century*, 2nd ed., A. K. Peters, Ltd., Wellesley, Massachusetts, 2008. **88**

[68] A. Bouhoula, E. Kounalis, and M. Rusinowitch, Automated mathematical induction, *J. Logic Comput.* **5** (1995), no. 5, 631–668. **18**

[69] N. Bourbaki, *The set theory*, Russian translation, MIR, Moscow, Edited V. A. Uspensky, p. 325, 1965. [*Theory of Sets*, is also published by Hermann, 1968, a translation of *Théorie des ensembles*, also published by Springer in 2004.] **12**

[70] C. B. Boyer *A History of Mathematics*, Revised by Uta C. Merzbach, John Wiley & Sons, 1989. **13**

[71] D. M. Bradley, More on teaching induction, letters to the editor, *MAA Focus* **28**, No. 7 (October 2008), p. 9. **103, 679**

[72] L. Brenton and A. Vasiliu, Znam's problem, *Math. Mag.* **75** (1) (Feb. 2002), 3–11. **177**

[73] C. Brezinski, *History of continued fractions and Padé approximants*, Springer-Verlag, New York, 1991. **179**

[74] D. S. Bridges, *Foundations of real and abstract analysis*, Graduate texts in mathematics **174**, Springer-Verlag, Inc., New York, 1998. **345, 739**

[75] R. L. Brooks, On colouring the nodes of a network, *Proc. Cambridge Phil. Soc.* **37** (1941), 194–197. **252**

[76] E. Brown and J. Tanton, A dozen hat problems, *Math Horizons*, April 2009, 22–25. [Published by the MAA.] **300**

[77] R. A. Brualdi, *Introductory combinatorics*, 2nd ed., Prentice-Hall Inc., Englewood Cliffs, NJ (USA), 1992. **17, 608**

[78] R. A. Brualdi, *Introductory combinatorics*, 3rd ed., Elsevier, New York, 1997. **17, 203**

[79] N. G. de Bruijn and P. Erdős, On a combinatiorial [sic] problem, *Indagationes Math.* **10** (1948), 421–423. **358, 788**

[80] N. G. de Bruijn and P. Erdős, A colour problem for infinite graphs and a problem in the theory of relations, *Indag. Math.* **13** (1951), 369–373. **253**

[81] C. Brumfiel, Using a game as a teaching device, *Mathematics Teacher* **67** (1974), 386–391. **300**

[82] R. C. Buck, Mathematical induction and recursive definitions, *Amer. Math. Monthly* **70** (2) (1963), 128–135. **18**

[83] F. Buckley and J. C. Molluzzo, *A first course in discrete mathematics*, Wadsworth Publishing Company, Belmont, CA, 1986. **16**

[84] R. T. Bumby, A problem with telephones, *SIAM Disc. Math.* **2** (1981), 13–18. **728**

[85] R. T. Bumby, F. Kochman, and D. B. West (eds.), Problems and solutions, *Amer. Math. Monthly* **102** (9) (1995), 843–844. **373, 801**

[86] B. H. Bunch, *Mathematical fallacies and paradoxes*, Van Nostrand Reinhold Company, 1982. **69**

[87] A. Bundy, The automation of proof by mathematical induction, in J. A. Robinson, A. Voronkov (eds), *Handbook of Automated Reasoning*, vol. 1, Elsevier and MIT press, Cambridge MA, 2001, pp. 845–912. **18**

[88] E. B. Burger and M. Starbird, *Coincidences, chaos, and all that math jazz*, W. W. Norton & Company, New York and London, 2005. **202**

[89] D. M. Burton, *The History of Mathematics—An Introduction*, Dubuque, Iowa, William C. Brown, 1988. **13**

[90] L. E. Bush, The William Lowell Putnam mathematical competition, *Amer. Math. Monthly* **68** (January 1961), 18–33. **258, 392, 700, 805**

[91] W. H. Bussey, Origins of mathematical induction, *Amer. Math. Monthly* **24** (1917), 119–207. **11, 13, 18, 126, 136, 320, 321, 396, 426, 428**

[92] F. Cajori, Origin of the name "mathematical induction", *Amer. Math. Monthly* **25** (5) (1918), 197–201. **13**

[93] D. Campbell, The computation of Catalan numbers, *Math. Mag.* **57** (1984), 195–208. **203**

[94] P. J. Cameron, *Combinatorics: topics, techniques, algorithms*, Cambridge University Press, 1994 (reprinted in 1996 with corrections). **17, 145**

[95] P. J. Cameron, *Sets, logic and categories*, Springer Undergraduate mathematics series, Springer-Verlag, New York, 1998. **16, 31, 56, 67, 310, 410, 415, 653, 655, 770, 815**

[96] C. Caratheodory, Über den Variabilitätsbereich der Koeffizienten von Potenzreihen, die gegebene Werts nicht annahmen, *Math. Ann.* **64** (1907), 95–115. **352**

[97] J.-D. Cassini, Une nouvelle progression de nombres, *Histoire de l'Académie Royale des Sciences*, Paris, Volume 1, (1733), 201. [Proceedings from 1680] **193**

[98] I. G. Chang and T. W. Sederberg, *Over and over again*, MAA, Washington D.C., 1997. **304**

[99] G. Chartrand, "The tower of Hanoi puzzle," in *Introductory graph theory*, Dover, New York, 1985 (pp. 135–139). **277**

[100] N. G. Chebotarëv, Several problems of the various divisions of algebra and number theory, *Uspekhi matematicheskikh nauk* [A Russian journal, "Successes of mathematical sciences", containing primarily surveys] (1938), issue 4, 284–286. **86**

[101] M.-D. Choi, Tricks or treats with the Hilbert matrix, *Amer. Math. Monthly* **90** No. 5, (May 1983), 301–312. **332**

[102] K.-M. Chong, An inductive proof of the A.M.-G.M. inequality, *Amer. Math. Monthly* **83** (1976), 369. **39**

[103] I. Z. Chorneyko and S. G. Mohanty, On the enumeration of certain sets of planted trees, *J. Combin. Theory Ser. B* **18** (1975), 209–221. **203**

[104] G. Chrystal, *Algebra II* (2nd. ed.), London, 1900. **39**

[105] W. Chu, A new combinatorial interpretation for generalized Catalan numbers, *Discrete Math.* **65** (1987), 91–94. **203**

[106] S. C. K. Chu and M-K. Siu, How far can you stick out your neck? *The College Mathematics Journal* **17** (1986), 122–132. **202**

[107] I.-P. Chu and R. Johnsonbaugh, Tiling deficient boards with trominoes, *Math. Mag.* **59** (no. 1) (1986), 34–40. **723, 724**

[108] V. Chvátal, A combinatorial theorem in plane geometry, *J. Combin. Theory Ser. A* **18** (1975), 39–41. **783**

[109] V. Chvátal and D. Sankoff, An upper-bound technique for lengths of common subsequences, Ch. 15 (pp. 353–357) in *Time warps, string edits, and macromolecules: the theory and practice of sequence comparison*, D. Sankoff and J. B. Kruskal eds., Addison-Wesley, Reading, Massachusetts, 1983. **237**

[110] R. Cleveland, The axioms of set theory, *Math. Mag.* **52** (1979), 256–257. [Used with kind permission from MAA: Copyright the Mathematical Association of America 2010. All rights reserved.] **817**

[111] D. H. Cohen, Conditionals, quantification, and strong mathematical induction, *J. Philos. Logic* **20** (1991), no. 3, 315–326. **16**

[112] P. J. Cohen, *Set theory and the Continuum Hypothesis*, W. A. Benjamin, Inc., New York, 1966. **816**

[113] A. M. Colman, Rationality assumptions of game theory and the backward induction paradox, pp. 353–371 in *Rational models of cognition*, M.l Oaksford and N. Chater, eds., Oxford University Press, 1998. **290**

[114] Consortium for Mathematics and Its Applications (U.S.), *For all practical purposes: introduction to contemporary mathematics*, W. H. Freeman and Company, New York, 1996. **38**

[115] J. H. Conway, *On numbers and games*, Academic Press, 1976. **290**

[116] J. H. Conway and R. K. Guy, *The book of numbers*, Springer-Verlag, New York, 1996. **8, 144, 200, 203, 211**

[117] C. Cooper and S. S. So, eds., Roots and coefficients of a quartic function, in Problems and solutions, *Coll. Math. J.* **40** (No.3, May 2009), 219. **320, 770**

[118] R. Cori and D. Lascar, *Mathematical logic: a course with exercises*, Part II, Recursion theory, Gödel's theorems, set theory, model theory, transl. by D. H. Pelletier, Oxford University Press, 2001. **322, 323**

[119] T. H. Cormen, C. E. Leiserson, R. L. Rivest, and C. Stein, *Introduction to algorithms*, 2nd ed., McGraw-Hill Book Company, and MIT Press, 2001. **278, 281, 282, 285, 286, 287**

[120] R. Courant and H. Robbins, *What is mathematics? An elementary approach to ideas and methods*, (2nd ed.) Oxford University Press, Oxford, England, 1996. [First edition published in 1941 has same references.] **17, 83**

[121] H. S. M. Coxeter, A problem of collinear points, *Amer. Math. Monthly* **55** (1948), 26–28. **358**

[122] H. S. M. Coxeter, *Introduction to geometry* (2nd ed.), Wiley, New York, 1969. **193, 784**

[123] H. S. M. Coxeter and S. L. Greitzer, *Geometry revisited*, Mathematical Association of America, Washington, D. C., 1967. **193**

[124] J. P. D'Angelo and D. B. West, *Mathematical thinking: problem solving and proofs*, Prentice-Hall, 2000. **17, 717**

[125] L. C. Dalton, A plan for incorporating problem solving throughout the advanced algebra curriculum, in *The secondary school mathematics curriculum, 1985 Yearbook*, (C. R. Hirsch, M. J. Zweng, eds.), National Council of Teachers of Mathematics, Inc., 1985. **207**

[126] L. Danzer, B. Grünbaum, and V. Klee, Helly's theorem and its relatives, pp. 101–180 in *Convexity*, Proc. Symposia in Pure Maths, vol. VII, Amer. Math. Soc., 1963. **354**

[127] A. Davidson, Forward thinking with "backward induction", article in *Asset securitization report*, 22 March 2004, Source Media, Inc., 2004. **290**

[128] W. Demopoulos, "Introduction", in *Frege's Philosophy*, (W. Demopoulos, ed.), Harvard University Press, Cambridge, Massachusetts, London, England, 1995. **14**

[129] J. Dénes and A. D. Keedwell, *Latin squares and their applications*, English Universities Press Limited, joint edition with Akadémiai Kiadó, Budapest, 1974. **333**

[130] J. Dénes and A. D. Keedwell, *Latin squares, (new developments in the theory and applications)*, Annals of Discrete Mathematics **46**, North-Holland, 1991. **333**

[131] N. Dershowitz and S. Zaks, Enumeration of ordered trees, *Disc. Math.* **31** (1980), 9–28. **203**

[132] B. Descartes, Solution to advanced problem No. 4526, *Amer. Math. Monthly* **61** (1954), 352. **384**

[133] J. M. Deshouillers,   G. Effinger, H. J. J. te Riel, and D. Zinoviev, New experimental results concerning the Goldbach conjecture, *Algorithmic number theory* (Proceedings of the 3rd international symposium, ANTS-III, held at Reed College, Portland OR, June 21–25, 1988, Ed. J. P. Buhler), Springer-Verlag (1990), 204–215. **83**

[134] W. Deuber, Developments based on Rado's dissertation "Studien zur Kombinatorik", *Surveys in Combinatorics, 1989*, J. Siemons ed., (Proc. of 12'th British Comb. Conf.), London Math. Soc. Lecture Notes **141**, Cambridge Univ. Press (1989), 52–74. **378**

[135] P. H. Diananda, A simple proof of the arithmetic mean-geometric mean inequality, *Amer. Math. Monthly*, **667** (1960), 1007. **39**

[136] L. E. Dickson, *History of the theory of numbers*, volume I (of three), Divisibility and Primality, AMS Chelsea Publishing, 1992. Reprinted from the first edition published by The Carnegie Institute of Washington, 1919. **608**

[137] R. Diestel, *Graph theory* (2nd ed.), Graduate texts in mathematics, Springer, 2000. **239**

[138] E. W. Dijkstra, Some beautiful arguments using mathematical induction, *Acta Inform.* **13** (1980), no. 1, 1–8. **18, 560**

[139] E. W. Dijkstra, Mathematical induction and computer science, *Nieuw Arch. Wisk.* (3) **30** (1982), no. 2, 117–123. **16, 18**

[140] M. V. Di Leonardo and T. Marino, Is the least integer principle equivalent to the principle of mathematical induction? (Italian, English, Italian summary), *Quad. Ric. Didatt.* no. 10 (2001), 102–114. **62**

[141] R. P. Dilworth, A decomposition theorem for partially ordered sets, *Ann. of Math.* (2) **51** (1950), 161–166. **224**

[142] J. H. Dinitz and D. R. Stinson, eds., *Contemporary design theory: a collection of surveys*, John Wiley & Sons, Inc., New York, 1992. **363**

[143] G. A. Dirac, Some theorems on abstract graphs, *Proc. Lond. Math. Soc.* **2** (1952), 69–81. **248, 676**

[144] D. Dolev, J. Halpern, and Y. Moses, Cheating Husbands and other stories, originally in an IBM research laboratory report of 1985, reprinted in *Proceedings of the Fourth ACM Conference on Principles of Distributed Computing*, 1985. **300**

[145] H. Doornbos, R. Backhouse, and J. van der Woude, A calculational approach to mathematical induction, *Theoret. Comput. Sci.* **179** (1997), no. 1–2, 103–135. **16**

[146] H. Dörrie, *100 great problems of elementary mathematics: their history and solutions*, Dover, New York, 1965. **41, 203**

[147] J. A. Dossey, A. D. Otto, L. E. Spence, and C. Vanden Eynden, *Discrete Mathematics*, 2nd ed., Harper Collins College Publishers, New York, 1993. **16, 723**

[148] H. Dubner and W. Keller, New Fibonacci and Lucas Primes, *Math. Comput.* **68** (1999), 417–427 and S1–S12. **193**

[149] V. Dubrovsky, Nesting puzzles, Part I: Moving oriental towers, *Quantum* **6** (1996), 53–57 (Jan.) and 49–51 (Feb.). **277**

[150] U. Dudley, *Elementary number theory*, 2nd ed., W. H. Freeman and Company, 1979. **18, 47, 82, 168, 423, 556, 557, 703**

[151] J. Dugundji, *Topology*, Allyn and Bacon, Inc., Boston, 1966. **665**

[152] D. S. Dummit and R. M. Foote, *Abstract algebra*, 2nd ed., John Wiley & Sons, Inc., New York, 1999. **334, 768**

[153] W. Dunham, *Journey through genius: the great theorems of mathematics*, Wiley, New York, 1990. **83**

[154] P. L. Duren, *Theory of H^p spaces*, Dover Publications, 2000. (Originally published by Academic Press, San Diego CA, 1970.) **148**

[155] W. L. Duren Jr., Mathematical induction in sets, *Amer. Math. Monthly* **64** (1957), no. 8, part II, 19–22. **18, 65**

[156] T. E. Easterfield, A combinatorial algorithm, *J. London Math. Soc.* **21** (1946), 219–226. **684**

[157] A. W. F. Edwards, A quick route to sums of powers, *Amer. Math. Monthly* **93** (1986), 451–455. **144**

[158] R. B. Eggleton and R. K. Guy, Catalan strikes again! How likely is a function to be convex? *Math. Mag.* **61** (1988), 211-219. **203**

[159] S. Elaydi, *An introduction to difference equations*, 3rd ed., Springer 2005. **271**

[160] H. B. Enderton, *Elements of set theory*, Academic Press, 1977. **53, 57, 87, 654, 655**

[161] A. Engel, *Problem-solving strategies*, Springer, 1998. **17, 94, 359, 363, 505, 514, 542, 549, 559, 560, 561, 586, 599, 629, 692, 695, 697, 712, 714, 720, 728, 743, 746, 764, 779, 789, 793**

[162] P. Erdős, Problems for solution, number 4065, *Amer. Math. Monthly* **50** (1943), 65. **358, 834**

[163] P. Erdős, On sets of distances of n points, *Amer. Math. Monthly* **53** (1946), 248–250. **349**

[164] P. Erdős, Graph theory and probability, *Canad. J. Math.* **11** (1959), 34–38. **384**

[165] P. Erdős, Extremal problems in graph theory, in *Theory of graphs and its applications*, (Proceedings of the Symposium held in Smolenice in June 1963), 29–36, Publishing House of the Czechoslovak Academy of Sciences, Prague, and Academic Press, New York, 1964. **256**

[166] P. Erdős, On the graph theorem of Turán (in Hungarian), *Mat. Lapok* **48** (1970), 249–251. **256**

[167] P. Erdős, Personal reminiscences and remarks on the mathematical work of Tibor Gallai, *Combinatorica* **2**(3)(1982), 207–212. **358**

[168] P. Erdős, Z. Füredi, R. Gould, and D. S. Gunderson, Extremal graphs for intersecting triangles, *J. Combin. Theory Ser. B* **64** (1995), 89–100. **693**

[169] P. Erdős and T. Gallai, On maximal paths and circuits of graphs, *Acta Math. Acad. Sci. Hung.* **10** (1959), 337–356. **241, 676**

[170] P. Erdős and A. Hajnal, On chromatic number of graphs and set systems, *Acta. Math. Acad. Sci. Hungar.* **17** (1966), 61–99. **384**

[171] P. Erdős and L. Lovász, Problems and results on 3-chromatic hypergraphs and some related results, *Infinite and finite sets* (Hajnal, A., Rado, R., and Sós, V. T., eds) *Coll. Math. Soc. J. Bolyai*, **11** (1975), 609–627. **397**

[172] P. Erdős and R. Steinberg, Three point collinearity, solution to problem 4065, [includes editor's note containing Gallai's solution as submitted by Erdős in original problem in [162].] *Amer. Math. Monthly* **51** (3) (1944), 169–171. **358**

[173] P. Erdős and G. Szekeres, A combinatorial problem in geometry, *Compositio Math.* **2** (1935), 463–470. **368, 370**

[174] P. Erdős and G. Szekeres, On some extremum problems in elementary geometry, *Ann. Univ. Sci. Budapest* **3** (1960), 53–62. **370**

[175] P. Ernest, Mathematical induction: a recurring theme, *Math. Gaz.* **66** (1982), no. 436, 120–125. **12, 14, 18, 162**

[176] J. Espinosa, *Mathematical Induction*, www.math.cl/induction.html. [Click on the .pdf link; some solutions are given by Naoki Sato] **175, 561, 583, 584, 592, 600, 601, 616, 617, 624, 707**

[177] L. Euler, Solutio problematis ad geometriam situs pertinentis, *Comment. Academiae Sci. I. Petropolitanae,* **8** (1736, though appeared in 1741), 128–140. **168**

[178] L. Euler, Methodus universalis series summandi ulterius promota, *Comment. Academiae Sci. I. Petropolitanae,* **8** (1736, though appeared in 1741), 147–158. [Reprinted in *Opera Omnia,* series 1, vol. 14, 124–137.] **209**

[179] L. Euler, Demonstratio Nonnullarum Insignium Proprietatum Quibus Solida Hedris Planis Inclusa Sunt Praedita, *Novi Comm. Acad. Sci. Imp. Petropol* **4** (1758), 140–160. **689**

[180] H. Eves, *An introduction of the history of mathematics,* 5th ed., Saunders College Publishing, New York, 1983. **12, 16, 73, 396**

[181] H. Eves, *Foundations and fundamental concepts of mathematics,* 3rd ed., Dover, 1997. [This is a reprint of 3rd edition originally published by PWS-Kent Publishing Company, Boston, 1990.] **16, 60, 651**

[182] P. Eymard and J.-P. Lafon, *The number π,* (Translated from French by Stephen S. Wilson) American Mathematical Society, Providence, Rhode Island, 2004. **186, 319**

[183] *Fantasia Mathematica,* edited by C. Fadiman, paperback, Simon and Schuster, New York, 1961. [Reprinted from 1958.] **818**

[184] R. Fagin, J. Y. Halpern, Y. Moses and M. Y. Vardi, *Reasoning about knowledge,* MIT Press, 1995. **295, 300**

[185] N. Falletta, *The Paradoxicon,* John Wiley & Sons, Inc., New York, 1983. **70, 415**

[186] J. Faulhaber, *Academia Algebra,* Darinnen die miraculosische Inventiones zu den höchsten Cossen weiters *continuirt* and *profitiert* werden. Augspurg, bey Hohann Ulrich Schönigs, 1631. [This book is rare; one copy is in the University of Cambridge library, and an annotated copy is at Stanford University.] **144**

[187] D. Fearnley-Sander, A LISP prover for induction formulae, *Internat. J. Math. Ed. Sci. Tech.* **18** (July-August 1987), 547–554. **18**

[188] T. Feder, *Stable networks and product graphs*, Mem. Amer. Math. Soc. vol. **555**, 1995, 223 pp. **252**

[189] R. P. Feynman, *Surely you're joking, Mr. Feynman! Adventures of a curious character*, Norton, New York, 1985. **115**

[190] R. P. Feynman, *QED: The strange theory of light and matter*, Princeton University Press, 1985 (first Princeton paperback printing, 1988). **118**

[191] L. Fibonacci, *Liber Abaci*, 1202. **603**

[192] S. Fisk, A short proof of Chvátal's watchman theorem, *J. Combin. Theory Ser. B* **24** (1978), 374. **783**

[193] P. Flajolet, J.C. Raoult, and J. Vuillemin, The number of registers required to evaluating arithmetic expressions, *Theoret. Comput. Sci.*, **9** (1979), 99–125. **277**

[194] M. E. Flahive and J. W. Lee, Some observations on teaching induction, *MAA Focus* **28** (5) (May/June 2008), 9–10. **17, 93**

[195] P. Fletcher, H. Hoyle, and C. W. Patty, *Foundations of discrete mathematics*, PWS-Kent Publishing Company, Boston, 1991. **16, 199**

[196] L. R. Foulds and D. G. Johnston, An application of graph theory and integer programming: Chessboard non-attacking puzzles, *Math. Mag.*, **57** (1984), 95–104. **749**

[197] D. Fowler, Could the Greeks have used mathematical induction? Critical remarks on an article by S. Unguru: "Greek mathematics and mathematical induction", *Physis Riv. Internaz. Storia Sci. (N. S.)* **31** (1994), no. 1, 273–289. **12, 14, 18, 162**

[198] J. B. Fraleigh, *A first course in abstract algebra*, 5th ed., Addison-Wesley Publishing Company, Reading Massachusetts, 1994. **768**

[199] M. M. France and G. Tenenbaum, *The prime numbers and their distribution*, American Mathematical Society, 2000, reprinted 2001 with corrections. [Originally published in French as Les Nombres Premiers, *Que sais-fe?*, no. 571, 1997 ed., Presses Universitaires de France, Paris, 1997. Translated from the French by Philip G. Spain.] **80**

[200] M. Fraňová and Y. Kodratoff, Predicate synthesis from formal specifications: using mathematical induction for finding the preconditions of theorems, *Nonmonotonic and inductive logic (Reinhardsbrunn Castle, 1991)*, 184–208, Lecture Notes in Comput. Sci. **659**, Springer, Berlin, 1993. **16**

[201] H. Freudenthal, Zur Geschichte der vollständigen Induktion (in German), *Arch. Internat. Hist. Sci.*, **6** (no. 22) (1953), 17–37. **11**

[202] S. H. Friedberg, A. J. Insel, and L. E. Spence, *Linear Algebra*, 4th ed., Prentice Hall (Pearson Education), Upper Saddle River, New Jersey, 2003. **774**

[203] J. W. Gaddum and E. A. Nordhaus, On complementary graphs, *Amer. Math. Monthly*, **63** (1956), 175–177. **687**

[204] D. Gale and L. S. Shapley, College admissions and the stability of marriage, *Amer. Math. Monthly* **69** (1962), 9–15. **251**

[205] G. Gamow, *One two three ... infinity*, Viking Press, 1947. (Also reprinted in paperback by Mentor Books, New York, 1953–1957.) **18, 277, 423**

[206] G. Gamow and M. Stern, *Puzzle-Math*, Viking Press, 1958. **296, 297**

[207] M. Gardner, *The Scientific American Book of Mathematical Puzzles and Diversions*, Scientific American, Inc., May 1957. **276**

[208] M. Gardner, Mathematical Games: about the remarkable similarity between the Icosian game and the Towers of Hanoi, *Scientific American* **196**, (May 1957), 150–156. **277**

[209] M. Gardner, "The Icosian game and the tower of Hanoi," in *The Scientific American Book of Mathematical Puzzles & Diversions*, Simon and Schuster, New York, 1959 (ch 6, pp 55–62). **277**

[210] M. Gardner, Catalan numbers: an integer sequence that materializes in unexpected places, *Scientific American* **234** (June 1976), 120–125. **203**

[211] M. Gardner, *Puzzles from other worlds*, Vintage, 1984. **300**

[212] M. Gardner, *The sixth book of mathematical games from Scientific American*, University of Chicago Press, Chicago, Il., 1984. **784**

[213] M. Gardner, Catalan numbers, *Time travel and other mathematical bewilderments*, W. H. Freeman, New York, 1988, pp. 253–266. **203**

[214] M. Gardner, *Penrose tiles to trapdoor ciphers*, Mathematical Association of America, Washington, DC, 1997. [Originally published by W. H. Freeman and Company, New York, 1989.] **17, 75, 77, 299, 300, 818**

[215] B. Garrison, Polynomials with large numbers of prime values, *Amer. Math. Monthly* **97** (1990), 316–317. **82**

[216] R. Gauntt, The irrationality of $\sqrt{2}$, *Amer. Math. Monthly* **63** (1956), 247. **48**

[217] K. Gödel, *The consistency of the Axiom of Choice and the Generalized Contin-uum Hypothesis with the axioms of set theory*, Annals of Mathematics Studies **3**, Princeton University Press, 1940. **816**

[218] S. W. Golomb, Checker boards and polyominoes, *Amer. Math. Monthly* **61** (no. 10, Dec.) (1954), 675–682. **294, 723**

[219] S. W. Golomb, *Polyominoes*, Scribner's, New York, 1965. **294**

[220] L. I. Golovina and I. M. Yaglom, *Induction in geometry*, Little mathematics library, MIR Publishers, Moscow, 1979. **16, 83, 363, 663, 791**

[221] I. J. Good, Normal recurring decimals, *J. London Math. Soc.* **21** (1946), 167–169. **257**

[222] E. G. Goodaire and M. M. Parmenter, *Discrete Mathematics with graph theory*, Prentice Hall, 1998. **16**

[223] R. L. Goodstein, On the Restricted Ordinal Theorem, *J. Symb. Logic* **9** (1944), 33-41. **88**

[224] H. W. Gould, *Bell & Catalan numbers: research bibliography of two special number sequences*, 6th ed., Math Monongliae, Morgantown, WV, 1985. **203**

[225] H. Gould and J. Kaucky, Evaluation of a class of binomial coefficient sum-mations, *J. Combin. Theory, Ser. A* **1** (1966), 233-247. **140**

[226] R. J. Gould, *Graph theory*, Benjamin/Cummings Publishing Company, Inc., 1988. **239, 693**

[227] W. T. Gowers, A new proof of Szemerédi's Theorem, *GAFA, Geom. Funct. Anal.* **11** (2001), 465–588. **372**

[228] R. L. Graham, *Rudiments of Ramsey theory*, Regional conference series in mathematics, No. 45, American Mathematical Society, 1981 (reprinted with corrections, 1983). [Published by AMS for the Conference Board of the Mathe-matical Sciences. Based on lectures presented at St. Olaf College, June 18–22, 1979.] **386**

[229] R. L. Graham and P. Hell, On the history of the minimum spanning tree problem, *Annals of the History of Computing* **7** (1) (1985), 43–57. **246**

[230] R. L. Graham, D. E. Knuth, and O. Patashnik, *Concrete mathematics: a foundation for computer science* (2nd ed.), Addison-Wesley, Reading, MA, 1994. **129, 144, 167, 193, 203, 210, 306, 360, 363, 619, 621, 728, 791**

[231] R. L. Graham, B. L. Rothschild, and J. H. Spencer, *Ramsey theory*, 2nd ed., Wiley-Interscience Ser. in Discrete Math., New York, 1990. **88, 322, 323, 365, 371, 372, 378, 383, 800**

[232] R. M. Grassl, Euler numbers and skew-hooks, *Math. Mag.* **66** (1993), 181–188. **213**

[233] F. Gray, *Pulse code communication*, U.S. Patent No. 2632058, March 15, 1953. **281**

[234] C. Greene and D. J. Kleitman, Strong versions of Sperner's theorem, *J. Combin. Theory Ser. A* **20** (1976), 80–88. **224**

[235] J. E. Greene, A new short proof of Kneser's conjecture, *Amer. Math. Monthly* **109** (2002), no. 10, 918–920. **221**

[236] G. E. Greenwood and A. M. Gleason, Combinatorial relations and chromatic graphs, *Canad. J. Math.* **7** (1955), 1–17. **368**

[237] A. Grigori, The principle of mathematical induction, (Spanish) *Bol. Mat.* **2** (1968), 1–5. **18, 62**

[238] R. P. Grimaldi, *Discrete and combinatorial mathematics: an applied introduction*, (4'th ed.), Addison Wesley Longman, Inc., 1999. **13, 16, 17, 322**

[239] R. P. Grimaldi, *Discrete and combinatorial mathematics: an applied introduction*, (5'th ed.), Pearson Addison Wesley, 2004. **704**

[240] B. Grünbaum and G. C. Shephard, Pick's theorem, *Amer. Math. Monthly* **100** (1993), 150–161. **784**

[241] E.-Y. Guma and M. B. Maschler, *Insights into game theory: an alternative mathematical experience*, Cambridge University Press, 2008. **252, 290, 686**

[242] D. S. Gunderson, *Induced Ramsey theory*, manuscript, 250 pages, 2006. **365**

[243] D. S. Gunderson and K. R. Johannson, On combinatorial upper bounds for van der Waerden numbers $W(3; r)$, *Congressus Numerantium* **190** (2008), 33–46. **372**

[244] R. K. Guy, Dissecting a polygon into triangles, Note 90, *Bull. Malayan Math. Soc.* **5** (1958), 57–60. **203**

[245] R. K. Guy, The strong law of small numbers, *Amer. Math. Monthly* **95** (October 1988), 697–712. **78, 79, 167, 424**

[246] R. K. Guy, *Fair game; how to play impartial combinatorial games*, COMAP, 1989. **291**

[247] R. K. Guy, The second strong law of small numbers, *Math. Mag.* **63** (1990), 3–20. **78, 79, 82, 198, 203, 204, 424, 644**

[248] R. K. Guy, *Unsolved problems in number theory*, (2nd ed.), Springer-Verlag, New York, 1994. **83, 419**

[249] R. Haas, Three-colorings of finite groups or an algebra of nonequalities, *Math. Mag.* **63** (1990), 211–225. **566**

[250] H. Hadwidger, H. Debrunner, and V. Klee, *Combinatorial geometry in the plane*, Holt, Rinehart and Winston, New York, 1964. **349, 353, 657, 779**

[251] J. Hagen, *Synopsis der Höheren Mathematik* **1** (1891), 64–68. [Berlin] **141**

[252] A. Hajnal, E. C. Milner, and E. Szemerédi, A cure for the telephone disease, *Canad. Math. Bull.* **15** (1976), 447–450. **728**

[253] A. W. Hales and R. I. Jewett, Regularity and positional games, *Trans. Amer. Math. Soc.* **106** (1963), 222–229. **375**

[254] M. Hall, An existence theorem for latin squares, *Bull. Amer. Math. Soc.* **51** (1944), 72. **333**

[255] M. Hall Jr., *Combinatorial Theory*, (2nd ed.), John Wiley and Sons, New York, 1986. **17**

[256] P. Hall, On representations of subsets, *J. London Math. Soc.* **10** (1935), 26–30. **249**

[257] P. R. Halmos and H. E. Vaughan, The marriage problem, *Amer. J. Math.* **72** (1950), 214–215. **684**

[258] J. Y. Halpern, *Epistemology: five questions* `www.cs.cornell.edu/home/halpern/papers/manifesto.pdf`, accessed Dec. 2009. **300**

[259] G. Hardy, J. E. Littlewood, and G. Pólya, *Inequalities*, 2nd ed., Cambridge Mathematical Library, Cambridge University Press, Cambridge, 1952. **38, 39, 320**

[260] V. C. Harris, On proofs of irrationality of $\sqrt{2}$, *Mathematics Teacher* **64** (1971), 19. **48**

[261] O. Haupt and P. Sengenhorst, Algebraic extensions of a field, Chapter 7 in *Fundamentals of mathematics; Volume 1, Foundations of Mathematics/Real number system and algebra*, (eds. F. Bachmann, H. Behnke, K. Fladt, and W. Suss) Translated by S. H. Gould, MIT Press, 1974. **340**

[262] S. Hayasaka, On the axiom of mathematical induction (Japanese, English summary), *Res. Rep. Miyagi Tech. College*, no. 2 (1965), 75–83. **18**

[263] E. Helly, Über Mengen konvexer Körper mit gemeinschaftlichen Punkten, *Jahrb. Deut. Math. Verein* **32** (1923), 175–176. **353**

[264] C. G. Hempel, Geometry and empirical science, in *The world of mathematics*, (J. R. Newman, ed.), vol 3, Simon and Schuster, New York, 1956. **22**

[265] L. Henkin, On mathematical induction, *Amer. Math. Monthly* **67** (4) (April 1960), 323–337. **14, 16, 18**

[266] J. Herman, R. Kučera, and J. Šimša, *Counting and configurations; Problems in combinatorics, arithmetic, and geometry*, CMS Books in Mathematics, Springer, 2003. **17, 371, 460, 609, 767, 783, 788**

[267] I. N. Herstein and I. Kaplansky, *Matters mathematical*, Harper and Row, 1974. **305, 306**

[268] D. Hilbert, Über die Irreduzibilatät ganzer rationaler Funktionen mit ganzzahligen Koeffizienten, [On the irreducibility of entire rational functions with integer coefficients] *J. Reine Angew. Math.* **110** (1892), 104–129. **371**

[269] D. Hilbert, *Die Grundlagen der Geometrie*, Leipzig, 1899 (published in *Festschrift zur Feier dr Enthüllung des Gauss-Weber-Denkmals in Göttingen*). **13, 22**

[270] P. Hilton and J. Pederson, Catalan numbers, their generalization and their uses, *Math. Int.* **13** (1991), 64–75. **203**

[271] N. Hindman, Ultrafilters and combinatorial number theory, Lecture Notes in Mathematics **751**, Proc. *Number Theory, Carbondale, 1979*, 119–184. **231**

[272] C. W. Ho and S. Zimmerman, On infinitely nested radicals, *Math. Mag.* **81** (Feb. 2008), 3–15. **159**

[273] K. Hoffman and R. Kunze, *Linear Algebra*, Prentice-Hall, Englewood Cliffs, New Jersey, 1961. **745**

[274] D. R. Hofstadter, *Gödel, Escher, Bach: an eternal golden braid*, Vintage Books, (a division of Random House), New York, 1979. **16, 21, 415**

[275] D. R. Hofstadter, *Metamagical themas: Questing for the essence of mind and pattern*, Basic Books, New York, 1985. **415**

[276] R. Honsberger, *Mathematical Gems I*, Mathematical Association of America, Washington, D. C., 1973. **203**

[277] R. Honsberger, *Mathematical Gems III*, Mathematical Association of America, Washington, D. C., 1985. **197, 203, 599, 723**

[278] R. Hoogerwoord, On mathematical induction and the invariance theorem, *Beauty is our business*, 206–211, Texts Monogr. Comput. Sci., Springer, New York, 1990. **18**

[279] W. A. Horn, Three results for trees, using mathematical induction, *J. Res. Nat. Bur. Standards Sect. B* **76B** (1972), 39–43. **243**

[280] D. Hrimiuc, Induction principle, π *in the sky*, The Pacific Institute for the Mathematical Sciences, (June, 2000), 24–26. **18, 514, 544, 551, 593, 597, 731**

[281] P. Howard and J. E. Rubin, *Consequences of the Axiom of Choice*, American Mathematical Society, Math. Surv. and Monographs, vol. 58, 1998. **57**

[282] *Hungarian Problem Book II*, New Mathematical Library, # 12, Random House, New York, 1963. **529**

[283] T. W. Hungerford, *Algebra*, Graduate texts in mathematics **73**, Springer-Verlag, 1974. **334**

[284] G. Hunter, *Metalogic: An introduction to metatheory of standard first order logic*, University of California Press, Berkeley, CA, 1971. **220**

[285] V. Ivanov, On properties of the coefficients of the irreducible equation for the partition of the circle, (Russian) *Uspekhi Matem. Nauk.* **9** (1941), 313–317. **86**

[286] B. W. Jackson and D. Thoro, *Applied combinatorics with problem solving*, Addison-Wesley Publishing Company, 1990. **205**

[287] I. M. Jaglom, L. I. Golovina, and I. S. Sominskij, *Die vollständige Induktion*, (German) [On mathematical induction] VEB Deutscher Verlag der Wissenschaften, Berlin, 1986. 183 pages. **16**

[288] R. D. James, Recent progress in the Goldbach problem, *Bulletin of the American Mathematics Society* **55** (1949), 125–136. **83**

[289] T. J. Jech, *Set theory*, Academic Press Inc., New York, 1978. **16, 32, 57, 60, 62, 63, 66, 653, 655, 815**

[290] D. M. Jiao, The method of mathematical induction in mathematical analysis, (Chinese) *Hanzhong Shiyuan Xuebao Ziran Kexue Ban* **1988**, no. 2, 68–70. **18**

[291] K. R. Johannson, *Variations of a theorem by van der Waerden*, Master's thesis, U. of Manitoba, 2007. Available at `http://mspace.lib.umanitoba.ca/dspace/handle/1993/321`. **372**

[292] R. Johnsonbaugh, *Discrete mathematics*, 5th ed., Prentice-Hall, Inc., 2001. **16, 17, 276, 294, 322, 703**

[293] B. Jonas, `http://en.wikipedia.org/wiki/User:B_jonas` **765**

[294] C. Jordan, Sur les assemblages de lignes, *J. Reine Angew. Math.* **70** (1869), 185–190. **680**

[295] Josephus Flavius, *Josephus: The complete works* (transl. by William Whiston, A. M.), Thomas Nelson Publishers, Nashville, 1998. **305**

[296] S. Jukna, *Extremal Combinatorics, with applications in computer science*, EATCS, Springer, 2001. **389, 799**

[297] Personal communication, University of Memphis, 20 March 2010. **225**

[298] S. A. Kalikow and R. McCutcheon, *An outline of ergodic theory*, Cambridge studies in advanced mathematics **122**, Cambridge University Press, to be released April 2010. **225**

[299] K. Kalmanson, *An introduction to discrete mathematics and its applications*, Addison-Wesley Publishing Company, Inc., 1986. **16**

[300] S. Kamiński, On the origin of mathematical induction, (Polish, Russian, English summary) *Studia Logica* **7** (1958), 221–241. **14, 18**

[301] M. Kaneko and J. J. Kline, nductive game theory: a basic scenario, Special issue on the Conferences at Coventry, Evanston, and Singapore, *Journal of Mathematical economics* **44** (12) (2008), 1332–3163. [Available on-line, `doi: 10.1016/jmateco.2008.07.009`.] **290**

[302] K. M. Kaplan, L. Burge, M. Garuba, and J. J. Kaplan, Mathematical Induction: The basis step of verification and validation in a modeling and simulation course, *Proceedings of the 2004 AMSE annual conference & exposition, session 1465*, American Society for Engineering Education, 2004. [Available at `http://me.nmsu.edu/~aseemath/1465_04_3.PDF`, accessed 1 Jan. 2009.] **18, 278**

[303] D. Kapur and H. Zhang, Automated induction: explicit vs. implicit (abstract), in *Proc. of Third International Symposium on artificial intelligence and mathematics* (Fort Lauderdale, Florida), 2004. **18**

[304] G. Károlyi, J. Pach, and G. Tóth, Ramsey-type results for geometric graphs I, *Discrete Comput. Geom.* **18** (1997), 247–255. **259, 700**

[305] E. Kasner and J. R. Newman, *Mathematics and the imagination*, Tempus Books, Redmond, WA, 1989. **277**

[306] G. O. H. Katona, Solution of a problem of A. Ehrenfeucht and J. Mycielski, *J. Combin. Theory Ser. A* **17** (1974), 265–266. **220**

[307] V. J. Katz, Combinatorics and induction in medieval Hebrew and Islamic manuscripts, in *Vita Mathematica: Historical Research and Integration with Teaching* (R. Calinger, ed.), 2000, M.A.A., 99–106. **13**

[308] J. L. Kelley, The Tychonoff product theorem implies the Axiom of Choice, *Fund. Math.* **37** (1950), 75-76. **667**

[309] L. M. Kelly, A resolution of the Sylvester–Gallai problem of J.-P. Serre, *Discrete Comput. Geom.* **1** (1986), 101–104. **358**

[310] J. Kepler, Letter to Joachim Tancke (12 May 1608), in his *Gesamelte Werke*, volume 16, 154–165. **193**

[311] A. Khinchin, *Continued fractions* (3rd ed.), Dover Publications, New York, 1997. **179, 184, 605**

[312] L. Kirby and J. Paris, Accessible independence results for Peano arithmetic, *Bull. London. Math. Soc.* **14** (1982), 285-93. **88**

[313] M. S. Klamkin, Solution to problem 1324, *Math. Mag.* **63** (June 1990), 193. **545**

[314] D. A. Klarner, Correspondences between plane trees and binary sequences, *Journal of Combinatorial Theory* **9** (1970), 401–411. **203**

[315] V. Klee, The Euler characteristic in combinatorial geometry, *Amer. Math. Monthly* **70** (1963), 119–127. **139, 473**

[316] S. C. Kleene, *Introduction to metamathematics*, Wolters-Noordhoff Publishing and North-Holland Publishing Company, 1971. **103, 322, 415, 654**

[317] D. J. Kleitman, and J. B. Shearer, Further gossip problems, *Discrete Math.* **30** (1980), 151–156. **728**

[318] A. B. Klionskiĭ, The method of mathematical induction and the theory of differential equations, (Russian), *Mathematical analysis*, 45–49, Leningrad. Gos. Ped. Inst., Leningrad, 1990. **18**

[319] H. Kneser, Eine direkte Ableitung des Zornschen Lemmas aus dem Auswahlaxiom, *Math Z.* **53** (1950), 110–113. **66**

[320] M. Kneser, Abschätzungen der asymptotischen Dichte von Summenmengen, *Math Z.* **58** (1953), 459–484. **371**

[321] M. Kneser, Aufgabe Nr. 300, *Jber. Deutsch. Math. Verein.* **58** (1955), 27. **221**

[322] M. Kneser, Aufgabe Nr. 360, *Jber. Deutsch. Math. Verein.* **59** (1956), 57.] **222**

[323] D. E. Knuth, *Sorting and searching*, volume 3 of *The art of computer programming*, Addison-Wesley, 1973. **278**

[324] D. E. Knuth, Convolution polynomials, *Mathematica J.* **2** (1992), 67-78. **141**

[325] D. E. Knuth, Johann Faulhaber and sums of powers, `arXiv:math/9207222v1[math.CA]`, 27 Jul 1992. **144**

[326] D. E. Knuth, *Stable marriage and its relation to other combinatorial problems: an introduction to the mathematical analysis of algorithms*, CRM Proceedings & Lecture Notes, vol. 10, A.M.S., 1997. **250, 252, 686**

[327] W. Kocay and D. L. Kreher, *Graphs, algorithms, and optimization*, Chapman & Hall/CRC, Boca Raton, 2005. **691**

[328] T. Koshy, *Fibonacci and Lucas numbers with applications*, Wiley, 2001. **635**

[329] M. Kraitchik, *Mathematical Recreations*, W. W. Norton & Company Inc., New York, 1942. **277**

[330] J. B. Kruskal, On the shortest spanning subtree of a graph and the traveling salesman problem, *Proc. Amer. Math. Soc.* **7** (1956), 48–50. **246**

[331] T. Kucera, *Personal communication*, University of Manitoba, 2005. **56**

[332] R. Kumanduri and C. Romero, *Number theory, with computer applications*, Prentice Hall, Upper Saddle River, NJ, 1998. **179, 180, 182, 186, 562**

[333] C. Kuratowski, Une méthode d'élimination des nombres transfinis des raisonnements mathématiques, *Fund. Math.* **3** (1922), 76–108. **63**

[334] R. Lafore, *Data structures and algorithms in Java*, 2nd ed., Sams publishing, Indianapolis, 2003. **281**

[335] J. Lagarias, he $3x + 1$ problem and its generalizations, *Amer. Math. Monthly* **92** (1985), 3–23. **263**

[336] I. Lakatos, *Proofs and refutations: the logic of mathematical discovery*, Cambridge University Press, Cambridge, 1976. **78, 79**

[337] B. M. Landman and A. Robertson, *Ramsey theory on the integers*, Student
      Mathematical Library **24**, American Mathematical Society, Providence, RI,
      2004. **372**

[338] H. G. Landau, On dominance relations and the structure of animal societies,
      III: The condition for score structure, *Bull. Math. Biophys.* **15** (1953), 143–
      148. **697, 699**

[339] E. Landau, *Grundlagen der Analysis*, (in German), Chelsea Publishing Com-
      pany, N. Y., 1960. **xxiii, 21**

[340] C. H. Langford and C. I. Lewis, History of symbolic logic, in *The world of
      mathematics*, Vol. 3, James R. Newman, Simon and Schuster, New York, 1956.
      (exerpted from *Symbolic Logic*, New York, 1932.) **22**

[341] M. E. Larsen, The eternal triangle—a history of a counting problem, *The
      College Math. J.* **20** (1989), 370–384. **357, 787, 788**

[342] L. Larson and the St. Olaf College Problem Solving Group, solutions to 1983
      IMO, *Math. Mag.* **57** (1984) (No. 2, March), 121–122. **701, 735**

[343] S. R. Lay, *Convex sets and their applications*, Dover Publications, Mineola,
      New York, 1982. **350, 351, 782**

[344] J. W. Layman, The Hankel transform and some of its properties, *J. Integer
      Sequences* **4** (2001), Article 01.1.5, 11 pages. [Available online] **765**

[345] C. G. Lekkerkerker, Voorstelling van natuurlijke getallen door een som van
      getallen van Fibonacci, *Simon Stevin* **29** (1952), 190–195. **197**

[346] H. R. Lewis and L. Denenberg, *Data structures and their algorithms*, Harper-
      Collins Publishers, 1991. **281, 285**

[347] S. Lipschutz, *Schaum's outline of theory and problems of set theory and related
      topics*, Schaum's Outline Series in Mathematics, McGraw-Hill Inc., New York,
      1964. **57, 60, 653**

[348] J. E. Littlewood, *A mathematician's miscellany*, Methuen, London, 1953. **75,
      295**

[349] J. E. Littlewood, *A mathematician's miscellany*, B. Bollobás ed., Cambridge
      University Press, 1986. **295, 300**

[350] L. Lopes, *Manuel d'Induction Mathématique*, (Manual of Mathematical In-
      duction), QED Texte, Boucherville, Quebec, 1998. **16, 467, 468, 469, 474,
      526, 528, 529, 531, 551, 574, 586, 588, 589, 643**

[351] M. Lothaire, *Combinatorics on words*, Cambridge University Press, 1997. ["M. Lothaire" is a pseudonym for a collection of authors, including Francois Perrot and Dominique Perrin.] **237**

[352] L. Lovász, On chromatic number of finite set systems, *Acta. Math. Acad. Sci. Hung.* **19** (1968), 59–67. **384**

[353] L. Lovász, Kneser's conjecture, chromatic number and homotopy, *J. Combin. Theory Ser. A* **25** (1978), 319–324. **221**

[354] L. Lovász, *Combinatorial problems and exercises* (2nd edition), Akadémiai Kiadó, Budapest 1993. **17, 481, 688, 689**

[355] L. Lovász, J. Pelikán, and K. Vesztergombi, *Discrete mathematics: elementary and beyond*, Springer, Undergraduate Texts in Mathematics, 2003. **16, 71, 72**

[356] L. Lovász and M. D. Plummer, *Matching Theory*, Annals of Discrete Mathematics **29**, North-Holland, 1986. **249**

[357] E. Lozansky and C. Rousseau, *Winning Solutions,* Problem books in mathematics, Springer-Verlag New York Inc., 1996. **17, 168, 562**

[358] É. Lucas, *Recreations mathematiques*, four volumes, Gauthier-Villars, Paris, 1891–1894. Reprinted by Albert Blanchard, Paris, 1960. (Tower of Hanoi discussed in Vol. 3, pp. 55–59.) **276**

[359] R. D. Luce and H. Raiffa *Games and decisions; introduction and critical survey*, Dover, 1989. [Copy of the 1957 original published by Wiley.] **289**

[360] S. H. Lui, *Lecture notes on linear algebra and its applications*, preliminary manuscript, 2008, 129 pages. **752, 754, 755**

[361] Z. Manna, S. Ness, and J. Vuillemin, Inductive methods for proving properties of programs, *Communications of the ACM* **16** (8) (1973), 491–502. **18**

[362] W. Mantel, Solution to Problem 28, by H. Gouwentak, W. Mantel, J. Teixeira de Mattes, F. Schuh, and W. A. Wythoff, *Wiskundige Opgaven* **10** (1907), 60–61. **255, 692**

[363] T. Marlowe, C. T. Ryan, and S. Washburn, *Discrete mathematics*, Addison Wesley Longman, Inc., 2000. **16, 199**

[364] M. Martelli and G. Gannon, Weighing coins: divide and conquer to detect a counterfeit, *The College Mathematics Journal* **28** (1997), 365–367. **303**

[365] G. Martin, *Polyominoes: Puzzles and problems in tiling*, Mathematical Association of America, Washington, D. C., 1991. **294**

[366] Problem A764 in "Quickies", *Math. Mag.* **63** (1990), 190, 198. **150**

[367] *Math. Mag.* **63** (1990), 204-5. **129**

[368] P. Matet, Shelah's proof of the Hales–Jewett theorem revisited, *European J. Combin.* **28** (2007), 1742–1745. **378**

[369] Iu. V. Matiiasevich, Diofantovost' perechislimykh mnozhestv, *Doklady Akademii Nauk SSSR* **191** (1970), 279–282. English translation, with amendments by the author, Enumerable sets are diophantine, *Soviet Mathematics— Doklady* **11** (1970), 354–357. **619**

[370] Y. Matijasevich, Solution to the tenth problem of Hilbert, *Mat. Lapok* **21** (1970), 83–87. **194**

[371] Y. Matiyasevich, *Hilbert's tenth problem*, The MIT Press, Cambridge, London, 1993. **194**

[372] J. Matoušek, *Lectures on discrete geometry*, Graduate Texts in Mathematics **212**, Springer, 2002. **222, 223, 259**

[373] H. F. Mattson, Jr., *Discrete mathematics with applications*, John Wiley & Sons, Inc., 1993. **16, 72, 738**

[374] S. Maurer, The king chicken theorems, *Math. Mag.* **53** (1980), 67–80. **698**

[375] S. B. Maurer and A. Ralston, *Discrete algorithmic mathematics*, Addison Wesley, 1991. **17**

[376] D. F. Maurolyci, Abbatis Messanensis, Mathematici Celeberrimi, *Arithmeticorum Libri Duo*, Venice, 1575. **11, 126**

[377] M. E. Mays and J. Wojciechowski, A determinant property of Catalan numbers, *Disc. Math.* **22** (2000), 125–133. **203**

[378] L. McGilvery, 'Speaking of paradoxes ...' or are we?, *Journal of Recreational Mathematics* **19** (1987), 15–19. **75, 300**

[379] J. D. McKinsey, *Introduction to the theory of games*, McGraw-Hill Book Company, Inc., New York, 1952. **289**

[380] D. G. McVitie and L. B. Wilson, Stable marriage assignment for unequal sets, *BIT* **10** (1970), 295–309. **252**

[381] D. G. McVitie and L. B. Wilson, The application of the stable marriage assignment to university admissions, *Oper. Res. Quart.* **21** (1970), 425–433. **252**

[382] M. T. Michaelson, A literature review of pedagogical research on mathematical induction, *Australian Senior Mathematics Journal* **22** (2) (2008), 57–62. **17**

[383] C. Milici, Le principe de l'induction mathématique dans le système formel de la théorie des ensembles fins [The principle of mathematical induction in the formal system of finite sets], (French, Romanian summary), *Proceedings of the Second symposium of Mathematics and its applications (Timişoara, 1987)*, 215–216, Res. Centre, Acad. SR Romania, Timişoara, 1988. **16**

[384] J. Miller, *Earliest known uses of some of the words of mathematics*, http:// jeff560.tripod.com/mathword.html, accessed 8 April 2010. **13**

[385] N. Miller and R. E. K. Rourke, *An advanced course in algebra*, MacMillan, Toronto, 1941 (Revised 1947). **457**

[386] L. Mirsky, *Transversal theory*, Academic Press, 1971. **660, 661**

[387] L. Mirsky, A dual of Dilworth's decomposition theorem, *Amer. Math. Monthly* **78** (1971), 876–877. **224, 661**

[388] D. S. Mitrinović, *Metod matematičke indukcije*, (Serbo-Croatian) 2nd ed. revised, Matematička Biblioteka, 4, Univerzitet u Beogradu, Belgrade, 1958. 63 pp. **15**

[389] D. S. Mitrinović, Matematička indukcija. Binomna formula. Kombinatorika. (Serbo-Croatian) (2nd revised and augmented edition), Matematička Biblioteka, No. 26, *Zavod za Izdavanje Udžbenika*, Socijalisticčke Republike Srbije, Belgrade, 1970. 101 pp. **15**

[390] J. M. Möllar, *General topology*, http://www.math.ku.dk/~moller/e03/3gt/ notes/gtnotes.pdf **665**

[391] M. Molloy and B. Reed, *Graph colouring and the probabilistic method*, (Algorithms and Combinatorics 23), Springer, 2002. **397**

[392] T. L. Moore, Using Euler's formula to solve plane separation problems, *The College Mathematics Journal* **22** (1991), 125–130. **360**

[393] O. Morgenstern and J. von Neumann, *Theory of games and economic behavior*, Princeton, 1944. [3rd ed. in 1953; 3rd ed. paper by John Wiley & Sons, New York, 1964.] **289**

[394] L. Moser, King paths on a chessboard, *Math. Gazette* **39** (1955), 54. **209**

[395] L. Moser and W. Zayachkowski, Lattice paths with diagonal steps, *Scripta Math.* **26** (1963), 223–229. **209**

[396] Th. Motzkin, The lines and planes connecting the points of a finite set, *Trans. Amer. Math. Soc.* **70** (1951), 451–464. **358**

[397] R. F. Muirhead, Proofs that the arithmetic mean is greater than the geometric mean, *Math. Gazette* **2** (1904), 283–287. **39**

[398] D. R. Musser, On proving inductive properties of abstract data types, *Proceedings of the 7th ACM SIGPLAN–SIGACT symposium on Principles of programming languages* (Jan. 28–30, Las Vegas, Nevada), pp. 154–162. **18, 281**

[399] J. Mycielski, Sur le coloriage des graphs, *Colloq. Math.* **3** (1955), 161–162. **383**

[400] E. Nagel and J. R. Newman, Goedel's proof, in vol. 3 of *The world of mathematics*, (J. R. Newman, ed.) Simon and Schuster, New York, 1956. **16, 415**

[401] M. B. Nathanson, *Additive number theory: inverse problems and the geometry of sumsets*, Graduate Texts in Mathematics**165**, Springer , 1996. **161, 371**

[402] R. B. Nelsen, Proof without words: sum of reciprocals of triangular numbers, *Math. Mag.* **64** (1991), 167. **462**

[403] R. B. Nelsen, *Proofs without words; exercises in visual thinking*, The Mathematical Association of America, Washington, D. C., 1993. **8, 125**

[404] R. B. Nelsen, *Proofs without words II; more exercises in visual thinking*, The Mathematical Association of America, Washington, D. C., 2000. **8, 125, 126**

[405] J. Nešetřil, Ramsey theory, in *Handbook of combinatorics*, R. L. Graham, M. Grötschel, L. Lovász eds., Vol 2, pp. 1331–1403, Elsevier Science B. V. (North Holland), Amsterdam, and The MIT Press, Cambridge Massachusetts, 1995. **386**

[406] J. Nešetřil and V. Rödl, A short proof of the existence of highly chromatic hypergraphs without short cycles, *J. Combin. Theory Ser. B* **27** (1979), 225–227. **384**

[407] J. Nešetřil and V. Rödl, Simple proof of the existence of restricted Ramsey graphs by means of a partite construction, *Combinatorica* **1** (2)(1981), 199–202. **384**

[408] J. Nešetřil and V. Rödl, The partite construction and Ramsey set systems, *Discrete Math.* **75** (1989), 327–334. **384**

[409] J. Nešetřil and V. Rödl, eds., *Mathematics of Ramsey theory*, Springer-Verlag, Berlin, 1990. **365, 384**

[410] K. Neu, Using random tilings to derive a Fibonacci congruence, *College Math. J.* **37** (2006), 44–47. **635**

[411] D. J. Newman, *A problem seminar*, Problem books in mathematics, Springer-Verlag New York Inc., 1982. **714, 744, 804, 806**

[412] J. R. Newmann, The Rhind papyrus, in *The world of mathematics*, vol 1, 170–178, Simon and Schuster, New York, 1956. **176**

[413] J. Nicod, *Foundations of geometry and induction*, (containing "Geometry in the sensible world" and "The logical problem of induction"), Harcourt, Brace and Company, New York, 1930. **79**

[414] A. Nilli, Shelah's proof of the Hales–Jewett theorem, in *Mathematics of Ramsey theory*, J. Nešetřil and V. Rödl, eds., pp. 150–151, Springer-Verlag, Berlin, 1990. **378**

[415] I. Niven and H. S. Zuckermann, Lattice points and polygonal area, *Amer. Math. Monthly* **74** (1967), 1195. **784**

[416] H. Noack and H. Wolff, Zorn's lemma and the high chain principle, Chapter 11 in *Fundamentals of mathematics; Volume 1, Foundations of Mathematics/Real number system and algebra*, (eds. F. Bachmann, H. Behnke, K. Fladt, and W. Suss) Translated by S. H. Gould, MIT Press, 1974. **57, 65, 770, 772**

[417] A. Nozaki, (illustrated by M. Anno) *Anno's hat tricks*, Philomel, 1985. **300**

[418] O. Ore, *The four-color problem*, (Pure and Applied Mathematics **27**) Academic Press, New York and London, 1967. **255**

[419] J. O'Rourke, *Art gallery theorems and algorithms*, Oxford University Press, 1987. **783**

[420] J. O'Rourke, *Computational geometry in C*, 2nd ed., Cambridge University Press, 1998. **287, 783**

[421] J. Pach and P. K. Agarwal, *Combinatorial geometry*, Wiley, New York, 1995. [This text is based on notes taken by Agarwal from lectures by Pach.] **222, 259, 780**

[422] L. Pacioli, *Summa de arithmetica, geometrica, proportioni et porportionalita* (briefly referred to as the *Sūma*), 1494. **396**

[423] R. Padmanabhan and S. Rudeanu, *Axioms for latttices and boolean algebras*, World Scientific Publishing Co. Pte. Ltd., Hackensack, NJ, 2008. **235**

[424] B. Pascal, *Oeuvres Complètes de Blaise Pascal*, Paris, 1889. **136**

[425] M. Paterson and U. Zwick, Overhang, *Amer. Math. Monthly* **116** (1), (2009), 19–44. **202**

[426] M. Paterson, Y. Peres, M. Thorup, P. Winkler, and U. Zwick, aximum overhang, *Amer. Math. Monthly* **116** (9), (2009), 763–787. **202**

[427] I. Peterson, A shortage of small numbers, *Science News*, 9 January 1988, p. 31. **78**

[428] I. Peterson, Prime conjecture verified to new heights, *Science News* **158** (7) (12 August 2000), p. 103. **83, 168**

[429] π *in the sky*, December 2000, The Pacific Institute for the Mathematical Sciences (PIMS). **147, 817**

[430] G. Pick, Geometrisches zur Zahlentheorie, *Lotos (Prague)* **19** (1899), 311–319. **356**

[431] A. D. Polimeni and H. J. Straight, *Foundations of discrete mathematics*, Brooks/Cole Publishing company, Monterey, CA, 1985. **17**

[432] J. L. Pollock, *Technical methods in philosophy*, Focus series, Westview Press, Boulder, San Francisco and London, 1990. **22**

[433] G. Pólya, *Induction and analogy in mathematics*, (Volume I: mathematics and plausible reasoning; Volume II: patterns of plausible inference), Princeton University Press, Princeton, New Jersey, 1954. **14, 17, 78, 89, 95, 114, 452, 783, 790, 792**

[434] G. Pólya, *How to solve it: a new aspect of mathematical method*, 2nd ed., Princeton University Press, Princeton, New Jersey, 1957. [Paperback printings 1971, 1973; first ed. 1945.] **17, 94**

[435] G. Pólya, *Mathematcal Discovery: on understanding, learning, and teaching problem solving*, two volumes, Wiley, 1962. [A combined edition was reprinted by John Wiley & Sons, New York, 1981.] **12, 17, 315, 741**

[436] D. G. Poole, The towers and triangles of Professor Claus (or, Pascal knows Hanoi), *Math. Mag.* **67** (1994), 323–344. **277**

[437] A. S. Posamentier and C. T. Salkind, *Challenging problems in algebra*, Dover Publications, Inc., New York, 1996. **273, 410, 547, 548, 549, 592, 714, 791**

[438] F. P. Preparata and S. J. Hong, Convex hulls of finite sets of points in two and three dimensions, *Commun. ACM* **20** (no. 2) (1977), 87-93. **287**

[439] R. C. Prim, Shortest connection networks and some generalizations, *Bell Systems Technical Journal* **36** (1957), 1389–1401. **246**

[440] H. J. Prömel, *Ramsey theory for discrete structures*, Habilitationsschrift, Bonn (1987), 354 pages. **365**

[441] V. Pták, Nondiscrete mathematical induction and iterative existence proofs, *Linear Algebra and Appl.* **13** (1976), no. 3, 223–238. **18**

[442] V. Pták, Nondiscrete mathematical induction, *General topology and its relations to modern analysis and algebra, IV, (Proc. Fourth Prague Topological Sympos., Prague, 1976), Part A*, pp. 166–178, Lecture notes in math., Vol. 609, Springer, Berlin, 1977. **18**

[443] W. V. O. Quine, *Mathematical Logic* (Revised ed.), Harvard University Press, Cambridge, 1951. **13**

[444] N. L. Rabinovitch, Rabbi Levi Ben Gershon and the origins of mathematical induction, *Archive for the History of Exact Sciences* **6** (1970), 237–248. doi: 10.1007/BF00327237 **13**

[445] J. Radon, Mengen konvexer Körper, die einen gemeinsamen Punkt enthalten, *Math. Ann.* **83** (1921), 113–115. **353**

[446] F. P. Ramsey, On a problem of formal logic, *Proc. London Math. Soc.*, **30** (1930), 264–286. **367**

[447] A. Rapoport, *Two-person game theory*, Dover Publications, 1966. **289**

[448] R. Rashed, *AHES* **9** (1972), 1–21. **11**

[449] L. Redei, Ein kombinatorischer Satz, *Acta. Lett. Sci. Szeged* **7** (1934), 39–43. **258**

[450] M. Renault, Four proofs of the ballot theorem, *Math. Mag.* **80** (5) (Dec. 2007), 345–351. **394, 395**

[451] M. D. Resnik, *Frege and the philosophy of mathematics*, Cornell University Press, 1980. **13**

[452] C. Reynolds and R. T. Yeh, Induction as the basis for program verification, *IEEE Transactions on Software Engineering*, SE-2 (4) (1976), 244–252. **18**

[453] I. Richards, Proof without words: Sum of integers, *Math. Mag.* **57** (1984) No. 2, 104. **8**

[454] J. Riordan, *Combinatorial identities*, John Wiley & Sons, Inc., 1968. **125, 140**

[455] F. S. Roberts, *Applied combinatorics*, Prentice-Hall Inc., Englewood Cliffs, New Jersey, 1984. **17, 266**

[456] J. Roberts, *Lure of the integers*, The Mathematical Association of America, 1992. **81**

[457] N. Robertson, D. Sanders, P. Seymour, and R. Thomas,   The four-colour theorem, *J. Combin. Theory Ser. B* **70** (1997), 2–44. **255**

[458] D. G. Rogers, Pascal triangles, Catalan numbers and renewal arrays, *Disc. Math.* **22** (1978), 301–310. **203**

[459] D. G. Rogers and L. Shapiro, Some correspondences involving the Schröder numbers, in *Combinatorial Mathematics: Proceedings of the international Conference, Canberra, 1977*, Springer-Verlag, New York, 1978, pp. 267–276. **209**

[460] H. Rogers, Jr., *Theory of recursive functions and effective computability*, McGraw-Hill Book Company, 1967. **261**

[461] N. Rogers, *Learning to reason: an introduction to logic, sets, and relations*, A Wiley-Interscience Publication, John Wiley and Sons, Inc., New York, 2000. **17**

[462] K. H. Rosen, *Discrete mathematics and its applications*, 6th ed., McGraw-Hill, New York, 2007. **17, 95, 252, 262, 290, 363, 416, 513, 542, 571, 663, 720, 784**

[463] K. A. Ross, *Informal introduction to set theory*, website manuscript, `www.uoregon.edu/~ross1/SetTheory.pdf`, accessed October 2001. **664, 666**

[464] K. A. Ross and C. R. B. Wright, *Discrete mathematics* (4th ed.), Prentice Hall, 1999. **17, 262, 285**

[465] P. Ross, A review of "A LISP prover for induction formulae", in *The College Mathematics Journal*, **20** (1989), 83. **447**

[466] S. M. Ross, *Stochastic processes*, 2nd ed., John Wiley & Sons, Inc., New York, 1996. **393, 394**

[467] S. M. Ross, *Introduction to Probability models*, 6th ed., Academic Press, 1997. **392, 805**

[468] H. Rothe, *Formulae de serierum reversione demonstratio universalis signis localibus combinatorio-analyticorum vicariis exhibita*, Leipzig, 1793. **141**

[469] J. R. Rotman, *A first course in abstract algebra*, 2nd ed., Prentice Hall, 2000. **770, 771**

[470] C. Rousseau and Y. Saint-Aubin, (transl. by C. Hamilton) *Mathematics and technology*, Springer undergraduate texts in mathematics and technology, Springer, 2008. **322**

[471] J. R. Royse, Mathematical induction in ramified type theory, *Z. Math. Logik Grundlagen Math.* **15** (1969), 7–10. **16**

[472] H. Rubin and J. E. Rubin, *Equivalents of the Axiom of Choice*, North-Holland Publishing Company, Amsterdam, 1970. **57**

[473] H. Rubin and J. E. Rubin, *Equivalents of the Axiom of Choice II*, North-Holland Publishing Company, Amsterdam, 1985. **57**

[474] A. D. Sands, On generalized Catalan numbers, *Disc. Math.* **21** (1978), 218–221. **203**

[475] N. Sauer, On the density of families of sets, *J. Combin. Theory Ser. A* **13** (1972), 145–147. **223**

[476] N. Sauer, *Homogeneous structures*, graduate course, University of Calgary, 1988. **378**

[477] A. Schach, Two forms of mathematical induction, *Math. Mag.* **32** (1958), 83–85. **37**

[478] P. H. Schoutte, De Ringen van Brahma, *Eigen Haard* **22** (1884), 274–276. **277**

[479] E. Schröder, Vier kombinatorische Probleme, *Z. Math. Phys.* **15** (1870), 361–376. **209**

[480] F. T. Schubert, *Nova Acta Acad. Petrop.*, II, ad annum 1793, 1798, mem., 174-7. **608**

[481] P. Schumer, The Josephus problem: once more around, *Math. Mag.* **75** (1) (Feb. 2002), 12–17. **306, 728**

[482] I. Schur, Uber die Kongruenz $x^m + y^m \equiv z^m$ (mod p), *Jber. Deutsch. Math.-Verein* **25** (1916), 114–116. **371**

[483] R. Séroul, Reasoning by induction, in *Programming for mathematicians*, Springer-Verlag, Berlin, 2000, pp. 22-25. **17**

[484] M. I. Shamos and D. Hoey, Closest-point problems, in *Proc. 16th Annual IEEE Symposium on Foundations of Computer Science*, 1975, 151–162. **286**

[485] D. Shanks, *Solved and unsolved problems in number theory*, Vol 1, Spartan Books, Washington, D. C., 1962. **562**

[486] J. Shearer, On a problem of Spencer, *Combinatorica* **5** (1985), 241–245. **399**

[487] S. Shelah, A combinatorial problem, stability and order for models and theories in infinitary languages, *Pacific J. Mathematics* **41** (1972), 247–261. **223**

[488] S. Shelah, Primitive recursive bounds for van der Waerden numbers. *J. Amer. Math. Soc.* **1** (3) (1988), 683–697. **372, 378, 379, 381, 383**

[489] K. Shirai, A relation between transfinite induction and mathematical induction in elementary number theory, *Tsukuba J. Math.* **1** (1977), 91–124. **18**

[490] W. E. Shreve, Teaching inductive proofs indirectly, Tips for Beginners, *The Mathematics Teacher* **61** (1963), 643–644. **17**

[491] H. S. Shultz, An expected value problem, em Two-year College Mathematics Journal **10** (Sept. 1979), 277-278. **392**

[492] T. Sillke, The gossip problem, `www.mathematik.uni-bielefeld.de/~sillke/PUZZLES/gossip` **728**

[493] R. Simson, An explication of an obscure passage in Albert Girard's commentary upon Simon Stevin's works, *Phil. Trans. Royal Society of London* **48** (1753), 368–376. **193**

[494] D. Singmaster, An elementary evaluation of the Catalan numbers, *Amer. Math. Monthly* **85** (1978), 366–368. **203**

[495] S. S. Skiena, *Implementing Discrete Mathematics: Combinatorics and Graph Theory with Mathematica*, Reading, MA: Addison-Wesley, 1990. **252**

[496] S. S. Skiena, *The algorithm design manual*, 2nd ed., Springer, 2008. **278, 281**

[497] N. J. A. Sloane, *A handbook of integer sequences*, Academic Press, 1973. **213, 424**

[498] I. S. Sominskiĭ, *The method of mathematical induction*, (Russian) Gosudarstv Izdat. Tehn.-Teor Lit., Moscow, 1956. 48 pp. **15, 16**

[499] I. S. Sominskiĭ, *The method of mathematical induction*, (Translated from the Russian by Halina Moss; translation editor Ian N. Sneddon), Pergamon Press, Oxford, 1961. (Also published by Blaisdell Publishing Co., a division of Random House, New York, London, 1961. vii+57 pp.) **16, 79, 139, 455, 485, 492, 493, 494, 496, 498, 500, 501, 509, 510, 543, 547, 705, 714, 739, 792**

[500] I. S. Sominski, *Matematiksel indüksiyon metodu*, (Turkish) [The method of mathematical induction] (translated by Bediz Asral) Turkish Mathematical Society Publications, No. 11, *Türk Matematik Derneği*, Istanbul, 1962. viii+72 pp. **16**

[501] I. S. Sominski, *Die Methode der vollständigen Induktion*, (German, 13th ed.) [The method of mathematical induction] (Translated from the Russian by W. Ficker and R-D Schröter) Kleine Ergänzungsreihe zu den Hochschulbüchern für Mathematik [Brief supplemental series to University Books for Mathematics], VEB Deutscher Verlag der Wissenschaften, Berlin, 1982, 55 pp. **16**

[502] I. S. Sominskii, *Método de inducción matemática* (Spanish, 2nd ed.) [The method of mathematical induction] (Translated from the Russian by Carlos Vega, with an epilogue by Yu. A. Gastev.) Lecciones Populares de Matemáticas, Mir, Moscow, 1985. 62 pp. **16**

[503] J. Spencer, *Ten lectures on the probabilistic method*, CBMS-NSF Regional Conf. Series in Applied Math. **52**, SIAM, 1987, 78 pages. **699**

[504] F. Spisani, Relative numbers, mathematical induction and infinite numbers, (Italian, with an English translation) *Internat. Logic Rev.* **29** (1984), 29–37. **18**

[505] R. P. Stanley, Hipparchus, Plutarch, Schröder, Hough, *Amer. Math. Monthly* **104** (1997), 344–350. **209**

[506] R. P. Stanley, *Enumerative combinatorics*, 2 volumes, Cambridge University Press, 1997, 1999. [Paperback editions, 1999, 2001.] **17, 203, 211, 245**

[507] J. Steiner, Einige Gesetze über die Theilung der Ebene und des Raumes, *J. Reine Angew. Math.* (also called "Crelle's journal") **1** (1826), 349–364. [Also reprinted in his *Gesammelte Werke*, Volume 1, 77–94.] **790**

[508] H. Steinhaus, *Mathematical snapshots*, 3rd ed., Dover, New York, 1999. [Paperback, with preface by Morris Kline published by Oxford University Press in 1983.] **5, 784**

[509] I. Stewart, *Galois Theory*, Chapman and Hall, New York, 1973. **340**

[510] I. Stewart, The riddle of the vanishing camel, in "Mathematical Recreations", *Scientific American* **266** (June 1993), 122–124. **177**

[511] I. Stewart, Mathematical recreations: Monks, blobs, and common knowledge, *Scientific American* **279** (August 1998), 96–97. **295**

[512] I. Stewart, *Galois Theory*, 3rd ed., Chapman and Hall/CRC, Boca Raton, 2004. **340**

[513] J. Stewart, *Single variable calculus, early transcendentals*, 4th ed., Brooks/Cole Publishing Company, 1999. **276**

[514] P. K. Stockmeyer, More on induction, Letters to the editor, *MAA Focus* **29** (No. 1), (January 2009), 28–29. **104**

[515] D. J. Struik, *A concise history of mathematics* (4th revised ed.), Dover Publications, Inc., New York, 1987. **11**

[516] G. J. Stylianides, A. J. Stylianides, and G. N. Philippou, G. N., Preservice teachers' knowledge of proof by mathematical induction. *J. Mathematics Teacher Education* **10** (2007), 145-166. **17**

[517] R. A. Sulanke, Bijective recurrences concerning Schröder paths, *Electronic J. Combinatorics* **5** (1998), No. 1, R47, 1–11. http://www.combinatorics.org **209**

[518] D. Sumner, The technique of proof by induction, //www.math.sc.edu/~sumner/numbertheory/induction/Induction.html, accessed 2010. **600**

[519] B. Sury, Mathematical induction—an impresario of the infinite, *Resonance*, February 1998, 69–76. **423**

[520] J. J. Sylvester, On the number of fractions contained in any 'Farey series' of which the limiting number is given, *The London, Edinburgh and Dublin Philosophical Magazine and Journal of Science*, series 5, **15** (1883), 251–257. Reprinted in his *Collected Mathematical papers*, volume 4, 101–109. **167**

[521] J. J. Sylvester, Mathematical Question No. 11851, *Educational Times* **59** (1893), p. 98. **358**

[522] J. J. Sylvester, *The collected papers of James Joseph Sylvester*, Vol. III, Cambridge University Press, London, 1904–1912. **603**

[523] H. Tahir, Pappus and mathematical induction, *Austral. Math. Soc. Gaz.* **22** (1995), no. 4, 166-167. **12, 14, 18**

[524] A. Takayama, *Mathematical economics* (2nd ed.), Cambridge University Press, Cambridge, 1985. **333**

[525] T. Tao, *Arithmetic Ramsey Theory*, http://www.math.ucla.edu/~tao/preprints/Expository/ramsey.dvi, accessed 14 December 2006. **372**

[526] Tao, T., *The blue-eyed islanders puzzle*, http://terrytao.wordpress.com/2008/02/05/the-blue-eyed-islanders-puzzle/ **300**

[527] T. Tao and V. Vu, *Additive combinatorics*, Cambridge Studies in Advanced Mathematics **105**, Cambridge University Press, Cambridge, 2006. **161, 371**

[528] E. Tardos, *Notes on CS684, algorithmic game theory*, written by Ara Hayrapetyan, May 7, 2004, http://www.cs.cornell/courses/cs684/2004sp/may7.ps, accessed 2010. **250**

[529] R. E. Tarjan, *Data structures and network algorithms*, SIAM, 1983. **246, 281**

[530] A. D. Taylor and W. S. Zwicker, *Simple games: desirability relations, trading, pseudoweightings*, Princeton University Press, 1999. **289, 290**

[531] R. Tidjeman, On a telephone problem, *Nieuw Arch. Wisk.* **3** (1971), 188–192. **728**

[532] J. Tierney, *Solution to 100-Hat Puzzle*, on-line *New York Times*, Tierney Lab, http:tierneylab.blogs.nytimes.com/2009/03/30/ **725**

[533] B. A. Trakhtenbrot, *Algorithms and automatic computing machines*, translated from the 1960 edition, in the series Topics in mathematics, D. C. Heath and Company, Boston, copyright by University of Chicago, 1963. **290, 291**

[534] D. Trim, *Calculus for engineers*, Prentice-Hall Canada, Inc., 1998. **17, 319, 503, 535, 596**

[535] J. K. Truss, *Discrete mathematics for computer scientists*, Addison-Wesley Publishers, Ltd., 1991. **17**

[536] P. Turán, Eine Extremalaufgave aus der Graphentheorie (in Hungarian), *Math. Fiz Lapook* **48** (1941), 436–452. **256**

[537] P. Turán, On the theory of graphs, *Colloquium Math.* **3** (1954), 19–30. **256**

[538] H. W. Turnbull, The great mathematicians, in, *The world of mathematics, Vol. 1* (J. R. Newmann, ed.), 75–168, Simon and Schuster, New York, 1956. **176**

[539] Zs. Tuza, Applications of the set-pair method in extremal hypergraph theory, in "Extremal problems for finite sets", (P. Frankl, Z. Füredi, G. Katona, D. Miklós, eds.) Proceedings of conference in Visegrád (Hungary), 16–21 June 1991, *Bolyai Society Mathematical Studies* **3** (1994), 479–514. **220**

[540] H. Tverberg, On Dilworth's decomposition theorem for partially ordered sets, *J. Combin. Theory* **3** (1967), 305–306. **224**

[541] S. Unguru, Greek mathematics and mathematical induction, *Physis Riv. Internaz. Storia Sci. (N. S.)* **28** (1991), no. 2, 273–289. **12, 14, 18, 162**

[542] S. Unguru, Fowling after induction. Reply to D. Fowler's comments: "Could the Greeks have used mathematical induction? Did they use it?" *Physis Riv. Internaz. Storia Sci. (N.S.)* **31** (1994), no. 1, 267–272. **12, 14, 18, 162**

[543] P. Urso and E. Kounalis, Sound generalizations in mathematical induction, *Theoret. Comp. Sci.* **323** (September 2004), 443–471. **18**

[544] G. Vacca, Maurolycus, the first discoverer of the principle of mathematical induction, *Bull. Amer. Math. Soc.* **16** (1909), 70–79. **11**

[545] R. van der Hofstad and M. Keane, An elementary proof of the hitting time theorem, *Amer. Math. Monthly* **115** (Oct. 2008), 753–756. **396**

[546] B. L. van der Waerden, Beweis einer Baudetschen Vermutung, *Nieuw. Arch. Wisk.* **15** (1927), 212–216. **372**

[547] B. L. van der Waerden, How the proof of Baudet's Conjecture was found, *Studies in Pure Mathematics*, L. Mirsky ed., Academic Press, London (1971), 251–260. **372**

[548] H. P. van Ditmarsch and B. P. Kooi, *The secret to my success*, Technical report OUCS-2004-10, `www.cs.atago.ac.nz/research/publications/oucs-2004-10.pdf`, accessed Dec. 2009. **300**

[549] R. H. van Lint and R. M. Wilson, *A course in combinatorics*, 2nd ed., Cambridge University Press, 2001. **649**

[550] E. P. Vance, *An introduction to modern mathematics*, (Chapter 11, Mathematical Induction), Addison-Wesley Publishing, Inc., 1968. [Also published separately in "Mathematical induction and conic sections", chapters 11 and 12 of the text, Addison-Wesley, 1971.] **15, 146, 485, 487, 488, 489, 491**

[551] V. Vapnik and A. Chervonenkis, On the uniform convergence of relative frequencies of events to their probabilities, *Theory of Probability and Its Applications* **16** (1971), 264–280. **223**

[552] I. Vardi, *Computer recreations in Mathematica*, Addison-Wesley, Reading, MA, 1991. **203, 209, 277**

[553] V. E. Vickers and J. Silverman, A technique for generating specialized Gray codes, *IEEE Transactions on Computers* **C-29** (1980), 329–331. **281**

[554] J. von Neumann, Zur Theorie der Gesellschaftsspicle, *Mathematische Annalen* **8** (1928), 295–320. **289**

[555] D. G. Wagner and J. West, Construction of uniform Gray codes, *Congressus Numerantium* **80** (1991) 217–223. **281**

[556] F. Waismann, *Introduction to mathematical thinking: the formation of concepts in modern mathematics*, Dover, 2003. [Reprint of the 4th printing (1966) of the Harper Torchbook edition (first published by Harper and Row, New York, 1959) of the Theodore J. Benac's translation of *Einführung in das mathematische Denken*; the translation was originally published by Frederick Ungar Publishing, New York, 1951.] **13, 14, 16, 27**

[557] C. Walther, Mathematical induction, in *Handbook of logic in artificial intelligence and logic programming, Vol. 2*, 127–228, Oxford Sci. Publ., Oxford Univ. Press, New York, 1994. **12, 18, 162**

[558] Wapedia *Wiki: Bernoulli number*, http://wapedia.mobi/en/ Bernoulli_number, accessed 2 January 2009. **141, 142**

[559] S. Warner, Mathematical induction in semigroups, *Amer. Math. Monthly* **67** (1960), 533–537. **334**

[560] R. Wasén, A note on the principle of mathematical induction, intuition, and consciousness in the light of ideas of Poincaré and Galois, *Chaos Solitons Fractals* **12** (2001), no. 11, 2123–2125. **17**

[561] D. S. Watkins, *Fundamentals of matrix computations*, 2nd ed., Wiley-Interscience, John Wiley & and Sons, 2002. **328**

[562] E. W. Weisstein, *CRC Concise encyclopedia of mathematics*, 2nd ed., Chapman & Hall/CRC, 2003. **211**

[563] D. G. Wells, *The Penguin dictionary of curious and interesting numbers*, Penguin, London, 1986. **203**

[564] D. G. Wells, *Hidden connections, double meanings*, Cambridge University Press, 1988. **80, 84**

[565] D. G. Wells, *The Penguin dictionary of curious and interesting geometry*, Penguin, London, 1991. **784**

[566] D. B. West, *Introduction to graph theory*, (2nd edition), Prentice Hall, Inc., 2001. **18, 239, 253, 674, 676, 680, 682, 684, 687, 689, 699**

[567] J. D. Weston, A short proof of Zorn's Lemma, *Arch. Math.* **8** (1957), 279. **66**

[568] H. Whitney, Congruent graphs and the connectivity of graphs, *Amer. J. Math.* **54** (1932), 150–168. [John Hopkins Univ. Press, Baltimore, MD.] **682**

[569] A. Wiles, Modular elliptic curves and Fermat's last theorem, *Ann. Math. (2)* **141** (1995), no. 3, 443–551. **47**

[570] H. S. Wilf, *Algorithms and complexity*, 2nd ed., A K Peters Ltd., Natick Massachusets, 2002. **278**

[571] S. Willard, *General topology*, Addison-Wesley Publishing Company, Ltd., 1970. **228, 666**

[572] J. D. Williams, *The compleat strategyst: being a primer on the theory of games of strategy*, revised edition, McGraw-Hill Book Company, 1966. [The first edition was 1954.] **289**

[573] R. Wilson, *Four colours suffice*, Penguin Books, 2002. **255**

[574] D. Wood, The towers of Brahma and Hanoi revisited, *Journal of Recreational Mathematics* **14** (1981), 17–24. **277**

[575] D. R. Woodall, Sufficient conditions for circuits in graphs, *Proc. Lond. Math. Soc.* **24** (1972), 739–755. **676**

[576] D. R. Woodall, Inductio ad absurdum? *Math. Gaz.* **59** (1975), 64–70. **248**

[577] J. Worpitzky, Studien über die *Bernoullischen* und *Eulerschen* Zahlen, *Journal für die reine und angewandte Mathematik* **94** (1883), 203–232. **212**

[578] Z. Xiao and P. Xiong, An equivalent form of the Dedekind axiom and its application. I. Also on the unity of the continuous induction, the mathematical induction, and the transfinite induction, (English), *Chinese Quart. J. Math.* **13** (1998) no. 3, 74–80. **18**

[579] Z. Xiao and P. Xiong, An equivalent form of the Dedekind axiom and its application. II. Also on the unity of the continuous induction, the mathematical induction, and the transfinite induction, (English), *Chinese Quart. J. Math.* **13** (1998) no. 4, 24–28. **18**

[580] D. Xie, Mathematical induction applied on Leontief systems, *Econom. Lett.* **39** (1992), no. 4, 405–408. **332, 763**

[581] M. Yadegari, The use of mathematical induction by Abū Kāmil Shujā'ibn Aslam (850–930), *Isis* **69** (1978), no. 247, 259–262. **14, 18**

[582] B. K Youse, *Mathematical Induction*, Prentice Hall, 1964. **16, 482, 508, 591, 595, 598, 709**

[583] B. K Youse, *Mathematical problems and theorems of enduring interest*, manuscript, 1998. **427, 512**

[584] T. Yuster, The reduced row echelon form of a matrix is unique: a simple proof, *Math. Mag.* **57** (1984), 93–94. **325, 745**

[585] E. Zeckendorf, Représentation des nombres naturels par une somme de nombres de Fibonacci ou de nombres de Lucas, *Bulletin de la Société Royale des Sciences de Liége* **41** (1972), 179–182. **197**

[586] E. Zermelo, Beweis, daßjede Menge wohlgeordnet werden kann (in German), *Math. Ann.* **59** (1904), 514–516. **60**

[587] E. Zermelo, Neuer Beweis für die Möglichkeit einer Wohlordnung (in German), *Math. Ann.* **65** (1908), 107–128. **60**

[588] H. Zhang, ed., *Automated mathematical induction*, Kluwer Academic, Dordrecht and Boston, 1996. [Reprinted from *Journal of automated reasoning* **16** (nos.1-2) (1996), 1–222.] **18**

[589] L. Zippin, *Uses of infinity*, Dover Publications, Mineola, New York, 2000. [First published by Random House, Inc., New York, 1962.] **17, 69, 190**

[590] M. Zorn, A remark on method in transfinite algebra, *Bull. Amer. Math. Soc.* **41** (1935), 667–670. **63**

[591] http://linas.org/mirrors.ltn.lv/2001.02.27/~podnieks/gt4.html [an account of Hilbert's tenth problem]. **194**

[592] http://logic.pdmi.ras.ru/Hilbert10/ [useful references, and article by YM on the history of Hilbert's tenth problem] **194**

Name index

Subject index

Milton Keynes UK
Ingram Content Group UK Ltd.
UKHW051926141024
449569UK00027B/1383